BARRON'S

AP®

PHYSICS C

WITH 4 PRACTICE TESTS

FIFTH EDITION

Robert A. Pelcovits, Ph.D.
Professor of Physics
Brown University
Providence, Rhode Island

Joshua Farkas, M.D.
Assistant Professor of Medicine
University of Vermont
Burlington, Vermont

Acknowledgments
I owe a tremendous debt of gratitude to my coauthor, Joshua Farkas, an exceptionally gifted former student of mine. Josh contributed the lion's share of work to the first edition, developing most of the examples and problems (especially the novel and creative ones), and offering numerous ideas and suggestions as to the best way to present difficult concepts and useful problem-solving strategies. I am very grateful for all he has done in making our book a success. While preparing the fifth edition, I was helped by yet another superb Brown undergraduate, Adam Scherlis. Adam reviewed all of the examples, problems, and practice tests, catching many errors and making many suggestions for improvements. I am very grateful for his expert help.

—RAP

I would first of all like to thank Robert Pelcovits for coauthoring this book with me and for his valiant efforts in its revision. To work with a man of his stature, both as a physicist and as a person, is indeed a privilege. This book never could have been possible without him. I would also like to thank Esta Farkas for heroically reviewing the initial drafts of this book for both content and English.

—JDF

About the Authors

Robert Pelcovits is a professor of physics at Brown University, where he has taught since 1979. He earned his B.A. and M.S. degrees at the University of Pennsylvania and his Ph.D. at Harvard University. The author of over 80 publications in theoretical condensed matter physics, he has lectured on his research at many universities throughout the world. He has taught a wide range of courses at Brown, from introductory to advanced graduate level. In 1999 he was awarded the Philip J. Bray Award for Teaching Excellence in the Physical Sciences by Brown. In 2008 he was named a Royce Family Professor of Teaching Excellence at Brown, and in 2011 the university awarded him the Harriet Sheridan Award for Distinguished Contribution to Teaching and Learning.

Having earned both Sc.B. and M.S. degrees from Brown University and an M.D. degree from Cornell University, *Joshua Farkas* is currently an assistant professor of medicine at the University of Vermont. He was also a National Merit Scholar, a Goldwater Scholar, and a Faculty Scholar at Brown University, with experience teaching college-level calculus and calculus-based physics. Having scored 5s on AP exams in BC Calculus, Biology, Chemistry, and Physics C, as well as 800s in SAT I Math and on SAT II exams in Math IIC, Chemistry, and Physics, he brought a student's perspective to this book.

Published by Kaplan, Inc.,
d/b/a Barron's Educational Series
750 Third Avenue
New York, NY 10017
www.barronseduc.com

ISBN: 978-1-4380-1285-8

10 9 8 7 6 5 4 3 2 1

Contents

As you review the content in this book to work toward earning that **5** on your AP PHYSICS C exam, here are five things that you **MUST** know above everything else:

1

Dimensional Analysis. Many AP Physics multiple-choice questions can be answered or have the number of possibly correct answers significantly reduced by using dimensional analysis to check the units of the answers.

- Even for free-response questions you should check your units. You may discover a mistake in your work by doing so.
- Whenever you solve a practice or homework problem, always check the units; this is a great habit to develop while mastering physics.

2

Vectors. Vectors are key quantities in physics, whether Newtonian mechanics or electricity and magnetism.

- Review the addition and subtraction laws for vectors in Chapter 1.
- Most important, review how to resolve vectors into components, including the relevant trigonometry.

3

Graphs. The AP exam will test your ability to understand physical information in graphical form, such as displacement of a moving object vs. time. You will need to draw graphs in the free-response questions and select correct graphs in the multiple-choice section.

- Review the use of graphs in the free-response questions. Also read "Graphical Analysis of Data" in Chapter 1.

Newton's Laws. Physics C Mechanics is all about the application of Newton's three laws of motion to mechanical systems.

- Review Chapter 4, paying close attention to the correct drawing of free-body diagrams and the proper application of Newton's third law to a pair of interacting objects.

5

Practice! Practice! Practice! The best (and maybe only) way to learn physics is to do practice problems. Reading a practice book or textbook over and over will not help you. Do as many practice problems as you can in this book, your textbook, and any online resources.

Introduction

HOW TO REVIEW FOR THE AP PHYSICS EXAM

Solving physics problems is like mastering any sport or musical instrument: If you want to do it well, you have to practice. Do not expect to understand everything immediately and do not be frustrated by the effort required to master some concepts and techniques. Contrary to popular conceptions of a vast divide between physics geniuses who can immediately solve problems and other people who simply cannot, learning physics requires hard work. However, a mastery of calculus-based physics will teach you analytical and mathematical skills that will prove extremely useful in a broad range of other disciplines.

GUIDE TO THIS REVIEW BOOK

Text

The text of the chapters works from the ground up without assuming extensive knowledge of physics on your part. Therefore, the text is appropriate at any stage of your mastery of AP physics. You may find it helpful to read the text along with your textbook when you are first learning the material and studying for course exams, or you may choose to read the chapters together as a review after you have completed most of your AP Physics course.

Questions

Because of space limitations, it is impossible for the questions to work from the ground up (i.e., start with very easy questions and progress to more difficult ones). Therefore, the questions are generally near the AP level and assume that you have some problem-solving experience. The questions are designed to raise you from a competent problem solver to an expert with extensive experience in solving AP problems. If you have difficulty with the questions, you may want to go back and solve some problems in your introductory physics textbook or AP physics textbook, as well as consider the advice in the next paragraph.

Difficulty of the Questions

The problems in this review book are generally slightly more difficult than actual AP exam questions. In order to maximize the amount that you learn from this book and your degree of comfort with AP physics questions, we have designed the questions to cover the material and problem-solving techniques of the AP Physics exam while being more challenging than questions you are actually likely to encounter on the exam. For example, our problems generally have more parts than standard AP problems in order to exhaustively explore various hypothetical situations. Therefore, we strongly urge you not to become discouraged if you are not able to solve all the problems on your first attempt. Instead, the best approach is to make your best effort at the problems, read (and study) the solutions of problems you got wrong, and later return to these problems and try them again. Because we have attempted to include all the common problem types that appear on the AP exam, after you have worked through all the problems in this guide correctly (even if not on your first or second attempt), you will probably be in excellent shape. Another approach that you might find useful for multistep problems is to do each part individually and check the answer before proceeding to the next part. This allows you to avoid wasting time by making a mistake in one part and then propagating it through the rest of the problem.

The advantage to this difficulty level is that after you obtain experience solving the problems in this book, you will be comfortable solving the problems on the AP exam.

Distribution of the Questions

Not all AP topics are created equal. Some topics (such as Gauss's and Faraday's laws) are tested extensively and are frequently the subject of entire free-response questions. Other topics (such as circuit analysis) are central to solving a range of questions but are generally not the sole topic of a free-response question. Finally, some topics (such as Maxwell's equations) are not tested very extensively.

The distribution of questions in this review book is based on a careful review of decades of AP examinations. Topics that have been tested frequently in the past are reviewed thoroughly, and topics tested infrequently are not reviewed as extensively. Additionally, more attention is paid to topics requiring more practice than others. (For example, although rotational kinematics is frequently tested, extensive practice with rotational kinematics is not crucial because of its similarity to linear kinematics.) We appreciate that the time you have to review for this exam is limited, so we have worked hard to provide you with exactly the material and questions you need to do well on the AP exam.

Derivations

From introductory physics courses, you may have received the impression that derivations aren't very important but rather that only the final results are relevant. This is not the case in Physics C, where the level of sophistication is high enough to require the derivation of most of the important equations. Since these derivations illustrate important principles and problem-solving approaches, we have included them here and urge you to make sure you understand them.

Problem-Solving Strategies to Watch For

As multiple-choice and free-response questions are presented and explained, we have tried to show, often in multiple approaches to the same problem, how various strategies can be applied. A quick way to answer a multiple-choice question, for example, often involves elimination of incorrect choices based on such considerations as dimensional analysis, behavior at extreme values of the system parameters, results of eliminating certain parameters, and the absence of parameters that by the nature of the problem must be included. In the free-response questions, results from previous parts frequently are used in later, more involved, parts; thus, notice how you can get valuable clues as to how to approach part (d) by rereading (and probably using results from) parts (a), (b), and (c). Remember the value of getting a good grasp of the situation described in a problem; using sketches, diagrams, and graphs; thinking of the relationships between given quantities; and relating situations to concepts that might be more familiar (as in rotational and linear motion, in electrical and gravitational fields, and in conservation of elastic and gravitational energy). As you carefully study our solutions, be sure to watch for these and the many other approaches that will improve your abilities to analyze and to understand physics problems and to develop your "physics intuition."

GRAPHICAL ANALYSIS OF DATA

The free-response section of the exam typically includes a question involving a lab experiment and data analysis. We have included such questions in this edition. Here are some general tips for successfully solving these questions.

- You will usually be asked to plot a relationship between quantities that is expected to be linear. You may need to evaluate a suitable function of one of the quantities in order to obtain a linear relation. For an example, see "Mechanics III" of the Diagnostic Test (page 19).
- If you are asked to plot quantity *A* as a function of quantity *B*, then *A* is on the vertical axis and *B* is on the horizontal axis. On the other hand, the problem may simply say to plot the data to test a linear relationship. In that case, it is your choice as to which quantity goes on which axis.
- You will need to examine the range of data points to determine the correct scale on the axes. Your data points should cover at least half of the grid. Your scale must be uniform along the axis; i.e., if neighboring major gridlines differ by 4 units, then this must be true of all neighboring major gridlines.
- Label the axes with the quantity plotted, and be sure to include the units.
- If you are asked to fit the data to a linear relationship, draw a straight line that best follows the trend of the data. Your line need not (and, in general, won't be able to) pass through all of the data points. Do not draw a jagged line that connects neighboring points! Your goal is to draw a single straight line that best approximates the data.

DESCRIPTION OF THE AP PHYSICS C EXAM

You are probably familiar with the concept of AP testing, which provides the opportunity to receive recognition for college-level coursework. While each college and university interprets AP scores (graded from a low of 1 to a high of 5) differently, good scores generally translate into college credits and advanced placement. (Note that you can miss many questions and still earn a score of 5.) Commonly, matriculation into many college courses without having had the experience of studying for the appropriate AP exam may place you at a severe disadvantage. Furthermore, college admissions committees tend to value applicants who have completed AP testing or have scheduled AP courses with the intention of taking the exam. (Note that it is permissable to take an AP exam after studying for it individually, even if your school system does not offer a corresponding course labeled AP.)

The Physics C exam is somewhat different from others you may have encountered. It is divided into two separate 90-minute sections—one on mechanics, the other on electricity and magnetism. These two sections are graded independently, and each is often considered the equivalent of a semester course in college. Although we have difficulty imagining how anyone could resist the beauty of either section, you have the option of taking either the mechanics or the electromagnetism test alone. This organization is reflected in this book, which is similarly divided between the two topics. Each half

NO HARM IN GUESSING!

There is no longer any penalty for wrong answers on the multiple-choice section, so you should answer all multiple-choice questions. Even if you have no idea of the correct answer, you should try to eliminate any obvious incorrect choices, and then guess!

of the book includes the many examples you will need of both question types, the multiple choice and the free response, as well as three complete model tests that mirror the question types and the distribution of topics likely to be encountered. Once the basic material is understood, careful study of the questions and solutions in this book will leave you well prepared for any questions on the actual exam.

Each of these two exams is divided evenly, by time and by contribution to grading, between a multiple-choice section and a free-response section. The 35-question multiple-choice section often lends itself to intuitive understanding, symbolic manipulation, and estimation by orders of magnitude. Both sections of the exams include a Table of Information indicating numerical values of physical constants and conversion factors. Both sections permit the use of calculators and do not require that memories be erased. An included equation table (reproduced at the end of this guide) can be used throughout the exam. While this reference information provides a handy memory check and a source of ideas for solutions, we do not recommend that you rely on it: Many equations are omitted, finding needed equations uses valuable time, and lack of mastery of the basics will hamper your thinking. Concepts from many areas of physics are often involved in each question. Carefully recording your work within the booklet (including the general laws and equations being used) is crucial for full or partial credit, as is the inclusion of units in final answers.

The topics covered on the Physics C exam and the percentage goals listed by the College Board are as follows (as you can take either the mechanics or electricity and magnetism exams separately, we have totaled each part to 100%):

I. Newtonian Mechanics

Topic	Percentage
Kinematics	18%
Newton's laws of motion	20%
Work, energy, and power	14%
Systems of particles, linear momentum	12%
Circular motion and rotation	18%
Oscillations and gravitation	18%

II. Electricity and Magnetism

Topic	Percentage
Electrostatics	30%
Conductors, capacitors, dielectrics	14%
Electric circuits	20%
Magnetic fields	20%
Induction and Maxwell's equations	16%

Further information about the AP Physics C exam can be found at the College Board website, which includes recent tests and an extensive guide to the exam containing sample questions.

OTHER USES OF THIS BOOK

Despite our careful modeling of this book to reflect the Physics C exam, it is also an invaluable aid for anyone taking courses in calculus-based mechanics or electromagnetism. Solving physics problems is what physics courses are generally about, and solving physics problems is what this book will teach you. Since it is hard to imagine any college physics exam that doesn't involve some combination of multiple-choice questions and more extensive problems, many test-taking strategies discussed in this guide should apply as well.

WHAT DO *YOU* THINK?

I am continually striving to improve this review book, so I welcome any questions or comments you might have, particularly those concerning errors or deficiencies in this edition. Please feel free to contact me either by ordinary mail or electronically. Your thoughts are greatly appreciated.

Robert A. Pelcovits
Department of Physics
Box 1843
Brown University
Providence, RI 02912
Robert_Pelcovits@brown.edu

Good luck!

ANSWER SHEET
Diagnostic Test

Mechanics

1.	Ⓐ Ⓑ Ⓒ Ⓓ Ⓔ	10.	Ⓐ Ⓑ Ⓒ Ⓓ Ⓔ	19.	Ⓐ Ⓑ Ⓒ Ⓓ Ⓔ	28.	Ⓐ Ⓑ Ⓒ Ⓓ Ⓔ
2.	Ⓐ Ⓑ Ⓒ Ⓓ Ⓔ	11.	Ⓐ Ⓑ Ⓒ Ⓓ Ⓔ	20.	Ⓐ Ⓑ Ⓒ Ⓓ Ⓔ	29.	Ⓐ Ⓑ Ⓒ Ⓓ Ⓔ
3.	Ⓐ Ⓑ Ⓒ Ⓓ Ⓔ	12.	Ⓐ Ⓑ Ⓒ Ⓓ Ⓔ	21.	Ⓐ Ⓑ Ⓒ Ⓓ Ⓔ	30.	Ⓐ Ⓑ Ⓒ Ⓓ Ⓔ
4.	Ⓐ Ⓑ Ⓒ Ⓓ Ⓔ	13.	Ⓐ Ⓑ Ⓒ Ⓓ Ⓔ	22.	Ⓐ Ⓑ Ⓒ Ⓓ Ⓔ	31.	Ⓐ Ⓑ Ⓒ Ⓓ Ⓔ
5.	Ⓐ Ⓑ Ⓒ Ⓓ Ⓔ	14.	Ⓐ Ⓑ Ⓒ Ⓓ Ⓔ	23.	Ⓐ Ⓑ Ⓒ Ⓓ Ⓔ	32.	Ⓐ Ⓑ Ⓒ Ⓓ Ⓔ
6.	Ⓐ Ⓑ Ⓒ Ⓓ Ⓔ	15.	Ⓐ Ⓑ Ⓒ Ⓓ Ⓔ	24.	Ⓐ Ⓑ Ⓒ Ⓓ Ⓔ	33.	Ⓐ Ⓑ Ⓒ Ⓓ Ⓔ
7.	Ⓐ Ⓑ Ⓒ Ⓓ Ⓔ	16.	Ⓐ Ⓑ Ⓒ Ⓓ Ⓔ	25.	Ⓐ Ⓑ Ⓒ Ⓓ Ⓔ	34.	Ⓐ Ⓑ Ⓒ Ⓓ Ⓔ
8.	Ⓐ Ⓑ Ⓒ Ⓓ Ⓔ	17.	Ⓐ Ⓑ Ⓒ Ⓓ Ⓔ	26.	Ⓐ Ⓑ Ⓒ Ⓓ Ⓔ	35.	Ⓐ Ⓑ Ⓒ Ⓓ Ⓔ
9.	Ⓐ Ⓑ Ⓒ Ⓓ Ⓔ	18.	Ⓐ Ⓑ Ⓒ Ⓓ Ⓔ	27.	Ⓐ Ⓑ Ⓒ Ⓓ Ⓔ		

Electricity and Magnetism

1.	Ⓐ Ⓑ Ⓒ Ⓓ Ⓔ	10.	Ⓐ Ⓑ Ⓒ Ⓓ Ⓔ	19.	Ⓐ Ⓑ Ⓒ Ⓓ Ⓔ	28.	Ⓐ Ⓑ Ⓒ Ⓓ Ⓔ
2.	Ⓐ Ⓑ Ⓒ Ⓓ Ⓔ	11.	Ⓐ Ⓑ Ⓒ Ⓓ Ⓔ	20.	Ⓐ Ⓑ Ⓒ Ⓓ Ⓔ	29.	Ⓐ Ⓑ Ⓒ Ⓓ Ⓔ
3.	Ⓐ Ⓑ Ⓒ Ⓓ Ⓔ	12.	Ⓐ Ⓑ Ⓒ Ⓓ Ⓔ	21.	Ⓐ Ⓑ Ⓒ Ⓓ Ⓔ	30.	Ⓐ Ⓑ Ⓒ Ⓓ Ⓔ
4.	Ⓐ Ⓑ Ⓒ Ⓓ Ⓔ	13.	Ⓐ Ⓑ Ⓒ Ⓓ Ⓔ	22.	Ⓐ Ⓑ Ⓒ Ⓓ Ⓔ	31.	Ⓐ Ⓑ Ⓒ Ⓓ Ⓔ
5.	Ⓐ Ⓑ Ⓒ Ⓓ Ⓔ	14.	Ⓐ Ⓑ Ⓒ Ⓓ Ⓔ	23.	Ⓐ Ⓑ Ⓒ Ⓓ Ⓔ	32.	Ⓐ Ⓑ Ⓒ Ⓓ Ⓔ
6.	Ⓐ Ⓑ Ⓒ Ⓓ Ⓔ	15.	Ⓐ Ⓑ Ⓒ Ⓓ Ⓔ	24.	Ⓐ Ⓑ Ⓒ Ⓓ Ⓔ	33.	Ⓐ Ⓑ Ⓒ Ⓓ Ⓔ
7.	Ⓐ Ⓑ Ⓒ Ⓓ Ⓔ	16.	Ⓐ Ⓑ Ⓒ Ⓓ Ⓔ	25.	Ⓐ Ⓑ Ⓒ Ⓓ Ⓔ	34.	Ⓐ Ⓑ Ⓒ Ⓓ Ⓔ
8.	Ⓐ Ⓑ Ⓒ Ⓓ Ⓔ	17.	Ⓐ Ⓑ Ⓒ Ⓓ Ⓔ	26.	Ⓐ Ⓑ Ⓒ Ⓓ Ⓔ	35.	Ⓐ Ⓑ Ⓒ Ⓓ Ⓔ
9.	Ⓐ Ⓑ Ⓒ Ⓓ Ⓔ	18.	Ⓐ Ⓑ Ⓒ Ⓓ Ⓔ	27.	Ⓐ Ⓑ Ⓒ Ⓓ Ⓔ		

Diagnostic Test

MULTIPLE-CHOICE QUESTIONS

Directions: Each multiple-choice question is followed by five answer choices. For each question, choose the best answer and fill in the corresponding circle on the answer sheet. You may refer to the formula sheet in the Appendix (pages 641–644).

Mechanics

1. A block lies on an inclined plane with angle of elevation θ, as shown in the figure. The inclined plane is frictionless, and the plane is accelerated to the left such that the block's height remains constant. What is the net force on the block?

(A) $mg\cot\theta$
(B) $mg\tan\theta$
(C) $mg\cos\theta$
(D) $mg\sin\theta$
(E) zero

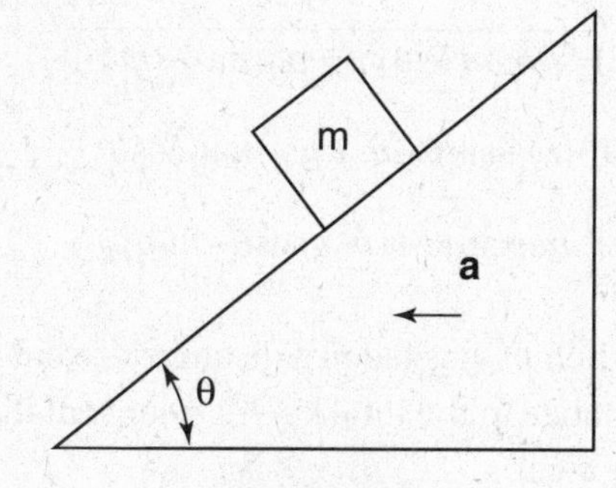

Question 1

2. Consider the Atwood machine shown in the figure, which consists of two masses connected to a pulley. If the pulley's mass is *not* negligible, which of the following is true of the acceleration a of each mass?

(A) $a > \left(\frac{m_1 + m_2}{m_1 - m_2}\right)g$

(B) $a < \left(\frac{m_1 + m_2}{m_1 - m_2}\right)g$

(C) $a > \left(\frac{m_1 - m_2}{m_1 + m_2}\right)g$

(D) $a = \left(\frac{m_1 - m_2}{m_1 + m_2}\right)g$

(E) $a < \left(\frac{m_1 - m_2}{m_1 + m_2}\right)g$

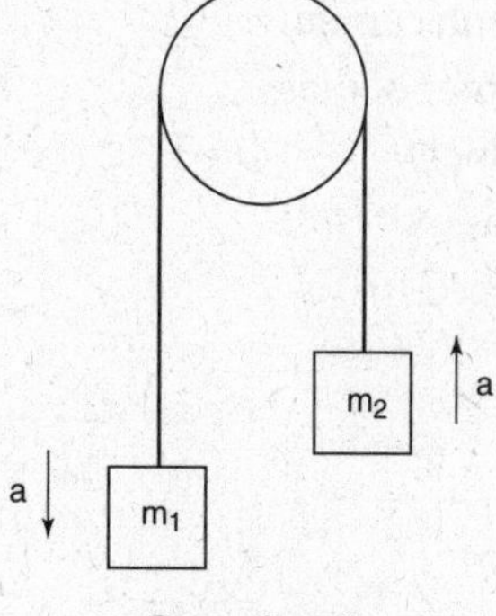

Question 2

GO ON TO THE NEXT PAGE

3. Consider the velocity and acceleration vectors shown in the figure for some object. In which would the object experience the greatest increase in speed over similar time intervals?

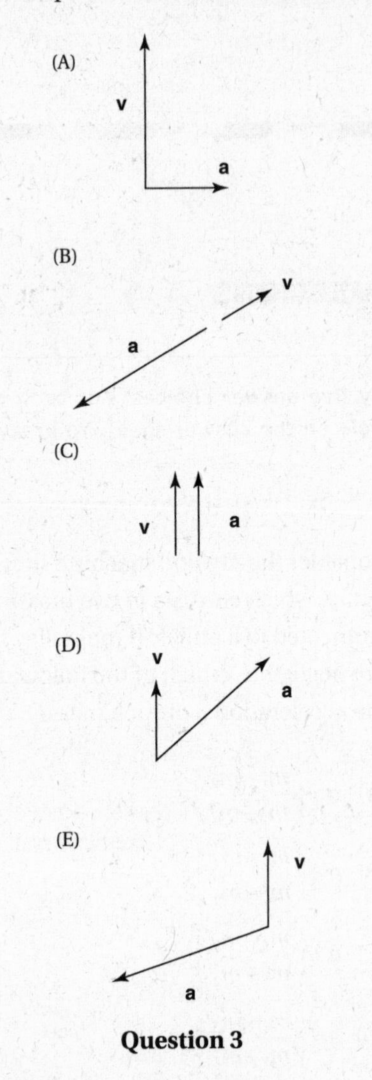

Question 3

4. The moment of inertia (rotational inertia), $\int r^2 dm$, is the rotational analog of

(A) displacement
(B) center of mass
(C) velocity
(D) mass
(E) force

5. A mass m connected to a spring with spring constant k oscillates at a frequency f. If the mass is increased by a factor of 4 and the spring constant is increased by a factor of 2, the new frequency will be

(A) $2f$
(B) $8f$
(C) $\sqrt{2}f$
(D) $f/\sqrt{2}$
(E) none of the above

Questions 6–8 refer to the figure below, which shows two masses connected by a massless pulley. Assume the incline is frictionless.

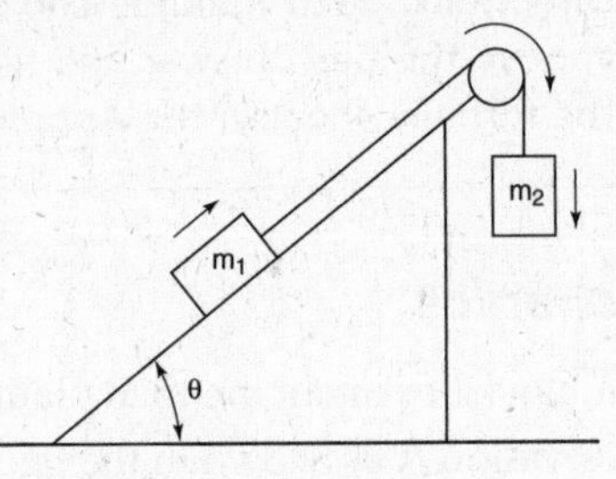

Questions 6–8

6. When the speed of each object is v, what is the total momentum of the system of two masses, m_1 and m_2? (Assume that the x-axis is horizontal and the y-axis is vertical.)

(A) $\frac{1}{2}(m_1 + m_2)v^2$

(B) $m_1 v(\cos\theta)\hat{i} + (m_1 v\sin\theta - m_2 v)\hat{j}$

(C) $\sqrt{[m_1 v\cos\theta]^2 + (m_1 v\sin\theta - m_2 v)^2}$

(D) $(m_1 v\sin\theta - m_2 v)\hat{i} + m_1 v(\cos\theta)\hat{j}$

(E) $m_1 v\sin\theta\hat{i} + (m_1 v\cos\theta - m_2 v)\hat{j}$

7. When m_2 has fallen a distance d, what is the change in the total kinetic energy of the system?

(A) $g(m_2 d + m_1 d\sin\theta)$
(B) $g(m_2 d - m_1 d\sin\theta)$
(C) $g(m_2 d + m_1 d\cos\theta)$
(D) $g(m_2 d - m_1 d\cos\theta)$
(E) Total kinetic energy is conserved.

GO ON TO THE NEXT PAGE

8. Now consider that the inclined plane, which has mass *M* and rests on a frictionless surface, can move as well. When both small masses start from rest, which of the following quantities evaluated for the system of the three masses is conserved?

(A) gravitational potential energy
(B) kinetic energy
(C) *x*-momentum
(D) *y*-momentum
(E) None of the above are conserved for this system.

9. Two objects with equal masses and speeds as shown collide and stick together. What is the final speed of the combined mass?

(A) $v/2$
(B) $v/\sqrt{2}$
(C) v
(D) $v\sqrt{2}$
(E) $2v$

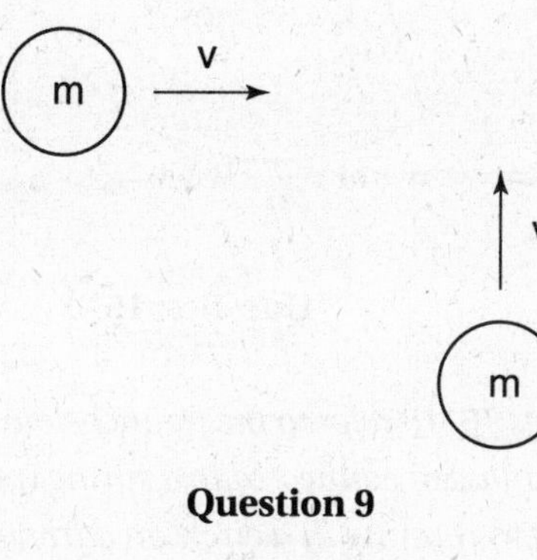

Question 9

10. An object moving with constant acceleration can have how many of the following path types?

I. a linear path
II. a circular path
III. a parabolic path
IV. an elliptical path
V. a spiral path

(A) none
(B) one
(C) two
(D) three
(E) four

11. A mass is pulled up a frictionless incline by a force applied parallel to the incline. The change in the object's kinetic energy is equal to which of the following?

(A) the negative of the change in potential energy
(B) the work done by gravity
(C) the work done by the applied force
(D) the work done by the applied force minus the work done by gravity
(E) the work done by the applied force plus the work done by gravity

12. Which of the following changes does not affect the period of a horizontal mass-spring system undergoing simple harmonic motion?

(A) doubling the mass
(B) attaching an identical spring in parallel (as in the figure)
(C) attaching an identical spring in series (as in the figure)
(D) the presence of friction between the block and the horizontal surface
(E) any manipulation of the masses and springs that does not affect the ratio of the maximum amplitude to the maximum velocity

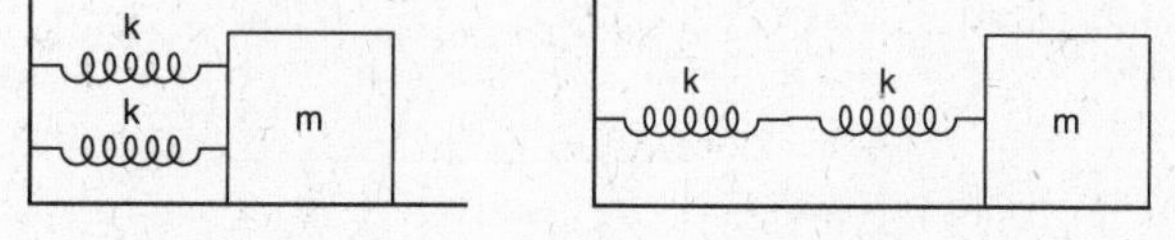

Question 12

GO ON TO THE NEXT PAGE

13. As a block moves up an incline at constant speed, the frictional force performs –5 J of work on the block. As the block moves down the same incline at constant speed over the same distance, how much work does the frictional force perform on the block?

(A) –5 J
(B) 0 J
(C) +5 J
(D) It depends on the angle of elevation of the incline.
(E) It depends on whether the applied force is the same in both situations.

14. The velocity as a function of time for an object undergoing simple harmonic motion is shown. Which of the following graphs shows the potential energy as a function of time?

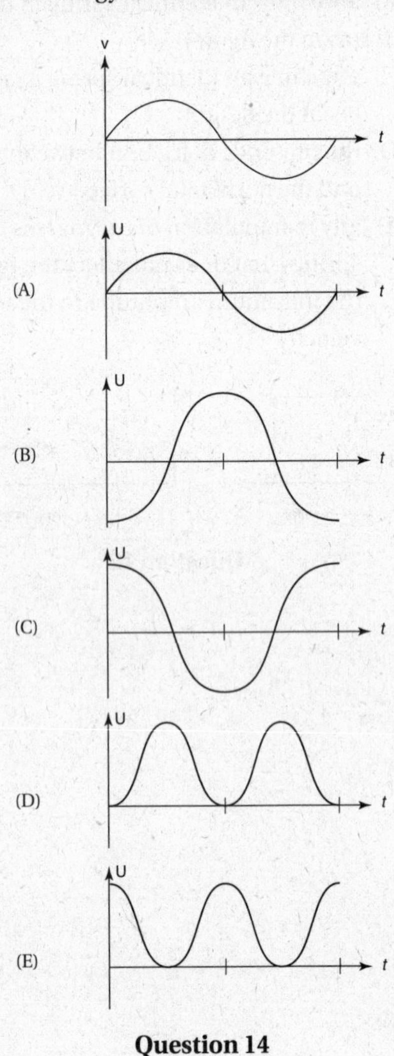

Question 14

15. Consider the two concentric cylinders in the figure, which are fastened to each other and suspended by a pin through their center, which exerts a force of F_1. Two additional forces, F_2 and F_3, act on the cylinders as shown. Which of the following equations is a necessary condition for the two-cylinder system to remain in static equilibrium?

(A) $F_1 - F_3 \sin\phi - F_2 \sin\theta = 0$
(B) $F_1 - F_3 \cos\phi + F_2 \cos\theta = 0$
(C) $F_2 \cos\theta - F_3 \cos\phi = 0$
(D) $F_2(\cos\theta) r_2 - F_3(\cos\phi) r_1 = 0$
(E) $F_2(\cos\theta) r_1 - F_3(\cos\phi) r_2 = 0$

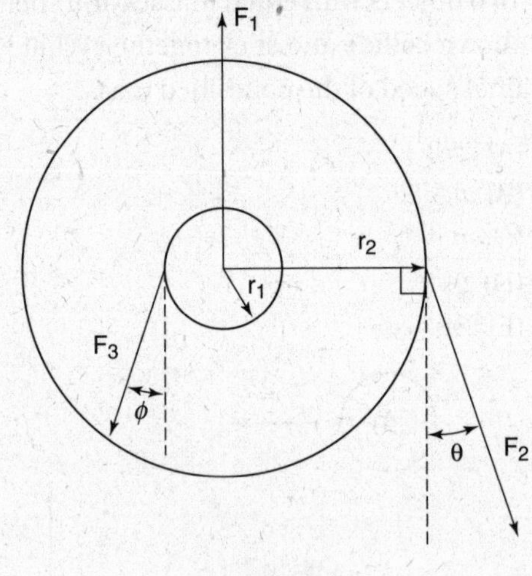

Question 15

Questions 16–18 refer to the figure below, which shows a mass m *connected to a spring (of spring constant* k*) rotating in a circle on a frictionless table at constant speed.*

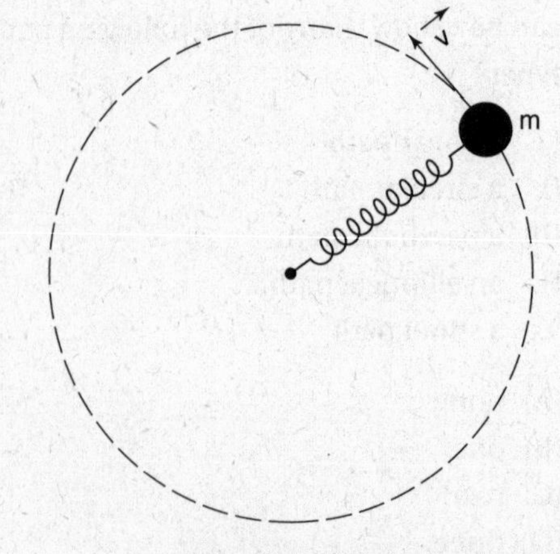

Questions 16–18

GO ON TO THE NEXT PAGE

16. If x is the extension of the spring from its unstretched length, v is the speed of the mass, and r is the radius of the circular path, which of the following equations is valid?

(A) $kx = mv^2/2$
(B) $kx^2/2 = mv^2/2$
(C) $kx^2/2 = mv^2/r$
(D) $kx = mv^2/r$
(E) None of the above equations are valid.

17. In terms of the mass m, the spring constant k, the extension of the spring x, and the radius of the circle r, what is the rate at which the spring does work on the mass?

(A) $kx\sqrt{kxr/m}$
(B) $kx\sqrt{m/kxr}$
(C) $kx(kxr/m)$
(D) $kx(m/kxr)$
(E) zero

18. Imagine that the spring in the figure is suddenly cut, causing the mass to continue along a path tangential to its motion at the instant the spring is cut. What is the change in the total energy of the system of the mass and spring caused by cutting the spring? (Express your answer in terms of the quantities given in question 16 and neglect the mass of the spring.)

(A) $-\frac{1}{2}kx^2$
(B) $-\frac{1}{2}kxr$
(C) $-\frac{1}{2}kx$
(D) $-\frac{1}{2}kx^3$
(E) No energy is lost.

19. Consider a disk that has uniform density and a hole as shown. Where is the center of mass of this disk located?

(A) point A
(B) point B
(C) point C
(D) point D
(E) point E

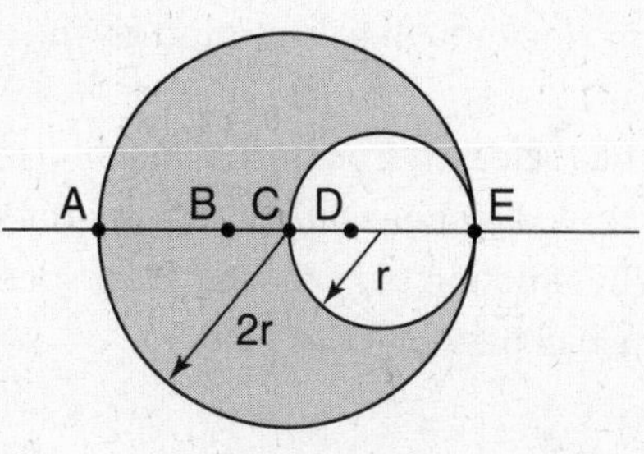

Question 19

20. A mass m is suspended by two ropes as shown. What is the tension in each of the ropes?

(A) $F_T = \frac{1}{2}mg\sin\theta$
(B) $F_T = \frac{1}{2}mg\csc\theta$
(C) $F_T = \frac{1}{2}mg\cos\theta$
(D) $F_T = \frac{1}{2}mg\sec\theta$
(E) $F_T = 2\,mg\cos\theta$

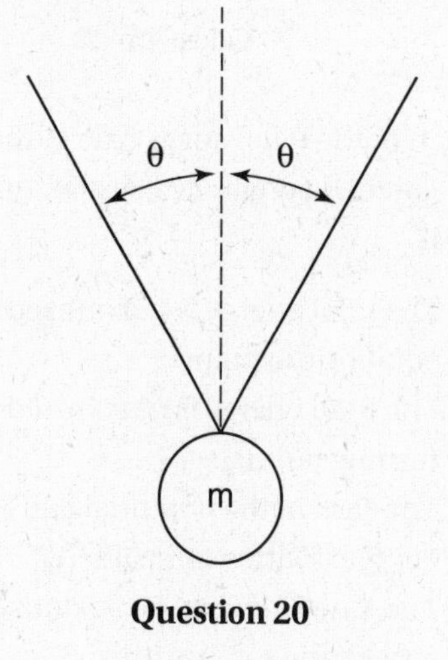

Question 20

GO ON TO THE NEXT PAGE

21. What is the maximum speed of a simple pendulum with length l whose maximum angle of displacement from the vertical is θ?

(A) $\sqrt{gl\cos\theta}$
(B) $\sqrt{2gl\cos\theta}$
(C) $\sqrt{gl(1-\cos\theta)}$
(D) $\sqrt{2gl(1-\cos\theta)}$
(E) Not enough information given.

22. A ball released from rest at height h rolls down a curved incline without slipping, as shown. What fraction of its final angular speed ω_f does the ball have at point P?

(A) $\omega_p = \omega_f / 4$
(B) $\omega_p = \omega_f / (2\sqrt{2})$
(C) $\omega_p = \omega_f / 2$
(D) $\omega_p = \omega_f / \sqrt{2}$
(E) $\omega_p = \omega_f$

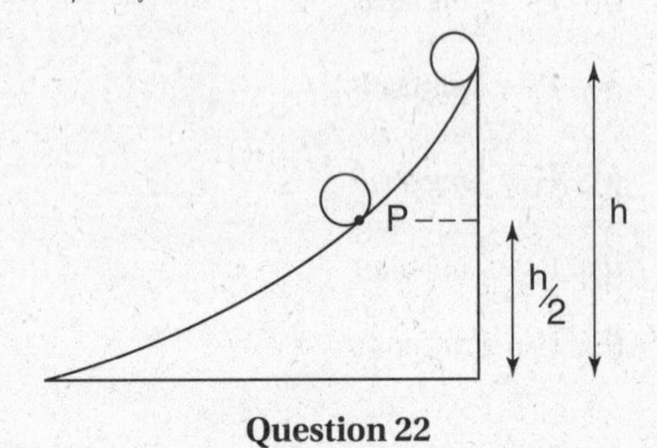

Question 22

23. Which of the following is true of the potential and kinetic energies in a system undergoing SHM?

(A) The total energy is maximized at the equilibrium point.
(B) The total energy is maximized at the turning points.
(C) The sum of the potential and kinetic energies varies sinusoidally.
(D) The kinetic energy is maximized when the acceleration is maximized.
(E) The maximum potential energy equals the maximum kinetic energy.

24. Consider a planet in an elliptical orbit around the sun as shown. Which of the following is the correct ratio of the magnitudes of the planet's acceleration at points A and B?

(A) $a_A / a_B = r_A / r_B$
(B) $a_A / a_B = r_B / r_A$
(C) $a_A / a_B = r_A^2 / r_B^2$
(D) $a_A / a_B = r_B^2 / r_A^2$
(E) $a_A / a_B = \sqrt{r_B / r_A}$

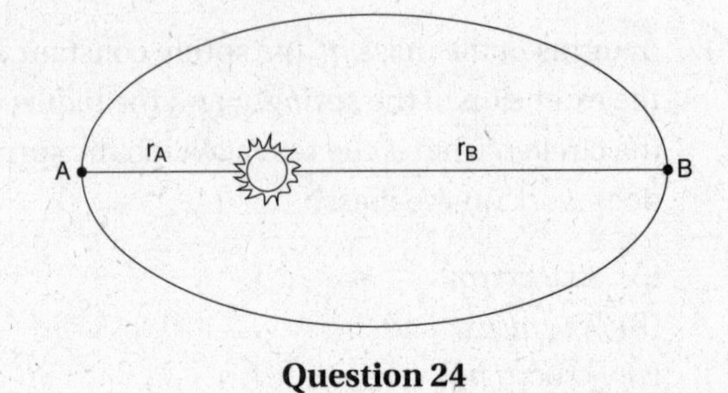

Question 24

25. If an object's position is given by the equation $x = vt$, the object's kinetic energy is given by the equation

(A) $\frac{1}{2}mv^2t^2$
(B) $\frac{1}{2}mx^2$
(C) $\frac{1}{2}mv^2$
(D) $\frac{1}{2}mx^3/t^2$
(E) none of the above

GO ON TO THE NEXT PAGE

26. In order to swim directly across a stream with water moving with speed v_{stream} with respect to the ground as shown, which of the following is required of the magnitude of the swimmer's speed with respect to the water, $v_{swimmer}$?

(A) $v_{swimmer} = v_{stream}$
(B) $v_{swimmer} > v_{stream}$
(C) $v_{swimmer} > \sqrt{2}v_{stream}$
(D) $v_{swimmer} > v_{stream}/\sqrt{2}$
(E) $v_{swimmer} > 0$

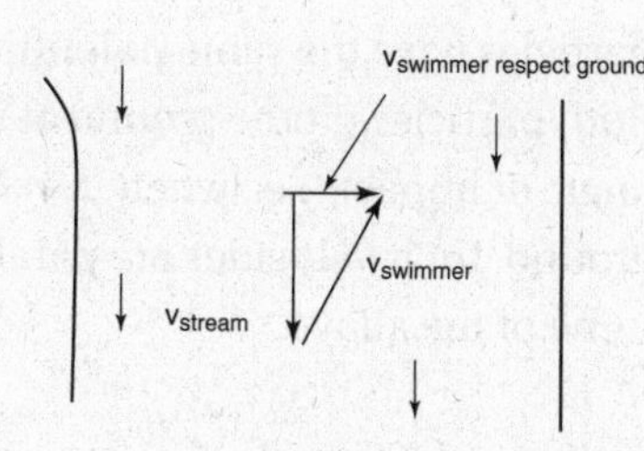

Question 26

27. What is the effective spring constant of two identical springs (each of spring constant k) attached in series (end-to-end—see question 12)?

(A) $k/2$
(B) $k/\sqrt{2}$
(C) k
(D) $k\sqrt{2}$
(E) $2k$

28. If the angular velocity of a disk is given in rads by the equation $\omega(t) = 9 + 3t$, what is the disk's angular displacement between $t = 0$ and $t = 1$ s?

(A) 0.75 rad
(B) 9 rad
(C) 8.25 rad
(D) 10.5 rad
(E) 18 rad

29. You open a 1.0 m-wide door by pushing on the edge farthest from the hinge. You apply a force of 2.0 N at right angles to the front of the door. If you rotate the door 0.5 radians in 0.5 s, at what rate have you done work on the door?

(A) 0.5 J/s
(B) 4 J/s
(C) 0.25 J/s
(D) −2.0 J/s
(E) 2.0 J/s

30. A projectile is launched vertically upward into the air. When is the rate of work done by gravity on the projectile maximized?

(A) at the top of the trajectory
(B) right after the particle is launched
(C) right before the particle hits the ground
(D) The rate is constant.
(E) The rate is zero.

31. A planet moves in an elliptical orbit around the sun (which is assumed to be stationary) as shown. What is the planet's speed v_2 at the point of closest approach to the sun?

(A) $v_2 = v_1$
(B) $v_2 = v_1 r_2 / r_1$
(C) $v_2 = (v_1 r_1 \sin\theta) / r_2$
(D) $v_2 = (v_1 r_1 \cos\theta) / v_2$
(E) $v_2 = (v_1 r_1 \tan\theta) / v_2$

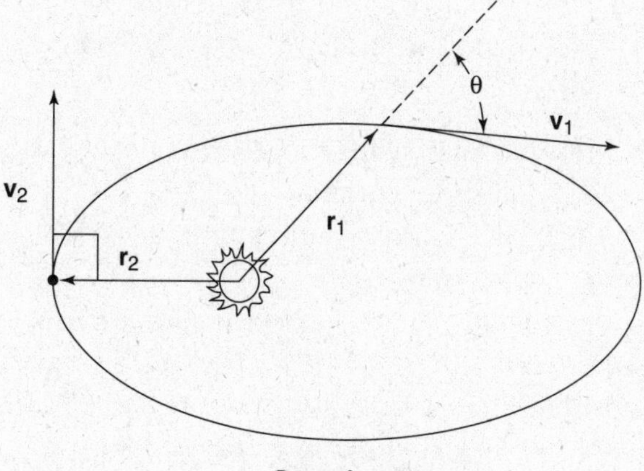

Question 31

GO ON TO THE NEXT PAGE

DIAGNOSTIC TEST

32. The potential function for an object in a region of space where there are no nonconservative forces is $U(x) = -ax^2 + bx - c$. What is the force on the object in this region?

(A) $F(x) = 2ax - b$
(B) $F(x) = -2ax + b$
(C) $F(x) = m(2ax - b)$
(D) $F(x) = m(-2ax + b)$
(E) $F(x) = (ax^3/3) - (bx^2/2) + cx$

33. If an object's position is given by the equation $x(t) = 2 + 4t - \frac{1}{2}t^2$ (where position is in meters and time is in seconds), what is the magnitude of the object's velocity at time $t = 1$?

(A) 1 m/s
(B) 2 m/s
(C) 3 m/s
(D) 4 m/s
(E) 6 m/s

34. Two particles are launched from the top of a building at the same time with the same initial speed. One particle is launched with an initial velocity pointing θ degrees *above* the horizontal, while the other particle is launched with an initial velocity pointing θ degrees *below* the horizontal, as shown. Discounting air resistance, which of the following is true?

(A) Both particles spend the same amount of time in the air.
(B) Both particles hit the ground the same distance from the building.
(C) At any particular instant in time, both particles have the same potential energy.
(D) Both particles hit the ground at the same angle of impact (i.e., when they hit the ground, their velocities are parallel).
(E) none of the above

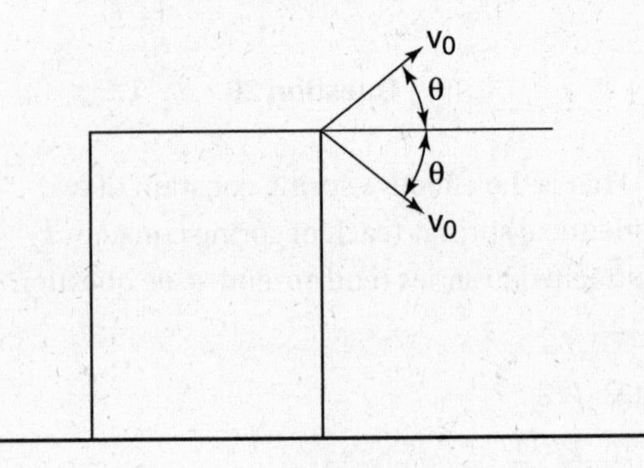

Question 34

GO ON TO THE NEXT PAGE

35. A figure skater is spinning without friction when he extends his arms, moving his hands farther from his axis of rotation and decreasing his angular velocity by a factor of 2. Which of the following statements is true?

(A) Both his angular momentum and kinetic energy are constant.
(B) His angular momentum decreases by a factor of 2, while his kinetic energy remains constant.
(C) His angular momentum remains constant, while his kinetic energy decreases by a factor of 4.
(D) His angular momentum remains constant, while his kinetic energy decreases by a factor of 2.
(E) His angular momentum remains constant, while his kinetic energy increases by a factor of 2.

DIAGNOSTIC TEST

GO ON TO THE NEXT PAGE

FREE-RESPONSE QUESTIONS

Directions: Use separate sheets of paper to write down your answers to the free-response questions. You may refer to the formula sheet in the Appendix (pages 641–644).

Mechanics

MECHANICS I

A pulley consists of two cylinders of radii r and $2r$ (see figure) with a total rotational inertia of I. Mass m_2 is sliding along a frictionless incline with an angle of elevation of θ. Assume that both m_1 and m_2 have the same mass m.

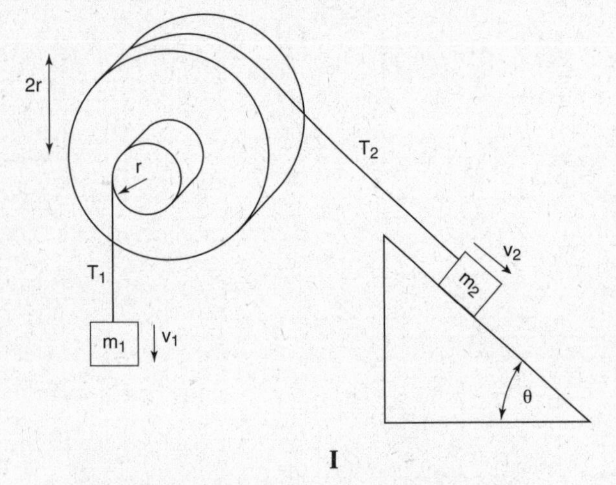

I

(a) Given that the angular acceleration of the pulley is zero (and assuming that the rope rolls without slipping over the pulley), what is the angle of elevation of the inclined plane?

(b) If the mass on the inclined plane is sliding down the ramp with a speed of v_2, what is the speed of the hanging mass?

Then the rope attaching m_2 to the pulley is cut, causing the hanging mass to accelerate.

(c) What is the acceleration of the hanging mass, m_1, under these conditions?

(d) What is the rate at which the kinetic energy of the pulley and m_1 is changing? (Your answer can include the angular speed of the pulley, ω.)

MECHANICS II

A roller coaster (mass m), starting from rest at a height h, goes around a circular loop of radius r as shown.

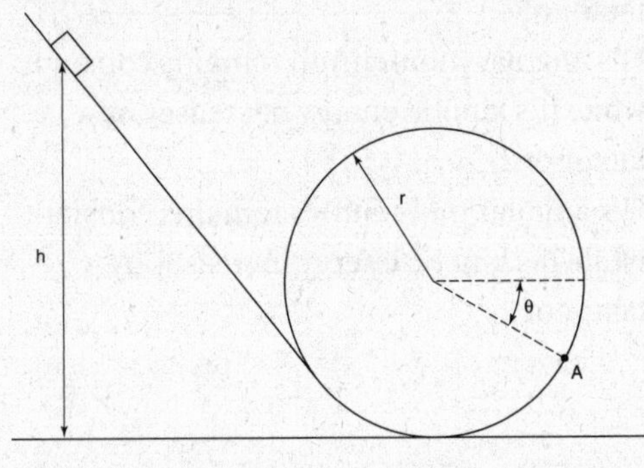

II

(a) If the normal force exerted on the roller coaster by the track at the top of the loop is zero, what is the roller coaster's speed at this point?

(b) Assuming that the track is frictionless, at what height h must the roller coaster have started?

(c) Draw a free-body diagram for the roller coaster as it passes point A.

(d) What is the object's speed as it passes through point A?

GO ON TO THE NEXT PAGE

MECHANICS III

A horizontal force F acts on a mass m sliding along an inclined plane with a coefficient of kinetic friction μ_k, as shown.

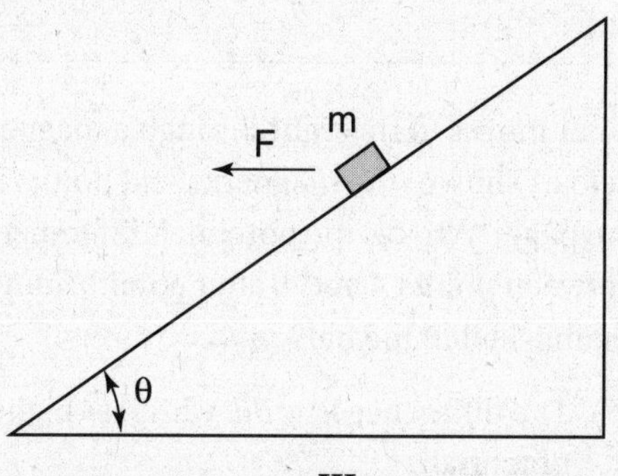

III

(a) Draw a free-body diagram indicating all the forces acting on the mass.

(b) Assuming that the mass slides downward along the incline, what is its acceleration?

(c) What is the maximum force F that can be applied without causing the block to lift off the incline?

(d) An experiment is carried out to check the prediction of part (c) for a mass of 1 kg. The angle of the incline is varied, and the value of the maximum force F that can be applied without causing the block to lift is measured using a spring scale. The following data are obtained:

θ (degrees)	20	30	45	60
F (N)	26.5	17	10	5.7

Which quantities should be plotted on the grid below to yield a straight line according to part (c)?

(e) Plot the data, fit the points as well as possible to a straight line, and check your prediction from part (c).

GO ON TO THE NEXT PAGE

MULTIPLE-CHOICE QUESTIONS

Directions: Each multiple-choice question is followed by five answer choices. For each question, choose the best answer and fill in the corresponding circle on the answer sheet. You may refer to the formula sheet in the Appendix (pages 641–644).

Electricity and Magnetism

1. Two metal spheres with radii r_1 and r_2 carrying positive charges Q_1 and Q_2 are separated by a very large distance. Suppose that a very thin wire connects the two spheres. Under what conditions will positive charge flow from sphere 1 to sphere 2?

   (A) $Q_1/4\pi\varepsilon_0 r_1^2 > Q_2/4\pi\varepsilon_0 r_2^2$
   (B) $Q_1/4\pi\varepsilon_0 r_1^2 < Q_2/4\pi\varepsilon_0 r_2^2$
   (C) $Q_1/4\pi\varepsilon_0 r_1 > Q_2/4\pi\varepsilon_0 r_2$
   (D) $Q_1/4\pi\varepsilon_0 r_1 < Q_2/4\pi\varepsilon_0 r_2$
   (E) $Q_1/4\pi\varepsilon_0 r_2 > Q_2/4\pi\varepsilon_0 r_1$

2. A charged particle is launched into a region of uniform magnetic field. Given that the particle's initial velocity has components parallel and perpendicular to the magnetic field, describe the particle's path qualitatively.

   (A) a straight line
   (B) a circle
   (C) a helical (corkscrew) path
   (D) a parabolic path
   (E) none of the above

3. A point charge $+Q$ is located at a fixed distance from a sphere. Compare the magnitude of the attractive force on the sphere if the sphere is (1) metal with net charge $-Q$, (2) metal with no net charge, (3) an insulator with net charge $-Q$, or (4) an insulator with no net charge.

   (A) charged metal = charged insulator > uncharged metal = uncharged insulator
   (B) charged metal > charged insulator > uncharged metal > uncharged insulator
   (C) charged insulator > charged metal > uncharged insulator > uncharged metal
   (D) charged insulator > charged metal > uncharged metal = uncharged insulator
   (E) charged insulator = charged metal > uncharged metal > uncharged insulator

4. A bar moves to the right through a magnetic field as shown (the magnetic field points into the page). What is the potential difference between points *A* and *B* after equilibrium is established in the bar?

   (A) The difference is v^2Bd, with *A* at higher potential.
   (B) The difference is v^2Bd, with *B* at higher potential.
   (C) The difference is vBd, with *A* at higher potential.
   (D) The difference is vBd, with *B* at higher potential.
   (E) Because there is no loop, there is no flux, so there is no potential difference.

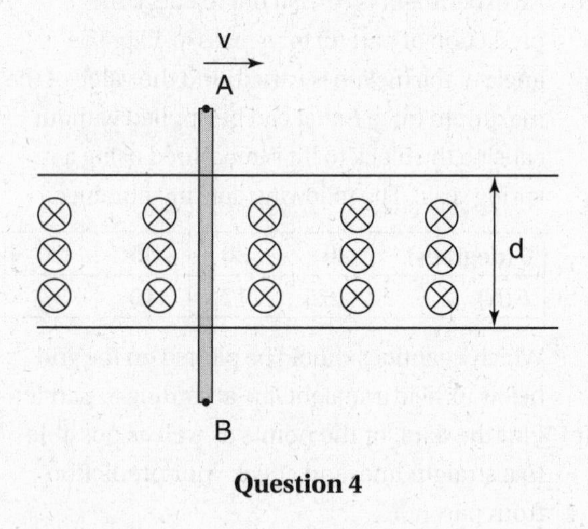

Question 4

GO ON TO THE NEXT PAGE

5. Suppose that the electric potential in a region of space is a constant. What can you conclude about the magnitude of the electric field in this region?

   (A) It is a nonzero constant.
   (B) It is zero.
   (C) It is proportional to the distance from the origin.
   (D) It is given by $E = \frac{1}{4\pi\varepsilon_0 r^2}$.
   (E) Not enough information is given.

Questions 6–8 refer to a circular region of radius a *with a uniform magnetic field pointing into the page that is increasing at a constant rate as shown. A circular loop of wire of radius* r *is centered on the middle of the region.*

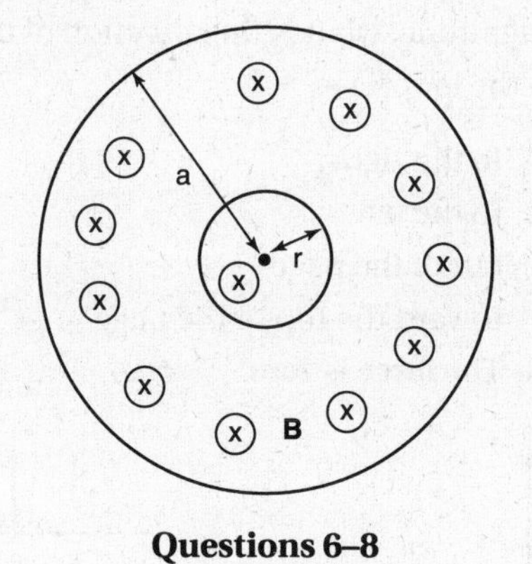

Questions 6–8

6. Which of the graphs plots the induced EMF in the circular loop of wire shown as a function of its radius r?

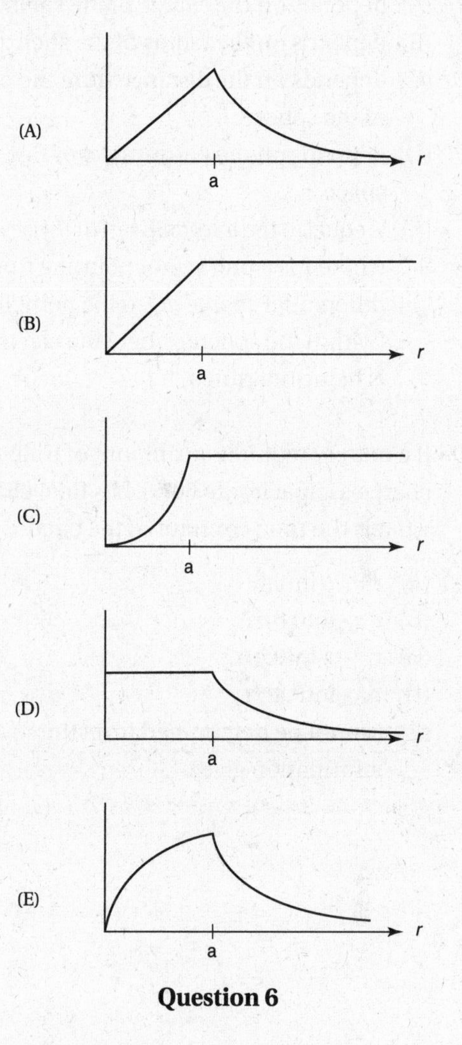

Question 6

7. Given that the resistance of the loop is proportional to the loop's circumference, which of the graphs above shows the induced current as a function of the radius of the loop?

8. Suppose that, instead of a magnetic field of increasing magnitude pointing into the page, a constant current of uniform current density flows into the page through a wire of radius a. Which graph above shows the magnitude of the magnetic field as a function of the radius, $B(r)$?

GO ON TO THE NEXT PAGE

9. The magnitude of the electric field inside a uniformly charged spherical shell

(A) depends on the charge of the shell
(B) depends on the radius of the shell
(C) depends on the distance from the center of the sphere, r
(D) depends on the permittivity of free space, ε_0
(E) is equal to the integral $\mathbf{E} = \int (dQ/4\pi\varepsilon_0 r^2)\hat{r}$, where $\hat{r}$ is a unit vector pointing from the differential charge dQ to the point in space (within the sphere) where the electric field is being measured

10. If a battery requires an amount of time t to charge a capacitor to 60% of its final charge, what is the time constant of the circuit?

(A) $\tau = -t/\ln(0.6)$
(B) $\tau = -\ln(0.6)/t$
(C) $\tau = -t/\ln(0.4)$
(D) $\tau = -\ln(0.4)/t$
(E) cannot be determined from the information given

11. Which of the following statements is true of the integral $\int_{\text{point } A}^{\text{point } B} \mathbf{E} \cdot d\mathbf{r}$?

(A) It is equal to the potential at point B minus the potential at point A.
(B) It is always independent of the path taken from point A to point B.
(C) It is the work per unit charge that the electric field performs as a charge moves from point A to point B.
(D) It is the work per unit charge that an external force performs to move a charge from point A to point B.
(E) It is the induced EMF.

12. A wire loop lies partially in a region where the magnetic field points out of the page, as shown. If the magnetic field is increasing in magnitude, what is the direction of the force on the loop?

(A) to the right
(B) to the left
(C) out of the page
(D) toward the top of the page
(E) The force is zero.

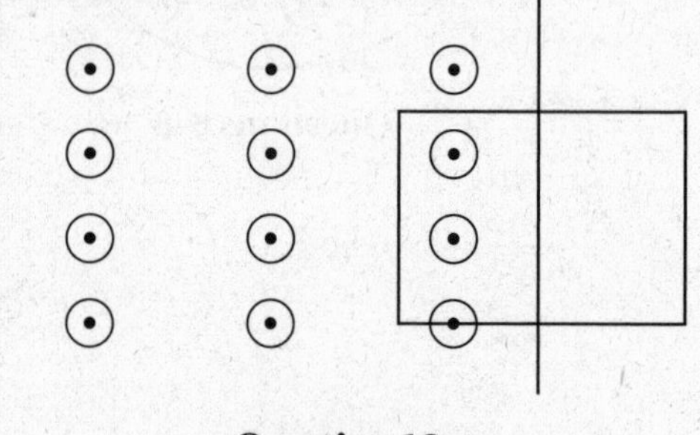

Question 12

GO ON TO THE NEXT PAGE

13. Suppose that, rather than an electron orbiting a proton as in atoms, an electron and a positron (a particle with the same mass as an electron and of opposite charge) orbit each other as shown. In terms of the magnitude of the charge on an electron e and the radius of the orbit r, what is the total kinetic energy of the system?

(A) $KE = e/4\pi\varepsilon_0 r$
(B) $KE = e^2/4\pi\varepsilon_0 r^2$
(C) $KE = -e^2/8\pi\varepsilon_0 r^2$
(D) $KE = -e^2/8\pi\varepsilon_0 r$
(E) $KE = e^2/16\pi\varepsilon_0 r$

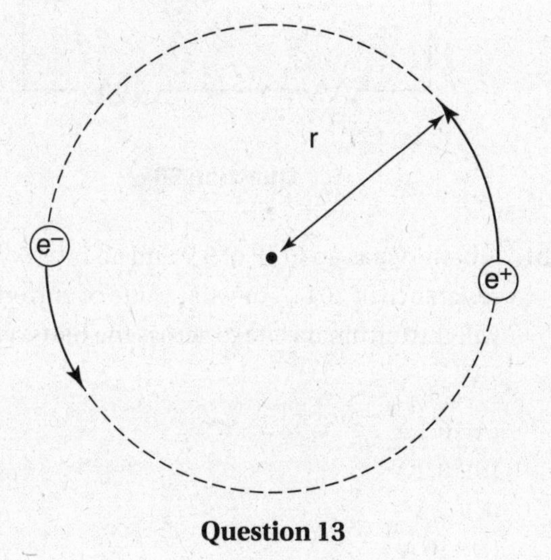

Question 13

14. Which one of Maxwell's equations states the experimental result that physicists have been unable to discover a magnetic monopole?

(A) $\oint \mathbf{E} \cdot d\mathbf{A} = Q_{\text{enclosed}}/\varepsilon_0$

(B) $\oint \mathbf{B} \cdot d\mathbf{s} = \mu_0 I_{\text{enclosed}} + \mu_0 \varepsilon_0 (d\Phi_E/dt)$

(C) $\oint \mathbf{B} \cdot d\mathbf{A} = 0$

(D) $\oint \mathbf{E} \cdot d\mathbf{s} = -d\Phi_B/dt = \text{EMF}$

(E) $\oint \mathbf{B} \cdot d\mathbf{s} = \mu_0 I + \mu_0 I_{\text{enclosed}}$

15. Two resistors with different resistances are connected in parallel. Which of the following statements is correct about this resistor network?

(A) The equivalent resistance is greater than the resistance of either of the resistors alone.
(B) The equivalent resistance is in between the resistance of either of the resistors alone.
(C) The equivalent resistance is smaller than the resistance of either of the resistors alone.
(D) The same amount of current passes through both resistors.
(E) The equivalent resistance depends on the potential difference across the resistor array.

16. The uniform electric field established between the parallel plates of a capacitor has magnitude E. A particle with a charge of magnitude q is released (initially at rest) somewhere between the plates and allowed to move a distance d under the influence of the capacitor's electric field. What is the change in the particle's electrostatic potential energy?

(A) $\Delta U = q^2 Ed$
(B) $\Delta U = -q^2 Ed$
(C) $\Delta U = qEd$
(D) $\Delta U = -qEd$
(E) none of the above

17. An isolated solid metal sphere contains a net charge of $+Q$. The magnitude of the electric field outside the sphere a distance d from the center of the sphere

(A) cannot be calculated because there isn't appropriate symmetry to apply Gauss's law
(B) is equal to $Q/4\pi\varepsilon_0 d^2$
(C) is equal to $Q/2\pi\varepsilon_0 d^2$
(D) is equal to $Q/2\sqrt{2}\pi\varepsilon_0 d^2$
(E) none of the above

GO ON TO THE NEXT PAGE

18. Point P is located within a cylindrical wire that carries current into the page with uniform current density as shown. What is the direction of the magnetic field at point P?

(A) The magnetic field is zero because of symmetry.
(B) toward the top of the page
(C) toward the bottom of the page
(D) into the page
(E) out of the page

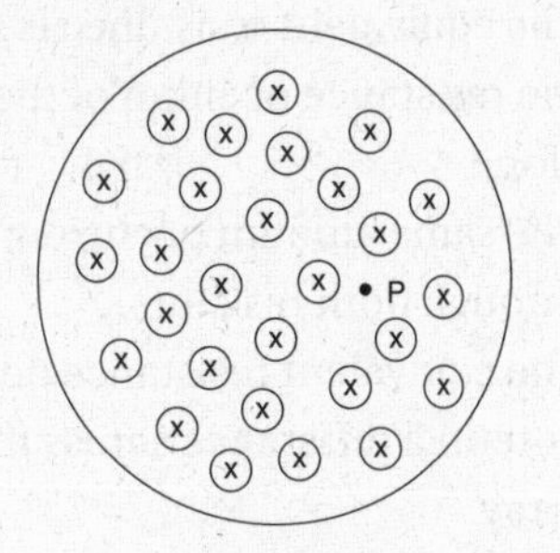

Question 18

19. A point charge $+q$ lies at the origin. Which of the following charge distributions could yield a zero net electric field (due to the $+q$ charge and the charge distribution) at the point (0, 1)?

I. a single charge $-Q$ with a smaller magnitude than $+q$, lying at some point along the x-axis
II. a single charge $-q$ with a magnitude equal to $+q$, lying at some point along the x-axis
III. a single charge $-Q$ with a larger magnitude than $+q$, lying at some point along the x-axis
IV. two charges both with charge $-Q$ and with larger magnitudes than $+q$, lying at points along the x-axis

(A) II
(B) III
(C) IV
(D) I and III
(E) II and IV

20. After closing the switch in the figure, the current I passing through the inductor increases from zero. What is the Kirchhoff's loop equation describing this situation?

(A) $V + IR + L(dI/dt) = 0$
(B) $V + IR - L(dI/dt) = 0$
(C) $V - IR + L(dI/dt) = 0$
(D) $V - IR - L(dI/dt) = 0$
(E) none of the above

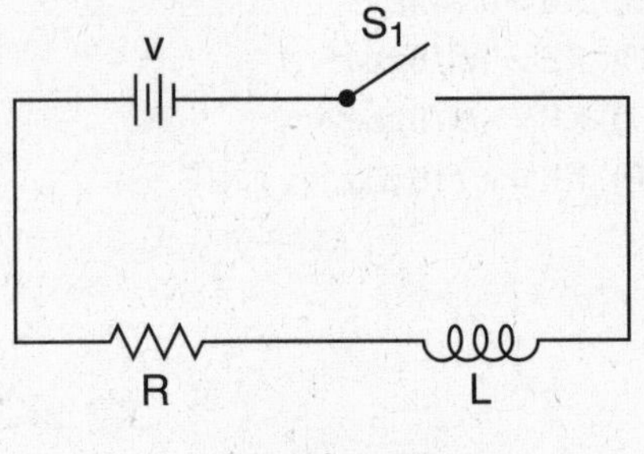

Question 20

21. A battery has an EMF of 9 V and an internal resistance of 20 Ω. For what value of current will the terminal voltage across the battery be 7 V?

(A) 0.1 A
(B) 0.2 A
(C) 5 A
(D) 10 A
(E) 40 A

22. Consider a parallel plate vacuum capacitor. If the plate area and the separation distance are doubled and a dielectric with dielectric constant $k_D = 2$ is inserted between the plates, completely filling the space, the capacitance will

(A) increase by a factor of 4
(B) increase by a factor of 2
(C) remain the same
(D) decrease by a factor of 2
(E) decrease by a factor of 4

GO ON TO THE NEXT PAGE

23. How much work is done by the electric field on a point charge of magnitude q moved from infinity into the center of a hollow spherical shell of radius R that has a uniform charge density and a net charge of $+Q$? (Assume there is a very small hole in the shell that allows the point charge to enter but does not disturb the spherical symmetry of the charge distribution.)

(A) $W = qQ/4\pi\varepsilon_0 R$
(B) $W = qQ/4\pi\varepsilon_0 R^2$
(C) $W = -qQ/4\pi\varepsilon_0 R$
(D) The work is undefined.
(E) The work is zero because of symmetry considerations.

24. Which unit is equal to one ohm divided by one henry?

(A) coulomb
(B) joule
(C) second
(D) second^{-1}
(E) farad

25. Suppose that when a charge is moved through a region of space with an electric field (which is produced by stationary charges), the field does no net work on the charge. Which of the following statements *must* be true?

(A) The electric force is nonconservative.
(B) The charge's path begins and ends along the same electric field line.
(C) The charge is moved along an equipotential line.
(D) The charge's path begins and ends on an equipotential line of the same value.
(E) Electric fields never do work on charges.

26. Consider a long solenoid with length l, radius r, and n turns per length. What effect will doubling the radius (while holding l and n constant) have on the *inductance* of the solenoid?

(A) Inductance will decrease by a factor of 4.
(B) Inductance will decrease by a factor of 2.
(C) Inductance will not change.
(D) Inductance will increase by a factor of 2.
(E) Inductance will increase by a factor of 4.

27. In the circuit shown, if the potential at point A is chosen to be zero, what is the potential at point B?

(A) –4 V
(B) +4 V
(C) –2.5 V
(D) 2.5 V
(E) none of the above

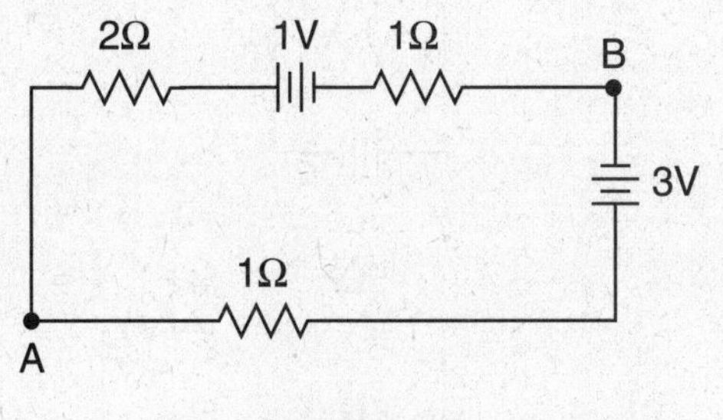

Question 27

28. What is the power dissipated in the resistors in the circuit introduced in question 27?

(A) 1 W
(B) 2 W
(C) 4 W
(D) 8 W
(E) 16 W

GO ON TO THE NEXT PAGE

Questions 29–31 refer to the circuits shown in the figure, in which all the batteries are identical.

(A) 3Ω

(B)

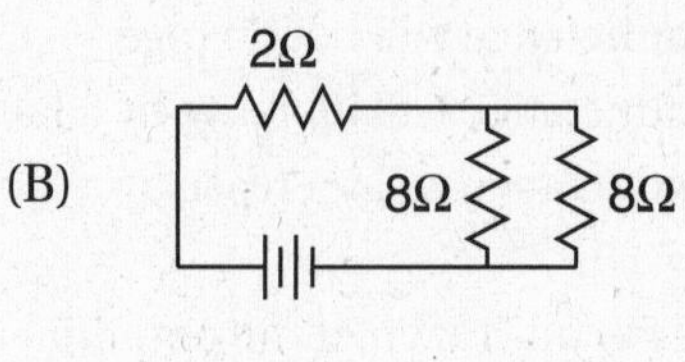

(C)

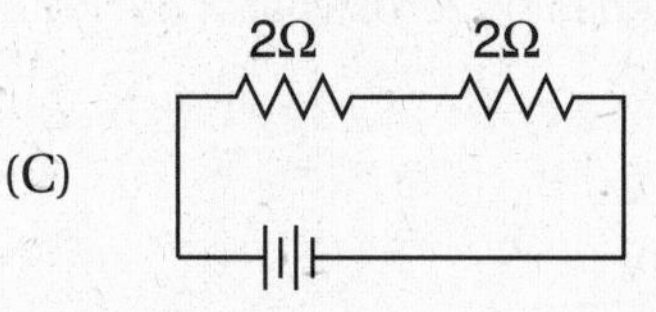

(D)

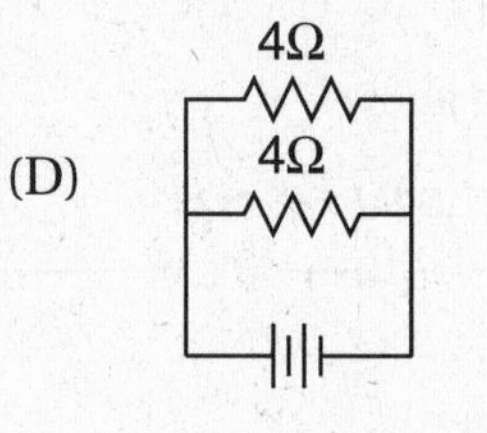

(E)

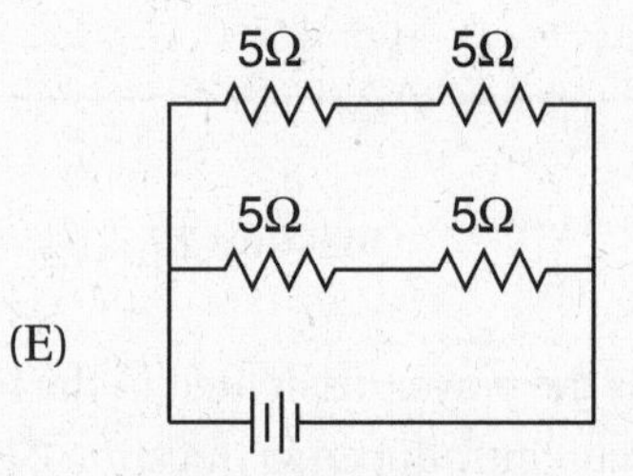

Questions 29–31

29. In which circuit is the equivalent resistance the greatest?

30. In which of the five circuits is the total power dissipated by all the resistors the greatest?

31. Of all of the circuits shown, which one contains the resistor with the smallest potential difference across it?

32. Which of the following statements is a necessary condition for the electric flux through a gaussian surface to be zero?

    (A) The electric field is zero along the entire gaussian surface.
    (B) The component of the field perpendicular to the gaussian surface is zero along the entire gaussian surface.
    (C) There is no charge enclosed within the gaussian surface.
    (D) There is no net charge enclosed within the gaussian surface.
    (E) Although the above conditions *would* cause the flux to be zero, they are not *required* for the flux to be zero.

33. Two moving particles of equal mass enter a region of uniform magnetic field pointing out of the page, causing them to move as shown. Which of the following is true about the charges on these particles?

    (A) $|Q_A| > |Q_B|$, $Q_A > 0$
    (B) $|Q_A| > |Q_B|$, $Q_A < 0$
    (C) $|Q_A| < |Q_B|$, $Q_A > 0$
    (D) $|Q_A| < |Q_B|$, $Q_A < 0$
    (E) none of the above

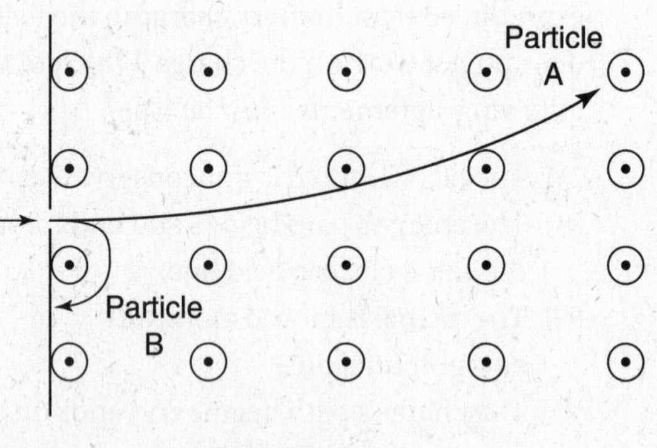

Question 33

GO ON TO THE NEXT PAGE

34. Given the electric field vector at a point in space, which of the following can be determined?

I. the direction of the electric force on any charged particle (of unspecified charge) at that point
II. the magnitude of the electric force on a particle with a charge of magnitude Q at that point
III. the rate of change of the potential with respect to position in the direction parallel to the field
IV. the rate of change of the potential with respect to position in the direction perpendicular to the field

(A) I and II
(B) III and IV
(C) I, III, and IV
(D) II, III, and IV
(E) I, II, III, and IV

35. In which of the situations shown is the force on the loop of wire greatest? (Assume that the dimensions of the figures and the strengths of the magnets are the same in all cases and that the magnet lies along the axis of the loop.)

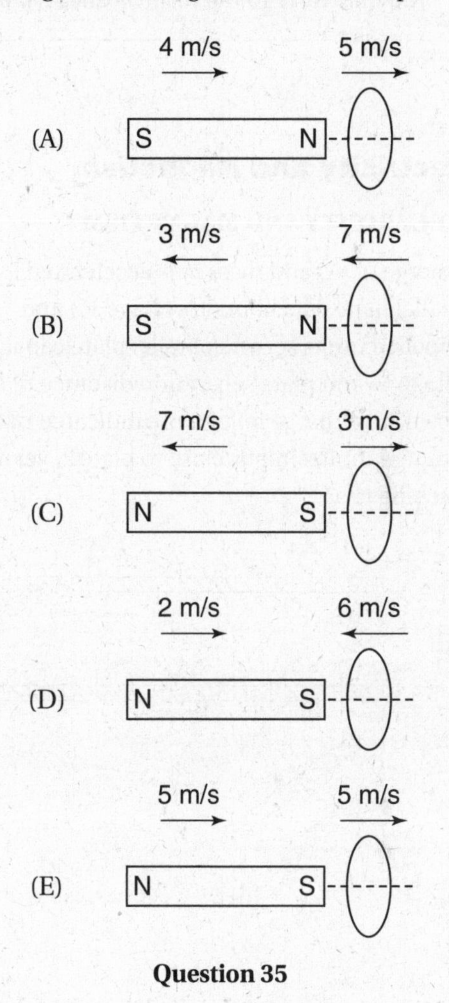

Question 35

GO ON TO THE NEXT PAGE

DIAGNOSTIC TEST

FREE-RESPONSE QUESTIONS

Directions: Use separate sheets of paper to write down your answers to the free-response questions. You may refer to the formula sheet in the Appendix (pages 641–644).

Electricity and Magnetism

ELECTRICITY AND MAGNETISM I

A charge of $+Q$ and mass m is accelerated by a parallel plate capacitor (at voltage V_1) and propelled into a second parallel plate capacitor (at voltage V_2 and plate separation distance d) as shown. The charge follows the indicated path, coming infinitesimally close to plate 2, yet never touching it.

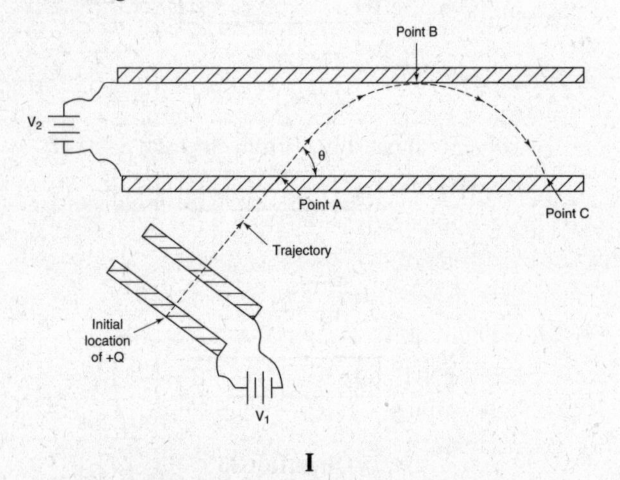

I

(a) What is the charge's initial velocity $\mathbf{v}_0$ as it enters the second capacitor?

(b) Describe the mathematical form of the charge's trajectory in the second capacitor.

(c) Calculate the charge's velocity vector at point C.

(d) Calculate the charge's velocity vector at point B. Use this vector to express V_2 in terms of V_1 and the angle θ.

ELECTRICITY AND MAGNETISM II

An infinitely long cylinder of radius a contains a volume charge density given by the function $\rho = b/r$, where b is a constant with units of coulombs·meters^{-2} (see figure).

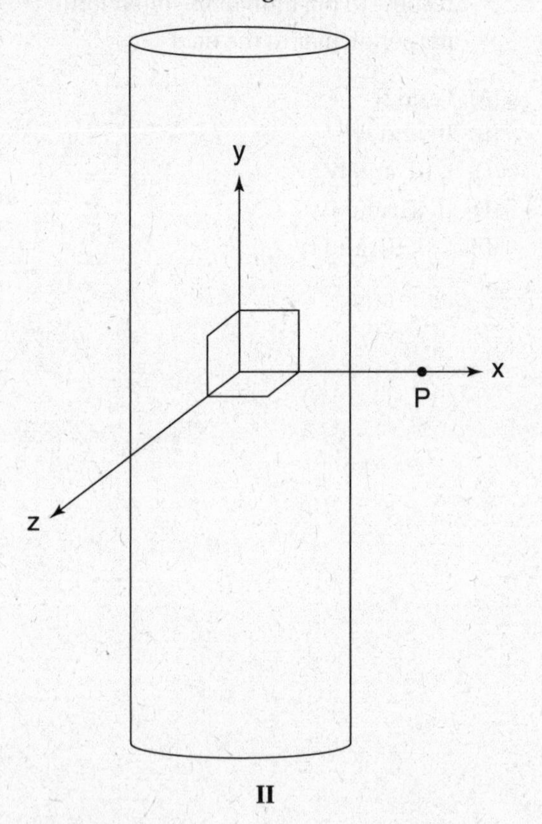

II

(a) Calculate the electric field in the following regions:

(i) $r < a$

(ii) $r > a$

(iii) $r = a$

GO ON TO THE NEXT PAGE

(b) Think of the cylinder as a combination of two semi-infinite identical cylinders, one above the xz-plane and one below the xz-plane, which are joined at the xz-plane. Discuss how the electric field vectors, because of these cylinders, add up to produce the net electric field at a point P along the x-axis.

Now imagine that the cylinder below the xz plane is removed so that only the cylinder above the xz-plane remains.

(c) Based on the insight gained from parts (a) and (b), describe the x-component of the field due to this new charge distribution (quantitatively, if possible).

Now consider a cylinder in which above the xz-plane the charge density is given by $\rho = b/r$, and below the xz-plane the charge density is given by $\rho = -b/r$.

(d) Describe the x-component of the field at point P (quantitatively, if possible).

(e) Taking the potential to be zero at $(\infty, 0)$, what is the potential along the x-axis, $V(x)$?

ELECTRICITY AND MAGNETISM III

Consider the circuit shown. The capacitor is initially uncharged when the switch S is closed at $t = 0$.

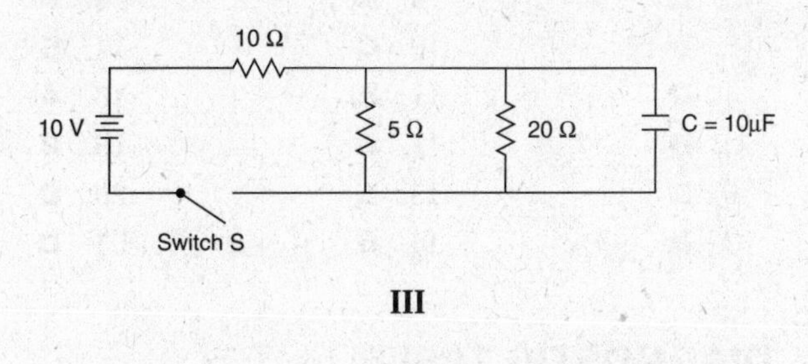

III

(a) What are the initial currents through each of the three resistors?

(b) Once the circuit has reached equilibrium, what are the final currents through each of the three resistors?

(c) At equilibrium, what is the voltage across the capacitor?

Then switch S is opened.

(d) Calculate the initial current through each of the resistors.

(e) The voltage across the capacitor is measured as a function of time, assuming that $t = 0$ when the switch is opened. To determine the time constant of the circuit, which two quantities would you plot to obtain a straight line?

(f) What is the predicted value of the time constant?

MECHANICS

ANSWER KEY

1. **B**
2. **E**
3. **D**
4. **D**
5. **D**
6. **B**
7. **B**
8. **C**
9. **B**
10. **C**
11. **E**
12. **E**
13. **E**
14. **E**
15. **D**
16. **D**
17. **E**
18. **A**
19. **B**
20. **D**
21. **D**
22. **D**
23. **E**
24. **D**
25. **C**
26. **B**
27. **A**
28. **D**
29. **E**
30. **C**
31. **C**
32. **A**
33. **C**
34. **D**
35. **D**

DIAGNOSTIC TABLE

See the appropriate chapter for additional explanations and practice.

Question	Chapter	Question	Chapter
1.	4	20.	4
2.	4	21.	5
3.	3	22.	5, 7
4.	8	23.	9
5.	9	24.	10
6.	6	25.	5
7.	5	26.	3
8.	6	27.	9
9.	6	28.	7
10.	3	29.	7
11.	5	30.	3, 5
12.	9	31.	8
13.	4, 5	32.	5
14.	9	33.	2
15.	8	34.	5
16.	3	35.	8
17.	5	Free response I	4, 5
18.	5	Free response II	4, 5
19.	6	Free response III	4

ANSWERS EXPLAINED

Multiple-Choice

1. **(B)** Consider the figure. Because the block accelerates horizontally, we choose horizontal (*x*) and vertical (*y*) coordinates rather than the usual tilted coordinates of inclined plane problems. The *y*-component of the normal force must balance gravity for the block's height to remain constant:

$$mg = F_N \cos\theta \Rightarrow F_N = \frac{mg}{\cos\theta}$$

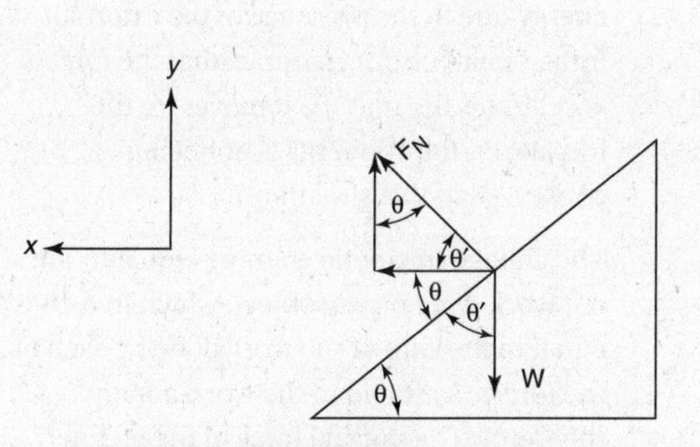

Question 1

Because the force in the *y*-direction is zero, the force in the *x*-direction is equal to the net force. Based on the diagram, this force is

$$F_{\text{net}} = F_x = F_N \sin\theta = \left(\frac{mg}{\cos\theta}\right)\sin\theta$$
$$= mg\tan\theta$$

For more practice with this type of problem, see Chapter 4.

2. **(E)** Consider the situation when the pulley is massless. In this case the tension throughout the rope is constant, and a free-body diagram can be drawn as shown. Applying Newton's second law yields

$$\begin{cases} m_1 a = m_1 g - T \\ m_2 a = T - m_2 g \end{cases}$$

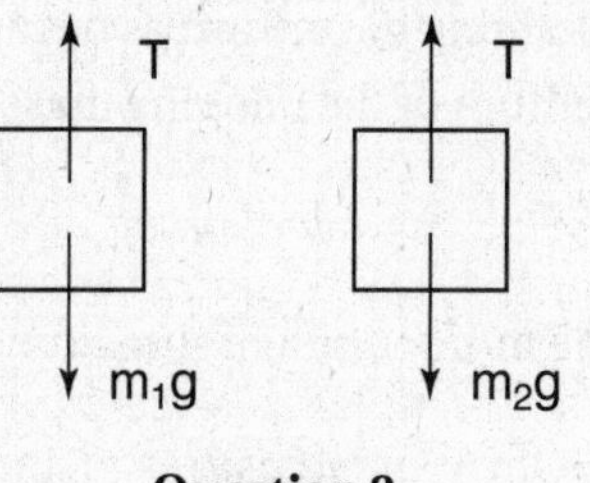

Question 2

Adding these equations and dividing by $m_1 + m_2$ yields

$$a = \left(\frac{m_1 - m_2}{m_1 + m_2}\right)g$$

Then, intuitively, if the pulley has mass, it has rotational inertia, slowing the rate of acceleration. Therefore, choice (E) is correct.

For more practice with this type of problem, see Chapter 4 (particularly Example 4.5 concerning the Atwood machine).

3. **(D)** The magnitude of the rate of change of an object's speed is equal to the magnitude of the component of the acceleration parallel to the velocity. When this component of acceleration is parallel to the velocity, the speed increases, whereas when the component is antiparallel to the object's velocity, the speed decreases. Because in choice (D) the acceleration has the greatest component in the same direction as the velocity, it experiences the greatest increase in speed.

For more information on this topic, see Chapter 3 (particularly Example 3.10).

4. **(D)** Mass is a measure of linear inertia. (For more information, see the definition of rotational inertia and the table comparing linear and angular quantities in Chapter 8.)

5. **(D)** According to the equation $f = (1/2\pi)\sqrt{(k/m)}$, quadrupling the mass and doubling the spring constant causes the frequency to decrease by a factor of $\sqrt{2}$. For more information about oscillating spring systems, see Chapter 9.

6. **(B)** Defining downward as negative, the momentum of the hanging mass is

$$\mathbf{p} = m\mathbf{v} = -m_2 v\hat{j}$$

and the momentum of the sliding mass is

$$\mathbf{p} = m\mathbf{v} = m_1\left[v(\cos\theta)\hat{i} + v(\sin\theta)\hat{j}\right]$$

Summing these two vectors to obtain the net momentum yields choice (B). For a review of momentum, see Chapter 6.

7. **(B)** The change in the potential energy is

$$\Delta U = \Delta U_{m1} + \Delta U_{m2} = m_1 g\Delta y_1 + m_2 g\Delta y_2$$
$$= m_1 gd\sin\theta - m_2 gd$$

Because of conservation of energy, the change in the kinetic energy must have the opposite sign of this change in the potential energy. More information about kinetic energy can be found in Chapter 5.

8. **(C)** As shown in question 6, the potential and kinetic energy both change, so choices (A) and (B) are not correct. There are no external forces acting in the x-direction, so the momentum in this direction must be conserved. There are external forces acting in the y-direction (gravity and the normal force exerted on the incline by the horizontal surface), so y-momentum is not conserved. Conservation of momentum is discussed in Chapter 6.

9. **(B)** This is an inelastic collision in two dimensions such that momentum (but not energy) is conserved. Thus, in the horizontal (x) direction we have the equation $mv + m(0) = (2m)v$; similarly, in the vertical (y) direction we have $m(0) + mv = (2m)v$. Therefore, the final velocity has components in both directions that are equal to $v/2$, so that the final velocity has a magnitude of

$$|v| = \sqrt{\left(\frac{v}{2}\right)^2 + \left(\frac{v}{2}\right)^2} = \frac{v}{\sqrt{2}}$$

Collision problems are discussed in Chapter 6.

10. **(C)** An object with constant acceleration can have a linear path (e.g., a projectile thrown directly upward) or a parabolic path (e.g., a projectile launched at an angle). Moving in circular, elliptical, or spiral paths all require centripetal forces that change in direction. Motion in multiple dimensions with a constant acceleration is essentially identical to projectile motion, which is discussed in Chapter 3.

11. **(E)** The kinetic energy is not equal to the negative of the change in the potential energy due to the presence of the external force. (Imagine, for example, that the force F accelerates the mass as it moves up the incline; certainly energy is not being conserved in this situation.)

    The change in kinetic energy is equal to the net work done on the object, which in turn is equal to the sum of the work done by each of the forces, as stated in the work-energy theorem. (The normal force of the incline does zero work.) Choice (D) is incorrect because although gravity *does* do negative work, this negative sign is contained within the phrase "the work done by gravity."

    The work-energy theorem is derived and discussed in Chapter 5.

12. **(E)** The period is given by the equation $T = 2\pi\sqrt{(m/k)}$. Therefore, doubling the mass, adding a spring in parallel (which doubles the effective spring constant), or adding a spring in series (which halves the effective spring constant) all affect the period. Adding friction complicates the motion such that it is no longer simple harmonic motion, and the period changes. Based on conservation of energy, $\frac{1}{2}mv_{max}^2 = \frac{1}{2}kA^2$, so that $m/k = A^2/v_{max}^2$. Therefore, any change in the parameters of the system that does not affect the ratio A/v_{max} does not affect the ratio m/k and thus does not affect the period. For further information on mass-spring systems, see Chapter 9.

13. **(E)** The fact that the net force is zero does *not* constrain the frictional force to a single value. Instead, the frictional force depends on the nature of the applied force and the resulting normal force. Consider, for example, the two cases in the figure that show two different ways to push a mass up an incline. In case 1, assume that the applied force exactly cancels the gravitational force; thus, both the normal and frictional forces are zero. In case 2, on the other hand, there are nonzero normal and frictional forces. (The component of the applied force parallel to the direction of motion cancels the frictional force such that the net force is zero.)

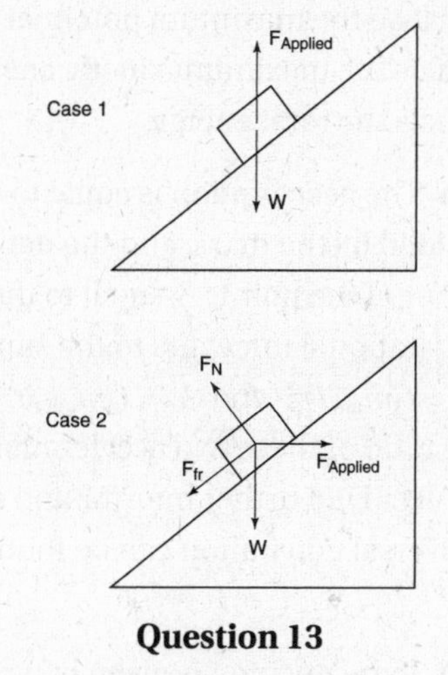

Question 13

Note: If the normal force *were* the same on the way up and on the way down, the frictional force would perform –5 J of work on the way down. The kinetic frictional force always opposes the motion and always performs negative work.

The friction force is discussed further in Chapter 4; work is discussed in Chapter 5.

14. **(E)** According to energy conservation, the potential energy plus the kinetic energy must equal a constant. Therefore, when the kinetic energy is minimized (when the velocity is zero), the potential energy must be maximized. The only graph that has maxima at all three points where the velocity is zero is choice (E).

For more information about how the velocity and potential energy of a system in simple harmonic motion change with time, see Chapter 9.

15. **(D)** The following equations are required for static equilibrium.

The net force in the *x*-direction must be zero:

$$F_2 \sin\theta - F_3 \sin\phi = 0$$

The net force in the *y*-direction must be zero:

$$F_1 - F_2 \cos\theta - F_3 \cos\phi = 0$$

The net torque about their centers must be zero:

$$F_3(\cos\phi)\,r_1 - F_2(\cos\theta)\,r_2 = 0$$

Inspection reveals that the only valid choice is (D), which is the negative of the torque condition above.

For further information about the static equilibrium of extended objects (objects that can rotate), see Chapter 8.

16. **(D)** The spring force provides the centripetal force, so it must equal the centripetal force.

The centripetal force is discussed in the section on uniform circular motion in Chapter 3.

17. **(E)** The rate of doing work is power, $P = \mathbf{F} \cdot \mathbf{v}$. Because the force is perpendicular to the velocity, the power is zero. Power is discussed further in Chapter 5.

18. **(A)** The kinetic energy of the mass remains constant throughout the entire action (both before, during, and after the spring is cut). However, the potential energy of the spring is lost (the spring reverts to its equilibrium state), and this decrease is equal to $-\frac{1}{2}kx^2$. (Since we are neglecting the mass of the spring, we can ignore the kinetic energy imparted to the spring.)

19. **(B)** There is more mass to the left of point *C* than to the right of point *C*, so the center of mass must lie to the left of point *C*. Because all the mass lies to the right of point *A*, point *A* cannot be the center of mass. Thus, by elimination, the center of mass must be point *B*. Note that the center of mass must lie on the horizontal line *AE* because of symmetry.

    Further information about calculating the center of mass of this object can be found in Example 6.15.

20. **(D)** Because of symmetry, the two tension forces will be equal in magnitude. Therefore, to keep the mass in static equilibrium requires $F_{\text{net},y} = 0 = 2F_T\cos\theta - mg$. Solving this equation for the tension force gives choice (D).

    More information about the static equilibrium of point masses can be found in Chapter 4 (particularly Example 4.3).

21. **(D)** At the maximum height of the pendulum, all the energy is gravitational potential energy. At the minimum height of the pendulum, all this potential energy has been converted to kinetic energy. Energy conservation between these extreme points yields

$$mgl(1-\cos\theta) = \frac{1}{2}mv^2$$

    (y is set equal to zero at the bottom of the pendulum's arc; see Example 9.1 for the geometry of the height difference.) Solving for velocity yields choice (D).

    Further information about the conservation of energy can be found in Chapter 5.

22. **(D)** Because the ball rolls without slipping, the relationship between the velocity of its center of mass and its angular velocity is $v = r\omega$. Therefore, the total kinetic energy of the ball is equal to

$$KE = \frac{1}{2}mv^2 + \frac{1}{2}I\omega^2 = \frac{1}{2}m(r\omega)^2 + \frac{1}{2}I\omega^2$$
$$= \omega^2\left(\frac{1}{2}mr^2 + \frac{1}{2}I\right)$$

    so that the kinetic energy is proportional to ω^2. The kinetic energy at the bottom of the incline is twice the kinetic energy at point *P* (because half of the gravitational potential energy has been converted to kinetic energy at point *P*). Therefore, the angular velocity is smaller by a factor of $\sqrt{2}$ at point *P*.

    Gravitational potential energy is discussed in Chapter 5 while rotational kinetic energy is covered in Chapter 7.

23. **(E)** Choices (A), (B), and (C) are false because the total energy is constant. Choice (D) is false because the kinetic energy is maximized at the equilibrium point, where the acceleration is zero. Choice (E) is correct because the maximum potential energy equals the maximum kinetic energy, which equals the total energy.

24. **(D)** The acceleration is equal to the net force divided by the mass, and the net force at either of the points is equal to the gravitational force due to the sun. Therefore, $a_A = Gm_{\text{sun}}/r_A^2$ and $a_B = Gm_{\text{sun}}/r_B^2$. Finding the ratio of these two accelerations yields choice (D). Further information about universal gravitation can be found in Chapter 10.

25. **(C)** If the position is given by $x = vt$, then the object is moving at constant velocity v, and the kinetic energy is given by $\frac{1}{2}mv^2$, choice (C).

26. **(B)** The relative velocity vectors are shown in the figure accompanying the question. The velocity of the swimmer with respect to the ground is equal to the velocity of the swimmer with respect to the water ($\mathbf{v}_{\text{swimmer}}$) plus the velocity of the water with respect to the ground ($\mathbf{v}_{\text{stream}}$). In order for the sum of these velocities to point directly across the stream, the swimmer's velocity $\mathbf{v}_{\text{swimmer}}$ must (1) cancel the velocity of the current $\mathbf{v}_{\text{stream}}$ and (2) provide a perpendicular component across the stream. For $\mathbf{v}_{\text{swimmer}}$ to have a

y-component equal in magnitude to $\mathbf{v}_{stream}$ and a nonzero x-component, its magnitude must be greater than $\mathbf{v}_{stream}$.

Imagine what would happen if $\mathbf{v}_{swimmer} = \mathbf{v}_{stream}$. If the swimmer swam in the direction opposite the current, the swimmer would remain stationary with respect to the shore. If the swimmer tried to cross the stream, the y-component of $\mathbf{v}_{swimmer}$ would be less than $\mathbf{v}_{stream}$, and the swimmer would be swept downstream. Although it would be possible to cross the stream, it would be impossible to cross the stream along a path perpendicular to the banks of the stream.

Further information about relative velocity can be found in Chapter 3 (in particular, see Example 3.8).

27. **(A)** Because the springs are identical, they are stretched the same amount. (This is required in order for the tension force to be constant throughout the two springs; if the tension in the springs were not equal, there would be a net force on the point where the springs were connected together, causing the extension of the springs to change until the two tension forces *were* equal.) Therefore, when a mass attached to the two springs is moved a distance Δx, each spring is displaced by an amount $\Delta x/2$ such that the tension force within each spring is $F = -k(\Delta x/2)$. This tension force is equal to the force the springs exert on the mass, so that the effective spring constant is $k_{effective} = -F/\Delta x = k/2$.

28. **(D)** The angular velocity is proportional to t, thus the angular acceleration is constant. The angular displacement then obeys (see Chapter 7)

$$\theta - \theta_0 = \omega_0 t + \frac{1}{2}\alpha t^2$$

with $\omega_0 = 9$ and $\alpha = 3$. Thus, the angular displacement is $(9)((1-0)) + 0.5(3)(1) =$ 10.5 rad.

29. **(E)** You exert a constant torque of (1.0 m) (2.0 N) sin 90° on the door. The rate at which you do work is given (for constant torque) by

$$\tau\frac{\Delta\theta}{\Delta t} = 2.0\ \text{N}\cdot\text{m}\frac{0.5\ \text{rad}}{0.5\ \text{s}} = 2.0\ \text{J/s}$$

choice (E).

30. **(C)** $\mathbf{P} = \mathbf{F} \cdot \mathbf{v}$. The force is the constant downward gravitational force, so the power is maximized when the downward velocity is maximized. This occurs just before the particle hits the ground, when it is moving with its maximum speed in the downward direction.

Projectile motion is discussed in Chapter 3, and power is discussed in Chapter 5.

31. **(C)** The angular momentum around the sun is constant. Therefore, using the equation $\mathbf{L} = \mathbf{r} \times \mathbf{p}$, $[|\mathbf{r}_1 \times m\mathbf{v}_1| = r_1 m v_{1,\perp} = r_1 m v_1 \sin\theta] = [|\mathbf{r}_2 \times m\mathbf{v}_2| = r_2 m v_2]$, solving for v_2 yields choice (C).

32. **(A)** Using $F(x) = -dU/dx$, we see that choice (A) is correct.

33. **(C)** Velocity is the time derivative of position, and taking the derivative yields $v(t) = 4 - t$. Substituting $t = 1$ yields choice (C).

34. **(D)** Conservation of energy requires that both particles have the same kinetic energy when they strike the ground. Because (on impact) they have the same x-component of velocity (their x-components of velocity are initially equal and remain unchanged) and the same initial speeds, they must have the same y-components of velocity (recall that $v = \sqrt{v_x^2 + v_y^2}$). Therefore, they have the same velocity vector on impact and must also share the same angle of impact.

35. **(D)** The ice exerts no torque on the skater. (The normal force acts essentially at the axis of rotation and exerts zero torque.) Therefore, the skater's angular momentum, $\mathbf{L} = I\omega$, must remain constant (such that if the angular velocity decreases by a factor of 2, the moment of inertia must increase by a factor of 2). Given these changes, the kinetic energy, $KE = \frac{1}{2}I\omega^2 = \left(\frac{1}{2}I\omega\right)\omega = (\text{constant})\omega$, decreases by a factor of 2.

Free-Response

MECHANICS I

(a) If the angular acceleration is zero, the acceleration of the two masses m_1 and m_2 must be zero (assuming that there's no slippage of the rope, so that the masses cannot accelerate while the pulley rotates at a constant angular speed). This means that the net forces on these objects must be zero, which requires that $T_1 = mg$ and $T_2 = mg\sin\theta$. Note that $m_1 = m_2 = m$.

Because the angular acceleration is zero, the net torque must also be zero. Designating counterclockwise torque as positive (and using the tension forces just calculated), we find that the net torque on the pulley is given by

$$\tau_{\text{net}} = mgr - mg(\sin\theta)(2r) = 0$$
$$\sin\theta = \frac{1}{2}$$
$$\theta = 30°$$

This part involves Newton's laws (Chapter 4) and their application to rotational motion.

(b) For both m_1 and m_2, the equation $v = r\omega$ (i.e., rolling without slipping of the rope) holds:

$$\omega = \frac{v_2}{2r} = \frac{|v_1|}{r}$$
$$v_1 = \frac{v_2}{2}$$

(c) Applying Newton's second law to the hanging mass (and designating downward acceleration as positive) yields

$$F_{\text{net}} = ma = mg - T_1$$

Applying Newton's second law to the pulley (and designating counterclockwise acceleration as positive, such that both the linear and angular accelerations are positive),

$$\tau_{\text{net}} = T_1 r = I\alpha$$

The final equation relates the linear acceleration of m_1 and the angular acceleration of the pulley (expressing the fact that the pulley rolls without slipping on the rope connecting it to m_1): $a = r\alpha$. Solving these three equations with three unknown variables for a yields

$$a = \frac{mg}{m + I/r^2}$$

(d) Because of conservation of energy, the rate at which the kinetic energy is changing is equal to the rate at which the potential energy is decreasing. Because the gravitational force is converting potential energy to kinetic energy, this rate is equal to the power exerted by gravity, given by the formula $P = \mathbf{F} \cdot \mathbf{v}$. When the pulley is rotating with a speed of ω, the speed of the falling mass is $v = r\omega$ (downward, parallel to gravity), so the power is $P = \mathbf{F} \cdot \mathbf{v} = gmr\omega$.

For further discussion of work, power, and translational KE, see Chapter 5.

MECHANICS II

(a) The normal force at the top of the loop is zero, so the radial force is equal to the gravitational force, which must equal the centripetal force:

$$F_{\text{net},y} = mg = F_{\text{cent}} = \frac{mv^2}{r}$$

Solving for the speed,

$$v = \sqrt{gr}$$

(b) Applying conservation of energy between the initial position of the roller coaster and the top of the loop (and using the previous result for the speed at the top of the loop),

$$mgh = mg(2r) + \frac{1}{2}mgr$$
$$h = 2.5r$$

(c) The forces are shown in the figure (the construction of free-body diagrams is discussed in Chapter 4).

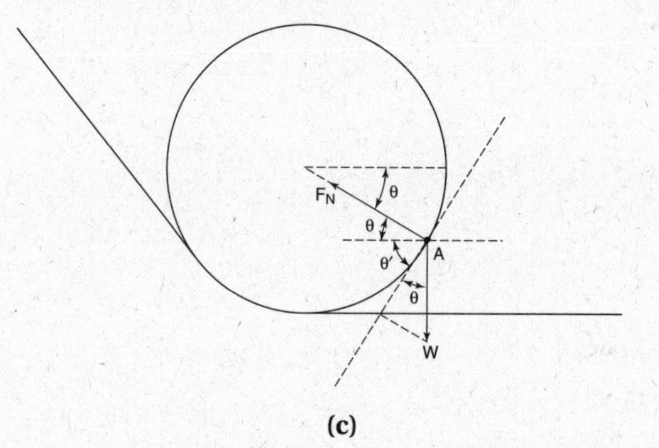

(c)

(d) Using energy conservation again [comparing the initial position of the roller coaster at height $2.5r$ with point A at height $r(1 - \sin\theta)$],

$$mg(2.5r) = mgr(1 - \sin\theta) + \frac{1}{2}mv^2$$
$$v = \sqrt{gr(3 + 2\sin\theta)}$$

MECHANICS III

(a) The free-body diagram is shown.

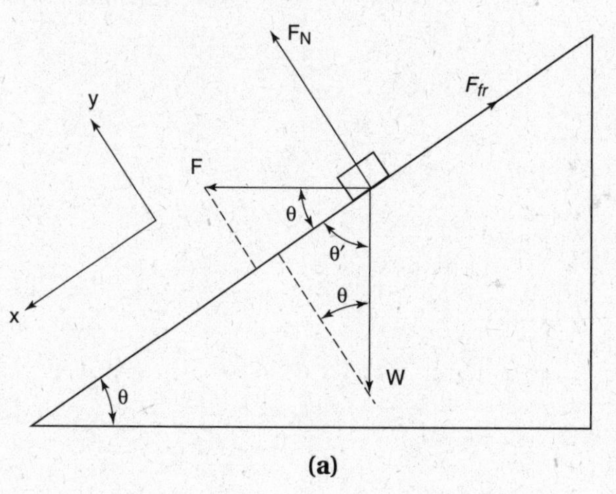

(a)

(b) Because the block remains on the incline, the acceleration and the net force in the y-direction are zero:

$$F_{\text{net},y} = F_N + F\sin\theta - mg\cos\theta = 0$$

The acceleration is parallel to the x-direction:

$$F_{\text{net},x} = ma = F\cos\theta + mg\sin\theta - F_{\text{fr}}$$

This yields two equations with three variables. The final equation is $F_{\text{fr}} = \mu_k F_N$. Solving these equations for the acceleration yields

$$a = \frac{F}{m}(\cos\theta + \mu_k \sin\theta) + g(\sin\theta - \mu_k \cos\theta)$$

(c) The block leaves the incline when it has a net force in the y-direction (and thus a net acceleration in the y-direction). Recall the equation for the net force in the y-direction:

$$F_{\text{net},y} = F_N + F\sin\theta - mg\cos\theta$$

Imagine increasing F from zero. As the applied force F increases, the normal force decreases such that the net force in the y-direction is zero and the block remains on the incline. However, the normal force cannot decrease indefinitely because it can never be negative. The maximum force that can be applied without causing the block to leave the incline occurs when the normal force is equal to zero:

$$F_{\text{net},y} = 0 = F\sin\theta - mg\cos\theta$$

Solving for F yields $F = mg\cot\theta$. Any force greater than this would produce a positive acceleration in the y-direction, causing the block to accelerate away from the incline.

For further information on the construction of free-body diagrams and solutions to inclined plane problems, see Chapter 4.

(d) According to part (c), the maximum force is given by $F = mg \cot \theta$. Thus, plot F on one axis and $\cot \theta$ on the other.

(e) The figure shows the plot and the straight-line fit. The slope of the line is given approximately by:

$$\text{slope} = \frac{(27 - 9.5)}{(2.75 - 1)} = 10\text{N}$$

The answer to part (c) predicts that this value should be $mg = (1)(9.8)\ \text{N} = 9.8\text{N}$.

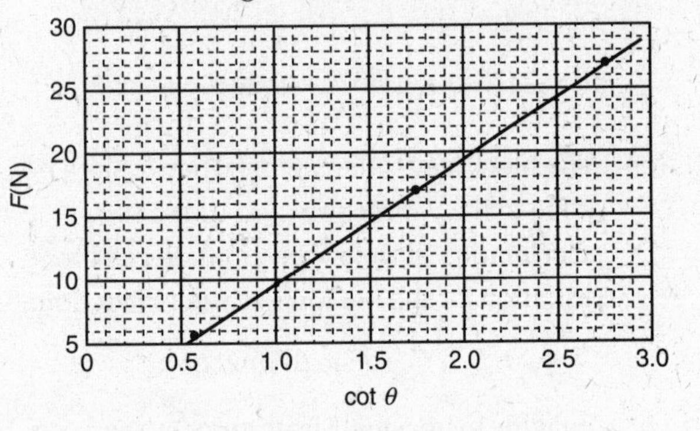

ELECTRICITY AND MAGNETISM

ANSWER KEY

1. **C**
2. **C**
3. **B**
4. **C**
5. **B**
6. **C**
7. **A**
8. **A**
9. **E**
10. **C**
11. **C**
12. **A**
13. **E**
14. **C**
15. **C**
16. **D**
17. **B**
18. **C**
19. **E**
20. **D**
21. **A**
22. **B**
23. **C**
24. **D**
25. **D**
26. **E**
27. **C**
28. **A**
29. **B**
30. **D**
31. **B**
32. **D**
33. **D**
34. **D**
35. **C**

DIAGNOSTIC TABLE

See the appropriate chapter for additional explanations and practice.

Question	Chapter	Question	Chapter
1.	11, 13	20.	19
2.	17	21.	14
3.	11	22.	15
4.	18	23.	12
5.	12	24.	19
6.	18	25.	12
7.	18	26.	19
8.	18	27.	14
9.	13	28.	14
10.	16	29.	14
11.	12	30.	14
12.	18	31.	14
13.	12	32.	13
14.	20	33.	17
15.	14	34.	11
16.	12	35.	18
17.	13	Free response I	11, 17
18.	17	Free response II	11, 12, 13
19.	11	Free response III	14

ANSWERS EXPLAINED

Multiple-Choice

1. **(C)** The potential of an isolated metal sphere with radius r and charge Q is $V = Q/4\pi\varepsilon_0 r$. Why? According to Gauss's law, the field outside the surface of the sphere depends only on the magnitude of the enclosed charge, so the field is the same as the field that would be produced if all of charge Q were concentrated at the center of the sphere. Therefore, the electric field outside the sphere is the same as the electric field due to a point charge Q, $|E| = Q/4\pi\varepsilon_0 r^2$. The magnitude of the potential at the surface of the sphere is then the integral of this electric field from infinity to the surface of the sphere, assuming the potential is zero at infinity. Because the electric field has the same form as the electric field of a point charge, the potential also has the same form as the potential of a point charge, $V = Q/4\pi\varepsilon_0 r$.

   For more information on Gauss's law and integrating an electric field to obtain potential, see Chapter 13. Positive charges move from higher potential to lower potential since the electric field points in the direction of decreasing potential, as discussed in Chapter 11.

2. **(C)** See the discussion of the motion of particles within magnetic fields in Chapter 17.

3. **(B)** This is a question about electrostatic induction. When a positive point charge is placed near a metal sphere (charged or uncharged), it attracts mobile electrons in the metal, causing the metal to become polarized such that more negative charge is on the side closer to the positive point charge and more positive charge is on the side farther away from the positive point charge. Because this charge redistribution results in moving negative charge *closer* to the positive point charge and moving positive charge *farther away* from the positive point charge, it increases the attractive force between the metal sphere and the positive point charge. Therefore, compared to an insulating sphere with the same net charge, a metal sphere always experiences a greater attractive force (such that charged metal > charged insulator and uncharged metal > uncharged insulator). Because this inductive effect is small compared to the effect of actually putting net charge $-Q$ on the sphere, charged insulator > uncharged metal.

4. **(C)** The charges within the bar are moving in a magnetic field, so they feel a magnetic force. Using the conventional current description, positive charges feel an upward force, causing positive charge to accumulate at the top end of the bar (leaving negative charge on the bottom end of the bar). This charge continues to accumulate until the electric force created by the charge imbalance cancels the magnetic force, mathematically $qE = qvB$. The magnitude of the potential difference between points A and B is then $|V| = |-\int \mathbf{E} \cdot d\mathbf{r}| = |vBd|$. The electric field points down (producing a force that opposes the magnetic force), and recalling that electric fields point from high to low potential, we see that point A is at the higher potential.

   Further information about motional EMF can be found in Example 18.14.

5. **(B)** The electric field is the negative derivative (with respect to the spatial coordinate) of the potential. Thus, with a constant potential the electric field must be zero.

6. **(C)** The induced EMF depends on the change in the enclosed flux. Because the magnetic field is uniform, and changing at a constant rate,

$$|\mathcal{E}| = \left|\frac{d\Phi_B}{dt}\right| = \left|\frac{d}{dt}\mathrm{A} \cdot \mathrm{B}\right| = A\left|\frac{dB}{dt}\right|$$

   Therefore, the induced EMF depends only on the area of the loop through which magnetic field lines pass. For $r < a$, the area within the loop exposed to the magnetic field is simply the entire area of the loop, which is a quadratic function, $A(r) = \pi r^2$. For $r > a$, increasing r does not increase the area exposed to magnetic field (because the magnetic field through the additional area is zero).

Therefore, for $r > a$ the induced EMF is constant. The only graph that is quadratic for small r and then constant is choice (C).

Calculating the induced EMF due to a magnetic field via Faraday's law is discussed in Chapter 18.

7. **(A)** The current is given by $I = V/R$. The resistance is proportional to the circumference of the loop, $2\pi r$. Therefore, for $r < a$, V is proportional to r^2 (as discussed in question 6), and $1/R$ is proportional to $1/r$, so the product $I = V/R$ is proportional to r (and thus its graph is a line passing through the origin).

   For $r > a$, V is constant and $1/R$ is proportional to $1/r$, so the product $I = V/R$ is inversely proportional to r. The only graph that satisfies these conditions is choice (A).

8. **(A)** According to Ampere's law, $\int \mathbf{B} \cdot d\mathbf{l} = \mu_0 I_{enclosed}$.

   Using circular Amperian paths shows that the magnitude of the magnetic field can be found from the equation $B = \mu_0 I_{enclosed}/2\pi r$. Compare this equation to the previous one, $I = V/R$, in question 7. The factor of $1/2\pi r$ in the equation for B is analogous to the factor $1/R$ in question 7 (both current and magnetic field are always inversely proportional to r), and $I_{enclosed}$ is analogous to V in question 7. (Both increase as a quadratic function for $r < a$ because they are proportional to area, and then remain constant.) Therefore, qualitatively the behavior of the ratio $B = \mu_0 I_{enclosed}/2\pi r$ in this problem is the same as the behavior of the ratio $I = V/R$ in question 7, and the shape of the graph is the same.

9. **(E)** According to Gauss's law (Chapter 13), the electric field inside a hollow sphere is zero because there is no enclosed charge. Therefore, the electric field is independent of the charge of the shell, the radius of the shell, the radial coordinate, and the permittivity of free space; it is zero inside the sphere independent of any of these factors.

   Note that the electric field is *always* given by the equation in choice (E).

10. **(C)** Setting the equation for $Q(t)$ equal to 60% of its final value yields

$$Q(t) = Q_f(1 - e^{-t/\tau}) = 0.6Q_f$$

$$1 - e^{-t/\tau} = 0.6$$

$$e^{-t/\tau} = 0.4$$

$$-\frac{t}{\tau} = \ln(0.4)$$

   Solving for τ then yields choice (C). Further information about RC circuits can be found in Chapter 16.

11. **(C)** Choice (A) is false: The potential at point B minus the potential at point A is given by $-\int_{\text{point } A}^{\text{point } B} \mathbf{E} \cdot d\mathbf{r}$. Choice (B) is false: The integral is path-independent only if there is no time-dependent magnetic field present. A time-dependent magnetic field produces a nonconservative electric force with no associated potential function. Choice (C) is true: The work per charge is the opposite of the change in potential energy per charge, which is equal to $\int_{\text{point } A}^{\text{point } B} \mathbf{E} \cdot d\mathbf{r}$ because of conservation of energy. Choice (D) is false: The work done by an arbitrary external force does not necessarily have any relationship with the electric field. Although choice (E) is true in the special case that the integral is a closed loop, that is, point A = point B, it is not true in general.

12. **(A)** The current induced in the loop moves in a clockwise direction to oppose the increasing outward flux through the loop according to Lenz's law. The movement of this current through the top and bottom of

the loop creates opposing forces that cancel each other, while the current flowing through the left side of the loop produces a force to the right, which equals the net force on the loop.

13. **(E)** Without doing a lot of work, we can eliminate most of the choices. First, kinetic energy is always positive, so choices (C) and (D) are incorrect. Second, recall the formula for the potential energy of two point charges separated by a distance r (ignoring signs for now):

$$U = q_1 V = q_1\left(\frac{q_2}{4\pi\varepsilon_0 R}\right) = \frac{q_1 q_2}{4\pi\varepsilon_0 R}$$

Comparing the form of this expression with our choices, allows us to eliminate choices (A), (B), and (C), simply on the basis of dimensional analysis. This leaves us only choice (E), which a detailed calculation shows is the correct answer.

14. **(C)** See Chapter 20 for a further explanation of Gauss's law for magnetism.

15. **(C)** The equivalent resistance is given by

$$R_{eq} = \frac{1}{(1/R_1)+(1/R_2)}$$

Because the denominator is greater than $1/R_1$ or $1/R_2$ alone, the entire fraction must be smaller than R_1 or R_2 alone. Choice (D) is incorrect: The voltage across the resistors must be the same because they are arranged in parallel. Therefore, the current through either resistor is given by the equation $I = V/R$. Because the resistors have different resistances, different amounts of current flow through them. The equivalent resistance of a resistor array is a physical property of the array itself (just as the resistance of a resistor is a physical property of the resistor), so it is independent of the voltage across the array (assuming, of course, that the resistors obey Ohm's law), eliminating choice (E).

16. **(D)** First, what is the sign of the change in energy? This is a tricky question because the actual sign of the charge and the direction of the electric field are not given. Recall that $F_x = -dU/dx$. Thus, a particle always feels a force pushing it in a direction that decreases the particle's potential energy. Therefore, whether the particle's charge is positive or negative, it always moves in such a way as to decrease its potential energy, so $\Delta U < 0$. This eliminates choices (A) and (C).

What, then, is the magnitude of the decrease in potential energy? The key equation here is $\Delta U = q\Delta V$ (based on the definition of potential). Because the particle moves parallel or antiparallel to the electric field (depending on the sign of its charge), $|\Delta V| = |-\int \mathbf{E}\cdot d\mathbf{r}| = Ed$. Therefore, the magnitude of the change in potential energy is $|\Delta U| = |q\Delta V| = qEd$, so choice (D) is correct.

17. **(B)** Based on Gauss's law, the enclosed charge term, $q_{enclosed}$, depends only on the magnitude of the charge and the symmetry properties of its distribution. Therefore, the field produced outside a charged sphere is the same as the field that would be produced if all the charge of the sphere were concentrated at a point at the sphere's center (which is easily calculated using Coulomb's law). Note that a charge placed on an *isolated* conducting sphere distributes itself uniformly across the surface of the sphere. Further information about Gauss's law and the distribution of a stationary charge along conductors is given in Chapter 13.

18. **(C)** According to Ampere's law and the symmetry of the situation, the only current that determines the magnetic field at point P is the current enclosed within the amperian path shown. (Thus, the current in region B has no effect on the magnetic field at point P.) The magnetic field produced at P due to the enclosed current is the same as the field produced by a thin wire along the axis of the cylinder, carrying the same current (since the magnetic field depends on only the enclosed current as long as it has cylindrical symmetry). The right-hand rule indicates

that this imaginary current-carrying wire would produce a downward magnetic field at point P, so the same downward magnetic field must be produced by the entire cylinder of current featured in this problem. For further information about calculating magnetic fields using Ampere's law, see Chapter 17.

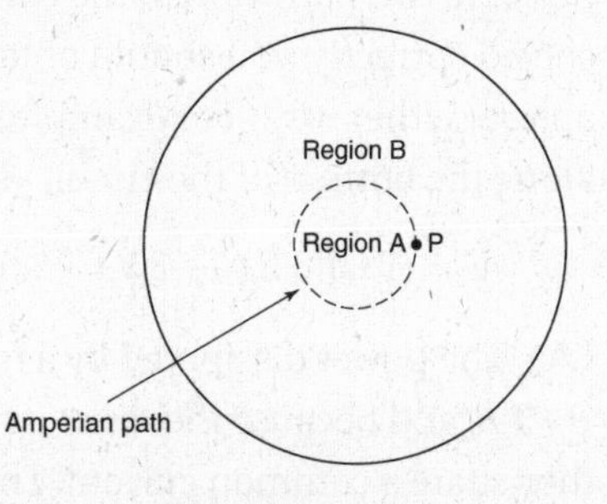

Question 18

19. **(E)** Distribution I will not be able to cancel the electric field due to $+q$ because it will never be able to achieve a high enough magnitude. Distribution II will work: Simply locate the charge $-q$ at the origin, and it will exactly cancel out the field due to $+q$ over all space. Distribution III will never be able to have an equal magnitude and opposite direction compared to the field due to $+q$; in order to have an opposite direction the negative charge must be located at the origin, but at this point it will produce a stronger magnitude field than the field due to $+q$. Distribution IV will definitely work. The charges must be located equal distances from the origin in order for the field due to both of the negative charges to be in the direction opposite the field due to $+q$. By moving the charges as far away from the origin as is necessary, the magnitude of the field can be adjusted so as to exactly equal the magnitude of the field due to $+q$.

20. **(D)** Tracing the loop in the counterclockwise direction: (1) The sign of the potential jump of the battery is positive because we are going from the negative terminal to the positive terminal. (2) The sign of the potential jump across the resistor is negative because we are going in the same direction as the current. (3) The sign of the potential jump across the inductor is negative. Why? The inductor opposes the increase in the current, so its potential jump must be of the opposite sign compared to that of the battery (which has a positive potential jump). Because current is increasing, $dI/dt > 0$, we must add a negative sign to the term $-L(dI/dt)$ in order for it to have an overall negative value as required.

21. **(A)** When the terminal voltage is 7 V, that means that there is a 2-V potential drop across the internal resistance. Applying Ohm's law to the internal resistance yields $I = V/R = 2\text{ V}/20\ \Omega = 0.1\text{ A}$.

22. **(B)** This problem relies on manipulation of the equation for the capacitance of a parallel plate capacitor with a dielectric, $C = \varepsilon_0 A k_D/d$. Increasing A, d, and k_D by factors of 2 results in a net increase of capacitance by a factor of 2.

23. **(C)** The potential at the surface of the sphere is the same as if all of the charge were concentrated at the center of the sphere, that is, $V = (1/4\pi\varepsilon_0)(Q/R)$ as discussed in question 1. The electric field inside the sphere is zero, so the potential at the center is the same as its value at the surface. The work done by the electric field to move the charge from infinity to this point is the *negative* of the change in the charge's potential energy: $W = -\Delta U = -q\Delta V = -q(Q/4\pi\varepsilon_0 R)$.

24. **(D)** The time constant of an RL circuit is $\tau = L/R$. Because the time constant has units of time, R/L has the units of inverse time.

    Alternative solution: Recall that the voltage drop across an inductor is $V = L(dI/dt)$ and the voltage drop across a resistor is $V = IR$. Equating the units of voltage in these two formulas, $V = IR = L(dI/dt)$. Manipulating this formula in terms of units yields $L/R = I(dt/dI) = t$. (Note that dt has the units of time and dI has the units of current.)

25. **(D)** Choice (C) would indeed result in no net work being done (no work is required to move a charge along an equipotential line).

However, this is not required for zero net work to be done: It is possible that the field might perform negative work along part of the charge's path and positive work along the remainder of the path, such that the net work is zero. The only condition that is *required* for zero net work to be done is for the charge to begin and end at the same voltage (so that the change in potential energy is zero, based on the definition of voltage), so choice (D) is correct. Choice (E) is true of magnetic fields, not electric fields, since the magnetic force is always perpendicular to the velocity. Choice (B) won't generally satisfy the requirement that no net work be done, except in the special case that the path begins and ends at the same point on the field line. In fact, any path that begins and ends at different points along an electric field line must have involved the electric field performing net work because electric field lines are parallel to the electric field and such movement causes a change in voltage. Choice (A) is false: Stationary charges always produce conservative electric forces.

26. **(E)** Based on the formula for the magnetic field within a solenoid (which can be quickly derived using Ampere's law), $B = \mu_0 nI$. Because the radius of the solenoid does not enter this equation, changing the radius will not affect the strength of the magnetic field within the solenoid. However, because increasing the radius will increase the area of each loop of the solenoid by a factor of 4 (because $A = \pi r^2$), it will increase the magnetic flux through each loop by a factor of 4 and thus increase the inductance by a factor of 4.

27. **(C)** First, constructing and solving a Kirchhoff's loop equation indicates that the current is 0.5 A flowing in a clockwise direction. More specifically, one valid loop equation is (choosing to designate positive current in the clockwise direction and traversing the loop in the clockwise direction):

$$-1\text{ V} - (1\ \Omega)\ I + 3\text{ V} - (1\ \Omega)\ I - (2\ \Omega)\ I = 0$$

(Alternatively, you could realize that the circuit has an equivalent resistance of 4 Ω and an equivalent voltage of 2 V, so applying Ohm's law indicates that the current is 0.5 A.)

Then, to calculate the potential at point *B*, begin at point *A* and sum potential jumps as you move to point *B*. (Use either of the possible two pathways. If the current is solved correctly, you should obtain the same answer either way.) For example, moving along the bottom of the circuit yields

$$V_B = (1\ \Omega)(0.5\text{ A}) - 3\text{ V} = -2.5\text{ V}$$

28. **(A)** The power dissipated by a resistor is $P = I^2R$, and because the resistors are in series, they share a common current. Therefore, the power dissipated by all the resistors is $P = \Sigma I^2 R = I^2 \Sigma R = \left(\frac{1}{2}\text{ A}\right)^2 (4\ \Omega) = 1\text{ W}$. (Note that this is equivalent to realizing that the equivalent resistance of the circuit is 4 Ω and plugging this value into the formula $P = I^2R$.)

29. **(B)** Application of the equivalent resistance formulas reveals that the equivalent resistance of circuits (A) through (E) are 3 Ω, 6 Ω, 4 Ω, 2 Ω, and 5 Ω.

30. **(D)** The most useful formula to use here for power is $P = V^2/R$. The circuit with the lowest equivalent resistance dissipates the greatest total power. (Remember, the voltage is the same for all circuits.)

31. **(B)** Elements in parallel have the same voltage, so the resistors in choices (A) and (D) all have a potential drop equal to the voltage of the battery, *V*. What are the potential drops across the resistors in circuits (C) and (E)? Because two identical resistors in series carry the same current and thus have the same potential drop, a Kirchhoff's loop rule quickly shows that all the resistors in these circuits have a potential drop of half the battery's voltage, *V*/2. The most complicated circuit is choice (B). Because the circuit has an equivalent resistance of 6 Ω, the current through the circuit is $I = V/6\ \Omega$. This current splits evenly between the two 8-Ω resistors,

such that their voltage is $V = IR = (V/12\ \Omega)(8\ \Omega) = \frac{2}{3}V$.

Comparison reveals that circuit (B) has the resistor ($R = 2\ \Omega$) with the lowest potential drop, $V/3$. (For information about applying Kirchhoff's and Ohm's laws to calculate potential drops, see Chapter 14.)

32. **(D)** According to Gauss's law, the electric flux is proportional to the net enclosed charge, so if the flux is zero, the net enclosed charge must be zero. The conditions in choices (A) and (B) would cause the *local flux* at any point on the gaussian surface to be zero, which would indeed cause the net flux to be zero. However, these are not required for the net flux to be zero: It is possible that positive flux through part of the gaussian surface cancels negative flux through another part of the gaussian surface such that the net flux is zero. Like choices (A) and (B), choice (C) would indeed result in the flux through the gaussian surface being zero. However, again this is not *required* for the net flux to be zero. It would be possible for a surface to enclose equal amounts of positive and negative charge such that the *net enclosed charge* and flux are zero.

    Further discussion of Gauss's law can be found in Chapter 13.

33. **(D)** The magnetic field provides the centripetal force for the circular motion shown, according to the equation $qvB = mv^2/r$. Because v, m, and B are fixed and shared by both particles, q is inversely proportional to r. Therefore, particle B, which follows a path of smaller radius, must have a greater magnitude of charge. The sign of the charge on particle A can be determined by the right-hand rule to be negative. (Remember that the centripetal force must always point toward the center of the arc.)

34. **(D)** I is false: The direction of the force depends on the sign of the charge on the particle, so it cannot be known for any charged particle. II is true: The magnitude of the electric force is simply $|F| = QE$. III is true: The rate of change in voltage with respect to position in the direction parallel to the field is simply E, with potential decreasing in the direction of the field. IV is true: The rate of change in potential with respect to position in the direction perpendicular to the field is always equal to zero.

35. **(C)** The loop feels a magnetic force caused by the external magnetic field if an induced current runs through the loop. (The magnetic field has no effect on stationary charges.) Because all the situations shown have the same geometry, the magnitude of the magnetic field and the length of current-carrying wire is the same in all cases, and the magnitude of the force depends only on the current. (A greater magnitude current produces a greater magnitude force.)

    The magnitude of the current depends on the rate at which the flux through the loop is changing, which depends on the relative velocity of the loop with respect to the magnet. In the frame of reference of the magnet, the velocities of the loops are +1 m/s, –4 m/s, +10 m/s, –8 m/s, and 0 m/s, respectively. Because the relative velocity of the loop in choice (C) is the greatest, this loop experiences the greatest induced current and thus the greatest magnetic force.

    For further information about induced currents and the forces they produce, see Chapter 18.

Free-Response

ELECTRICITY AND MAGNETISM I

(a) Conservation of energy gives us the magnitude of velocity (electrical potential energy is converted to kinetic energy). Based on the definition of voltage, the potential energy of the point charge is simply the value of the charge multiplied by the voltage across the capacitor.

$$QV_1 = \frac{1}{2}mv_0^2$$

$$v_0 = \sqrt{\frac{2QV_1}{m}}$$

Now, we know that the velocity has an angle of elevation of θ, so we can express its velocity vector in terms of $\hat{i}$ and $\hat{j}$ using vector geometry:

$$\mathbf{v}_0 = \sqrt{\frac{2QV_1}{m}}(\cos\theta)\hat{i} + \sqrt{\frac{2QV_1}{m}}(\sin\theta)\hat{j}$$

(b) The particle experiences a constant force in the $-y$-direction. Mathematically, this is analogous to gravity near the Earth's surface, such that the charge moves like a projectile. Therefore, the form of the trajectory is a parabola (as proven in Example 3.3).

(c) Because of the symmetry of parabolas (or energy conservation, if you prefer), the particle has the same speed at point C as at point A. The x-component of the velocity remains unchanged (because there is no force in the x-direction), so in order for the speed to remain the same, the magnitude of the y-component of the velocity must be the same (although clearly the y-component of velocity changes sign). Again, this is entirely analogous to a projectile under the influence of gravity.

$$\mathbf{v}_f = \sqrt{\frac{2QV_1}{m}}(\cos\theta)\hat{i} - \sqrt{\frac{2QV_1}{m}}(\sin\theta)\hat{j}$$

(d) We will use energy conservation here. Because we wish to express our answer in terms of V_1, we invoke energy conservation starting with the initial position of the charge in capacitor 1. There are two key points to note:

1. At the peak of its trajectory, when the charge is very close to B, the y-component of velocity is zero and the x-component of velocity has the same value as it had at point A (again, this can be understood in terms of an analogy with projectile motion).
2. The electric potential energy at point B equals the potential drop across the second capacitor multiplied by $+Q$ [as in part (a)].

We begin by equating the charge's energy before the trajectory begins (when the charge is stationary at the positive plate of capacitor 1) to the charge's energy at point B.

$$QV_1 = QV_2 + \frac{1}{2}mv_x^2$$

$$= QV_2 + \frac{1}{2}m\left[\frac{2QV_1}{m}(\cos^2\theta)\right]$$

The charge cancels:

$$V_1 = V_2 + V_1\cos^2\theta$$

Solving for V_2,

$$V_2 = V_1(1 - \cos^2\theta) = V_1\sin^2\theta$$

ELECTRICITY AND MAGNETISM II

(a)

(i) This is a standard application of Gauss's law with cylindrical symmetry (see Chapter 13). The only tricky part is that, because of the varying charge density, we must integrate to obtain the enclosed charge. To mirror the cylindrical symmetry, we select a cylindrical gaussian surface (length l, radius r) and integrate using cylindrical shells (length l, radius r, thickness dr).

$$dQ = \rho dV = \left(\frac{b}{r}\right)(2\pi rl\,dr) = 2\pi bl\,dr$$

$$Q = \int dQ = \int_0^r 2\pi bl\,dr = 2\pi blr$$

Units check: Because b has units of C/m^2, the above equation [of the form b (length^2)] has the correct unit, coulombs.

Applying Gauss's law (the electric field points perpendicularly away from the center of the cylinder, so $\mathbf{E}\cdot d\mathbf{A} = |E||dA|$ (for the curved sides).

$$\Phi = \oint \mathbf{E}\cdot d\mathbf{A} = |E|\oint dA = 2\pi rl|E|$$

$$= \left(\frac{Q_{enc}}{\varepsilon_0} = \frac{2\pi blr}{\varepsilon_0}\right)$$

Solving for the electric field,

$$E = \frac{b}{\varepsilon_0}$$

Units check: b has the same units as surface charge density (C/m^2). Therefore, this equation has the same dimensions as the field because of an infinitely large sheet of charge, $E = \sigma / 2\varepsilon_0$.

(ii) Again, we need to integrate to obtain the enclosed charge. (This time our gaussian surface encloses a cylinder of charge with radius a.)

$$Q = \int dQ = \int_0^a 2\pi bl\, dr = 2\pi abl$$

$$\Phi = \oint \mathbf{E} \cdot d\mathbf{A} = |E| \oint dA = 2\pi rl |E|$$

$$= \left[\frac{Q_{enc}}{\varepsilon_0} = \frac{2\pi abl}{\varepsilon_0} \right]$$

$$E = \frac{ab}{r\varepsilon_0}$$

(iii) Evaluating either of the above equations yields $E = b / \varepsilon_0$. (Because the volume charge density is never infinite (i.e., there are no surface or point charges), the electric field is continuous.)

(b) The x-components of the electric fields add together and directly contribute to the net electric field in the x-direction. The y-components of the electric field cancel, causing the net field in the y-direction to be zero. (This is an example of the second type of understanding symmetry.)

(c) In the infinitely long cylinder, both halves contribute equally to the field in the x-direction without any cancellation. Therefore, E_x due to half a cylinder is simply half of E_x due to a complete cylinder:

On the interval $r < a$:

$$E_x = \frac{b}{2\varepsilon_0}$$

On the interval $r > a$:

$$E_x = \frac{ab}{2r\varepsilon_0}$$

(d) Here we can use the superposition of electric fields. In part (c), we calculated the field due to a semi-infinite cylinder. To calculate the field due to two such cylinders, we can simply superpose the fields due to each. Because the two semi-infinite cylinders produce fields of equal magnitude and opposite direction along the x-axis (because they are oppositely charged), the net field in the x-direction at point P is zero.

(e) Recall that $V(x) = V(x) - V(\infty) = -\int_\infty^x E_x dx$

From part (d) the x-component of the electric field is zero everywhere along the x-axis, so this integral is zero and the potential is zero everywhere along the x-axis.

ELECTRICITY AND MAGNETISM III

(a) Initially the capacitor acts as a short circuit, so the current through the 5- and 20-Ω resistors is zero. The entire potential jump is across the 10-Ω resistor, causing a current of $I = 10\text{ V}/10\ \Omega = 1\text{ A}$.

(b) At equilibrium, the capacitor acts as an open circuit. Therefore, the equivalent resistance of the entire circuit is

$$R_{eq} = 10\ \Omega + \frac{1}{(1/5\ \Omega) + (1/20\ \Omega)} = 14\ \Omega$$

Therefore, the current flowing through the battery and the 10-Ω resistor is

$$I_{10} = \frac{10\text{ V}}{14\ \Omega} = \frac{5}{7}\text{ A}$$

Thus, the potential drop across the 10-Ω resistor is $V_{10} = I_{10}R_{10} = (0.71\text{ A})(10\ \Omega) = 7.1\text{ V}$, and the potential drop across the 5- and 20-Ω resistors is $10\text{ V} - 7.1\text{ V} = 2.9\text{ V}$. Applying Ohm's law across the 5- and 20-Ω resistors yields

$$I_5 = \frac{2.9\text{ V}}{5\ \Omega} = 0.57\text{ A}$$

$$I_{20} = \frac{2.9\text{ V}}{20\ \Omega} = 0.14\text{ A}$$

Answer check: 0.57 A + 0.14 A = 0.71 A, as required by the Kirchhoff node rule. Note that in calculating the currents through the 5- and 20-Ω resistors, we have been careful to keep sufficient numbers of digits (in the voltages) so that our answers are correct to two significant figures.

(c) At equilibrium, the capacitor has the same potential jump as the 5- and 20-Ω resistors (because they are in parallel), which is 2.9 V, as calculated in part (b).

(d) Initially, the capacitor acts as a battery with 2.9 V. Therefore, as in part (b),

$$I_5 = \frac{2.9\text{ V}}{5\,\Omega} = 0.57\text{ A}$$

$$I_{20} = \frac{2.9\text{ V}}{20\,\Omega} = 0.14\text{ A}$$

and $I_{10} = 0$.

(e) The voltage across a discharging capacitor obeys $V(t) = V_0 e^{-t/RC}$ (see Chapter 16), where V_0 is the voltage at $t = 0$ and R is the equivalent resistance of the circuit. The time constant is given by $\tau = RC$. Thus, if you plot ln V versus t, you expect a straight line whose slope is given by $-1/\tau$.

(f) The equivalent resistance of the parallel combination of the two resistors is $\frac{(5)(20)}{5+20}\,\Omega = 4\,\Omega$ and $\tau = (4\,\Omega)(10^{-5}\,F) = 4 \times 10^{-5}$ s. It is not practical to carry out this experiment as stated. You would need to replace at least one of the two resistors with a "megohm" resistor, i.e., a resistor with resistance of the order $10^6\,\Omega$, which would yield a time constant about 1 second long.

For further information on calculating currents within circuits, see Chapter 14.

Background

1

→ VECTORS AND SCALARS
→ ADDITION, SUBTRACTION, AND MULTIPLICATION OF VECTORS
→ UNIT ANALYSIS

This chapter is intended to provide you with a review of some background topics that are particularly important to Physics C, as well as some general tips on approaching questions on the AP exam. Because of space limitations, it is impossible to review all the math required for the AP exam. Therefore, we urge you to review topics with which you are having difficulty in appropriate math textbooks.

VECTORS

Scalars specify magnitude and no direction. For example, speed, which indicates how fast an object is moving but not in what direction, is a scalar. Other scalars include time and mass.

Vectors specify both magnitude and direction. For example, velocity, which indicates both how fast an object is moving and in what direction, is a vector. Other vectors include acceleration and force. Scalars can be equated with scalars, and vectors can be equated with vectors, but *scalars can never be equated with vectors.*

Graphical Vector Manipulation

A vector is generally represented by an arrow whose direction is in the direction of the vector and whose length is proportional to the vector's magnitude (Figure 1.1).

- *Vector addition* (using the tip-to-tail method): For example, to add **a** + **b**, the tip of **a** is placed at the tail of **b**. Then, **a** + **b** is the vector pointing from the tail of **a** to the tip of **b**.
- *Scalar multiplication*: To multiply a vector by a positive scalar, simply multiply the vector's magnitude by the scalar. (For example, to multiply by 2 simply double the vector's magnitude.) To multiply a vector by a negative scalar, change the vector's magnitude *and* reverse the direction of the vector.
- *Vector subtraction*: To subtract two vectors, **a** – **b**, add the first vector to the negative of the second vector: **a** – **b** = **a** + (–**b**). The negative of the second vector is obtained by reversing its direction (an example of scalar multiplication), and then the sum of these two vectors is obtained using the tip-to-tail method (vector addition).

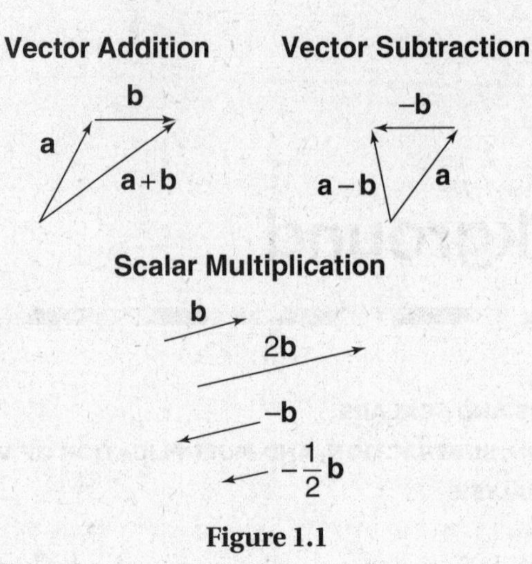

Figure 1.1

Polar[1] and Cartesian Coordinates

There are generally two ways to write vectors: using polar coordinates and using Cartesian coordinates. For example, consider the vector pointing from the origin to point (3, 4) in Figure 1.2.

- *Polar coordinates*: We can specify the magnitude and direction of this vector by saying that it has a magnitude of 5 (i.e., a length of 5) and makes an angle of 53.1 degrees with the $+x$-axis.
- *Cartesian coordinates*: Unit vectors $\hat{i}$, $\hat{j}$, and $\hat{k}$ are defined as vectors of length 1 that point parallel to the x-, y-, and z-axes, respectively. Therefore, using vector addition (and scalar multiplication), we can specify the magnitude and direction of this vector by saying that it is equal to $3\hat{i} + 4\hat{j}$.

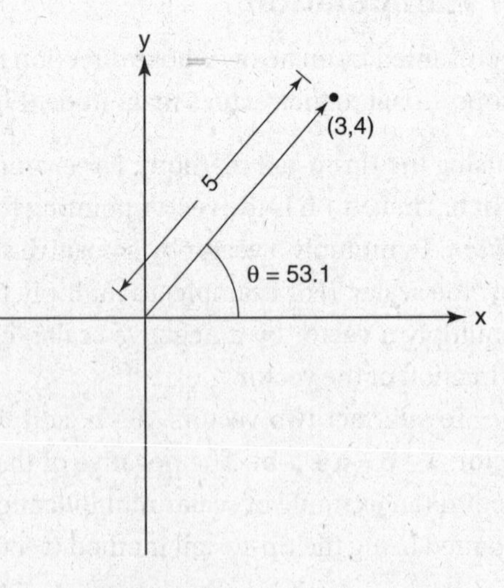

Figure 1.2

[1]In polar coordinates, the angle of the vector is generally specified in a *counterclockwise* sense starting at the $+x$-axis. However, in Physics C the direction of vectors is often indicated in terms of other angles. Therefore, we use the term "polar coordinates" here somewhat loosely to mean any vector whose direction is specified by some sort of angle.

CONVERSIONS BETWEEN CARTESIAN AND POLAR COORDINATES

Because it is easier to manipulate vectors in Cartesian coordinates, we generally solve problems using Cartesian coordinates. However, we must still be able to convert between polar and Cartesian coordinates.

- *Conversion from Cartesian coordinates*: The magnitude of the vector can be calculated using the Pythagorean theorem: $a = \sqrt{a_x^2 + b_y^2 + a_z^2}$. The direction of the vector must be obtained using trigonometry. (For example, in Figure 1.2 the angle could be calculated using the formula $\theta = \tan^{-1}\frac{4}{3}$.)
- *Conversion from polar to Cartesian coordinates*: This involves using trigonometry. For example, in Figure 1.2, $a_x = 5\cos 53.1°$ and $a_y = 5\sin 53.1°$.

ALGEBRAIC MANIPULATION OF VECTORS IN CARTESIAN COORDINATES

- *Scalar multiplication*: Multiply each component of a vector by the scalar. For example, if $\mathbf{a} = (2\hat{i} - 1\hat{j})$, then $2\mathbf{a} = (4\hat{i} - 2\hat{j})$ and $-\mathbf{a} = (-2\hat{i} + 1\hat{j})$.
- *Vector addition*: Add the *x*-, *y*-, and *z*-components together. For example, if $\mathbf{a} = (a_x\hat{i} + a_y\hat{j})$ and $\mathbf{b} = (b_x\hat{i} + b_y\hat{j})$, then $\mathbf{a} + \mathbf{b} = (a_x + b_x)\hat{i} + (a_y + b_y)\hat{j}$.
- *Vector subtraction*: Subtract the *x*-, *y*-, and *z*-components. For example, if $\mathbf{a} = (a_x\hat{i} + a_y\hat{j})$ and $\mathbf{b} = (b_x\hat{i} + b_y\hat{j})$, then $\mathbf{a} - \mathbf{b} = (a_x - b_x)\hat{i} + (a_y - b_y)\hat{j}$.

Dot Product

The dot product of two vectors, $\mathbf{a} \cdot \mathbf{b}$, is equal to the product of the length of either vector and the length of the component of the other vector parallel to the first vector. We are familiar with resolving vectors into components parallel to the *x*-, *y*-, and *z*-axes when converting vectors in polar coordinates to Cartesian coordinates. Computing the length of the component of a vector parallel to another vector is similar. For example, when calculating the length of the component of **b** parallel to **a** in Figure 1.3, we resolve the **b** vector into components that are parallel ($b_{\parallel}$) and perpendicular ($b_{\perp}$) to **a** as shown. When we show the components of a vector on the same diagram as the original vector, we add a squiggly line to the original vector to indicate that we are replacing the vector by its components; otherwise, we will have duplicate information about the vector. The length of the component of **b** parallel to **a**, $b_{\parallel}$, is equal to $b\cos\theta$.

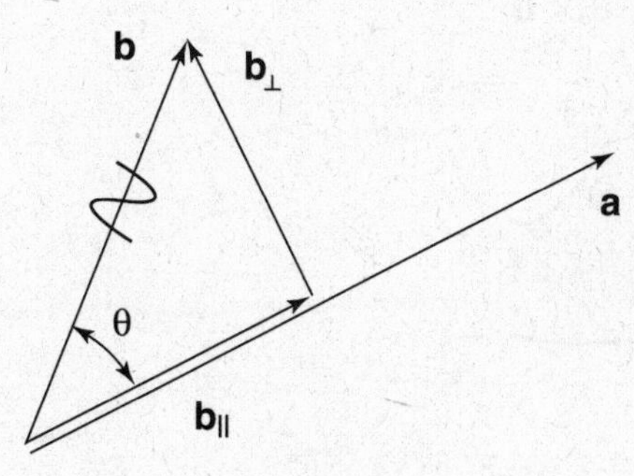

Figure 1.3

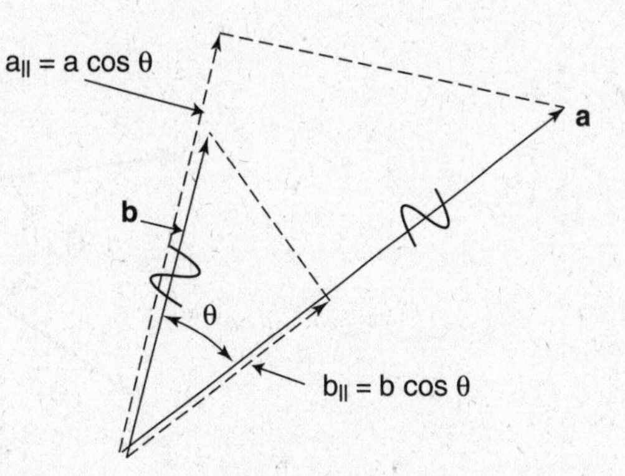

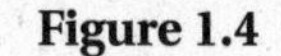

Figure 1.4

Thus, we can restate our definition of the dot product in two ways (Figure 1.4):

$$\mathbf{a} \cdot \mathbf{b} = ab_{\parallel} = ab\cos\theta$$

$$\mathbf{a} \cdot \mathbf{b} = ba_{\parallel} = ba\cos\theta$$

Therefore, if two vectors are perpendicular, their dot product will equal zero [$\cos(\pi/2) = 0$]; if two vectors are parallel, their dot product will equal the product of their magnitudes ($\cos 0 = 1$); and if two vectors are antiparallel, their dot product will be the negative of the product of their magnitudes [$\cos(\pi) = -1$].

In Cartesian coordinates, the dot product of two vectors $\mathbf{a} = a_x\hat{i} + a_y\hat{j} + a_z\hat{k}$ and $\mathbf{b} = b_x\hat{i} + b_y\hat{j} + b_z\hat{k}$ is given by the formula $\mathbf{a} \cdot \mathbf{b} = a_xb_x + a_yb_y + a_zb_z$. The dot product is also known as the *scalar product* because the final result is a scalar.

Cross-Product

Unlike the dot product, the cross-product of two vectors yields a third vector. (Thus, the cross-product is also known as the *vector product.*)

The magnitude of the cross-product of two vectors, $|\mathbf{a} \times \mathbf{b}|$, is equal to the magnitude of either vector multiplied by the length of the component of the second vector perpendicular to the first vector. As in the case of the dot product, we can write the vector product in two equivalent ways:

$$|\mathbf{a} \times \mathbf{b}| = ab_{\perp} = ab\,|\sin\theta|$$

$$|\mathbf{a} \times \mathbf{b}| = ba_{\perp} = ba\,|\sin\theta|$$

Therefore, if two vectors are perpendicular, their cross-product is equal to the product of their magnitudes [$\sin(\pi/2) = 1$]; and if two vectors are parallel or antiparallel, their cross-product is zero ($\sin 0 = \sin\pi = 0$).

The direction of the cross-product is *perpendicular* to the plane containing vectors **a** and **b**. Therefore, in determining the direction, first imagine a plane that contains the two vectors. There are *two* antiparallel directions perpendicular to this plane, but which one points in the correct direction? To determine this, you must use the right-hand rule. Imagine placing the palm of your right hand along the plane so that your fingers point parallel to **a**. If your fingers curl in the direction of **b**, your thumb will indicate the correct perpendicular direction (Figure 1.5).

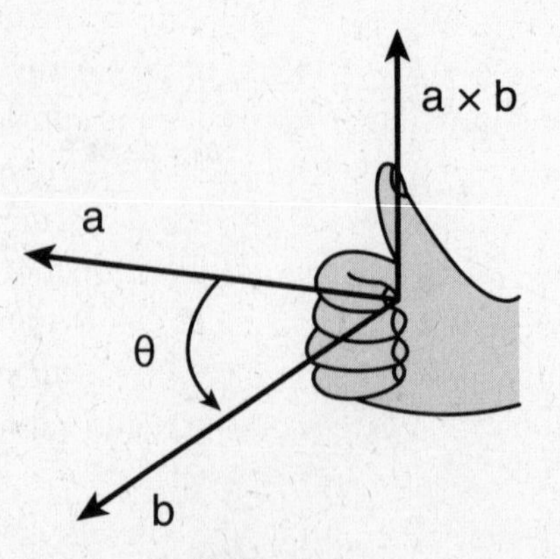

Figure 1.5

An alternative method of finding the direction of the cross-product in the special case that **a** and **b** are perpendicular is as follows. Point the index finger of your right hand out and extend your middle finger as shown in Figure 1.6. If your index finger points parallel to **a** and your middle finger points parallel to **b**, the cross-product will point parallel to your thumb.

Make sure to use your right hand!

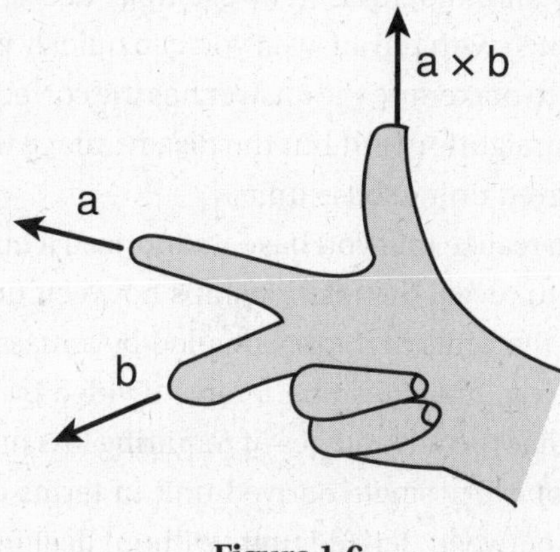

Figure 1.6

UNIT ANALYSIS

All measurements and observable quantities have units; otherwise they would be meaningless. For example, to specify a length it is not sufficient to say that it is "4": Some unit must also be indicated, such as "4 inches" or "4 miles." Additionally, each quantity must be specified in units of the appropriate dimension: We can specify length in terms of inches, centimeters, or miles, but we cannot say a length is "4 seconds." This is the key to unit analysis: *If an answer has the wrong units, it must be wrong.*

Rules of Unit Analysis

The following rules govern the ways in which units combine in formulas and equations.

- *Multiplication and division*: Units are multiplied and divided just as variables are. For example, if a has units of seconds and b has units of meters, b/a has units of meters/second and b/a^2 has units of m/s^2.
- *Addition and subtraction*: The sum or difference of two quantities with the same units has those same units (e.g., 1 m + 3 m = 4 m). It is always incorrect to add or subtract quantities with different units! For example, it is meaningless to say that a certain quantity is equal to 1 meter plus 1 second. Therefore, if your answer to a problem is the sum of two quantities with different units (i.e., it has mixed units), then it must be incorrect.
- *Exponential functions*: The argument x of an exponential function, such as e^x, must be dimensionless, such as the ratio of two lengths. For example, $e^{10\text{m}}$ is meaningless.
- *Trigonometric functions*: Arguments of trigonometric functions, such as $\sin x$ and $\tan^{-1} x$, also must be dimensionless (generally one that is the ratio of two lengths).

Two Approaches to Unit Analysis

The best single way to check the results of a problem that yields a complicated formula is to analyze its units and make sure they are consistent with the type of quantity specified by the formula. There are two basic strategies for memorizing and manipulating units in order to do this. The first strategy is to memorize how every derived unit is expressed in terms of base units: meters, seconds, and kilograms. (For example, acceleration is m/s^2 and force is $kg \cdot m/s^2$.) Then when presented with the answer, you can quickly express everything in terms of base units and then check to make sure the answer has the correct base units. This approach has the advantage that it is straightforward but the disadvantage that it requires memorizing the relationship of every derived unit to base units.

TIP

Unit analysis is one of the best ways to check your answers. Even Nobel laureates in physics use unit analysis to check their work.

The second approach is to realize that you have memorized a number of formulas and that you can use these formulas to reveal the relationships between units. For example, suppose you are trying to determine the units of a force divided by a mass. In that case, comparison to Newton's second law, $\mathbf{F} = m\mathbf{a}$, indicates that a force divided by a mass must have units of acceleration. This approach has the advantages of requiring less memorization (you need not memorize the relationship of every single derived unit in terms of base units) and of often allowing for quick shortcuts between derived units without dealing with base units.

PROBLEM-SOLVING TECHNIQUES

Other Ways to Check Symbolic Answers (Answers that Are Expressions of Variables)

- Make sure each variable is related to the final answer you would expect based on qualitative intuition. For example, you would expect that an object's speed after falling a certain distance would increase if gravity increased. Therefore, if the answer has g in the denominator, this should be an indication that the answer might be wrong.
- Make sure that the solution is consistent with various limiting situations. For example, in the situation of a falling mass, if you imagine that g becomes infinitely big, the final speed of the mass should approach infinity.

These answer checks will be illustrated throughout the text. Because of the ease of checking symbolic answers by unit analysis and by the methods described above, it is always wise to solve a multistep problem symbolically and then substitute in values at the end (rather than substituting in values as you proceed through the problem). This allows you to check your answer more thoroughly and gives you insight into general physical statements beyond the problem at hand.

Strategy for Multiple-Choice Questions

The scoring of the multiple-choice questions is designed such that random guessing does not affect your score. However, if you are able to eliminate even one choice, educated guessing on the average increases your score. Therefore, even if you cannot obtain the correct answer to a multiple-choice question, it is worth your time to eliminate as many choices as you can and then guess the correct solution. Many methods for eliminating choices for multiple-choice questions (e.g., unit analysis and the symbolic answer checks discussed above) are explained in the solutions to our model multiple-choice questions.

Strategy for Free-Response Questions

These questions are graded by physics teachers using detailed, partial-credit rubrics, so it's generally best to do as much as you can and show all your work. One useful technique in approaching free-response questions is to realize that the parts within each question often build on each other. Therefore, it may be helpful to read all the parts before you start working, to get an idea of where the questions are leading. If you do not know the answer to one part of a question and the answer is needed in subsequent parts, it is advisable to make up a value for the unknown answer using appropriate units (do not use the values 0, 1, or 10). Then use this value in the later parts of the question. This approach will demonstrate the full extent of your knowledge and maximize the partial credit you will receive.

One-Dimensional Kinematics

2

→ **INSTANTANEOUS SPEED, VELOCITY, AND ACCELERATION**

→ **AVERAGE SPEED, VELOCITY, AND ACCELERATION**

→ **UNIFORMLY ACCELERATED MOTION: FREELY FALLING OBJECTS**

Based on the following three definitions, we will derive all the equations for one-dimensional kinematics and uniformly accelerated motion (UAM), or constant acceleration, using calculus. All the derivations are posed as questions, and we suggest you attempt them on your own before checking our solutions.

DEFINITIONS

Instantaneous velocity:

$$v(t) = \lim_{\Delta t \to 0} \frac{\Delta x}{\Delta t} = \frac{dx}{dt} \Leftrightarrow \Delta x = \int dx \equiv \int_{t_1}^{t_2} v(t)dt$$

Instantaneous speed:

$$|v(t)|$$

Instantaneous acceleration:

$$a(t) = \lim_{\Delta t \to 0} \frac{\Delta v}{\Delta t} = \frac{dv}{dt} \equiv \frac{d^2x}{dt^2} \Leftrightarrow \Delta v = \int dv \equiv \int_{t_1}^{t_2} a(t)dt$$

Note: These are the one-dimensional definitions of velocity, speed, and acceleration. In the next chapter, we will introduce two-dimensional definitions. For now, no formal vector notation is needed; a value's sign (+/–) is sufficient to specify direction along one dimension.

Velocities and speeds are expressed in length/time (SI units of m/s), and accelerations are expressed in length/time^2 (SI units of m/s^2).

Instantaneous speed is defined to be the magnitude of instantaneous velocity (velocity is a vector, and speed is a scalar). For example, if I told you that a bullet was traveling at 100 m/s, that would be the speed of the bullet. Alternatively, if I told you that a bullet was traveling at 100 m/s toward a target (magnitude and direction), that information would describe the velocity of the bullet.

Derivation of Average Velocity For an arbitrary $x(t)$ function over some time interval (t_1 to t_2), what is the average velocity?

Solution

Recall from calculus the formula for the average value of a function.

$$\text{Average value} = \frac{\int_{x_1}^{x_2} f(x)\,dx}{x_2 - x_1}$$

We need to apply this formula to the velocity function:

$$\text{Average velocity} = \bar{v} = \frac{\int_{t_1}^{t_2} v(t)\,dt}{t_2 - t_1}$$

However, we want to express this average velocity using the position function $x(t)$, so we must relate velocity to position.

$$v(t) = \frac{dx}{dt} \Rightarrow v(t)\,dt = \frac{dx}{dt}\,dt = dx$$

$$\text{Average velocity} = \bar{v} = \frac{\int_{t_1}^{t_2} dx}{t_2 - t_1} = \frac{[x]_{t_1}^{t_2}}{t_2 - t_1} = \frac{x(t_2) - x(t_1)}{t_2 - t_1} = \frac{\Delta x}{\Delta t}$$

The numerator, Δx, is called the *displacement*. Displacement is the net difference in the location of an object independent of how the object got there. For example, if you run 30 complete laps at a track, your displacement is precisely zero.

Average quantities are often denoted as the variable with a bar over it ($\bar{v}$, $\bar{a}$).

Graphical interpretation (Figure 2.1): The *average* velocity between any two points on an $x(t)$ graph is the slope of the line segment connecting the two points. Note that this involves displacement and therefore is path independent (e.g., the two paths shown have the same average velocity between points A and B). Alternatively, the *instantaneous* velocity at any point is the slope of the $x(t)$ graph at that particular point.

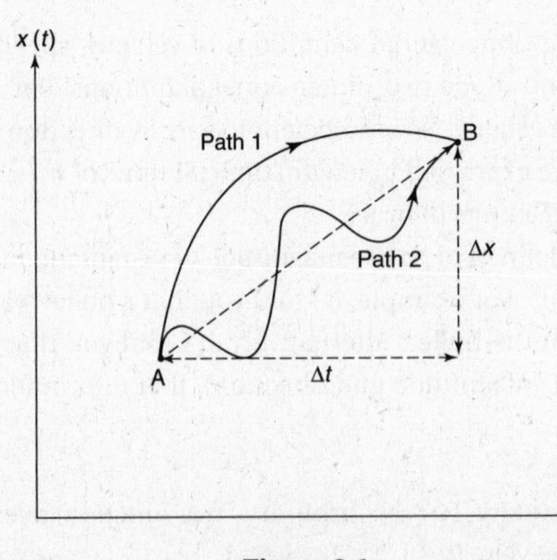

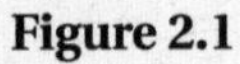
Figure 2.1

Derivation of Average Speed For an arbitrary $x(t)$ function over some time interval (t_1 to t_2), what is the average speed?

Solution

We'll use the same approach (the formula for the average value from calculus):

$$\text{Average speed} = \frac{\int_{t_1}^{t_2} |dx|}{t_2 - t_1} = \frac{\text{total distance}}{\Delta t}$$

Unlike displacement, total distance is the distance traveled irrespective of direction. For example, if you make 30 complete laps around a track, although your net displacement is zero, the total distance is equal to 30 times the circumference of the track.

Average speed is *not* the magnitude of average velocity.

This difference can be understood in terms of the fact that velocities and displacements are vectors, whereas speeds and distances are scalars. Consider the integral for displacement: $\Delta x = \int v dt$. Since vdt is a vector (v is a vector, and dt is a scalar, so the product is a vector), it can be positive or negative and thus can cancel itself during the integration. Alternatively, when integrating total distance, $|dx|$ is always positive, so it cannot cancel itself during the integration. Therefore, it makes sense that for any object's path during any time interval,

$$\text{Total distance} \geq \text{displacement}$$

If the path is in a single direction along a straight line, the total distance equals the displacement. Otherwise, the total distance is greater than the displacement.

Derivation of Average Acceleration For an arbitrary $v(t)$ function over some time interval (t_1 to t_2), what is the average acceleration?

Solution

Same approach as above:

$$\text{Average acceleration} = \bar{a} = \frac{\int_{t_1}^{t_2} a(t)dt}{t_2 - t_1} = \frac{\int_{t_1}^{t_2} \frac{dv}{dt}dt}{t_2 - t_1} = \frac{\int_{t_1}^{t_2} dv}{t_2 - t_1} = \frac{v\big|_{t_1}^{t_2}}{t_2 - t_1} = \frac{v(t_2) - v(t_1)}{t_2 - t_1} = \frac{\Delta v}{\Delta t}$$

The graphical interpretation is completely analogous to the interpretation of average velocity: The average acceleration between any two points (a and b) on a $v(t)$ graph is the slope of the line segment connecting them, which is path independent. Alternatively, the instantaneous acceleration at any point is the slope of the $v(t)$ graph at that particular point.

Derivation of Velocity as a Function of Time for UAM For an object with uniform acceleration a and initial velocity v_0 at $t = 0$, calculate the velocity as a function of time, $v(t)$.

Solution

The following fundamental schematic is a guide to conversion between position, velocity, and acceleration. (This is nothing new but rather simply a compact way to write down the definitions of position, velocity, and acceleration introduced at the beginning of the chapter.)

TIP

How to relate position, velocity, and acceleration

$$x(t)\underset{\text{integrate}}{\overset{\text{differentiate}}{\rightleftarrows}}v(t)\underset{\text{integrate}}{\overset{\text{differentiate}}{\rightleftarrows}}a(t)$$

In this case, we are converting from acceleration to velocity, so we need to integrate once:

$$a=\frac{dv}{dt}$$

Separating variables,

$$dv=adt$$

Integrating both sides of the equation,

$$\int dv=\int a\,dt$$

$$v=at+C$$

where C is the constant of integration. From the initial conditions, it is clear that $v_0=C$, so that

$v=v_0+at$ **valid only for uniformly accelerated motion**

Quantities with zeros as subscripts (for example, x_0, v_0, a_0, and t_0) are interpreted to mean the initial value of that parameter, the value at time $t=0$.

Alternative solution: Whenever it is possible to perform an indefinite integral and solve for the constant of integration, it is equally valid to simply use a definite integral.

$$\int_{v_0}^{v_f} dv=\int_{t=0}^{t=t_f} a\,dt \Rightarrow v_f-v_0=at_f \Rightarrow v_f=v_0+at_f$$

Derivation of Position as a Function of Time for UAM For an object with uniform acceleration a, initial velocity v_0, and position x_0 at $t=0$, calculate the position as a function of time, $x(t)$.

Solution

Again referring to the fundamental schematic,

$$x(t)\underset{\text{integrate}}{\overset{\text{differentiate}}{\rightleftarrows}}v(t)\underset{\text{integrate}}{\overset{\text{differentiate}}{\rightleftarrows}}a(t)$$

We see that to get from $a(t)=a$ to $x(t)$, we need to perform two integrations. The first integration is identical to that in the derivation of velocity, with the same final result:

$$v=\frac{dx}{dt}=v_0+at$$

Separating variables,

$$dx=v_0dt+atdt$$

Integrating both sides,

$$\int dx=\int v_0dt+\int at\,dt$$

$$x=v_0t+\frac{1}{2}at^2+C$$

From the initial conditions, the constant of integration C must equal x_0.

$$x = x_0 + v_0 t + \frac{1}{2}at^2$$ **valid only for uniformly accelerated motion**

Alternative solution: Again, a definite integral can also be used.

$$\int_{x_0}^{x_f} dx = \int_{t=0}^{t=t_f} v_0\,dt + \int_{t=0}^{t=t_f} at\,dt$$

$$x_f - x_0 = v_0 t + \frac{1}{2}at^2$$

Derivation: Relating Velocity to Position for UAM We've just worked out the relationship between velocity and time, $v(t)$, and between position and time, $x(t)$. Now we want to combine these two equations to get an equation that relates velocity directly to position.

Solution

The time variable can be eliminated using substitution. Starting with $v(t)$,

$$v = v_0 + at$$

Solving for t,

$$t = \frac{v - v_0}{a}$$

Substituting into the $x(t)$ equation,

$$x = \frac{1}{2}at^2 + v_0 t + x_0 = \frac{1}{2}a\left(\frac{v - v_0}{a}\right)^2 + v_0\left(\frac{v - v_0}{a}\right) + x_0$$

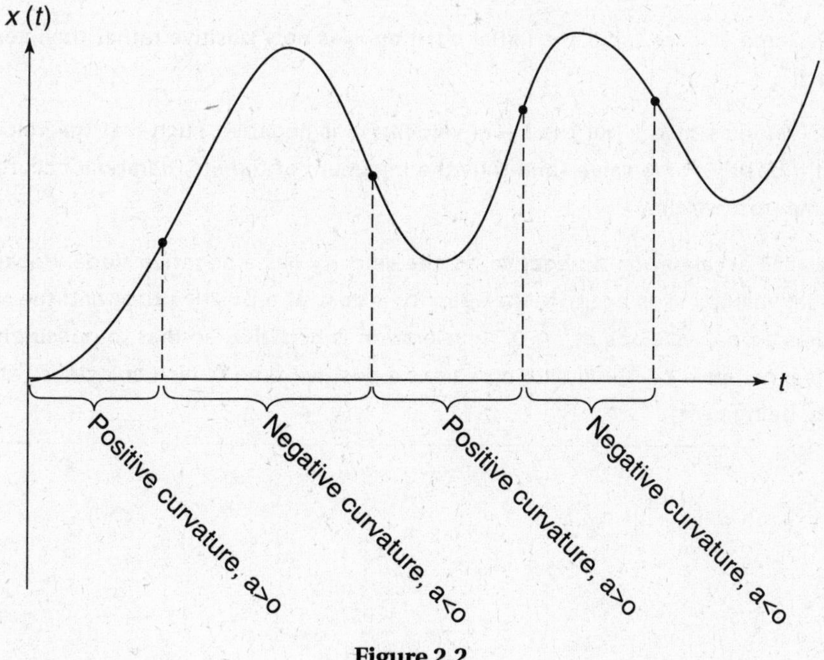

Figure 2.2

Simplifying,

$$v^2 = v_0^2 + 2a(x - x_0) \quad \text{valid only for uniformly accelerated motion}$$

In general, the position, velocity, and acceleration graphs are successive derivatives of each other. Their graphs reflect this and should remind you of graphs of $f(x)$, $f'(x)$, and $f''(x)$ encountered in calculus courses. Recall that the sign of the acceleration [the second derivative of an $x(t)$ curve] can be determined based on the *curvature*: Positive acceleration corresponds to positive curvature (concave up), whereas negative acceleration corresponds to negative curvature (concave down), as shown in Figure 2.2.

For graphs for uniformly accelerated motion, the following are true statements:

1. Acceleration is constant, so the $a(t)$ graph is a horizontal line.
2. $v = v_0 + at$. The $v(t)$ graph is a line that may have a nonzero slope (if there is nonzero acceleration). A positive slope corresponds to positive acceleration, and a negative slope to negative acceleration.
3. $x = \frac{1}{2}at^2 + v_0 t + x_0$. The $x(t)$ graph is a parabola if there is nonzero acceleration, a line with nonzero slope if there is zero acceleration but nonzero v_0, and a line with zero slope if a and v_0 are both zero.

EXAMPLE 2.1

A few of these cases are shown in Figure 2.3. You should attempt to understand these situations before reading the following descriptions.

Case 1: The acceleration is positive, so the velocity graph has a positive slope. Because the velocity line passes through the x-axis at the origin, the parabola has a horizontal tangent at that point [the $x(t)$ function's minimum].

Case 2: Same as case 1, but the initial position x_0 is now positive rather than zero as in case 1.

Case 3: Same as case 1, but the initial velocity v_0 is negative, such that the velocity is zero at a positive time value [and thus the minimum of the $x(t)$ parabola occurs at this same positive time].

Case 4: The acceleration is negative, so the velocity has a negative slope. Because the initial velocity v_0 is positive, the velocity is zero at a positive time and the $x(t)$ curve has a positive slope at $t = 0$. Acceleration is negative, so that the position curve is a negative parabola with a peak at a positive time (which coincides with velocity being zero).

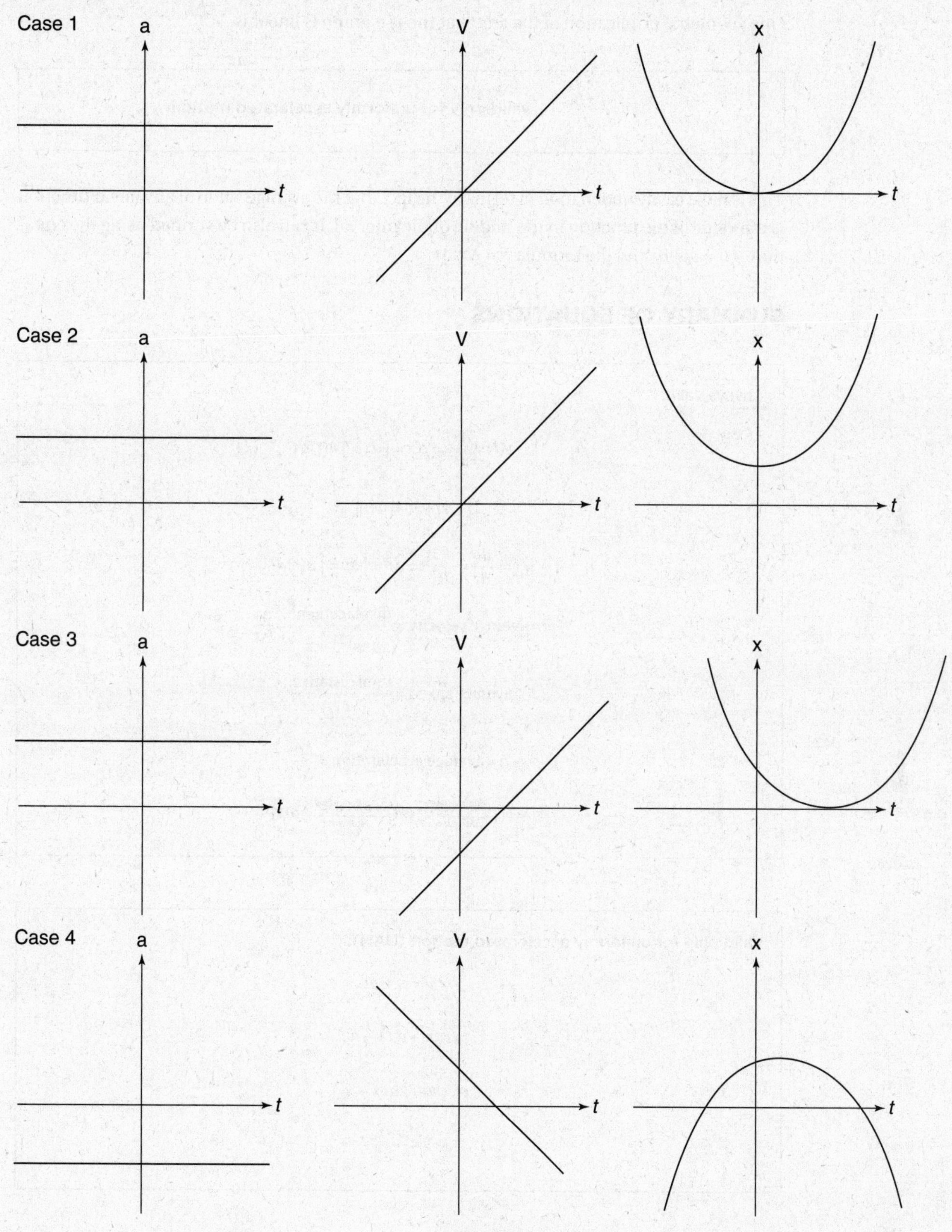

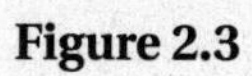

Figure 2.3

One geometric implication of the fact that the $v(t)$ graph is linear is

$$\bar{v} = \frac{v + v_0}{2}$$ **valid only for uniformly accelerated motion**

This is most easily understood in terms of the fact that the average value of any linear function is the value of the function in the middle of the interval. It can also be verified using the equation $v(t) = v_0 + at$ and the formula $\bar{v} = \Delta x / \Delta t$.

SUMMARY OF EQUATIONS

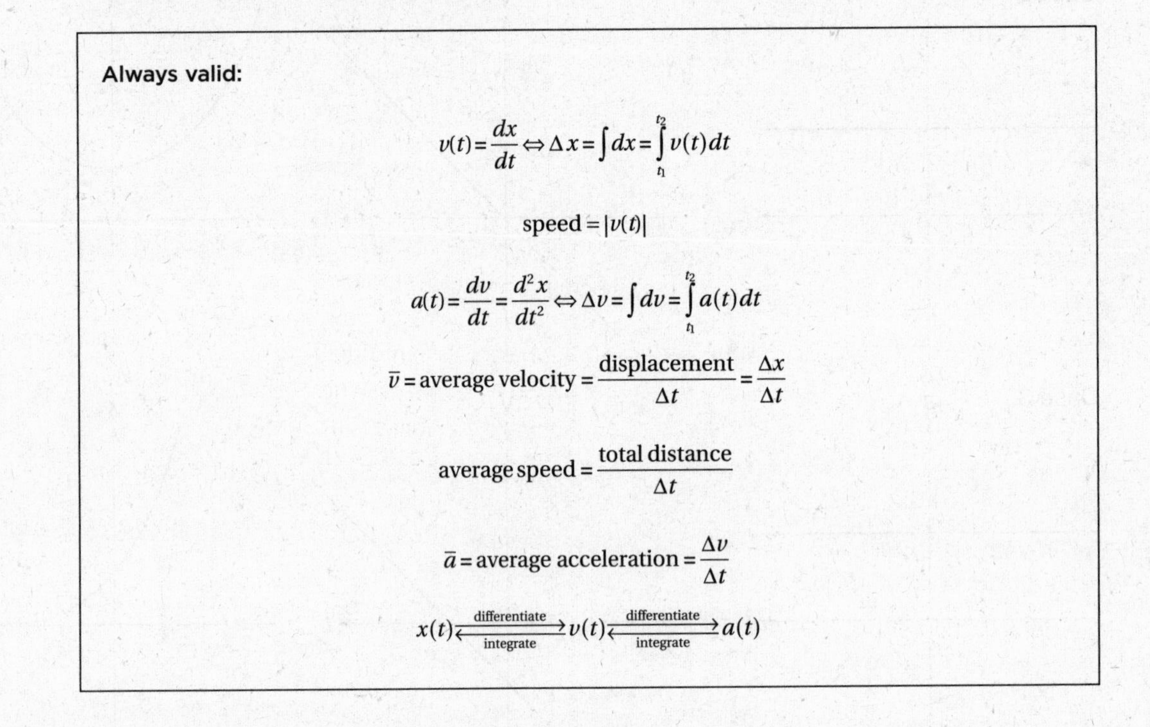

Always valid:

$$v(t) = \frac{dx}{dt} \Leftrightarrow \Delta x = \int dx = \int_{t_1}^{t_2} v(t)\,dt$$

$$\text{speed} = |v(t)|$$

$$a(t) = \frac{dv}{dt} = \frac{d^2x}{dt^2} \Leftrightarrow \Delta v = \int dv = \int_{t_1}^{t_2} a(t)\,dt$$

$$\bar{v} = \text{average velocity} = \frac{\text{displacement}}{\Delta t} = \frac{\Delta x}{\Delta t}$$

$$\text{average speed} = \frac{\text{total distance}}{\Delta t}$$

$$\bar{a} = \text{average acceleration} = \frac{\Delta v}{\Delta t}$$

$$x(t) \underset{\text{integrate}}{\overset{\text{differentiate}}{\rightleftarrows}} v(t) \underset{\text{integrate}}{\overset{\text{differentiate}}{\rightleftarrows}} a(t)$$

Valid only for uniformly accelerated motion (UAM):

$$v = v_0 + at$$

$$x = x_0 + v_0 t + \frac{1}{2}at^2$$

$$v^2 = v_0^2 + 2a(x - x_0)$$

$$\bar{v} = \frac{v + v_0}{2}$$

PROBLEM SOLVING

The two sets of equations above define roughly two basic types of problems:

1. *Nonuniform accelerated motion problems*: These generally involve conversion between position, velocity, and acceleration via differentiation or integration. (Always remember the constant of integration if you are using indefinite integrals.) You've probably seen similar or identical problems in calculus.
2. *UAM problems*: These problems give you a set of values (such as x_0, v_0, a, t_1, and t_2) and ask you to calculate other values from them. They generally involve algebraic manipulations of the equations valid only for UAM listed above.

Tips for Solving Free-Fall Problems

One of the most common types of UAM problems involve objects falling only under the force of gravity (air resistance ignored), such that their acceleration is g. The following tricks are useful in solving this type of problem:

1. At the peak of any trajectory, the object's y-component of velocity is zero. This is often used to calculate the peak height (by setting the velocity equal to zero).
2. Because of conservation of energy (which will be discussed later), and as reflected mathematically in the symmetry of parabolas, an object's *speed* as it passes a certain height is exactly the same on the way up as it is on the way down.
3. Also because of the symmetry of parabolas, the magnitude of the time difference between when the object is at the peak of its trajectory (t_{peak}) and when it is at a particular y-position below its peak is the same whether the object is ascending or descending at the latter position.
4. You should memorize that g is 9.8 m/s^2.

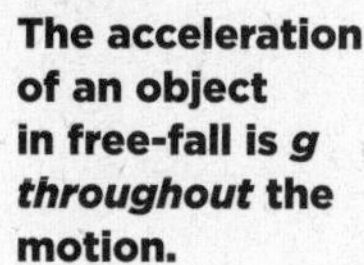

The acceleration of an object in free-fall is *g* *throughout* the motion.

CHAPTER SUMMARY

Motion along a straight line (one dimension) is described in terms of the position $x(t)$ of the object at time t with respect to the origin, the instantaneous velocity $v = \frac{dx}{dt}$, and the instantaneous acceleration $a = \frac{dv}{dt}$. Speed is the absolute value of the velocity. Average velocity and acceleration are defined in terms of the net change of the displacement and velocity, respectively. When the acceleration is constant (UAM), the average and instantaneous accelerations are equal, and the position and instantaneous velocity of the object can be found in terms of the acceleration, and the initial position and velocity. An important physical example of UAM is "free-fall," the motion of an object under the influence of gravity alone near the surface of the Earth.

PRACTICE EXERCISES

Multiple-Choice Questions

1. The position of an object is given by the equation $x(t) = 2 + 4t - t^2$, where position is measured in meters and time in seconds. What is the particle's average acceleration from $t = 0$ to $t = 2$?

   (A) $-4\ \text{m/s}^2$
   (B) $-2\ \text{m/s}^2$
   (C) $0\ \text{m/s}^2$
   (D) $2\ \text{m/s}^2$
   (E) $4\ \text{m/s}^2$

2. A 400-kg car can accelerate from rest to a final speed v over a distance d. To what speed can the car accelerate within a distance of $2d$ (again, starting from rest)? Assume the same value of acceleration in both cases.

   (A) $\sqrt{2}v$
   (B) $2v$
   (C) $2\sqrt{2}v$
   (D) $4v$
   (E) $8v$

3. Figure 2.4 compares the $v(t)$ curves of two objects. The area under both curves from $t = 0$ to $t = t_f$ is the same. Which of the following quantities is the same from $t = 0$ to $t = t_f$?

   (A) average position
   (B) average velocity
   (C) average acceleration
   (D) total displacement
   (E) both (B) and (D) are correct

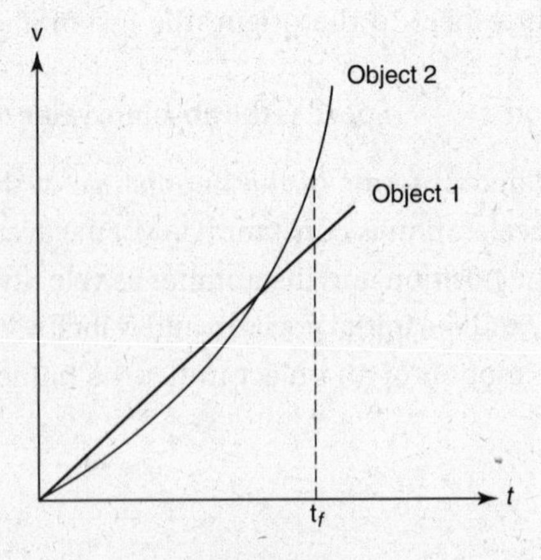

Figure 2.4

4. In the situation introduced in question 3, if both objects start at $x = 0$ at $t = 0$, which of the following statements is correct?

(A) Object 1 will be ahead of object 2 for the entire interval, $0 < t < t_f$.
(B) Object 2 will be ahead of object 1 for the entire interval, $0 < t < t_f$.
(C) Object 1 will initially be ahead but will lose its lead part way through.
(D) Object 2 will initially be ahead but will lose its lead part way through.
(E) The objects switch positions in the lead twice.

5. The velocity of a particle moving in one dimension is given by the equation $v(t) = 4 + 3t^2$, where velocity is in m/s and time is in seconds. What is the average value of the velocity in the interval $t = 0$ to $t = 2$?

(A) 4 m/s
(B) 8 m/s
(C) 12 m/s
(D) 16 m/s
(E) 20 m/s

6. An object initially at rest at position $x = 0$ starts moving with constant acceleration. After 1 s, the object is located at $x = 2$. What is the object's velocity at $t = 2$ s?

(A) 0.5 m/s
(B) 1 m/s
(C) 2 m/s
(D) 4 m/s
(E) 8 m/s

7. The velocity function for an object falling from rest under the influence of gravity and air resistance is shown in Figure 2.5. Which of the following position functions shown in the figure is consistent with this velocity function?

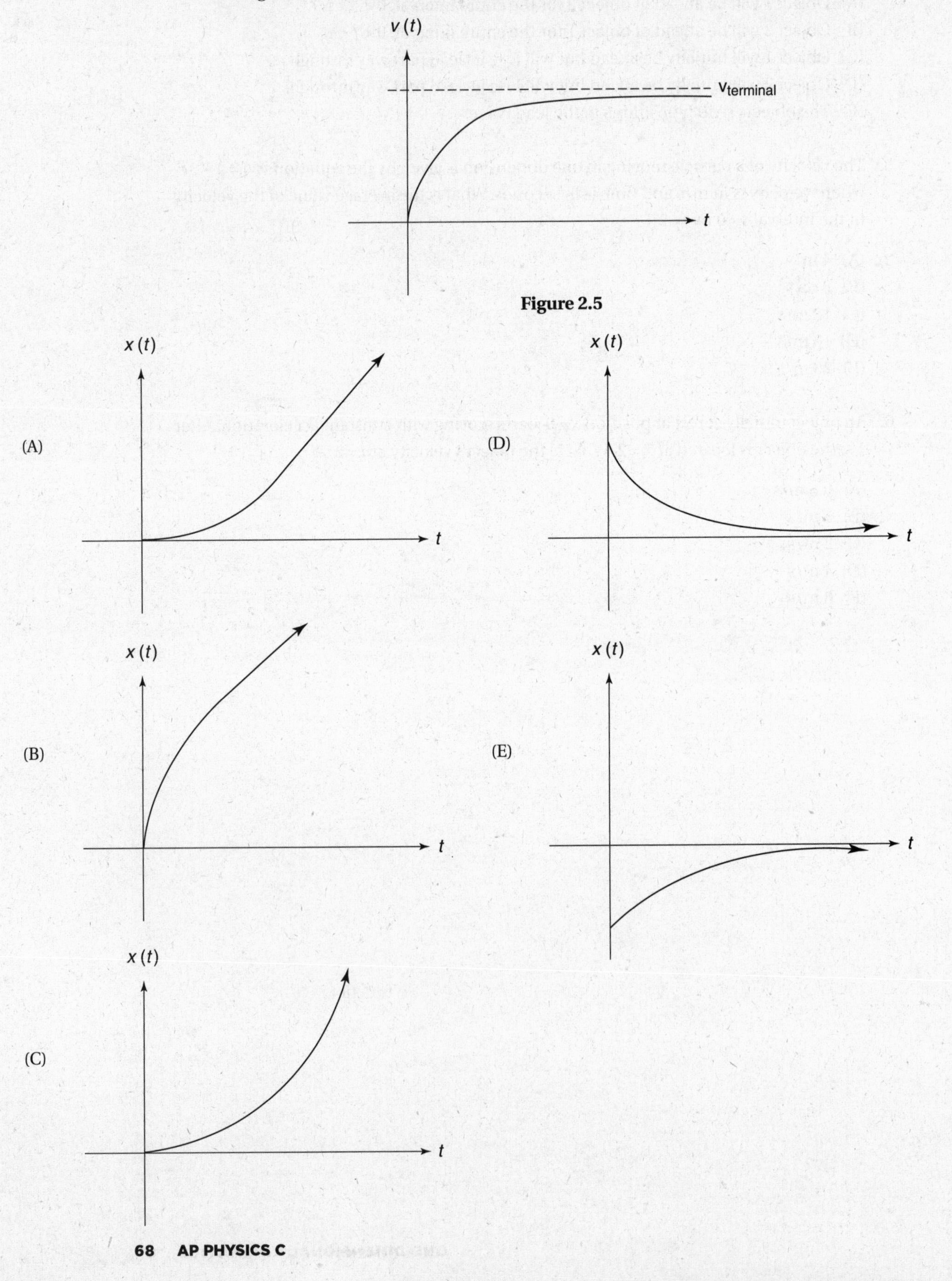

Figure 2.5

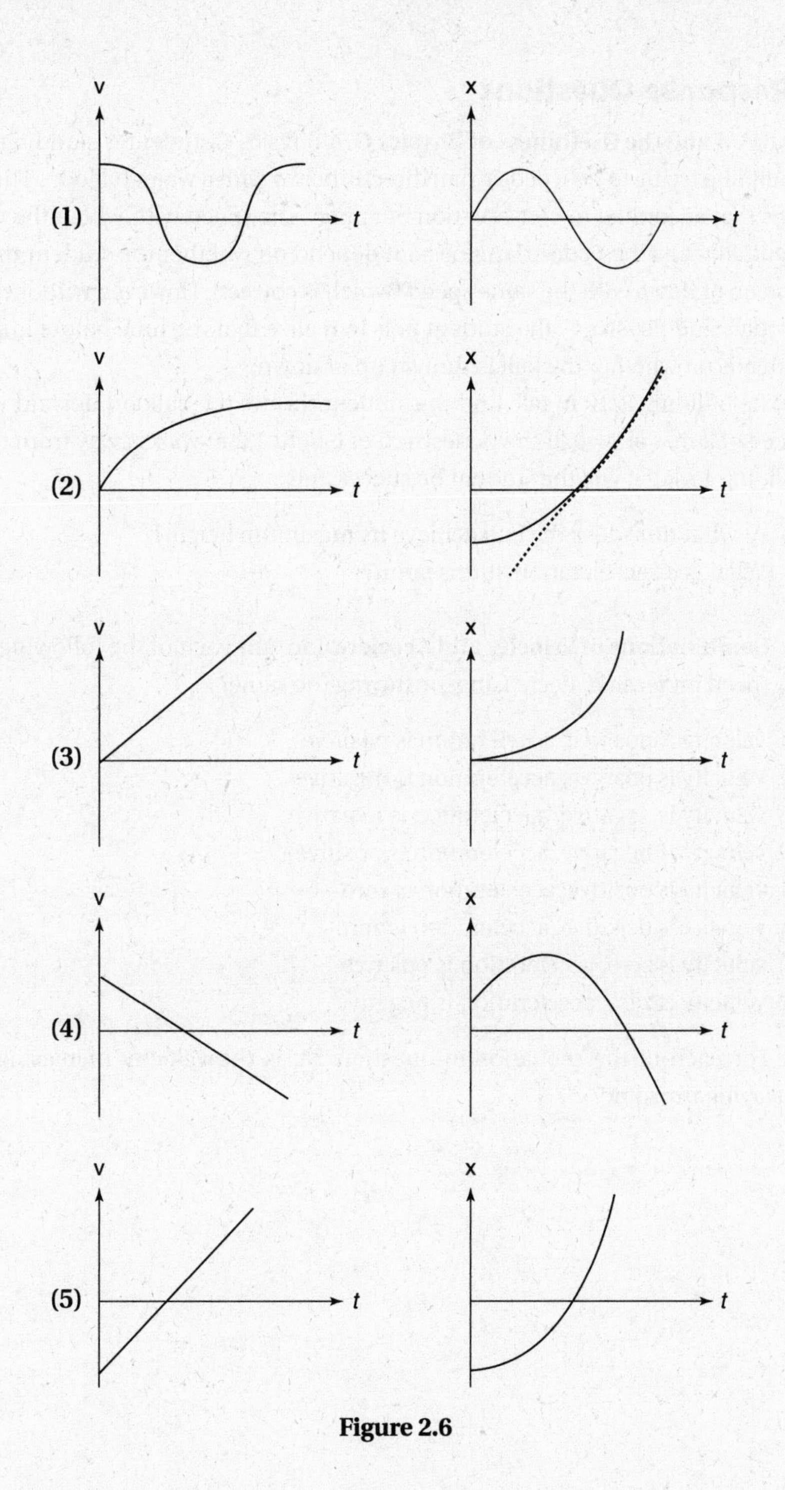

Figure 2.6

8. Shown in Figure 2.6 are five pairs of velocity and position functions. How many of the pairs are qualitatively consistent with each other?

   (A) one
   (B) two
   (C) three
   (D) four
   (E) five

Free-Response Questions

1. (a) **UAM and the Usefulness of Physics C** A Physics C student is standing on the top of a building trying to hit a pedestrian directly below with a water balloon. The student reasons that according to conservation of energy, the speed with which the water balloon eventually hits the pedestrian does not depend on whether the student throws the balloon up or down with the same speed (which is correct). However, without knowledge of calculus and Physics C, the student fails to realize that the time before impact critically depends on whether the ball is thrown up or down.

   The building is 10 m tall, and the student throws the balloon upward with an initial speed of 2 m/s at $t = 0$. If the pedestrian of height 1.2 m walks away from the side of the building 1 s later, will the student be successful?

   (b) At what time does the ball achieve its maximum height?

   (c) What is its acceleration at this point?

2. (a) **Combinations of Velocity and Acceleration** For each of the following situations, is the speed increasing, decreasing, or staying the same?

   (A) velocity is positive, acceleration is positive
   (B) velocity is positive, acceleration is negative
   (C) velocity is negative, acceleration is negative
   (D) velocity is negative, acceleration is positive
   (E) velocity is positive, acceleration is zero
   (F) velocity is negative, acceleration is zero
   (G) velocity is zero, acceleration is positive
   (H) velocity is zero, acceleration is negative

   (b) For each of the scenarios in question 3a, is the velocity increasing, decreasing, or staying the same?

3. **Graphs for Uniformly Accelerated Motion** Match the $v(t)$ graphs in the left column with the corresponding $x(t)$ graphs in the right column in Figure 2.7.

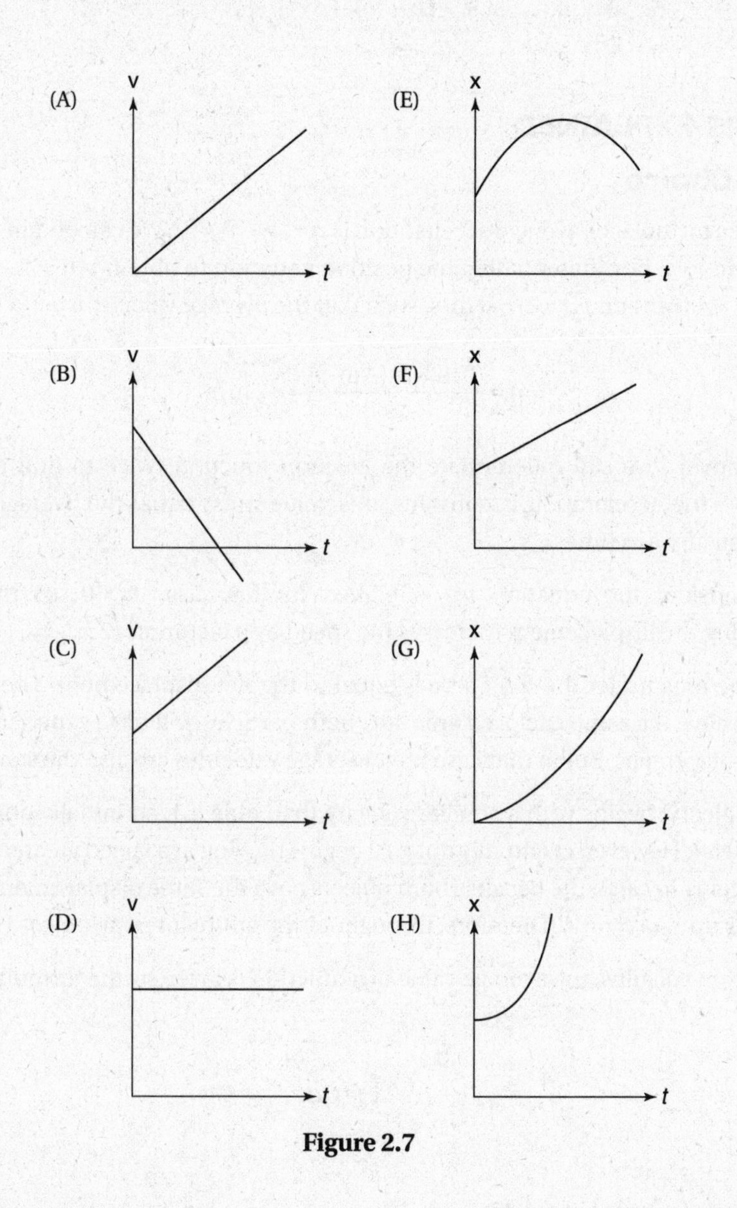

Figure 2.7

4. **The UAM Chase Problem** James Bond is racing Forrest Gump. After the race begins, it takes Forrest 3 s to remove his metal leg braces before he starts running. If Gump accelerates with a constant 3 m/s^2 once he begins to run and Bond accelerates with a constant 1.5 m/s^2, what will be the difference in their speeds when Forrest passes Bond?

CHALLENGE

5. **The Classic Well Problem**

   (a) A rock is dropped into a well, and the splash is heard 20 s later. What is the depth of the well? (Take the speed of sound to be 340 m/s and g to be -9.8 m/s^2.)

   (b) What is the speed of the rock when it hits the water?

   (c) What is the acceleration of the rock just before it hits the water?

ANSWER KEY

1. **B**	4. **A**	7. **A**
2. **A**	5. **B**	8. **D**
3. **E**	6. **E**	

ANSWERS EXPLAINED

Multiple-Choice

1. **(B)** The formula for average acceleration is $\bar{a} = (v_f - v_0)/\Delta t$. We can obtain the velocity at $t = 0$ and $t = 2$ by differentiating the position equation to obtain $v(t) = 4 - 2t$. Therefore, $v(t = 0) = 4$ m/s and $v(t = 2) = 0$ m/s, such that the average velocity is

$$\bar{a} = \frac{(0\text{ m/s}) - (4\text{ m/s})}{2\text{s}} = -2\text{ m/s}^2$$

Alternatively, we can differentiate the position function twice to find that $a(t) = -2$. Because the acceleration is constant, this value must equal the average acceleration between any two times.

2. **(A)** Consider the equation $v_f^2 = v_0^2 + 2a\Delta x$. In this case, $v_0^2 = 0$, so that $v_f^2 = 2a\Delta x$. Doubling the displacement increases the speed by a factor of $\sqrt{2}$.

3. **(E)** The area under the $v(t)$ curve is equal to the net displacement. Therefore, if both curves have the same enclosed area at t_f, both particles will have experienced the same net displacement. But in that case their average velocities are also the same.

4. **(A)** Object 2 begins with a smaller velocity than object 1, so initially object 1 must be in the lead. However, eventually object 2 begins moving at a speed greater than object 1, so it begins to catch up. Because both objects have the same displacement at t_f, object 2 catches up *exactly* at t_f. Therefore, throughout the entire interval, object 1 is in the lead.

5. **(B)** From calculus, the average value of a function is given by the formula

$$\bar{f} = \frac{\int_{t_1}^{t_2} f(t)\,dt}{t_2 - t_1}$$

Using this formula here yields

$$\bar{f} = \frac{\int_0^2 (4 + 3t^2)\,dt}{2} = 8$$

(In the language of kinematics, this is equivalent to integrating $v(t)$ to get the displacement, Δx, and dividing the displacement by the time increment to obtain the average velocity.)

6. **(E)** For constant acceleration, the position as a function of time is given by $x(t) = x_0 + v_0 t + \frac{1}{2}at^2$. Substituting $x = 2$ at $t = 1$, we find that the acceleration is 4 m/s^2. The velocity as a function of time is then given by the equation $v(t) = v_0 + at = 0 + (4)t$. Substituting into this equation, we find that the object's velocity at $t = 2$ s is 8 m/s.

7. **(A)** The velocity is the derivative of the position function. Therefore, if the velocity approaches a constant positive value, the position will approach a constant positive slope. This leaves choices (A) and (B). The slope of the $v(t)$ curve is initially positive, meaning that the acceleration is positive and thus the second derivative of the position is positive. Because the second derivative is positive, the position curve will initially be concave upward, which is true only of choice (A).

8. **(D)** $v = dx/dt$. This is really a calculus problem: How many of the derivatives correctly match their functions? In choice (A), the position is a positive sine function, so the derivative is a positive cosine function with the same period. In choice (B), the position function's slope increases from zero to a constant value and then remains at the constant value. In choice (C), the position function is a concave-upward parabola whose slope is initially zero. Because the position function is quadratic, the velocity is linear (beginning at zero and increasing at a constant rate). In choice (D), the position function is a concave-downward parabola whose slope is initially positive. Because the position function is quadratic, the velocity is linear (beginning at a positive value and decreasing at a constant rate). In choice (E), the initial value of $v(t)$ is negative, but the slope of $x(t)$ is initially zero (rather than being negative, as it should be). Therefore, this pair does not match.

Free-Response

1. (a) Let's call the balloon's direction of motion y, set the $+y$ direction pointing upward, and let the ground lie at $y = 0$. In this case, we have $y_0 = 10$ m, $v_0 = 2$ m/s, and $a = -9.8$ m/s^2. We can plug this into the $x(t)$ equation for UAM:

$$y = y_0 + v_0 t + \frac{1}{2}at^2 = 10\,\text{m} + (2\,\text{m/s})t - (4.9\,\text{m/s}^2)t^2$$

We can solve for the time when the balloon will reach the height of the pedestrian's head by setting $y(t) = 1.2$ m:

$$1.2\,\text{m} = 10\,\text{m} + (2\,\text{m/s})t - (4.9\,\text{m/s}^2)t^2$$

Solving this equation with the quadratic formula yields $t = 1.56$ s. Because this value is greater than 1 s, the Physics C student misses the pedestrian. Note that when we use the quadratic formula, we also obtain a negative solution, which we discard (clearly this solution is meaningless).

Note: What happens if we choose $y = 0$ at the top of the building? This is equally valid so long as we are consistent about our conventions. For example, if we set $y = 0$ at the top of the building, in order to calculate the time when the balloon will be at pedestrian height, we must solve for $y(t) = -10.0 + 1.2 = -8.8$ m.

(b) This is a very common problem. The key is to realize that when the ball is at its maximum height, the y-component of the instantaneous velocity is zero. Therefore, we can solve for the time when the ball is its maximum height by solving for when $v = 0$ (since there is no x-component of velocity in this problem):

$$v = v_0 + at = 2\,\text{m/s} - (9.8\,\text{m/s}^2)t = 0$$

This yields $t = 0.20$ s.

(c) What is the acceleration at this point? The velocity is instantaneously zero, but the acceleration is *always* $-9.8\,\text{m/s}^2$, even at the peak of the trajectory.

2. (a) Imagine a car undergoing each of these scenarios: (A) accelerating to the right (speed increases), (B) moving to the right and braking (speed decreases), (C) accelerating to the left (speed increases), (D) moving to the left and braking (speed decreases), (E) moving to the right with constant speed, (F) moving to the left with constant speed, (G) accelerating to the right from rest (speed increases), (H) accelerating to the left from rest (speed increases).

   Note: If the velocity is parallel to the acceleration, the speed increases; if the velocity is antiparallel to the acceleration, the speed decreases. In Chapter 5 we will see that this can be understood in terms of power.

   (b) This is much simpler than the last example: Velocity increases whenever acceleration is positive, decreases whenever acceleration is negative, and stays the same when acceleration is zero. Thus, velocity increases for (A), (D), and (G), decreases for (B), (C), and (H), and stays the same for (E) and (F).

3. (A) and (H), (B) and (E), (C) and (G), (D) and (F). In graph (A) the object starts with zero velocity and then moves to the right. The slope of the graph is constant (e.g., the object's acceleration is uniform), so that $x(t)$ is a parabola with zero slope at $t = 0$ [graph (H)]. In graph (B) the object moves first with decreasing speed to the right and then switches direction (where the curve crosses the axis) and moves to the left. This motion corresponds to what is shown in graph (E) (a parabola whose slope is initially positive at $t = 0$ and then becomes negative). Graph (C) looks similar to graph (A), with the exception that the initial velocity is nonzero. Thus, it corresponds to graph (G), where the parabola has a positive slope at $t = 0$. Finally, in graph (D), the velocity is constant, so that the $x(t)$ graph is a straight line [graph (F)].

4. As a general approach, define two equations for $x(t)$, one for Bond and another for Gump, and solve for the time when they are equal (the time when Forrest catches up with Bond). Substituting this time into the derivative of each of the position equations yields the velocity of each runner at that time, and then subtraction gives the difference in their speeds.

   Let $t = 0$ when the race begins. Then, the $x(t)$ equation for Bond is

$$x_{\text{Bond}} = x_0 + v_0 t + \frac{1}{2}at^2 = \frac{1}{2}(1.5\,\text{m/s}^2)t^2$$

Let $t' = 0$ when Gump begins to run. Then, $x(t')$ for Gump is

$$x_{\text{Gump}} = x_0 + v_0 t' + \frac{1}{2} a t'^2 = \frac{1}{2}(3\,\text{m/s}^2) t'^2$$

We have to define x_{Bond} and x_{Gump} in terms of the same time parameter. This can be done using the relation $t = t' + 3$ s because Gump starts running 3 s after Bond. Substituting this into x_{Gump} yields

$$x_{\text{Gump}} = \frac{1}{2}(3\,\text{m/s}^2) t'^2 = \frac{1}{2}(3\,\text{m/s}^2)(t - 3\,\text{s})^2$$

Looking ahead, we don't bother expanding, but immediately solve for $x_{\text{Bond}} = x_{\text{Gump}}$:

$$x_{\text{Gump}} = \frac{1}{2}(3\,\text{m/s}^2)(t - 3s)^2 = x_{\text{Bond}} = \frac{1}{2}(1.5\,\text{m/s}^2) t^2$$

Taking the square root yields a linear equation that can be quickly solved to yield the time when Gump passes Bond, $t_{\text{pass}} = 10.2$ s. Bond's velocity at this time is

$$v_{\text{Bond}} = \frac{dx_{\text{Bond}}}{dt} = \frac{d}{dt}\left[\frac{1}{2}(1.5\,\text{m/s}^2) t^2\right] = (1.5\,\text{m/s}^2) t$$

$$v_{\text{Bond}}(t = 10.2\,\text{s}) = 15.3\,\text{m/s}$$

Gump's velocity at this time is

$$v_{\text{Gump}} = \frac{dx_{\text{Gump}}}{dt} = \frac{d}{dt}\left[\frac{1}{2}(3\,\text{m/s}^2)(t - 3s)^2\right] = (3\,\text{m/s}^2)(t - 3)$$

$$v_{\text{Gump}}(t = 10.2\,\text{s}) = 21.6\,\text{m/s}$$

Therefore, Gump is running faster than Bond by 21.6 m/s − 15.3 m/s = 6.3 m/s. Intuitively, it makes sense that Gump is running faster (otherwise, how could he possibly catch up?).

5. (a) In time interval t_{down}, the rock falls and hits the water. In time interval t_{up}, the sound travels back up the well. We need to express both t_{down} and t_{up} in terms of the depth of the well, set their sum equal to 20 s, and solve for the depth of the well.

CHALLENGE

Calculating t_{down} (time from dropping the rock until it hits the water): Setting $y = 0$ at the top of the well and $+y$ pointing down,

$$y = y_0 + v_0 t + \frac{1}{2} g t^2 = \frac{1}{2}(9.8\,\text{m/s}^2) t^2$$

$$y_{\text{well}} = (4.9\,\text{m/s}^2) t^2_{\text{down}} \Rightarrow t_{\text{down}} = \sqrt{\frac{y_{\text{well}}}{4.9\,\text{m/s}^2}}$$

Calculating t_{up}: Sound moves at constant speed, so we can simply divide length by velocity to calculate the time (mathematically, if $a = 0$, then $\Delta x = v\Delta t$):

$$t_{up} = \frac{y_{well}}{340\text{ m/s}}$$

Summing and setting the equation equal to 20 s,

$$20\text{ s} = t_{down} + t_{up} = \sqrt{\frac{y_{well}}{4.9\text{ m/s}^2}} + \frac{y_{well}}{340\text{ m/s}}$$

We can get rid of the square root by isolating it and squaring:

$$\left(20\text{ s} - \frac{y_{well}}{340\text{ m/s}}\right)^2 = \left(\sqrt{\frac{y_{well}}{4.9\text{ m/s}^2}}\right)^2$$

$$400\text{ s}^2 + \frac{y^2_{\ well}}{115{,}600\text{ m}^2/\text{s}^2} - \frac{2y_{well}}{17} = \frac{y_{well}}{4.9\text{ m/s}}$$

We can solve with a quadratic equation, obtaining $y_{well} = 1287$ m and 35,904 m. By squaring, we have introduced an extra mathematical solution. Plugging these two solutions into the equation *before* squaring indicates that only the solution 1287 m works. From the result we find $t_{down} = 16.2$ s.

(b) The velocity is given by $v = v_0 + at_{down}$, which yields $|v| = 0 + (9.8)(16.2) = 159$ m/s.

(c) The acceleration of an object in free-fall is always $g = 9.8\text{ m/s}^2$.

3 Two-Dimensional Kinematics

→ INSTANTANEOUS VELOCITY, SPEED, AND ACCELERATION IN TWO DIMENSIONS
→ UNIFORMLY ACCELERATED MOTION (UAM) INCLUDING PROJECTILE MOTION
→ RELATIVE POSITION, VELOCITY, AND ACCELERATION
→ UNIFORM CIRCULAR MOTION (UCM)

Velocity and Acceleration Are Really Vectors As discussed at the beginning of the last chapter, velocity and acceleration are actually vectors (although in one dimension, there is no need for formal vector notation, so this is easily confused). In two dimensions, more formal vector notation is required.

Defining the Position Vector A natural way to describe the position of an object in more than one dimension is to define a position vector that points from the origin to the location of the object. Its magnitude gives the distance from the origin to the object's location. Mathematically, in two dimensions,

$$\text{Position vector} = \mathbf{r} = x\hat{i} + y\hat{j}$$

Complete Definitions of Velocity, Speed, and Acceleration Again, we'll start by defining velocity, speed, and acceleration.

Instantaneous velocity:

$$\mathbf{v}(t) \equiv \lim_{\Delta t \to 0} \frac{\Delta \mathbf{r}}{\Delta t} = \frac{d\mathbf{r}}{dt} \Leftrightarrow \Delta \mathbf{r} = \int d\mathbf{r} \equiv \int_{t_1}^{t_2} \mathbf{v}(t)\,dt$$

Instantaneous speed:

$$\text{speed} \equiv \mathrm{v} = \sqrt{v_x^2 + v_y^2}$$

Instantaneous acceleration:

$$\mathbf{a}(t) \equiv \lim_{\Delta t \to 0} \frac{\Delta \mathbf{v}}{\Delta t} = \frac{d\mathbf{v}}{dt} \equiv \frac{d^2\mathbf{r}}{dt^2} \Leftrightarrow \Delta \mathbf{v} = \int d\mathbf{v} \equiv \int_{t_1}^{t_2} \mathbf{a}(t)\,dt$$

Although these definitions look very similar to the one-dimensional definitions, the subtle difference is that the definitions of velocity and acceleration are two-dimensional vector equations, each of which is equivalent to a set of two one-dimensional equations. For example, consider the definition of velocity:

$$\mathbf{v} = \frac{d\mathbf{r}}{dt}$$

We can represent a two-dimensional vector equation by a set of one-dimensional equations if we write the vectors in terms of their components:

$$\mathbf{v} = v_x\hat{i} + v_y\hat{j}$$

$$\frac{d\mathbf{r}}{dt} = \frac{dx}{dt}\hat{i} + \frac{dy}{dt}\hat{j}$$

Therefore,

$$v_x\hat{i} + v_y\hat{j} = \frac{dx}{dt}\hat{i} + \frac{dy}{dt}\hat{j}$$

Geometrically, the components in one direction cannot "mix" with one another (e.g., an $\hat{i}$-component can never equal or affect a $\hat{j}$-component). Therefore, for the two vectors to be equal, their *x*-components, and *y*-components, must separately be equal. This allows us to write the single two-dimensional equation $\mathbf{v} = d\mathbf{r}/dt$ as two one-dimensional equations:

$$\begin{cases} v_x = \dfrac{dx}{dt} \\ v_y = \dfrac{dy}{dt} \end{cases}$$

Can you see how an analogous argument could be used to convert the two-dimensional equation $\mathbf{a} = d\mathbf{v}/dt$ to the following set of two one-dimensional equations?

$$\begin{cases} a_x = \dfrac{dv_x}{dt} \\ a_y = \dfrac{dv_y}{dt} \end{cases}$$

Speed Speed is, as always, the magnitude of the velocity vector. As explained above, the velocity vector can be written in terms of its components:

$$\mathbf{v} = v_x\hat{i} + v_y\hat{j}$$

Speed, the magnitude of this vector, is calculated using the Pythagorean theorem:

$$\text{Speed} = \text{v} = \sqrt{v_x^2 + v_y^2}$$

GENERALIZATION OF ONE-DIMENSIONAL KINEMATICS EQUATIONS

The one-dimensional kinematics equations derived in Chapter 2 are still valid and can be viewed as special cases of two-dimensional kinematics (when one of the components is zero). Additionally, it shouldn't be too surprising that the one-dimensional kinematics equations derived in the *x*-direction are valid in the *y*- as well, if we simply change the variables.

Mathematically, you can see this from the fact that the two-dimensional vector definitions of velocity and acceleration can be broken down into one-dimensional definitions, which (in any given direction) are *identical* to the original definitions of velocity and acceleration given in the last chapter.

Because these original definitions are valid (for any particular direction), all the equations derived from them must also be valid (for that particular direction). For example, the following UAM equation was derived for one-dimensional motion in Chapter 2:

$$v = v_0 + at$$

According to the argument above, this is true in two dimensions for any motion with uniform acceleration:

$$\begin{cases} v_x = v_{x0} + a_x t \\ v_y = v_{y0} + a_y t \end{cases}$$

$\mathbf{v} = \mathbf{v}_0 + \mathbf{a}t$ **valid only for uniformly accelerated motion**

THE PHYSICAL SIGNIFICANCE OF THE VECTORS

1. For the instantaneous velocity vector, the magnitude is the speed (a scalar), and its direction is in the direction of motion (tangent to the path of the object). The direction can be understood in terms of the fact that the displacement vector, $\mathbf{r}_2 - \mathbf{r}_1$, approaches the tangent to the trajectory as $\Delta t \to 0$, as illustrated in Figure 3.1.

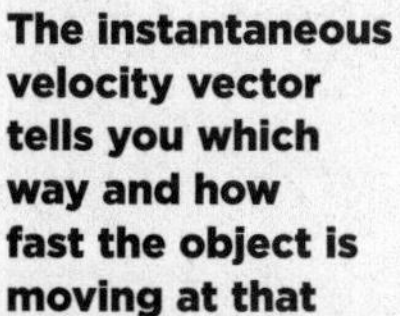

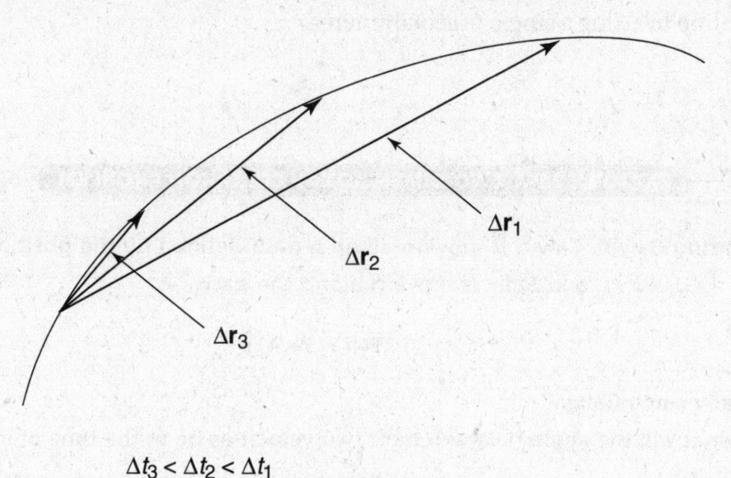

Figure 3.1

2. In the previous chapter, we saw that displacement is a vector (such that if an object starts and ends of the same position, positive and negative displacements sum to zero). In contrast, total distance can never be negative and can be zero only if the object does not move at all.

$\Delta\mathbf{r} = \Delta x\hat{i} + \Delta y\hat{j}$ **Definition of displacement**

Therefore, the displacement vector points from the object's initial position to its final position. As in one dimension, displacement is independent of the path taken between the initial and final points (whereas total distance is path dependent). As before, displacement ≤ total distance.

3. The average velocity during a given time interval is parallel to the displacement vector, as can be seen from the following equation (recall that multiplying a vector by a positive scalar produces a parallel vector).

$$\overline{\mathbf{v}}_{\text{average}} = \left(\frac{1}{\Delta t}\right)\Delta\mathbf{r}$$

4. The direction, magnitude, and significance of the acceleration vector will be discussed later in this chapter.

GENERAL APPROACH TO SOLVING KINEMATICS PROBLEMS IN TWO DIMENSIONS

Two-dimensional kinematics problems are solved by treating each dimension using the methods in Chapter 2.

Because two-dimensional vector equations reduce to independent one-dimensional equations, the different components of motion are entirely independent of each other (other than being time synchronized). Therefore, kinematics problems in two dimensions can be solved by applying one-dimensional kinematics equations to each particular dimension and coordinating the motion by using a single time parameter.

EXAMPLE 3.1 A TWO-DIMENSIONAL COLLISION PROBLEM

Sally is chasing Calvin. Calvin is moving along a path defined by the position vector $\mathbf{r}_{\text{Calvin}} = (t-1)\hat{i} + (11-t^2)\hat{j}$, and Sally is moving along the path

$$\mathbf{r}_{\text{Sally}} = (5-5t)\hat{i} + (5+5t)\hat{j}$$

(a) Will Sally catch Calvin?

(b) If so, what will the angle between their two velocities be at the time of impact?

SOLUTION

(a) For two point objects to collide, they must have equal *x*- and *y*-components at a particular instant of time. Mathematically speaking, we write down the simultaneous equations $x(t)_{\text{Calvin}} = x(t)_{\text{Sally}}$ and $y(t)_{\text{Calvin}} = y(t)_{\text{Sally}}$ and see if there are any solutions to these two equations with a common value of *t*. We find

$$x(t)_{\text{Calvin}} = t - 1 = x(t)_{\text{Sally}} = (5 - 5t) \Rightarrow t = 1$$

$$y(t)_{\text{Calvin}} = (11 - t^2) = y(t)_{\text{Sally}} = (5 + 5t) \Rightarrow \text{valid at } t = 1$$

Therefore, there is a collision at *t* = 1; both Sally and Calvin are at (0, 10).

(b) First we must calculate their velocities by differentiation:

$$\mathbf{v}_{\text{Calvin}} = \frac{d\mathbf{r}_{\text{Calvin}}}{dt} = \frac{d}{dt}\left[(t-1)\hat{i} + (11-t^2)\hat{j}\right] = \hat{i} - 2t\hat{j}$$

$$\mathbf{v}_{\text{Sally}} = \frac{d\mathbf{r}_{\text{Sally}}}{dt} = \frac{d}{dt}\left[(5-5t)\hat{i} + (5+5t)\hat{j}\right] = -5\hat{i} + 5\hat{j}$$

Now, plugging into these formulas to calculate the velocity at t = 1,

$$\mathbf{v}_{\text{Calvin}} = \hat{i} - 2\hat{j} \text{ and } \mathbf{v}_{\text{Sally}} = -5\hat{i} + 5\hat{j}$$

One way to calculate the angle between these vectors is to sketch them, use trigonometry to calculate the angle between them and the coordinate axes, and then sum various angles to compute the angle between them. An easier method involves one of the definitions of the dot product:

$$\mathbf{v}_{\text{Calvin}} \cdot \mathbf{v}_{\text{Sally}} = v_{\text{Calvin}} v_{\text{Sally}} \cos\theta \rightarrow \theta = \cos^{-1}\left(\frac{\mathbf{v}_{\text{Calvin}} \cdot \mathbf{v}_{\text{Sally}}}{v_{\text{Calvin}} v_{\text{Sally}}}\right)$$

The dot product is easily computed using the formula $(a\hat{i} + b\hat{j}) \cdot (c\hat{i} + d\hat{j}) = ac + bd$, and the magnitudes of the velocity vectors can be computed using the Pythagorean theorem for a final result of θ = 162°.

EXAMPLE 3.2 POSITION, VELOCITY, AND ACCELERATION IN TWO DIMENSIONS

A particle is moving in two dimensions with a velocity given by $\mathbf{v} = 4t^3\hat{i} - (\cos t)\hat{j}$ with velocity in units of m/s.

(a) What is its acceleration as a function of time?

(b) What is its displacement between t = 0 and t = 3?

SOLUTION

We need to differentiate to get from velocity to acceleration, and integrate to get from velocity to displacement:

(a) $\mathbf{a} = \frac{d\mathbf{v}}{dt} = \frac{d}{dt}\left(4t^3\hat{i} - (\cos t)\hat{j}\right) = 12t^2\hat{i} + (\sin t)\hat{j}$

(b) $\Delta\mathbf{r} = \int \mathbf{v}dt = \int_0^3 \left(4t^3\hat{i} - (\cos t)\hat{j}\right)dt = \left[t^4\hat{i} - (\sin t)\hat{j}\right]_0^3 = 81\hat{i} - (\sin 3)\hat{j}$

PROJECTILE MOTION

Conversion of Velocity Between "Polar" and "Rectangular" Form: A Review of Some Vector Algebra and Trigonometry The velocity of a projectile can be written in either polar form (a magnitude and an angle of elevation) or rectangular form, $\mathbf{v} = v_x\hat{i} + v_y\hat{j}$. As you will see, when actually solving these problems it is generally most useful to work with the rectangular form of velocity. However, you should be comfortable converting between polar form and rectangular form as follows.

Conversion from rectangular form to polar form:

$$v = \sqrt{v_x^2 + v_y^2}$$

$$\theta = \tan^{-1}\left(\frac{v_y}{v_x}\right)$$

Conversion from polar form to rectangular form:

$$v_x = v\cos\theta$$

$$v_y = v\sin\theta$$

Standard Assumptions Unless told otherwise, you can assume that there is no air resistance and that the only force acting on the object is gravity, which pulls the object downward causing an acceleration of 9.8 m/s^2, which we denote by g.

Equations Relevant to Projectile Motion x-Direction: There are no forces acting in the x-direction, so $a_x = 0$ and v_x is constant:

$$v_x(t) = v_{x0}$$

$$x(t) = x_0 + v_x t$$

TIP

The key equations for projectile motion

y-Direction: The acceleration in the y-velocity is $-g$ (designating upward as positive).

$$a_y = -g$$

$$v_y = v_{y0} - gt$$

$$y = y_0 + v_{y0}t - \frac{1}{2}gt^2$$

$$v_y^2 = v_{y0}^2 - 2g(y - y_0)$$

PROBLEM-SOLVING TIPS

This list is an extension of the tips for one-dimensional free-fall problems.

1. At the peak of the trajectory, $v_y = 0$ (discussed in Chapter 1).
2. Because of conservation of energy (reflected in the symmetry of parabolas), the speed at any two points on the object's trajectory with the same y-coordinate must be the same. Because v_x is constant and the y-velocity reverses direction, this means that the y-components of the velocity at the two points are equal and opposite in sign; that is, $v_{y1} = -v_{y2}$.
3. As will be proven in Example 3.3, the trajectory $y(x)$ is parabolic for all projectile motion. (Recall, as discussed in Chapter 2, that $y(t)$ is also parabolic; however, $y(x)$ gives the actual physical path followed by the object). Because of the symmetry of these parabolas, the magnitude of the time differences between when the object is at the peak of its trajectory (t_{peak}) and when it is at any two points of equal height (or, equivalently, an equal x-distance away from the location of the peak) must be equal.
4. It's generally easiest to resolve any vectors into x- and y-components and solve the problem using the one-dimensional equations for motion listed above. The final answer generally results from linking the two sets of equations by a common time parameter.
5. If in doubt, writing down the position equations [$x(t)$ and $y(t)$] is often a good way to start the problem (because most quantities can be calculated from these fundamental equations).

EXAMPLE 3.3 (DERIVATION)

Show that $y(x)$ (the trajectory) is parabolic for projectile motion.

SOLUTION

We choose our coordinate system such that the object's initial position is at the origin, $x_0 = y_0 = 0$. Thus,

$$x(t) = v_x t$$

$$y(t) = v_{y0}t - \frac{1}{2}gt^2$$

We can convert this parametric equation to the form $y(x)$ by solving the $x(t)$ equation for t in terms of x and then substituting into the other equation to eliminate t.

$$t = \frac{x}{v_x}$$

Substituting,

$$y(x) = v_{y0}\left(\frac{x}{v_x}\right) - \frac{1}{2}g\left(\frac{x}{v_x}\right)^2$$

Simplifying,

$$y(x) = \left(\frac{v_{y0}}{v_x}\right)x + \left(\frac{-g}{2v_x^2}\right)x^2$$

This is the equation of a parabola that passes through the origin (as it should, based on the initial conditions). As long as $v_x \neq 0$, $y(x)$ will be parabolic. (If $v_x = 0$, then the path will simply be a one-dimensional line up and down, which you can consider a limiting case of a parabola.)

***Symbolic answer check*: (1) The answer has correct units; for the second term in the $y(x)$ equation:**

$$\frac{\text{m/s}^2}{\text{m}^2/\text{s}^2}\text{m}^2 = \text{m}$$

(2) The initial slope of the path (dy/dx) at $x = 0$ is v_{y0}/v_{x0} (which is the slope of $\mathbf{v}_0$). (3) If g is made larger artificially, then y will be smaller everywhere along the trajectory.

EXAMPLE 3.4 A STANDARD PROJECTILE MOTION PROBLEM

An object is thrown off a cliff 50 m high with an initial velocity of 25 m/s and an angle of elevation of 30°. How far away from the base of the cliff will it land? (Take g = 10 m/s².)

SOLUTION

Let's start by writing down the position equations $x(t)$ and $y(t)$. Taking the origin to be the base of the cliff,

$$x = x_0 + v_{x0}t = v_0(\cos\theta)t = (25\text{ m/s})(\cos 30°)t = (21.7\text{ m/s})t$$

$$y(t) = y_0 + v_{y0}t - \frac{1}{2}gt^2 = y_0 + v_0(\sin\theta)t - \frac{1}{2}gt^2$$
$$= 50\text{ m} + (25\text{ m/s})(\sin 30°)t - \frac{1}{2}(10\text{ m/s}^2)t^2$$

$$y(t) = 50\text{ m} + (12.5\text{ m/s})t - (5\text{ m/s}^2)t^2$$

We can calculate the time of impact by solving for y = 0 and then substituting this time into the equation for $x(t)$ to compute the distance away from the cliff the object lands.

$$y(t_{\text{impact}}) = 50\text{ m} + (12.5\text{ m/s})t_{\text{impact}} - (5\text{ m/s}^2)t_{\text{impact}}^2 = 0$$

Using the quadratic equation to solve (and ignoring the negative solution) yields t_{impact} = 4.65 s.

$$x(t_{\text{impact}}) = (21.7\text{ m/s})(4.65\text{ s}) = 101\text{ m}$$

EXAMPLE 3.5 MAXIMUM PROJECTILE RANGE

In order to calculate the maximum range of a rocket, you fire the rocket straight up and record the time it takes for it to return to the ground, $t_{\text{trajectory}}$. Based on this single piece of data,

(a) What is the speed with which the rocket is launched?
(b) Compute the rocket's range as a function of the angle of elevation θ and the initial speed v_0.
(c) At what angle is the range maximized?
(d) What is the maximum range?

SOLUTION

(a) This is a one-dimensional kinematics problem. We start with the $v_y(t)$ equation:

$$v_y = v_{y0} - gt$$

We solve for the time at which the rocket reaches its peak height, t_{peak}, by setting v_y equal to zero:

$$v_y = 0 = v_{y0} - gt_{\text{peak}} \Rightarrow t_{\text{peak}} = \frac{v_{y0}}{g}$$

Because of the symmetry of parabolas, the total trajectory time is exactly twice the time spent going up:

$$t_{\text{trajectory}} = 2t_{\text{peak}} = \frac{2v_{y0}}{g} \Rightarrow v_0 = v_{y0} = \frac{t_{\text{trajectory}}\, g}{2}$$

Alternate solution 1: Based on the symmetry of parabolas, the *y*-velocity reverses direction during the trajectory so that $v_{y,\text{final}} = -v_{y0}$. Using this equation along with $v_y(t) = v_{y0} - gt$ yields the same answer for $t_{\text{trajectory}}$ as shown above.

Alternate solution 2: Simply write out the *y*(*t*) equation and solve for *y*(*t*) = 0:

$$y(t) = v_{y0}t - \frac{1}{2}gt^2 = t\left(v_{y0} - \frac{1}{2}gt\right) = 0$$

This equation reveals that the object is on the ground at time $t = 0$ and $t = 2v_{y,0}/g$; solving the second equation for the initial velocity yields $v_{y0} = t_{\text{trajectory}}g/2$.

Symbolic answer check: (1) The answer has the correct units s(m/s²) = m/s = *velocity*. (2) When the trajectory time increases, the initial velocity increases, which makes sense (a greater initial velocity would be required to keep the rocket in the air for a longer amount of time). Additionally, if the value of *g* increases, the initial velocity will increase, which also makes sense (if gravity were pulling down harder, the rocket would need a larger initial speed to stay up in the sky for the same length of time).

(b) We can use the same approach as in part (a) to obtain the equation:

$$t_{\text{trajectory}} = \frac{2v_{y0}}{g} = \frac{2v_0 \sin\theta}{g}$$

The range (the *x*-displacement) is simply the *x*-velocity multiplied by the time in the air:

$$\text{Range} = \Delta x = v_x t_{\text{trajectory}} = (v_0 \cos\theta)\frac{2v_0 \sin\theta}{g} = \frac{2v_0^2 \sin\theta\cos\theta}{g}$$

$$\text{Range} = \frac{2v_0^2 \sin\theta\cos\theta}{g}$$

Symbolic answer check: (1) The answer has the correct units

$$\frac{(\text{m/s})^2}{\text{m/s}^2} = \text{m}$$

(2) If the initial velocity increases, the range will increase, whereas if gravity increases, the range will decrease (both of which make sense). In terms of the angle, we can simplify the equation using the identity $2\sin\theta\cos\theta = \sin 2\theta$ to obtain:

$$\text{Range} = \frac{v_0^2 \sin 2\theta}{g}$$

Thus, the maximum range occurs for $\theta = 45°$, whereas the range is zero for $\theta = 90°$ (the object goes straight up or down and never changes the x-coordinate) and also zero for $\theta = 0°$ (the object immediately falls to the ground and never covers any x-displacement).

(c) To maximize the range function,

$$\text{Range} = \frac{v_0^2 \sin 2\theta}{g}$$

we note that the only part of this function that depends on θ is $\sin 2\theta$, which is maximized when $\theta = \pi/4$. It makes sense that the maximum range occurs when the initial velocity is 45° above the horizontal.

Alternative solution: If you don't spot the $2\sin\theta\cos\theta = \sin 2\theta$ shortcut, you can always maximize the function in the customary calculus manner by setting its first derivative equal to zero:

$$\text{Range} = \frac{2v_0^2}{g} \sin\theta \cos\theta$$

$$\frac{d(\text{Range})}{d\theta} = 0 = \frac{2v_0^2}{g}(\cos^2\theta - \sin^2\theta) \Rightarrow \sin\theta = \cos\theta$$

Because of the physical constraints of the situation, $0 \le \theta \le \pi/2$. The only solution to the above equation on this interval is $\theta = \pi/4$. In a calculus course you might want to verify that this is a maximum by checking that the second derivative is negative. In this problem, however, we expect the extremum to be a maximum, and it makes intuitive sense that it occurs at an angle of elevation of 45°, so it isn't necessary to take the second derivative.

(d) The maximum range can be obtained by plugging $\theta = \pi/4$ into the range function to obtain

$$\text{Range}_{\text{max}} = \frac{v_0^2}{g}$$

From part (a) we know that

$$v_0 = \frac{t_{\text{trajectory}}\, g}{2}$$

Substituting,

$$\text{Range}_{\text{max}} = \frac{g t_{\text{trajectory}}^2}{4}$$

Symbolic answer check: (1) The answer has the correct units $(\text{m/s}^2)\text{s}^2 = \text{m}$. (2) Increasing the trajectory time increases the maximum range.

EXAMPLE 3.6

A friendly naturalist is trying to shoot a monkey with a tranquilizer dart. She aims the dart directly at the monkey. At the instant that the naturalist shoots, the monkey notices and lets go of the branch (free-fall starting from rest). Prove that the monkey will be hit by the dart (assuming that the dart reaches the *x*-position of the monkey before the monkey or the dart lands on the ground).

SOLUTION

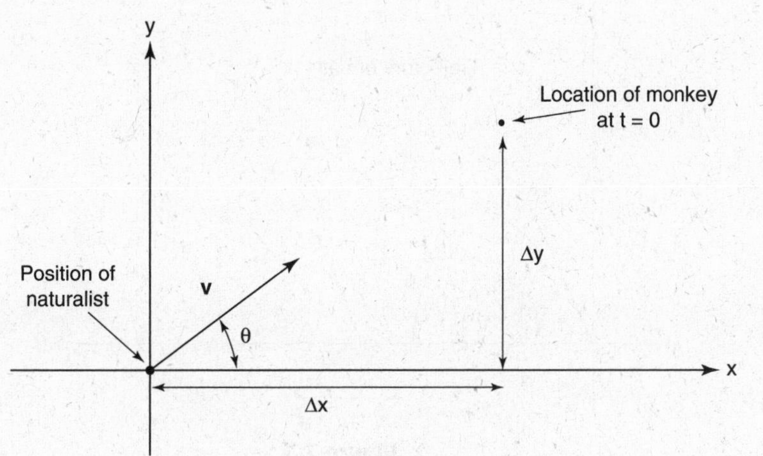

Figure 3.2

We start by making a sketch, choosing an origin, and defining a few basic quantities (Δx, Δy, v, θ). Now we can write down the position equations:

$$x_{\text{dart}} = v(\cos\theta)t$$

$$y_{\text{dart}} = v(\sin\theta)t - \frac{1}{2}gt^2$$

$$x_{\text{monkey}} = \Delta x$$

$$y_{\text{monkey}} = \Delta y - \frac{1}{2}gt^2$$

TIP

If a problem has two objects, you will need two sets of kinematic equations, one set for each object.

To prove that two objects collide, we must show that at some point in time, $x_{\text{monkey}} = x_{\text{dart}}$ and $y_{\text{monkey}} = y_{\text{dart}}$. Therefore, our general approach is to solve for the time when $x_{\text{monkey}} = x_{\text{dart}}$ and insert this time into the $y(t)$ equations to show that $y_{\text{monkey}} = y_{\text{dart}}$.

$$x_{\text{dart}} = v(\cos\theta)t = x_{\text{monkey}} = \Delta x \Rightarrow t_{\text{collision?}} = \frac{\Delta x}{v(\cos\theta)}$$

$$y_{\text{monkey}}(t_{\text{collision?}}) = \Delta y - \frac{1}{2}gt^2_{\text{collision?}} \stackrel{?}{=} y_{\text{dart}}(t_{\text{collision?}}) = v(\sin\theta)t_{\text{collision?}} - \frac{1}{2}gt^2_{\text{collision?}}$$

Simplifying,

$$\Delta y \stackrel{?}{=} v(\sin\theta)t_{\text{collision?}}$$

Substituting for $t_{\text{collision?}}$,

$$\Delta y \stackrel{?}{=} v(\sin\theta)\frac{\Delta x}{v(\cos\theta)} = \Delta x(\tan\theta)$$

Geometrically, based on the figure this is indeed true. Therefore, the dart and the monkey will collide.

EXAMPLE 3.7 THE HILL PROBLEM

As shown in Figure 3.3, a ball is thrown up a hill with an initial speed of v_0 at an angle of elevation of θ. The hill is inclined at angle of ϕ. At what time will the ball land?

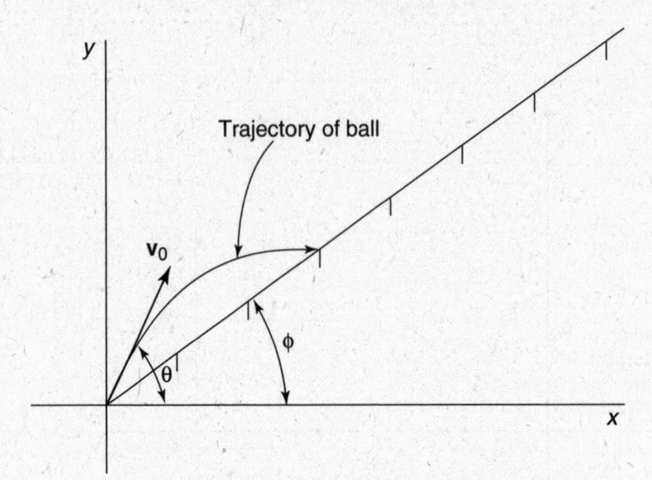

Figure 3.3

SOLUTION

We begin by writing down the position equations:

$$x = v_0(\cos\theta)t$$

$$y = v_0(\sin\theta)t - \frac{1}{2}gt^2$$

From Figure 3.3 the object will hit the hill when $y/x = \tan\phi$. We can use this to solve for the time of collision.

$$\frac{y}{x} = \frac{v_0(\sin\theta)t_{\text{collision}} - \frac{1}{2}gt_{\text{collision}}^2}{v_0(\cos\theta)t_{\text{collision}}} = \tan\phi \Rightarrow t_{\text{collision}} = \frac{2v_0}{g}(\sin\theta - \cos\theta\tan\phi)$$

Symbolic answer check: (1) The correct units are

$$\frac{\text{m/s}}{\text{m/s}^2} = \text{s}$$

(2) When we increase v_0, $t_{\text{collision}}$ increases. (3) When we increase g, $t_{\text{collision}}$ decreases. (4) When we decrease ϕ, $t_{\text{collision}}$ increases. These all make intuitive sense.

RELATIVE POSITION, VELOCITY, AND ACCELERATION

Statements of position, velocity, and acceleration are always made with respect to a frame of reference. For example, the position $(3m\hat{i} + 4m\hat{j})$ is meaningless unless the origin and orientation of the coordinate system are specified. How can we transform information from one frame of reference to another when the two frames are in relative motion?

MAIN DERIVATION

Consider Figure 3.4. Vector addition relates the position of an object relative to two different frames of reference:

$$\mathbf{r}_{P \text{ relative to } B} = \mathbf{r}_{P \text{ relative to } A} + \mathbf{r}_{A \text{ relative to } B}$$

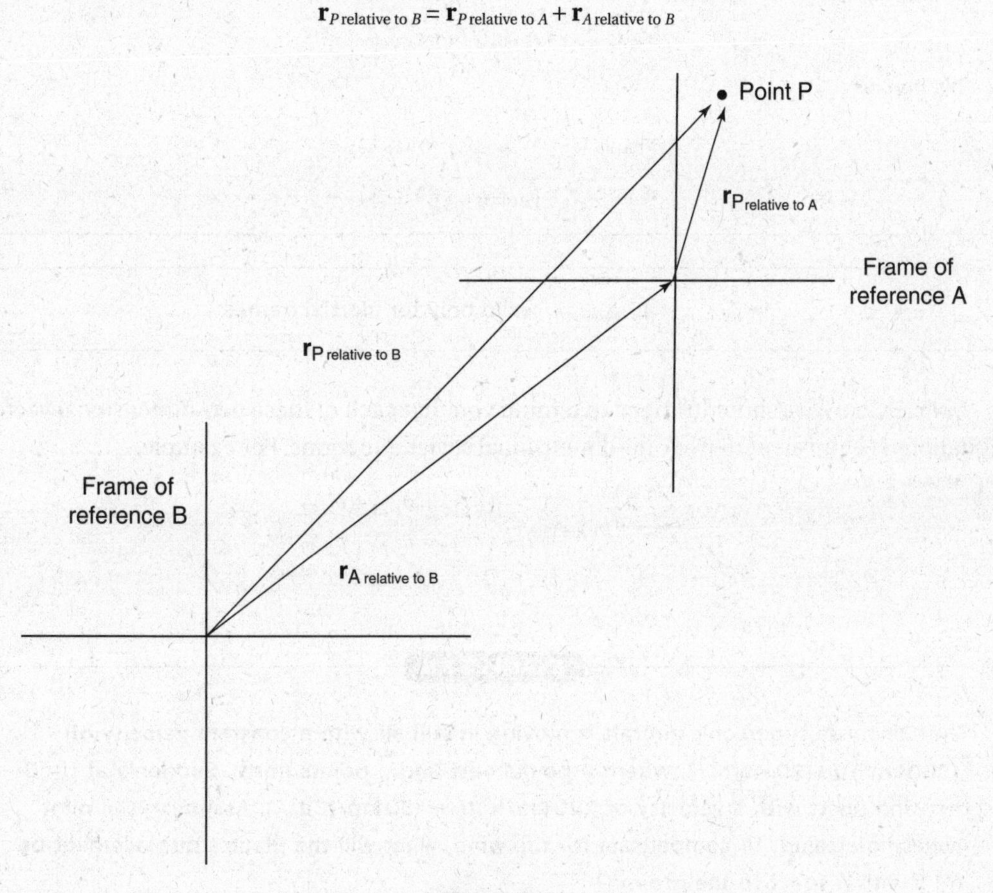

Figure 3.4

To convert this position equation to an equation of relative velocities, we must differentiate:

$$\frac{d\mathbf{r}_{P \text{ relative to } B}}{dt} = \frac{d\mathbf{r}_{P \text{ relative to } A}}{dt} + \frac{d\mathbf{r}_{A \text{ relative to } B}}{dt}$$

$$\mathbf{v}_{P \text{ relative to } B} = \mathbf{v}_{P \text{ relative to } A} + \mathbf{v}_{A \text{ relative to } B}$$

Another derivative yields an equation for relative accelerations:

$$\frac{d\mathbf{v}_{P \text{ relative to } B}}{dt} = \frac{d\mathbf{v}_{P \text{ relative to } A}}{dt} + \frac{d\mathbf{v}_{A \text{ relative to } B}}{dt}$$

$$\mathbf{a}_{P \text{ relative to } B} = \mathbf{a}_{P \text{ relative to } A} + \mathbf{a}_{A \text{ relative to } B}$$

For most problems that you will have to deal with, the two reference frames will be *inertial reference frames*, which means that they move with a constant velocity with respect to each other. This implies that $\mathbf{a}_{B\,\text{relative to}\,A} = 0$, making the acceleration the same in both inertial reference frames.

Many physics books omit the "relative to" when they write these vectors, such that our $\mathbf{v}_{P\,\text{relative to}\,B}$ is written as $\mathbf{v}_{PB}$.

SUMMARY OF EQUATIONS

By inspection of Figure 3.4, vector algebra shows that

$$\mathbf{r}_{P\,\text{relative to}\,B} = \mathbf{r}_{P\,\text{relative to}\,A} + \mathbf{r}_{A\,\text{relative to}\,B}$$

This implies

$$\mathbf{v}_{P\,\text{relative to}\,B} = \mathbf{v}_{P\,\text{relative to}\,A} + \mathbf{v}_{A\,\text{relative to}\,B}$$

$$\mathbf{a}_{P\,\text{relative to}\,B} = \mathbf{a}_{P\,\text{relative to}\,A} + \mathbf{a}_{A\,\text{relative to}\,B}$$

$\mathbf{a}_{P\,\text{relative to}\,B} = \mathbf{a}_{P\,\text{relative to}\,A}$ **valid only for inertial frames**

Note: By now, we shouldn't need to remind you that each of these two-dimensional vector equations is equivalent to two one-dimensional scalar equations. For example,

$$\mathbf{v}_{PB} = \mathbf{v}_{PA} + \mathbf{v}_{AB} \Leftrightarrow \begin{cases} v_{x,PB} = v_{x,PA} + v_{x,AB} \\ v_{y,PB} = v_{y,PA} + v_{y,AB} \end{cases}$$

EXAMPLE 3.8

The Concorde supersonic aircraft is moving in still air with a constant velocity of $[(200\,\text{km/h})\hat{i} + (20\,\text{km/h})\hat{j}]$, where $\hat{i}$ points east and $\hat{j}$ points north. Suddenly at $t = 0$ the wind gusts with a velocity of $[(20\,\text{km/h}^2)t\hat{i} - (30\,\text{km/h}^3)t^2\hat{j}]$. Assuming the pilot makes no attempt to compensate for the wind, what will the plane's displacement be in 1 h with respect to the ground?

SOLUTION

$$\mathbf{v}_{\text{plane relative to air}} = (200\text{ km/h})\hat{i} + (20\text{ km/h})\hat{j}$$

$$\mathbf{v}_{\text{air relative to ground}} = (20\text{ km/h}^2)t\hat{i} - (30\text{ km/h}^3)t^2\hat{j}$$

$$\mathbf{v}_{\text{plane relative to ground}} = \mathbf{v}_{\text{plane relative to air}} + \mathbf{v}_{\text{air relative to ground}}$$

$$\mathbf{v}_{\text{plane relative to ground}} = (200\text{ km/h} + 20t\text{ km/h}^2)\hat{i} + (20\text{ km/h} - 30t^2\text{ km/h}^3)\hat{j}$$

$$\Delta\mathbf{r} = \int \mathbf{v}\,dt = \int_{t=0\text{ h}}^{t=1\text{ h}} \left[(200\text{ km/h} + 20t\text{ km/h}^2)\hat{i} + (20\text{ km/h} - 30t^2\text{ km/h}^3)\hat{j}\right]dt$$

$$\Delta\mathbf{r} = \left[(200t\text{ km/h} + 10t^2\text{ km/h}^2)\hat{i} + (20t\text{ km/h} - 10t^3\text{ km/h}^3)\hat{j}\right]_{t=0\text{ h}}^{t=1\text{ h}}$$

$$\Delta\mathbf{r} = (210\text{ km})\hat{i} + (10\text{ km})\hat{j}$$

EXAMPLE 3.9 THE FASTEST WAY TO CROSS A STREAM

Two ways to cross a stream are (1) to compensate for the current and swim directly across the stream (from point *A* to point *B* as shown in Figure 3.5); and (2) to make no effort to compensate for the current and swim perpendicularly to the banks of the stream, causing drifting downstream, and then swim directly upstream (from point *A* to point *C* to point *B*). Assuming you can swim at speed *s*, the current flows at speed *c*, and the distance between the banks of the stream is *d*, how long does it take to cross the stream via each path?

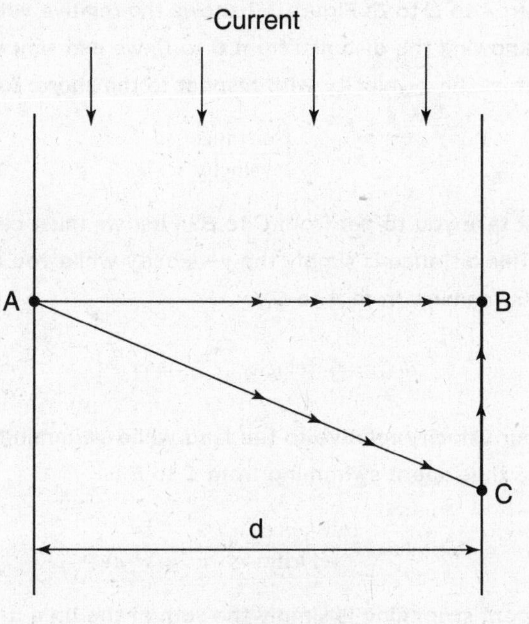

Figure 3.5

SOLUTION

Direct path from *A* to *B*: The relative velocity vectors and vector algebra are shown in Figure 3.6. By simple vector geometry, $v_{\text{you relative to land}} = \sqrt{s^2 - c^2}$. Therefore (in the land frame of reference), the time it takes to get across (moving at constant velocity) is simply

$$\Delta t = \frac{\text{distance}}{\text{velocity}} = \frac{d}{(s^2 - c^2)^{1/2}}$$

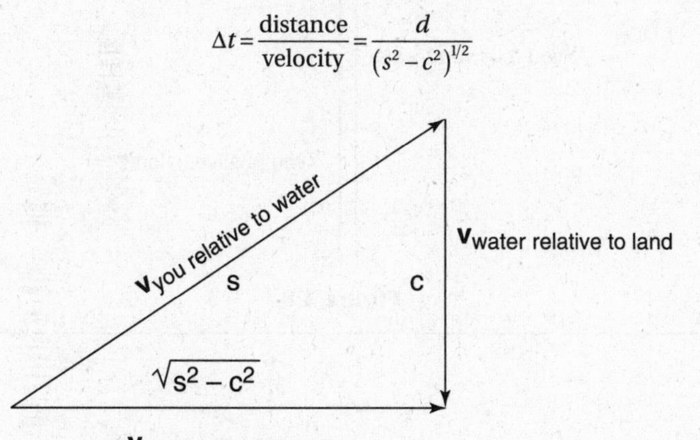

Figure 3.6

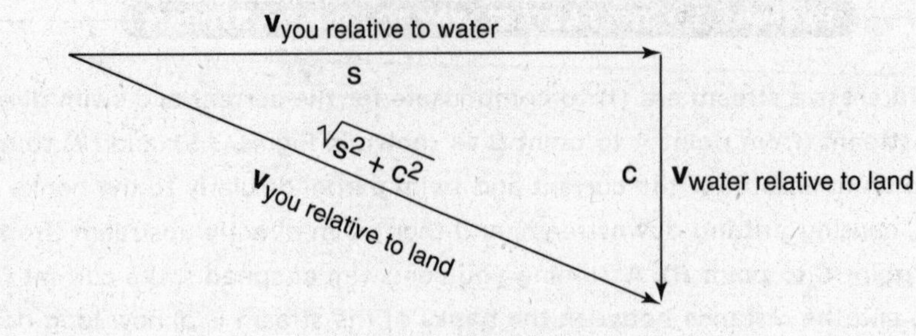

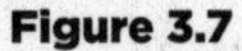

Figure 3.7

Indirect path from *A* to *C* to *B*: Figure 3.7 shows the relative velocity vectors in this situation. Without knowing the distance from *B* to *C*, we can simply say that the current does not affect the *x*-velocity with respect to the shore, so that

$$t_{A\,\text{to}\,C} = \frac{x\text{-distance}}{x\text{-velocity}} = \frac{d}{s}$$

How long does it take you to get from *C* to *B*? First we must calculate the distance between *C* and *B*. This distance is simply the *y*-velocity while you drift multiplied by t_{AC} (the time spent swimming from *A* to *C*):

$$d_{C\,\text{to}\,B} = (y\text{-velocity})(t_{A\,\text{to}\,C}) = c\left(\frac{d}{s}\right)$$

From Figure 3.8 your velocity relative to the land while swimming from *C* to *B* is $s - c$. Therefore, the time spent swimming from *C* to *B* is

$$t_{C\,\text{to}\,B} = \frac{\text{distance}}{\text{velocity}} = \frac{cd/s}{s-c} = \frac{cd}{s(s-c)}$$

The total time spent swimming is simply the sum of the time it takes to swim from *A* to *C* and from *C* to *B*:

$$t_{\text{total}} = t_{A\,\text{to}\,C} + t_{C\,\text{to}\,B} = \frac{d}{s} + \frac{cd}{s(s-c)} = \frac{d}{s-c}$$

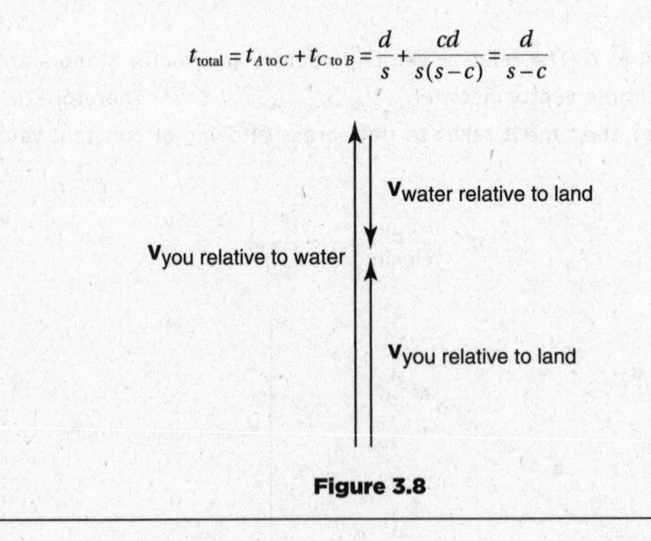

Figure 3.8

UNIFORM CIRCULAR MOTION

The Direction and Significance of the Acceleration Vector It is useful to divide the acceleration vector into a component parallel to the velocity vector ($a_{\parallel}$) and a component perpendicular to the velocity vector ($a_{\perp}$).

- $a_{\parallel}$ (the *tangential acceleration*) affects only the *magnitude* of the velocity vector (the speed). Mathematically, $a_{tan} = dv/dt$.
- $a_{\perp}$ (the *radial acceleration*) affects only the *direction* of the velocity vector.

A few special types of motion can be classified along these lines:

1. When $a_{\perp}(t) = 0$ and $a_{\parallel}(t) \neq 0$, the direction of the motion never changes: The path of the particle lies along a line (rectilinear motion), but its speed changes.
2. When $a_{\parallel}(t) = 0$ and $a_{\perp}(t) \neq 0$, the magnitude of the velocity is constant and only the direction changes. A special case of this is uniform circular motion (UCM), which occurs whenever $a_{\parallel} = 0$ and $a_{\perp}$ is constant.
3. Both $a_{\parallel}(t) \neq 0$ and $a_{\perp}(t) \neq 0$: This most general situation corresponds to more complicated motions such as projectile motion or nonuniform circular motion.

When an object moves at constant speed in uniform circular motion, its acceleration is not zero!

Calculation of the Acceleration Required to Keep an Object in Uniform Circular Motion (UCM) Mathematically, what is a general equation for uniform circular motion? As shown in Figure 3.9, the position vector of a point along a circle centered at the origin, with an angle of θ swept out counterclockwise from the $+x$-axis, is

$$\mathbf{r} = r(\cos\theta)\hat{i} + r(\sin\theta)\hat{j}$$

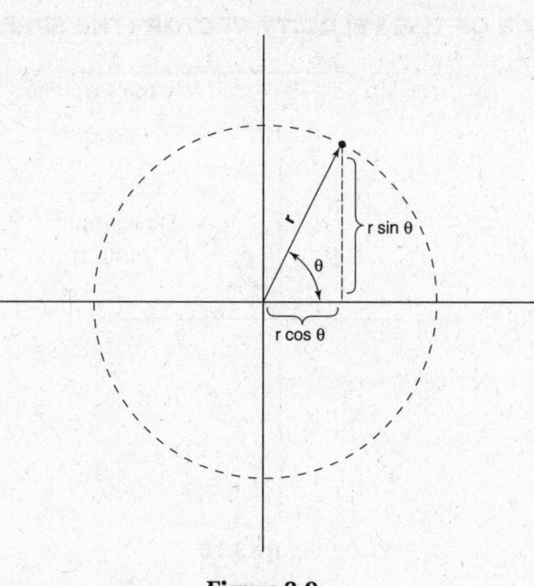

Figure 3.9

To express this position equation in terms of time for an object moving at constant speed, we make the substitution $\theta = \omega t + \phi$:

$$\mathbf{r}(t) = r\left[\cos(\omega t + \phi)\hat{i} + \sin(\omega t + \phi)\hat{j}\right]$$

Because the motion is uniform, the angle θ changes at a constant rate, which is given by ω (the angular velocity) according to the equation $\omega = d\theta/dt$. Thus, the sign of ω determines the sense of rotation (by convention, $\omega > 0$ refers to counterclockwise rotation) and the magnitude of ω determines how quickly the $\mathbf{r}(t)$ vector rotates. The *phase shift angle*, ϕ, is a parameter that determines the initial angle and thus the initial position.

Now that we have $\mathbf{r}(t)$, we can simply differentiate twice to calculate the acceleration:

$$\mathbf{r}(t) = r\left[\cos(\omega t + \phi)\hat{i} + \sin(\omega t + \phi)\hat{j}\right]$$

Differentiating once,

$$\mathbf{v}(t) = \omega r\left[-\sin(\omega t + \phi)\hat{i} + \cos(\omega t + \phi)\hat{j}\right]$$

Differentiating again,

$$\mathbf{a}(t) = -\omega^2 r\left[\cos(\omega t + \phi)\hat{i} + \sin(\omega t + \phi)\hat{j}\right]$$

Now we have expressions for $\mathbf{r}(t)$ and $\mathbf{a}(t)$, but we still must figure out the meaning of these expressions.

1. THE DIRECTION OF THE VELOCITY VECTOR

The dot product $\mathbf{r} \cdot \mathbf{v} = \left(r_x\hat{i} + r_y\hat{j}\right) \cdot \left(v_x\hat{i} + v_y\hat{j}\right) = r_x v_x + r_y v_y = 0$, indicating that the velocity vector is perpendicular to the position vector (and thus tangent to the circle). This is not surprising because the velocity vector is *always* tangent to the path of the particle, which in this case is tangent to the circle and thus perpendicular to the position vector as shown in Figure 3.10.

2. THE MAGNITUDE OF THE VELOCITY VECTOR (THE SPEED)

$$v = \sqrt{r^2\omega^2 \sin^2(\omega t + \phi) + r^2\omega^2 \cos^2(\omega t + \phi)} = r\omega$$

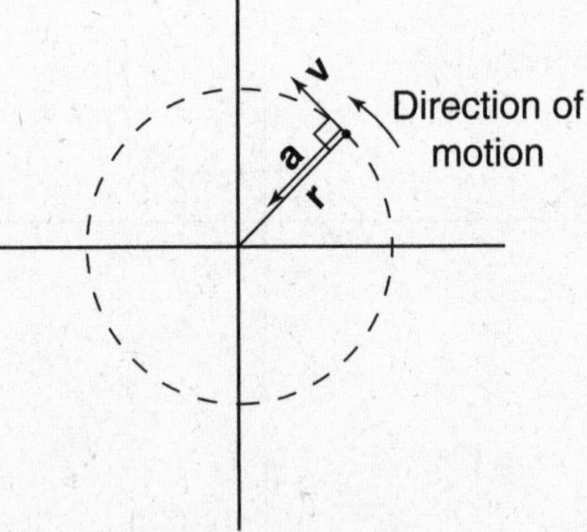

Figure 3.10

3. THE DIRECTION OF THE ACCELERATION VECTOR

From inspection, we can see that $\mathbf{a}(t) = -\omega^2\mathbf{r}(t)$. Because ω^2 is positive, this equation implies that the acceleration vector is antiparallel to the position vector. Because $\mathbf{r}(t)$ points from the origin outward to the particle's location, the acceleration must point inward, toward the center of the rotation. For this reason, the acceleration is called *centripetal acceleration.*

4. THE MAGNITUDE OF THE ACCELERATION VECTOR

Applying the Pythagorean theorem,

$$a(t) = \sqrt{\omega^4 r^2 \cos^2(\omega t + \phi) + \omega^4 r^2 \sin^2(\omega t + \phi)} = \omega^2 r$$

By using the equation derived above that relates velocity to angular velocity:

$$v = r\omega$$

we can also write:

$$a(t) = \frac{v^2}{r}$$

The significance of this acceleration, and the associated centripetal force, will be discussed in the next chapter.

SUMMARY OF RESULTS FOR UCM

1. **The direction of the velocity vector is tangent to the circle in the direction of motion.**
2. **The magnitude of the velocity vector (the speed) is related to the angular velocity by the equation** $v = r\omega$.
3. **The direction of the acceleration vector is always toward the center of the circle.**
4. **The magnitude of the acceleration vector is given by the equation** $a(t) = v^2/r$.

ACCELERATION IN NONUNIFORM CIRCULAR MOTION

Nonuniform circular motion refers to motion in a circular path with nonconstant velocity (e.g., a roller coaster that moves in a vertical circle). In nonuniform circular motion, the acceleration does not point directly toward the center of the circle. Instead, it has both radial and tangential components. As in UCM, the radial component describes how the velocity vector changes *direction* such that the particle remains on a circular path. The new component of acceleration, the tangential component, describes the change in the object's *speed*.

$$a_{\text{radial}} = a_{\text{centripetal}} = \frac{v^2}{r}$$

$$a_{\tan} = \frac{dv}{dt}$$ **Acceleration for nonuniform circular motion**

Note: We will defer further discussion of related problems and the derivation of the above equations until Chapter 7 when we will be equipped to solve more interesting problems.

EXAMPLE 3.10

Figure 3.11 shows three objects with labeled velocity and acceleration vectors. Discuss these objects in terms of whether their speed and/or direction of motion is constant.

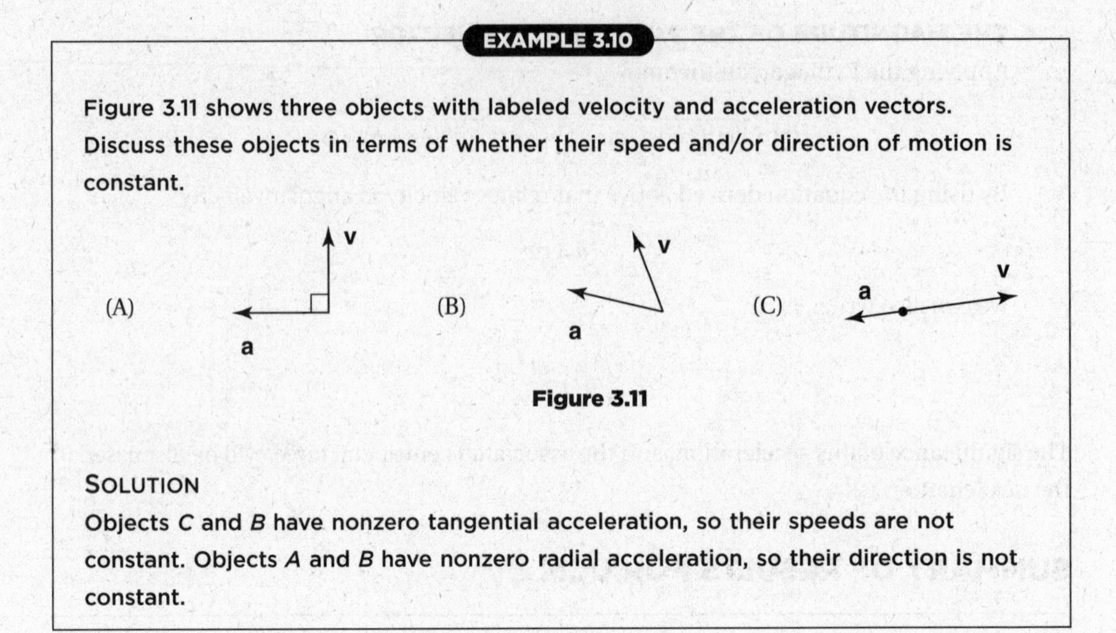

Figure 3.11

SOLUTION

Objects *C* and *B* have nonzero tangential acceleration, so their speeds are not constant. Objects *A* and *B* have nonzero radial acceleration, so their direction is not constant.

CHAPTER SUMMARY

Using vector quantities the position, velocity, and acceleration of an object moving in two dimensions can be quantified. Other than being time synchronized, the different vector components of the motion are independent of each other. Projectile motion provides an example of UAM in two dimensions. All kinematic quantities are defined with respect to a frame of reference. Observations in different frames of reference can be related in terms of the relative velocity of the two frames.

An object undergoes uniform circular motion (UCM) when it moves in a circle at constant *speed.* However, it does undergo acceleration, with the acceleration vector directed radially inward. If the motion is nonuniform, the acceleration vector also has a component tangent to the circle.

PRACTICE EXERCISES

Multiple-Choice Questions

1. The paths of three projectiles are shown in Figure 3.12. Which projectile spends the most time in the air?

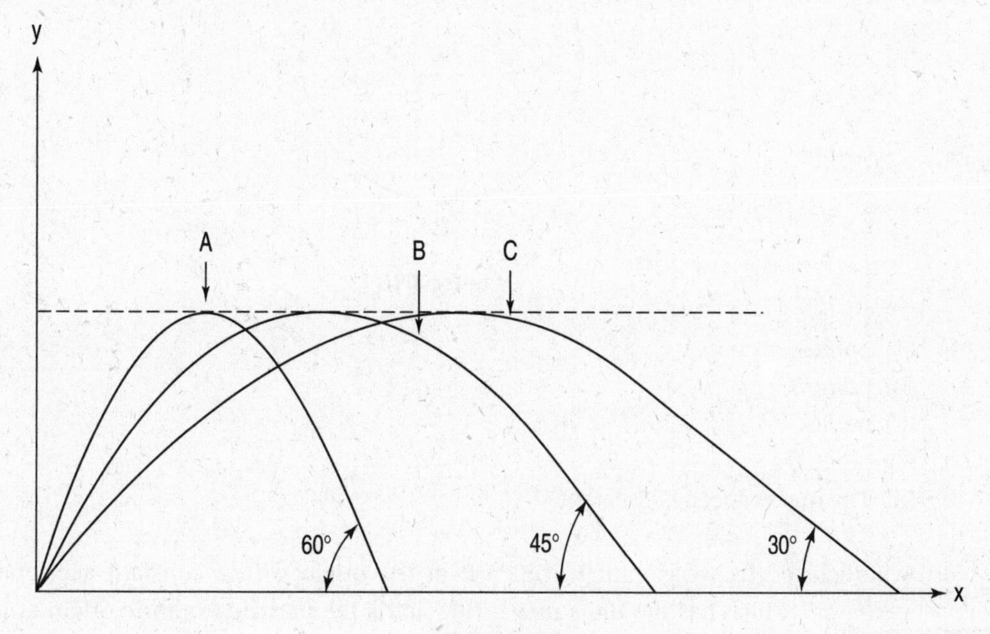

Figure 3.12

(A) projectile *A*
(B) projectile *B*
(C) projectile *C*
(D) projectiles *A* and *C*
(E) All three projectiles spend the same amount of time in the air.

2. Two boats initially next to each other begin moving with velocities $\mathbf{v}_1 = (\hat{i} + 6\hat{j})$ and $\mathbf{v}_2 = (-2\hat{i} + 2\hat{j})$ (where velocities are measured in m/s). What is the rate at which the distance between the boats is increasing?

(A) $3\hat{i} + 4\hat{j}$
(B) $-3\hat{i} - 4\hat{j}$
(C) 5 m/s
(D) $\sqrt{17}$ m/s
(E) 25 m/s

3. A mass tethered to a string rotates in a vertical circle as shown in Figure 3.13. At what point is the mass's speed increasing at the greatest rate?

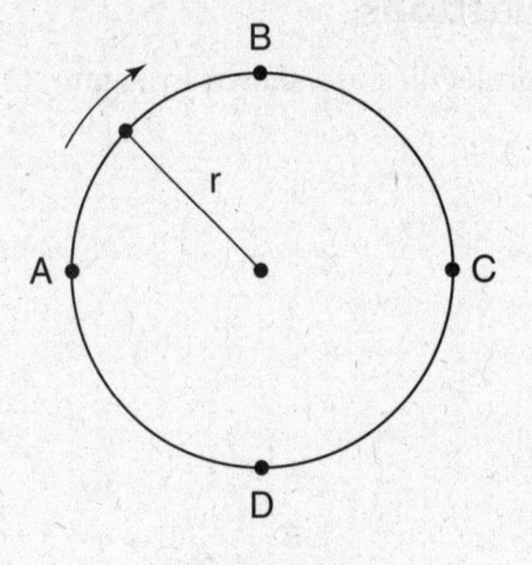

Figure 3.13

(A) point A
(B) point B
(C) point C
(D) point D
(E) The mass's speed is constant.

4. A particle begins accelerating from rest at the origin with a constant acceleration $\mathbf{a} = 2\hat{i} - 4\hat{j}$ (which has the units m/s²). How far is the particle from the origin at time $t = 1$?

(A) 1 m
(B) 2 m
(C) 3 m
(D) 5 m
(E) none of the above

5. A particle is launched from the ground with a speed v and an angle of elevation θ. What is the difference between the object's minimum and maximum speeds?

(A) $\sqrt{v^2 - v^2 \sin^2 \theta}$
(B) $\sqrt{v^2 - v^2 \cos^2 \theta}$
(C) $v(1 - \cos\theta)$
(D) $v(1 - \sin\theta)$
(E) $v\sin\theta$

6. If a projectile is launched from the ground with an initial speed v and angle of elevation θ, which of the following is true? (Designate downward angles of elevation as negative.)

(A) At the peak of the trajectory, the velocity has a magnitude v with an angle of elevation of 0°.
(B) When the object hits the ground, its velocity has a magnitude v with an angle of elevation of θ.

(C) During the course of the projectile's motion, the velocity is never perpendicular to the acceleration.
(D) During the course of the projectile's motion, the velocity is never perpendicular to the initial velocity.
(E) During the course of the projectile's motion, the velocity can be perpendicular to the initial velocity depending on the value of θ.

7. Consider an object moving along a curved path through space. The acceleration vector has two components, one tangential to the path and the other perpendicular to the path. All the following statements are true *except*

   (A) The perpendicular acceleration always points toward the concave side of a curve (e.g., toward the center of a circular path).
   (B) The tangential acceleration always points tangent to the curve in the same direction as the velocity ($\mathbf{a}_{\text{tangential}} = c\mathbf{v}$, $c > 0$).
   (C) The magnitude of the tangential acceleration is equal to the magnitude of the derivative of the speed: $a_{\text{tangential}} = dv/dt$.
   (D) When moving in a circular path, $a_{\text{centripetal}} = v^2/r$ even if the speed is *not* constant.
   (E) The perpendicular component of acceleration for a projectile is not zero.

8. Two particles are simultaneously launched off the top of a cliff with the same initial speeds and angles as indicated in Figure 3.14. What is the velocity of the top projectile with respect to the bottom projectile as a function of time?

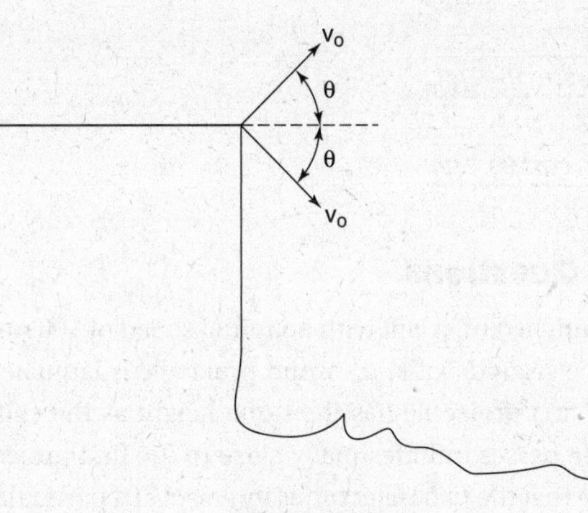

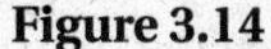
Figure 3.14

   (A) $v_0(\sin\theta)\hat{j}$
   (B) $v_0(\sin\theta + gt)\hat{j}$
   (C) $2v_0(\sin\theta)\hat{j}$
   (D) $2v_0(\sin\theta + 2gt)\hat{j}$
   (E) none of the above: the relative velocity also has an x-component

9. A piece of metal breaks off the top of a truck and falls in a linear path with respect to the truck, as indicated by the solid line with the arrow in Figure 3.15. Which of the following is true?

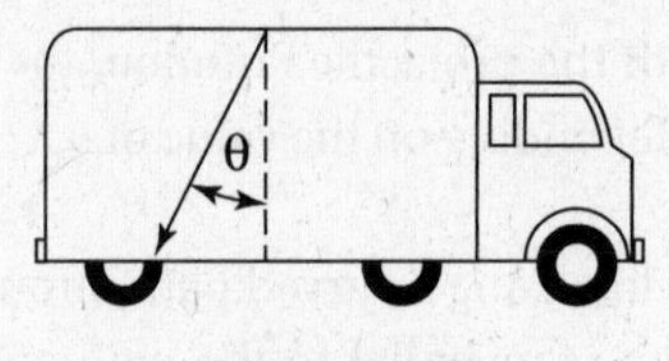

Figure 3.15

(A) The truck is accelerating to the right with constant acceleration $g\cot\theta$.
(B) The truck is accelerating to the left with constant acceleration $g\cot\theta$.
(C) The truck is accelerating to the right with constant acceleration $g\tan\theta$.
(D) The truck is accelerating to the left with constant acceleration $g\tan\theta$.
(E) The truck's acceleration is not constant.

10. A mass is launched off a cliff of height h with an initial speed of v_0 and an angle of elevation of θ above the horizontal. How long is the mass in the air?

(A) $t = \dfrac{2v_0 \sin\theta}{g}$

(B) $t = (v_0 \sin\theta)/g$

(C) $\dfrac{v_0 \sin\theta + \sqrt{v_0^2 \sin^2\theta + 2gh}}{g}$

(D) $\dfrac{v_0 \sin\theta - \sqrt{v_0^2 \sin^2\theta + 2gh}}{g}$

(E) $\dfrac{v_0 \cos\theta - \sqrt{v_0^2 \cos^2\theta + 2gh}}{-g}$

Free-Response Questions

1. Projectile 1 is launched off a cliff with an initial speed of 100 m/s and an angle of elevation of 45°. Three seconds later, a second projectile is launched off the same cliff. The instant that the first projectile has the same height as the cliff (on its way down), the second projectile passes infinitesimally close to the first projectile without touching it (you can assume that the two trajectories intersect at a particular time, but do not worry about a physical collision between the two objects).

   Sketch the paths of the two projectiles and answer the following questions *qualitatively.* (You can choose to answer some of the following questions before finishing the sketch.)

   (a) Compare the initial y-components of the velocities of the two projectiles when they are launched.
   (b) Compare the maximum heights of the two projectiles and the total time spent by each projectile in the air.
   (c) Compare the x-components of the positions of the *peaks* of the trajectories of the two projectiles.
   (d) Compare the x-components of the velocities of the two projectiles.

(e) Compare the angles of elevation (at any particular time) of the velocity vectors of the two projectiles.

(f) Compare the x- and y-components of the velocities of the two projectiles as they pass each other.

(g) Which projectile has the longer range?

2. A particle is moving along the curve $y = x^3 - 3x^2 + 1$ as shown in Figure 3.16 with a constant x-component of velocity given by $dx/dt = v_x$. Express all your answers in terms of x and v_x.

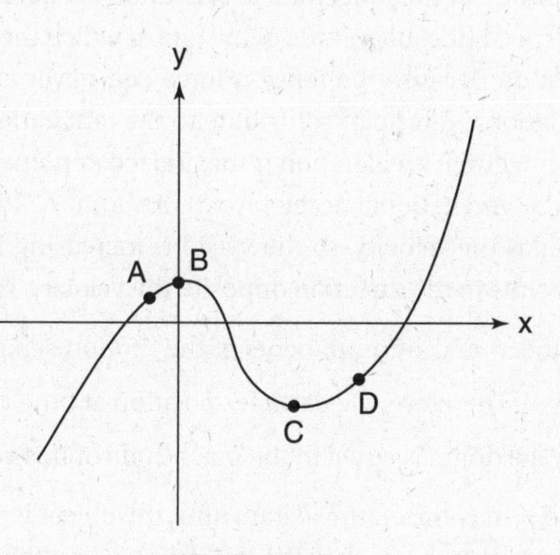

Figure 3.16

(a) Sketch the velocity and acceleration vectors at points C and D.

(b) Calculate the y-component of velocity v_y as a function of x.

(c) Where is v_y equal to zero? What is the qualitative significance of this (these) point(s)?

(d) Calculate the y-component of acceleration a_y as a function of x.

(e) Where is a_y equal to zero? What is the qualitative significance of this (these) point(s)?

(f) Now suppose that a particle moves along the curve with constant *speed*. Sketch the acceleration and velocity vectors at points C and D for this situation.

ANSWER KEY

1. **E**	4. **E**	7. **B**	10. **C**
2. **C**	5. **C**	8. **C**	
3. **C**	6. **E**	9. **C**	

ANSWERS EXPLAINED

Multiple-Choice

1. **(E)** The key to solving this problem is realizing that we can treat the x-motion and the y-motion of the projectile separately. Because all the projectiles have the same maximum height, they all have the same initial y-component of velocity, have the same $y(t)$ function, and spend the same amount of time in the air. The initial x-component of velocity, which determines how far the projectiles travel in the x-direction, has no impact on motion in the y-direction.

2. **(C)** The rate at which the magnitude of a distance changes is a *speed*, not a velocity, eliminating choices (A) and (B) (Why? The derivative of a scalar, in this case distance, must be another scalar, in this case speed). We use the equation for relative velocities, $\mathbf{v}_{\text{boat 1 respect water}} = \mathbf{v}_{\text{boat 1 respect boat 2}} + \mathbf{v}_{\text{boat 2 respect water}}$ to calculate the relative velocity of boat 2 with respect to boat 1 to be $\mathbf{v}_{\text{boat 1 respect boat 2}} = \mathbf{v}_{\text{boat 1 respect water}} - \mathbf{v}_{\text{boat 2 respect water}} = 3\hat{\imath} + 4\hat{\jmath}$. Thus, an observer standing on boat 2 sees boat 1 moving away with a velocity of $(3\hat{\imath} + 4\hat{\jmath})$. The rate at which boat 2 is moving away from boat 1 is equal to the magnitude of this velocity, 5 m/s.

3. **(C)** The magnitude of the tangential acceleration is equal to the magnitude of the derivative of the speed (the magnitude of the rate at which the speed changes). Mathematically, $a_{\text{tan}} = dv/dt$. Because the tension force can never exert tangential acceleration, only the gravitational force can contribute to the tangential acceleration, and the magnitude of the tangential acceleration is maximized at points *A* and *C* (where the velocity is parallel to the gravitational acceleration). At point *C*, the acceleration points in the same direction as the velocity, so the speed is increasing; alternatively, at point *A*, the acceleration points in the direction opposite the velocity, so the speed is decreasing.

4. **(E)** Treating the *x*- and *y*-components of the motion separately yields $x(t) = \frac{1}{2}a_x t^2 = 1$, $y(t) = \frac{1}{2}a_y t^2 = -2$. Therefore, the particle's position at time $t = 1$ is $\mathbf{r} = \hat{\imath} - 2\hat{\jmath}$. The particle's distance from the origin is equal to the magnitude of this vector: $r = \sqrt{(1)^2 + (-2)^2} = \sqrt{5}$.

5. **(C)** The speed is maximized the instant after the object is launched (when it is equal to v) and is minimized at the peak of the trajectory (when the *y*-component of the velocity is zero, such that the speed is equal to the *x*-component of velocity, $v\cos\theta$). Therefore, the difference between the maximum and minimum speeds is $v - v\cos\theta$.

6. **(E)** Consider the projectile trajectory shown in Figure 3.17 and the accompanying comparison of the velocity vectors at various points during the trajectory. In choice (A), for any trajectory the velocity at the peak is parallel to the ground and thus has an angle of elevation of 0° (with no *y*-component). Because there is no force in the *x*-direction, the velocity in the *x*-direction must remain constant. Therefore, if the initial velocity is $\mathbf{r}_0 = v_x\hat{\imath} + v_y\hat{\jmath}$, the velocity at the peak of the trajectory will be $v_{\text{peak}} = v_x\hat{\imath}$. This velocity's magnitude is *less* than v. In choice (B), when the object hits the ground, its velocity has a magnitude of v, but its angle of elevation $-\theta$ (pointing downward rather than upward). In choice (C), at the peak of the trajectory, the velocity is horizontal and the acceleration (as always) is straight down, so the velocity *is* perpendicular to the acceleration. In choices (D) and (E), the initial angle of elevation is θ and during the course of the motion the angle of elevation decreases to $-\theta$. Therefore, the maximum difference in direction between the velocity at any time and the initial velocity is 2θ. If $\theta < 45°$, the velocity will never be perpendicular to the initial velocity. However, if $\theta \geq 45°$, it will be.

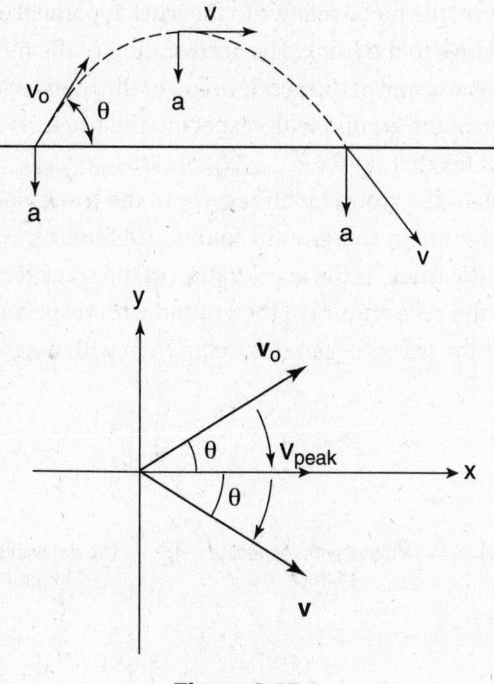

Figure 3.17

7. **(B)** The tangential acceleration may point *parallel* or *antiparallel* to the velocity: $\mathbf{a}_{tangential} = c\mathbf{v}$ or $\mathbf{a}_{tangential} = -c\mathbf{v}$. In choice (E), a perpendicular component of acceleration is involved whenever an object's direction changes, not only in the case of circular motion.

8. **(C)** Both projectiles have the same *x*-component of velocity, so the velocity of the top particle relative to the bottom has only a *y*-component. The *y*-component of the velocity of the top particle with respect to the ground is $v_{\text{top respect ground},y} = v_0 \sin\theta - gt$, and the *y*-component of the velocity of the bottom particle with respect to the ground is $v_{\text{bottom respect ground},y} = -v_0 \sin\theta - gt$. Applying the equation for relative velocities yields $\mathbf{v}_{\text{top relative ground}} = \mathbf{v}_{\text{bottom relative ground}} + \mathbf{v}_{\text{top relative bottom}}$. Therefore,

$$v_{\text{top relative bottom},y} = v_{\text{top relative ground},y} - v_{\text{bottom relative ground},y}$$
$$= (v_0 \sin\theta - gt) - (-v_0 \sin\theta - gt)$$

Alternative solution: The initial relative velocity of the top particle with respect to the bottom particle is

$$\mathbf{v}_{\text{top relative bottom}} = \mathbf{v}_{\text{top relative ground}} - \mathbf{v}_{\text{bottom relative ground}}$$
$$= [(v_0 \sin\theta) - (-v_0 \sin\theta)]\,\hat{j}$$

Then, because both particles experience the same acceleration, their velocities are changed by the same amount such that their relative velocity is unaffected (i.e., their *relative acceleration* is zero, so that their relative velocity does not change).

9. **(C)** The motion of the metal relative to the truck is parallel to the net acceleration vector of the metal relative to the truck. The acceleration of the metal with respect to the truck is equal to the vector sum of the acceleration of the metal with respect to the ground and the acceleration of the ground with respect to the truck, as shown in Figure 3.18. Vector geometry indicates that $\tan\theta = a_{\text{ground respect truck}}/a_{\text{metal respect ground}}$. Solving this equation for the acceleration of the ground with respect to the truck yields $a_{\text{ground respect truck}} = g\tan\theta$ to the left. Since we are in the ground frame of reference, what we commonly call "the acceleration of the truck" is the acceleration of the truck relative to the ground, which is the *negative* of the acceleration of the ground with respect to the truck. Therefore, from our perspective the truck accelerates to the *right* with magnitude $g\tan\theta$.

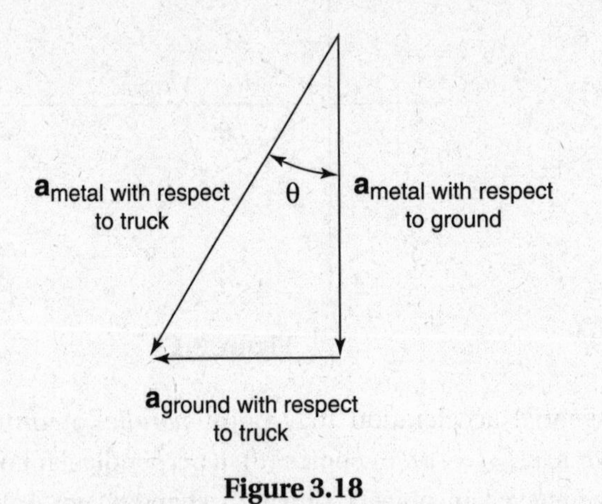

Figure 3.18

10. **(C)** Designating $y = 0$ at the base of the cliff, the y-position of the projectile is given by the equation $y(t) = y_0 + v_{y,0}t - \frac{1}{2}gt^2 = h + v_0(\sin\theta)t - \frac{1}{2}gt^2$. Setting this equation equal to zero (the position at which the mass hits the ground) and solving using the quadratic equation formula yields

$$t = \frac{v_0 \sin\theta \pm \sqrt{v_0^2 \sin^2\theta + 2gh}}{g}$$

Selecting the positive solution yields choice (C).

Free-Response

1. The projectiles' trajectories are shown in Figure 3.19.

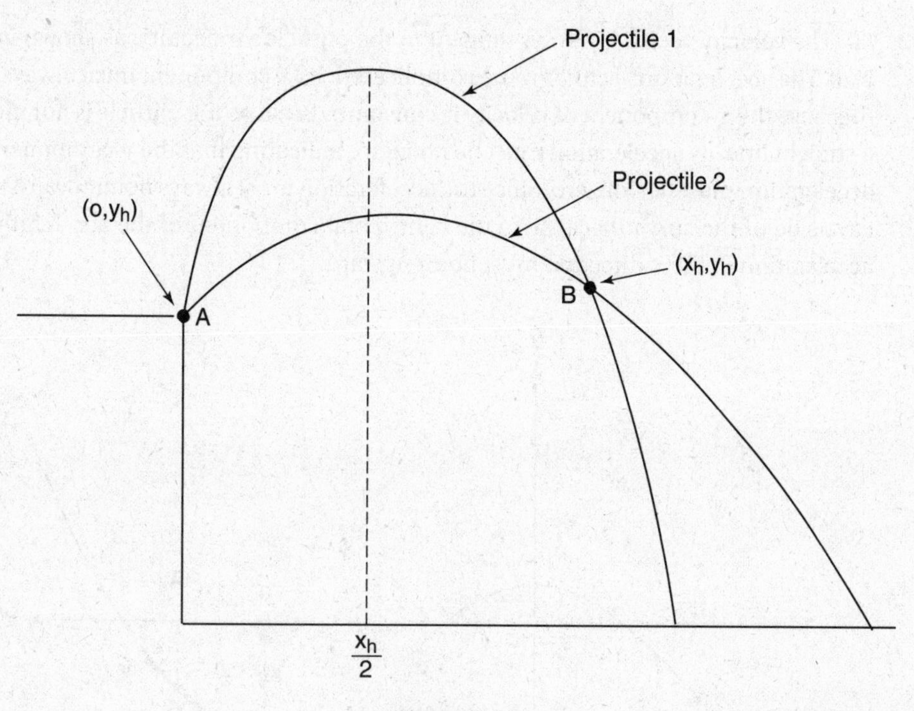

Figure 3.19

(a) The amount of time it takes for a projectile to fall to the height at which it was launched depends only on its *y*-component of velocity v_y: A projectile with a larger initial value of v_y will require longer to fall to its original height. Because it takes projectile 2 three fewer seconds to fall to its original height (i.e., to go from point *A* to point *B*), it must have a smaller initial v_y.

(b) Because projectile 2 has a smaller initial v_y, it has a smaller maximum height and spends less time in the air.

(c) Because of the symmetry of parabolas, the *x*-component of the position of the peak of the trajectories of both projectiles occurs at exactly $x_h/2$ (Figure 3.19).

(d) Because projectile 2 can cover the same *x*-displacement as projectile 1 (the distance between points *A* and *B*) in three fewer seconds, projectile 2 has a greater v_x.

(e) Because projectile 2 has a smaller initial v_y and a greater initial v_x, it has a smaller angle of elevation at any time during its flight. (If in doubt, a quick sketch should confirm this.)

(f) Because point *B* is at the same height as point *A*, v_y for each of the particles is the negative of their initial values (because of the symmetry of parabolas). Therefore, for projectile 1 v_y still has a greater magnitude. The values of v_x remain unchanged, such that projectile 2 still has the greater *x*-component of velocity.

(g) At point *B*, projectile 1 is falling with a greater magnitude v_y, so after passing point *B* it spends less time in the air than projectile 2. Because projectile 1 spends less time in the air and has a *smaller* value of v_x, it has a smaller range.

2. (a) The velocity vector is always tangent to the particle's trajectory, as shown in Figure 3.20. The acceleration vector is more complicated. Its *x*-component must always be zero (because the *x*-component of velocity is constant). Because the particle is not moving in a straight line, its acceleration must be nonzero, indicating that the *y*-component of the acceleration must be nonzero. Since the acceleration must always point toward the concave side of the curve (because of the centripetal component of the acceleration), the acceleration in the *y*-direction must point upward.

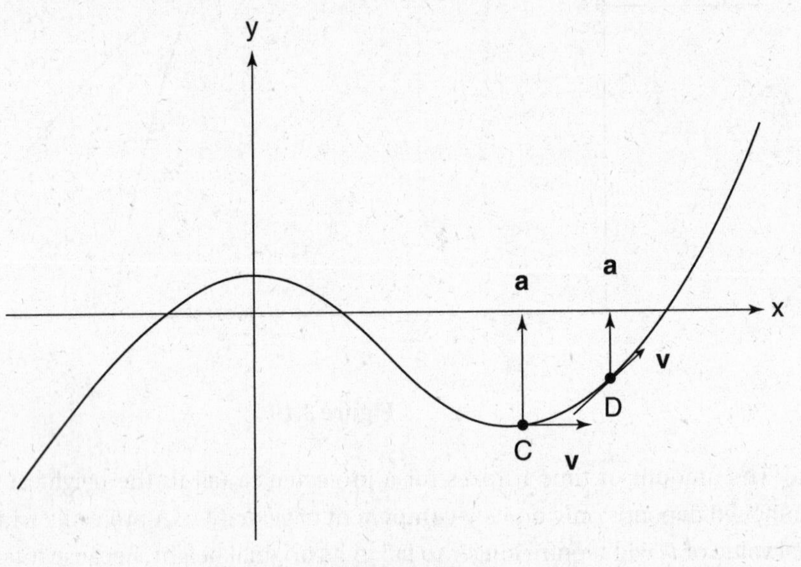

Figure 3.20

Alternative solution: Because the *x*-component of velocity is constant, the *x*-position is proportional to time, so we can substitute *x* with $v_x t$ to obtain $y(t)$. This function [like the original $y(x)$] is concave upward, and thus a_y must be positive.

(b) $v_y = \frac{dy}{dt} = \frac{dy}{dx}\frac{dx}{dt} = (3x^2 - 6x)(v_x)$

(c) Setting the above equation equal to zero, we find that the *y*-component of velocity is equal to zero at $x = 0$ and $x = 2$, locations at which the slope of the $y(x)$ curve [which is qualitatively similar in shape to the $y(t)$ curve] is zero.

(d) $a_y = \frac{d^2y}{dt^2} = \frac{d}{dt}[(3x^2 - 6x)(v_x)] = \frac{d(3x^2 - 6x)(v_x)}{dx}\frac{dx}{dt} = (6x - 6)v_x^2$

(e) Setting the above equation equal to zero, we find that a_y is zero at $x = 1$. This is the inflection point of the $y(x)$ and $y(t)$ curves, the point where the second derivative is zero. This makes sense: It is where the curvature changes from negative to positive, and thus at this point the curvature is zero. Because the curvature is zero, the acceleration is zero as well.

(f) If the speed is constant, the tangential acceleration must be zero. The acceleration vector is then zero or perpendicular to the object's trajectory. At points C and D, because the path is curved, the acceleration must be nonzero such that the acceleration vector must point toward the center of the path, as shown in Figure 3.21.

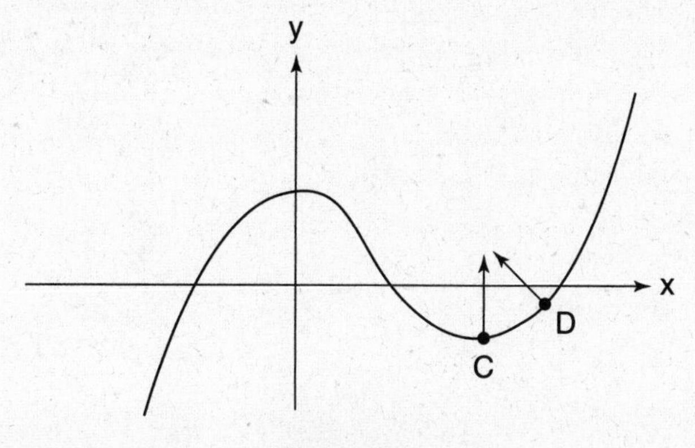

Figure 3.21

Newton's Laws

4

→ NEWTON'S THREE LAWS OF MOTION
→ MASS VS WEIGHT
→ APPLICATION OF NEWTON'S LAWS

Newton's first law:

$$\mathbf{F}_{\text{net}} = 0 \Rightarrow \mathbf{a} = 0$$

Stated in words: When the net force acting on a body is zero, its acceleration must be zero, meaning that the velocity remains constant. This corresponds to two physical situations: (1) the object can either remain at rest ($\mathbf{v} = 0$) or (2) continue along a straight-line path at constant velocity ($\mathbf{v} \neq 0$).

Superposition of Forces What do we mean by net force? What happens if we exert several forces on a body at the same time? We find experimentally that the body behaves as if a single force is acting on it equal to the vector sum of all the individual forces (we call this vector sum the net force, $\mathbf{F}_{\text{net}}$).

$$n \text{ individual forces} \xleftrightarrow{\text{equivalent}} \mathbf{F}_{\text{net}} = \sum_{i=1}^{i=n} \mathbf{F}_i$$

The superposition of forces also works the other way, letting us treat each force as the sum of its components:

$$\mathbf{F} = F_x\hat{i} + F_y\hat{j}$$

When problem solving, we often combine these two approaches to add various forces together component-wise, obtaining a set of one-dimensional equations such as the following (which can then be related to the one-dimensional equations for uniformly accelerated motion).

$$\mathbf{F}_{\text{net}} = \sum_{i=1}^{i=n} \mathbf{F}_i \Leftrightarrow \begin{cases} F_{\text{net},x} = \sum_{i=1}^{i=n} F_{x,i} \\ F_{\text{net},y} = \sum_{i=1}^{i=n} F_{y,i} \end{cases}$$

Concept of Inertia Inertia refers to how much an object resists a change in its velocity and is measured by *mass*. For example, it is easy to drastically change the velocity of an object with little mass or inertia (e.g., flicking a paperclip), but it is harder to change the velocity of an object with lots of mass or inertia (e.g., throwing a bowling ball).

Corollary to Newton's first law:

$$\mathbf{v} = 0 \text{ and/or } \mathbf{v} = \text{constant} \quad \Leftrightarrow \quad \mathbf{a} = 0 \quad \Leftrightarrow \quad \mathbf{F}_{\text{net}} = 0$$

What this equation means is that if an object is at rest (*static equilibrium*) or in motion with a constant velocity (*dynamic equilibrium*), it must have zero acceleration and zero net force. This equation is the basis of solving all static equilibrium problems involving particles.

Newton's Laws Are Valid Only in Inertial Reference Frames If you accelerate your car too quickly, everything on the dashboard goes flying back and lands in your lap. Consider the situation from the frame of reference of the car: These objects are accelerating with respect to the car, yet there is no net force acting on them. Is this a violation of Newton's first law? No. Newton's first law (as well as his second and third laws) are valid only in inertial reference frames. The first law postulates that such frames exist in the real world.

Recall from Chapter 3 that inertial reference frames are reference frames that move at a constant velocity with respect to other inertial reference frames. The most familiar inertial frame is Earth (at a local level, neglecting its spin and revolution). Very rarely in Physics C do we encounter noninertial reference frames, which include frames attached to objects that accelerate linearly (such as the car in the example above), objects that experience centripetal acceleration (such as a merry-go-round), and objects that experience both linear and radial acceleration (such as a car speeding up as it goes around a curve).

What happens if there is a net force on an object? Newton's second law provides the answer:

> **Newton's second law:**
>
> $$\mathbf{F}_{\text{net}} = m\mathbf{a}$$

This law reveals that *force is a vector* parallel to the acceleration. Force is measured in pounds or newtons (N), the SI unit of force defined as 1 newton ≡ 1 kg m/s^2.

TIP

Newton's third law applies to a pair of objects, never a single object alone.

> **Newton's third law: For every force exerted by one object on another, there is another force equal in magnitude and opposite in direction that is exerted back by the second object on the first.**

For example, if you push on a wall, the wall exerts an equal and opposite force on you. Note that the third law involves *two* objects, never just one. Newton's third law is less obvious with long-range forces than with contact forces but is still always valid. For example, if Earth is exerting a downward force on you, you are exerting an upward force on Earth. (Because Earth is much more massive than you, it does not accelerate very much because of this force, as explained by Newton's second law.)

MASS VS WEIGHT

- *Mass* is a measure of inertia (the degree to which an object resists changes in its velocity) and is the proportionality constant that relates force to acceleration in Newton's second law, $\mathbf{F}_{\text{net}} = m\mathbf{a}$. The SI unit for mass is kilograms (kg).
- *Weight* is the magnitude of the force exerted on an object by the closest nearby planet (typically Earth) according to the formula

$$w = mg$$

Thus, weight has units of force [newtons (SI) or pounds]. (Note that this is simply a special case of the formula $F = ma$, where F is weight and a is acceleration due to gravity.)

An object has the same mass anywhere in the universe, but not the same weight.

It's easy to lose *weight*—just move to a planet with a smaller g or move to a higher altitude (with a smaller local g). However, if you want to lose *mass*, you'll have to actually exercise.

Various Forces Solving Newton's law problems requires familiarity with the forces that you are likely to encounter. Some of the following may seem obvious now, but in the context of complicated problems it's easy to confuse things.

1. **GRAVITATIONAL FORCE**
   For now, we will be dealing with gravity near Earth's surface.
   Magnitude: $F = w = mg$
   Direction: Down toward the center of Earth (not necessarily *perpendicular* to the surface that the object sits on, as in inclined plane problems).
2. **NORMAL FORCE**—which we will denote F_N.
   Magnitude: Determined by Newton's second law.
   Direction: *Always* perpendicular (hence the term *normal*) to the surface the object is on and pointing away from the surface toward the object.
   Example: While a book of mass m is resting on a table, the table exerts a normal force on the book that points up and has a magnitude mg. It is Newton's second law that tells us that the normal force must have magnitude mg because the acceleration is zero. (Newton's third law tells us that the book exerts an equal and opposite normal force on the table.) If we exert an additional downward force of magnitude F on the book, then the normal force exerted on the book by the table will still point up but will have a *larger* magnitude of $mg + F$, so that the net force on the book will remain zero.
3. **FRICTIONAL FORCE**
   Magnitude: The magnitude of the frictional force can be determined from simple experiments. Slowly apply an increasingly strong horizontal force to an object on your desk. What do you observe? Up to a certain magnitude of force, the object doesn't budge, but after a threshold force is reached, the object suddenly begins to accelerate. This can be explained in terms of Figure 4.1, which plots the frictional force as a function of the applied horizontal force.

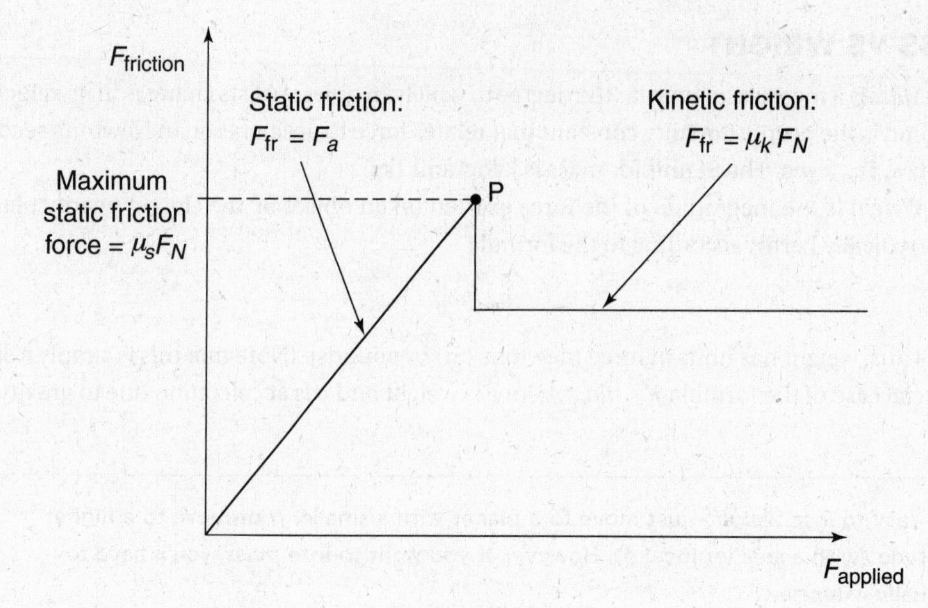

Figure 4.1

For values of the applied force below threshold, the frictional force is large enough to balance the applied force and the object does not move; the magnitude of the frictional force is exactly equal to the applied force and is "whatever it takes to prevent motion" (a variable magnitude similar to the magnitude of the normal force). This frictional force, present when the objects are not sliding relative to each other, is called *static friction* and has a variable magnitude with a *maximum* magnitude of $F_{\text{maximum static friction}} = \mu_s F_N$ (where μ_s is the coefficient of static friction, a dimensionless number that is a property of the two materials in contact).

However, above this threshold value (to the right of point *P* in Figure 4.1) friction is unable to balance the applied force, and the object moves. This frictional force, present when the objects are sliding relative to each other, is called *kinetic friction*. Experimentally, its magnitude has a constant value, $F_{\text{kinetic friction}} = \mu_k F_N$ (where μ_k is the coefficient of kinetic friction, similar to the coefficient of static friction and always lower in magnitude). Thus, the kinetic frictional force is always lower than the maximum static frictional force.

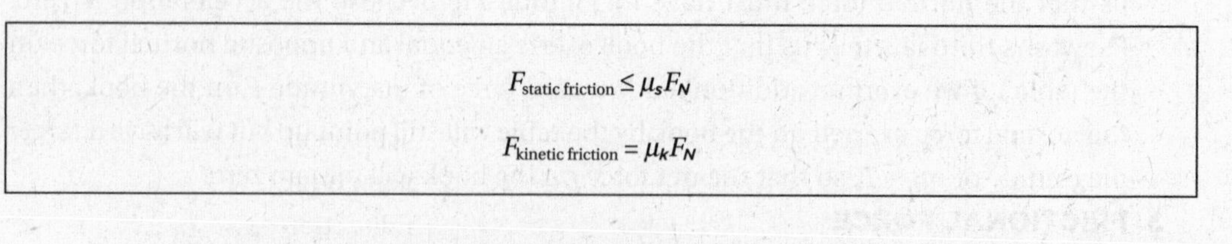

$$F_{\text{static friction}} \leq \mu_s F_N$$

$$F_{\text{kinetic friction}} = \mu_k F_N$$

Direction: The frictional force points *parallel* to the plane of contact between the two objects and in a direction that opposes the motion or incipient motion of the object. (Thus, to determine the direction of static friction, ask the question: In what direction would the object slip if there were no friction? The frictional force points in the *opposite* direction.)

EXAMPLE 4.1

A ladder is sliding down the side of the building. Sketch all the forces on the ladder.

SOLUTION

The forces are shown in Figure 4.2. Note how the normal forces are perpendicular to the surfaces of contact and how friction opposes the motion of each end of the ladder.

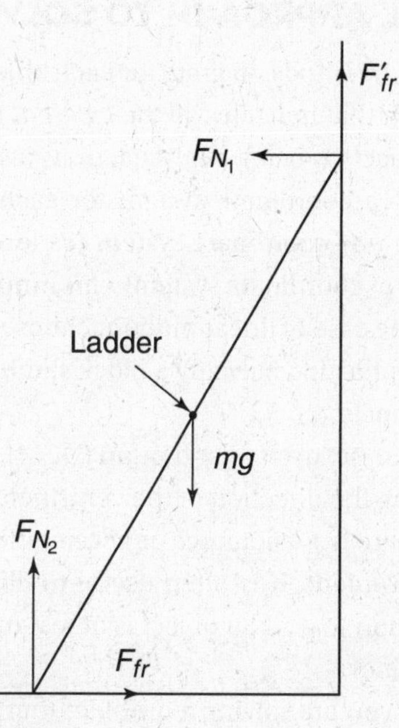

Figure 4.2

EXAMPLE 4.2

Suppose you apply the minimum force required to move an object initially at rest on a horizontal surface. If you continue to apply this force, once the object moves, what will be the acceleration of the object? Express your answer in terms of m, g, μ_s, and μ_k.

SOLUTION

The minimum force required to make the object move is $F_{\text{applied}} = \mu_s F_N = \mu_s mg$. (Given that the vertical acceleration is zero, the normal force must cancel the gravitational force and have a magnitude of mg.) However, once the block begins to move, there are kinetic frictional forces such that (taking $+x$ to be the direction of motion)

$$F_{\text{net},x} = F_{\text{applied},x} + F_{\text{frictional},x}$$

$$F_{\text{net},x} = \mu_s mg - \mu_k mg$$

$$a_{\text{net},x} = \frac{F_{\text{net},x}}{m} = (\mu_s - \mu_k)g$$

4. TENSION FORCE

Magnitude: Variable magnitude determined by Newton's second law. Generally ropes are approximated as massless, such that the tension at every point along the rope is the same.

Direction: The tension force is always parallel to the rope or string. The rope or string always *pulls* on the objects to which it's connected, and thus at each end the tension force points toward the middle of the rope or string.

GENERAL APPROACH TO SOLVING NEWTON'S LAW PROBLEMS

TIP

Follow this step-by-step approach, and you will solve every Newton's laws problem successfully.

1. Draw a free-body diagram for each object in the problem. A free-body diagram is simply a sketch that indicates all the *external forces* exerted on the object. If the motion of several objects is being analyzed, draw a separate free-body diagram for each object.
2. Choose a coordinate system for each object. Although it is mathematically valid to choose *any* coordinate system (as long as it is an inertial reference frame), a careful choice of coordinate systems can simplify problems dramatically.
   a. In the case of linear motion, choose one axis parallel to the direction of motion. (For example, in analyzing a block sliding down an incline, one axis should be parallel to the incline.)
   b. For uniform circular motion (UCM), choose one axis parallel to the radial coordinate, along the direction of the centripetal acceleration.
   c. If there is no net force or acceleration, use your judgment as to which axes are most convenient. It is often useful to align one axis parallel to the direction of *possible* motion (e.g., if an object is at rest on an inclined plane, align one axis parallel to the incline).

   If you are solving a problem that contains multiple bodies moving with the same acceleration (e.g., connected by a rope), make sure that the accelerations of the objects are related appropriately (i.e., positive acceleration of one object in its coordinate system corresponds to positive acceleration of a connected object in *its* coordinate system).
3. Resolve all the forces in the free-body diagram into components along the axes of your coordinate system. Add the components in each direction separately and apply Newton's second law.

$$\mathbf{F}_{\text{net}} = m\mathbf{a} \Leftrightarrow \begin{cases} \sum F_x = ma_x \\ \sum F_y = ma_y \end{cases}$$

4. Solve these simultaneous equations.
   a. Static equilibrium (object at rest) $\Rightarrow F_{\text{net},x} = F_{\text{net},y} = 0$ (the corollary of Newton's first law, discussed above).
   b. Circular motion. The centripetal force is the net force required for circular motion. It can be provided by any number of forces such as tension, normal force, gravity, or friction. Therefore, the centripetal force should never appear on a free-body diagram but should rather be set equal to the net force when solving Newton's second-law equations.

In any uniform circular motion problem, the net force must equal the centripetal force, that is, $F_{net} = mv^2/r$, and this net force must point toward the center of the circle.

c. You may also need to use the equations that define the magnitudes of the forces (e.g., the equations for static and kinetic friction or gravity).

d. The accelerations of any two objects attached by a string of fixed length have equal magnitudes (as discussed in Example 4.5, assuming the string is massless and taut).

EXAMPLE 4.3 STATIC EQUILIBRIUM

A mass *m* is suspended by two ropes as shown in Figure 4.3 and remains at rest. Calculate the tension in each rope in terms of *m*, *g*, θ, and ϕ.

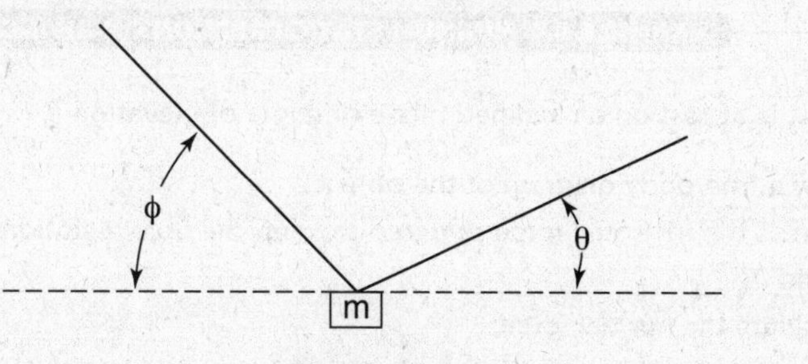

Figure 4.3

SOLUTION

A free-body diagram with a coordinate system is shown in Figure 4.4. Because there is no net acceleration and thus no net force, we are free to choose whichever axes are most convenient. Here we choose the *x*-axis horizontal and the *y*-axis vertical. Summing the components of forces along these directions yields

$$F_{net,x} = 0 = T_1 \cos\theta - T_2 \cos\phi$$

$$F_{net,y} = 0 = T_1 \sin\theta + T_2 \sin\phi - mg$$

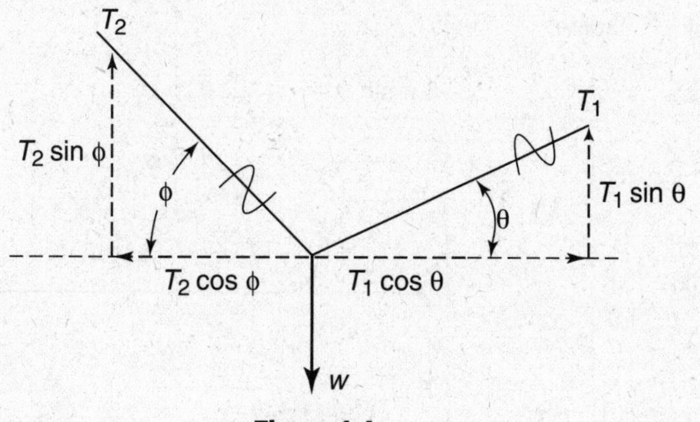

Figure 4.4

We now have two linear equations and two variables (T_1 and T_2), so we can solve for T_1 and T_2, obtaining

$$T_1 = \frac{mg}{\sin\theta + \cos\theta\tan\phi} = \frac{mg\sec\theta}{\tan\phi + \tan\theta}$$

$$T_2 = \frac{mg\sec\phi}{\tan\phi + \tan\theta}$$

Symbolic answer check: (1) The units are correct. The trigonometric functions are dimensionless, so the units of the tensions are the same as those of the weight (both having units of force). (2) Increasing the value of g yields increased tension forces, which makes sense. (3) If we allow $\theta \to 0$, then $T_2 = mg/\sin\phi \Rightarrow mg = T_2\sin\phi$. That is, if the right rope is horizontal, the entire gravitational force will be provided by the y-component of T_2. The same is true if $\phi \to 0$; then $T_1 = mg/\sin\phi \Rightarrow mg = T_1\sin\phi$. (4) If $\theta \to 90°$, then T_1 can support the mass on its own and T_2 approaches zero. Vice versa, it is also true that if $\phi \to 90°$, then T_1 approaches zero.

EXAMPLE 4.4 MINIMUM STATIC FRICTION ON AN INCLINE

An object is at rest on an inclined plane of angle of elevation θ.

(a) Draw a free-body diagram of the object.

(b) Express the frictional force required to keep the object stationary in terms of m, g, and θ.

(c) Calculate the normal force.

(d) What is the minimal coefficient of static friction required to keep the object from slipping?

SOLUTION

(a) The free-body diagram is shown in Figure 4.5.

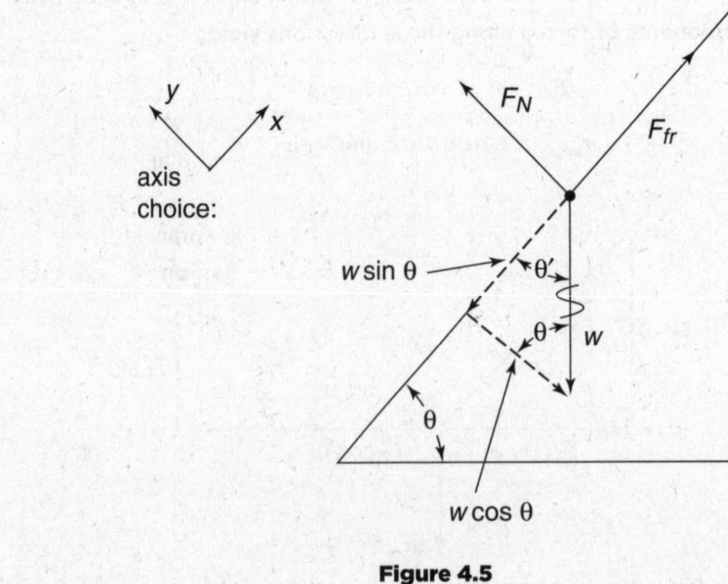

Figure 4.5

Angle Calculations: The following is a method for keeping track of angles that will prove useful when dealing with more complicated problems. Designate θ' as $90 - \theta$, which is known as the *complement of* θ. Therefore, if θ is one angle in a right triangle, the other angle must be θ' (see Figure 4.6). Similarly, if θ divides a right angle, the other angle inscribed in the right angle must be θ'.

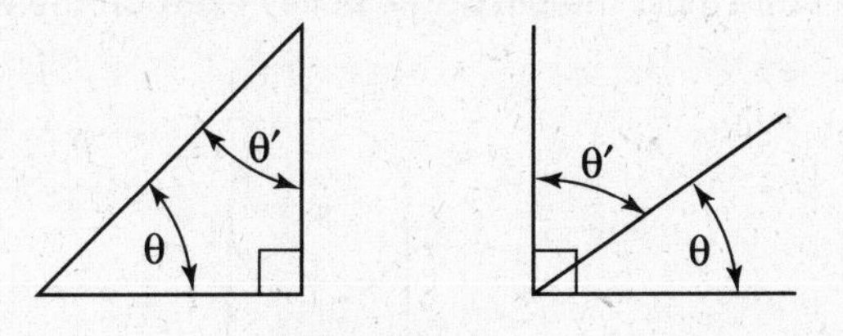

Figure 4.6

How is this useful? The key is that just as θ' is the complement of θ, θ is the complement of θ'. Therefore, once you identify certain angles as θ', you can then identify *their* complements as θ. This often allows you to quickly label all the angles in a problem as either θ or θ'. Finally, $\sin\theta = \cos\theta'$ and $\cos\theta = \sin\theta'$, so it is easy to work with the trigonometric functions of these angles.

(b) Calculating the net force in the *x*-direction,

$$F_{\text{net},x} = F_{\text{fr}} - mg\sin\theta = 0 \Rightarrow F_{\text{fr}} = mg\sin\theta$$

(c) Calculating the net force in the *y*-direction,

$$F_{\text{net},y} = 0 = F_N - mg\cos\theta \Rightarrow F_N = mg\cos\theta$$

(d) We can calculate the coefficient of static friction using the following equation. (The coefficient of friction is *minimized* if the force exerted is the *maximum* possible force for that coefficient of friction.)

$$F_{\text{maximum static friction}} = \mu_s F_N$$

$$\mu_s = \frac{F_{\text{fr}}}{F_N} = \frac{mg\sin\theta}{mg\cos\theta} = \tan\theta$$

Symbolic answer check: (1) The units are correct (coefficients of friction are dimensionless numbers). (2) For $\theta = 0$, no friction is needed. As θ approaches 90°, the needed amount of friction approaches infinity.

Note: Make sure you are comfortable with the free-body diagram and geometry used in this problem. Inclined plane problems are very common on AP exams.

EXAMPLE 4.5 THE ATWOOD MACHINE

The Atwood machine consists of two masses attached to a pulley as shown in Figure 4.7.

(a) **What is the acceleration of the system consisting of the two masses and the rope? (Treat the rope and pulley as frictionless and massless.)**

(b) **What vertical force must the pin of the pulley exert on the wheel of the pulley to hold it up?**

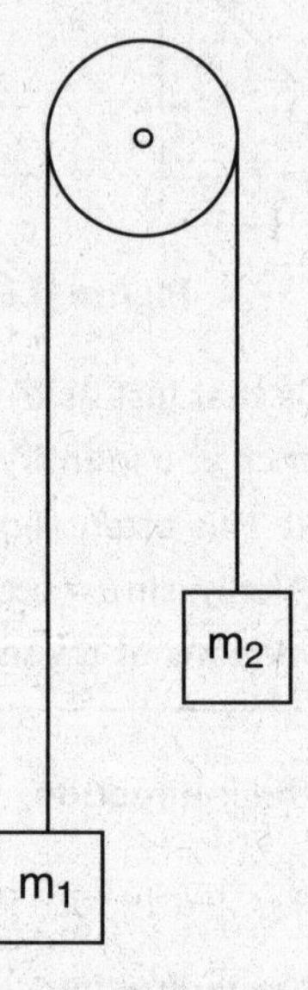

Figure 4.7

SOLUTION

(a) The free-body diagrams for the two masses are shown in Figure 4.8. Calculating the net force on each mass yields (with "up" as the positive direction for each mass):

$$F_{\text{net,m1}} = F_T - m_1 g = m_1 a_1$$

$$F_{\text{net,m2}} = F_T - m_2 g = m_2 a_2$$

We have two equations in three variables (F_T, a_1, and a_2). The last equation comes from realizing that because the masses are connected by the rope, their velocities must have equal magnitudes (but in opposite directions):

$$\mathbf{v}_1 = -\mathbf{v}_2$$

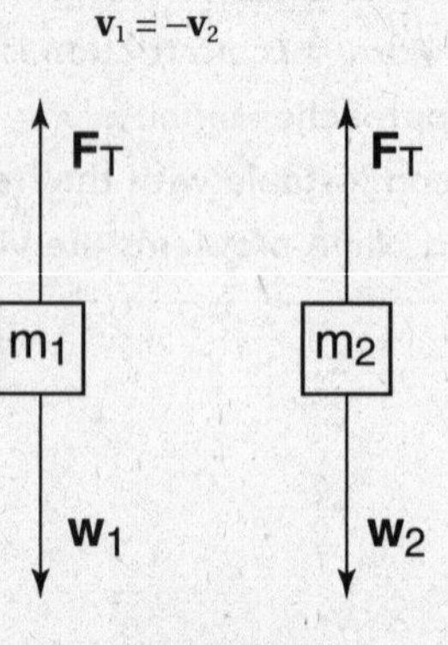

Figure 4.8

Differentiating this yields the last equation. (This demonstrates the assertion above that masses attached by ropes experience accelerations with equal magnitudes.)

$$\mathbf{a}_1 = -\mathbf{a}_2$$

Note that we could have chosen two different coordinate systems for the masses, with y positive *up* for m_1 and y positive *down* for m_2. Then Newton's second law for m_2 would read

$$F_{\text{net},m_2} = m_2 g - F_T = m_2 a_2$$

and the two velocities would be *equal*, without a negative sign, and thus,

$$\mathbf{a}_1 = \mathbf{a}_2$$

However, the solution for the tension and the magnitudes of the accelerations are the same as in our original choice of coordinates.

Solving the equations in our first choice of coordinates yields

$$F_T = \left(\frac{2m_1 m_2}{m_1 + m_2}\right) g$$

$$a_1 = g\left(\frac{m_2 - m_1}{m_1 + m_2}\right) \quad \text{and} \quad a_2 = g\left(\frac{m_1 - m_2}{m_1 + m_2}\right)$$

***Symbolic answer check*: (1) The units are correct (both accelerations have units of g, which is also an acceleration). (2) As m_1 approaches infinity or m_2 approaches zero, a_1 approaches $-g$ and a_2 approaches $+g$. This corresponds to m_2 being too light to make much difference, such that m_1 is essentially in free-fall, taking m_2 for the ride. The reverse situation occurs if m_2 approaches infinity or m_1 approaches zero.**

(b) The free-body diagram for the wheel of the pulley is shown in Figure 4.9. For the pulley not to fall down, the pin must exert an upward force exactly canceling the downward tension forces. (Note that although the pulley is rotating, it is not translating, so its net force is still zero. The requirements for static equilibrium in extended objects, which accounts for the rotation being zero, are addressed in Chapter 8.)

$$F_{\text{net},y} = F_{\text{pin}} - 2F_T \Rightarrow F_{\text{pin}} = 2F_T = \frac{4m_1 m_2 g}{m_1 + m_2}$$

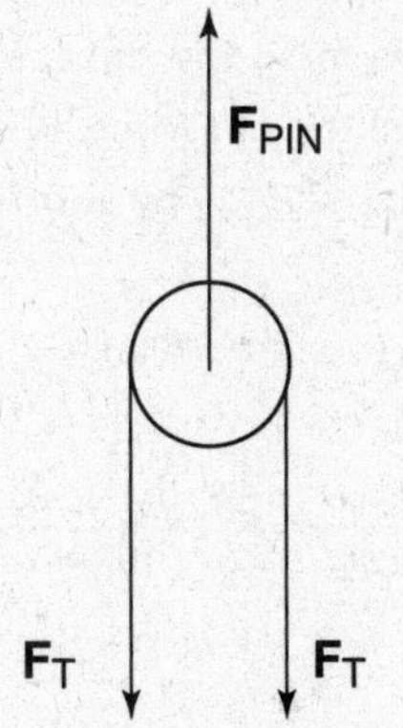

Figure 4.9

***Symbolic answer check*: (1) The units are correct, $m \cdot g$. (2) If either mass approaches zero, F_{pin} approaches zero (this corresponds to one mass in free-fall and the massless string sliding limply over the pulley should read pulley). (3) If the masses are equal and at rest, the F_T simply equals their combined mass, 2 mg.**

EXAMPLE 4.6 THE MULTIPLE-BLOCK PROBLEM

A force F is pushing three adjacent blocks along a frictionless surface as shown in Figure 4.10.

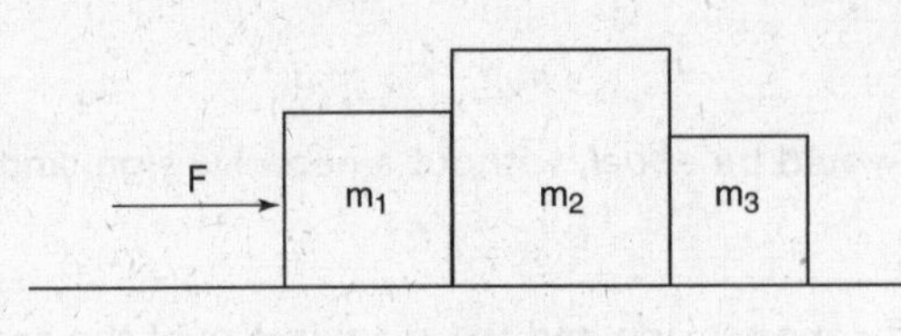

Figure 4.10

(a) What is the acceleration of each block?

(b) What is the net force on m_1?

(c) What is the magnitude of the force that m_1 exerts on m_2?

(d) What is the magnitude of the force that m_2 exerts on m_3?

SOLUTION

(a) Because all the blocks remain in contact, they move together and accelerate at the same rate. We can thus treat them as one larger object, ignore complicated internal forces, and easily find their common acceleration.

When several objects move together and accelerate at the same rate, we can treat them as a single object to find their common acceleration.

$$a = \frac{F}{m_{\text{total}}} = \frac{F}{m_1 + m_2 + m_3}$$

(b) To calculate the net force on m_1, we can simply substitute into Newton's second law (because we already know the acceleration of m_1).

$$F_{\text{net},m1} = m_1 a = m_1\left(\frac{F}{m_1 + m_2 + m_3}\right) = \frac{m_1 F}{m_1 + m_2 + m_3}$$

(c) By Newton's third law, the contact force that m_1 exerts on m_2 is equal and opposite to the force that m_2 exerts on m_1. We can find the force m_2 exerts on m_1 by drawing a free-body diagram for m_1 (as shown in Figure 4.11) and applying Newton's second law. The force F pushes m_1 to the right, while m_2 exerts a force on m_1. The sum of these forces must equal the net force as calculated in part (b), allowing us to calculate the magnitude of $F_{2\text{ on }1} = F_{1\text{ on }2}$ by subtraction:

$$F_{\text{net},x\text{ on }m1} = F - F_{2\text{ on }1} = \frac{m_1 F}{m_1 + m_2 + m_3}$$

$$F_{2\text{ on }1} = F - \frac{m_1 F}{m_1 + m_2 + m_3} = F\left(1 - \frac{m_1}{m_1 + m_2 + m_3}\right) = F\left(\frac{m_2 + m_3}{m_1 + m_2 + m_3}\right)$$

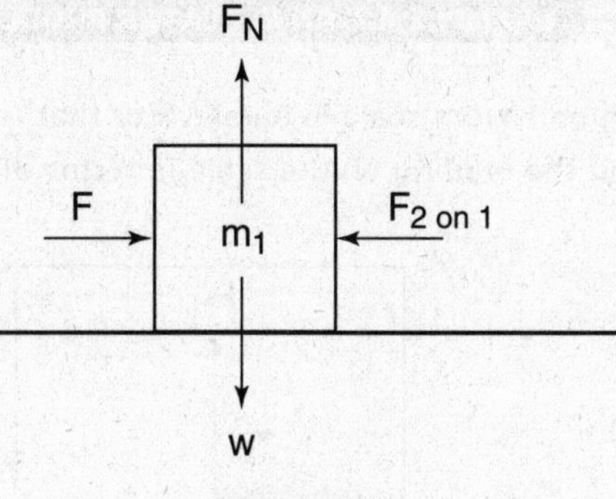

Figure 4.11

Alternative solution: The contact force of m_1 on m_2 must equal the net force on the combined masses of m_2 and m_3, which by Newton's second law is given by $(m_2 + m_3)a$ (treating these two masses as a unit and ignoring internal forces). Substituting, we obtain

$$F_{\text{net},m_1} = (m_2 + m_3)\left(\frac{F}{m_1 + m_2 + m_3}\right)$$

Symbolic answer check: (1) The units are correct (the masses cancel, leaving units of force). (2) The sign is correct: We calculate the magnitude of $F_{2\text{ on }1}$ to be positive, as it should be. (3) As m_1 approaches infinity, the majority of the applied force F is needed to accelerate m_1 (since it has the greatest inertia), so $F_{2\text{ on }1}$ approaches zero. (4) If m_2 or m_3 approaches infinity, it will require most of the applied force F and $F_{2\text{ on }1}$ will approach F.

(d) The only force acting on m_3 parallel to the direction of motion is the force that m_2 exerts on m_3. Applying Newton's second law,

$$F_{\text{net},x,m_3} = F_{2\text{ on }3} = m_3 a = m_3\left(\frac{F}{m_1 + m_2 + m_3}\right) = \frac{m_3 F}{m_1 + m_2 + m_3}$$

Symbolic answer check: (1) The units are correct. (2) If m_3 approaches infinity, m_3 will absorb all the force and $F_{2\text{ on }3}$ will equal F. (3) If m_1 or m_2 approaches infinity, it will absorb all the force and $F_{2\text{ on }3}$ will approach zero.

EXAMPLE 4.7 AN ELEVATOR PROBLEM

You are standing on a bathroom scale in an elevator that is accelerating upward with an acceleration *a*. Find the reading on the scale in terms of your mass *m* and *g*.

Figure 4.12

SOLUTION

The free-body diagram is shown in Figure 4.12. Note that the scale measures the normal force you exert on it (which compresses a spring within the scale); by Newton's third law, this is equal to and opposite the normal force that the scale exerts on you.

Even though you feel "weightless" when the elevator is in free-fall, you are not "massless."

Summing the forces along the *y*-direction and applying Newton's second law,

$$F_{\text{net},y} = F_N - mg = ma \Rightarrow F_N = mg + ma = m(g + a)$$

Thus, the scale gives a "weight" of $m(g + a)$.

Symbolic answer check: (1) The units are correct, $m \cdot a$. (2) If $a = 0$, the elevator is at rest or has constant velocity, and the scale will simply read *mg*. (3) Increasing the acceleration *a* pushes you harder against the scale, increasing your normal force and the reading of the scale. (4) If $a < 0$, the elevator is accelerating downward, causing the normal force to be less than *mg* (and thus causing the scale to read a weight *less* than *mg*). If the elevator is in free-fall, $a = -g$ and the scale reads zero: You feel "weightless" in the frame of reference of the elevator. The scale reads the normal force.

EXAMPLE 4.8 A HANGING ACCELEROMETER

A pendulum bob hanging in a truck is deflected through an angle of θ as the truck accelerates, as shown in Figure 4.13. The truck accelerates at a constant rate, such that θ is constant.

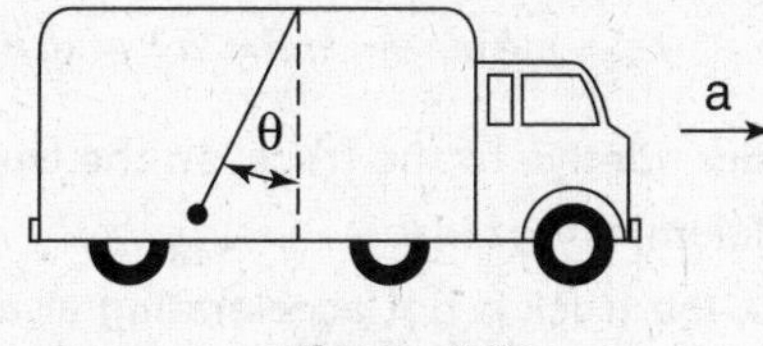

Figure 4.13

(a) Draw a free-body diagram of the forces acting on the pendulum bob.

(b) Solve for the magnitude of the tension force in terms of the gravitational force and θ.

(c) Express the net force on the bob in terms of m, g, and θ.

(d) What is the acceleration of the truck in terms of g and θ?

SOLUTION

(a) The free-body diagram is shown in Figure 4.14.

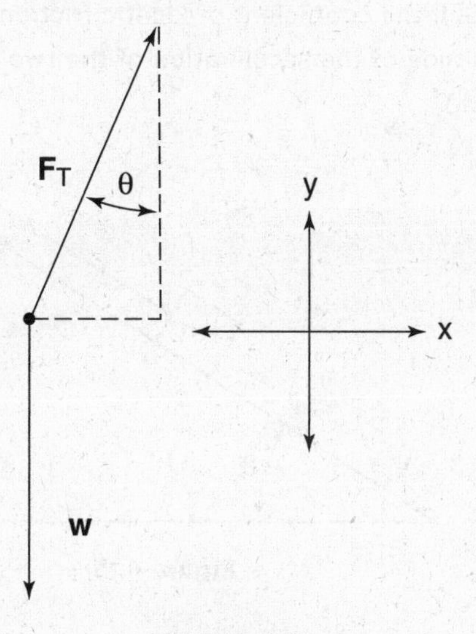

Figure 4.14

(b) The y-acceleration is zero (the ball accelerates along with the truck to the right; its vertical position does not change). Therefore,

$$F_{\text{net},y} = F_T \cos\theta - mg = 0 \Rightarrow F_T = \frac{mg}{\cos\theta}$$

(c) As discussed in part (b), there is no net acceleration in the y-direction, so the net force is simply the x-component of the tension force.

$$F_{\text{net}} = F_T(\sin\theta)$$

Substituting for F_T as calculated above,

$$F_{\text{net}} = \left(\frac{mg}{\cos\theta}\right)(\sin\theta) = mg(\tan\theta)$$

(d) Applying Newton's second law,

$$F_{\text{net}} = mg(\tan\theta) = ma \Rightarrow a = g\tan\theta$$

The pendulum is stationary relative to the truck, so the truck must have the same acceleration as the pendulum, $a = g\tan\theta$.

Answer check: If $\theta = 0$, the truck is not accelerating at all. As θ approaches $\pi/2$ (i.e., the pendulum swings to a horizontal position), the acceleration of the truck approaches infinity.

EXAMPLE 4.9 A CLASSIC INCLINED PLANE PROBLEM

Two masses (m_1 and m_2) are connected by a rope hanging from a pulley on an incline as shown in Figure 4.15. If the coefficient of kinetic friction between m_1 and the plane is μ_k, what is the magnitude of the acceleration of the two masses? (Assume that m_2 is accelerating downward.)

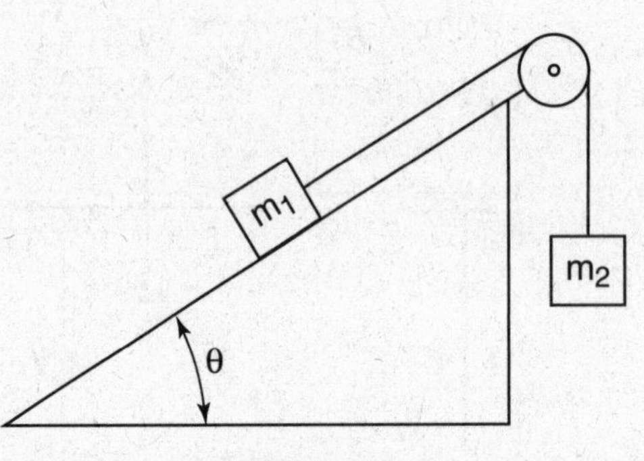

Figure 4.15

SOLUTION

Free-body diagrams and coordinate axes are shown in Figure 4.16. Note that the coordinate axes are chosen so that the common acceleration a is positive for each object within its own coordinate system. The only force not already parallel to a coordinate axis is gravity, which is divided into components as in Example 4.4.

For the sliding block,

$$F_{\text{net},x} = F_T - F_{\text{fr}} - m_1 g\sin\theta = m_1 a$$

$$F_{\text{net},y} = F_N - m_1 g\cos\theta = 0$$

The kinetic friction on the sliding block is

$$F_{\text{fr}} = \mu_k F_N$$

For the falling block,

$$F_{\text{net},y} = m_2 g - F_T = m_2 a$$

We can choose different coordinate systems for the two masses in this example.

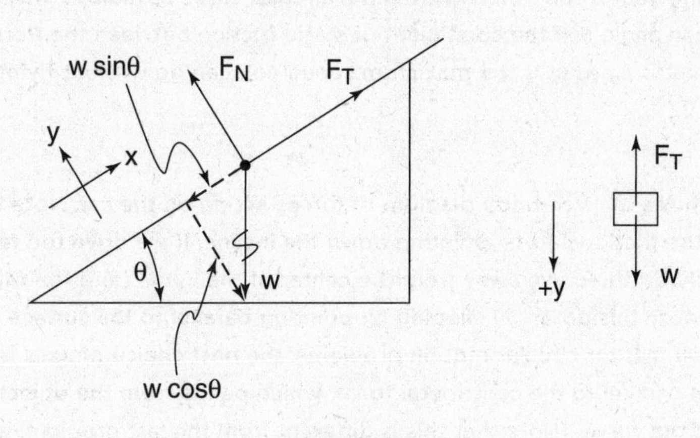

Figure 4.16

We have four variables (F_T, F_{fr}, a, and F_N) and four equations. (Note that because we are assuming that the rope is massless, the magnitude of the tension forces acting on both blocks is the same.) It's not too hard to solve these equations by standard methods for simultaneous linear equations (successive substitution and/or addition of equations to eliminate variables), obtaining

$$a = \frac{(m_2 - \mu_k m_1 \cos\theta - m_1 \sin\theta)\,g}{m_1 + m_2}$$

Symbolic answer check: (1) All the terms in the numerator are forces, and all the terms in the denominator are masses, so the ratio is an acceleration. (2) Mathematically, from the above formula the limit of a as m_2 approaches infinity is g. Alternatively, as m_1 approaches zero, the acceleration approaches g. This makes sense—in both cases, the effect of m_1 on m_2 is negligible, and m_2 is effectively in free-fall. (3) As the value of g increases, the acceleration increases. (4) As μ_k increases, the acceleration decreases. (5) Careful! You might be tempted, as we are, to make the statement, "According to the above equation, as m_1 increases, the acceleration becomes negative (the block slides *down* the incline), which makes sense." The problem with this statement is that if the block is sliding down the incline, the friction force *changes* directions, and we will need to rewrite our free-body diagram and re-solve our equations. The above equation for acceleration is valid only if m_2 is going downward. (Incidentally, although our quantitative results do not apply to the situation where m_1 is sliding down, our qualitative conclusions are correct—increasing m_1 will eventually make the block slide down the incline.)

EXAMPLE 4.10 A CLASSIC BANKED CURVE PROBLEM

You are driving your turbo Porsche around a circular curve of radius *r* where the road is banked at an angle θ. If the coefficient of static friction between the Porsche's tires and the asphalt is μ_s, what is the maximum speed you can go without flying off the road?

SOLUTION

Figure 4.17 shows the free-body diagram of forces acting on the car. Note that we have drawn the frictional force pointing *down* the incline. If we drive too fast, the car will tend to fly off the curve away from the center of the curve (and the frictional force will oppose this potential slipping by pointing parallel to the surface *down* the incline). Recall that for circular motion problems, the best choice of axes is generally with one axis parallel to the centripetal force, which points from the object toward the center of the curve. (Note that this is different from the last problem, where we chose axes parallel and perpendicular to the incline.)

Angle calculations are performed using the θ and θ' method introduced in Example 4.4. Resolving all the forces into components along the axes, summing, and applying Newton's second law yields

$$F_{\text{net},y} = 0 = F_N \cos\theta - mg - F_{\text{fr}} \sin\theta$$

$$F_{\text{net},x} = \frac{mv^2}{r} = F_{\text{fr}} \cos\theta + F_N \sin\theta$$

TIP

Here we do not tilt our coordinate system, because the acceleration is horizontal.

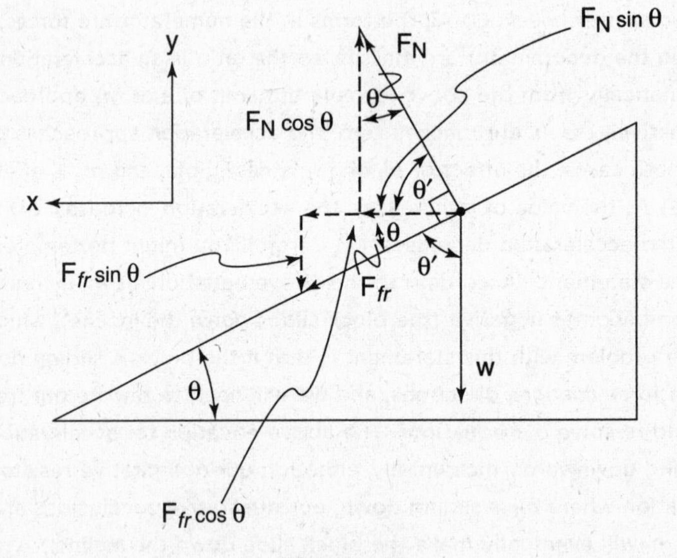

Figure 4.17

Note the form of these equations: Each of the forces that had to be resolved (F_N and F_{fr}) shows up twice—once with a sine factor and the other time with a cosine factor. Once you have had enough practice with these types of problems, a set of force equations without matched sine and cosine factors will leap out at you as likely being incorrect.

The maximum speed corresponds to the maximum frictional force:

$$F_{\text{maximum, static friction}} = \mu_s F_N$$

Now we have three equations with four variables (F_N, F_{fr}, v, and m). What shall we do? After pausing to make sure we didn't neglect any equations, we bravely forge ahead, hoping that mass will cancel out somewhere. Happily, it does, and repeated substitution yields

$$v = \sqrt{\frac{gr(\mu_s \cos\theta + \sin\theta)}{\cos\theta - \mu_s \sin\theta}}$$

Symbolic answer check: (1) g has units of m/s^2, r has units of m, and the sines, cosines, and coefficient of friction are dimensionless. Therefore, gr has units of m^2/s^2, and the square root of this term has units of m/s, the correct units for a velocity. (2) Increasing gravity holds the car down on the road better, increasing the maximum velocity. (3) Increasing the radius, making the curve less tight, increases the maximum velocity. (4) Increasing the coefficient of friction decreases the car's tendency to slip, increasing the maximum velocity. (5) Note that our solution will be undefined if θ gets too large, in particular if $\tan\theta > 1/\mu_s$ (the denominator will then be negative, and the square root will yield an imaginary number). In that case, the bank is too steep for the car to remain on the road, and it begins to slide *down* rather than *up* the incline. We must then reconsider the problem with the frictional force pointing up the bank.

It may seem surprising that the friction between the tire and the road is *static*—after all, the car is in motion. However, if you really consider the interaction of the rubber and the road, during normal driving the rubber does not slide relative to the road—each portion of the rubber briefly touches the road. Kinetic friction between the tires and the road would correspond to skidding, where the rubber actually rubs against the road.

EXAMPLE 4.11 MULTIPLE CHOICE

Two masses are connected by a pulley as shown in Figure 4.18. A force F is applied to m_2 as shown. Given that the horizontal surface is frictionless and the pulley and string are massless, for what range of values of F does the tension in the string equal zero?

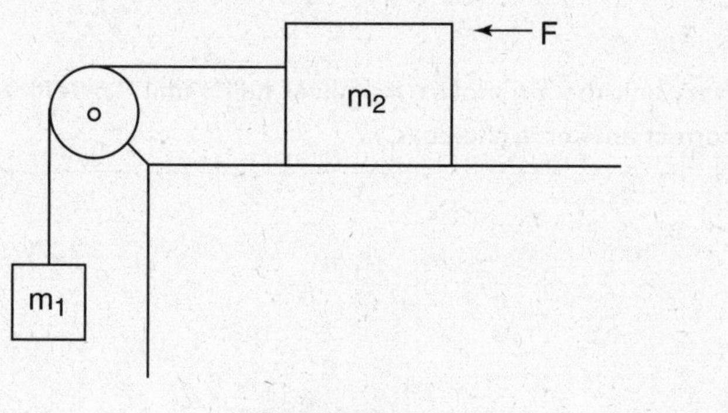

Figure 4.18

(A) $F \geq \frac{m_1 m_2}{m_1 + m_2} g$

(B) $F \geq (m_1 + m_2)g$

(C) $F \geq m_2 g$

(D) $F \geq m_1 g$

(E) none of the above

(C) If there is no tension in the string, m_1 will experience free-fall. Thus, for m_2 to "keep up" with m_1, it must experience an acceleration to the left of g. If the acceleration to the left is greater than this, the tension will still be zero (the slack in the string will grow). If the acceleration is less than this, the string will become taut and m_1 will pull on m_2. Using these observations along with Newton's second law, we see that the answer is choice (C).

EXAMPLE 4.12 MULTIPLE CHOICE

In the system shown in Figure 4.19, what is the relationship between a_1 and a_2?

(A) $a_1 = a_2$

(B) $a_1 = 2a_2$

(C) $a_2 = 2a_1$

(D) $a_1 = \sqrt{2}a_2$

(E) There is no simple relationship between the two accelerations.

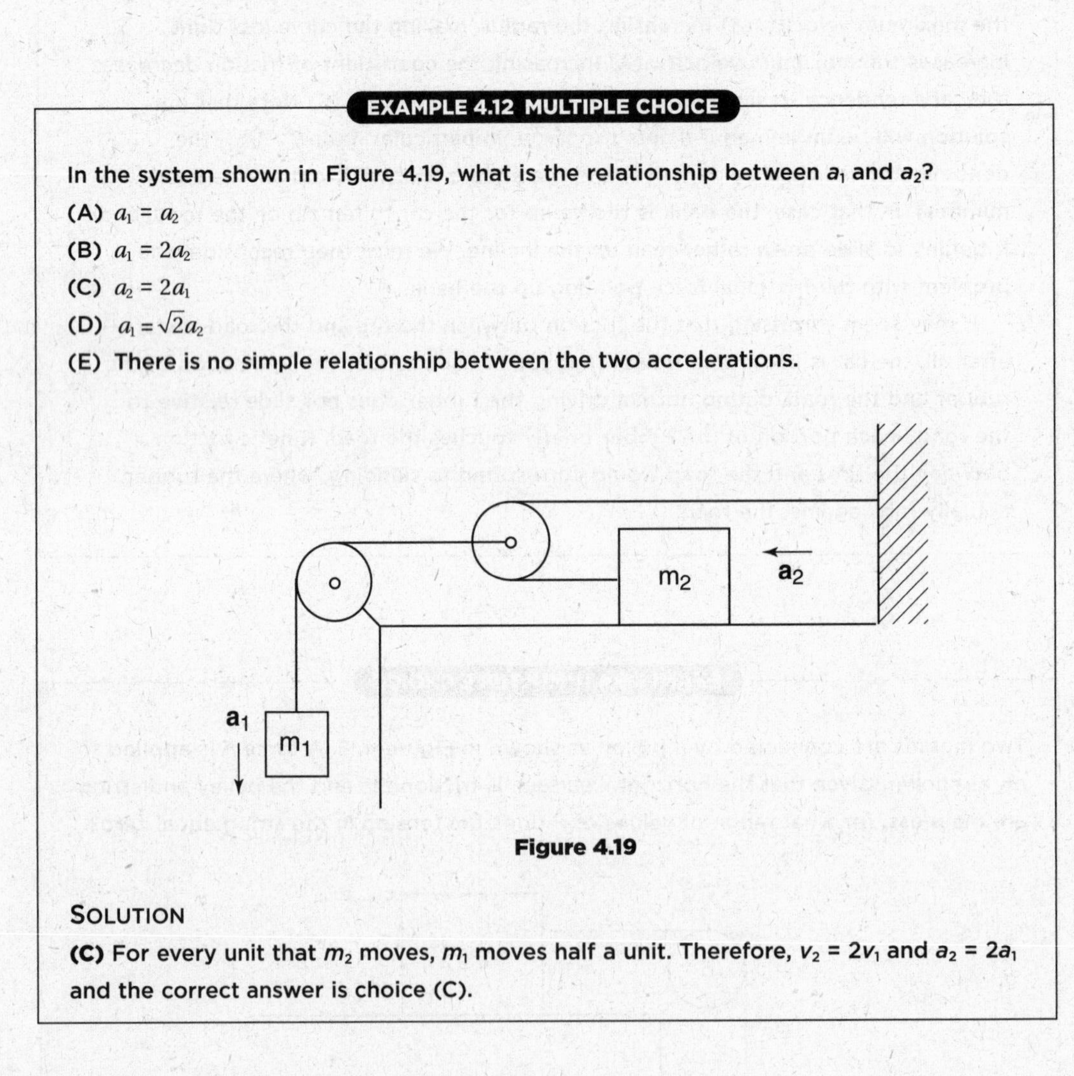

Figure 4.19

SOLUTION

(C) For every unit that m_2 moves, m_1 moves half a unit. Therefore, $v_2 = 2v_1$ and $a_2 = 2a_1$ and the correct answer is choice (C).

CHAPTER SUMMARY

Newton's three laws of motion describe how an object responds to interactions with other objects that exert forces on it. If the net force on the object is zero, it will have zero acceleration. A nonzero force produces acceleration in proportion to the mass of the object. The weight of an object is the gravitational force exerted on it by Earth. If object A exerts a force on object B, then B exerts an equal and opposite force on A.

PRACTICE EXERCISES

Multiple-Choice Questions

1. Consider a mass moving up the incline shown in Figure 4.20 under the influence of an applied force *F*. If the block's mass is *m* and the magnitude of its acceleration is *a*, what is the magnitude of the friction force?

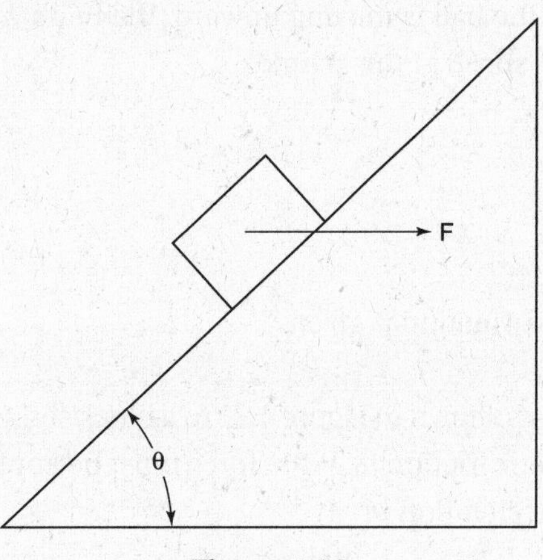

Figure 4.20

(A) $F_{fr} = F\cos\theta - mg\sin\theta - ma$
(B) $F_{fr} = F\sin\theta - mg\cos\theta - ma$
(C) $F_{fr} = F\cos\theta + mg\sin\theta - ma$
(D) $F_{fr} = F\cos\theta - mg\sin\theta + ma$
(E) $F_{fr} = F\cos\theta - mg\cos\theta - ma$

2. What is the coefficient of kinetic friction in the situation introduced in question 1?

(A) $\mu_k = \dfrac{F_{fr}}{F\sin\theta + mg\sin\theta}$

(B) $\mu_k = \dfrac{F_{fr}}{F\sin\theta + mg\cos\theta}$

(C) $\mu_k = \dfrac{F_{fr}}{F\cos\theta + mg\sin\theta}$

(D) $\mu_k = \dfrac{F_{fr}}{F\cos\theta + mg\cos\theta}$

(E) $\mu_k = \dfrac{F_{fr}}{F\tan\theta + mg\cos\theta}$

3. Consider Figure 4.21, in which all the blocks have the same mass and the same acceleration a and the coefficient of kinetic friction between the blocks and the horizontal surface is μ_k. What is the tension in the string between mass i ($i = 1, \ldots, N$) and the mass to its right?

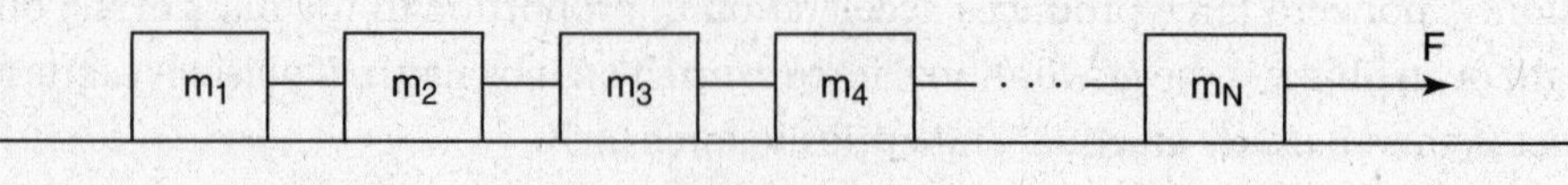

Figure 4.21

(A) $T = i(ma - \mu_k mg)$
(B) $T = i(ma + \mu_k mg)$
(C) $T = (i + 1)(ma + \mu_k mg)$
(D) $T = (i + 1)(ma - \mu_k mg)$
(E) $T = N(ma - \mu_k mg)$

4. A ball attached to a string of length r rotates in a vertical circle. When the string is parallel to the ground and the ball is moving upward, the ball's velocity is v. What is the rate of change in the ball's speed at this point?

(A) $\sqrt{g^2 + v^4/r^2}$
(B) $-\sqrt{g^2 + v^4/r^2}$
(C) $+g$
(D) $-g$
(E) Not enough information is given.

5. A force F is applied as shown in Figure 4.22 to a block sliding on a horizontal surface with a coefficient of kinetic friction μ_k. What force must be applied to cause the mass to move to the right with acceleration a?

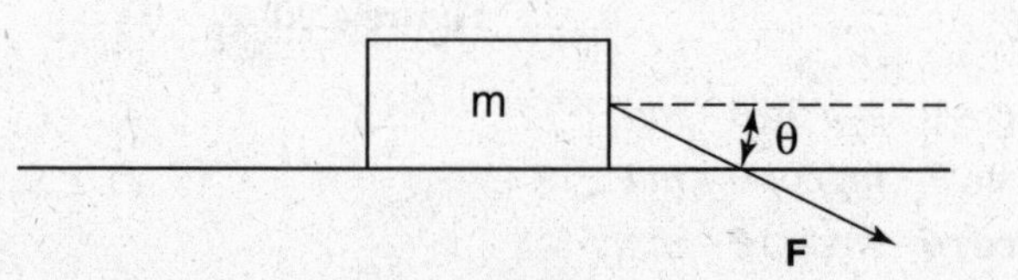

Figure 4.22

(A) $F = ma \sec\theta$

(B) $F = \dfrac{ma + \mu_k mg}{\cos\theta}$

(C) $F = \dfrac{ma + \mu_k mg}{\sin\theta}$

(D) $F = \dfrac{ma + \mu_k mg}{\sin\theta - \mu_k \cos\theta}$

(E) $F = \dfrac{ma + \mu_k mg}{\cos\theta - \mu_k \sin\theta}$

6. A ball is dropped and bounces several times, each time making a partially inelastic collision with the ground. Which of the graphs in Figure 4.23 shows the ball's velocity as a function of time? (Upward velocity is designated positive.)

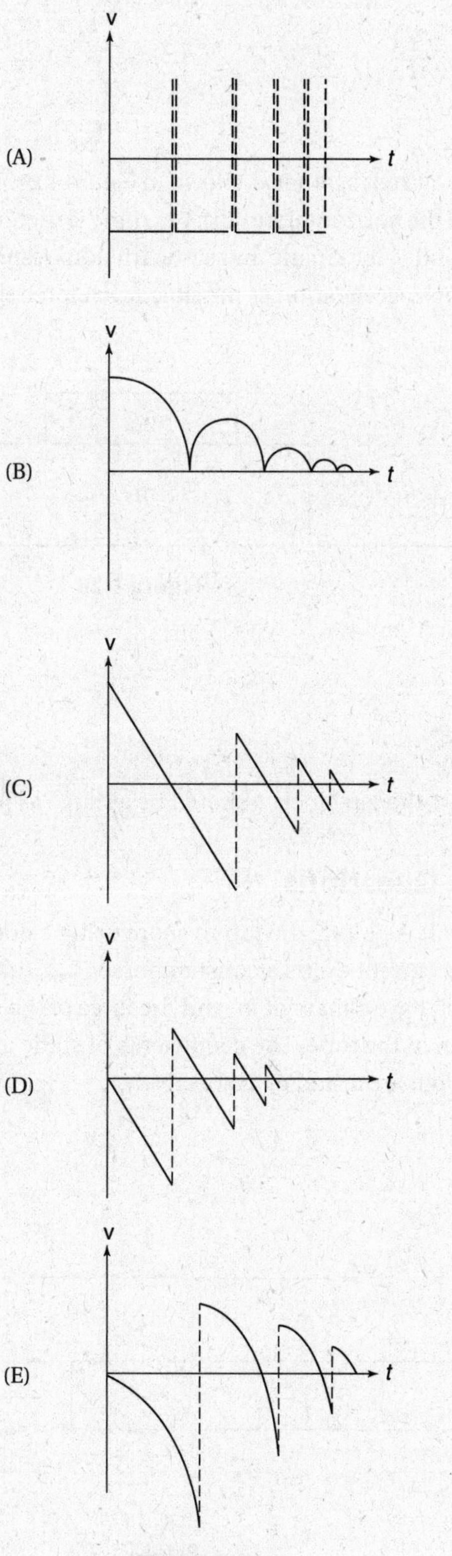

Figure 4.23

7. A variable force exerted during a time interval Δt changes a mass m's velocity from $\mathbf{v}_0$ to $\mathbf{v}_f$. What is the average force exerted during this time interval?

(A) $(\mathbf{v} - \mathbf{v}_0)/\Delta t$
(B) $m\mathbf{v}_f - m\mathbf{v}_0$
(C) $m(v_f^2 - v_0^2)/2$
(D) $m(v_f^2 - v_0^2)/2\Delta t$
(E) $m(\mathbf{v}_f - \mathbf{v}_0)/\Delta t$

8. Two blocks are stacked together as shown in Figure 4.24. There is no friction between the lower block and the horizontal surface, but there is friction between the two blocks, with coefficients μ_k and μ_s for kinetic and static friction, respectively. In order to provide the top block as much acceleration as possible, what force should be applied to the bottom block?

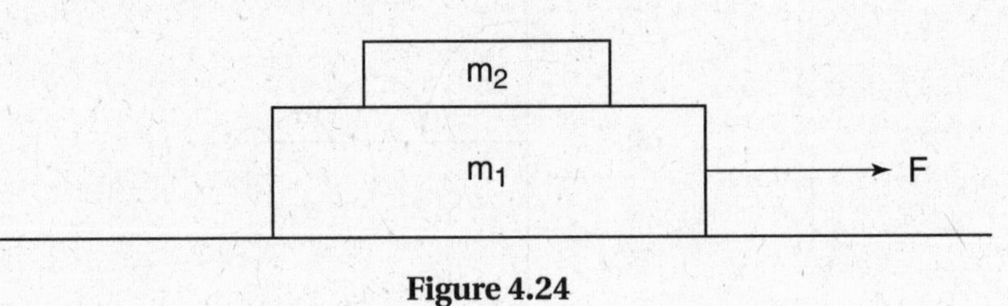

Figure 4.24

(A) $\mu_s m_1 g$
(B) $\mu_k m_1 g$
(C) $\mu_s(m_1 + m_2)g$
(D) $\mu_k(m_1 + m_2)g$
(E) To accelerate the top block, F should be as large as possible.

Free-Response Questions

1. A person on a raft is pulling the raft to shore using a double pulley system as shown in Figure 4.25. The current exerts a constant force F_{current} on the raft to the left. The person and the raft each have a mass of m, and the force of the current is greater in magnitude than the tension in the rope. The coefficients of static and kinetic friction between the person and the raft are μ_s and μ_k, respectively.

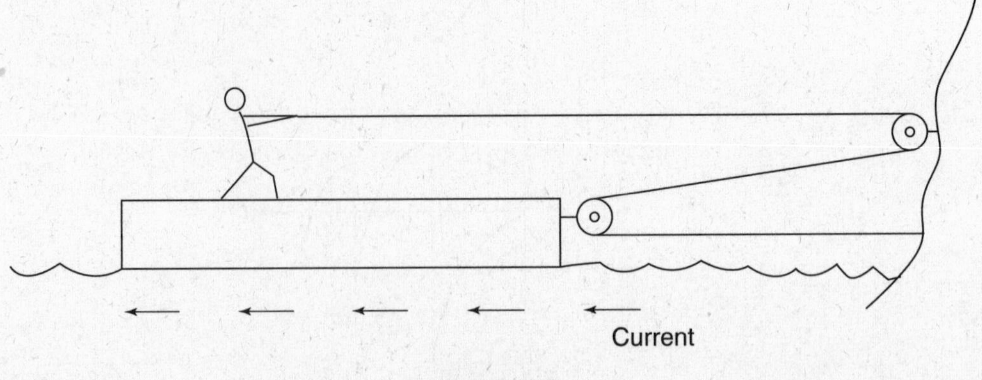

Figure 4.25

(a) Draw free-body diagrams for the person, the raft, and the raft-person system (i.e., treating both the raft and the person as a single object).

(b) Compute the acceleration of the raft relative to the shore as a function of the tension force the person exerts on the rope, given that the person does not slip relative to the raft.

(c) What is the greatest tension force the person can exert without slipping?

(d) After the person begins to slip, what is the acceleration of the raft relative to the shore as a function of the tension force the person exerts on the rope?

2. A common amusement park ride is the rotor. It consists of a cylindrical room in which people stand with their backs against a rough wall. The cylinder is rotated, and after it reaches a certain speed the floor is retracted, yet the riders do not fall down.

(a) Draw a free-body diagram of a person in the rotor.

(b) In terms of the radius R of the rotor, the linear speed of the passengers v, and the gravitational acceleration g, express the minimum coefficient of static friction required to keep the passengers from slipping.

In another design of the rotor, new *frictionless* walls are angled like a funnel, as shown in Figure 4.26.

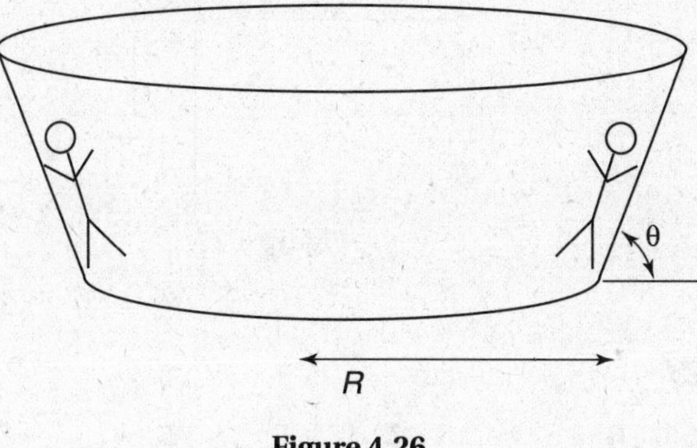

Figure 4.26

(c) At what linear speed must the passengers travel so that they move neither up nor down the walls of the funnel when the floor is removed?

3. A car of mass m is driving around a circular racetrack that has a radius of curvature r and is banked at an angle of elevation θ. The coefficient of static friction satisfies $\mu_s < \tan\theta$.

(a) At what speed is there no frictional force between the car and the banked curve?

(b) What is the maximum speed the car can go without flying off the banked curve?

(c) What is the minimum speed the car can go around the curve without slipping down the incline?

4. Experimentally testing Newton's second law requires accurate measurements of mass, acceleration, and force. The first two of these quantities can be measured accurately using a scale and photobridges, respectively. Measuring force accurately is more difficult; spring scales are very crude force-measuring devices. However, a simple two-mass system with a pulley can do the job. Place one mass, m_1, onto a horizontal airtrack, and connect it to a second mass, m_2, by a string that hangs over a lightweight pulley (see Figure 4.27).

   (a) Write Newton's second law for each mass.
   (b) Combine the two equations in part (a) to obtain Newton's second law for the combined system of the two masses.
   (c) Explain how the force appearing in the second law statement of part (b) can be accurately measured.
   (d) What approximations have been made in deriving the answer to part (b)? Which masses have been neglected?
   (e) In an actual experiment, you keep $m_1 + m_2$ fixed but transfer small pieces of mass from m_1 to m_2. Sketch a plot of the acceleration of the system as a function of the weight of m_2. What is the value of the slope of the line plotted?

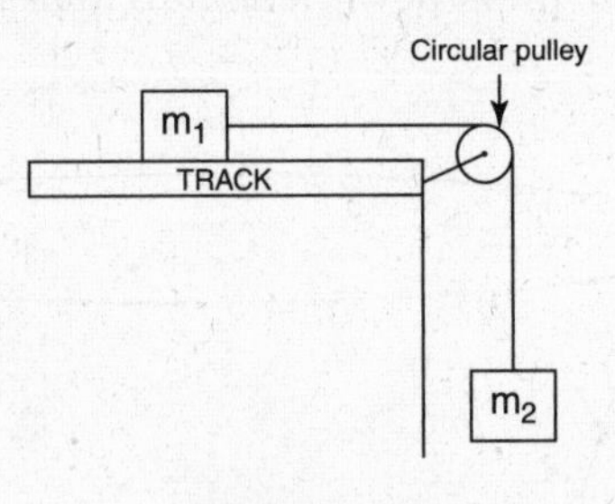

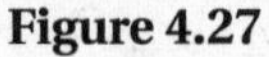
Figure 4.27

ANSWER KEY

1. **A**	3. **B**	5. **E**	7. **E**
2. **B**	4. **D**	6. **D**	8. **C**

ANSWERS EXPLAINED

Multiple-Choice

1. **(A)** Consider the free-body shown in Figure 4.28. Resolving all forces into components parallel and perpendicular to the direction of motion yields the following equation for the forces along the inclined plane:

$$F_{net} = ma = F\cos\theta - F_{fr} - mg\sin\theta$$

Solving this equation for the frictional force yields choice (A).

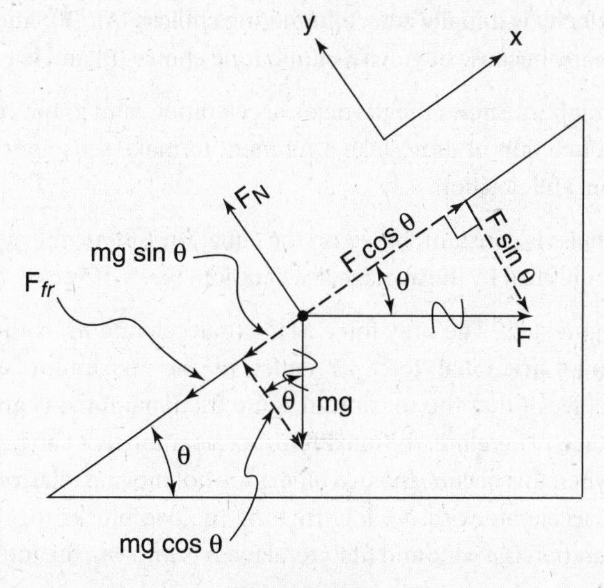

Figure 4.28

2. **(B)** Referring to the free-body diagram, summing the net force in the y-direction yields

$$F_{\text{net},y} = 0 = F_N - F\sin\theta - mg\cos\theta$$

This allows us to solve for the normal force:

$$F_N = F\sin\theta + mg\cos\theta$$

The coefficient of friction is then the ratio of the frictional force and the normal force:

$$F_{\text{fr}} = \mu_k F_N \Rightarrow \mu_k = \frac{F_{\text{fr}}}{F_N}$$

3. **(B)** The applied force required to overcome friction and give any single block an acceleration of a is $F_{\text{applied}} = ma + \mu_k mg$. (To obtain this equation, sketch a free-body diagram of a single block and apply Newton's second law to obtain $F_{\text{net}} = ma = F_{\text{applied}} - F_{\text{friction}}$.) Because the tension force in the string to the right of mass i must supply this force to the i blocks to its left, it must have i times this magnitude.

4. **(D)** The rate at which the ball's speed changes is equal in magnitude to its tangential acceleration. When the string is parallel to the ground, the tangential force is equal to the gravitational force, and thus the magnitude of the tangential acceleration is g. Since the acceleration is in the direction opposite the velocity, the ball is slowing down. (As explained in Chapter 5, this can also be understood in terms of gravitational potential energy.)

5. **(E)** The net force in the y-direction is zero: $F_{\text{net},y} = 0 = F_N - mg - F\sin\theta$, allowing us to solve for the normal force: $F_N = mg + F\sin\theta$. Applying Newton's second law in the x-direction, $F_{\text{net},x} = ma = F\cos\theta - F_{\text{fr}}$. Substituting for $F_{\text{fr}} = \mu_k F_N$ yields $F_{\text{net},x} = ma = F\cos\theta - \mu_k(mg + F\sin\theta)$. Solving this equation for F gives choice (E).

6. **(D)** The velocity is initially zero, eliminating choices (A), (B), and (C). The velocity decreases linearly because of gravity, eliminating choice (E) and leaving only choice (D).

   Note: Graph (A) shows the particle's acceleration, and graph (B) shows the particle's height as a function of time. Take a moment to make sense of these plots of velocity, acceleration, and position.

7. **(E)** Since mass is constant, based on the equation $F = ma$, the average force is equal to the mass multiplied by the average acceleration, $(\mathbf{v}_f - \mathbf{v}_0)/\Delta t$.

8. **(C)** See Figure 4.29. The only force that can accelerate m_2 is the frictional force, and the maximum frictional force is equal to the maximum static frictional force, $F_{fr} = \mu_s mg$. (Recall that the maximum static frictional force is greater than the kinetic frictional force.) Therefore, the maximum acceleration that can be given to the top block is $a = \mu_s g$. When this occurs, the two blocks are not moving relative to each other, so they must *both* accelerate with $a = \mu_s g$. Treating the two blocks together as one system, if their acceleration is $a = \mu_s g$ and their total mass is $m_1 + m_2$, the total external force, equal to F, must be $F = m_{total}a = (m_1 + m_2)(\mu_s g)$.

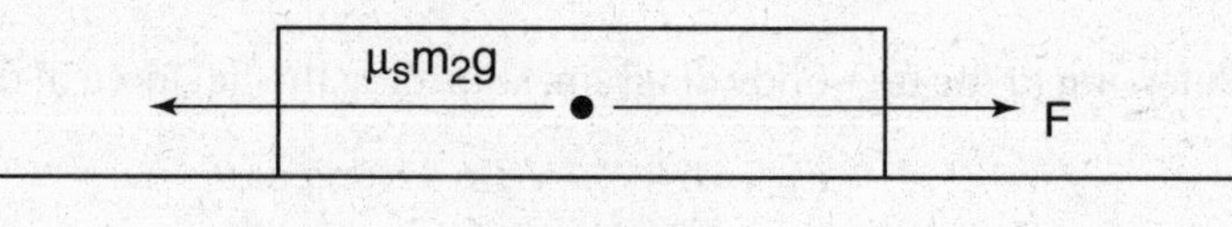

Figure 4.29

   Alternative solution: Without using the shortcut of treating the two blocks together in the last step, the applied force F can be determined using the free-body diagram in the figure. (*Note*: The free-body diagram ignores forces in the y-direction for clarity.) Two forces are exerted on the lower block: the applied force F and the frictional force that m_2 exerts on m_1 (which is equal to and opposite the frictional force that m_1 exerts on m_2). The net acceleration must be $a = \mu_s g$, so we can apply Newton's second law: $[F_{net} = m_1 a = m_1\mu_s g] = F - \mu_s m_2 g$. Solving this equation for the applied force gives choice (C).

Free-Response

1. (a) The free-body diagrams are shown in Figure 4.30.

   Free-body diagram of the person: What is the direction of the frictional force on the person? First, imagine what would happen in the absence of friction. The net force on the person would be F_T to the right, and the net force on the raft would be $2F_T - F_{current}$ to the right. Since $F_{current} > F_T$ (as indicated in the question), the rightward force on the person would be greater than the rightward force on the raft, causing the person to slip to the right relative to the raft. The force of friction on the person must oppose this potential slipping to the right by pointing to the left.

   Free-body diagram of the raft: Because the friction force and the normal force are forces exerted between the person and the raft, the raft must experience equal and opposite normal and friction forces as shown.

Free-body diagram of the raft-person system: When treating the raft and the person as a single object, internal forces (namely, the frictional force and the normal forces between the raft and the person) can be ignored. Note that the gravitational force in this diagram refers to the gravitational force on both the raft and the person combined, $w = (2m)g$.

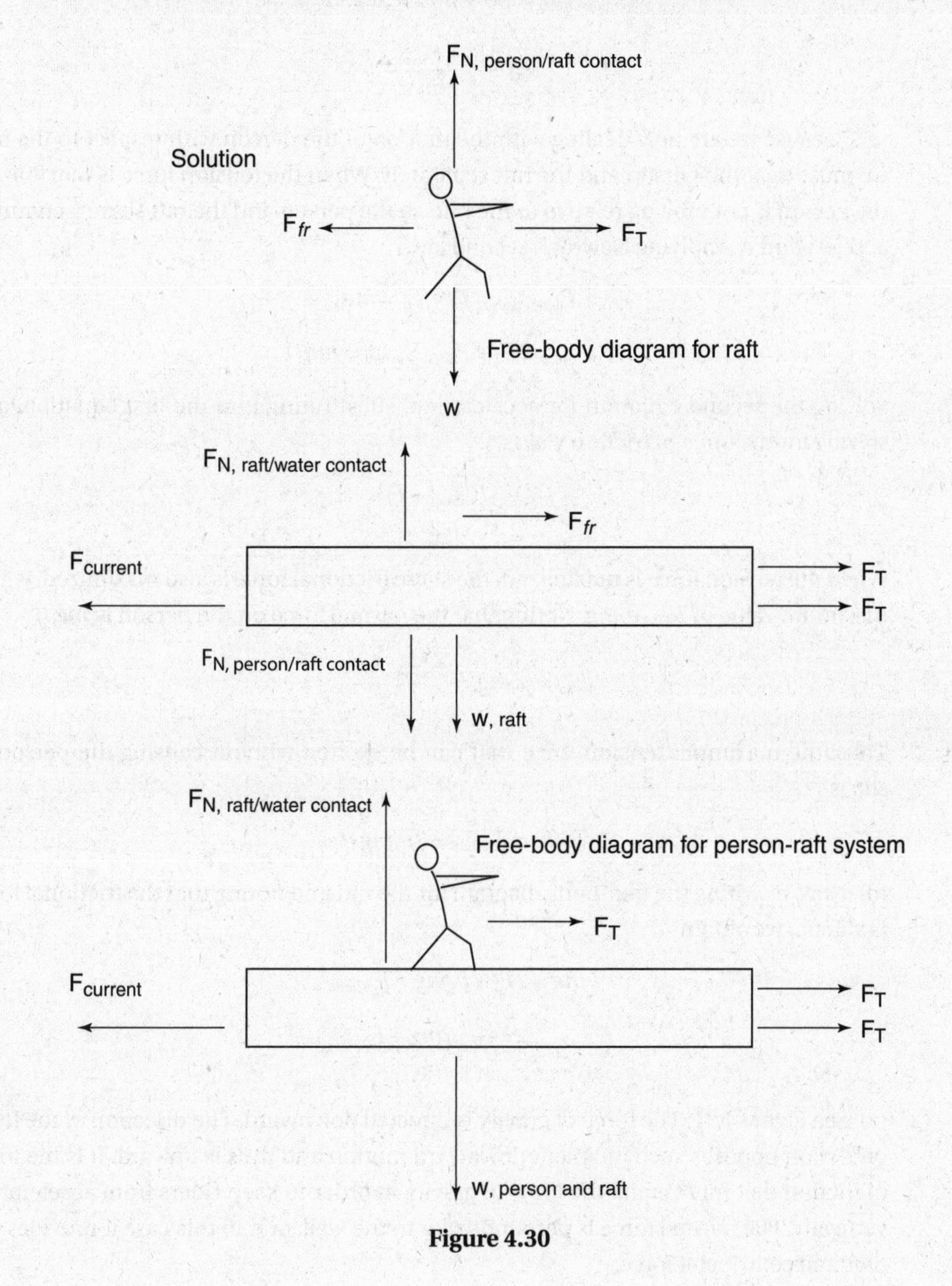

Figure 4.30

(b) When the person does not slip relative to the raft, we can treat the person and the raft as a single system, as in the previous free-body diagram. Applying Newton's second law,

$$F_{\text{net,person-raft}} = (m + m)a = 3F_T - F_{\text{current}}$$

$$a = \frac{3F_T - F_{\text{current}}}{2m}$$

(c) Because we are now dealing with the motion of the person with respect to the raft, we must treat the person and the raft separately. When the tension force is maximized, the person is not moving relative to the raft, so the person and the raft share a common acceleration *a*. Applying Newton's second law,

$$F_{\text{net,person}} = F_T - F_{\text{fr}} = ma$$

$$F_{\text{net,raft}} = 2F_T + F_{\text{fr}} - F_{\text{current}} = ma$$

Solving the second equation for acceleration, substituting into the first equation, and solving for the force of friction yields

$$F_{\text{fr}} = \frac{F_{\text{current}} - F_T}{2}$$

When the tension force is maximized, the static frictional force is also maximized, with a maximum value of $F_{\text{fr}} = \mu_s mg$. Noting that the normal force on the person is *mg*,

$$F_{\text{fr}} = \mu_s mg = \frac{F_{\text{current}} - F_T}{2}$$

Thus, the maximum tension force that can be exerted without causing the person to slip is

$$F_T = F_{\text{current}} - 2\mu_s mg$$

(d) Now, rewriting the free-body diagram for the raft and noting that the frictional force is kinetic, we obtain

$$ma = 2F_T + \mu_k mg - F_{\text{current}}$$

$$a = \frac{2F_T + \mu_k mg - F_{\text{current}}}{m}$$

2. (a) See Figure 4.31. The force of gravity is directed downward. The direction of the force of friction opposes such incipient downward motion and thus is upward. It is the force of friction that must equal the force of gravity in order to keep riders from accelerating vertically. The normal force is perpendicular to the wall, and in this case it provides the required centripetal force.

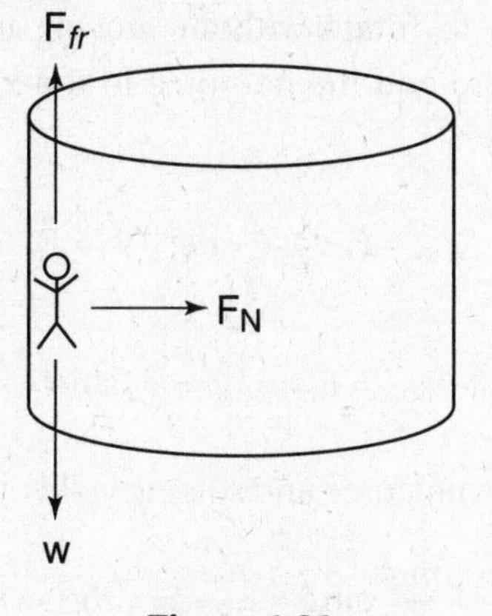

Figure 4.31

(b) The key to this problem is to realize that the net force is equal to the centripetal force: $F_{\text{net}} = F_{\text{net},x} = F_{\text{centripetal}} = F_N = mv^2/R$. The friction force must cancel the gravitational force, so it must have the same magnitude as the gravitational force, $F_{\text{fr}} = mg$. The necessary coefficient of friction is then the ratio of the frictional force to the normal force:

$$\mu_s = \frac{F_{\text{fr}}}{F_N} = \frac{mg}{mv^2/R} = \frac{gR}{v^2}$$

Symbolic answer check: If g increases, the required μ_s will increase. (More friction is needed to hold the passengers up.) Alternatively, if v^2/R, the centripetal acceleration, decreases, the normal force will decrease, so a greater coefficient of friction will be required.

(c) Again, this is an example of uniform circular motion, with $F_{\text{net}} = F_{\text{centripetal}}$. Even though the walls of the ride are no longer vertical, the motion is circular in a horizontal plane and the net force is also horizontal, directed toward the center of rotation. Thus, it is convenient to choose one coordinate axis in the horizontal direction as shown in Figure 4.32. Labeling the angles θ and θ' (complement of θ) makes clear how to resolve F_N into components along the axes, as shown in the figure. (A quick way to check whether you have done your geometry correctly is to see if the resolution of the forces agrees with the case considered above when $\theta = \pi/2$. For example, the y-component of the normal force is $F_N \cos\theta$ according to the figure. This reduces to zero, as expected, when $\theta = \pi/2$.)

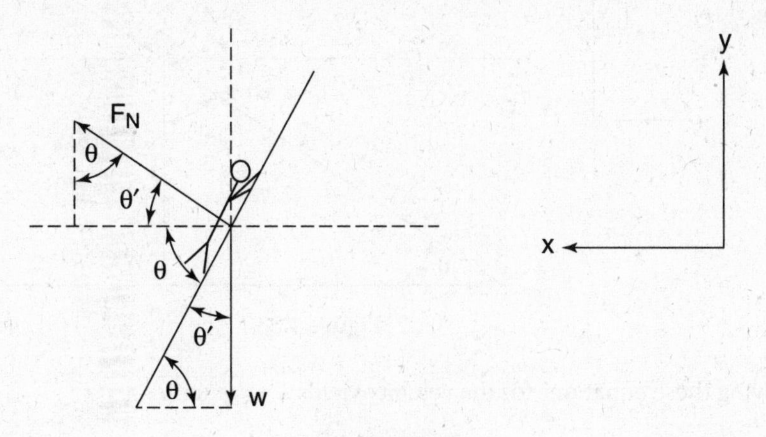

Figure 4.32

For the passengers to rotate without moving up or down, the net force in the y-direction must be zero and the net force in the x-direction must be the centripetal force.

$$F_{\text{net},y} = F_N \cos\theta - mg = 0 \Rightarrow F_N = \frac{mg}{\cos\theta}$$

$$F_{\text{net},x} = F_{\text{centripetal}} \Rightarrow F_N \sin\theta = \frac{mv^2}{R}$$

Substituting for the normal force and solving yields

$$\frac{mg}{\cos\theta}\sin\theta = \frac{mv^2}{R} \Rightarrow \sqrt{Rg\tan\theta} = v$$

Symbolic answer check: (1) As θ approaches 90° (i.e., the walls of the funnel approach the vertical), the required velocity increases to infinity. This makes sense: If the walls are vertical, no matter how quickly the rotor is spinning, the person will slip if there is no friction (since there is no source of upward force to cancel gravity. (2) As θ approaches zero degrees (i.e., the walls of the funnel approach the horizontal), the required velocity decreases to zero. This corresponds to riders who, lying on their backs, require no velocity at all in order to remain motionless on the now horizontal surface.

3. (a) The forces acting on the car in this situation are shown in Figure 4.33. Note that the axes are chosen such that one axis (the x-axis) is parallel to the centripetal force. The net force in the y-direction must be zero, and the net force in the x-direction must be the centripetal force:

$$F_{\text{net},y} = 0 = F_N \cos\theta - mg$$

$$\left[F_{\text{net},x} = F_{\text{cent}} = \frac{mv^2}{r}\right] = F_N \sin\theta$$

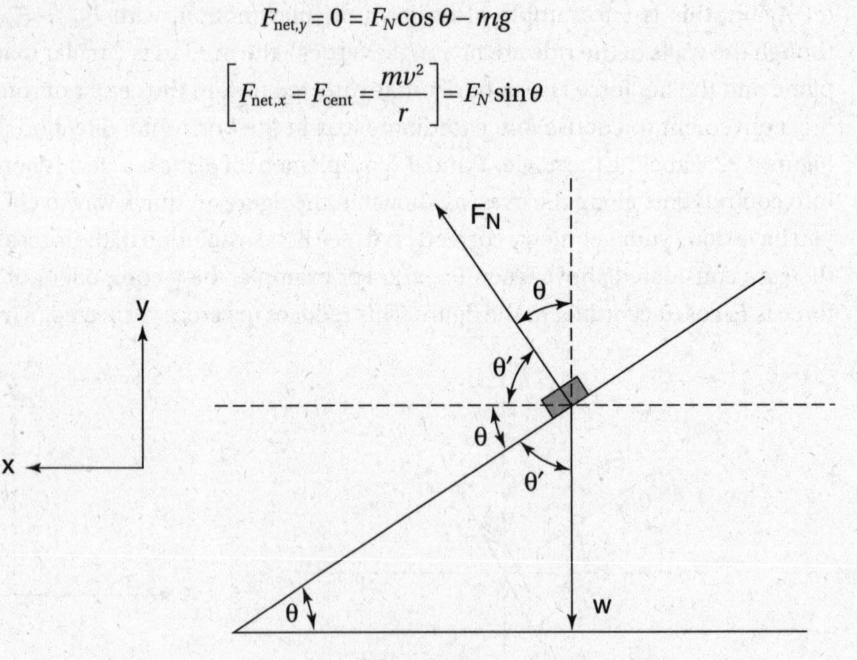

Figure 4.33

Solving these equations for the velocity yields $v = \sqrt{gr\tan\theta}$.

(b) The free-body diagram for this situation is shown in Figure 4.34. Note that because the car has a tendency to slip *up* the incline, the frictional force points in the opposite direction, *down* the incline. Again, the net force in the *y*-direction must be zero, and the net force is the centripetal force:

$$F_{net,y} = 0 = F_N \cos\theta - F_{fr}\sin\theta - mg$$

$$\left[F_{net,x} = F_{cent} = \frac{mv^2}{r}\right] = F_{fr}\cos\theta + F_N\sin\theta$$

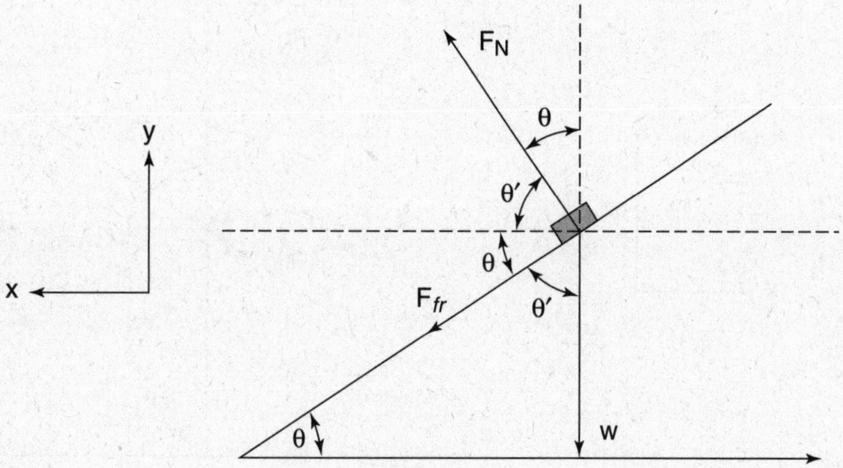

Figure 4.34

At this point we have two equations with three variables (F_{fr}, F_N, and v). The third equation relates the frictional force to the normal force: $F_{fr} = \mu F_N$ (note that we are using the maximum value of the static friction force so that we can find the maximum speed possible). Solving these equations for the velocity yields

$$v = \sqrt{\frac{gr(\mu\cos\theta + \sin\theta)}{\cos\theta - \mu\sin\theta}}$$

Note: A more detailed solution is shown in Example 4.10.

(c) The free-body diagram for this situation is shown in Figure 4.35. Note that because the car has a tendency to slip *down* the incline, the frictional force now points in the opposite direction, *up* the incline. As before, the net force in the y-direction must be zero, and the net force in the x-direction must be the centripetal force:

$$F_{\text{net},y} = 0 = F_N \cos\theta + F_{\text{fr}} \sin\theta - mg$$

$$\left[F_{\text{net},x} = F_{\text{cent}} = \frac{mv^2}{r}\right] = F_N \sin\theta - F_{\text{fr}} \cos\theta$$

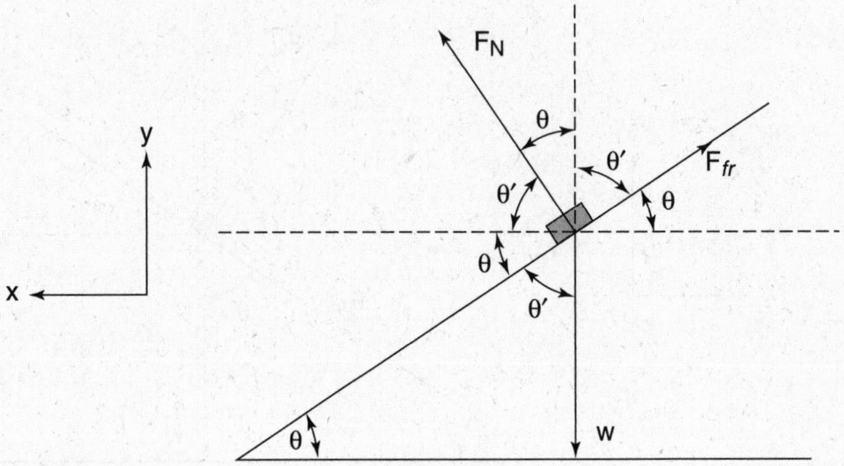

Figure 4.35

Again, the third equation required to solve for v is $F_{\text{fr}} = \mu F_N$. Solving these equations for the velocity yields

$$v = \sqrt{\frac{gr(\sin\theta - \mu\cos\theta)}{\cos\theta + \mu\sin\theta}}$$

4. (a) If T is the tension in the string, then Newton's second law reads $T = m_1 a$ for mass m_1 and $m_2 g - T = m_2 a$ for mass m_2.

(b) By adding the two equations (which eliminates T), we find $m_2 g = (m_1 + m_2)a$.

(c) The net force in the second law statement of part (b) is the weight of m_2, which is easily measured with an accurate lab scale. Measuring tension in the string is much more difficult.

(d) Any friction present in the airtrack has been neglected. The masses of the pulley and the string have also been neglected.

(e) Let $M = m_1 + m_2$. M is held fixed in the experiment. Thus, a plot of the acceleration a of the system versus $m_2 g$ will be a straight line with slope $\frac{1}{M}$; see part (b).

5 Work, Energy, and Power

- → DEFINITION OF WORK, KINETIC AND POTENTIAL ENERGY, AND POWER
- → WORK-ENERGY THEOREM
- → CONSERVATIVE AND NONCONSERVATIVE FORCES
- → HOOKE'S LAW AND ELASTIC POTENTIAL ENERGY
- → CONSERVATION OF MECHANICAL ENERGY
- → ENERGY DIAGRAMS

WORK

The work done on a body by a single external force **F** is defined by the equation

$$dW = \mathbf{F} \cdot d\mathbf{r}$$

The definition of work by one force

What if *n* external forces are acting on an object at the same time? Because of the principle of superposition of forces, the net work is

$$dW = \mathbf{F}_{\text{net}} \cdot d\mathbf{r}$$

Work due to multiple forces

When this expression is expanded, we find that the net work is equal to the algebraic sum of the work done by each individual force. (Recall that $\mathbf{A} \cdot \mathbf{B} + \mathbf{C} \cdot \mathbf{B} = (\mathbf{A} + \mathbf{C}) \cdot \mathbf{B}$.)

$$dW = \left(\sum \mathbf{F}_i\right) \cdot d\mathbf{r} = \sum (\mathbf{F}_i \cdot d\mathbf{r}) = \sum dW_i$$

Work due to multiple forces

Thus, to calculate the net work done by multiple forces, we can (1) calculate the work done by each force and sum these individual works, or (2) calculate the net force and compute the work done by the net force using the equation $dW = \mathbf{F}_{\text{net}} \cdot d\mathbf{r}$.

Work (and energy) has units of newton · meter, which is defined to be a joule ($\text{J} = \text{N} \cdot \text{m} = \text{kg} \cdot \text{m}^2/\text{s}^2$).

Because work is equal to the dot product of the net force and the displacement vector, it is a scalar whose magnitude and sign are determined by the component of the net force *parallel* to the direction of motion. This can be shown explicitly by resolving the force into components parallel and perpendicular to the displacement vector and expanding the dot product:

$$dW = \mathbf{F} \cdot d\mathbf{r} = (\mathbf{F}_{\perp} + \mathbf{F}_{\parallel}) \cdot d\mathbf{r} = \mathbf{F}_{\perp} \cdot d\mathbf{r} + \mathbf{F}_{\parallel} \cdot d\mathbf{r} = 0 + \mathbf{F}_{\parallel} \cdot d\mathbf{r}$$

There are three sign possibilities for work, which correspond to three different situations discussed below. This following discussion can be understood based on the fact that the displacement vector $d\mathbf{r}$ is parallel to the velocity and on our discussion of tangential acceleration (which affects an object's speed) and centripetal acceleration (which affects an object's direction) from Chapter 3.

Case 1 (work is negative): This occurs when the force has a component that is antiparallel to the displacement vector $d\mathbf{r}$ and thus antiparallel to the velocity. In this case, there is a tangential acceleration that causes the object's speed to decrease. (Later in the chapter we will see that this decrease can also be understood in terms of the conservation of energy.)

TIP

The work done by a force on an object can be positive, negative, or even zero.

Case 2 (work is zero): This occurs when the force has no component parallel to the displacement vector $d\mathbf{r}$ and thus no component parallel to the velocity and no tangential acceleration. Since there is no tangential acceleration, the object's speed does not change.

Case 3 (work is positive): This is the opposite of case 1: The force has a component parallel to the displacement vector $d\mathbf{r}$ and the velocity, producing a tangential acceleration parallel to the velocity and increasing the object's speed.

For *constant* forces, we can easily evaluate the work integral:

$$\int dW = \int \mathbf{F} \cdot d\mathbf{r}$$

Because of the distributive property of the dot product $[\mathbf{A} \cdot \mathbf{B} + \mathbf{A} \cdot \mathbf{C} = \mathbf{A} \cdot (\mathbf{B} + \mathbf{C})]$, we can put the constant force in front of the integral sign to obtain

$$\int dW = \mathbf{F} \cdot \left(\int d\mathbf{r}\right)$$

$$W = \mathbf{F} \cdot \Delta\mathbf{r}$$

Work due to a constant net force

Graphical Significance of the Definition of Work

For motion in one dimension (e.g., in the x-direction), $W = \int \mathbf{F} \cdot d\mathbf{r} = \int \mathbf{F} \cdot dx\hat{i} = \int F_x dx$. In this case, work is equal to the area underneath the $F_x(x)$ curve.

EXAMPLE 5.1

A mass is dragged a distance Δx along a rough surface by a force F applied at an angle of elevation θ as shown in Figure 5.1. The coefficient of friction is μ_k.

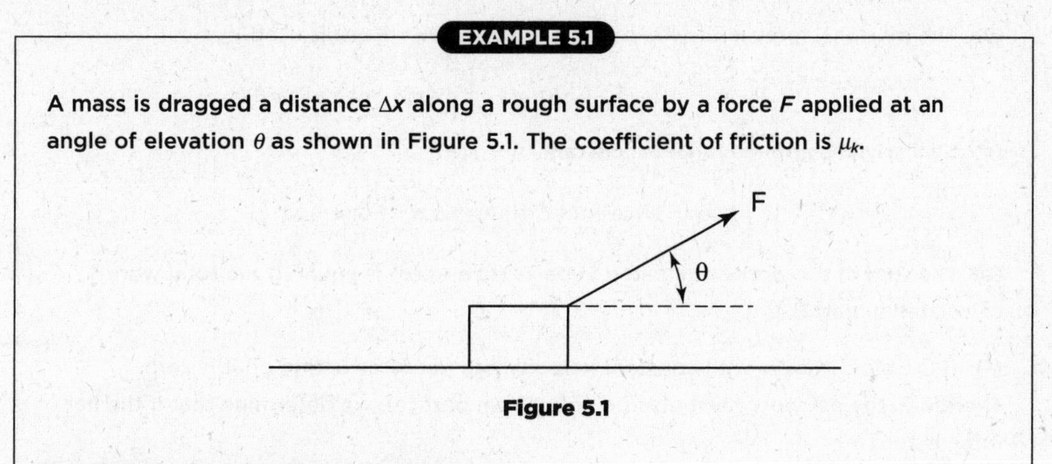

Figure 5.1

(a) What is the net force on the block?
(b) What is the net work done on the block?
(c) What is the work done by friction?
(d) What is the work done by the applied force?
(e) Show that $W_{\text{net}} = \Sigma W_i$.
(f) If the velocity of the mass remains constant, what is the net work done on the mass? What is the value of F in this case?

SOLUTION

(a) See the free-body diagram in Figure 5.2. The net force in the y-direction is zero, allowing us to solve for the normal force:

$$F_{\text{net},y} = 0 = F_N + F\sin\theta - mg = 0 \Rightarrow F_N = mg - F\sin\theta$$

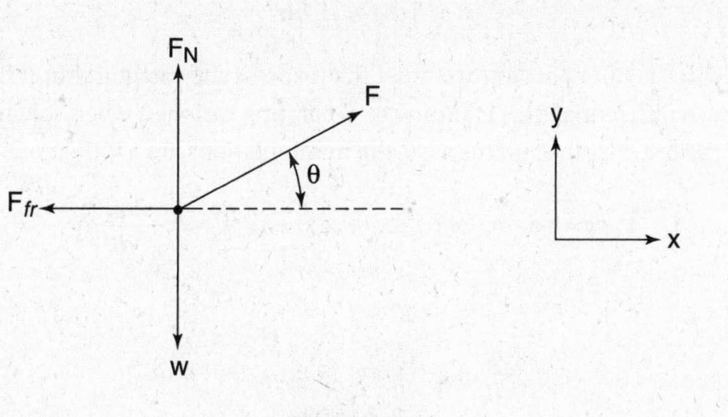

Figure 5.2

From the normal force, the frictional force can be calculated:

$$F_{\text{fr}} = \mu F_N = \mu(mg - F\sin\theta)$$

Given the frictional force, the net force in the x-direction is

$$F_{\text{net},x} = F\cos\theta - F_{\text{fr}} = F\cos\theta - \mu(mg - F\sin\theta) = F(\cos\theta + \mu\sin\theta) - \mu mg$$

(b) Calculating the work done by the net force,

$$W = \mathbf{F}_{\text{net}} \cdot \Delta\mathbf{r} = [F(\cos\theta + \mu\sin\theta) - \mu mg]\hat{i} \cdot (\Delta x)\hat{i} = [F(\cos\theta + \mu\sin\theta) - \mu mg]\Delta x$$

(c) The frictional force is constant so the equation $W = \mathbf{F} \cdot \Delta\mathbf{r}$ is used.

$$W = \mathbf{F} \cdot \Delta\mathbf{r} = -\mu(mg - F\sin\theta)\hat{i} \cdot (\Delta x)\hat{i} = \Delta x(-\mu mg + \mu F\sin\theta)$$

(d) Similarly, the applied force is constant,

$$W = \mathbf{F} \cdot \Delta\mathbf{r} = \left[F(\cos\theta)\hat{i} + F(\sin\theta)\hat{j}\right] \cdot \Delta x\hat{i} = F(\cos\theta)\Delta x$$

(e) The sum of the works calculated in parts (c) and (d) is equal to the total work calculated in part (b).

(f) If the block moves with constant velocity, the net force acting on it is zero. Therefore, the net work must also be zero. From part (a), we determine that if the net force is zero,

$$F = \frac{\mu mg}{(\cos\theta + \mu\sin\theta)}$$

The Relationship Between Work and Speed

We have discussed qualitatively how the sign of work is related to the change in an object's speed (forces that exert positive work increase speed, forces that exert negative work decrease speed, and forces that exert zero work do not affect speed). Now we would like to define this relationship mathematically. We start with the definition of work and integrate to obtain the total work:

$$\int dW = \int \mathbf{F} \cdot d\mathbf{r}$$

If the net force is not constant, we must first express the integrand in terms of one variable and its own differential (this is similar to separating variables when solving air resistance problems in the last chapter). In this case, the manipulations are a little more tricky:

$$\mathbf{F} \cdot d\mathbf{r} = m\mathbf{a} \cdot d\mathbf{r} = m(a_x dx + a_y dy) = m\left(\frac{dv_x}{dt}dx + \frac{dv_y}{dt}dy\right)$$

Rearranging,

$$= m\left(\frac{dx}{dt}dv_x + \frac{dy}{dt}dv_y\right) = m(v_x dv_x + v_y dv_y)$$

Now the right side is ready to integrate: It is composed of two integrands, each of which is defined by a different variable and its own differential.

$$\int dW = \int m(v_x dv_x + v_y dv_y)$$

$$W = m\left[\frac{v_x^2}{2} + \frac{v_y^2}{2}\right]_{v_0}^{v_F}$$

$$W = \frac{m}{2}\left[(v_{x,F}^2 + v_{y,F}^2) - (v_{x,0}^2 + v_{y,0}^2)\right]$$

Given the definition of speed as $v = \sqrt{v_x^2 + v_y^2}$, this can be rewritten as

$$W = \frac{mv_F^2}{2} - \frac{mv_0^2}{2}$$

We are thus led to define a new quantity, the kinetic energy:

$$KE = \frac{mv^2}{2}$$

Definition of kinetic energy

Similar to work and other forms of energy, kinetic energy has units of joules ($J = N \cdot m = kg \cdot m^2/s^2$).

Work-Kinetic Energy Theorem

Now we can appreciate the physical significance of work rather than viewing it simply as a mathematical definition: Work is the change in an object's kinetic energy due to the action of a given force.

$$W_{net} = \Delta KE$$

Work-kinetic energy theorem

This is the first and most fundamental of a number of equations we will soon derive relating various types of work and energy. This theorem is a restatement of Newton's laws, and it is always valid.

TIP

This theorem is another way of stating Newton's second law.

EXAMPLE 5.2

A force $\mathbf{F} = (4\ kg/s^2)x\hat{i}$ acts on an object of mass 2 kg as it moves from $x = 0$ to $x = 5$ m. Given that the object is at rest at $x = 0$,

(a) Calculate the net work.

(b) What is the final speed of the object?

SOLUTION

(a) $W = \int \mathbf{F} \cdot d\mathbf{r} = \int_0^5 (4\ kg/s^2)x\hat{i} \cdot dx\hat{i} = \int_0^5 (4\ kg/s^2)x\,dx = 50\ J$

(b) $50\ J = \Delta KE = KE_f - 0 = \frac{1}{2}(2\ kg)v_f^2$

$v_f = \sqrt{50}\ m/s$

CONSERVATIVE AND NONCONSERVATIVE FORCES

The Law of Conservation of Energy

Energy is never created or destroyed; it merely changes form. The total energy of an isolated system is constant when all forms of energy are accounted for (e.g., sound, light, heat, KE).

According to the law of conservation of energy, energy can only be converted between different forms (e.g., gravitational potential energy can be converted to kinetic energy during free-fall). At a macroscopic level, not all energy conversions are reversible. For example, when a block comes to rest under the influence of friction, kinetic energy is converted to heat. However, the reverse is impossible. (After the object comes to rest, can it draw heat out of the room and accelerate to its original speed?) Alternatively, many conversions of energy are reversible. For example, when you throw an object into the air, kinetic energy is converted to gravitational potential energy on the way up, and that conversion is exactly reversed on the way down.

Conservative forces (e.g., gravity) are involved in reversible energy conversions, where we can get our kinetic energy back (thus, the energy at our disposal is "conserved"). While the kinetic energy is temporarily gone, the total energy is stored in the form of potential energy, energy due to the position or configuration of the system. Alternatively, nonconservative forces (e.g., kinetic friction) are involved in irreversible energy conversions; although total energy is always conserved, energy is converted to forms (most notably sound and heat) from which we cannot recover it.

Summarizing,

Conservative forces: kinetic energy ⇒ potential energy ⇒ kinetic energy

Nonconservative forces: kinetic energy ⇒ sound, heat, etc. ⇒ cannot be easily recovered

When conservative forces perform negative work, removing kinetic energy from an object, the energy is stored in the form of potential energy. This is reflected mathematically in that every conservative force is associated with a potential energy function. Indeed, one definition of a conservative force is any force for which a potential energy function, $U(x, y, z)$, can be defined such that the work done moving the object between any two points is the difference in the potential energy at those points:

TIP

Potential energy can only be defined for conservative forces.

$$W = -\Delta U = -U(x_2, y_2, z_2) + U(x_1, y_1, z_1)$$

General definition of a potential energy function

This directly reveals the following properties of conservative forces:

1. The work done by a conservative force when an object moves between two points is independent of the path taken by the object and depends only on the location of the initial and final points (i.e., the work is *path independent*).
2. Based on the above equation, when the starting and ending points of the motion are the same (i.e., the path is a closed loop), the net work must always be zero.

GRAVITATIONAL POTENTIAL ENERGY

To derive an expression for gravitational potential energy, we must calculate the work involved in an arbitrary displacement and use the equation $W = -\Delta U$ to determine the change in the potential energy. Because the net force is constant, the following equation applies:

$$W = m\mathbf{g} \cdot \Delta\mathbf{r}$$

Unless stated otherwise, the $+y$-direction is generally taken to point up when dealing with gravitational potential energy. Therefore, the work done by gravity as the object moves from y_1 to y_2 is

$$W_{\text{gravity}} = -mg\hat{j} \cdot \Delta\text{r} = -mg\hat{j} \cdot \left(\Delta x\hat{i} + \Delta y\hat{j}\right) = mgy_1 - mgy_2 = -\Delta(mgy)$$

Comparing this equation with the definition of potential energy, $W = -\Delta U$, reveals that

$$W_{\text{gravity}} = -\Delta U_{\text{gravitational}} = -\Delta(mgy)$$

Note that our definition of potential energy involves only the *difference* in height between the two points. As we will continue to observe, it is only differences in potential energy (and, in this case, differences in height) that are significant. Therefore, we are free to choose any reference point where we define $U = 0$. For the case of gravity we choose our reference point to be $y = 0$, such that

$$U_{\text{gravitational}} = mgy$$

Gravitational potential energy

$$W_{\text{gravity}} = -\Delta U_{\text{gravitational}} = -\Delta(mgy)$$

Work done by the gravitational force

What happens if we set $y = 0$ at a different location? This changes the value of U at a particular point, but recall that it is the difference in potential energy that yields the work and ultimately the change in speed.

Only differences in potential energy are significant. The actual magnitude of potential energy at any particular point is arbitrary and depends on the choice of coordinate systems.

Potential energy, like work and kinetic energy, is a scalar with units of joules ($\text{J} = \text{Nm} = \text{kg}\cdot\text{m}^2/\text{s}^2$).

To review various sign conventions, consider an object falling toward Earth under the influence of gravity. The gravitational force is parallel to the velocity, so gravity performs positive work on the object, increasing its speed (and thus its kinetic energy). The y-coordinate decreases, so that $U = mgy$ decreases and thus the potential energy decreases. You can verify that these signs are consistent with the equation within the equation $W_{\text{gravity}} = -\Delta U_{\text{gravitational}} = -\Delta(mgy) = \Delta\text{KE}$.

EXAMPLE 5.3

You lift an object at constant velocity. What are the *signs* of (a) the work you do on the object, (b) the work gravity does on the object, (c) the net work on the object, (d) the change in kinetic energy of the object, and (e) the change in potential energy of the object?

SOLUTION

A simple free-body diagram reveals that to lift the object with constant velocity (i.e., with zero acceleration and thus zero net force), you must exert an upward force on the object equal to the gravity force. Therefore, (a) you exert a force parallel to the displacement, performing positive work, and (b) gravity exerts a force antiparallel to the displacement, so gravity does negative work. (c) The net force is zero, so the net work is zero. You can also obtain this answer from the work-energy theorem, in that (d) because the velocity is constant, the change in kinetic energy is zero and thus the net work must be zero. (e) Because the y-coordinate increases, the potential energy increases. Although the work you do does not increase the object's kinetic energy, it does increase the object's potential energy.

SPRING FORCE (HOOKE'S LAW)

We now turn to a classic example of a nonconstant force where we can see how to use our general definitions of potential energy and work. Experimentally it is found that when a spring is stretched, it opposes the stretching with a force given by Hooke's law:

$$F_x = -kx$$

Force due to a spring

In the above equation, and generally when we are dealing with springs, x is defined as the spring's displacement from its equilibrium, unstretched length.

- The magnitude of the spring force is proportional to the displacement from the spring's equilibrium position. The proportionality constant k, called the *spring constant* or *force constant*, is a property of the particular spring. It is a measure of the stiffness of the spring (a spring with a larger spring constant produces larger forces at the same displacement).
- The direction of the spring force is opposite the displacement, as indicated by the negative sign. (Because the force always tries to restore the spring to its relaxed state, whether it has been stretched or compressed, it is called a *restoring force*. We will encounter other restoring forces in Chapter 9.)

EXAMPLE 5.4 (UNIT ANALYSIS)

What are the units of the spring constant k?

SOLUTION

$F_x = -kx$. Writing this equation in terms of units,

$$\text{N} = k\,\text{m} \Rightarrow k = \frac{\text{N}}{\text{m}} = \frac{\text{kg} \cdot \text{m/s}^2}{\text{m}} = \frac{\text{kg}}{\text{s}^2}$$

ELASTIC POTENTIAL ENERGY

In order to obtain an expression for the elastic potential energy stored in a spring, we can take the same approach as we did with gravity: calculate the work as a function of position and then apply the equation $W = -\Delta U$. Suppose that $\mathbf{F}_{\text{net}} = -kx\hat{i}$ such that the spring lies along the x-axis with its equilibrium position at the origin. How much work is done by the spring force during a given displacement? The force varies, so we must use the differential definition of work this time:

$$dW = \mathbf{F}_{\text{net}} \cdot d\mathbf{r}$$

$$dW = -kx\hat{i} \cdot dx\hat{i} = -kx\,dx$$

$$W_{\text{spring}} = \int -kx\,dx = \left.\frac{-kx^2}{2}\right|_{x_1}^{x_2} = \frac{kx_1^2}{2} - \frac{kx_2^2}{2} = -\Delta\left(\frac{1}{2}kx^2\right)$$

Comparison with the definition of potential energy as $W = -\Delta U$ reveals that

$$U_{\text{elastic}} = \frac{kx^2}{2} \quad \text{(given that } x = 0 \text{ at the equilibrium position)}$$

Elastic potential energy

$$W_{\text{spring}} = -\Delta U_{\text{elastic}} = -\Delta\left(\frac{1}{2}kx^2\right)$$

The work done by the spring force

Note that the above expression for U_{elastic} requires that $x = 0$ corresponds to the unstretched state of the spring. Unlike the case of gravitational potential energy where we could choose $y = 0$ wherever we liked, here we must choose $x = 0$ at the equilibrium position.

EXAMPLE 5.5

A mass attached to a spring starts from rest at equilibrium. A variable external force is exerted on the mass, resulting in stretching the spring by an amount *a* and accelerating the mass to a speed of v_f.

(a) What is the work done by the spring force?

(b) What is the change in the object's kinetic energy?

(c) What is the work done by the external force?

SOLUTION

(a) $W_{\text{spring}} = -\Delta\left(\frac{1}{2}kx^2\right) = -\frac{1}{2}ka^2$

Note that the work done is negative. (As the spring is stretched, the restoring force is antiparallel to the displacement, so the work is negative.)

(b) $\Delta\text{KE} = \Delta\left(\frac{1}{2}mv^2\right) = \left(\frac{1}{2}mv^2\right)_{\text{final}} - \left(\frac{1}{2}mv^2\right)_{\text{initial}} = \frac{1}{2}mv_f^2$

(c) According to the work-energy theorem,

$$W_{\text{net}} = W_{\text{external}} + W_{\text{spring}} = \Delta\text{KE} = \frac{1}{2}mv_f^2$$

Solving for W_{external} yields

$$W_{\text{external}} = \frac{1}{2}mv_f^2 - W_{\text{spring}} = \frac{1}{2}mv_f^2 - \left(-\frac{1}{2}ka^2\right) = \frac{1}{2}mv_f^2 + \frac{1}{2}ka^2$$

UNIFICATION OF GRAVITATIONAL AND ELASTIC POTENTIAL ENERGY

What if we have a situation where multiple conservative forces are acting (e.g., gravity *and* the spring force)? The total work is simply the sum of the work performed by each conservative force, which is equal to the sum of the negative difference in potential energies:

$$W_{\text{net}} = \sum W_i = \sum(-\Delta U_i)$$

Now if we let W and U represent the *net* work and *net* potential energy, respectively, we can write this as

$$W_{\text{net}} = -\Delta U_{\text{net}} \quad \textbf{Valid only if all forces are conservative}$$

We have just derived a special case of the law of conservation of energy. The total mechanical energy of an object is defined as the sum of its kinetic and potential energies:

$$E_{\text{mech}} = \text{KE} + U$$

In the absence of nonconservative forces, this total energy is constant, as we now show. (Recall that according to the work-energy theorem, $W = \Delta \text{KE}$.)

$$\Delta E_{\text{mech}} = \Delta(\text{KE} + U) = \Delta\text{KE} + \Delta U = W + \Delta U = 0$$

From the above equation, it is easy to see that as KE increases, PE decreases, and vice versa, such that the total change in mechanical energy is always zero.

What if some of the forces are nonconservative? Nonconservative forces do not have associated potential energy functions, so the work they do is *not* accounted for in the $-\Delta U$ term, and the above equation is *not* valid. Thus, in the presence of nonconservative forces, $W = \Delta\text{KE} \neq -\Delta U$ and the total mechanical energy, $E_{\text{mech}} = \text{KE} + U$, is *not* constant. To account for energy added or removed by nonconservative forces, we must modify our energy conservation equation as follows.

$$\Delta E_{\text{mech}} = W_{\text{nonconservative}}$$

This equation should make sense. Friction, which can perform only negative work (because the force is antiparallel to the velocity) decreases the total mechanical energy. Alternatively, if you stick your hand into the system and increase the speed of an object by doing positive work (exerting a force parallel to the velocity), the net mechanical energy increases. The most useful and general form of the energy conservation equation is

$$\text{KE}_1 + U_1 + W_{\text{nonconservative}} = \text{KE}_2 + U_2$$

Work-energy theorem taking account of nonconservative forces (most useful version, always valid and applicable)

EXAMPLE 5.6

As shown in Figure 5.3, a mass *m* is dropped from a height *h* above the equilibrium position of a spring. Set up the equation that determines the spring's compression *d* when the object is instantaneously at rest.

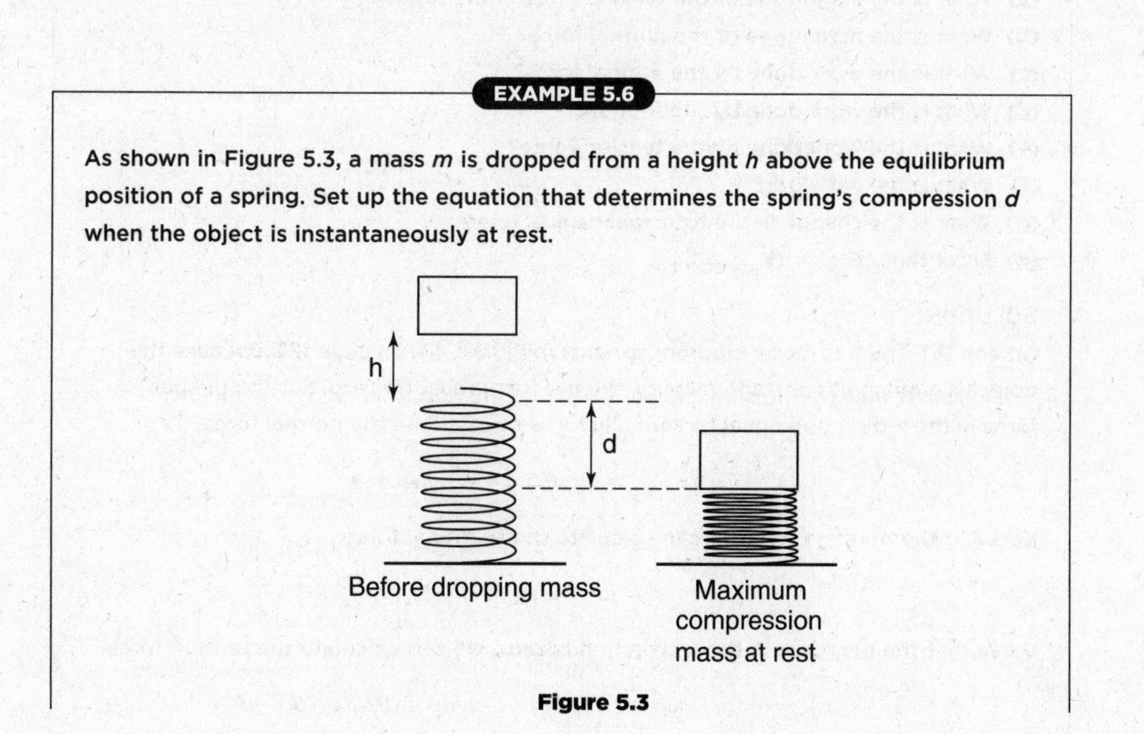

Figure 5.3

SOLUTION

We use the equation $KE_1 + U_1 + W_{nonconservative} = KE_2 + U_2$. Because all the forces involved (gravity and the spring force) are conservative, $W_{nonconservative} = 0$. The object starts and ends up at rest, so $KE_1 = KE_2 = 0$. Designating $y = 0$ at the height where the spring is in equilibrium, $U_{gravitational,1} = mgh$ and $U_{gravitational,2} = -mgd$. Initially, the energy stored in the spring is zero, and when the spring is maximally compressed, the potential energy stored in the spring is $U_{spring,2} = \frac{1}{2}kd^2$. Recalling that the total potential energy at any point is equal to the sum of the potential energy stored in different forms, $U = \Sigma U_i$, we can finally set up our conservation of energy equation:

$$KE_1 + U_1 + W_{nonconservative} = KE_2 + U_2$$

$$KE_1 + U_{gravitational,1} + U_{spring,1} + W_{nonconservative} = KE_2 + U_{gravitational,2} + U_{spring,2}$$

$$0 + mgh + 0 + 0 = 0 + (-mgd) + \frac{1}{2}kd^2$$

This could be solved using the quadratic formula to calculate d, the maximal compression of the spring.

EXAMPLE 5.7

A mass m is pulled a distance d up an incline (angle of elevation θ) at constant speed using a rope that is parallel to the incline. The coefficient of friction is μ_k.

(a) What is the magnitude of the tension force in the rope?
(b) What is the magnitude of the normal force?
(c) What is the work done by the normal force?
(d) What is the work done by the friction?
(e) What is the work done by the tension force?
(f) What is the net work?
(g) What is the change in the total mechanical energy?
(h) Show that $\Delta E_{mech} = W_{nonconservative}$.

SOLUTION

(a) and **(b)** The free-body diagram appears in Figure 4.17 on page 126. Because the object is moving at constant velocity, the net force must be zero. Setting the net force in the y-direction equal to zero allows us to calculate the normal force:

$$F_{net,y} = 0 = F_N - mg\cos\theta \quad \Rightarrow \quad F_N = mg\cos\theta$$

Knowing the normal force, we can calculate the frictional force,

$$F_{fr} = \mu F_N = \mu mg\cos\theta$$

Given that the net force in the x-direction is zero, we can calculate the tension force:

$$F_{net,x} = 0 = F_T - mg\sin\theta - F_{fr} \quad \Rightarrow \quad F_T = mg(\sin\theta + \mu\cos\theta)$$

(c) The work done by the normal force is zero because the displacement is perpendicular to the normal force.

(d) and (e) Since these forces are constant, the work can be calculated using the equation $W = \mathbf{F} \cdot \Delta\mathbf{r}$:

$$W_{\text{fr}} = \mathbf{F}_{\text{fr}} \cdot \Delta\mathbf{r} = (\mu mg\cos\theta)(d\cos 180°) = -\mu mg(d\cos\theta)$$

$$W_T = \mathbf{F}_T \cdot \Delta\mathbf{r} = mg(\sin\theta + \mu\cos\theta)d\cos 0° = mgd(\sin\theta + \mu\cos\theta)$$

(f) Because the speed is constant, the change in kinetic energy is zero and thus the net work is zero (according to the work-energy theorem, $W = \Delta\text{KE}$). We can verify this mathematically as follows.

$$W_{\text{net}} = W_{\text{gravity}} + W_T + W_{\text{fr}} + W_{\text{normal}}$$

We have already calculated the work done by each of these forces except for the gravitational force.

$$W_{\text{gravity}} = -\Delta U_{\text{gravity}} = -\Delta(mgy) = -mgd\sin\theta$$

Substituting this and the expressions for W_{tension}, W_{fr}, and W_{normal} from parts (b)-(e) into the equation for the net work yields

$$W_{\text{net}} = -mgd\sin\theta + mgd(\sin\theta + \mu\cos\theta) - \mu mgd\cos\theta + 0 = 0$$

(g) Evaluating the change in the total mechanical energy directly,

$$\Delta E_{\text{mech}} = \Delta\text{KE} + \Delta U = 0 + mgd\sin\theta$$

(h) The friction force and the tension force are nonconservative. Therefore, the net nonconservative work is

$$W_{\text{nonconservative}} = W_{\text{tension}} + W_{\text{fr}} = mgd(\sin\theta + \mu\cos\theta) - \mu mgd\cos\theta = mgd\sin\theta$$

POWER

Power is the rate at which a force does work on a system. Therefore, power is the rate at which energy is added or removed from a system (by a nonconservative force) or the rate at which energy is converted between different forms (by a conservative force).

The average power over a time interval is given by:

$$\bar{P} = \frac{W}{\Delta t}$$

Average power

For constant forces

$$P = \mathbf{F} \cdot \frac{d\mathbf{r}}{dt} = \mathbf{F} \cdot \mathbf{v}$$ **valid only for constant forces**

Power has units of J/s, which is defined as a watt $\left(W = \frac{J}{s} = \frac{(kg \cdot m^2/s^2)}{s} = \frac{kg \cdot m^2}{s^3}\right)$.

The Relationship Between Force and Potential Energy in the Absence of Nonconservative Forces So far, we have related force to work and work to potential energy. What, then, is the relationship between force and potential energy? We start by relating force and work:

$$dW = \mathbf{F}_{net} \cdot d\mathbf{r}$$

We will deal only with the relationship between force and potential energy for objects moving along a line (rectilinear motion). If we are moving along the x-axis, the above equation can be rewritten as follows.

$$dW = \mathbf{F}_{net} \cdot d\mathbf{r} = \left(F_x\hat{i} + F_y\hat{j}\right) \cdot \left(dx\hat{i}\right) = F_x dx$$

If there are no nonconservative forces present, $W = -\Delta U = -(U - U_0)$. Differentiating this equation yields

$$dW = -dU = F_x dx$$

Rearranging this equation, we have

$$F_x = -\frac{dU}{dx}$$

Relationship between force and potential energy for rectilinear motion along the *x*-axis in the absence of nonconservative forces

Note: There's nothing special about the x-axis. For example, if we were moving along the y-axis instead of along the x-axis, the following analogous equation would be valid: $F_y = -dU/dy$.

A good way to understand this equation is in the context of energy diagrams, which are plots of potential energy vs position, $U(x)$, for a conservative force. Converting a $U(x)$ graph to an $F(x)$ graph is exactly the same as converting position to velocity or velocity to acceleration (by taking the derivative), except for the negative sign. One helpful way to conceptualize $U(x)$ curves is to imagine a ball rolling around on top of the $U(x)$ curve; the ball experiences a force whose magnitude is proportional to the slope of the $U(x)$ curve and whose direction points toward decreasing potential energy.

As shown in Figure 5.4, the total mechanical energy is often shown in energy diagrams in addition to the $U(x)$ curve. In the absence of nonconservative forces the mechanical energy is constant: E_{mech} = constant = KE + U. Thus, the kinetic energy is the difference between the total mechanical energy and the potential energy. Graphically, at any particular point (e.g., point D), the distance between the x-axis and the $U(x)$ curve is the potential energy and the difference between the $U(x)$ curve and the mechanical energy is the KE. The object cannot move to positions where the $U(x)$ curve is greater than the mechanical energy: This would require that it have a negative kinetic energy, which is impossible $\left(\frac{1}{2}mv^2 \geq 0\right)$. Therefore, the motion of the object in this example is bounded. For example, if an object with the indicated total energy were released between points A and B, it would never be able to escape—instead, it would forever oscillate between points A and B. At these points (the *turning points*), all the energy is potential energy, so the object is temporarily at rest (like a ball rolling back and forth between two hills). At point C, the potential energy is minimized, causing the kinetic energy to be maximized, such that the object is traveling at its maximum speed.

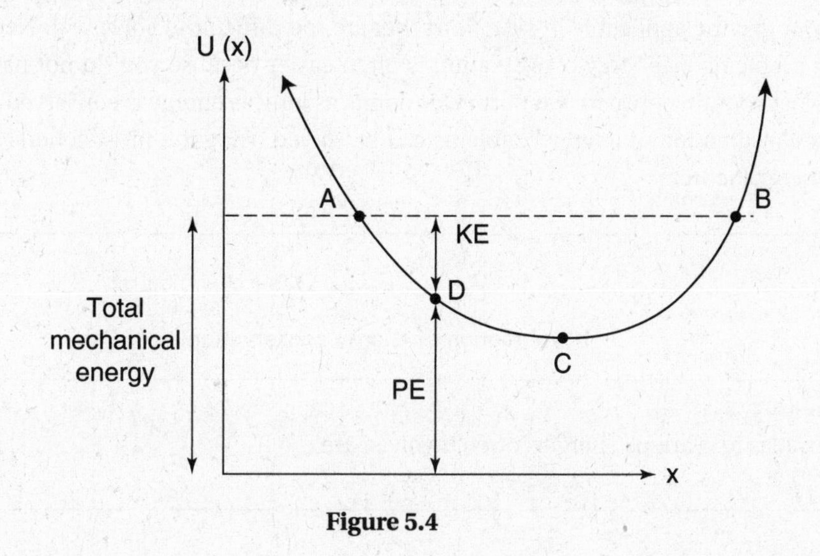

Figure 5.4

Points at which the force is zero [i.e., the tangent of the $U(x)$ curve is horizontal] are called *equilibrium points* because if a particle were placed at these points with zero kinetic energy, it could remain there forever at rest. There are three types of equilibrium points, which are all illustrated in Figure 5.5. *Stable equilibrium points* (e.g., point C) are points where, if the object is given a small nudge, it will remain close to the equilibrium point, oscillating within an "energy well" (much like a ball at rest in a gully). *Unstable equilibrium points* (points A and B) are just the opposite: If the object if given a small displacement, it will end up far away from the equilibrium point (much like a ball at rest on the top of a hill). You may be confused about point A: If given a nudge to the right, it will go very far in the $+x$ direction, whereas if given a nudge to the left, it will soon reverse direction. However, *after* it reverses direction, it will then go very far in the $+x$ direction; in either case, it will not end up anywhere near point A, so this is a point of unstable equilibrium. *Neutral equilibrium points* are locations where the $U(x)$ curve is completely flat, so if the particle is given a little nudge, it will continue to move with a constant velocity (similar to a ball at rest on a flat field).

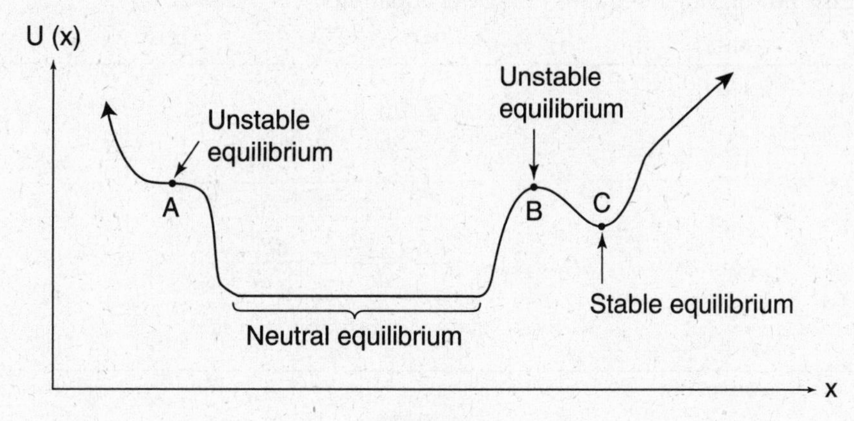

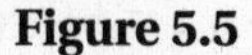

Figure 5.5

SOLVING CONSERVATION OF ENERGY PROBLEMS (EQUATION SUMMARY)

Although results described in this chapter were derived from Newton's laws, energy conservation provides a novel way to solve problems that is often easier than applying Newton's laws directly or may be applicable in situations that are too difficult to solve with Newton's laws. Solving problems by energy conservation is often easier because you do not have to worry about what goes on between two particular points as long as energy is conserved.

Most conservation of energy problems can be solved using the most general form of the work-energy theorem:

$$\mathrm{KE}_1 + U_1 + W_{\text{nonconservative}} = \mathrm{KE}_2 + U_2$$

Key equation of energy conservation!

The formulas for various energies often involved are

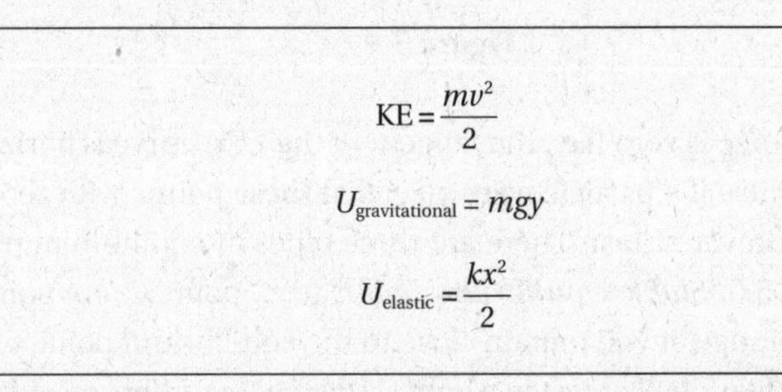

$$\mathrm{KE} = \frac{mv^2}{2}$$

$$U_{\text{gravitational}} = mgy$$

$$U_{\text{elastic}} = \frac{kx^2}{2}$$

If you need to compute the work done by an arbitrary force, you may need to integrate the equation

$$dW = \mathbf{F}_{\text{net}} \cdot d\mathbf{r}$$

Although power generally is not a useful way to solve problems, you may be required to calculate or use power, which requires the power equations:

$$P = \frac{dW}{dt}$$

$$\bar{P} = \frac{W}{\Delta t}$$

$$P_{\text{nonconservative}} = \frac{dW_{\text{nonconservative}}}{dt} = \frac{dE_{\text{total}}}{dt}$$

$$P = \mathbf{F} \cdot \mathbf{v} \quad \text{valid only for constant forces}$$

Force can be related to potential energy in the absence of nonconservative forces:

$$F_x = -\frac{dU}{dx}$$ **valid only in absence of nonconservative forces**

And of course, do not forget the spring force:

$$F_x = -kx$$

Force due to a spring

CHAPTER SUMMARY

Defining "work" in a particular way allows us to recast Newton's second law in a very convenient form: the work-energy theorem. If the work done by a force on an object is independent of the path followed by the object, a "potential energy" associated with this force can be defined. Two examples of such conservative forces are gravity and the spring force. In the presence of conservative forces, an object's total mechanical energy, kinetic plus potential, is conserved.

PRACTICE EXERCISES

Multiple-Choice Questions

1. In Figure 5.6 the person pulls the mass (initially at rest) all the way to the top of the incline and then allows it to slide halfway down the incline before exerting enough tension force to stop the mass at this point. If the coefficient of kinetic friction is μ_k and the length of the incline is *d*, what is the total work done by the person?

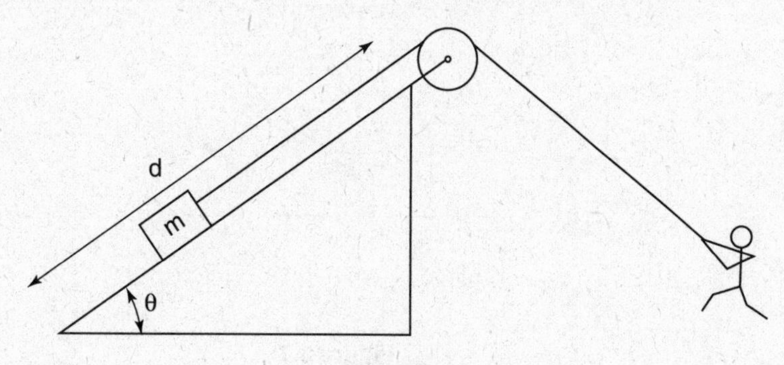

Figure 5.6

(A) $mgd\sin\theta$

(B) $\frac{1}{2}mgd\sin\theta$

(C) $\frac{1}{2}mgd\sin\theta + \frac{3}{2}\mu_k mgd\cos\theta$

(D) $\frac{1}{2}mgd\sin\theta - \frac{3}{2}\mu_k mgd\cos\theta$

(E) $\frac{1}{2}mgd\sin\theta + \frac{1}{2}\mu_k mgd\cos\theta$

2. If a projectile thrown directly upward reaches a maximum height h and spends a total time in the air of T, the average power of the gravitational force during the trajectory is

(A) $P = 2mgh/T$
(B) $P = -2mgh/T$
(C) $P = 0$
(D) $P = mgh/T$
(E) $P = -mgh/T$

3. Given only the constant net force on an object and the object's net displacement, which of the following quantities can be calculated?

(A) the net change in the object's velocity
(B) the net change in the object's mechanical energy
(C) the average acceleration
(D) the net change in the object's kinetic energy
(E) the net change in the object's potential energy

4. A particle is launched with initial speed v and angle of elevation θ. What is the maximum height that the particle will reach during its trajectory?

(A) $y_{max} = (v^2/2g) \sin^2 \theta$
(B) $y_{max} = (v^2/2g) \cos^2 \theta$
(C) $y_{max} = (2v^2/g) \sin \theta \cos \theta$
(D) $y_{max} = (v^2/2g) \sin \theta \cos \theta$
(E) $y_{max} = (v/2g) \cos \theta$

5. Consider the potential energy function shown in Figure 5.7. Assuming that no nonconservative forces are present, if a particle of mass m is released from position x_0, what is the maximum speed it will achieve?

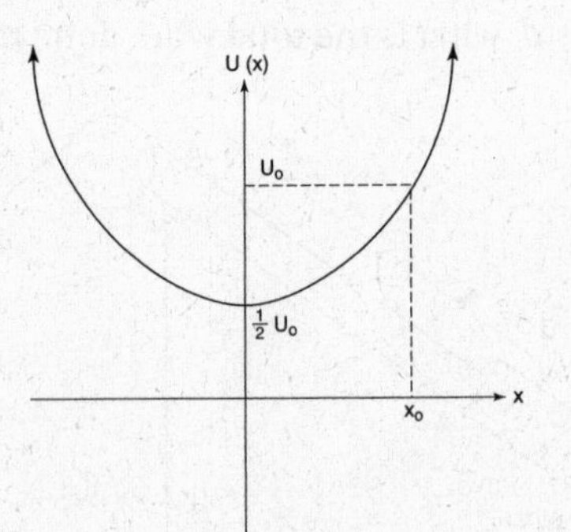

Figure 5.7

(A) $\sqrt{4U_0/m}$
(B) $\sqrt{2U_0/m}$
(C) $\sqrt{U_0/m}$
(D) $\sqrt{U_0/2m}$
(E) The particle will achieve no maximum speed but instead will continue to accelerate indefinitely.

6. Which of the following is the most accurate description of the system introduced in question 5?

   (A) stable equilibrium
   (B) unstable equilibrium
   (C) neutral equilibrium
   (D) a bound system
   (E) There is a linear restoring force.

7. If the only force acting on an object is given by the equation $F(x) = 2 - 4x$ (where force is measured in newtons and position in meters), what is the change in the object's kinetic energy as it moves from $x = 2$ to $x = 1$?

   (A) +4 J
   (B) −4 J
   (C) +2 J
   (D) −2 J
   (E) +8 J

8. Consider a mass falling toward Earth, as gravitational potential energy is converted to kinetic energy. Which of the following statements is true?

   (A) The mass and Earth experience equal and opposite accelerations.
   (B) The change in Earth's velocity is greater than the change in the mass's velocity.
   (C) The mass and Earth experience equal forces.
   (D) The kinetic energy of both the mass and Earth increase by the same amount.
   (E) The increase in the kinetic energy of the mass is greater than the increase in Earth's kinetic energy.

9. A 10-kg mass is sliding to the right with a speed of 4 m/s. How much work would be required to reverse the mass's direction such that it moves to the left with a speed of 4 m/s?

   (A) +160 J
   (B) +80 J
   (C) 0 J
   (D) −80 J
   (E) −160 J

10. If the mass shown in Figure 5.8 is moved a distance x from its equilibrium position, what will be the magnitude of the spring force it experiences?

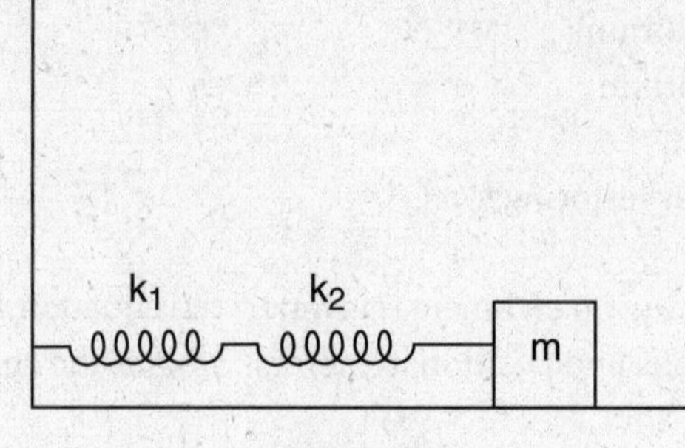

Figure 5.8

(A) $x(k_1 + k_2)$
(B) $xk_1k_2/(k_1 + k_2)$
(C) $xk_1k_2/(k_1 - k_2)$
(D) $xk_1k_2/(k_2 - k_1)$
(E) $x(k_1 + k_2)/k_1k_2$

11. A pendulum bob of mass m is released from rest as shown in Figure 5.9. What is the tension in the string as the pendulum swings through the lowest point of its motion?

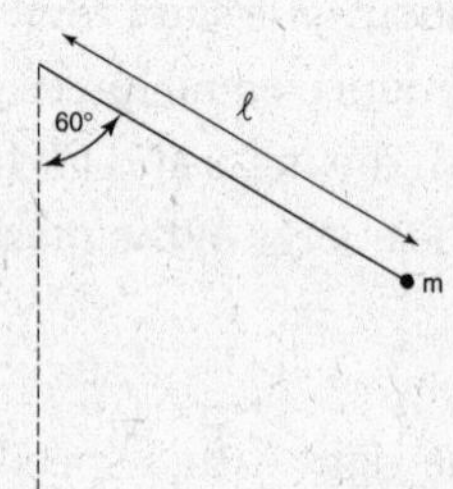

Figure 5.9

(A) $T = \frac{1}{2}mg$
(B) $T = mg$
(C) $T = \frac{3}{2}mg$
(D) $T = 2mg$
(E) none of the above

Free-Response Questions

1. Consider the potential energy function shown in Figure 5.10. A particle is released at the origin with the given total energy, and in the absence of any nonconservative forces.

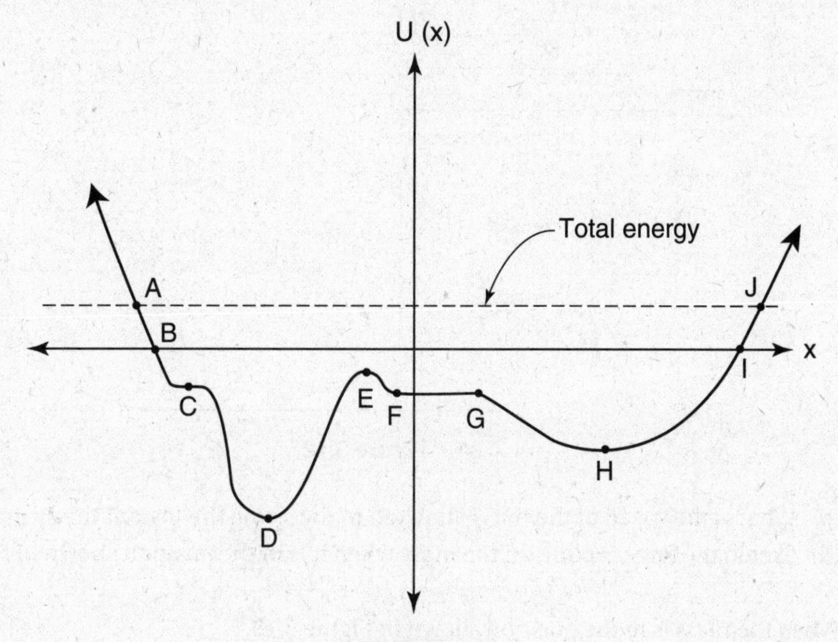

Figure 5.10

(a) Identify all points or intervals of stable, unstable, and neutral equilibrium.
(b) Where is the object moving with the greatest speed? With zero speed?
(c) Is the motion of the object bounded?
(d) In what region or point is the *magnitude* of the force maximized? Minimized?
(e) Within which region(s) is the force negative?

2. A mass m is placed on an incline of angle θ at a distance d from the end of a spring as shown in Figure 5.11. The coefficient of kinetic friction between the mass and the plane is μ.

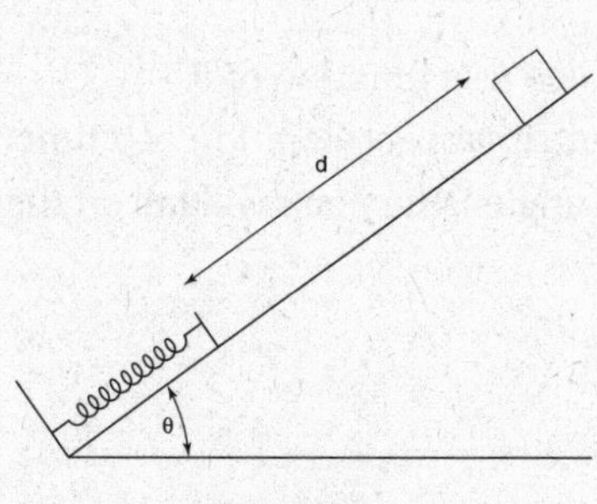

Figure 5.11

(a) The mass is released from rest at the position shown. Using Newton's laws, calculate the block's speed when it reaches the spring.
(b) Using energy conservation, calculate the block's speed when it reaches the spring.
(c) The spring has spring constant k. At what value x of the compression of the spring does the object reach its maximum speed?

3. A mass m attached to a string of length $2r$ swings, starting at rest when the string is horizontal, until the string is vertical. At the instant the string is vertical, the mass makes contact with a horizontal surface, the string is cut, and the mass continues along a frictionless track as shown in Figure 5.12.

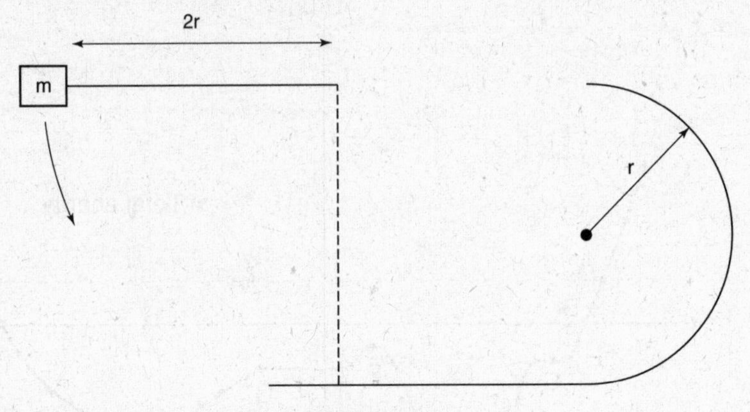

Figure 5.12

(a) What is the speed of the mass attached to the string the instant the string is cut?

(b) Sketch the forces acting on the mass when it is in the position shown in Figure 5.13.

When the mass is in the position shown in Figure 5.13,

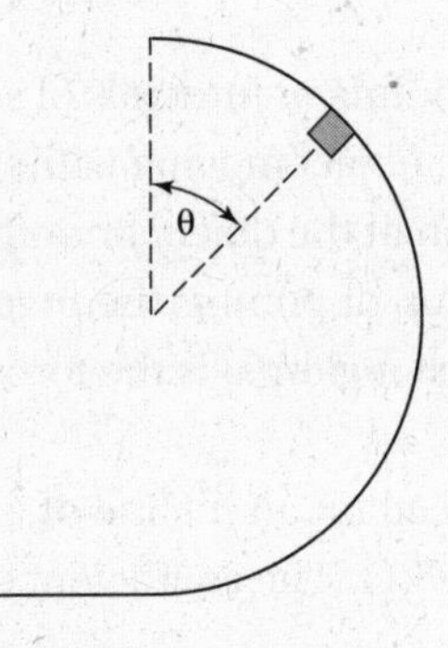

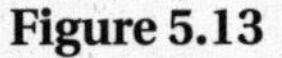

Figure 5.13

(c) find the object's speed as a function of θ.

(d) find the object's centripetal acceleration as a function of θ.

(e) determine at what angle θ the mass will fall off the track.

4. A rope of length l, mass m, and uniform mass density hangs vertically from a ceiling, as shown in Figure 5.14. The bottom end of the rope is raised very slowly until it is at the same height as the top of the rope.

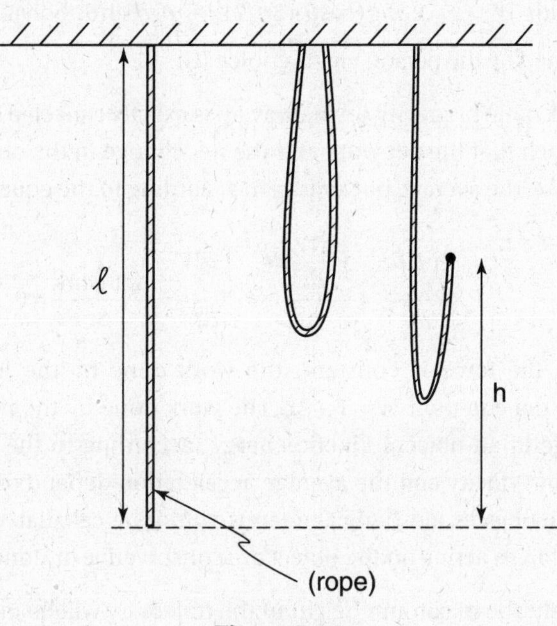

Figure 5.14

(a) Based on the change in the position of the center of mass of the rope, what is the change in the rope's potential energy?

(b) What vertical force applied at the end of the rope is required to hold the end a distance h above the bottom of the hanging rope, as shown in the figure?

(c) Integrate to evaluate the minimum work required to lift the bottom of the rope all the way to the top of the rope.

ANSWER KEY

1. **C**	4. **A**	7. **A**	10. **B**
2. **C**	5. **C**	8. **E**	11. **D**
3. **D**	6. **D**	9. **C**	

ANSWERS EXPLAINED

Multiple-Choice

1. **(C)** When in doubt, use the most complete equation for energy conservation:

$$KE_0 + U_0 + W_{nonconservative} = KE_f + U_f$$

The initial and final kinetic energies are zero because the mass is at rest in both cases. Setting $y = 0$ at the bottom of the incline, $U_0 = 0$ and $U_f = \frac{1}{2}mgd\sin\theta$. There are two nonconservative forces, the force exerted by the person and the frictional force, so $W_{nonconservative}$ is the sum of these two forces. What is the work done by the frictional force? The frictional force has a magnitude of $\mu_k mg \cos\theta$, pointing in a direction that always opposes the direction of motion. Therefore, the frictional force always performs negative work (both going up *and* going down), with the total amount of work equal to the frictional force multiplied by the total distance (not the displacement): $W_{\text{friction}} = -\mu_k mg\cos\theta(3d/2)$. Substituting all these terms into the conservation of energy equation yields $W_{\text{person}} - \mu_k mg(\cos\theta)(3d/2) = \frac{1}{2}mgd\sin\theta$. Solving this equation for the work performed by the person yields choice (C).

2. **(C)** The work done by gravity on the way up is exactly canceled by the work done on the way down, such that the net work and the net change in the object's kinetic energy are zero. Therefore, the average power is zero according to the equation

$$\bar{P} = \frac{\int_{\Delta t} P\,dt}{\Delta t} = \frac{\int_{\Delta t} \frac{dW}{dt}dt}{\Delta t} = \frac{\int_{\Delta t} dW}{\Delta t} = \frac{\text{net work}}{\Delta t} = 0$$

3. **(D)** Because the force is constant, the work done by the force can be calculated according to the equation $W = \mathbf{F} \cdot \Delta\mathbf{r}$. The work done by the net force is always equal to the change in an object's kinetic energy (according to the work-energy theorem). The change in velocity and the average acceleration depend on the object's mass. The change in the object's mechanical energy cannot be calculated without knowledge of whether the forces acting on the object are conservative or nonconservative.

4. **(A)** Intuitively, the maximum height of the trajectory will be maximized when the particle is launched straight up ($\theta = 90°$). This eliminates choices (B), (C), and (D), which are all zero at $\theta = 90°$. Dimensional analysis can then be used to eliminate choice (E), which has units of seconds, to arrive at choice (A) by elimination.

A more straightforward solution is to use energy conservation to equate the initial kinetic energy to the kinetic energy at the peak of the trajectory (note that the velocity at the peak of the trajectory is the x-component of velocity, equal to $v\cos\theta$):

$$\frac{1}{2}mv^2 = mgh + \frac{1}{2}m(v\cos\theta)^2$$

Solving for h and applying the identity $1 - \cos^2\theta = \sin^2\theta$ yields choice (A).

Note: This problem demonstrates that kinetic energy can be broken up into components:

$$\text{KE} = \frac{1}{2}mv^2 = \frac{1}{2}m(v_x^2 + v_y^2) = \frac{1}{2}mv_x^2 + \frac{1}{2}mv_y^2 = \text{KE}_x + \text{KE}_y$$

Utilizing this property of the kinetic energy, we could have solved the problem simply by equating the initial y-component of the kinetic energy to the gravitational potential energy:

$$\frac{1}{2}mv_y^2 = \frac{1}{2}m(v\sin\theta)^2 = mgy$$

Alternative solution: This problem can also be solved using kinematics by calculating the time to reach the maximum height (using the equation $v_y = v_{y,0} - gt = 0$) and then substituting this expression for time into $y(t) = y_0 + v_{y,0}t - \frac{1}{2}gt^2$.

5. **(C)** Imagine placing a ball on top of the potential energy curve. The ball rolls back and forth, as if within a gully. The maximum speed is achieved at the minimum potential energy, which occurs at the bottom of the gully, where $x = 0$ and $U(x) = U_0/2$. Conservation of energy between the initial point and this point yields the following equation.

$$U_0 = \frac{U_0}{2} + \frac{1}{2}mv^2$$

Solving this equation for v yields choice (C).

6. **(D)** The system is bound: The total energy is equal to U_0, so the particle cannot go beyond the interval $-x_0 < x < x_0$ (as this would require having negative kinetic energy). Although if the particle were at rest at $x = 0$, it *would* be in stable equilibrium, the particle is *not* at rest, so it is not in stable equilibrium. Although there is a restoring force that always directs the particle toward $x = 0$, without knowing that the $U(x)$ curve is parabolic it cannot be concluded that the restoring force is linear.

7. **(A)** The force is in the negative direction (because the function $F(x) = 2 - 4x$ is negative on the interval $1 < x < 2$), and the displacement is in the negative direction. Therefore, the force does positive work on the object, speeding it up and increasing its kinetic energy. The fact that $\Delta KE > 0$ eliminates choices (B) and (D).

   The change in the kinetic energy is equal to the net external work (according to the work-energy theorem)

$$\Delta KE = W = \int F_x\,dx = \int_2^1 (2-4x)dx = 4\text{ J}$$

8. **(E)** In choice (A), Earth and the mass experience equal and opposite *forces.* Using Newton's second law, $|F| = m_1a_1 = m_2a_2$, we see that because Earth has a greater mass, the magnitude of its acceleration is smaller. In choice (B), since the mass has a greater acceleration (as discussed above), it will have a greater change in its velocity. In choice (C), the mass and Earth experience forces that are equal in magnitude but *opposite* in direction. In choices (D) and (E), the mass and Earth experience forces with the same magnitude. Based on the equation for power, $P = \mathbf{F} \cdot \mathbf{v}$, the rate at which the gravitational force increases the kinetic energy is larger for the mass because it has a greater velocity parallel to the force.

9. **(C)** The work is equal to the change in kinetic energy, which is zero because the initial and final speeds are the same. This can be understood as follows: While slowing down the mass, the force must exert negative work. To accelerate the mass in the opposite direction, positive work must be done. These negative and positive works cancel each other out exactly.

10. **(B)** Designate x_1 as the extension of spring 1 from its equilibrium length, and x_2 as the extension of spring 2 from its equilibrium length. The two springs share the same spring force, so $x_1k_1 = x_2k_2$. Additionally, conservation of length requires that the net extension of the two springs together, x, be equal to the sum of the extensions of each of the individual springs: $x = x_1 + x_2$.

    Solving $x_1k_1 = x_2k_2$ for x_2 yields $x_2 = k_1x_1/k_2$. Substituting this into the conservation of length equation yields $x = x_1 + k_1x_1/k_2$. Solving this equation for x_1 yields

$$x_1 = \frac{x}{1 + k_1/k_2}$$

The force on mass m is equal to the tension in either of the springs (recall that these tensions are the same), which is equal to

$$F = k_1 x_1 = k_1\left(\frac{x}{1 + k_1/k_2}\right) = \frac{x k_1 k_2}{k_1 + k_2}$$

Alternative solution: Choice (E) can be eliminated because it has incorrect units (position divided by spring constant). Choices (C) and (D) can be eliminated because they do not have the appropriate equivalence symmetry: The relative location of the two springs doesn't affect their tension forces or the net force on mass m. Therefore, because the two springs are functionally equivalent, the qualities of each spring should have equal standing in the final result, such that switching all the subscripts (1s and 2s) yields the same result. This is true for the actual answer, but not choices (C) and (D).

Finally, choice (A) would be the force if both springs were individually stretched a distance x. Qualitative understanding of the problem (i.e., realizing that each spring is stretched a distance less than x and that they share a common tension force) allows us to eliminate this choice. The only remaining choice is (B), which has correct units and equivalence symmetry.

11. **(D)** Based on the geometry of the 30–60–90 right triangle (Figure 5.15), the vertical height of the pendulum decreases by $l/2$ as the pendulum swings down to the lowest point of its motion. Conservation of energy then requires that $mg(l/2) = \frac{1}{2}mv^2$. Rearranging this equation (by multiplying by 2 and dividing by l) yields $mv^2/l = mg$. Therefore, the centripetal force, which is equal to the net force, must equal mg pointing *upward.* (Note that this is not saying that gravity provides the centripetal force, but rather that numerically the magnitude of the centripetal force at this point equals the weight of the object.) Designating upward as positive,

$$[F_{net,y} = F_{cent} = +mg] = [F_{tension} - F_g = F_{tension} - mg]$$

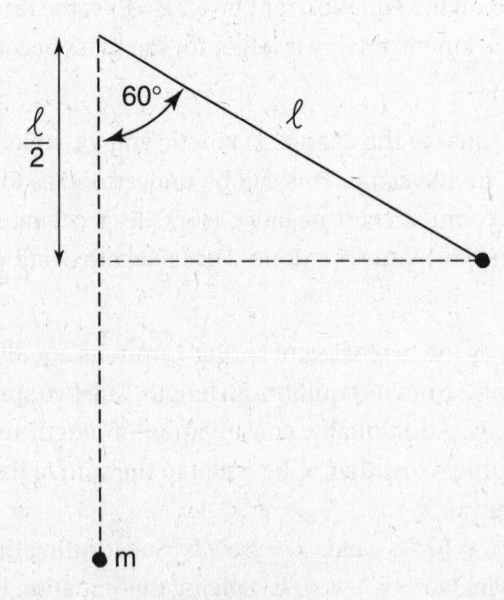

Figure 5.15

Solving for the tension force gives $T = 2mg$.

Free-Response

1. (a) Equilibrium points are points where there is no force on the object [i.e., the slope of the $U(x)$ curve is zero], such that an object could theoretically remain at rest (in *equilibrium*) at these points forever. Stable equilibrium occurs in energy wells, such that small nudges result in small-amplitude oscillations back and forth within the energy well; points D and H are points of stable equilibrium. Unstable equilibrium occurs at points where, if given a small nudge, the particle will move away from the equilibrium point, similar to a ball balanced on the top of a hill; points C and E are points of unstable equilibrium. Neutral equilibrium occurs at points where, if given a small nudge, the particle will continue to move with constant velocity, similar to a ball on a horizontal field; the interval between points F and G is a region of neutral equilibrium.

   (b) KE + PE = E_{total}. The object's speed (and thus its kinetic energy) is maximized when PE is minimized, at point D. Alternatively, the speed is zero when the potential energy is equal to the total energy, at points A and J.

   (c) Yes, the object is bounded within the region between points A and J. (Beyond this region, the potential energy is greater than the total energy; this would require the kinetic energy to be negative, which is impossible.)

   (d) $F(x) = -dU/dx$. Therefore, the magnitude of the force is maximized when the magnitude of the slope is the greatest, which occurs along the steep drop between points C and D. The magnitude is zero when the slope is zero, which occurs at all the equilibrium points discussed in parts (a) through (c).

   (e) Based on the equation $F(x) = -dU/dx$, the force is negative when the slope is positive, in segments DE and HJ. (Intuitively, if an imaginary ball were released on the potential energy surface within these regions, it would roll in the $-x$-direction).

2. (a) See the free-body diagram in Figure 5.16. Using Newton's second law in the y-direction, $F_{net,y} = 0 = F_N - mg\cos\theta$, we find that the normal force is $F_N = mg\cos\theta$, such that the frictional force is $F_{fr} = \mu F_N = \mu mg\cos\theta$. Using Newton's second law in the x-direction,

$$F_{net,x} = ma = mg\sin\theta - F_{fr} = mg\sin\theta - \mu mg\cos\theta$$

   we find that the acceleration $a = g(\sin\theta - \mu\cos\theta)$. To calculate the final velocity given the acceleration and the displacement, we use the equation from one-dimensional kinematics $v_f^2 = v_0^2 + 2a\Delta x$:

$$v_f^2 = 0 + 2dg(\sin\theta - \mu\cos\theta) \Rightarrow v_f = \sqrt{2dg(\sin\theta - \mu\cos\theta)}$$

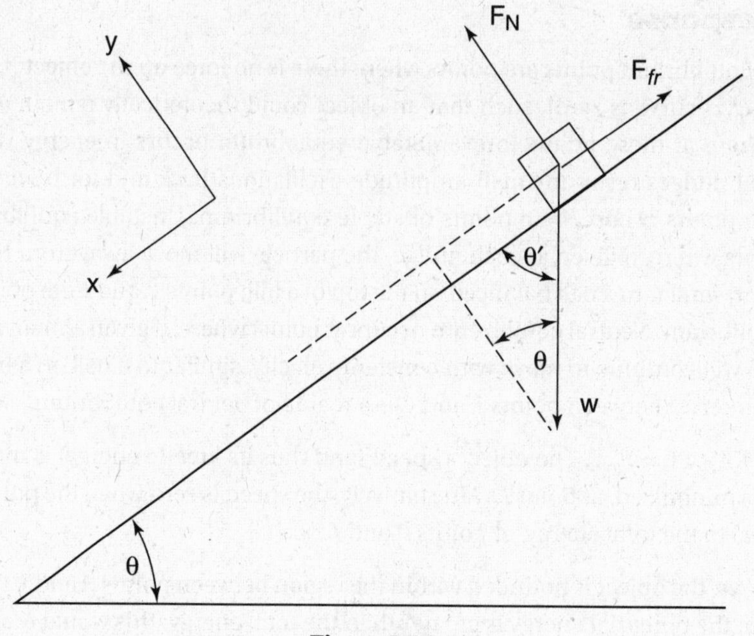

Figure 5.16

(b) Let $y = 0$ when the mass reaches the spring and apply the modified conservation of energy equation,

$$KE_1 + U_1 + W_{nonconservative} = KE_2 + U_2$$

where point 1 is the initial position of the mass and point 2 is where the mass contacts the spring. The term $W_{nonconservative}$ is the work done by friction:

$$W = \mathbf{F} \cdot \mathbf{d} = F_{fr} d \sin 180° = -(\mu mg \cos \theta)(d)$$

The energy equation then reads

$$0 + mg(d \sin\theta) - \mu mgd \cos\theta = \frac{1}{2} mv_b^2 + 0$$

$$v_b = \sqrt{2dg(\sin\theta - \mu\cos\theta)}$$

Symbolic answer check: (1) The units are correct. (2) The answer is reasonable because speed increases with increasing d and g and decreases with increasing coefficient of friction.

(c) Before the object hits the spring, it is accelerating downward with constant acceleration $a = g(\sin\theta - \mu\cos\theta)$. As the spring is compressed, the spring force increases from zero (on impact, when the spring is uncompressed), eventually growing large enough to cause the object to slow down and come to rest. The maximum speed occurs at the instant the object stops speeding up and begins to slow down; this corresponds to a net acceleration and net force of zero:

$$F_{net,x} = 0 = mg \sin\theta - \mu mg \cos\theta - F_{spring}$$

$$F_{spring} = mg \sin\theta - \mu mg \cos\theta = kx$$

$$x = \frac{mg(\sin\theta - \mu\cos\theta)}{k}$$

Symbolic answer check: (1) The units are correct. (2) Increasing m or g or decreasing μ increases x. (Greater downward force causes a greater compression of the spring.) Increasing k decreases x. (Greater spring constant decreases the compression.)

3. (a) Conservation of energy (gravitational potential energy is converted to kinetic energy):

$$2rmg = \frac{1}{2}mv^2$$

$$v = 2\sqrt{rg}$$

(b) Consider the free-body diagram in Figure 5.17.

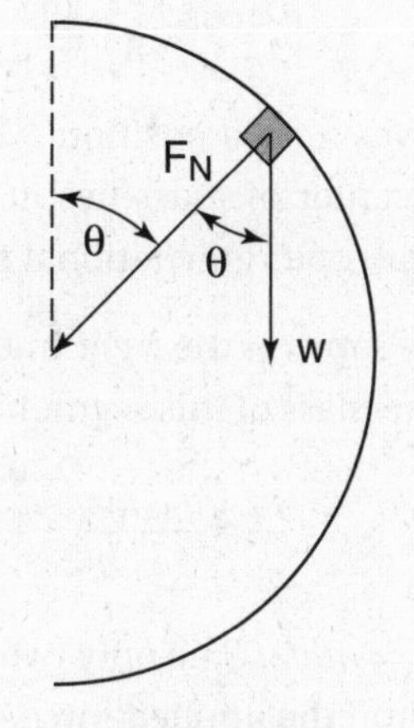

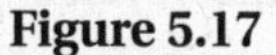

Figure 5.17

(c) Again using conservation of energy,

$$2rmg = mg(r + r\cos\theta) + \frac{1}{2}mv^2$$

$$v = \sqrt{2gr(1-\cos\theta)}$$

(d) Substituting the above velocity into the equation for the centripetal acceleration yields

$$a_{\text{cent}} = \frac{v^2}{r} = 2g(1-\cos\theta)$$

(e) We write down the radial component of Newton's second law. The normal force is purely radial, while gravity has a radial component given by $mg\cos\theta$. Thus,

$$F_{\text{net,radial}} = mg\cos\theta + F_N = F_{\text{cent}} = 2mg(1-\cos\theta)$$

When the component of the force in the radial direction is *greater* than the centripetal force, the mass moves in a *tighter* curve than the circular track, causing it to fall toward the center of the track. As the mass moves up the track, its speed decreases (decreasing the necessary centripetal force), and the component of the gravitational force in the radial direction increases. As this happens, the normal force decreases because less force is required to maintain the centripetal force. However, the normal force cannot decrease beyond zero (it cannot be negative), so beyond a certain point even if the normal force is zero, the radial force is greater than the centripetal force, causing the mass to fall. Therefore, the instant that the mass begins to fall is when the normal force is zero (the point at which it cannot decrease further). Setting $F_N = 0$ in Newton's second law we find

$$2mg(1 - \cos\theta) = mg\cos\theta$$

$$\cos\theta = \frac{2}{3}$$

$$\theta = \cos^{-1}\frac{2}{3} = 48°$$

4. (a) The center of mass moves from a position $l/2$ beneath the ceiling to a position $l/4$ beneath the ceiling. [The center of mass lies at the middle of the rope segment(s).] Therefore, the increase in the rope's gravitational potential energy is $\Delta U = mg\Delta y = mgl/4$.

(b) The vertical segment of rope on the right in the figure has length $h/2$. Because the mass density is constant, the mass of this segment can be calculated using the equation

$$\lambda = \frac{m}{l} = \frac{m_{\text{hanging segment}}}{h/2}$$

with the result $m_{\text{hanging segment}} = mh/2l$. The only two forces acting on this segment of rope are the gravitational force and the applied force, which balance each other. Therefore, the magnitude of the applied force is given by $F_{\text{applied}} = m_{\text{hanging segment}}\, g = mgh/2l$.

(c) The minimum work is the work required to lift the rope infinitesimally slowly (such that no kinetic energy is produced). The force required to do this was calculated in part (b), so we integrate to calculate the required work:

$$W = \int \mathbf{F}\cdot d\mathbf{r} = \int_0^l \frac{mgh}{2l}(dh) = \frac{mgl}{4}$$

This answer is consistent with the result from part (a), as it must be.

Linear Momentum and Center of Mass

6

- → DEFINITION OF LINEAR MOMENTUM
- → NEWTON'S SECOND LAW IN TERMS OF MOMENTUM; IMPULSE
- → CONSERVATION OF MOMENTUM IN THE ABSENCE OF A NET EXTERNAL FORCE
- → ELASTIC AND INELASTIC COLLISIONS
- → CENTER OF MASS
- → MOTION OF THE CENTER OF MASS

MOMENTUM

$$\text{Momentum} = \mathbf{p} = m\mathbf{v}$$

Linear momentum is a vector parallel to velocity. Momentum obeys superposition, such that the net momentum of a collection of objects is the vector sum of the momentum of each object.

The Relationship Between Force and Momentum

$$\mathbf{F}_{\text{net}} = m\mathbf{a} = \frac{md\mathbf{v}}{dt} = \frac{d}{dt}(m\mathbf{v}) = \frac{d\mathbf{p}}{dt}$$

Thus, when the net force on an object or system of objects is zero, the net momentum remains constant (it is conserved). Recall from Chapter 4 that internal forces cannot contribute to the net force because they always cancel each other out. (This is a direct consequence of Newton's third law: Any internal force produces an equal and opposite internal force that cancels it.) Therefore, in the absence of a net external force ($F_{\text{net}} = 0$), the total momentum of a body or system of bodies is conserved (even if mechanical energy is not conserved). Mathematically,

If the net external force on a system of objects is zero, then the total momentum is conserved.

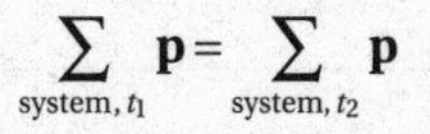

$$\sum_{\text{system}, t_1} \mathbf{p} = \sum_{\text{system}, t_2} \mathbf{p}$$

Impulse is defined as follows.

$$\text{Impulse} = \mathbf{J} = \int_{t_1}^{t_2} \mathbf{F}dt$$

If the force is constant, we can put it in front of the integral sign:

$$\mathbf{J} = \int_{t_1}^{t_2} \mathbf{F}dt = \mathbf{F}\int_{t_1}^{t_2} dt = \mathbf{F}(t_2 - t_1) = \mathbf{F}\Delta t \quad \textbf{valid only for constant force}$$

The Relationship Among Impulse, Force, and Time (Impulse-Momentum Theorem)

Recall from above that

$$\mathbf{F} = \frac{d\mathbf{p}}{dt} \Rightarrow d\mathbf{p} = \mathbf{F}\,dt$$

Therefore,

$$\mathbf{J} = \int_{t_1}^{t_2} \mathbf{F}\,dt = \int d\mathbf{p} = \Delta\mathbf{p}$$

Relationship between impulse and force: the impulse-momentum theorem

Of course, this vector equation can also be written as a set of scalar equations:

$$\begin{Bmatrix} J_x = \int_{t_1}^{t_2} F_x dt = \Delta p_x = mv_{x,2} - mv_{x,1} \\ J_y = \int_{t_1}^{t_2} F_y dt = \Delta p_y = mv_{y,2} - mv_{y,1} \end{Bmatrix}$$

The Relationship Between Impulse and the Average Force

Applying the calculus formula to calculate the average value of a function,

$$\mathbf{F}_{\text{average}} = \frac{\int_{t_1}^{t_2} \mathbf{F}\,dt}{t_2 - t_1} = \frac{\mathbf{J}}{t_2 - t_1} = \frac{\mathbf{J}}{\Delta t}$$

The graphical significance of impulse and average force is shown in Figure 6.1. The equation $\mathbf{J} = \int_{t_1}^{t_2} \mathbf{F}\,dt$ tells us that the impulse is the area under the $F(t)$ graph. The average force is this area divided by Δt, which equals the height of a rectangle of width Δt that contains the same area as the integral $\mathbf{J} = \int_{t_1}^{t_2} \mathbf{F}\,dt$.

EXAMPLE 6.1

During a collision with a wall lasting from t = 0 to t = 2 s, the force acting on a 2-kg object is given by the equation $\mathbf{F} = (4\,\text{kg}\cdot\text{m/s}^4)t(2s - t)\hat{i}$.

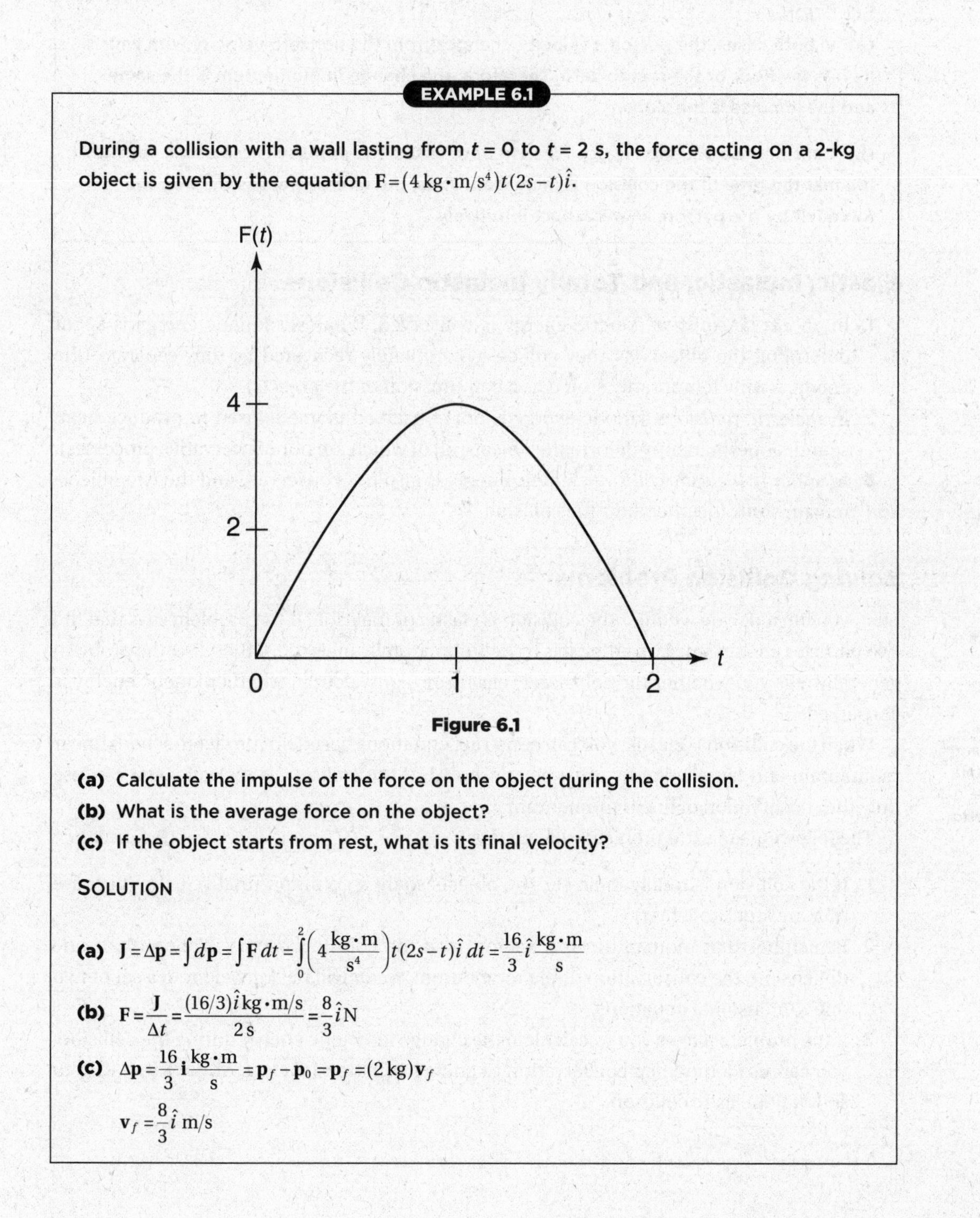

Figure 6.1

(a) Calculate the impulse of the force on the object during the collision.
(b) What is the average force on the object?
(c) If the object starts from rest, what is its final velocity?

SOLUTION

(a) $\mathbf{J} = \Delta\mathbf{p} = \int d\mathbf{p} = \int \mathbf{F}\,dt = \int_0^2 \left(4\frac{\text{kg}\cdot\text{m}}{\text{s}^4}\right) t(2s - t)\hat{i}\,dt = \frac{16}{3}\hat{i}\,\frac{\text{kg}\cdot\text{m}}{\text{s}}$

(b) $\mathbf{F} = \frac{\mathbf{J}}{\Delta t} = \frac{(16/3)\hat{i}\,\text{kg}\cdot\text{m/s}}{2\,\text{s}} = \frac{8}{3}\hat{i}\,\text{N}$

(c) $\Delta\mathbf{p} = \frac{16}{3}\mathbf{i}\frac{\text{kg}\cdot\text{m}}{\text{s}} = \mathbf{p}_f - \mathbf{p}_0 = \mathbf{p}_f = (2\,\text{kg})\mathbf{v}_f$

$\mathbf{v}_f = \frac{8}{3}\hat{i}\,\text{m/s}$

EXAMPLE 6.2

Compare a person falling on a bare gym floor with a person falling on a mat. Assume that the mat increases the time it takes for the person to come to rest. Compared to the collision without a mat, how does the mat influence (a) the impulse and (b) the average force on the person?

SOLUTION

(a) In both cases, the person's velocity changes from the nonzero velocity with which he hits the floor or the mat to zero. Therefore, the change in momentum is the same and the impulse is the same.

(b) Consider the equation $\mathbf{F}_{\text{average}} = \mathbf{J}/\Delta t$. In both cases, the impulse is the same, yet with the mat the time of the collision is increased. Thus, the mat decreases the average force felt by the person, as we expect intuitively.

Elastic, Inelastic, and Totally Inelastic Collisions

1. In an *elastic collision,* kinetic energy is conserved. Whatever kinetic energy is spent deforming the objects as they collide is completely recovered as they separate (the energy is only temporarily stored as a compression of the objects).
2. In *inelastic collisions,* kinetic energy is not conserved (some is used to produce heat, sound, or permanently deform the objects, all of which are nonconservative processes).
3. In a *totally inelastic collision,* kinetic energy is also not conserved, and the two objects remain stuck together after the collision.

Solving Collision Problems

Always use conservation of momentum when solving a collision problem, and conservation of energy if the collision is elastic.

First, you must decide whether the collision is elastic or inelastic. If the problem says that the two particles end up stuck together, it is by definition totally inelastic. Otherwise, the problem generally tells you whether the collision is elastic or (equivalently) whether kinetic energy is conserved.

When the collision is elastic, you can construct equations based on the fact that both linear momentum and kinetic energy must be conserved. If the collision is inelastic, you can use only the conservation of linear momentum.

The following are extra problem-solving tips:

1. If the collision is totally inelastic, the objects share a common final velocity (because they are stuck together).
2. Remember that momentum is a vector, so if the collision occurs in more than one dimension, the conservation of the momentum *vector* will be equivalent to a set of two one-dimensional equations.
3. If the problem allows you to calculate the change in kinetic energy during the collision, you can use a modified conservation of energy equation (such as $\Delta KE = KE_{fr} - KE_0$) to solve an inelastic collision.

4. When you think of a situation where linear momentum is conserved, you probably think of a collision involving contact forces. However, momentum is conserved even when forces act at a distance (such as gravitational forces or electrostatic forces), as long as there are no net external forces. For example, the momentum of a system of planets, which attract each other gravitationally, is conserved.

EXAMPLE 6.3

A bullet of mass 0.005 kg moving at a speed of 100 m/s lodges within a 1-kg block of wood resting on a frictionless surface and attached to a horizontal spring of k = 50 N/m.

(a) What is the velocity of the block the instant after the bullet strikes it?

(b) What is the maximum compression of the spring?

SOLUTION

(a) This is a totally inelastic collision, so linear momentum (but not energy) is conserved. Noting that both bodies are moving with a common speed after the collision,

$$m_1 v_1 + m_2 v_2 = (m_1 + m_2) v_f$$

$$(0.005\text{ kg})(100\text{ m/s}) + (1\text{ kg})(0) = (1.005\text{ kg}) v_f \Rightarrow v_f = 0.498\text{ m/s}$$

(b) At this point, we have a standard mass-and-spring problem: A mass of 1.005 kg moves with an initial velocity of 0.5 m/s when the spring of k = 50 N/m is at its equilibrium position (assuming that the bullet embeds itself so quickly in the block that it is brought to rest relative to the block while the spring is still in its equilibrium position). Energy is conserved during the compression of the spring, and we calculate the maximum compression by converting all the kinetic energy to potential energy:

$$\frac{1}{2} m v_0^2 = \frac{1}{2} k x_f^2$$

$$\frac{1}{2}(1.005\text{ kg})(0.5\text{ m/s})^2 = \frac{1}{2}(50\text{ N/m}) x_f^2 \Rightarrow x_f = 0.07\text{ m}$$

Common error: The initial kinetic energy of the bullet cannot be equated to the final energy stored in the spring because mechanical energy is not conserved during the collision; mechanical energy is conserved *after* the bullet lodges in the block.

EXAMPLE 6.4

A variable force $F(t)$ acts on an object of mass m that is initially at rest for a time interval t.

(a) Find an expression for the final velocity by calculating the impulse and relating it to the momentum.

(b) Calculate the final velocity using Newton's second law ($\mathbf{F} = m\mathbf{a}$).

SOLUTION

(a) Recall the impulse-momentum theorem:

$$\mathbf{J} = \int_0^t \mathbf{F}(t)dt = \mathbf{p}_f - \mathbf{p}_0$$

Because the mass is initially at rest, $\mathbf{p}_0 = 0$.

$$\int_0^t \mathbf{F}(t)dt = \mathbf{p}_f = m\mathbf{v}_f$$

$$\mathbf{v}_f = \frac{1}{m}\int_0^t \mathbf{F}(t)dt$$

(b)

$$\mathbf{a} = \frac{\mathbf{F}(t)}{m}$$

$$\Delta\mathbf{v} = \mathbf{v}_f - \mathbf{v}_0 = \mathbf{v}_f = \int_0^t \mathbf{a}(t)dt = \int_0^t \frac{\mathbf{F}(t)dt}{m}$$

Simplifying,

$$\mathbf{v}_f = \frac{1}{m}\int_0^t \mathbf{F}(t)dt$$

This example illustrates that the two solutions are fundamentally identical; both are rooted in Newton's laws, so they differ only in notation.

EXAMPLE 6.5

Particle C collides and sticks to particle S. If the particles have masses m_C and m_S and initial velocities $\mathbf{v}_{S,0} = v_{S,x}\hat{i} + v_{S,y}\hat{j}$ and $\mathbf{v}_{C,0} = v_{C,x}\hat{i} + v_{C,y}\hat{j}$, what is the final velocity of the two particles?

SOLUTION

This is a totally inelastic collision, so conservation of momentum yields

$$m_C\mathbf{v}_C + m_S\mathbf{v}_S = (m_C + m_S)\mathbf{v}_f$$

$$\mathbf{v}_f = \frac{m_C\mathbf{v}_C + m_S\mathbf{v}_S}{m_C + m_S}$$

In terms of components, this can be written

$$\begin{cases} v_{f,x} = \dfrac{v_{C,x}m_C + v_{S,x}m_S}{m_C + m_S} \\ v_{f,y} = \dfrac{v_{C,y}m_C + v_{S,y}m_S}{m_C + m_S} \end{cases}$$

EXAMPLE 6.6 AN ELASTIC COLLISION IN ONE DIMENSION

A mass of 3 kg moving to the right at a speed of 5 m/s collides elastically with a 10-kg mass moving at 1 m/s to the left. What is the velocity of the 10-kg mass after the collision?

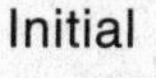

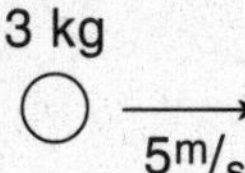

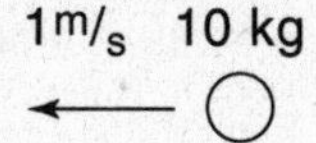

Figure 6.2

SOLUTION

The initial and final situations are shown in Figure 6.2. How are the final directions of the velocity vectors determined? The directions shown in the figure were selected arbitrarily. If the chosen direction is incorrect, the calculated magnitude will be negative (so it won't be a problem).

Conservation of momentum:

$$(3\text{ kg})(5\text{ m/s}) + (10\text{ kg})(-1\text{ m/s}) = (3\text{ kg})(v_1) + (10\text{ kg})(v_2)$$

Conservation of energy:

$$\frac{1}{2}(3\text{ kg})(5\text{ m/s})^2 + \frac{1}{2}(10\text{ kg})(1\text{ m/s})^2 = \frac{1}{2}(3\text{ kg})(v_1^2) + \frac{1}{2}(10\text{ kg})(v_2^2)$$

Solving the first equation for v_1 yields

$$v_1 = \frac{5\text{ kg m/s} - (10\text{ kg})v_2}{3\text{ kg}}$$

Inserting this into the second equation and solving via the quadratic equation yields two answers: $v_2 = -1$ m/s and $v_2 = 1.769$. Do we choose the second solution because we are expecting the 10-kg mass to move to the right? Maybe. We suspect that the 10-kg mass will move to the right, but we aren't sure. Instead, we work backward to see which solution makes physical sense. Using the above equation that defines v_1 in terms of v_2, we calculate

$$v_2 = -1\text{ m/s} \Rightarrow v_1 = +5\text{ m/s}$$

$$v_2 = 1.769\text{ m/s} \Rightarrow v_1 = -4.23\text{ m/s}$$

Note that the first solution for v_2 corresponds to the initial situation. This certainly makes sense: Everything is conserved between the initial situation and itself. Thus, the second solution is the one we are interested in: The 10-kg mass has a final velocity of 1.769 m/s to the right.

EXAMPLE 6.7 THE CLASSIC POOL PROBLEM

The six ball, traveling at 10 m/s on a pool table, collides elastically with the eight ball (initially at rest), sending the eight ball off in a different direction as shown in Figure 6.3. What is the speed and direction of the six ball after the collision? The two balls have identical mass.

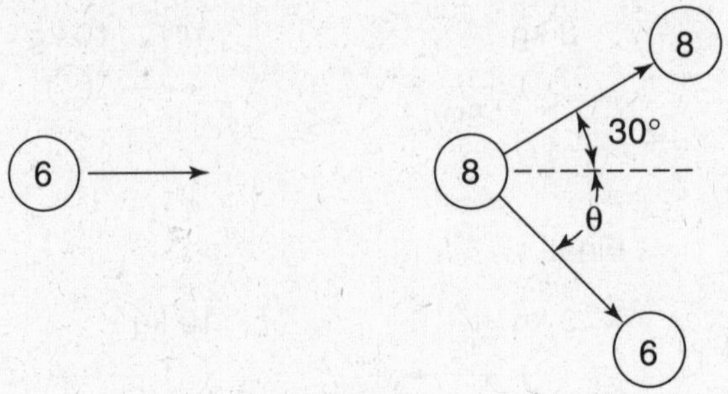

Figure 6.3

SOLUTION

This is an elastic collision, so we have conservation of both momentum and energy. Because we are working in two dimensions, the conservation of vector momentum corresponds to two scalar equations (conservation of *x*- and *y*-momentum).

Conservation of kinetic energy (letting v_8 and v_6 be the magnitudes of the final velocities of the eight and six balls, respectively):

TIP

Momentum is a vector. You must apply conservation of momentum to each vector component independently.

$$\frac{1}{2}m(10\,\text{m/s})^2 = \frac{1}{2}mv_8^2 + \frac{1}{2}mv_6^2$$

Conservation of *x*-momentum:

$$m(10\text{ m/s}) = mv_8\cos 30 + mv_6\cos\theta$$

Conservation of *y*-momentum:

$$0 = mv_8\sin 30 - mv_6\sin\theta$$

Rewriting these equations,

$$\begin{cases} (10\,\text{m/s})^2 = v_8^2 + v_6^2 \\ 10\,\text{m/s} = \dfrac{\sqrt{3}}{2}v_8 + v_6\cos\theta \\ 0 = \dfrac{v_8}{2} - v_6\sin\theta \end{cases}$$

One way to solve a set of equations like this one is to isolate the sine and cosine terms, square these equations, and apply the identity $\sin^2\theta + \cos^2\theta = 1$ to eliminate the unknown angle.

Isolating the cosine and sine terms,

$$\cos\theta = \frac{10\ \text{m/s} - \left(\sqrt{3}/2\right)v_8}{v_6}$$

$$\sin\theta = \frac{v_8}{2v_6}$$

Squaring these two equations we have

$$\cos^2\theta = \frac{100\ \text{m}^2/\text{s}^2 - (10\ \text{m/s})\sqrt{3}v_8 + \frac{3}{4}v_8^2}{v_6^2}$$

$$\sin^2\theta = \frac{v_8^2}{4v_6^2}$$

We add these two equations and use the identity $\sin^2\theta + \cos^2\theta = 1$

$$\frac{100\ \text{m}^2/\text{s}^2 - (10\ \text{m/s})\sqrt{3}v_8 + \frac{3}{4}v_8^2}{v_6^2} + \frac{v_8^2}{4v_6^2} = 1$$

Simplifying,

$$v_8^2 - (10\ \text{m/s})\sqrt{3}v_8 + 100\ \text{m}^2/\text{s}^2 = v_6^2$$

Recall from above the conservation of energy equation:

$$(10\ \text{m/s})^2 = v_8^2 + v_6^2$$

Substituting for v_6^2,

$$100\ \text{m}^2/\text{s}^2 = v_8^2 + \left[v_8^2 - (10\ \text{m/s})\sqrt{3}v_8 + 100\ \text{m}^2/\text{s}^2\right]$$

Because the term 100 m^2/s^2 appears on both sides of the equation, it cancels, leaving us with a linear equation that is easily solved:

$$0 = 2v_8^2 - (10\ \text{m/s})\sqrt{3}v_8 \Rightarrow v_8 = 8.66\ \text{m/s}$$

Substitution into the previous equations allows us to recover the other values:

$$v_8^2 - (10\ \text{m/s})\sqrt{3}v_8 + 100\ \text{m}^2/\text{s}^2 = v_6^2 \Rightarrow v_6 = 5\ \text{m/s}$$

$$\sin\theta = \frac{v_8}{2v_6} \Rightarrow \theta = 60°$$

Note that the angle between the paths of the two balls after the collision is 90°. This is a very general result. The angle between the paths of two balls of equal mass after an elastic collision where one ball was initially at rest will always be 90°.

CENTER OF MASS

The center of mass is defined to be the *weighted average* of the location of mass in a system. Mathematically,

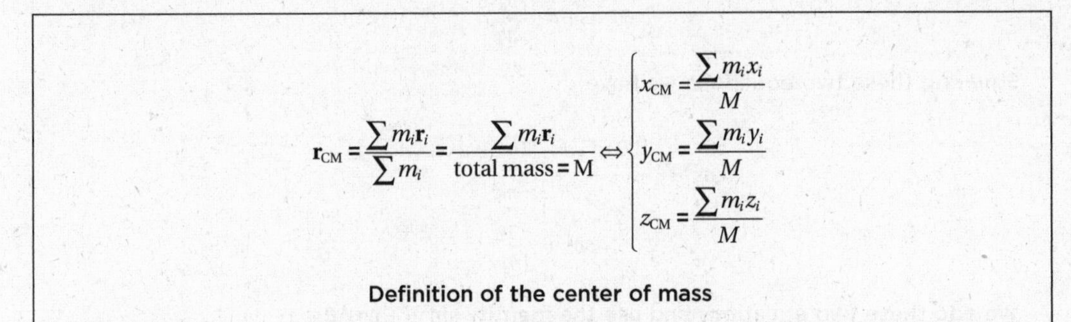

$$\mathbf{r}_{CM} = \frac{\sum m_i \mathbf{r}_i}{\sum m_i} = \frac{\sum m_i \mathbf{r}_i}{\text{total mass} = M} \Leftrightarrow \begin{cases} x_{CM} = \frac{\sum m_i x_i}{M} \\ y_{CM} = \frac{\sum m_i y_i}{M} \\ z_{CM} = \frac{\sum m_i z_i}{M} \end{cases}$$

Definition of the center of mass

EXAMPLE 6.8 COMPUTING THE CENTER OF MASS OF A GROUP OF POINT MASSES

Three masses are aligned in a coordinate system as follows: mass 1 (1.5 kg) at (1 m, 2 m, 4 m); mass 2 (4 kg) at (2 m, 1 m, 3 m); mass 3 (2 kg) at (0.5 m, 5 m, 0 m). Calculate the center of mass.

SOLUTION

Computing the center of mass of a group of point masses is straightforward: simply plug into the above formulas:

$$x_{CM} = \frac{(1.5\,\text{kg})(1\,\text{m}) + (4\,\text{kg})(2\,\text{m}) + (2\,\text{kg})(0.5\,\text{m})}{1.5\,\text{kg} + 4\,\text{kg} + 2\,\text{kg}} = 1.4\,\text{m}$$

$$y_{CM} = \frac{(1.5\,\text{kg})(2\,\text{m}) + (4\,\text{kg})(1\,\text{m}) + (2\,\text{kg})(5\,\text{m})}{1.5\,\text{kg} + 4\,\text{kg} + 2\,\text{kg}} = 2.3\,\text{m}$$

$$z_{CM} = \frac{(1.5\,\text{kg})(4\,\text{m}) + (4\,\text{kg})(3\,\text{m}) + (2\,\text{kg})(0\,\text{m})}{1.5\,\text{kg} + 4\,\text{kg} + 2\,\text{kg}} = 2.4\,\text{m}$$

Therefore, the center of mass lies at (1.4 m, 2.3 m, 2.4 m).

Numeric answer check: We note that each of these coordinates lies somewhere between the corresponding coordinates of the masses (e.g., in the *x*-direction, 1.4 m is between 0.5 m and 2 m). If this was not the case (e.g., if we computed x_{CM} = 20 m), then we would clearly be mistaken.

Computing the Center of Mass of Continuous Mass Distributions

How can we compute the center of mass of a continuous mass distribution (e.g., a wire, a basketball, or a semicircular plate)? How can we apply our point-mass CM definition to extended objects? This involves a common theme in Physics C: moving to a differential level. To calculate the CM of a baseball bat we imagine dividing the bat into infinitesimally small pieces and then applying our point-mass CM equation to all these pointlike pieces. Mathematically,

$$\mathbf{r}_{CM} = \lim_{n \to \infty} \frac{\sum_{i=1}^{i=n} m_i \mathbf{r}_i}{M_{\text{total}}}$$

We recognize an integral (the mass of the pieces shrinks to zero and becomes a differential, whereas the position vector remains fixed).

$$\mathbf{r}_{\text{CM}} = \frac{\int \mathbf{r}\, dm}{M} \Leftrightarrow \begin{cases} x_{\text{CM}} = \dfrac{\int x\, dm}{M} \\ y_{\text{CM}} = \dfrac{\int y\, dm}{M} \\ z_{\text{CM}} = \dfrac{\int z\, dm}{M} \end{cases}$$

Symmetry shortcut: Any plane of symmetry, mirror line, axis of rotation, or point of inversion must contain the center of mass. Why? If you think of the center of mass as a weighted average, it makes sense: If there are identical mass distributions on either side of a plane, axis, or point, the weighted average must lie on that plane, axis, or point.

EXAMPLE 6.9 SYMMETRY ELEMENTS

Use symmetry considerations to find the center of mass of uniform planar mass distributions in the shapes of (a) a rectangle and (b) an equilateral triangle.

SOLUTION

See Figure 6.4. In both cases, at least two lines of mirror symmetry can be drawn (a line of mirror symmetry means that if a mirror were placed along the line perpendicular to the page, it could generate the entire figure by combining one half of the figure with its mirror image). Because the CM must lie on *both* of these lines, it must lie at their intersection. This type of argument can often be used to locate the CM at the object's geometric center.

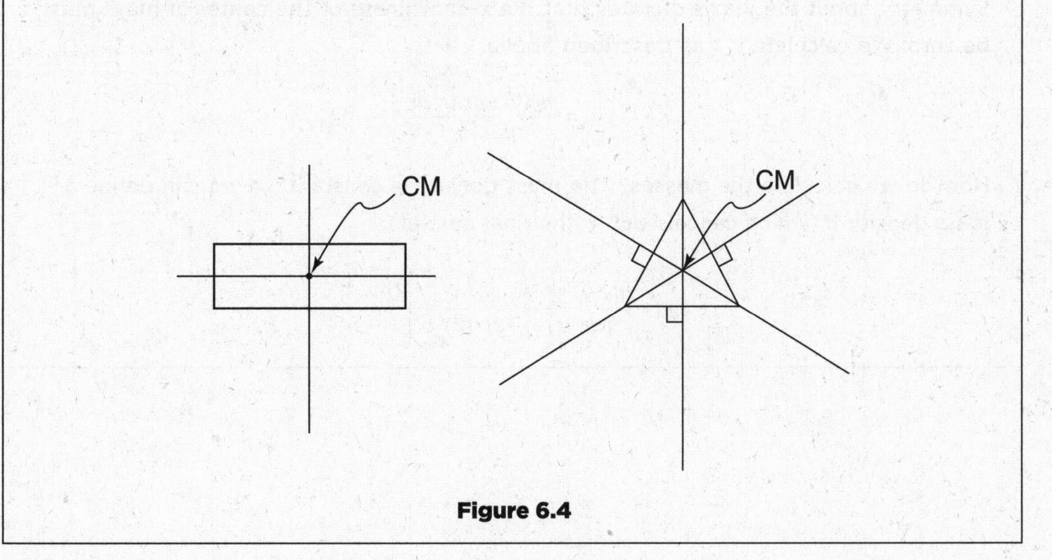

Figure 6.4

Calculating the Center of Mass Using Negative Mass

Consider calculating the center of mass of the object shown in Figure 6.5, which has uniform density. The smaller circle has no mass. The easiest way to do this calculation is to consider the object the sum of a complete circle with mass density σ and a smaller circle with mass density $-\sigma$ such that, in the area of the hole, the positive and negative mass densities cancel, leaving a region of zero mass. The "negative mass" has no physical significance; this is a purely mathematical trick (which is nonetheless very useful).

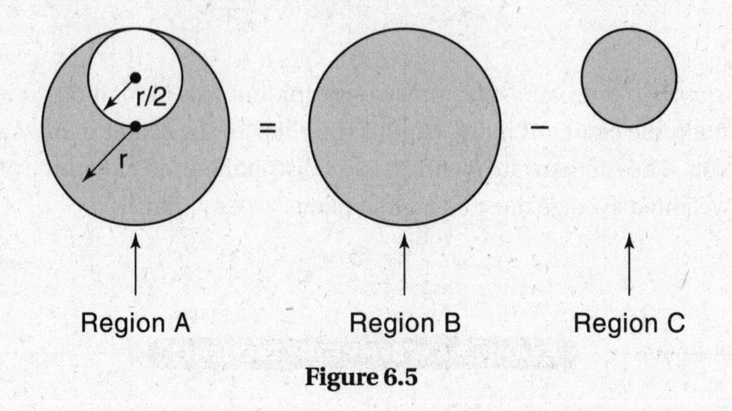

Figure 6.5

EXAMPLE 6.10

Calculate the center of mass of this object, with the center of the larger circle as the origin of your coordinate system. The object has uniform mass density and a radius of *r* (the radius of the smaller circle is half the radius of the larger circle).

SOLUTION

Symmetry about the *y*-axis dictates that the *x*-coordinate of the center of mass must be zero. We calculate y_{CM} as described above:

$$y_{CM} = \frac{m_b(0) + m_c(r/2)}{m_b + m_c}$$

How do we calculate the masses? The mass density is constant, so we can define a mass density σ (which cancels out in the final answer):

$$y_{CM} = \frac{m_b(0) + \left[-\pi(r/2)^2\sigma\right](r/2)}{(\pi r^2\sigma) + \left[-\pi(r/2)^2\sigma\right]} = \frac{-r}{6}$$

The Velocity and Acceleration of the Center of Mass

We have determined the position of the center of mass:

$$\mathbf{r}_{CM} = \frac{\sum m_i \mathbf{r}_i}{M}$$

What about its velocity and acceleration? We can calculate these quantities by differentiating the equation for $\mathbf{r}_{CM}$ with respect to time:

$$\mathbf{v}_{CM} = \frac{d\mathbf{r}_{CM}}{dt} = \frac{d}{dt}\left(\frac{\sum m_i \mathbf{r}_i}{M}\right) = \frac{1}{M}\frac{d}{dt}\sum m_i \mathbf{r}_i = \frac{\sum m_i (d\mathbf{r}_i/dt)}{M} = \frac{\sum m_i \mathbf{v}_i}{M}$$

Noting that $\Sigma m_i \mathbf{v}_i$ is the sum of the momentum of each object, the total momentum $\mathbf{P}_{net}$ of the system is given by

$$\mathbf{P}_{net} = M\mathbf{v}_{CM}$$

Differentiating (recall that $\mathbf{F} = d\mathbf{P}/dt$),

$$\frac{d\mathbf{P}_{net}}{dt} = \mathbf{F}_{net} = M\frac{d\mathbf{v}_{CM}}{dt} = M\mathbf{a}_{CM}$$

Summarizing,

$$\mathbf{v}_{CM} = \frac{\sum m_i \mathbf{v}_i}{M}$$

$$\mathbf{P}_{net} = M\mathbf{v}_{CM}$$

$$\mathbf{F}_{net} = \frac{d\mathbf{P}_{net}}{dt} = M\mathbf{a}_{CM}$$

What does this mean?

1. The center of mass moves as if it were a point of mass M acted on by the net force.
2. This is an alternative derivation of the conservation of momentum for a system of particles: If the net force on a system of particles is zero, the total momentum of the system remains constant.

EXAMPLE 6.11 USING THE CENTER OF MASS TO SOLVE PROBLEMS

A satellite in deep space is stationary at the origin of our coordinate system. It explodes, producing two pieces, one of 200 kg and one of 500 kg. When the 500-kg piece is located at (30 m, 100 m), where is the 200-kg piece?

SOLUTION

There is no net external force acting on the satellite, so the acceleration of its center of mass is zero and the center of mass must remain at the origin. Therefore,

$$\begin{cases} x_{CM} = 0 = \dfrac{(500\text{ kg})(30\text{ m}) + (200\text{ kg})(x_{200\text{ kg}})}{200\text{ kg} + 500\text{ kg}} \\ y_{CM} = 0 = \dfrac{(500\text{ kg})(100\text{ m}) + (200\text{ kg})(y_{200\text{ kg}})}{200\text{ kg} + 500\text{ kg}} \end{cases}$$

It is easy to solve these equations to obtain the location of the 200-kg fragment of the satellite at (–75 m, –250 m).

EXAMPLE 6.12

Explain the results for Example 6.5 on page 178 (an inelastic collision in two dimensions) from the point of view of center of mass.

SOLUTION

Because there is no net external force, the velocity of the center of mass remains constant. When the two particles stick together, they act like a point mass and therefore must move with velocity $\mathbf{v}_{CM}$. Therefore, the final velocity is simply a calculation of the $\mathbf{v}_{CM}$. Examination of the calculated final velocity in Example 6.5 reveals that it is indeed equal to $\mathbf{v}_{CM} = \Sigma m_i \mathbf{v}_i / M$.

CHAPTER SUMMARY

The momentum of an object of mass m and velocity $\mathbf{v}$ is defined by $\mathbf{p} = m\mathbf{v}$. Newton's second law then reads $\mathbf{F} = \frac{d\mathbf{p}}{dt}$. If the net force is zero, then momentum is conserved. If we have a system of objects subject to zero external force, then the total momentum of the objects is conserved even though the objects exert forces on each other. In an elastic collision of two or more objects, the total kinetic energy is conserved, which is not the case for inelastic collisions. In a totally inelastic collision, the objects stick together after the collision. The center of mass (CM) of a system of objects or a single object of finite extent can be defined so that the CM moves as if it were a pointlike object subject only to external forces.

PRACTICE EXERCISES

Multiple-Choice Questions

1. Two masses moving along the coordinate axes as shown in Figure 6.6 collide at the origin and stick to each other. What is the angle θ that the final velocity makes with the x-axis?

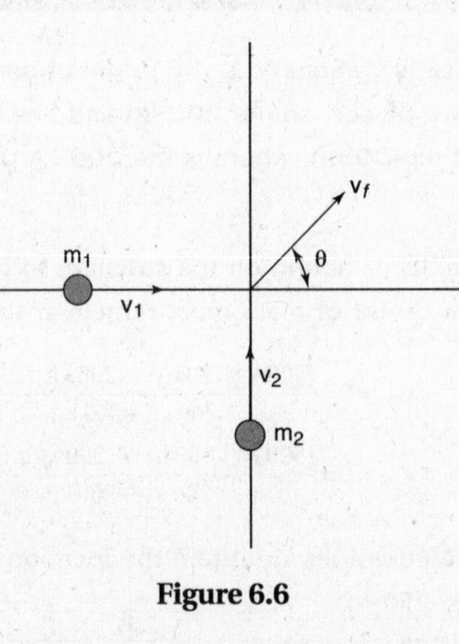

Figure 6.6

(A) $\tan^{-1}(v_2/v_1)$
(B) $\tan^{-1}[m_1v_1/(m_1+m_2)]$
(C) $\tan^{-1}(m_1v_2/m_2v_1)$
(D) $\tan^{-1}(m_2v_2^2/m_1v_1^2)$
(E) $\tan^{-1}(m_2v_2/m_1v_1)$

2. Stating that the net force on an object is zero is *exactly equivalent* to which of the following statements?

(A) The object's speed is constant.
(B) The object is in static equilibrium.
(C) The object's linear momentum is constant.
(D) There are no forces acting on the object.
(E) none of the above

3. Mass m_1 moving with initial speed v_0 has a glancing collision with a second mass m_2 initially at rest, as shown in Figure 6.7. Which of the following statements is necessarily true?

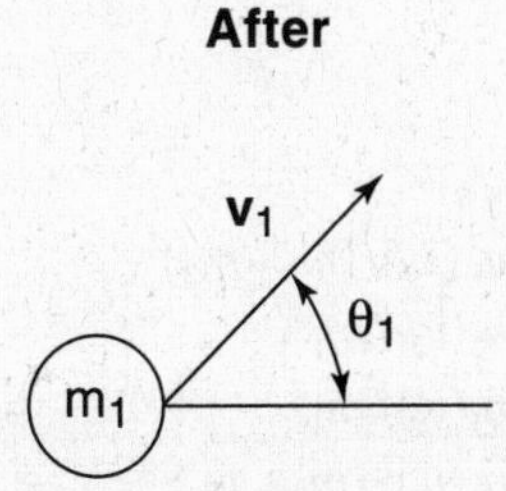

Before

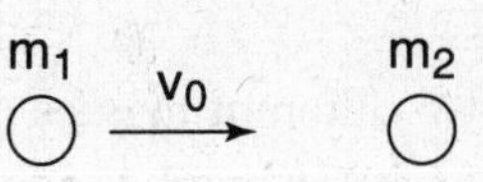

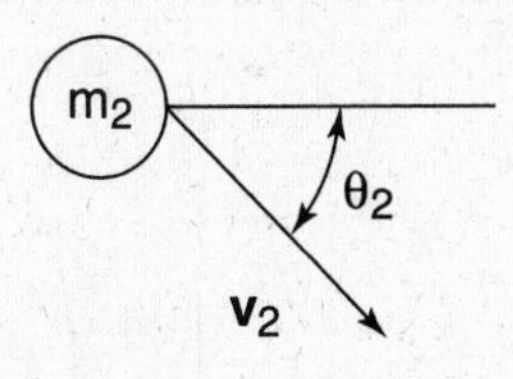

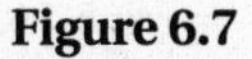

Figure 6.7

(A) $\frac{1}{2}m_1v_0^2 = \frac{1}{2}m_1v_1^2 + \frac{1}{2}m_2v_2^2$.
(B) If $v_1 < v_2$, then $\theta_1 < \theta_2$.
(C) If $m_1v_1 < m_2v_2$, then $\theta_1 < \theta_2$.
(D) If $m_1v_1 < m_2v_2$, then $\theta_2 < \theta_1$.
(E) None of the above are true.

4. A mass traveling in the $+x$-direction collides with a mass at rest. Which of the following statements is true?

   (A) After the collision, the two masses will move with parallel velocities.
   (B) After the collision, the masses will move with antiparallel velocities.
   (C) After the collision, the masses will both move along the x-axis.
   (D) After the collision, the y-components of the velocities of the two particles will sum to zero.
   (E) none of the above

5. A mass m_1 initially moving at a speed v_0 collides with and sticks to a spring attached to a second, initially stationary mass m_2 (Figure 6.8). The two masses continue to move to the right on a frictionless surface as the length of the spring oscillates. At the instant that the spring is maximally extended, the velocity of the first mass is

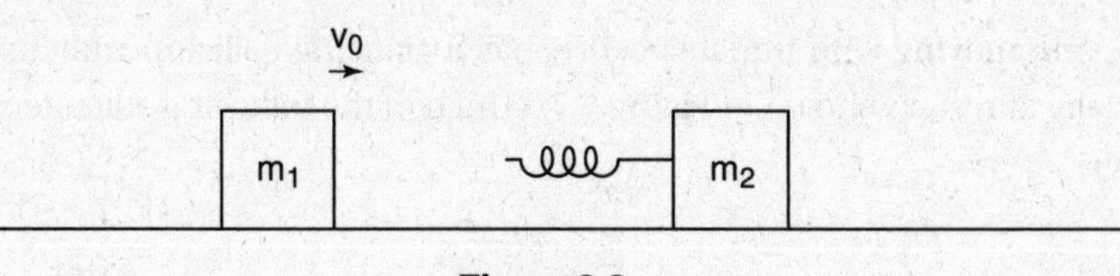

Figure 6.8

   (A) v_0
   (B) $m_1^2(v_0)/(m_1+m_2)^2$
   (C) $m_2(v_0)/m_1$
   (D) $m_1(v_0)/m_2$
   (E) $m_1(v_0)/(m_1+m_2)$

6. The force as a function of time experienced by two different masses during two collisions is shown in Figure 6.9. The total area under each $F(t)$ graph is equal. How many of the following quantities are the same for the two collisions?

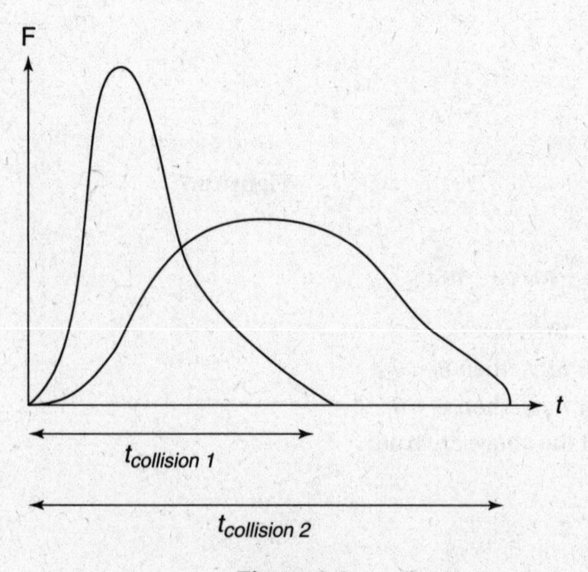

Figure 6.9

I. average force during the collision
II. net change in velocity during the collision
III. net change in momentum during the collision
IV. net displacement during the collision

(A) None of the quantities are the same.
(B) One of the quantities is the same.
(C) Two of the quantities are the same.
(D) Three of the quantities are the same.
(E) Four of the quantities are the same.

7. A variable net force acts on an initially stationary 2-kg mass as shown in Figure 6.10. What is the mass's final velocity?

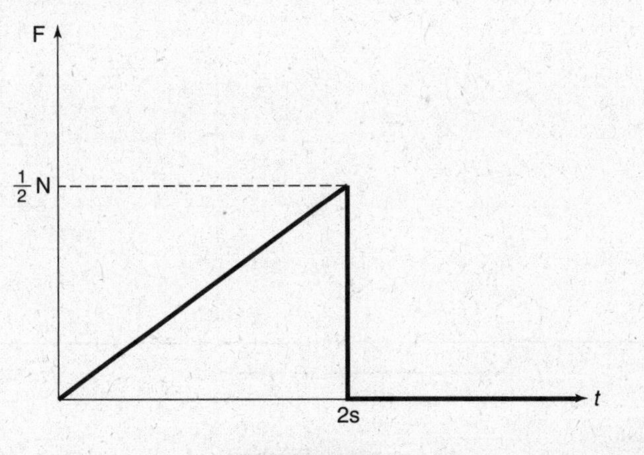

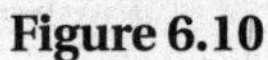
Figure 6.10

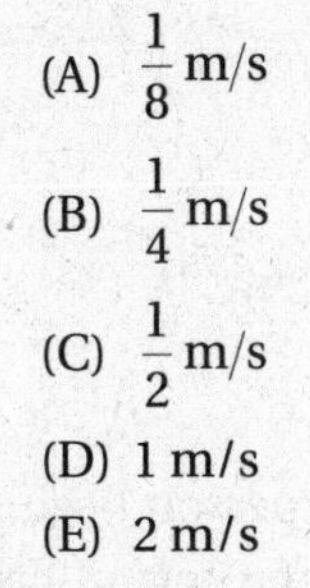
(A) $\frac{1}{8}$ m/s
(B) $\frac{1}{4}$ m/s
(C) $\frac{1}{2}$ m/s
(D) 1 m/s
(E) 2 m/s

8. A mass is falling vertically toward the ground when it explodes into two fragments of masses m_1 and m_2, which strike the ground at the same time (Figure 6.11). If the first mass lands a distance d_1 from the place it would have landed had the explosion not occurred (see the figure), what is the final distance between the two masses after they land?

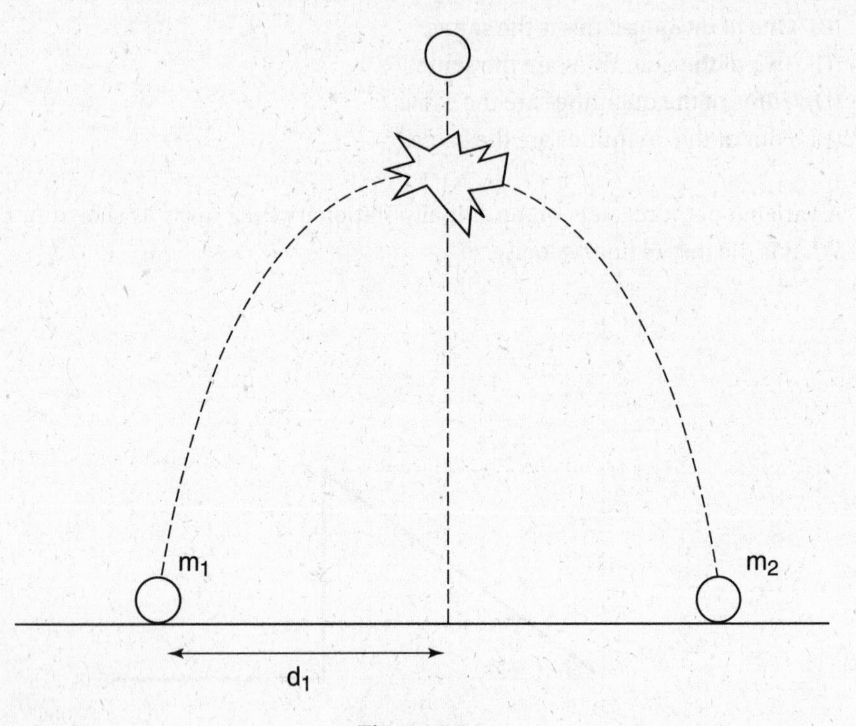

Figure 6.11

(A) $d_1(1+m_1^2/m_2^2)$
(B) $d_1(1+m_2^2/m_1^2)$
(C) $d_1(1+m_1/m_2)$
(D) $d_1(1+m_2/m_1)$
(E) Not enough information is given.

9. A person is at rest in a canoe that is motionless in still water. If the person begins to move with speed v_p relative to the water, what is the kinetic energy of the person-canoe system? (The mass of the person and the canoe are m_p and m_c, respectively.)

(A) $\mathrm{KE}=\frac{1}{2}m_p v_p^2$

(B) $\mathrm{KE}=\frac{1}{2}(v_p+v_c)v_p^2$

(C) $\mathrm{KE}=\frac{1}{2}m_c v_p^2$

(D) $\mathrm{KE}=m_p^2 v_p^2/2m_c$

(E) $\mathrm{KE}=\frac{1}{2}m_p v_p^2(1+m_p/m_c)$

10. If an object is moving with a constant acceleration, which of the following is true?

   (A) It must be under the influence of a single force equal to $m\mathbf{a}$.
   (B) In each second, its speed changes by an amount equal in value to a.
   (C) Its kinetic energy increases at a rate equal to its speed multiplied by $m\mathbf{a}$.
   (D) In each second, its momentum changes by an amount equal in value to $m\mathbf{a}$.
   (E) The object moves along a parabolic path.

Free-Response Questions

1. A large mass M with a semicircular cutout lies on a frictionless horizontal surface as shown in Figure 6.12. A mass m is released from rest at the right side of the semicircular cutout (which is also frictionless).

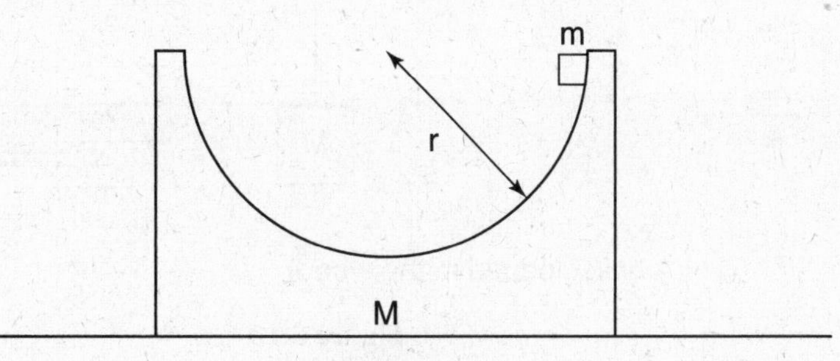

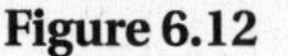
Figure 6.12

   (a) Qualitatively describe the motion of the two masses.
   (b) When the small mass m is at the bottom of the semicircle, what is its speed with respect to the table?
   (c) If M were fixed to the table, would this increase or decrease the maximum speed of m?
   (d) When m has reached the left side of the semicircular cutout, how far has M been displaced horizontally?

2. A projectile is fired from the edge of a cliff 100 m high with an initial speed of 60 m/s at an angle of elevation of 45°.

   (a) Write equations for $x(t)$, $y(t)$, v_x, and v_y. Choose the origin of your coordinate system at the particle's original location.
   (b) Calculate the location and velocity of the particle at time $t = 5$ s.

   Suppose the projectile experiences an internal explosion at time $t = 4$ s with an internal force purely in the y-direction, causing it to break into a 2-kg and a 1-kg fragment.

   (c) If the 2-kg fragment is 77 m above the height of the cliff at $t = 5$ s, what is the y-coordinate of the position of the 1-kg piece?
   (d) If the speed of the 2-kg fragment is 46 m/s and the fragment is falling at $t = 5$ s, what is the y-component of the velocity of the 1-kg fragment?

3. The Ballistic Pendulum. To determine the muzzle speed of a gun, a bullet is shot into a mass M hanging from a string as shown in Figure 6.13, causing M to swing upward through a maximum angle of θ.

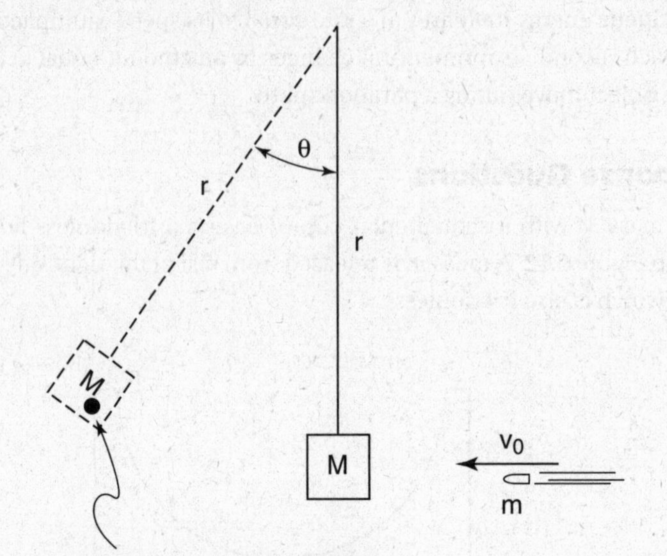

Figure 6.13

(a) What is the speed of M the instant after the bullet lodges in it?
(b) What is the speed of the bullet before it hits M?
(c) What is the tension in the string at the highest point of the pendulum's swing (when the string makes an angle of θ with the vertical as shown)?
(d) Make a sketch of the speed of the bullet, v_0, as a function of the angle θ, $0 < \theta < \frac{\pi}{2}$, before the bullet hits M. Pay careful attention to the correct curvature of your plot.

ANSWER KEY

1. **E**	4. **E**	7. **B**	10. **D**
2. **C**	5. **E**	8. **C**	
3. **D**	6. **B**	9. **E**	

ANSWERS EXPLAINED

Multiple-Choice

1. **(E)** This is an inelastic collision, so we use conservation of momentum (essentially the only way to solve inelastic collision problems).

Conservation of the x-component of momentum:

$$m_1 v_1 = (m_1 + m_2)\, v_{fx}$$

Conservation of the y-component of momentum:

$$m_2 v_2 = (m_1 + m_2)\, v_{fy}$$

Finally, using trigonometry to obtain the angle the final velocity makes with the x-axis,

$$\theta = \tan^{-1}\left(\frac{v_{f,y}}{v_{f,x}}\right) = \tan^{-1}\left(\frac{\dfrac{m_2 v_2}{m_1 + m_2}}{\dfrac{m_1 v_1}{m_1 + m_2}}\right) = \tan^{-1}\left(\frac{m_2 v_2}{m_1 v_1}\right)$$

Alternative solution: Choice (A) can be eliminated because it does not refer to the masses, which intuitively should definitely enter into the final answer. (For example, if $m_1 << m_2$, then $\theta \approx 90°$, whereas if $m_1 >> m_2$, then $\theta \approx 0°$.) Choice (B) can be eliminated because the quantity inside the arctan function isn't a dimensionless number. Choice (C) can be eliminated because, according to this answer, if $m_1 << m_2$, then θ approaches zero, which doesn't make sense. The only way to choose between choices (D) and (E) is to do the algebra, although one might guess the correct solution is choice (E); because we are using conservation of momentum here, it might be expected that the velocity would enter the answer raised to the first power.

2. **(C)** In choice (A), an object moving in uniform circular motion has constant speed but nonzero net force. In choice (B), objects in dynamic equilibrium (i.e., objects moving with constant velocity) are not in static equilibrium but do experience zero net force. (Additionally, as we will see later, static equilibrium for an extended object requires that *both* the net force and the net torque be zero.) In choice (C), $\mathbf{F}_{net} = d\mathbf{p}/dt$, so zero net force is equivalent to zero momentum. In choice (D), an object with zero net force can be acted on by several forces that sum to a net force of zero.

3. **(D)** Conservation of the y-component of momentum requires that $m_1 v_1 \sin\theta_1 = m_2 v_2 \sin\theta_2$ such that, if $m_1 v_1 < m_2 v_2$, this must be compensated by $\sin\theta_1 > \sin\theta_2$, which in turn requires that $\theta_2 < \theta_1$. Choice (A) is not necessarily correct because the problem does not indicate whether energy is conserved.

4. **(E)** If it is a glancing collision as in the previous problem, then (A), (B), and (C) will not be true. As for choice (D), it is true that the sum of the y-components of the *momentum* of the two particles after the collision will be zero. However, if the two particles have different *masses*, the sum of the two y-components of velocity will *not* be zero. None of the choices (A) through (D) will necessarily be correct.

5. **(E)** Consider the spring. As the spring extends, the distance between the two masses increases. As the spring compresses, the distance between the two masses decreases. At the instant that the spring is at its *maximum* extension or compression, the distance between the two masses is not changing, and the masses are moving at the same velocity.

   Linear momentum is conserved in this situation. (The spring force is an internal force, which does not change the net linear momentum.) Therefore, $m_1 v_0 = v_1(m_1 + m_2)$. Solving this equation for v_1 yields choice (E).

6. **(B)** In statement I, the average force, according to calculus, is given by the equation $\bar{F} = \int \mathbf{F} dt / \int dt$. The numerator, $\int \mathbf{F} dt$, is the area under the graph, so it is the same for both collisions. However, the denominator, which is equal to the duration of the collision, is different for the two collisions. Therefore, the ratio, which is the average force, is not the same for both collisions. In statement III, the net change momentum is equal to the impulse, defined as $\int F\, dt$ [the area under the $F(t)$ curve], which is the same for both

collisions. In statement II, because the change in *momentum* is the same for both particles, the particle with less mass experiences a greater change in velocity. Therefore, the net velocity difference is not the same for the two masses. In statement IV, while the impulse is the same for the two masses, this does not require that the displacement during the collision is the same. Why? Imagine two collisions, each of which has the same impulse (net change in momentum). In one collision, suppose a very large force is exerted for a small time, almost instantaneously imparting momentum to the particle (such as a collision between two hard objects). Because the collision is extremely brief, there is little time for the particle to move appreciably. In a second, slower collision with the same impulse, because the particle has more time to accelerate, it has a greater displacement.

7. **(B)** The area under the $F(t)$ graph is the impulse, which is equal to $\frac{1}{2}(\text{base})(\text{height}) = \frac{1}{2}(2\text{ s})(0.5\text{ N}) = 0.5\text{ N}\cdot\text{s}$. This is equal to the change in the object's momentum: $\Delta p = m\Delta v = mv_f - mv_0 = mv_f$. Solving these equations for the final velocity $v_f = \text{impulse}/m$ yields choice (B).

8. **(C)** Internal forces do not affect the position of the center of mass of the m_1-m_2 system. Therefore, the center of mass of the two particles continues to move straight downward. At the instant the two particles hit the ground, the center of mass is on the ground, exactly where the mass would have landed if the explosion hadn't taken place. This location of the center of mass is set as the origin as shown in Figure 6.14, yielding the following equation for the center of mass: $\text{CM} = 0 = m_2d_2 - m_1d_1$. Solving this equation for d_2 yields $d_2 = d_1m_1/m_2$. The final distance between the two masses is equal to

$$d_1 + d_2 = d_1(1 + m_1/m_2)$$

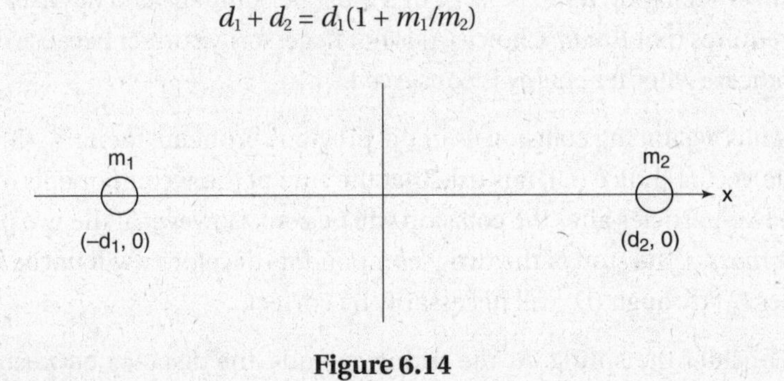

Figure 6.14

9. **(E)** Conservation of momentum requires that the canoe move in the opposite direction (with respect to the water) with a magnitude given by the equation $m_pv_p = m_cv_c$. Solving for the velocity of the canoe yields $v_c = v_pm_p/m_c$. The total kinetic energy is then the total energy of the person and the canoe:

$$\text{KE} = \frac{1}{2}m_pv_p^2 + \frac{1}{2}m_cv_c^2 = \frac{1}{2}m_pv_p^2 + \frac{1}{2}m_c\left(\frac{v_pm_p}{m_c}\right)^2$$

Simplification yields choice (E).

10. **(D)** In choice (A), the *net* force must equal $m\mathbf{a}$, but this force could be the sum of several forces. Therefore, it is not true that the object must be under the influence of a *single* force equal to $m\mathbf{a}$. Choice (B) is true only if the acceleration is parallel to the velocity such that the acceleration equals the tangential acceleration. (Consider, for example, a projectile at the peak of its parabolic path. The change in the object's speed is zero, even though a constant acceleration is acting on it.) In choice (C), the rate at which its kinetic energy increases, equal to the power exerted by the force, is given by $P = m\mathbf{a}\cdot\mathbf{v}$, *not* simply the ordinary product of ma and the speed v. This is the same sort of error made by choice (B): It ignores the possibility that the force is not parallel to the velocity. Choice (D) is true. $\mathbf{F} = d\mathbf{p}/dt$, and for constant forces $\mathbf{F} = \Delta\mathbf{p}/\Delta t$. Therefore, the change in the momentum during 1 s is $\Delta\mathbf{p} = \mathbf{F}\Delta t$. Choice (E) is not necessarily true. If the object has no component of velocity in the direction perpendicular to the force, the path will be linear (e.g., the trajectory of an object thrown straight up).

Free-Response

1. (a) As the mass m slides down the semicircular cutout, it speeds up. Because there are no external forces on the two-mass system in the x-direction, the x-momentum must remain constant. Therefore, as m speeds up toward the left, M must accelerate to the right. Then, as m begins climbing the left side of the groove, its speed and the speed of M both decrease. When m reaches the top of the groove, it (and M) are instantaneously both at rest (such that the net momentum is equal to zero, as it initially was). Conservation of energy requires that the final height of m be the same along the sides of the semicircle. This pattern is repeated as both masses oscillate.

(b) Based on the qualitative discussion above, m moves to the left with speed v_m, while M moves to the right with speed v_M. Conservation of momentum requires that $mv_m = Mv_M$. Conservation of energy requires that

$$mgr = \frac{1}{2}mv_m^2 + \frac{1}{2}Mv_M^2$$

Substituting $v_M = mv_m/M$ and solving for v_m yields

$$v_m = \sqrt{\frac{2gr}{1+m/M}}$$

(c) If M were fixed, then *all* of the gravitational potential energy would be converted to m's kinetic energy (rather than the potential energy being converted to the kinetic energy of both masses). This would transfer more kinetic energy to m, giving it a larger velocity.

(d) Since there are no forces in the x-direction, the x-component of the velocity of the center of mass must remain constant. Because this velocity is initially zero, the x-coordinate of the center of mass must remain fixed. Consider Figure 6.15. Based on the symmetry of the initial and final positions, the distance M moves is equal to twice the distance between its center of mass and the center of mass of the system.

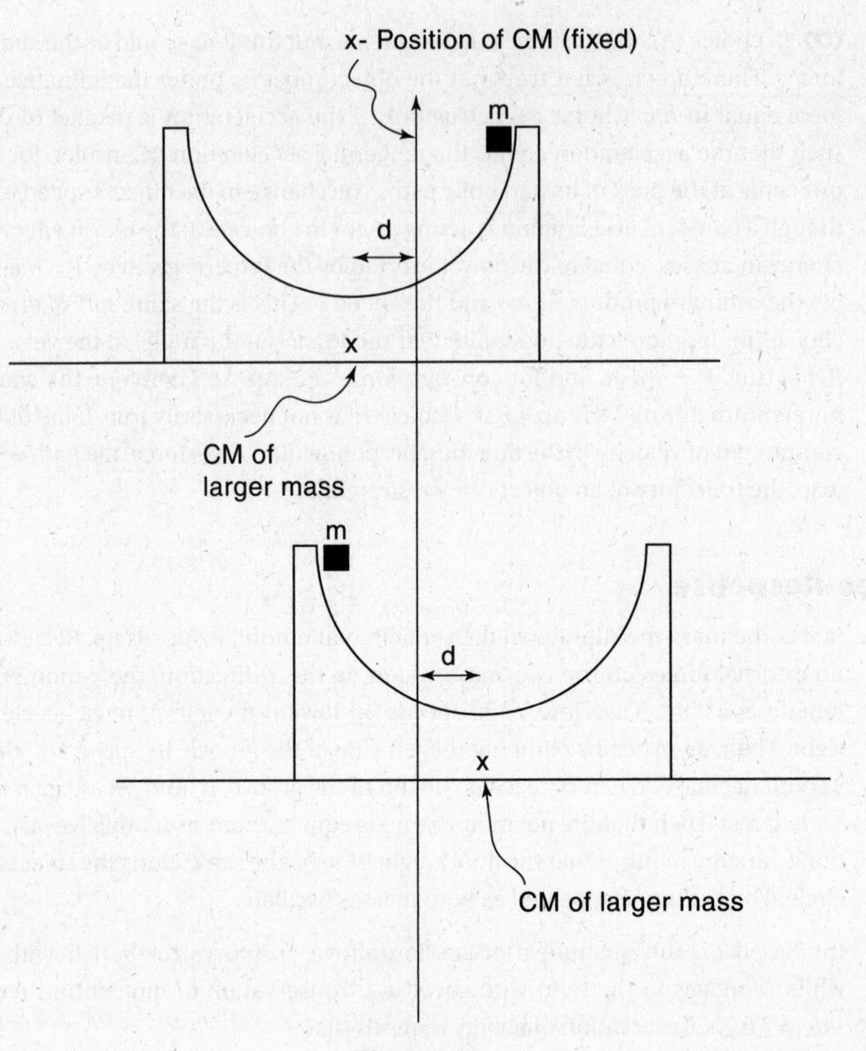

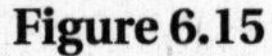
Figure 6.15

What is the distance between the center of mass of M and the center of mass of the system? Consider Figure 6.16, where the y-axis passes through the center of mass of the larger mass. Based on this figure,

$$d = x_{CM} = \frac{M(0) + m(r)}{M + m} = \frac{mr}{m + M}$$

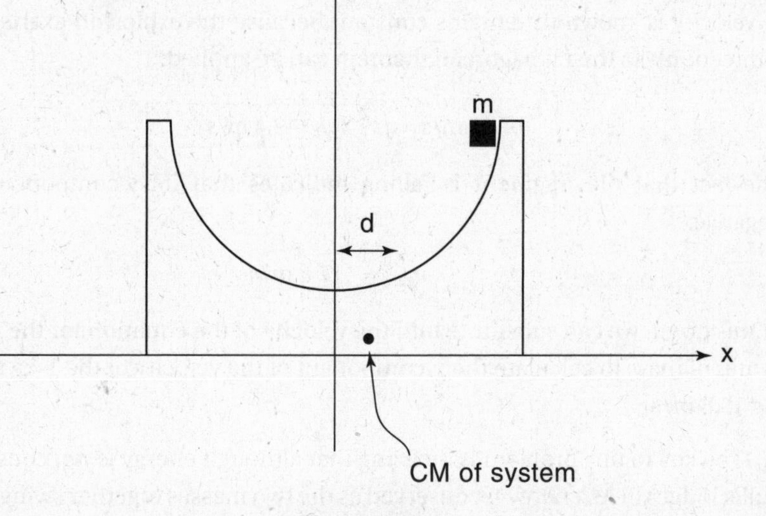

Figure 6.16

(Note that the x-component of the center of mass of M lies at its geometric center.) The distance that the center of mass of M travels is twice this distance, $2mr/(m+M)$.

2. (a) Using the standard equations for projectile motion from Chapter 3,

$$x = (42.4\ \text{m/s})t$$

$$v_x = \frac{dx}{dt} = 42.4\ \text{m/s}$$

$$y = (42.4\ \text{m/s})t - (4.9\ \text{m/s}^2)t^2$$

$$v_y = \frac{dy}{dt} = 42.4\ \text{m/s} - (9.8\ \text{m/s}^2)t$$

(b) Plugging $t = 5$ into these equations yields a position of (212 m, 89.5 m) with respect to the edge of the cliff and a velocity $\mathbf{v} = (42.4\ \text{m/s})\hat{i} - (6.6\ \text{m/s})\hat{j}$.

(c) Because there is no external force, the center of mass continues to move along the original trajectory. Therefore, the y-position of the center of mass must still be equal to 89.5 m as calculated above.

$$89.5\ \text{m} = \frac{(2\ \text{kg})(77\ \text{m}) + (1\ \text{kg})(y_{1\,\text{kg}})}{3\ \text{kg}}$$

Solving for the y-coordinate of the 1-kg fragment yields $y_{1\text{kg}} = 115$ m.

(d) Because there is no external force, the velocity of the center of mass is unchanged by the explosion, and its y-component at $t = 5$ is given by

$$-6.6\ \text{m/s} = \frac{(2\ \text{kg})v_{y,2\,\text{kg}} + (1\ \text{kg})v_{y,1\,\text{kg}}}{3\ \text{kg}}$$

However, the problem gives us the total *speed* rather than the y-component of velocity of the 2-kg fragment. How can we calculate the y-component of velocity? The x-component of velocity is known (it remains constant because the explosion exerts no force in the x-direction), so the Pythagorean theorem can be applied:

$$46\text{ m/s} = \sqrt{v_{y,2\text{ kg}}^2 + (42.4\text{ m/s})^2}$$

The fact that the fragment is falling indicates that its y-component of velocity is negative:

$$v_{y,2\text{kg}} = -17.8\text{ m/s}$$

At this point, we can substitute into the velocity of the equation for the y-velocity of the center of mass to calculate the y-component of the velocity of the 1-kg fragment (yielding 15.8 m/s).

3. (a) The key to this problem is realizing that although energy is *not* conserved while the bullet lodges in M, energy *is* conserved as the two masses together swing upward. Therefore, the speed of the two masses after the collision can be calculated using conservation of energy (equating the energy of the masses the instant after the collision equal to their energy at the top of the pendulum's swing). Designating v_f as the speed of M and the embedded bullet the instant after the bullet hits it,

$$\frac{1}{2}(m+M)v_f^2 = (m+M)gr(1-\cos\theta)$$

Solving for v_f,

$$v_f = \sqrt{2gr(1-\cos\theta)}$$

(b) Momentum is conserved during the inelastic collision of the bullet and the block (designating v_0 as the initial speed of the bullet).

$$mv_0 = (M+m)v_f$$

Substituting for v_f from above and solving for v_0,

$$v_0 = \frac{m+M}{m}(v_f) = \frac{m+M}{m}\sqrt{2gr(1-\cos\theta)}$$

(c) The free-body diagram for this part is shown in Figure 6.17. Because the mass is instantaneously at rest, its centripetal acceleration is zero and the only acceleration present is tangential acceleration (in the x-direction). Therefore, the net force in the y-direction must be zero:

$$F_{\text{net},y} = 0 = F_T - (m+M)g\cos\theta$$

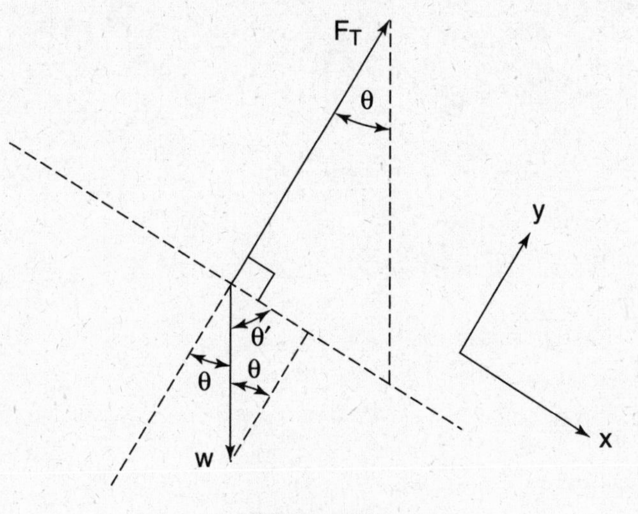

Figure 6.17

Rearranging this equation indicates that $F_T = (m + M)g\cos\theta$.

(d) You are plotting the answer to part (b) as a function of θ. An easy way to do this is to use the trig identity $\sin\frac{\theta}{2} = \sqrt{\frac{1-\cos\theta}{2}}$, in which case the answer to part (b) can be written as $v_0 = \frac{m+M}{m}\sqrt{gr}\sin\frac{\theta}{2}$. Thus, the velocity is a concave downward function of θ (see Figure 6.18). Note that the velocity is not maximized for $\theta = \frac{\pi}{2}$, so the curve does not flatten at that point. If you considered the pendulum rising to $\theta = \pi$, you would encounter the maximum value.

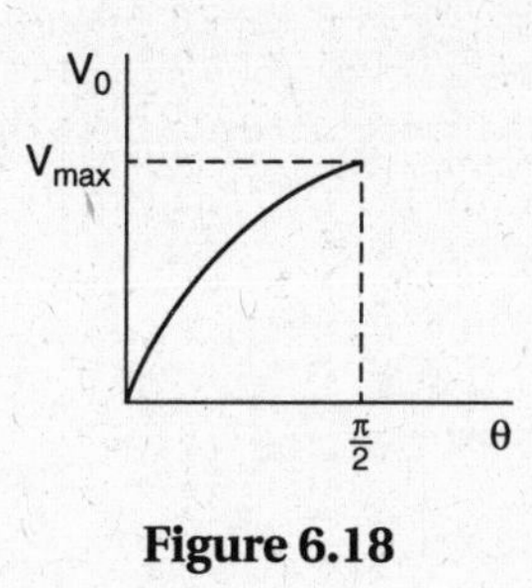

Figure 6.18

Rotation I: Kinematics, Force, Work, and Energy 7

→ DEFINITIONS OF ANGULAR POSITION, VELOCITY, AND ACCELERATION
→ RELATIONSHIPS BETWEEN ANGULAR AND TRANSLATIONAL KINEMATIC QUANTITIES
→ ROTATIONAL KINETIC ENERGY; ROTATIONAL INERTIA
→ TORQUE
→ ROTATIONAL ANALOG OF NEWTON'S SECOND LAW
→ WORK AND POWER IN ROTATIONAL MOTION

This chapter explores the rotation of rigid bodies about fixed axes. Newton's laws are valid for rotational motion just as they apply to linear motion. However, it will help to recast kinematics and Newton's second law into a more useful form for dealing with rotational motion.

THE MOTIVATION FOR DEVELOPING NEW PARAMETERS

When an object is undergoing pure translational motion, every particle in the object has the same displacement, velocity, and acceleration. When a rigid body is rotating about an axis, every particle in the object moves in a circular path about the axis, but the particles can have *different* velocities and accelerations (e.g., particles farther away from the axis move faster), making it impossible to assign a single velocity or acceleration to the object. However, every particle rotates through the same angle in a given time interval, thus sharing a common angular displacement ($\Delta\theta$), angular velocity ($d\theta/dt$), and angular acceleration ($d^2\theta/dt^2$). Because these angular quantities apply to every particle in the object, they are useful in characterizing rotational motion.

Angular Position, θ

All parts of a rotating rigid body have a common angular velocity and acceleration.

1. The angle θ does not merely refer to an object's current position relative to some reference position (i.e., it does not "reset" after every rotation). For example, if we define a wheel's angle to be $\theta = 0$ and then rotate the wheel three times counterclockwise, its angular position is $\theta = 6\pi$, not $\theta = 0$.
2. Angle (as measured in radians) is defined to be the ratio of a circular arc to its radius according to the equation θ = arc length/radius. This equation illustrates that θ is *dimensionless.* (However, angles are often tagged with the label "radians" simply as a reminder that they are angles.) The conversion between radians, degrees, and revolutions is

$$1 \text{ revolution} = 2\pi \text{ radians}^1 = 360 \text{ degrees}$$

Conversion between units for θ

3. Angular displacement, given by the equation

$$\Delta\theta = \theta_{\text{final}} - \theta_{\text{initial}}$$

Angular displacement

is analogous to linear displacement (e.g., if an object is rotated one revolution clockwise and one revolution counterclockwise, the net angular displacement is zero).

Angular Velocity and Acceleration

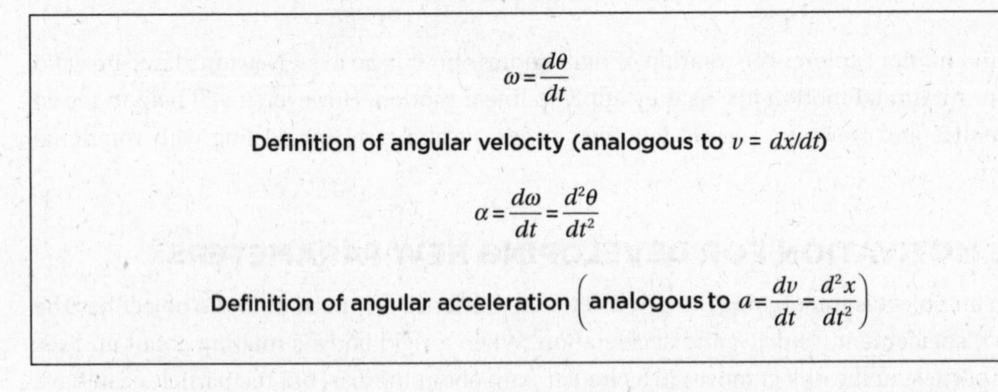

From the above equations we can see that angular velocity ω has units of s^{-1} and angular acceleration α has units of s^{-2}.

Sign Conventions Concerning Angular Position, Velocity, and Acceleration Because of the relationship $\alpha = d\omega/dt = d^2\theta/dt^2$, angular position ($\theta$), velocity ($\omega$), and acceleration ($\alpha$) must all follow the same sign convention.

Convention 1 (right-hand rule): When you point your right thumb along the axis of rotation such that your fingers curl in the direction of angular displacement, velocity, or acceleration, your right thumb indicates the vector *direction* of angular displacement, velocity, or acceleration (respectively). This provides a way of associating a vector with the sense of rotation of the object (i.e., clockwise or counterclockwise).

Convention 2 (special case of the right-hand rule in two dimensions): In two dimensions, counterclockwise rotation is generally taken to be positive, and clockwise to be negative (as in polar coordinates). This is simply a special case of the right-hand rule in two dimensions; from Figure 7.1, you can see that counterclockwise rotation in the *xy*-plane corresponds to an angular velocity in the $+\hat{k}$ direction, whereas clockwise rotation corresponds to an angular velocity in the $-\hat{k}$ direction.

[1]One revolution corresponds to a complete circle. Because the circumference of a circle is $2\pi r$, the number of radians in a circle is θ = arc length/radius $= 2\pi r/r = 2\pi$.

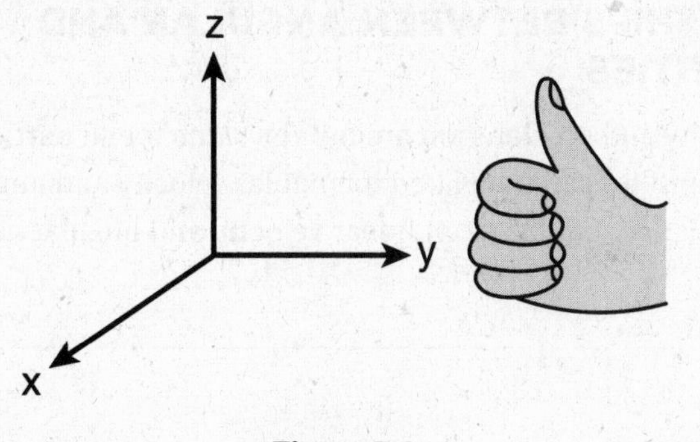

Figure 7.1

PROBLEM SOLVING AND DIRECTIONS

When solving problems, you can deviate from these conventions and simply designate a sense of rotation (e.g., counterclockwise or clockwise) as positive for convenience and keep track of signs that way. (This is the rotational equivalent of choosing convenient axes.)

FUNDAMENTAL EQUATIONS OF ANGULAR KINEMATICS

The definitions of angular position, angular velocity, and angular acceleration are mathematically analogous to the relationship among linear position, linear velocity, and linear acceleration. Thus, the same relationships apply between the angular quantities as between the linear quantities: The same proofs can be written, simply by replacing linear variables with their rotational relatives. We will not rewrite these proofs (if you wish to refresh your memory, see Chapter 2) but will rather simply list the important results.

Fortunately, the similarities between angular motion and one-dimensional motion are so strong that we can apply the problem-solving techniques of one-dimensional motion to angular motion.

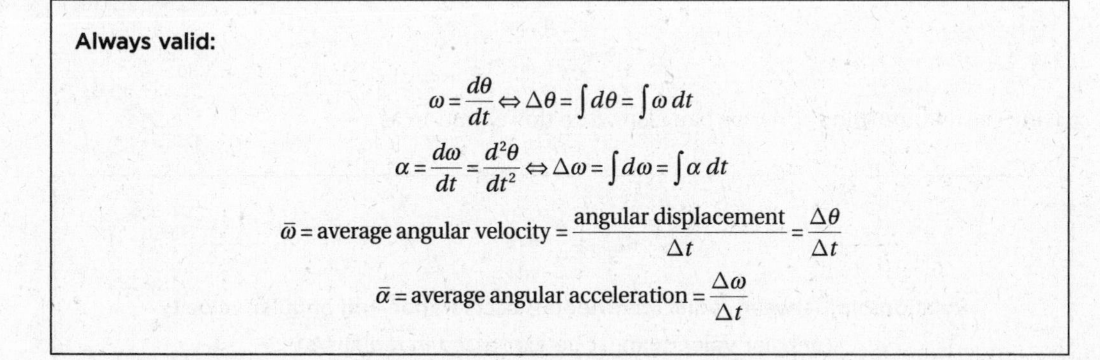

Always valid:

$$\omega = \frac{d\theta}{dt} \Leftrightarrow \Delta\theta = \int d\theta = \int \omega\, dt$$

$$\alpha = \frac{d\omega}{dt} = \frac{d^2\theta}{dt^2} \Leftrightarrow \Delta\omega = \int d\omega = \int \alpha\, dt$$

$$\bar{\omega} = \text{average angular velocity} = \frac{\text{angular displacement}}{\Delta t} = \frac{\Delta\theta}{\Delta t}$$

$$\bar{\alpha} = \text{average angular acceleration} = \frac{\Delta\omega}{\Delta t}$$

Valid only for uniformly accelerated motion (UAM):

$$\omega = \omega_0 + \alpha t$$

$$\theta = \theta_0 + \omega_0 t + \frac{1}{2}\alpha t^2$$

$$\omega^2 = \omega_0^2 + 2\alpha(\theta - \theta_0)$$

$$\bar{\omega} = \frac{\omega + \omega_0}{2}$$

TIP

These equations are analogous to those for translational UAM in Chapter 2.

THE RELATIONSHIPS BETWEEN ANGULAR AND LINEAR QUANTITIES

Although linear velocity and acceleration are not the same for all particles in a rigid rotating body, these linear quantities can be related to angular velocity and angular acceleration at a given radius r. The relationship between linear velocity and angular velocity was derived in Chapter 3:

$$v = r\omega$$

Relationship between speed and angular velocity (a scalar equation)
(angular speed must be expressed in radians/s)

Recalling that the tangential acceleration is the time derivative of the speed (discussed in Chapter 3), we can differentiate again and take the absolute value to relate the tangential linear acceleration to the rotational acceleration:

$$a_{\text{tan}} = \frac{dv}{dt} = \frac{d}{dt} r\omega = r\alpha$$

$$a_{\text{tan}} = r\alpha$$

Calculating the magnitude of tangential acceleration (a scalar equation)
(angular acceleration must be expressed in radians/s^2)

In Chapter 3 we derived the relationship among radial acceleration, velocity, and radius for an object undergoing *uniform* circular motion:

$$a_{\text{radial}} = \frac{v^2}{r}$$

Using the relationship $v = r\omega$, we can also write this equation as

$$a_{\text{radial}} = \frac{v^2}{r} = r\omega^2$$

Relationship between radial (centripetal) acceleration and angular velocity
(angular velocity must be expressed in radians/s)

ROTATIONAL KINETIC ENERGY

What is the kinetic energy of a rigid body due to its rotation? We cannot plug directly into the formula for the kinetic energy, $KE = \frac{1}{2}mv^2$, because different parts of the body are at different radii from the axis of rotation and thus are moving at different velocities (recall that $v = r\omega$). As with center of mass, this can be solved only by moving to a differential level. If we imagine dividing the object into infinitesimally small pieces, we can apply the KE equation to these pieces to obtain the total KE by summing the KE of each piece:

$$KE_{total} = \lim_{n \to \infty} \sum_{i=1}^{n} \frac{1}{2} m_i v_i^2$$

We recognize an integral (the mass of the pieces shrinks to zero and becomes a differential, whereas the velocity does not):

$$KE = \int \frac{1}{2} v^2 dm$$

This can be expressed in angular terms using the relationship $v = r\omega$:

$$KE = \int \frac{1}{2} r^2 \omega^2 dm$$

Putting the constants in front (all the differential masses share the same angular velocity),

$$KE = \frac{1}{2} \omega^2 \int r^2 dm$$

This equation is very similar to the equation for translational KE, $KE = \frac{1}{2}mv^2$. The linear squared velocity v^2 is analogous to the angular squared velocity ω^2. The integral, $\int r^2\, dm$, is the angular analog of mass. Just as mass is a measure of translational inertia, this integral is a measure of rotational inertia (often called the *moment of inertia*).

$$I = \int r^2 dm$$

Definition of rotational inertia

TIP

Rotational inertia measures a body's resistance to rotation.

The next chapter will describe how to evaluate this integral for a variety of simple objects. We can now rewrite the rotational kinetic energy:

$$KE_{rotational} = \frac{1}{2} I \omega^2$$

Rotational kinetic energy $\left(\text{analogous to } KE = \frac{1}{2}mv^2\right)$

The rotational KE is not a "new" type of kinetic energy: It is simply a useful way to sum the translational kinetic energies of the particles that make up a rotating object.

TORQUE

In words, *torque* is the ability of a force to cause an object to accelerate angularly (i.e., to rotate at a nonconstant angular velocity). Torque is the angular analog of force. Mathematically,

$$\tau = \mathbf{r} \times \mathbf{F}$$

$$\tau = rF \sin \theta$$

Definition of torque

Here **r** is defined as a relative position vector pointing from the axis of rotation to the point where the force is applied (which requires that torque be calculated about some specified axis). The angle θ is defined as the angle between the position and force vectors when they are placed tail to tail. It makes sense that this cross-product determines the rotation that the force imparts:

1. Clearly the rotation produced should be proportional to the magnitude of the force.
2. It makes sense that the torque is proportional to the magnitude of r. Forces acting at greater radial distances from the axis (greater "lever arms") exert more torque. If you don't believe this, try closing a door by pushing on it close to the hinges.
3. The torque also depends on the angle between **F** and **r**; torque is maximized when the force is perpendicular to **r** and zero when it is parallel or antiparallel (which also makes sense: Try closing a door by pushing or pulling the door away from its hinges).

The units of torque are force • length, generally newton meters (N•m).

Computing Torque in Problems

1. *Magnitude:* Recall from Chapter 1 that the magnitude of the cross-product is equal to the length of either vector multiplied by the projection of the other vector perpendicular to the first vector. In practice, it is often easiest to calculate the magnitude of the torque this way (designating $F_\perp$ the component of the force perpendicular to **r**, and $r_\perp$ the component of **r** perpendicular to **F**).

$$\tau = rF_\perp = r_\perp F$$

Alternative formula for computing the magnitude of torque

Based on the situation, it may be easier to compute $F_\perp$ and r or $r_\perp$ and F.

2. *Direction:* To be compatible with our definitions of angular position, velocity, and acceleration, the direction of torque is governed by the same right-hand rule convention. Point your right thumb along the axis of rotation such that your fingers curl in the direction of the rotation caused by the torque, and your right thumb will point in the direction of the torque. In two-dimensional situations, you can use the convention that torques that rotate the object in the counterclockwise direction are positive and clockwise torques are negative. If, in solving a problem, you deviate from the sign

conventions and define a certain direction as positive, you must be consistent and assign signs to torques according to the same rules.

3. *Superposition of torques and net torque:* In part as a consequence of the principle of superposition for forces (a body being acted on by several forces behaves as if one net force, equal to the vector sum of the individual forces, is acting on it), torques obey the same principle. That is, if an object is being acted on by several torques, it behaves as if it were being acted on by one net torque equal to the vector sum of each of the individual torques.

ROTATIONAL ANALOG OF NEWTON'S SECOND LAW

If you are getting used to the concept of rotational and translational analogs, you may be expecting a rotational analog of $\mathbf{F} = m\mathbf{a}$: Perhaps torque (the rotational analog of force) should be related in some way to rotational inertia (the rotational analog of mass) and angular acceleration.

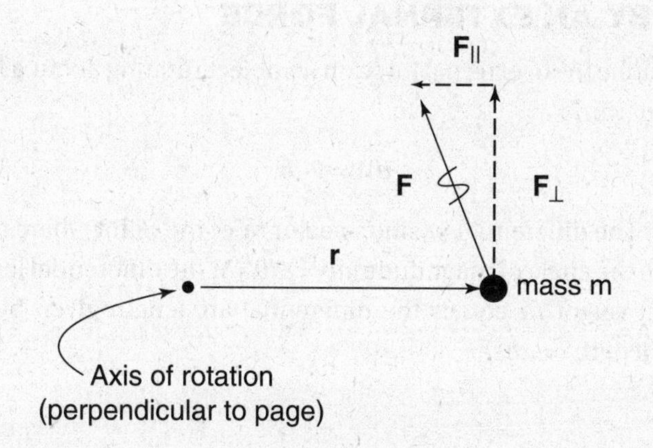

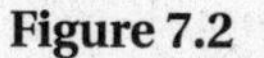

Figure 7.2

Consider Figure 7.2. Let's start by discussing the torque exerted by a force **F** on a point mass m about an arbitrary axis. Calculating the magnitude of the torque,

$$\tau = rF_{\perp} = rF_{\text{tangential}}$$

From above we know that $a_{\tan} = r\alpha$, such that (applying Newton's second law)

$$F_{\text{tangential}} = ma_{\tan} = mr\alpha$$

Substituting,

$$\tau = rF_{\text{tangential}} = rma_{\tan} = rmr\alpha = mr^2\alpha$$

We recognize rotational inertia $I = mr^2$, as we were expecting and obtain[2]

$$\tau = I\alpha$$

Because torque and angular acceleration follow the same sign conventions (the right-hand rule), their signs agree, so we can ignore the absolute value notation. Based on the principle of superposition for torques, if several torques are involved, they will act as if they exert a single torque equal to the net torque:

[2]Recall the definition of rotational inertia as $I = \int r^2 dm$. For a point particle, all the mass is the distance r from the axis of rotation, so that $I = \int r^2 dm = r^2 \int dm = mr^2$.

$$\tau_{net} = I\alpha$$

Newton's second law for rotational motion (analogous to F = *m*a)

We derived this equation for a point mass—is it also valid for extended objects? We can simply say that the equation is true for each particle in the extended object (on a differential level):

$$d\tau = \alpha dI$$

Then, by summing both sides (recall that the angular acceleration is constant over an extended body), we recover our initial statement on a macroscopic level:

$$\left[\int d\tau = \tau_{net}\right] = \left[\int \alpha dI = \alpha \int dI = \alpha I\right]$$

WORK DONE BY AN EXTERNAL FORCE

How much work is done by an external force on an object rotating about a fixed axis? We start with the definition of work:

$$dW = \mathbf{F} \cdot d\mathbf{r}$$

Consider Figure 7.3. The differential position vector $d\mathbf{r}$ of the point where the force is exerted is a vector tangent to the circle of magnitude $|d\mathbf{r}| = rd\theta$. (At the differential level, the magnitude of the displacement vector $d\mathbf{r}$ equals the differential arc length given by the equation arc length $= r\theta \Rightarrow d(\text{arc length}) = rd\theta$.)

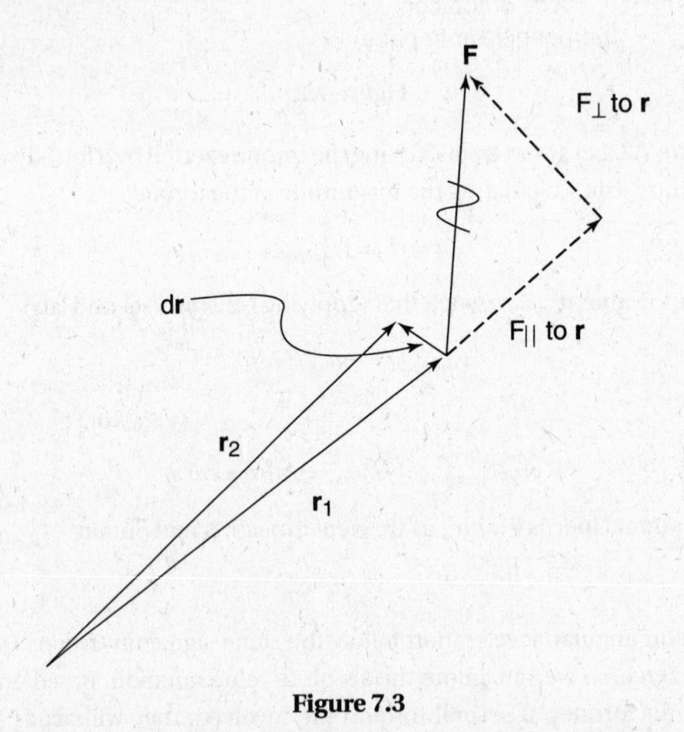

Figure 7.3

Recall that the dot product is equal to the magnitude of either of the vectors multiplied by the component of the second vector parallel to the first. Thus, we can evaluate the magnitude of the dot product $dW = \mathbf{F} \cdot d\mathbf{r}$ by multiplying the magnitude of the $d\mathbf{r}$ vector by the component of the force parallel to $d\mathbf{r}$ (in keeping with the notation used so far, we designate $F_\perp$ as the force perpendicular to the position vector **r**, *not* $d\mathbf{r}$. Recall that $d\mathbf{r}$ is perpendicular to **r**). Thus,

$$dW = \mathbf{F}\cdot d\mathbf{r} = (F_\perp)(r\,d\theta)$$

For the signs to work, $F_\perp$ must obey the same sign convention as other angular values (such that if $F_\perp$ is parallel to $d\theta$ the dot product is positive, whereas if $F_\perp$ is antiparallel to $d\theta$ the dot product is negative). We can then make the substitution $\tau = F_\perp r$:

$$dW = \tau\,d\theta \Leftrightarrow W = \int \tau\,d\theta$$

Calculation of work in rotational systems (analogous to $dW = \mathbf{F}\cdot d\mathbf{r}$)

The work is positive if the torque and angular velocity[3] both have the same sign causing the object to speed up, and negative if torque and angular velocity have opposite signs causing the object to slow down. This is the rotational equivalent of the sign possibilities for translational work ($W > 0$ if $\mathbf{F}_{net}$ has a component parallel $d\mathbf{r}$ causing the object to speed up, whereas $W < 0$ if $\mathbf{F}_{net}$ has a component in the direction opposite $d\mathbf{r}$ causing the object to slow down).

If the torque is constant, it can be put in front of the integral sign:

$$W = \int \tau\,d\theta = \tau \int d\theta = \tau\Delta\theta$$

Work done by a constant torque (analogous to $W = \mathbf{F}\cdot\Delta\mathbf{r}$ for constant forces)

Relationship Between Torque and Potential Energy When Energy Is Conserved

In the absence of nonconservative forces, energy is conserved:

$$\text{Total energy} = \text{KE} + U = \text{constant}$$

The differential form of this statement is

$$d(\text{total energy}) = 0 = d\text{KE} + dU$$

Recall that work is equal to the change in kinetic energy, $W = \Delta\text{KE} = \text{KE} - \text{KE}_0$, which in a differential form reads $dW = d\text{KE}$. Thus, the relationship between work and potential energy when the total energy remains constant is

$$dW = d\text{KE} = -dU$$

We have already related torque to work in the equation

$$dW = \tau\,d\theta$$

Using the above arguments, we can relate torque to potential energy as well:

$$\tau = \frac{dW}{d\theta} = -\frac{dU}{d\theta}$$

[3]Because $\omega = d\theta/dt$ and time is always increasing ($dt > 0$), such that ω has the same sign as $d\theta$.

$$\tau = -\frac{dU}{d\theta}$$

Relationship between torque and potential energy valid *only* if the torque is produced by conservative forces (analogous to $F_x = -dU/dx$, which is valid only if the force is due to conservative forces)

POWER ASSOCIATED WITH A TORQUE

Recall the definition of power: $P = dW/dt$. Inserting the expression for dW in a rotational system yields

$$P = \frac{dW}{dt} = \frac{\tau\, d\theta}{dt} = \tau\frac{d\theta}{dt} = \tau\omega$$

$$P = \tau\omega$$

Calculating power in rotational systems (analogous to $P = \mathbf{F}\cdot\mathbf{v}$)

Because we are assuming rotation about a fixed axis, the torque and the angular velocity lie along the same axis (though they may be antiparallel), and thus the expression for power does not include a dot product.

EQUATION SUMMARY WITH TRANSLATIONAL ANALOGS

At the beginning of this chapter we noted how the parallels between translational and rotational kinematics and dynamics may help you remember equations and solve problems. Here is a summary of the important equations presented in this chapter and their translational analogs (note that because we are assuming fixed-axis rotation, the rotational vector quantities have only a single component; were we to consider more general rotation, such as for a precessing gyroscope, three-dimensional vectors would be required):

Variables

	Rotational	Translational
Position	θ	x, y, z
Velocity	ω	$\mathbf{v}$
Acceleration	α	$\mathbf{a}$
Inertia	$I = \int r^2 dm$	Mass
"Force"	$\boldsymbol{\tau} = \mathbf{r} \times \mathbf{F}$	$\mathbf{F}$

Equations

Rotational	Translational
$\omega = \frac{d\theta}{dt} \Leftrightarrow \Delta\theta = \int d\theta = \int \omega\, dt$	$\mathbf{v} = \frac{d\mathbf{r}}{dt} \Leftrightarrow \Delta\mathbf{r} = \int d\mathbf{r} = \int \mathbf{v}\, dt$
$\alpha = \frac{d\omega}{dt} = \frac{d^2\theta}{dt^2} \Leftrightarrow \Delta\omega = \int d\omega = \int \alpha\, dt$	$\mathbf{a} = \frac{d\mathbf{v}}{dt} = \frac{d^2\mathbf{r}}{dt^2} \Leftrightarrow \Delta\mathbf{v} = \int d\mathbf{v} = \int \mathbf{a}\, dt$
$\bar{\omega} = \frac{\text{angular displacement}}{\Delta t} = \frac{\Delta\theta}{\Delta t}$	$\bar{\mathbf{v}} = \frac{\text{linear displacement}}{\Delta t} = \frac{\Delta\mathbf{r}}{\Delta t}$
$\bar{\alpha} = \frac{\Delta\omega}{\Delta t}$	$\bar{\mathbf{a}} = \frac{\Delta\mathbf{v}}{\Delta t}$
$\omega = \omega_0 + \alpha t$ (UAM)	$\mathbf{v} = \mathbf{v}_0 + \mathbf{a}t$ (UAM)
$\theta = \theta_0 + \omega_0 t + \frac{1}{2}\alpha t^2$ (UAM)	$\mathbf{r} = \mathbf{r}_0 + \mathbf{v}_0 t + \frac{1}{2}\mathbf{a}t^2$ (UAM)
$\omega^2 = \omega_0^2 + 2\alpha(\theta - \theta_0)$(UAM)	$v_f^2 = v_0^2 + 2a\,\Delta\, x$(UAM)
$\bar{\omega} = \frac{\omega_{\text{final}} + \omega_0}{2}$(UAM)	$\bar{\mathbf{v}} = \frac{\mathbf{v} + \mathbf{v}_0}{2}$(UAM)
$\text{KE}_{\text{rotational}} = \frac{1}{2}I\omega^2$	$\text{KE}_{\text{translational}} = \frac{1}{2}mv^2$
$\tau = I\alpha$	$\mathbf{F} = m\mathbf{a}$
$dW = \tau d\theta$	$dW = \mathbf{F} \cdot d\mathbf{r}$
$W = \tau\Delta\theta$ (constant torque)	$W = \mathbf{F} \cdot \Delta\mathbf{r}$ (constant force)
$\tau = -\frac{dU}{d\theta}$	$F_x = -\frac{dU}{dx}$
$P = \tau\omega$	$P = \mathbf{F} \cdot \mathbf{v}$

PROBLEM SOLVING

Pure Rotation

The similarities between rotational and translational motion continue into problem solving, so all our well-honed problem-solving techniques will carry over.

Combined Rotation and Translation

One very common AP problem involves calculating the acceleration of a system including both rotation (e.g., the rotation of a pulley) and translation (e.g., the vertical motion of masses connected by a string passing over the pulley). For each mass, there is generally one Newton's second-law-type equation that can be written. In the case of masses and a pulley, the masses can be described by the translational second law, while the rotational second law describes the pulley. The rotational motion can be related to the translational using the equation $a_{\tan} = r\alpha$, which is valid in the case of a pulley if the string does not slip over the pulley.

TIP

The tension in the string on either side of a pulley is *not* the same when we include the mass of the pulley.

Caution: When solving a pulley problem taking into account the rotation of the pulley (and thus its mass), the tension in the string is not the same on either side of the pulley as it was when we neglected the mass of the pulley. If it were, there would be no net torque on the pulley (the two forces of tension on either side of the pulley would exert equal and opposite torques) and the pulley would not rotate.

CHAPTER SUMMARY

The angular position, velocity, and acceleration of a rigid body rotating about a fixed axis can be defined in analogy to the translational quantities defined in Chapter 2. If the rotational acceleration is constant, these definitions yield equations for the angular position and velocity analogous to those in Chapter 2 for UAM. By examining the kinetic energy of a rotating object, the rotational analog of mass, the rotational inertia, can be defined. The rotational analogs of Newton's second law, work and power, all take the forms expected by straightforward analogies.

PRACTICE EXERCISES

Multiple-Choice Questions

1. The constant torque required to accelerate a disk (of rotational inertia I) from rest to a speed of f rotations per second in a time interval of t is

   (A) $2\pi ft/I$
   (B) $2\pi I/ft$
   (C) $fI/2\pi t$
   (D) $2\pi fI/t$
   (E) $2\pi f/t$

2. Figure 7.4 shows five graphs of angular velocity as a function of time $\omega(t)$. In which graph does the particle have the greatest angular displacement?

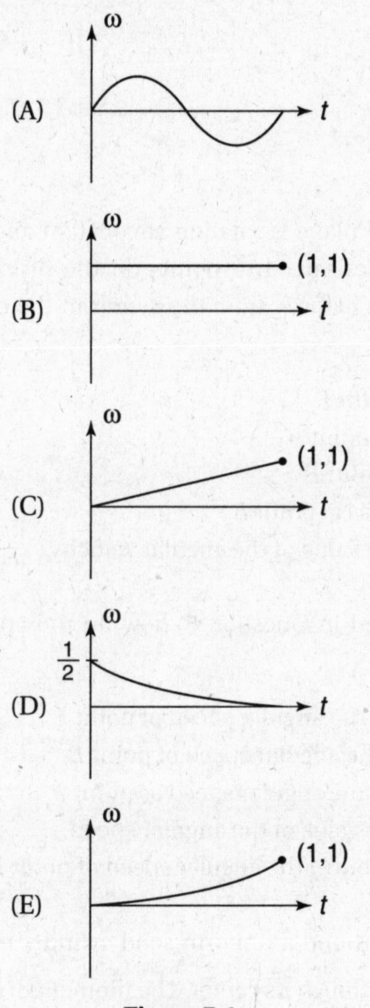

Figure 7.4

3. Of the graphs introduced in question 2, which has the greatest average angular acceleration $\bar{\alpha}$?

(A) graph *A*
(B) graph *B*
(C) graph *C*
(D) graph *E*
(E) graphs *C* and *E*

4. Of the graphs introduced in question 2, which has the smallest average angular velocity?

(A) graph *A*
(B) graph *B*
(C) graph *C*
(D) graph *D*
(E) graphs *A* and *B*

5. Imagine a truck driving on the XY plane in the $+y$-direction. If it is slowing down, in what direction is the angular acceleration of its wheels?

(A) the $+x$-direction
(B) the $-x$-direction
(C) the $+y$-direction
(D) the $-y$-direction
(E) the $+z$-direction

6. A disk in a horizontal plane is rotating at constant angular speed about a vertical axis through its center. Consider two points on the disk: point A, which is at the edge, and point B, which is halfway from the center to the edge. What is the linear speed of point A?

(A) the same as at point B
(B) twice as big as at point B
(C) half as big as at point B
(D) four times as big as at point B
(E) It depends on the value of the angular velocity.

7. For the disk described in Question 6, how do the angular speeds of points A and B compare?

(A) Point A has twice the angular speed of point B.
(B) Point A has half the angular speed of point B.
(C) Point A has the same angular speed as point B.
(D) It depends on the value of the angular speed.
(E) Point A has one-fourth the angular speed of point B.

8. A cable is wrapped around a uniform solid cylinder of mass M and radius R that can rotate about an axis through its center. The moment of inertia of the cylinder is $\frac{1}{2}MR^2$. If the cable is pulled with a force F and if the direction of F is tangent to the cylinder, what is the linear acceleration of a point on the cable?

(A) F/M
(B) $2F/M$
(C) FR/M
(D) $2FR/M$
(E) M/F

9. Two forces, F_1 and F_2, are applied to the bar shown in Figure 7.5. What is the net torque about the left end of the bar?

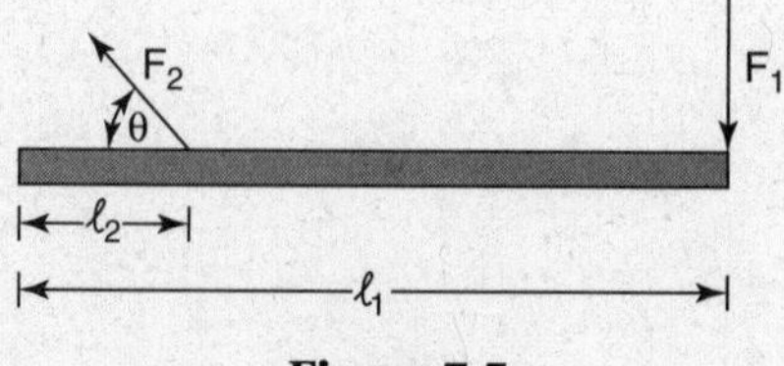

Figure 7.5

(A) $F_2 - F_1$
(B) $F_1 - F_2$
(C) $F_2\ell_2 - F_1\ell_1$
(D) $F_2\ell_2 \sin\theta - F_1\ell_1$
(E) $F_1\ell_1 - F_2\ell_2 \sin\theta$

10. The rotor of a centrifuge (originally at rest) is accelerated with an angular acceleration α for a time interval Δt_1, left at constant speed for a time interval Δt_2, and then slowed down at a constant angular acceleration $-\alpha$ until it stops rotating. What is the total angular displacement during this process?

(A) $\Delta\theta = \frac{1}{2}\alpha(\Delta t_1)^2 + \alpha\Delta t_1\,\Delta t_2 + \frac{1}{2}\alpha(\Delta t_2)^2$

(B) $\Delta\theta = \frac{1}{2}\alpha(\Delta t_1)^2 + \alpha\Delta t_1\,\Delta t_2$

(C) $\Delta\theta = \alpha(\Delta t_1)^2 + \alpha\Delta t_1\Delta t_2$

(D) $\Delta\theta = \frac{1}{2}\alpha(\Delta t_1)^2 + 2\alpha\Delta t_1\,\Delta t_2$

(E) zero

Free-Response Questions

1. A rod of uniform mass density, length L, rotational inertia $\frac{1}{3}mL^2$ (about the axle shown), and mass m is pivoted without friction about an axle as shown in Figure 7.6. The rod is allowed to fall from a horizontal to a vertical position under the influence of gravity.

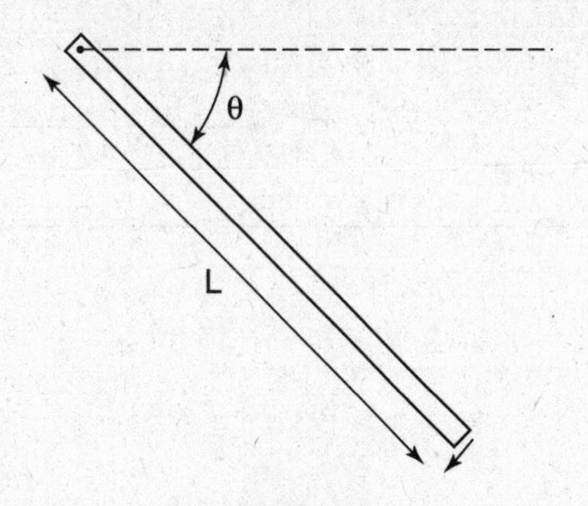

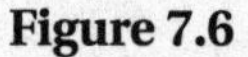
Figure 7.6

(a) Express the differential work performed by gravity in terms of m, g, and θ.
  (i) Use the formula $dW = \mathbf{F} \cdot d\mathbf{r}$.
  (ii) Use the formula $dW = \tau d\theta$.
(b) Integrate your expression for dW to obtain the total work done by gravity while the object swings from the horizontal to the vertical position.
(c) Demonstrate that energy has been conserved.
(d) What is the angular velocity of the rod when it is vertical?

2. The pulley shown in Figure 7.7 is a uniform cylinder of radius R with a moment of inertia $I = \frac{1}{2}MR^2$, and the incline is frictionless. Assume that the rope rolls without slipping along the pulley connecting the two masses. Express your answers to the following questions in terms of m_1, m_2, M, R, and g.

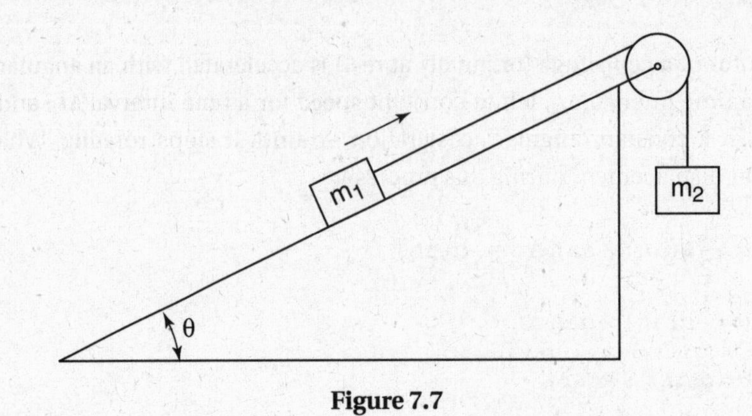

Figure 7.7

(a) Draw free-body diagrams for the two masses and the pulley.
(b) Write Newton's second-law equations for each mass and the pulley.
(c) Calculate the acceleration of m_2.

3. A toy boat m_1 is attached to a hanging mass m_3 by a cylindrical massless pulley (radius r, rotational inertia $\frac{1}{2}m_2r^2$) as in Figure 7.8. The boat experiences a frictional force $F_{fr} = -bv$ (where b is a constant with units of kg/s).

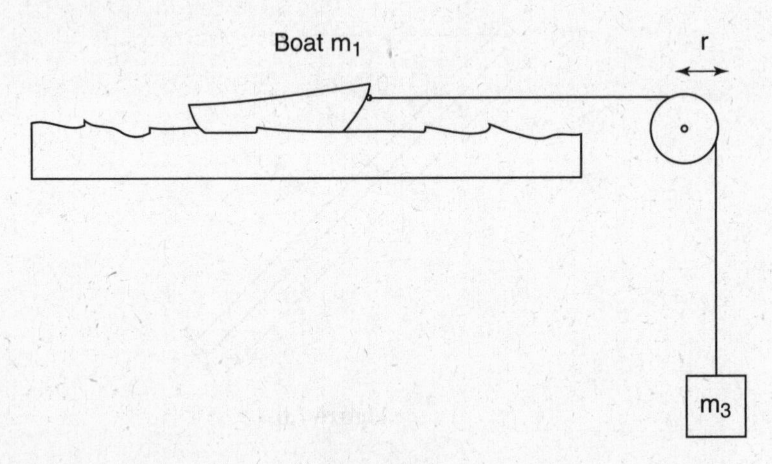

Figure 7.8

(a) Write down Newton's second-law equations for each mass (setting the direction of motion in the positive direction).
(b) Solve these equations for the linear acceleration of the system as a function of velocity.

4. A rod of mass m, length l, and uniform density is pivoted about its right end as shown in Figure 7.9. The rod is released from rest in the horizontal position and allowed to swing downward under the influence of gravity. The rotational inertia of the rod about its end is $\frac{1}{3}ml^2$. The instant after the rod is released,

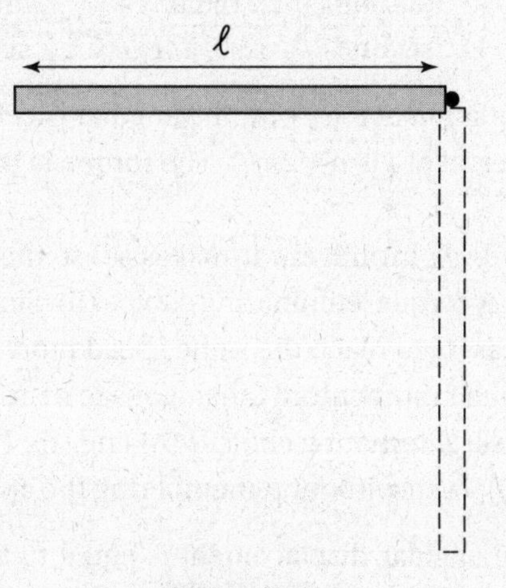

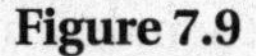
Figure 7.9

(a) What is the net torque on the rod about its pivot? (Note that in terms of torque, the gravitational force acts as if it is being exerted at the center of mass of an object.)

(b) What is the acceleration of the rod's center of mass?

(c) What is the force exerted by the pivot on the rod?

The instant that the rod is vertical,

(d) What is the net torque on the rod about its pivot?

(e) What is the acceleration of the rod's center of mass?

ANSWER KEY

1. **D**	4. **A**	7. **C**	10. **C**
2. **B**	5. **A**	8. **B**	
3. **E**	6. **B**	9. **E**	

ANSWERS EXPLAINED

Multiple-Choice

1. **(D)** The final angular speed is

$$\left(f\frac{\text{rotations}}{\text{second}}\right)\left(\frac{2\pi\ \text{radians}}{\text{rotation}}\right)=2\pi f\frac{\text{radians}}{\text{second}}$$

According to the equation $\omega_f = \omega_0 + \alpha t$, the angular acceleration required to achieve this speed in a time interval of t is $\alpha = 2\pi f/t$. The torque is then given by the equation $\tau = I\alpha$ to be $\tau = 2\pi f I/t$.

Alternative approach: Intuitively, it makes sense that increasing the final speed increases the necessary torque, eliminating choice (B). Similarly, if t is decreased, a greater torque will be necessary to reach the same speed more quickly, eliminating choice (A). The greater the angular momentum (analogous to mass), the greater the torque needed to accelerate the disk, eliminating choices (A) and (E). This leaves us with a 50-50 guess between (C) and (D), even without remembering the appropriate formulas.

2. **(B)** $\Delta\theta = \int \omega dt$. The angular displacement is equal to the time integral of the angular velocity, which is equal to the area under the $\omega(t)$ curve. Graph (B) has the greatest enclosed area, so it has the greatest angular displacement.

3. **(E)** The average angular acceleration is given by the equation $\bar{\alpha} = (\omega_f - \omega_0)/\Delta t$. Graphs C and E tie for the largest net increase in the angular velocity, $\omega_f - \omega_0$, so they have the greatest average angular acceleration.

4. **(A)** $\bar{\omega} = \Delta\theta/\Delta t = \int \omega\, dt/\Delta t$. Therefore, the graph with the smallest area under the curve has the smallest average angular velocity. Graph A has zero area under its curve (positive area under the first half of the graph cancels negative area under the second half of the graph), so it has the smallest average angular velocity, namely, zero.

5. **(A)** As shown in Figure 7.10, when the truck is moving in the $+y$-direction, its wheels have an angular velocity pointing in the $-x$-direction. An angular acceleration that decreases this angular velocity must point in the opposite direction: in the $+x$-direction.

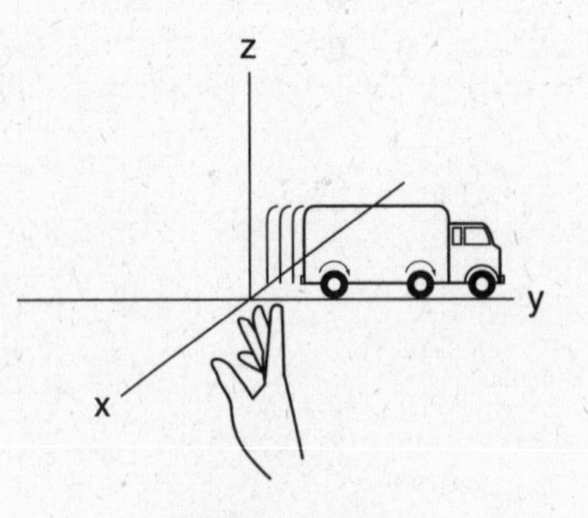

Figure 7.10

6. **(B)** The linear speed v of a point on the disk is related to the angular speed ω by $v = r\omega$, where r is the distance of the point from the rotation axis. Since A is twice the distance from the axis compared to B, the linear speed of A is twice as big as that of B.

7. **(C)** All points on a rigid rotating body have the same angular speed and angular velocity.

8. **(B)** The torque exerted by F is $\tau = RF$. The angular acceleration of the cylinder is then $\alpha = \frac{\tau}{I} = \frac{2RF}{MR^2} = \frac{2F}{MR}$, and the linear acceleration of a point on the rope (which is the same as a point on the surface of the cylinder) is $a = R\alpha = 2F/M$.

9. **(E)** Force F_1 is perpendicular to the rod. So the torque exerted by F_1 has magnitude $F_1\ell_1$ and acts to cause a counterclockwise rotation; hence, it's a positive torque in our convention. The torque exerted by F_2 has magnitude $F_2\ell_2 \sin\theta$ and acts to cause a clockwise rotation; hence, it's a negative torque.

10. **(C)** Figure 7.11 provides a sketch of the angular velocity as a function of time. The total angular displacement is equal to the *area* enclosed by this curve. What this graph reveals is that the displacement while the rotor is accelerating is equal to the displacement while the rotor is decelerating because the magnitudes of the acceleration and deceleration are equal. (Geometrically, the areas of the two triangles are equal.)

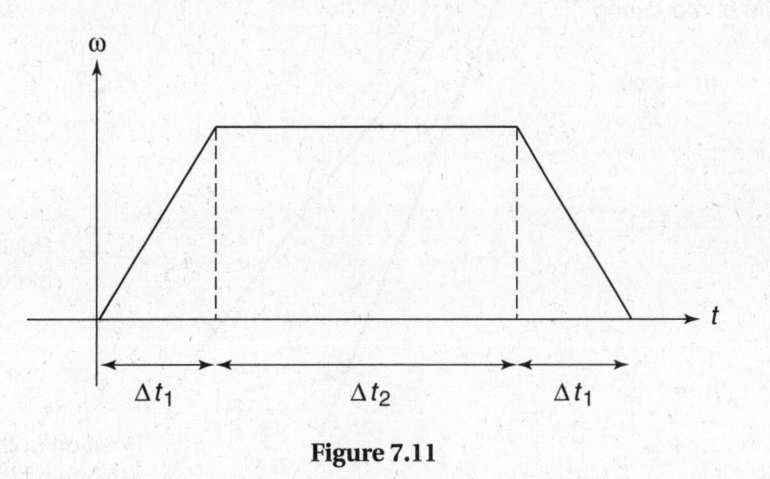

Figure 7.11

The displacement while the rotor is accelerating is given by the equation $\Delta\theta = \theta_f - \theta_0 = \omega_0\Delta t + \frac{1}{2}\alpha(\Delta t)^2 = \frac{1}{2}\alpha(\Delta t_1)^2$. The constant velocity of the intermediate interval is given by $\omega = \omega_0 + \alpha t = \alpha\Delta t_1$, so the angular displacement during this interval (with zero angular acceleration) is $\Delta\theta = \theta_f - \theta_0 = \omega_0\Delta t + \frac{1}{2}\alpha(\Delta t)^2 = (\alpha\Delta t_1)\Delta t_2 + 0$. Finally, summing twice the angular displacement while the rotor is accelerating (we multiply by 2 to account for the deceleration time interval) plus the angular displacement while the speed is constant yields choice (C).

Free-Response

1. (a) Consider Figure 7.12. Note that the gravitational force acts as if it were exerted at the center of mass of the rod (in this case, the middle of the rod).

(i) In order to express $d\mathbf{r}$ in terms of θ, we note that $d\mathbf{r}$ is perpendicular to the rod, such that for very small changes in θ it is approximately equal to the arc length swept out by an arc of radius $L/2$ and angle $d\theta$.

$$dW = F_{\parallel}dr = mg(\cos\theta)\left(\frac{L}{2}d\theta\right)$$

Because the gravitational force has a component in the same direction as the displacement, this work is positive.

$$dW = \frac{1}{2}mgL\cos\theta d\theta$$

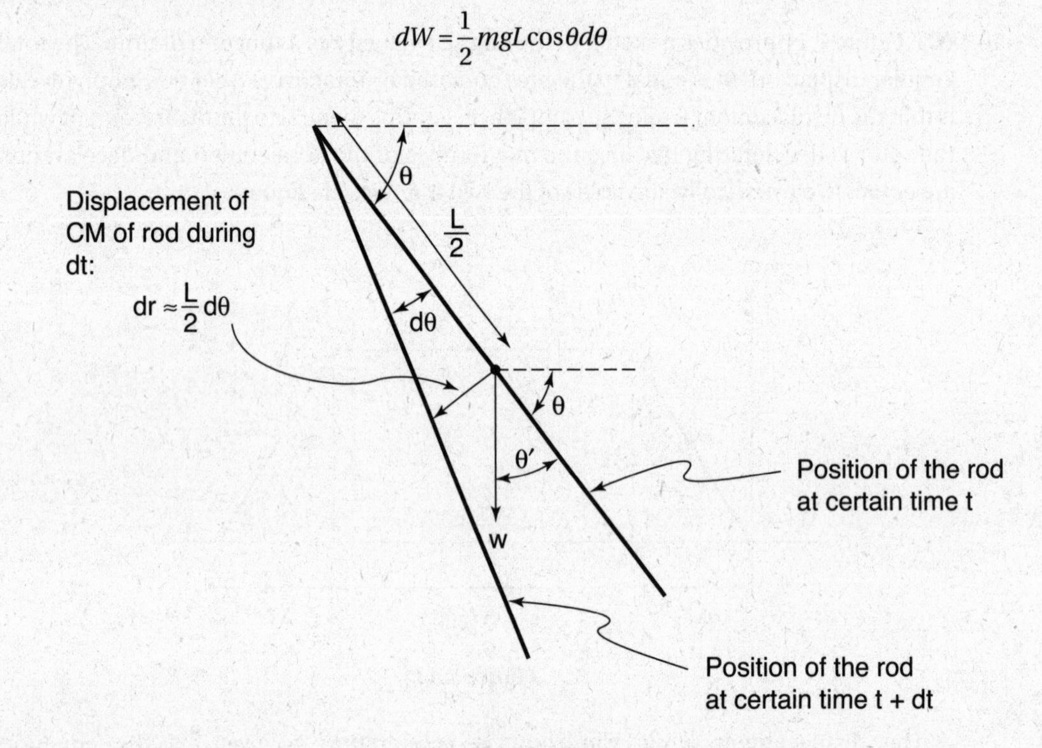

Figure 7.12

(ii)

$$|\tau| = |F_{\perp}||r| = (mg\cos\theta)\left(\frac{L}{2}\right)$$

$$|dW| = \tau d\theta = (mg\cos\theta)\left(\frac{L}{2}\right)d\theta$$

Again, qualitatively we can determine that this work is positive, yielding the same answer as in part (i).

(b) $W = \int dW = \int_0^{\pi/2} \frac{1}{2}mgL\cos\theta d\theta = \frac{1}{2}mgL$

(c) The change in potential energy is $\Delta U = mg\,\Delta h = -\frac{1}{2}mgL$. According to the work-energy theorem, the change in the kinetic energy is $\Delta KE = work_{net} = \frac{1}{2}mgL$. Since the change in the potential energy is the negative of the change in the kinetic energy, the total mechanical energy (which is the sum of the kinetic and potential energies) is constant.

(d) Using conservation of energy (gravitational PE has been converted to rotational KE),

$$\frac{1}{2}mgL = \frac{1}{2}\left(\frac{mL^2}{3}\right)\omega^2$$

Solving for the angular velocity yields

$$\omega = \sqrt{\frac{3g}{L}}$$

2. (a) The free-body diagrams are shown below in Figure 7.13.

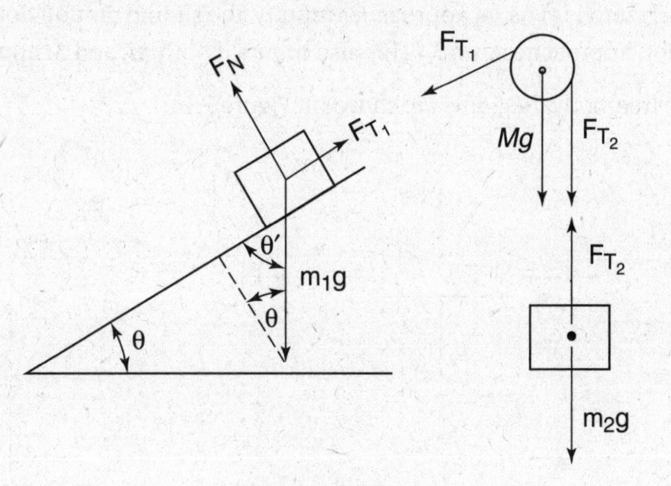

Figure 7.13

(b) For m_1 we have (using a tilted coordinate system with $+x$ pointing up the incline)

$$F_{net,x,M1} = m_1 a = T_1 - m_1 g\sin\theta$$

Note that for the torque equation for the pulley, we let the direction of net torque (clockwise) be positive, consistent with the positive, upward motion of m_1. The Newton's second-law equation for the pulley is then

$$\tau_{net} = I\alpha = \left(\frac{1}{2}MR^2\right)\alpha = T_2R - T_1R$$

For m_2, Newton's second law (with the positive direction chosen down) reads

$$F_{net,y,M2} = m_2 a = m_2 g - T_2$$

(c) We currently have three equations with four unknowns (a, T_1, T_2, and α). The fourth equation we need comes from relating linear and angular acceleration due to the rolling without slipping condition (noting that both are positive such that their signs agree).

$$a = R\alpha$$

Now we have four equations in four variables. Solving this system of equations is straightforward (although a little tedious) by repeated substitution to eliminate unwanted variables. The final result is

$$a = \frac{2(m_2 - m_1 \sin\theta)}{M + 2m_2 + 2m_1} g$$

Symbolic answer check: There are many limiting conditions that can be used to check the answer. (1) The acceleration increases if g increases. (2) As M approaches infinity, the pulley becomes impossible to turn and the acceleration approaches zero. (This also occurs if both m_2 and m_1 approach zero.) (3) As m_2 approaches infinity, the free-fall motion of m_2 dominates and the acceleration approaches g. (This also occurs if both m_1 and M approach zero.) (4) As m_1 approaches infinity, the sliding motion dominates and the acceleration approaches $g\sin\theta$. (This also occurs if both m_2 and M approach zero.)

3. (a) The free-body diagrams are shown in Figure 7.14.

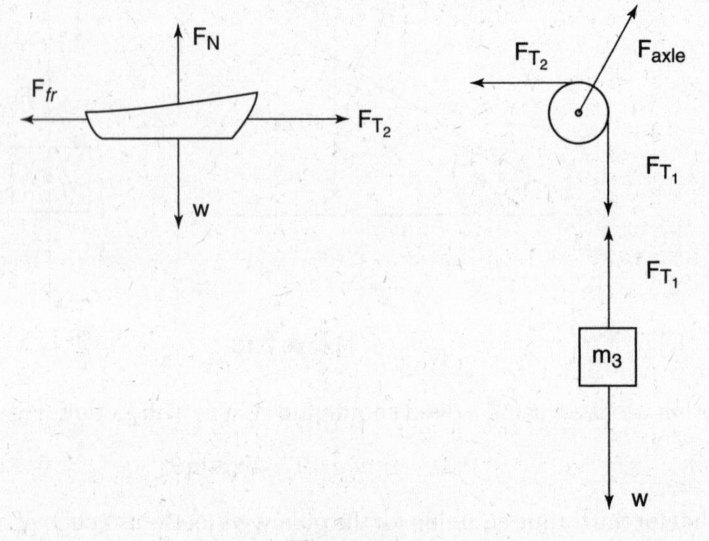

Figure 7.14

Mass 3: $m_3 g - T_1 = m_3 a$

Mass 1: $T_2 - bv = m_1 a$

Mass 2 (quickly eliminating the rotational acceleration using the substitution $\alpha = a/r$):

$$\tau_{net} = T_1 r - T_2 r = I\alpha = \left(\frac{1}{2} m_2 r^2\right)\left(\frac{a}{r}\right)$$

which simplifies to

$$T_1 - T_2 = \frac{1}{2} m_2 a$$

(b) Solving three equations in three variables is tedious but straightforward. (One quick way to solve these equation is to solve the net force equations for mass 1 and 3 for T_1 and T_2 (respectively), substitute into the equation for mass 2, and solve for the acceleration.)

$$a=\frac{m_3 g-bv}{m_1+\frac{m_2}{2}+m_3}$$

4. (a) The only force exerting a torque on the rod is the gravitational force. The gravitational force acts as if it is being exerted at the rod's center of mass, which is a distance $l/2$ from the pivot. Because in the horizontal position the gravitational force is perpendicular to the radial vector from the pivot to the rod's center of mass, the magnitude of the torque is $\tau = rF_\perp = (l/2)mg$. The torque causes the rod to rotate counterclockwise.

   (b) The angular acceleration of the rod is

$$\alpha=\frac{\tau}{I}=\frac{\frac{l}{2}mg}{\frac{1}{3}ml^2}=\frac{3g}{2l}$$

   The acceleration of the center of mass is equal to $a = r\alpha = (l/2)(3g/2l) = 3g/4$.

   (c) Recall that the net force is equal to the mass multiplied by the acceleration of the center of mass. Because the center of mass is accelerating downward with an acceleration *less* than g, the pivot must exert an upward force. Designating downward as positive,

$$\left[F_{\text{net}}=ma=\frac{3mg}{4}\right]=mg-F_{\text{pivot}}$$

   Therefore, the pivot must exert a force of $mg/4$ upward.

   (d) The net torque is zero because the gravitational force is parallel to $\vec{r}$ when the rod is vertical.

   (e) At the instant the rod becomes vertical, its angular speed is $\omega=\sqrt{3g/l}$ and not changing. Thus, the center of mass does not experience any tangential acceleration. However, the center of mass must experience a centripetal acceleration of $a=v^2/r=\omega^2 r=\frac{3}{2}g$ upward for it to remain in circular motion.

Rotation II: Inertia, Equilibrium, and Combined Rotation/Translation

8

→ CALCULATION OF MOMENT OF INERTIA FOR COMPLEX OBJECTS
→ PARALLEL AXIS THEOREM
→ ANGULAR MOMENTUM
→ NEWTON'S SECOND LAW FOR ROTATIONAL MOTION IN TERMS OF ANGULAR MOMENTUM
→ CONSERVATION OF ANGULAR MOMENTUM
→ ROLLING WITHOUT SLIPPING
→ STATIC EQUILIBRIUM FOR EXTENDED OBJECTS

Recall the definition of rotational inertia (where r is the distance between each differential mass and the axis of rotation).

$$I = \int r^2 dm$$

Definition of rotational inertia

Calculations of the rotational inertia for point masses, rings rotating about their axes, and cylindrical shells rotating about their axes (as shown in Figure 8.1) are straightforward because all the differential masses share a common r.

$$I = \int r^2 dm = r^2 \int dm = r^2 M$$

Rotational inertia for point masses, rings, and cylindrical shells rotating about their axes

What about more general objects where every differential mass does not share a common distance from the axis of rotation? In this more general case, we must proceed as we did with center-of-mass calculations for continuous masses: Divide the object into differential masses, calculate the rotational inertia of each differential mass, and sum these contributions to the total rotational inertia in an integral.

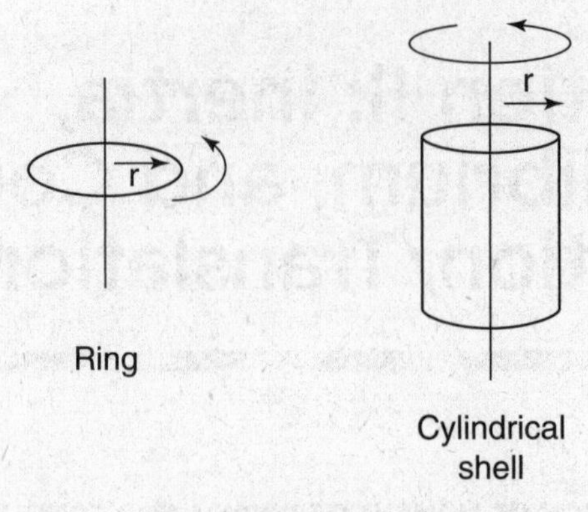

Figure 8.1

For one-dimensional objects, the differential regions consist of points (differential line segments). While it would also be perfectly valid to divide two- or three-dimensional objects into pointlike differential areas, summing these points would require double or triple integrals. As in center-of-mass calculations, to avoid the use of multiple integrals, we can divide two- or three-dimensional mass distributions into carefully chosen two- or three-dimensional differential regions. We must choose these two- or three-dimensional differential regions such that every particle of mass in a given differential region shares a common radius from the axis of rotation, allowing us to calculate the rotational inertia of the differential regions using the equation $dI = r^2 dm$.

Stepwise Procedure (clarified in the examples below)

1. Determine the geometry of the differential regions you will use. Your choice of differential region will be guided by the requirement that every particle of mass inside the differential region must share a common distance to the axis of rotation.
   Figure 8.2 illustrates the common differential regions described below.
   a. One-dimensional mass ⇒ differential line segments
   b. Two-dimensional mass rotating about an axis perpendicular to its plane ⇒ *rings*
   c. Two-dimensional mass rotating about an axis contained in its plane ⇒ *rectangular strips*
   d. Three-dimensional mass ⇒ *cylindrical shells* whose axes lies along the axis of rotation

   Note: If the object is already shaped like one of these regions, that is, if all its particles already share a common distance from the axis of rotation, you can immediately apply the macroscopic equation $I = Mr^2$ to obtain the rotational inertia.
2. Choose a variable in terms of which dm and r are defined. This is generally the variable whose differential is one of the dimensions of the differential region.
3. Express the differential length, area, or volume of the differential region in terms of the chosen variable, its differential, and various constants. The following equations may be helpful:
   a. Area of a ring = $2\pi r\, dr$. (Imagine unwrapping an infinitesimally thin ring; you will be left with a very skinny rectangle of dimensions $2\pi r \times dr$.)

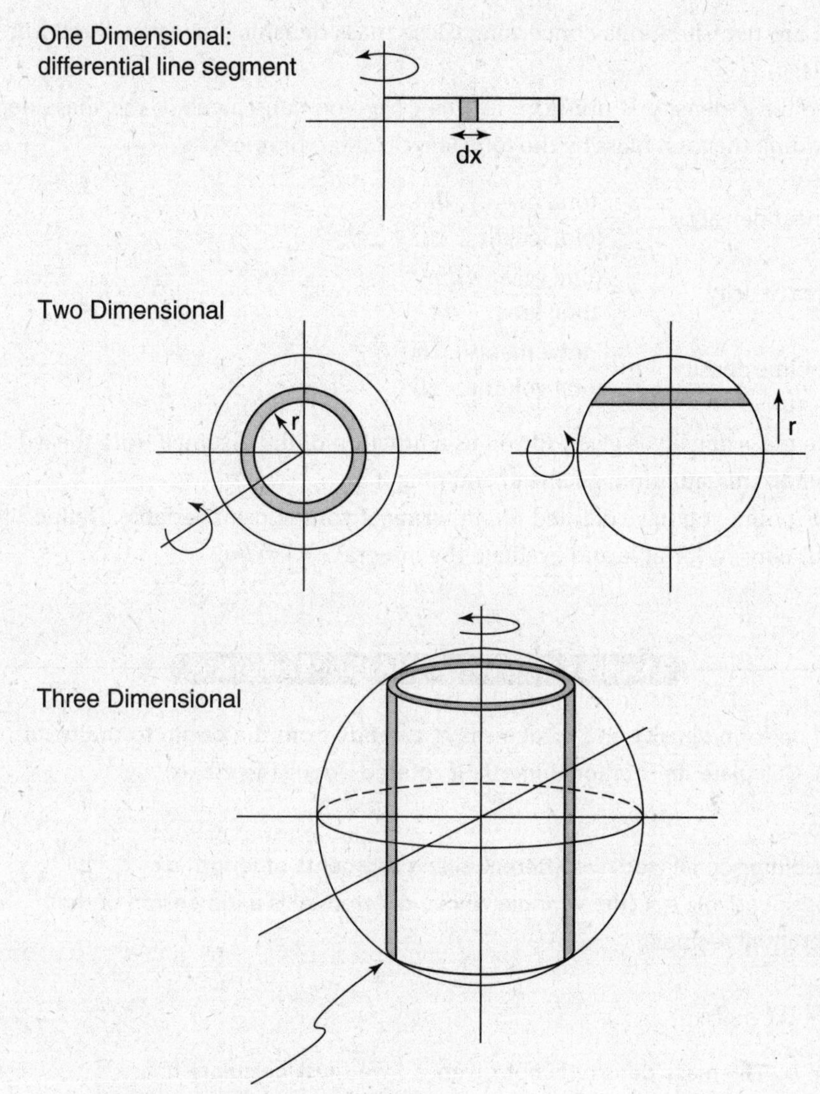

Figure 8.2

b. Volume of a cylindrical shell = $(2\pi r\ dr)h$. (Imagine unwrapping an infinitesimally thin cylindrical shell; you will be left with a very thin rectangular solid of dimensions $2\pi r \times dr \times h$.)

4. Multiply the differential length, area, or volume of the differential mass by a density factor to obtain the differential mass. As with center of mass, linear (λ), area (σ), and volume (ρ) densities are defined as follows.

Linear density: $\lambda = \dfrac{dm}{d\ell} \Rightarrow dm = \lambda d\ (\text{length})$

Area density: $\sigma = \dfrac{dm}{dA} \Rightarrow dm = \sigma d\ (\text{area})$

Volume density: $\rho = \dfrac{dm}{dV} \Rightarrow dm = \rho d\ (\text{volume})$

There are two situations concerning these mass densities that you should be able to handle:

a. The mass density is uniform. In this case, you can calculate the mass density by dividing the total mass by the total length/area/volume:

Linear density: $\lambda = \dfrac{\text{total mass}}{\text{total length}} = \dfrac{dm}{d\ell}$

Area density: $\sigma = \dfrac{\text{total mass}}{\text{total area}} = \dfrac{dm}{dA}$

Volume density: $\rho = \dfrac{\text{total mass}}{\text{total volume}} = \dfrac{dm}{dV}$

b. The mass density is given to you as a function of the distance from the axis of rotation. In this situation, go ahead and plug it in.

5. At this point, you have defined dm in terms of your chosen variable. Define r in terms of your chosen variable and evaluate the integral $I = \int r^2 dm$.

EXAMPLE 8.1 A LINEAR MASS DISTRIBUTION

A rod of uniform density and total mass *M* extends from the origin to the point (*L*, 0, 0). Calculate its rotational inertia if rotated about the *z*-axis.

SOLUTION

1. **One-dimensional (rod) ⇒ differential line segments of length *dx*.**
2. **Chosen variable = *x* (the variable whose differential is a dimension of the differential regions).**
3. $dl = dx$
4. $dm = \lambda dl = \lambda dx$

Situation b: The mass density is not given, so we must calculate it:

$$\lambda = \frac{\text{total mass}}{\text{total length}} = \frac{M}{L} \Rightarrow dm = \frac{M}{L}dx$$

5. $I = \int r^2 dm = \int_{x=0}^{x=L} (x^2)\left(\frac{M}{L}dx\right) = \frac{ML^2}{3}$

EXAMPLE 8.2 A CIRCULAR MASS DISTRIBUTION ROTATING ABOUT AN AXIS PERPENDICULAR TO ITS PLANE

A planar circular mass distribution of radius R is rotating about an axis perpendicular to its plane passing through its center. The density is given by the equation $\sigma = ar^2$, where a is a constant with units of kg/m^4. Calculate the rotational inertia.

SOLUTION

1. This is a two-dimensional mass rotating about an axis perpendicular to its plane, so we use rings of radius r and thickness dr.
2. Choose the variable r (the variable whose differential is a dimension of the differential region).
3. The area of the ring is $dA = 2\pi r\,dr$.
4. We are given the mass density function, so we simply insert it into the equation:

$$dm = \sigma dA = (ar^2)(2\pi r dr) = 2\pi a r^3 dr$$

5. $I = \int r^2 dm = \int_{r=0}^{r=R} r^2 (2\pi a r^3\, dr) = \frac{\pi a R^6}{3}$

EXAMPLE 8.3 A CIRCULAR MASS DISTRIBUTION ROTATING ABOUT AN AXIS CONTAINED IN ITS PLANE

A planar circular mass distribution of radius R centered at the origin (and lying in the xy-plane) is rotating about the x-axis. The object's mass M is distributed uniformly. Calculate the rotational inertia.

SOLUTION

1. This is a two-dimensional mass rotating about an axis contained within its plane, so we use rectangular strips as shown in Figure 8.3.

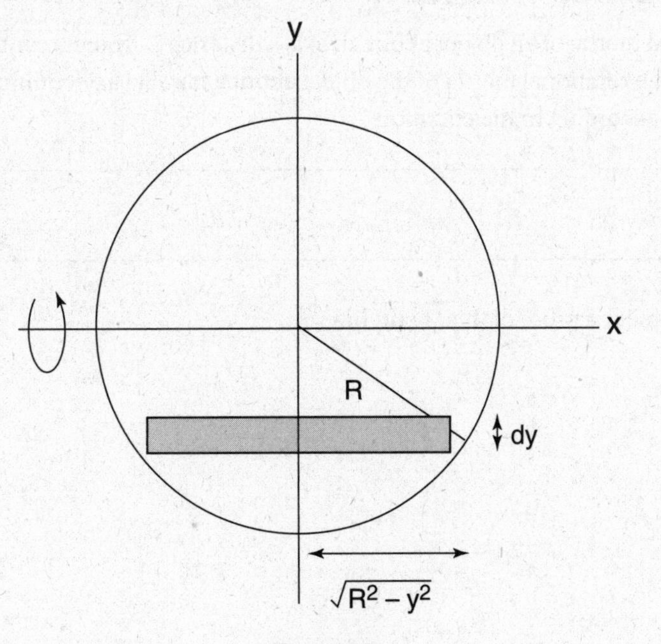

Figure 8.3

2. Chosen variable = y (the variable whose differential is a dimension of the differential region).
3. As shown above, $dA = 2\sqrt{R^2 - y^2}dy$.
4. Because the density is not given, we must calculate area mass density σ = total mass/total area = $M/\pi R^2$:

$$dm = \sigma dA = \left(\frac{M}{\pi R^2}\right)\left(2\sqrt{R^2 - y^2}dy\right) = \frac{2M}{\pi R^2}\sqrt{R^2 - y^2}dy$$

5. $I = \int r^2\, dm = \int_{y=-R}^{y=R} (y^2)\left(\frac{2M}{\pi R^2}\sqrt{R^2 - y^2}dy\right) = \frac{2M}{\pi R^2}\int_{y=-R}^{y=R} y^2\sqrt{R^2 - y^2}\, dy$

The easiest way to solve this integral is to perform a trigonometric substitution using

$$\begin{cases} y = R\sin\theta \\ dy = R\cos\theta\, d\theta \end{cases}$$

This yields

$$I = \frac{2M}{\pi R^2}\int_{-\pi/2}^{\pi/2} R^4 \sin^2\theta \cos^2\theta d\theta$$

Using the identity $\sin 2\theta = 2\sin\theta\cos\theta$,

$$I = \frac{MR^2}{2\pi}\int_{-\pi/2}^{\pi/2} \sin^2 2\theta d\theta$$

Using the identity $\sin^2 x = (1 - \cos 2x)/2$,

$$I = \frac{MR^2}{2\pi}\int_{-\pi/2}^{\pi/2} \frac{1 - \cos 4\theta}{2} d\theta = \frac{MR^2}{4\pi}\left[\theta - \frac{\sin 4\theta}{4}\right]_{-\pi/2}^{\pi/2} = \frac{MR^2}{4}$$

PARALLEL AXIS THEOREM

TIP

Once you know the rotational inertia for an axis through the CM, you can use this theorem to find *I* through a parallel axis.

The rotational inertia of an object about an axis a distance D from its center of mass ($I_{\text{parallel axis}}$) is related to the rotational inertia of the object about a parallel axis running through its center of mass (I_{CM}) according to the equation

$$I_{\text{parallel axis}} = I_{\text{CM}} + MD^2$$

We do not present a proof of this theorem.

EXAMPLE 8.4 USE OF THE PARALLEL AXIS THEOREM

In Example 8.1, it was found that the rotational inertia about the z-axis of a uniform rod of mass M reaching from the origin along the x-axis to $x = L$ is $I = ML^2/3$. Calculate the moment of inertia of the rod about an axis parallel to the z-axis passing through the middle of the rod.

SOLUTION

See Figure 8.4.

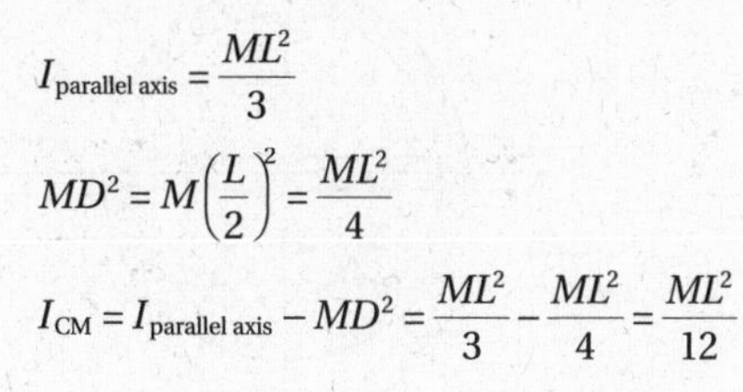

$$I_{\text{parallel axis}} = \frac{ML^2}{3}$$

$$MD^2 = M\left(\frac{L}{2}\right)^2 = \frac{ML^2}{4}$$

$$I_{\text{CM}} = I_{\text{parallel axis}} - MD^2 = \frac{ML^2}{3} - \frac{ML^2}{4} = \frac{ML^2}{12}$$

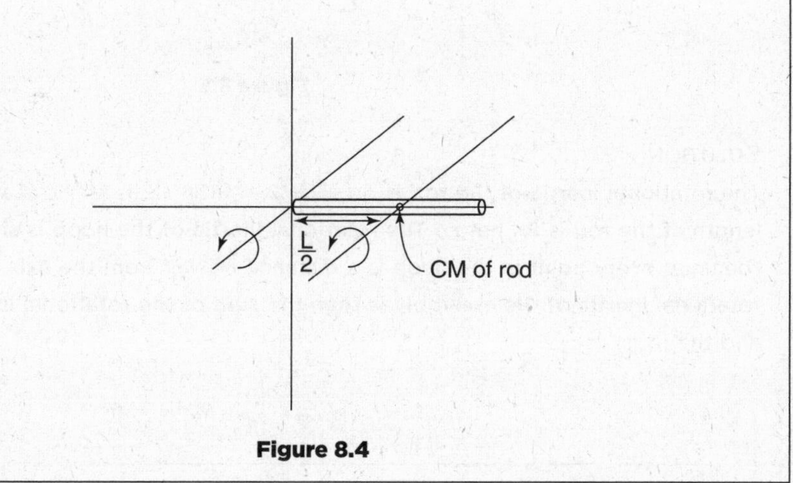

Figure 8.4

Assemblies of Piecewise Smooth Objects or Points Suppose two piecewise smooth objects are fixed together; what is the rotational inertia of this assembly?

Based on the property of the integral, $\int_a^c f(x)dx = \int_a^b f(x)dx + \int_b^c f(x)dx$,

$$I = \int_{\text{both objects}} r^2 dm = \int_{\text{object 1}} r^2 dm + \int_{\text{object 2}} r^2 dm = I_{\text{object 1}} + I_{\text{object 2}}$$

Thus, the rotational inertia of any assembly of objects is simply the sum of the rotational inertias of the individual objects.

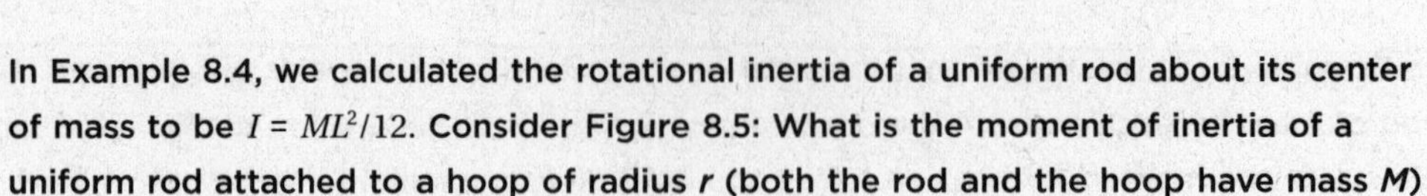

EXAMPLE 8.5

In Example 8.4, we calculated the rotational inertia of a uniform rod about its center of mass to be $I = ML^2/12$. Consider Figure 8.5: What is the moment of inertia of a uniform rod attached to a hoop of radius r (both the rod and the hoop have mass M) about an axis perpendicular to the rod and the hoop?

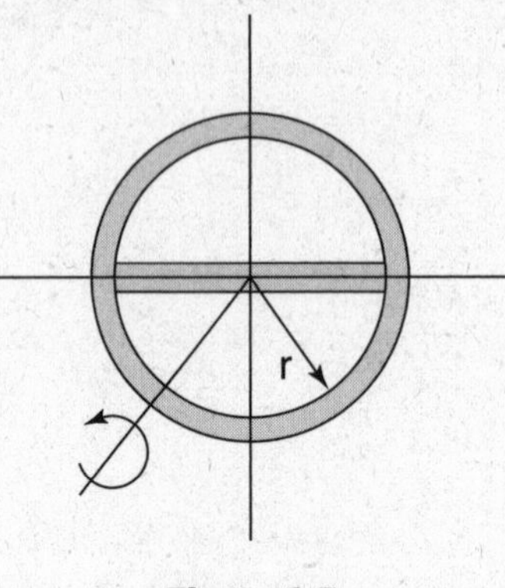

Figure 8.5

SOLUTION

The rotational inertia of the rod is $I = ML^2/12 = M(2r)^2/12 = Mr^2/3$. (Careful! The total length of the rod is $2r$, not r.) The rotational inertia of the hoop is simply Mr^2 (because every point in the hoop is a distance r away from the axis of rotation). The rotational inertia of the assembly is then the sum of the rotational inertia of the hoop and the rod:

$$I_{\text{assembly}} = I_{\text{rod}} + I_{\text{hoop}} = \frac{Mr^2}{3} + Mr^2 = \frac{4}{3}Mr^2$$

ANGULAR MOMENTUM

In this section, we introduce yet another useful quantity associated with rotational motion: angular momentum. As usual, we begin with a mathematical definition whose utility we will soon appreciate.

The angular momentum of a particle is defined as follows (where $\mathbf{p}$ is the linear momentum and $\mathbf{r}$ is a relative position vector pointing from the axis of rotation to the object).

$$\mathbf{L} = \mathbf{r} \times \mathbf{p}$$

Definition of angular momentum for a point particle

This is very similar to the definition of torque:

$$\boldsymbol{\tau} = \mathbf{r} \times \mathbf{F}$$

$$\mathbf{L} = \mathbf{r} \times \mathbf{p}$$

Similarities include the following.

1. The angular momentum depends on the magnitude of the translational momentum, the distance from the axis of rotation, and the angle between **r** and **p** (so that only components of **p** perpendicular to **r** contribute to the angular momentum).
2. Like torque, the value of the angular momentum depends on the choice of axis of rotation.
3. The angular momentum of a particle can therefore be calculated using the techniques used to calculate torque or other cross-products; for example, the magnitude of the angular momentum is $L = rp_{\perp} = pr_{\perp}$. The sign can be chosen using the counterclockwise/clockwise convention or the right-hand rule (point your right thumb along the axis of rotation and curl your fingers in the direction of **p** and your thumb will point in the direction of **L**).
4. Angular momenta add as vectors such that the angular momentum of a system of particles is the vector sum of the angular momentum of each of the particles.

$$\mathbf{L}_{\text{system}} = \sum \mathbf{L}_i$$

Net angular momentum of a system of particles

Note that a particle need not *rotate* about the axis to have angular momentum (all that is needed is a component of velocity perpendicular to **r**). Therefore, an object translating past an axis of rotation has angular momentum.

Relationship Between Rotational Inertia and Angular Momentum

How can we calculate the angular momentum of an extended object? Do we divide the object into differential pieces, calculate the differential angular momentum $d\mathbf{L} = \mathbf{r} \times d\mathbf{p}$ of each piece, and sum these differential angular momenta? Although this would be valid, it is not necessary; instead, there is a simple relationship between rotational inertia and angular momentum, $\mathbf{L} = I\boldsymbol{\omega}$, so that once we have calculated the rotational inertia (which *may* involve integration), we can calculate the angular momentum without further integrations. (This relationship is valid only when the object rotates about a symmetry axis, but we will consider only situations of this type.)

Our goal here is to prove the relationship $\mathbf{L} = I\boldsymbol{\omega}$. We start by proving this relationship on a differential level (with a particle) and then generalize to extended masses and systems of particles by integration.

For an object or particle rotating about a fixed axis, $v = r\omega$ and this velocity is perpendicular to the radial vector **r**. Therefore,

$$L = rp_{\perp} = r(mv) = r(mr\omega) = mr^2\omega$$

L and **ω** are governed by the same sign conventions, so we can ignore the absolute value signs:

$$\mathbf{L} = mr^2\boldsymbol{\omega} = I\boldsymbol{\omega}$$

Note that this equation is valid only if both the angular momentum and the rotational inertia are calculated about the same axis.

Now consider an extended object rotating about the z-axis. For a thin slice of the object lying in the xy-plane we can generalize the above particle equation to the entire slice by saying that it is valid for each tiny particle (on a differential level) that makes up the slice. Therefore, by summing both sides of the equation, the macroscopic sums will also equal each other:

$$\mathbf{L}_{\text{thin slice}} = \int d\mathbf{L} = \int dI\omega = \omega \int dI = \omega I$$

When we try to extend this calculation to the rest of the object, that is, to the thin slices that lie parallel but not on the *xy*-plane, we encounter a complication. The position vectors **r** have *z*-components, which yield a component of angular momentum perpendicular to the *z*-axis, the axis of rotation. However, if the *z*-axis is an axis of symmetry for the object, it can be shown that the net *z*-component of angular momentum is zero. Thus, in this case we obtain

TIP

Note the analogy between $\mathbf{L} = I\omega$ and $\mathbf{p} = m\mathbf{v}$.

$$\mathbf{L} = I\omega$$

Relationship between angular momentum and angular inertia for an extended object rotating about a symmetry axis

Relationship Between Torque and Angular Momentum

Given the parallels between torque and angular momentum, we might expect the relationship between torque and angular momentum to be similar to the relationship between their translational analogs, force and translational momentum ($\mathbf{F}_{\text{net}} = d\mathbf{p}/dt$). This is indeed the case:

$$\tau_{\text{net}} = I\alpha = I\frac{d\omega}{dt} = \frac{d}{dt}(I\omega) = \frac{d\mathbf{L}}{dt}$$

$$\tau_{\text{net}} = \frac{d\mathbf{L}}{dt}$$

Relationship between net torque and angular momentum

This equation was derived for one rotating object with one angular acceleration. However, we can generalize to a system of objects simply by summing both sides of the equation:

$$\tau_{\text{net, system}} = \sum \tau_i = \sum \frac{d\mathbf{L}_i}{dt} = \frac{d\sum \mathbf{L}_i}{dt} = \frac{d\mathbf{L}_{\text{system}}}{dt}$$

Conservation of Angular Momentum

Recall how the equation $\mathbf{F}_{\text{net,system}} = d\mathbf{p}_{\text{system}}/dt$ implies that linear momentum is conserved if $\mathbf{F}_{\text{net}} = 0$. (The latter condition is satisfied whenever the net external force is zero; internal forces always cancel each other out and do not contribute to $\mathbf{F}_{\text{net}}$.) The analogous statement, $\tau_{\text{net,system of particles}} = d\mathbf{L}_{\text{system}}/dt$, requires angular momentum to be conserved when $\tau_{\text{net}} = 0$ (which is satisfied whenever the net external torque is zero and the internal forces act along the line joining one particle with another; in this case the internal torques cancel and do not contribute to τ_{net}).

If the net external torque acting on an object or system of objects is zero, the total angular momentum is constant.

EXAMPLE 8.6

A person is sitting on a rotating lab stool holding a top that is free to rotate, as shown in Figure 8.6. The top is initially pointing vertically and rotating counterclockwise (as viewed from above) so that its angular momentum vector of magnitude *L* points in the +*z*-direction. The person turns the top (which continues to rotate about its axis with a constant speed) to a final angle θ as shown. If the person is initially at rest, what is the person's final angular velocity with respect to the laboratory? (The person-top system has a rotational inertia of *I* about the *z*-axis.)

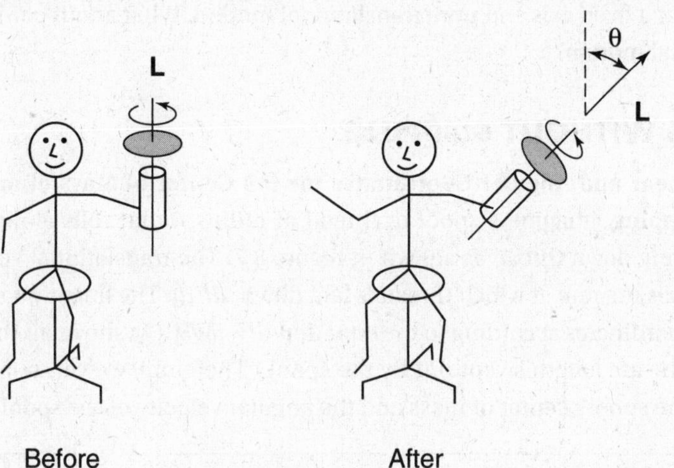

Figure 8.6

SOLUTION

We start by defining the system as consisting of the person and the top. Neglecting any friction in the axle of the stool, there are no external torques exerted in the *z*-direction. Therefore, L_z is conserved. Conservation of L_z requires that

$$L_{z,0} = L_{z,\text{top,final}} + L_{z,\text{person,final}}$$

Because the top continues to rotate about its axis with a constant angular velocity about its axis, the magnitude of its angular momentum remains constant. We can use geometry to calculate the component of this angular momentum in the *z*-direction:

$$L_{z,\text{top,final}} = L\cos\theta$$

Therefore,

$$L = L\cos\theta + L_{z,\text{person}} = L\cos\theta + I\omega_{\text{person}}$$

$$\omega_{\text{person}} = \frac{L(1-\cos\theta)}{I}$$

The direction of this rotation must be parallel to the original rotation of the top, which is counterclockwise (looking downward from above the person).

Review of Analogous Linear and Angular Equations

Rotational Analog	Linear Analog
$\mathbf{L} = I\boldsymbol{\omega}$ (rotational inertia)	$\mathbf{p} = m\mathbf{v}$ (translational inertia)
$\tau_{net} = \frac{d\mathbf{L}}{dt} = I\alpha$	$\mathbf{F} = \frac{d\mathbf{p}}{dt} = m\mathbf{a}$ (Newton's second law)
Angular momentum is conserved if the external torque is zero.	Linear momentum is conserved if the external force is zero.

Combined Rotational and Translational Motion So far we have discussed pure rotational motion about a fixed axis and pure translational motion. What about combined translational and rotational motion?

ROLLING WITHOUT SLIPPING

Relating Linear and Angular Coordinates for the Center of Mass of an Object that Rolls Without Slipping. Imagine a spool of thread of radius R that rolls along a surface without slipping, laying down thread as shown in Figure 8.7. The translational velocity of the center of mass equals the rate at which thread is laid down, dl/dt. The linear quantity dl is related to rotational coordinates according to the equation $dl = R|d\theta|$. (As shown in the figure, this length is equal to the arc length swept out by the spool.) Therefore, we can relate the translational velocity of the spool's center of mass and the angular velocity of the spool:

TIP

How translation and rotation are synchronized when an object rolls without slipping

$$v_{center} = \frac{dl}{dt} = \frac{Rd\theta}{dt} = R\frac{d\theta}{dt} = R\omega$$

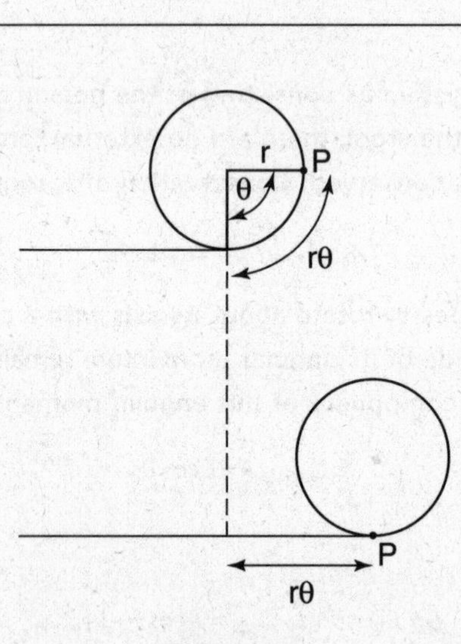

Figure 8.7

Another way to understand this equation is to imagine viewing the spool in a frame of reference where the axle is stationary. In this case, the length of unwound thread increases with a magnitude equal to the velocity of the edge of the spool, which is $|\mathbf{v}| = r|\boldsymbol{\omega}|$.

Differentiation yields the relationship between linear and rotational acceleration:

$$a_{center} = R\alpha$$

Computing the Motion of Other Particles in the Object and the Kinetic Energy

A useful way to analyze rolling without slipping is to consider it as pure rotation about the point of contact with the ground. The beauty of this approach is that it side-steps the tricky issue of dealing with combined translation and rotation (about the object's center) by converting the situation to pure rotation (about the point of contact).

We begin by showing that the point of contact of the object with the ground has zero translational velocity if the object rolls without slipping. Consider a particle on the edge of a wheel that is about to make contact with the ground as the wheel rotates. Before the particle touches the ground, it is moving vertically toward the ground, and afterward it is moving vertically away from the ground. As the velocity of this point goes from downward to upward, it must instantaneously be zero. It can also be seen in Figure 8.8 that combined rotational and translational motion can be considered a superposition of rotational and translational motion. At the point of contact, the sum of the translational velocity and the rotational velocity is zero. Thus, as shown above, we can consider the object to undergo pure rotation about this (instantaneously) stationary point of contact with the ground.

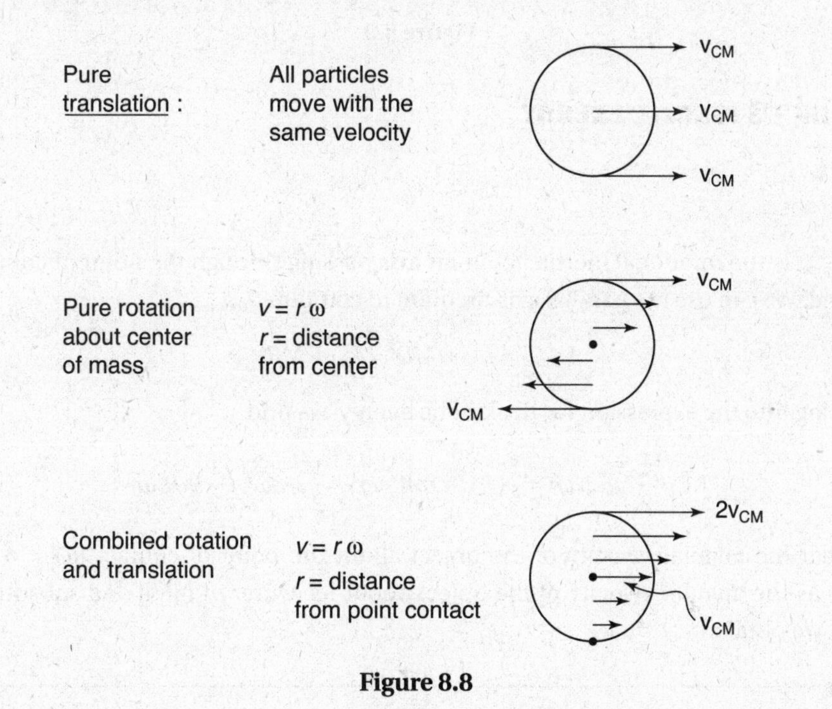

Figure 8.8

THE INSTANTANEOUS VELOCITY OF OTHER PARTICLES IN THE OBJECT

We know that the center of mass is moving with v_{CM} and that it is a distance R from our instantaneously stationary axis of rotation. Therefore, the angular velocity about the contact point is

$$\omega = \frac{v_{CM}}{R}$$

We can now calculate the velocity of any particle in the object by applying the formula $|\mathbf{v}| = r|\boldsymbol{\omega}|$ and noting that the velocity is perpendicular to the vector pointing from the axis to the particle, as shown in Figure 8.9. For example, we can easily calculate the speed of a particle at the top of the object to be

$$v_{top} = r\omega = (2R)\left(\frac{v_{CM}}{R}\right) = 2v_{CM}$$

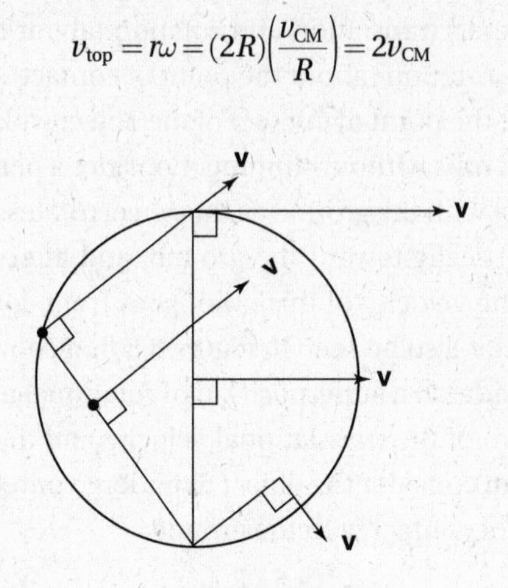

Figure 8.9

THE OBJECT'S KINETIC ENERGY

$$KE = \frac{1}{2} I_{contact}\omega^2$$

where $I_{contact}$ is the rotational inertia about an axis passing through the point of contact with the ground. We can use the parallel axis theorem to compute $I_{contact}$:

$$I_{contact} = I_{CM} + MD^2 = I_{CM} + MR^2$$

Substituting into the expression for the kinetic energy, we find

$$KE = \frac{1}{2} I_{contact}\omega^2 = \frac{1}{2}\left(I_{CM} + mR^2\right)\omega^2 = \frac{1}{2} I_{CM}\omega^2 + \frac{1}{2} mR^2\omega^2$$

Noting that the angular velocity of the object about the point of contact ($|\boldsymbol{\omega}| = |\mathbf{v}_{CM}|/R$) is the same as the angular velocity of the object about its center of mass and substituting for $v_{CM} = R\omega$, we obtain

$$KE = \frac{1}{2} I_{CM}\omega_{CM}^2 + \frac{1}{2} Mv_{CM}^2 = KE_{\text{pure rotation}} + KE_{\text{pure translation}}$$

The kinetic energy of combined rotation and translation is simply the sum of the KE of pure rotation and pure translation.

Although we do not prove it here, the kinetic energy of a rolling object is *always* equal to the sum of the kinetic energies of pure rotation and pure translation (even if the motion is not rolling without slipping).

The Force Required for Rotation Without Slipping

Recall that if an object rolls without slipping the following equation is valid.

$v_{CM} = R\omega$ **condition equivalent to rolling without slipping**

By differentiating, we can make the statement that

$$\text{If } v_{CM} = R\omega, \quad \text{then} \quad a_{center} = R\alpha$$

$a_{center} = R\alpha$

condition necessary but not sufficient for rolling without slipping

What physical insight is there to be gained from these equations? Consider a few situations.

***Case 1*:** If a circular object rolls without slipping at a constant translational speed, the translational acceleration is zero ($a_{CM} = 0$), and consequently the angular acceleration is zero ($\alpha = 0$) and the net torque is zero. Thus *no net torque* is required to keep an object rolling without slipping at a constant speed. (Conservation of angular momentum is enough to keep the object turning at constant speed.) An object that rolls without slipping over a frictionless surface continues to do so, and an object that rolls without slipping at a constant velocity over a surface with friction experiences no frictional force (recall that static friction is variable; you get as much as you need to prevent slipping, up to the maximum value of $\mu_s F_N$). A free-body diagram of this situation is shown in Figure 8.10, and you can observe that neither the normal force nor gravity exerts a torque; the net torque is zero.

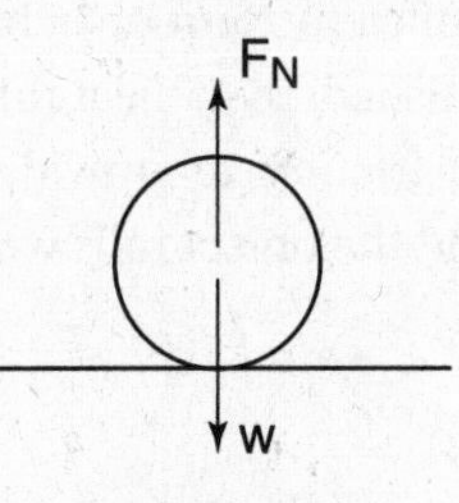

Figure 8.10

***Case 2*:** Consider Figure 8.11. What happens if a force $F_{applied}$ is applied to the center of mass as shown in the absence of friction? The velocity v_{CM} changes at a rate of $\mathbf{a}_{CM} = \mathbf{F}_{applied}/m$. However, neither the applied force nor the normal or gravity forces exert any torque. Thus, the angular velocity of the object does *not* change, and the condition for rolling without slipping, $v_{CM} = r\omega$, is soon violated (v_{CM} changes but ω does not). Thus, the object begins sliding.

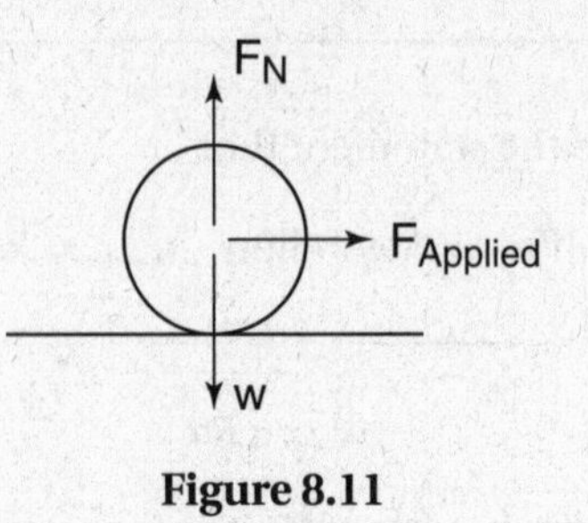

Figure 8.11

***Case 3*:** Consider Figure 8.12. What happens if a force $F_{applied}$ is applied to the center of mass as shown in the presence of friction? The center of mass accelerates with $\mathbf{a}_{CM} = (\mathbf{F}_{applied} - \mathbf{F}_{fr})/m$. There is now a torque exerted by the frictional force. There are two possible cases.

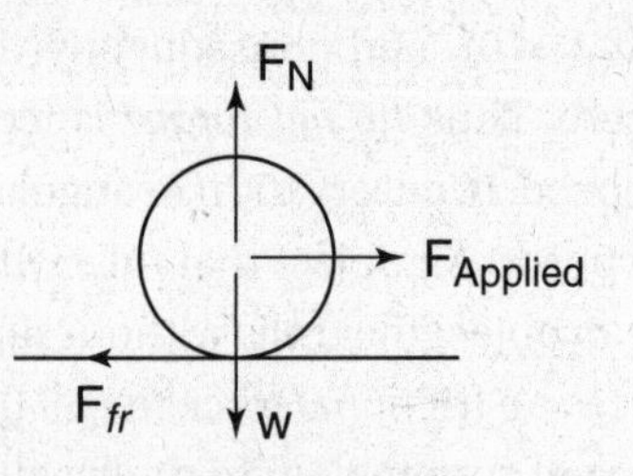

Figure 8.12

a. There is sufficient static friction to prevent the object from sliding. In this case, the frictional force provides sufficient torque such that $a_{center} = r\alpha$. Because the initial condition $v_0 = r\omega_0$ is also satisfied, the object rolls without slipping.
b. There is insufficient static friction to provide enough torque for the condition $a_{center} = r\alpha$ to be satisfied; and the object rolls while slipping.

EXAMPLE 8.7

A cylinder $\left(I = \frac{1}{2}mR^2\right)$ is at rest on a flat surface. When a horizontal force $\mathbf{F}_{app}$ is exerted on the cylinder's axle, what is the minimum coefficient of static friction μ_s required to keep the cylinder from slipping?

SOLUTION

The minimum coefficient of friction occurs when the object is on the verge of slipping (i.e., the frictional force must be the maximum static frictional force $F_{fr} = \mu_s F_N = \mu_s mg$). The free-body diagram is the same as in case 3.

$$F_{net} = ma_{CM} = F_{app} - F_{fr} = F_{app} - \mu_s mg \Rightarrow a_{CM} = \frac{F_{app} - \mu_s mg}{m}$$

In calculating the torque, the frictional force is already perpendicular to the **r** vector. As for signs, the frictional force clearly exerts a torque that helps the object to rotate in the direction to promote rolling without sliding (clockwise), so we need worry only about magnitudes.

$$\tau = F_{fr}R = \mu_s mgR = I\alpha = \frac{1}{2}mR^2\alpha \Rightarrow \alpha = \frac{2\mu_s g}{R}$$

Now, for the object to roll without slipping, the equation $a_{center} = R\alpha$ must be satisfied:

$$\left[a_{CM} = \frac{F_{app} - \mu_s mg}{m}\right] = [R\alpha = 2\mu_s g]$$

$$\mu_s = \frac{F_{app}}{3mg}$$

Symbolic answer check: (1) Correct units (the coefficient of friction is dimensionless). (2) Increasing gravity increases the normal force, causing the necessary coefficient of friction to decrease. (3) Increasing the applied force increases the necessary frictional torque, increasing the necessary coefficient of friction. (4) Increasing the mass increases the normal force, which would be expected to decrease the coefficient of static friction. However, increasing the mass *also* increases the rotational inertia (increasing the necessary torque needed to achieve a given rotational acceleration) and decreases the translational acceleration). The relationship between μ_s and m is thus too complicated to be useful as a quick answer check.

STATIC EQUILIBRIUM FOR EXTENDED OBJECTS

The general term "equilibrium" refers to a system that is not accelerating. For a point mass, equilibrium is defined by the requirement that $\mathbf{F}_{\text{net}} = 0$. As discussed in Newton's first law, the two cases of a point mass in equilibrium are a mass that remains at rest (*static equilibrium)* or travels at constant velocity *(dynamic equilibrium).*

In this chapter, we are interested in static equilibrium. For point masses, the requirements for static equilibrium are the only requirement for equilibrium, $\mathbf{F}_{\text{net}} = 0$, in addition to the initial condition that the object is at rest. What about extended objects? Consider the extended object shown in Figure 8.13. Although the condition for translational static equilibrium is satisfied ($\mathbf{F}_{\text{net}} = 0$, $v_0 = 0$), the object clearly does not remain at rest but instead rotates. This demonstrates the necessity of adding another requirement for static equilibrium of extended objects: that $\tau_{\text{net}} = 0$ about any axis. Therefore, the requirement for static equilibrium for extended objects is that both $\mathbf{F}_{\text{net}} = 0$ and $\tau_{\text{net}} = 0$ (in addition to the fact that the object is *initially* at rest, mathematically that $v_{\text{CM}} = 0$ and $\omega_0 = 0$).

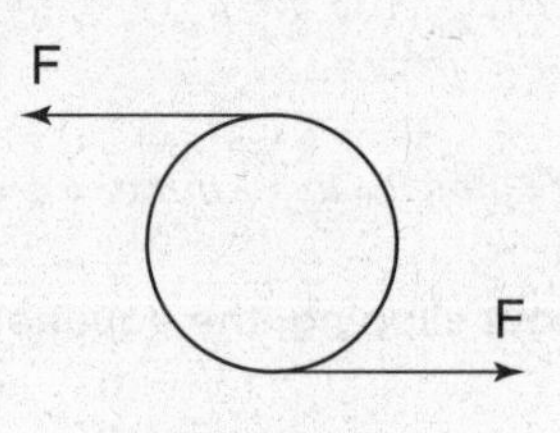

Figure 8.13

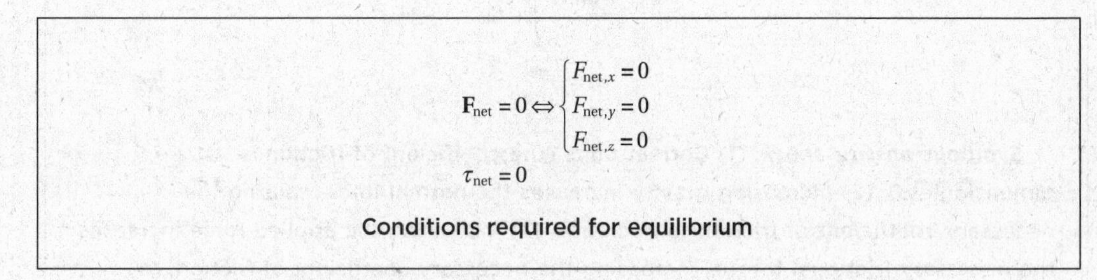

$$\mathbf{F}_{\text{net}} = 0 \Leftrightarrow \begin{cases} F_{\text{net},x} = 0 \\ F_{\text{net},y} = 0 \\ F_{\text{net},z} = 0 \end{cases}$$

$$\tau_{\text{net}} = 0$$

Conditions required for equilibrium

Solving Static Equilibrium Problems

If there are several objects involved, you need to perform the following steps on each object or group of objects, depending on the nature of the problem.

1. Satisfying $\mathbf{F}_{\text{net}} = 0$: Choose a convenient coordinate system and sketch all forces acting on the object (or assembly of objects, for example, a child and a seesaw). Resolve these forces into components along the axes, sum the forces acting along each axis, and set these sums equal to zero.
2. Satisfying $\tau_{\text{net}} = 0$: The following statements are equivalent if the net force is zero. Torque about an axis is zero ⇔ Object is not rotating in the plane perpendicular to that axis ⇔ torque about any parallel axis is zero.

(AP problems deal only with forces acting in two dimensions, so that you only have to worry about rotation in a single plane about axes perpendicular to that plane. This statement is not valid in three dimensions.)

The point here is that to satisfy $\tau_{net} = 0$, it is sufficient that the torque about any particular "chosen axis" perpendicular to the plane where the forces are acting is zero. This guarantees that the torque about *all* parallel axes is zero (so long as the net force is zero), making it redundant to repeat the torque calculation about another axis.

Make sure to select your "chosen axis" wisely. Based on the definition of torque as $\boldsymbol{\tau} = \mathbf{r} \times \mathbf{F}$, if you choose your axis to pass through the point where a particular force is acting, the torque exerted by that force is zero since $\mathbf{r} = 0$. You can use this strategically to avoid calculating the torque due to unknown or complicated forces.

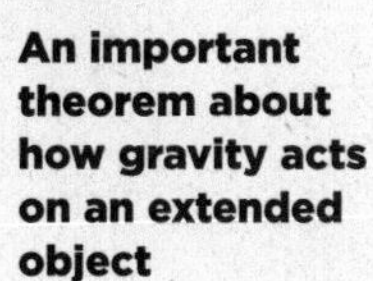

TIP

An important theorem about how gravity acts on an extended object

The torque exerted by gravity acts as if the gravitational force is exerted at the center of mass. *Proof*: Consider the torque exerted by gravity about the *z*-axis on an arbitrary mass distributed along an *xy*-plane. Because the points making up the mass lie at different **r** from the origin, the net torque must be calculated by integration:

$$d\boldsymbol{\tau} = \mathbf{r} \times d\mathbf{F} = \mathbf{r} \times \mathbf{g}dm$$

Using the properties of the cross-product that $c\mathbf{A} \times \mathbf{B} = \mathbf{A} \times c\mathbf{B}$ and that $\mathbf{A} \times \mathbf{C} + \mathbf{B} \times \mathbf{C} = (\mathbf{A} + \mathbf{B}) \times \mathbf{C}$,

$$\tau = \int d\tau = \int \mathbf{r} \times \mathbf{g}dm = \int \mathbf{r}\,dm \times \mathbf{g} = \left(\int \mathbf{r}\,dm\right) \times \mathbf{g}$$

Recall the definition of center of mass:

$$\mathbf{r}_{CM} = \frac{\int \mathbf{r}dm}{M}$$

We can then rewrite the torque as

$$\boldsymbol{\tau} = M\mathbf{r}_{CM} \times \mathbf{g}$$

This is the same torque that would be caused if the net gravitational force were exerted at the center of mass ($\mathbf{r}_{CM}$).

EXAMPLE 8.8 CLASSIC STATIC EQUILIBRIUM PROBLEM

A ladder of length *L* and mass *M* rests on a *frictionless* wall at an angle of θ as shown in Figure 8.14. Calculate the magnitude of the force that the ground exerts on the ladder (assume there is a frictional force where the ladder contacts the ground).

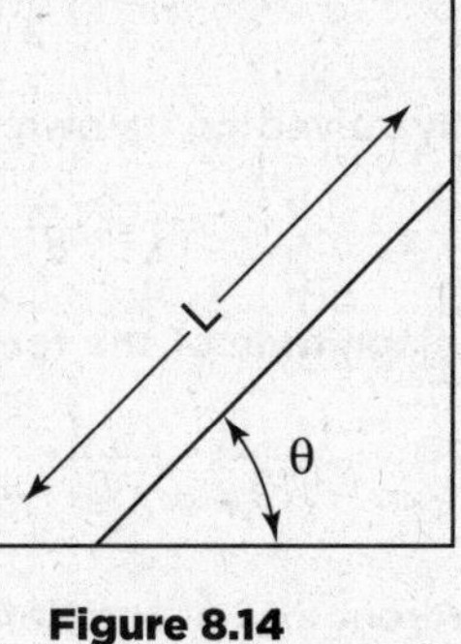

Figure 8.14

SOLUTION

1. **Choose a coordinate system and draw a free-body diagram showing all the forces acting on the ladder, as shown in Figure 8.15. Note how the frictional force on the**

ground opposes the slipping of the ladder. Because the wall is frictionless, it exerts only a horizontal normal force. Setting the net force on the ladder to zero yields

$$\begin{cases} F_{\text{net},x} = F_{\text{fr}} - F'_N = 0 \\ F_{\text{net},y} = F_N - mg = 0 \end{cases}$$

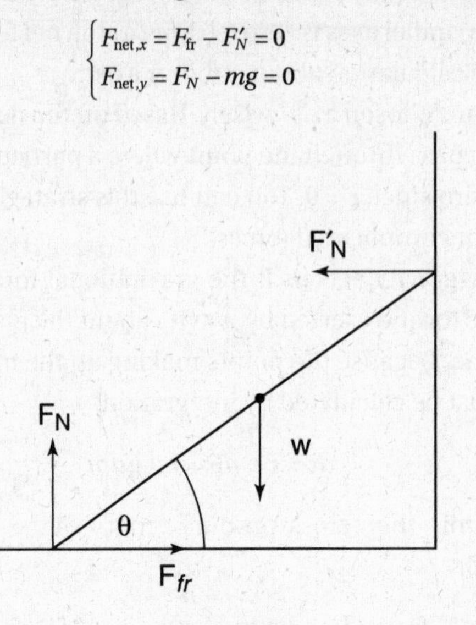

Figure 8.15

2. Now we have three unknowns (F_{fr}, F'_N, and F_N) and two equations, so we need one more equation, which is provided by the condition that the torque be zero.

Selection of axis: If we choose the axis to pass through the point where the ladder contacts the ground, the torque due to F_N and F_{fr} will be zero, giving us a simple equation:

$$\tau_{\text{net}} = +(F'_N \sin\theta)L - (mg\cos\theta)\frac{L}{2} = 0$$

Note: The magnitude of the torques was calculated using the formula $\tau = F_\perp r$, and signs were assigned based on the convention that counterclockwise is positive and clockwise is negative.

We now have three equations in three variables, so we can solve for all the variables. This is straightforward because of our choice of axis: The last equation can be immediately solved for F'_N, which can then be plugged into the first equation to calculate F_{fr}:

$$F_{\text{fr}} = F'_N = \frac{mg\cot\theta}{2}$$

The second equation is easily solved on its own:

$$F_N = mg$$

The problem asked for the magnitude of the force exerted by the ground:

$$F_{\text{ground}} = \sqrt{F^2_{N,\text{ground}} + F^2_{\text{fr}}} = mg\sqrt{1 + \frac{\cot^2\theta}{4}}$$

Note: If you choose a different axis of rotation (as long as it is perpendicular to the *xy*-plane), you will still obtain the correct answer (although the math might be more complicated).

CHAPTER SUMMARY

The rotational inertia of any extended object, no matter how complex in shape, can be calculated in principle using the integral formulation, dividing the object up into pointlike constituents. Once the rotational inertia has been calculated for rotation about an axis, it can be easily recalculated for a second axis parallel to the first by using the parallel axis theorem.

In analogy with linear momentum, the angular momentum of a rotating object can be defined and used to recast Newton's second law for rotational motion. Angular momentum is conserved if the net external torque on a rotating object is zero. If an object is in static equilibrium, neither translating nor rotating, then both the net torque and force on the object must be zero.

PRACTICE EXERCISES

Multiple-Choice Questions

1. A figure skater spins on the ice, holding two weights a distance d from the axis about which he is rotating (Figure 8.16). At a certain time, he pulls these weights toward his body as shown, causing his rotational inertia to decrease by a factor of 2. How does this affect the maximum speed with which any part of his body moves?

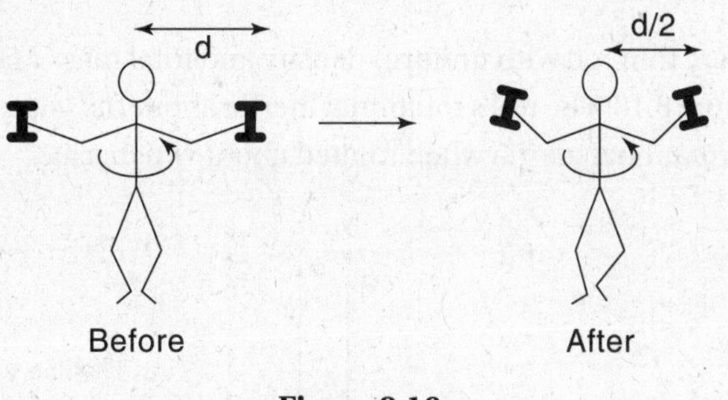

Figure 8.16

(A) The maximum speed decreases by a factor of 2.
(B) The maximum speed decreases by a factor of $\sqrt{2}$.
(C) The maximum speed remains the same.
(D) The maximum speed increases by a factor of $\sqrt{2}$.
(E) The maximum speed increases by a factor of 2.

2. Consider the massless balance (which is at rest) shown in Figure 8.17. What is the weight of m?

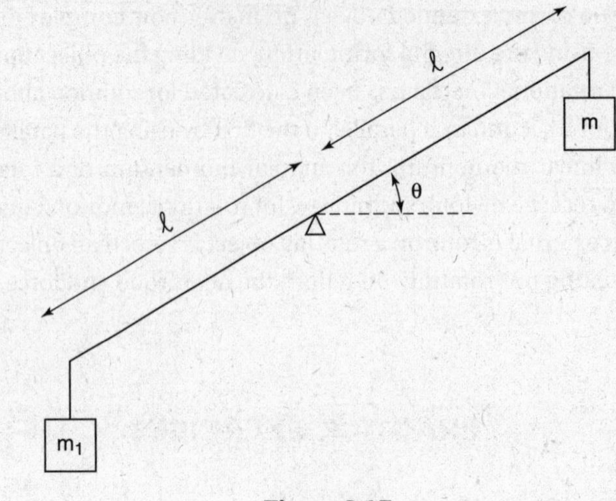

Figure 8.17

(A) m_1
(B) m_1g
(C) $m_1g\cos\theta$
(D) $m_1g\sin\theta$
(E) $m_1g\sec\theta$

3. Consider a very thin rod with uniform density and total mass M lying along the x-axis as shown in Figure 8.18. The rod's rotational inertia about the y-axis is $ML^2/12$. The rod has the greatest rotational inertia when rotated about which axis?

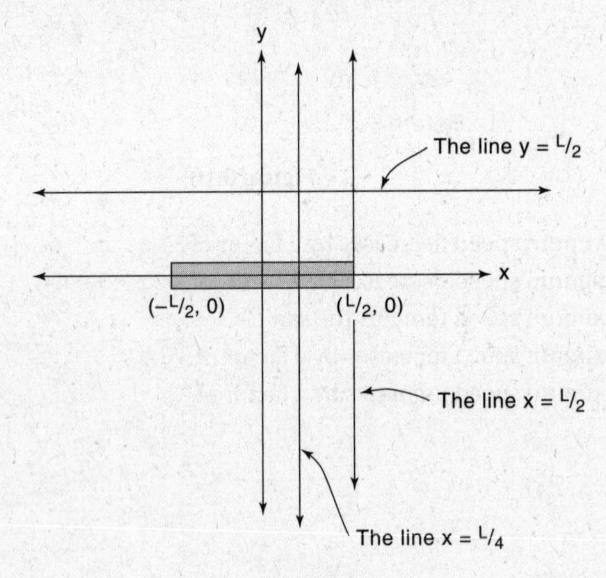

Figure 8.18

(A) the y-axis
(B) the x-axis
(C) the vertical line $x = L/4$
(D) the vertical line $x = L/2$
(E) the horizontal line $y = L/2$

4. A ball rolls along a horizontal frictionless surface without slipping toward an incline. The ball moves up the incline a certain distance before reversing direction and moving back down the incline. In which of the following situations does the ball move the greatest distance up the incline before turning around?
   (A) The incline is frictionless.
   (B) There is sufficient static friction between the ball and the incline to prevent the ball from slipping.
   (C) There is insufficient static friction to prevent slipping, but there is some kinetic friction.
   (D) The amount of friction is irrelevant.
   (E) The situation described in this problem is impossible: A ball cannot roll along a horizontal frictionless surface without slipping.

5. A point mass moves with constant velocity in the y-direction as shown in Figure 8.19. Which of the graphs plots the magnitude of the mass's angular momentum, $\mathbf{L} = \mathbf{r} \times \mathbf{p}$, as a function of the particle's y-position?

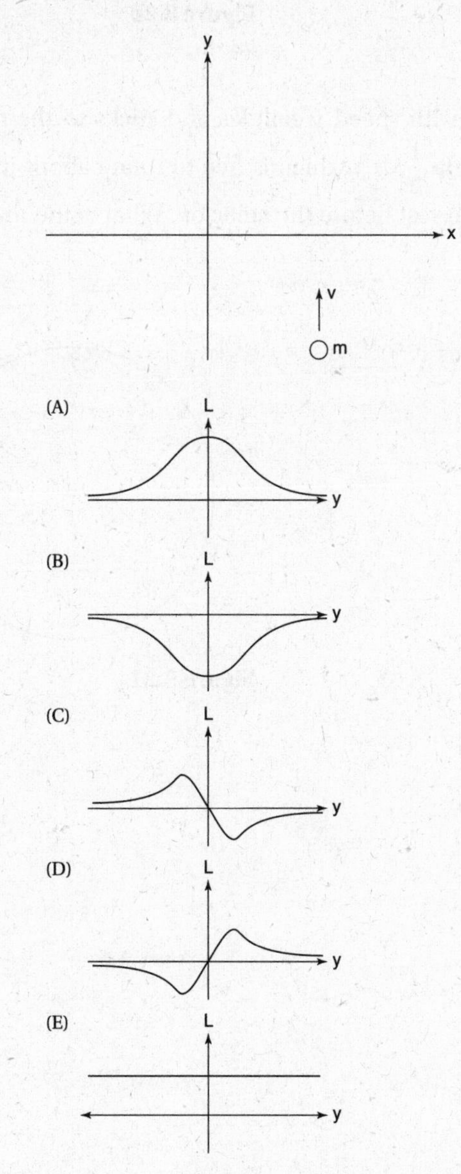

Figure 8.19

6. A disc with uniform density is supported by a pivot and string as shown in Figure 8.20. What is the direction of the force exerted by the pivot on the disc?

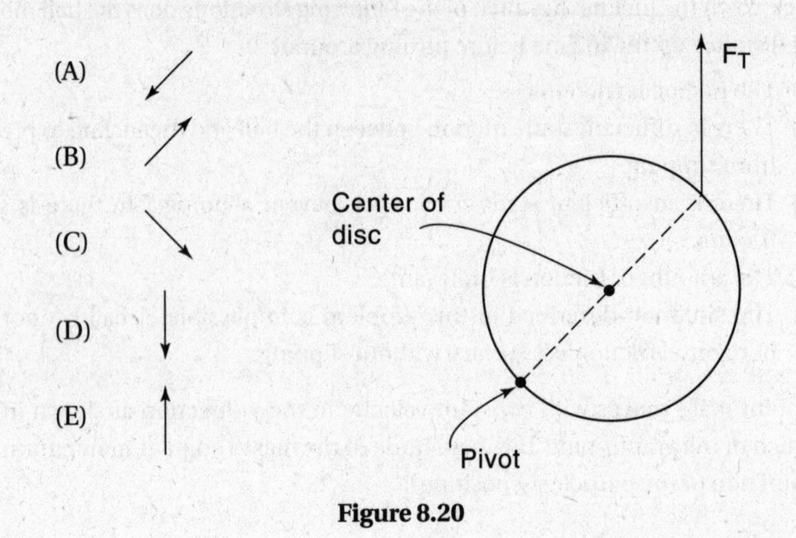

Figure 8.20

7. A mass m moving with speed v collides and sticks to the rim of a wheel of radius r and rotational inertia $\frac{1}{2}Mr^2$, which is free to rotate about its axle, as shown in Figure 8.21. The wheel is at rest before the collision. What is the mass's linear speed after the collision?

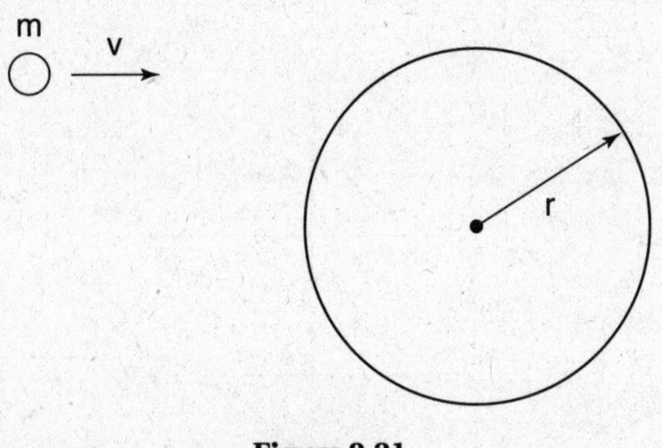

Figure 8.21

(A) $mvr\Big/\left(m+\frac{1}{2}M\right)$

(B) $mv\Big/\left(m-\frac{1}{2}M\right)$

(C) $mv\Big/\left(m+\frac{1}{2}M\right)r$

(D) $2mv/M$

(E) $mv\Big/\left(m+\frac{1}{2}M\right)$

8. A wheel rolls along a frictionless surface with a constant angular velocity as shown in Figure 8.22. A force applied to the axle of the wheel causes the wheel's center of mass to accelerate to the right with a magnitude a (without affecting the wheel's angular velocity). What is the acceleration of a point P on the wheel?

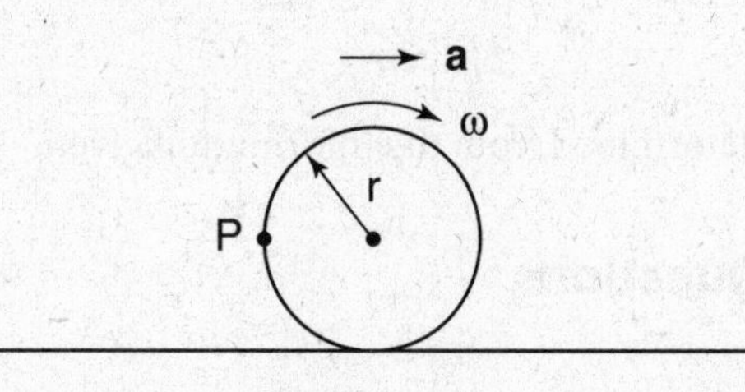

Figure 8.22

(A) $a\hat{i}$
(B) $(a-\omega^2 r)\hat{i}$
(C) $(a+\omega^2 r)\hat{i}$
(D) $a\hat{i}+\omega^2 r\hat{j}$
(E) $(a+\omega^2 r)\hat{i}-g\hat{j}$

9. As shown in Figure 8.23, four identical point masses are arranged at the vertices of a rectangle that is rotated about a perpendicular axis passing through the midpoint of one of the line segments of the rectangle. Calculate the moment of inertia of this system.

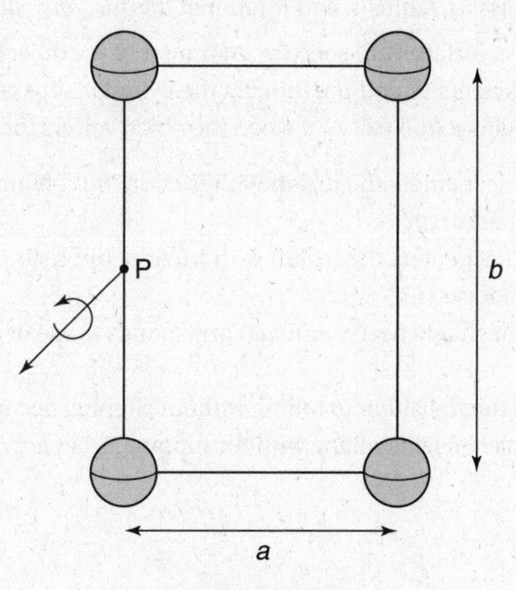

Figure 8.23

(A) $ma^2/2$
(B) $(ma^2/2)+2mb^2$
(C) $m(2a^2+b^2)$
(D) $m(a^2+2b^2)$
(E) none of the above

10. A wheel with radius R, mass M, and rotational inertia I rolls without slipping across the ground. If its translational kinetic energy is E, what is its rotational kinetic energy?

(A) E
(B) $E/2$
(C) $E(I/MR^2)$
(D) $E(MR^2/I)$
(E) It cannot be determined from the information given.

Free-Response Questions

1. A cylinder of radius r, mass m, and rotational inertia $\frac{1}{2}mr^2$ rolls without slipping down an incline of length d and angle of elevation θ (starting at rest at the top of the incline).

   (a) Draw a free-body diagram, being careful to indicate the point of action of all the forces.
   (b) Construct three Newton's second-law equations, two for the components of the net force and one for the net torque.
   (c) Solve these equations for the linear acceleration of the cylinder.
   (d) What is the minimum coefficient of friction required to prevent slipping? Is this kinetic or static friction?

2. A cylinder of mass m, radius r, and rotational inertia $\frac{1}{2}mr^2$ slides without rolling along a flat, frictionless surface with speed v_0. At time $t = 0$ the object enters a region with friction (with coefficients μ_k and μ_s). Initially the cylinder slips relative to the surface, but eventually it begins to roll. Set $t = 0$ when the object enters the region with friction.

   (a) After the object enters the region with friction, but before it begins rolling without slipping, what is $v(t)$?
   (b) After the object enters the region with friction, but before it begins rolling without slipping, what is $\omega(t)$?
   (c) What relationship between v and ω corresponds to the beginning of rolling without slipping?
   (d) When does the transition to rolling without slipping occur?
   (e) After the object begins rolling without slipping, what are $v(t)$ and $\omega(t)$?

3. A 4-kg projectile is launched with an initial velocity $\mathbf{v} = (2\,\text{m/s})\hat{i} + (2\,\text{m/s})\hat{j}$. On impact, it lands on the very top of a sphere of radius 3 m and rotational inertia 3 kg·m², which is pivoted about an axle (at its center) coming out of the page as shown in Figure 8.24. The sphere is initially at rest, the collision is elastic, and the projectile rebounds with a speed of 2 m/s.

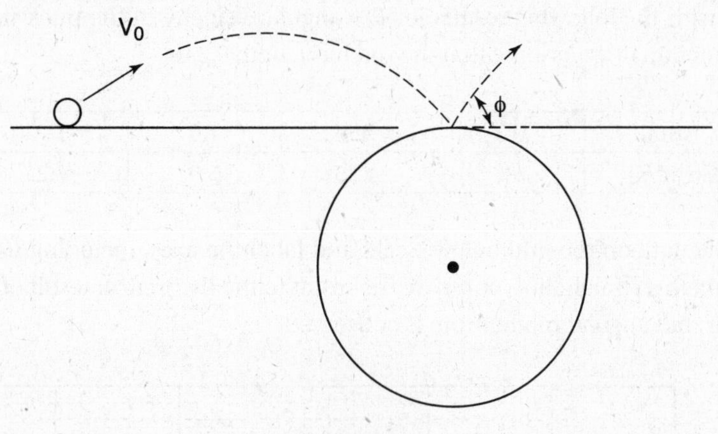

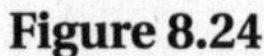
Figure 8.24

(a) What is the velocity of the projectile an instant before impact with the sphere?
(b) What is the angular velocity of the sphere after the collision?
(c) What is the angle of elevation ϕ of the projectile as it bounces off the sphere?

4. In the two collisions shown in Figure 8.25, a puck of mass m sliding along a frictionless table with speed v_0 collides and sticks to a rod of total mass m with uniform linear mass density. In the first case, the puck sticks to the middle of the rod, whereas in the second case the puck sticks to the end of the rod. The rotational inertia of the rod about its center of mass is $\frac{1}{12}ml^2$.

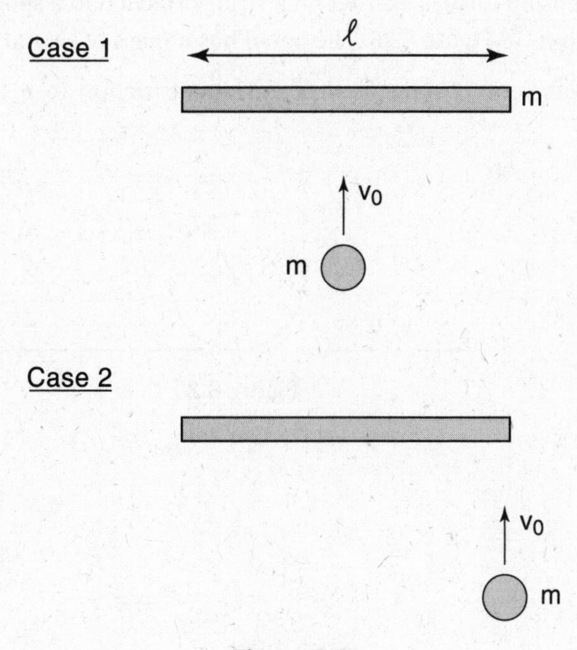

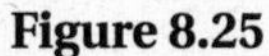
Figure 8.25

(a) In case 1, what is the final velocity of the rod?
(b) In case 2, what is the final velocity of the center of mass of the puck-mass system?
(c) In case 2, one point in the puck-mass system moves in a straight line. Where is this point? (Specify in terms of the distance from the end of the rod that the puck hits.)
(d) In case 2, what is the final angular velocity of the puck-mass system?
(e) An experiment is carried out to test conservation of angular momentum in case 2 with the following results for the angular velocity of the puck-mass system as a function of v_0. Assume that the rod has length 0.2 m.

v_0 (m/s)	0.5	0.8	1.0	1.2	1.4
ω (rad/s)	3.1	4.75	5.9	7.2	8.5

Plot the data on the grid below. Scale and label the axes, including units. After fitting the data to a straight line, compare the result to the theoretical result of part (d). Does it appear that angular momentum is conserved?

5. An upward tension force is exerted on a string attached to a spool of thread resting on a table as shown in Figure 8.26. The spool has a mass of m and a rotational inertia of $\frac{1}{2}mr^2$; the coefficients of static friction and kinetic friction are μ_s and μ_k, respectively.

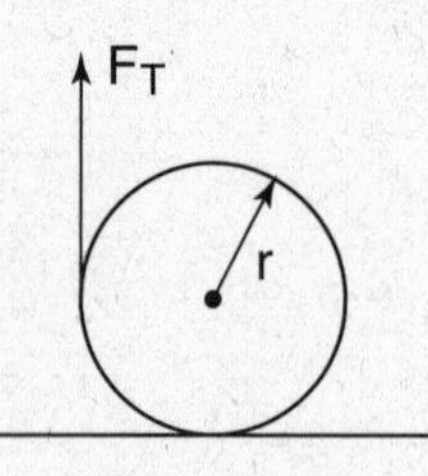

Figure 8.26

(a) Draw a free-body diagram of the spool.
(b) What is the maximum tension force that can be exerted without causing the spool to slip as it rolls along the table? Write down, but *do not solve*, the system of equations that could be used to answer this question.

Then the tension force is increased beyond this limit to a force F_T, so that the spool slips as it rolls along the table.

(c) What is the magnitude of the angular acceleration?
(d) What is the magnitude of the linear acceleration of the center of mass of the spool?

6. A mass m attached to a spring with spring constant k rotates in a circle of radius r on a horizontal frictionless surface as shown in Figure 8.27. The mass moves with speed v_0.

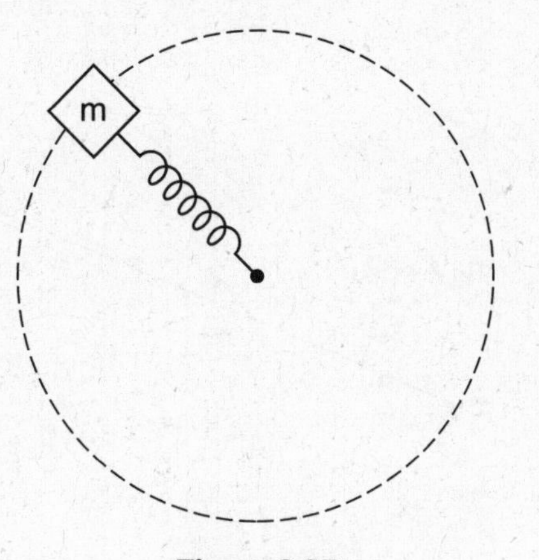

Figure 8.27

(a) What is the equilibrium length of the spring when not attached to the mass?

Then the mass is given a small tap in the radial direction, which causes the length of the spring to oscillate as the mass revolves around the circle.

(b) In terms of the length of the spring, l, at any particular point during this oscillation, what is the mass's tangential velocity?
(c) The instant after the spring is tapped its radial speed is v_r. What is the spring's maximum extension from its unstretched length? Write down the equations that could be used to answer this question but do not solve them.

ANSWER KEY

1. **C**	4. **B**	7. **E**	10. **C**
2. **B**	5. **E**	8. **C**	
3. **D**	6. **E**	9. **C**	

ANSWERS EXPLAINED

Multiple-Choice

1. **(C)** Angular momentum $\mathbf{L} = I\omega$ is conserved, so that decreasing the rotational inertia by a factor of 2 causes ω to increase by a factor of 2. The maximum speed with which any part of the skater's body moves is equal to $v = \omega r_{max}$, where r_{max} is the maximum distance that any part of the skater's body is from the axis of rotation. Based on Figure 8.16, r_{max} decreases by a factor of 2. Finally, because ω increases by a factor of 2 and r_{max} decreases by a factor of 2, the product $v = \omega r_{max}$ remains the same.

2. **(B)** For the balance to be at rest (in static equilibrium), the net torque on it must be zero, in Figure 8.28. Calculating torque via $\tau = Fr_{\perp}$, this is equivalent to

$$mgl\cos\theta = m_1 gl\cos\theta \Rightarrow m = m_1$$

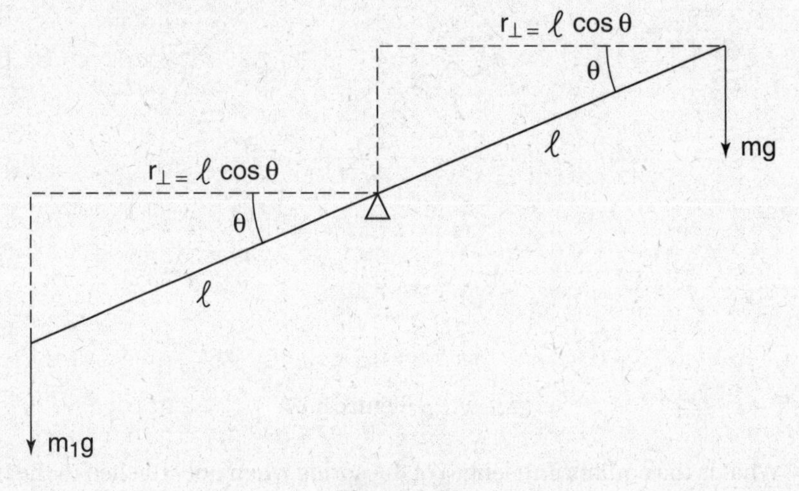

Figure 8.28

The key here is that rotating the balance affects the $r_{\perp}$ of both sides in exactly the same way, so that the two masses are equal. (This is tricky because it is counterintuitive; we are used to the idea that the heavier mass rests lower on the balance. This is true, but in real life the heavier mass tips the balance *all the way* until an obstruction such as a table is encountered.)

Finally, recalling that weight is the gravitational force on an object, the weight of m is equal to $mg = m_1 g$, choice (B).

3. **(D)** In choice (A), the rotational inertia about the y-axis is given to be $ML^2/12$. In choice (B), because all the mass lies on the x-axis (a distance of zero from the x-axis), the rotational inertia about the x-axis is zero. In choice (C), according to the parallel axis theorem, the rotational inertia about the vertical line $x = L/4$ is

$$I_{\text{parallel axis}} = I_{\text{CM}} + MD^2 = \frac{ML^2}{12} + M\left(\frac{L}{4}\right)^2$$

In choice (D), as in choice (C),

$$I_{\text{parallel axis}} = I_{\text{CM}} + MD^2 = \frac{ML^2}{12} + M\left(\frac{L}{2}\right)^2$$

In choice (E), because all of the mass lies a distance $L/2$ from the horizontal line $y = L/2$, the inertia about this line is simply $MR^2 = M(L/2)^2$.

4. **(B)** If the incline is frictionless as in choice (A), it exerts no torque on the ball, so that the ball's angular velocity is constant. In this situation, the translational kinetic energy is converted to gravitational potential energy at the top of the ball's path, but the rotational energy remains constant and is not converted to gravitational potential energy.

If there is sufficient static friction to prevent the ball from slipping as in choice (B), at the peak of the incline when the ball's translational velocity is zero, its angular velocity $\omega = v/r$ must also be zero. In this situation, both translational and kinetic energy are zero at the summit of the path, at which point they *both* have been converted to gravitational potential energy.

Choice (C) is an intermediate between these cases. At the summit of the trajectory the ball is rotating but with less than its original angular speed. Therefore some, but not all, of the rotational kinetic energy has been converted to kinetic energy. Additionally some energy has been lost to friction.

Because the most energy is converted to gravitational potential energy in choice (B), the ball reaches the highest point on the incline in choice (B) (recall that $U = mgy$).

In choice (E), no torque is required for a ball to roll with constant angular velocity, so no friction is needed for the ball to roll steadily along a horizontal floor with *any* particular angular velocity. (Of course, at some point, something must have accelerated the ball, but no torque is required to keep it rolling.)

5. **(E)** There is no net force acting on the mass and thus no torques (note that for an extended object, it is possible to have a nonzero torque while having zero net force), so its angular momentum about any axis is constant. This can also be understood in terms of the equation $L = pr_{\perp}$: In this case, the linear momentum remains constant and $r_{\perp}$, the perpendicular component of the position vector, also remains constant. (It is equal to the x-intercept of the particle's trajectory.)

6. **(E)** Consider Figure 8.29. In order for the net torque about the pivot to be zero, the tension force must be exactly half of the gravitational force. (This is because the tension force acts at twice the distance as the gravitational force, and both forces are parallel so the fraction perpendicular to **r** is the same.) Thus, in order for the net force to be zero, the pivot must exert a force of $\frac{1}{2}mg$ upward.

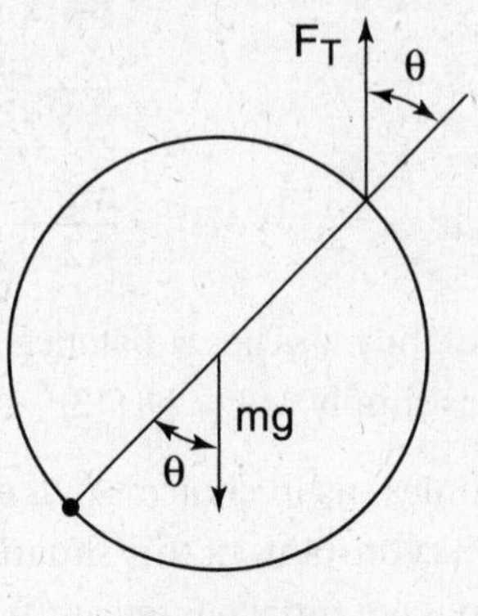

Figure 8.29

7. **(E)** This is an inelastic rotational collision, so the solution is based on conservation of angular momentum:

$$mvr = \left(mr^2 + \frac{1}{2}Mr^2 \right)\omega$$

This equation allows for calculation of the final angular velocity, from which the speed can be calculated using the equation $v = r\omega$. The initial angular momentum was calculated using the formula $L = pr_\perp$. (Based on Figure 8.30 and the fact that the mass is on a collision course with the rim of the wheel, $r_\perp$ equals the radius of the wheel, r.) The final rotational inertia of the mass-wheel combination is equal to the rotational inertia of the mass, mr^2, plus the rotational inertia of the wheel, $\frac{1}{2}Mr^2$.

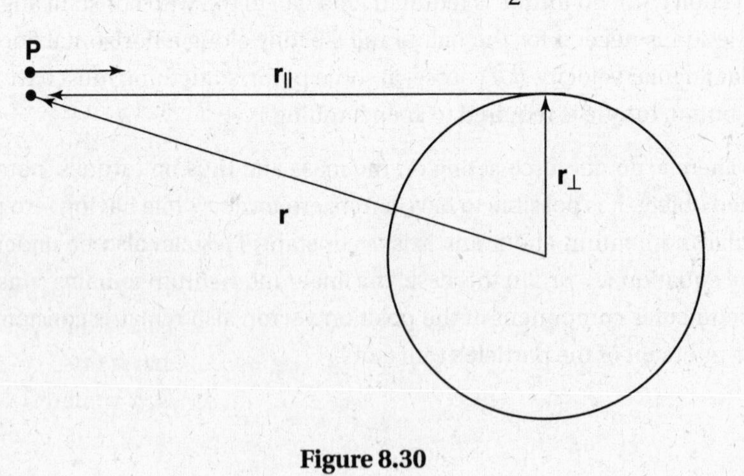

Figure 8.30

8. **(C)** Consider Figure 8.31, which shows how the position of point P can be considered the sum of the position of the center of mass with respect to some stationary origin and the position of P with respect to the center of mass:

$$\mathbf{r}_{P\text{ respect origin}} = \mathbf{r}_{\text{CM respect origin}} + \mathbf{r}_{P\text{ respect CM}}$$

Taking the second derivative of this equation reveals that the accelerations add similarly:

$$\mathbf{a}_{P\text{ respect origin}} = \mathbf{a}_{\text{CM respect origin}} + \mathbf{a}_{P\text{ respect CM}}$$

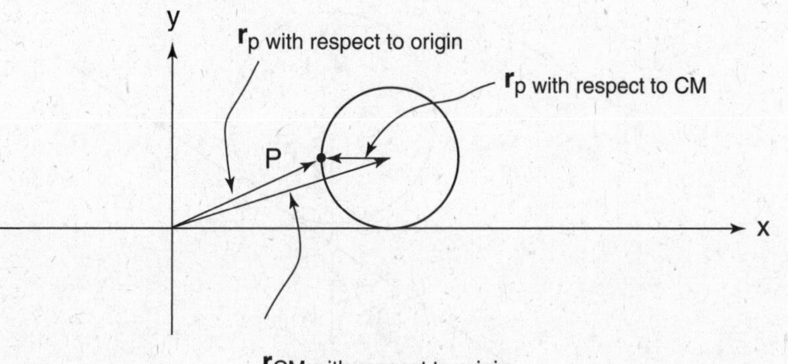

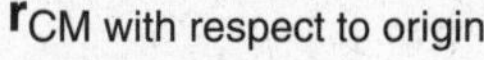

Figure 8.31

The acceleration of the center of mass with respect to the origin is given in the problem to be $a\hat{i}$. The acceleration of point P with respect to the center of mass is equal to the centripetal acceleration (pointing toward the center of the circle), $\mathbf{a}_{P\text{ respect CM}} = v^2/r\hat{i} = \omega^2 r\hat{i}$. Taking the sum of these accelerations yields the net acceleration, choice (C).

9. **(C)** The rotational inertia of a system of point masses is given by the equation $I = \sum m_i r_i^2$:

$$I = 2\left[m\left(\frac{b}{2}\right)^2\right] + 2\left[m\left(\left(\frac{b}{2}\right)^2 + a^2\right)\right] = m(2a^2 + b^2)$$

10. **(C)** Based on the equation $E = \frac{1}{2}mv_{\text{CM}}^2$, the velocity of the center of mass is $v_{\text{CM}} = \sqrt{2E/m}$. This makes the angular velocity of the wheel that is rotating without slipping equal to $\omega = v_{\text{CM}}/r = (1/r)\sqrt{2E/m}$. Finally, plugging this expression for the angular velocity into the equation for rotational kinetic energy, $\text{KE}_{\text{rot}} = \frac{1}{2}I\omega^2$, yields choice (C).

Free-Response

1. (a) The free-body diagram is shown in Figure 8.32. How is the direction of the frictional force determined? The frictional force is the only force that exerts a torque on the cylinder. Because we know intuitively that the cylinder rotates in a counterclockwise sense as it rolls without slipping down the incline, the frictional force must exert a counterclockwise torque by pointing up the slope.

   (b) Net force:

$$F_{\text{net},x} = ma = mg\sin\theta - F_{\text{fr}}$$

$$F_{\text{net},y} = 0 = F_N - mg\cos\theta$$

Net torque:

$$\tau_{\text{net}} = F_{\text{fr}} r = I\alpha = \left(\frac{1}{2}mr^2\right)\left(\frac{a}{r}\right)$$

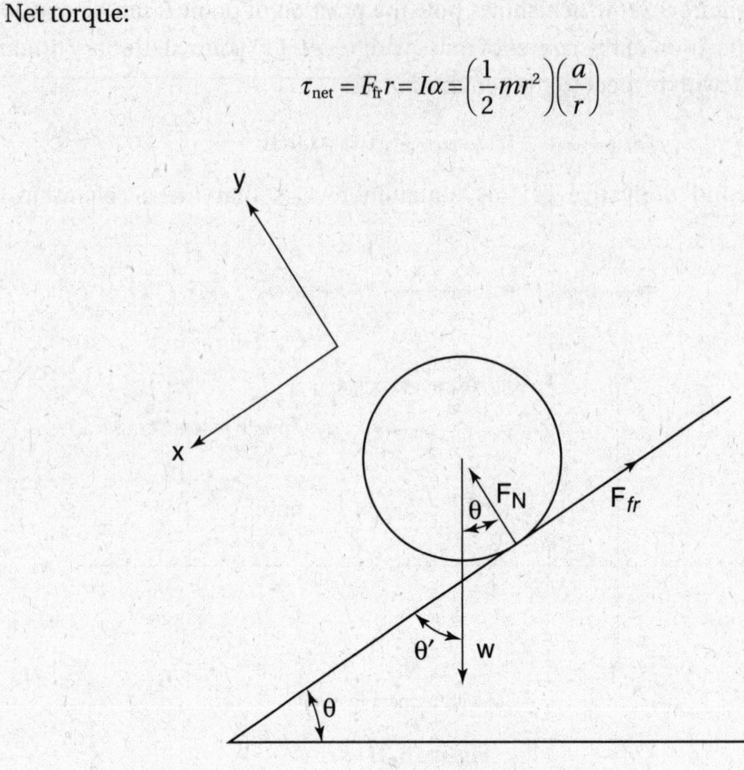

Figure 8.32

(c) The net torque equation can be rearranged to yield $F_{\text{fr}} = \frac{1}{2}ma$. Substituting this equation into the equation for the net force in the x-direction,

$$F_{\text{net},x} = ma = mg\sin\theta - F_{\text{fr}} = mg\sin\theta - \frac{1}{2}ma$$

Solving this equation for the acceleration yields

$$a = \frac{2}{3}g\sin\theta$$

(d) From part (c), the frictional force required to prevent slipping is

$$F_{\text{fr}} = \frac{1}{2}ma = \frac{1}{2}m\left(\frac{2}{3}g\sin\theta\right) = \frac{1}{3}mg\sin\theta$$

Also from part (c), the normal force is $F_n = mg\cos\theta$. Therefore, the minimum coefficient of friction needed is

$$\mu_s = \frac{F_{\text{fr}}}{F_N} = \frac{\frac{1}{3}mg\sin\theta}{mg\cos\theta} = \frac{1}{3}\tan\theta$$

This is static friction because the surface of the cylinder does not slide along the incline. (Although the cylinder is moving, the point at the bottom of the cylinder in contact with the incline is instantaneously at rest.)

2. (a) While the object is sliding, the net force on the object is the kinetic frictional force, $F_{fr} = -\mu_k mg$, and so the acceleration is $a = -\mu_k g$. Therefore, the translational velocity as a function of time is $v(t) = v_0 + at = v_0 - \mu_k gt$.

(b) While the object is rolling and sliding, the net torque on the object is due to the frictional force, $\tau = F_{fr}r = \mu_k mgr$. Applying the rotational analog of Newton's second law, $\alpha = \tau/I = \mu_k mgr \Big/ \frac{1}{2}mr^2 = 2\mu_k g/r$. Therefore, its angular velocity as a function of time is $\omega(t) = \omega_0 + \alpha t = 2\mu_k gt/r$ (designating clockwise rotation as positive).

(c) When the cylinder rolls without slipping, $v = r\omega$.

(d) Setting $v(t) = r\omega(t)$ yields $v_0 - \mu_k gt = 2\mu_k gt$. Solving for time yields $t = v_0/3\mu_k g$.

(e) After the object begins rolling without slipping, its translational and angular velocities remain constant, as explained in case 1. Substituting the value of t calculated above into the equation for $\omega(t)$ allows us to calculate the constant angular velocity of this rolling motion:

$$\omega(t) = \frac{2\mu_k g}{r}\left(\frac{v_0}{3\mu_k g}\right) = \frac{2v_0}{3r}$$

Multiplying by the radius then yields the constant translational velocity of the rolling motion: $v(t) = \frac{2}{3}v_0$.

Answer check: The final velocity is less than the initial velocity due to loss of translational kinetic energy to friction and the conversion of some translational kinetic energy to rotational kinetic energy.

3. (a) Because of the symmetry of parabolas (or conservation of energy, if you prefer), the x-component of velocity remains constant while the y-component changes sign so that the velocity an instant before impact is $\mathbf{v} = (2\text{m/s})\hat{i} - (2\text{m/s})\hat{j}$.

(b) The collision is elastic, so energy is conserved.

$$\frac{1}{2}mv_0^2 = \frac{1}{2}mv_f^2 + \frac{1}{2}I\omega^2$$

$$\frac{1}{2}(4\text{kg})\left(2\sqrt{2}\text{m/s}\right)^2 = \frac{1}{2}(4\text{kg})(2\text{m/s})^2 + \frac{1}{2}(3\text{kg}\cdot\text{m}^2)\omega^2$$

$$\omega = \frac{4}{\sqrt{3}}\text{s}^{-1}$$

(c) At the instant of the collision, there is no external torque on the projectile-sphere system such that its angular momentum is conserved. Designating the clockwise direction as positive and choosing the origin of the coordinate system at the center of the sphere,

$$\mathbf{r} \times \mathbf{p}_0 = \mathbf{r} \times \mathbf{p}_f + I\boldsymbol{\omega}$$

Note: In computing $\mathbf{r} \times \mathbf{p}$, the sign can be obtained by the right-hand rule, and the magnitude is found by multiplying $|\mathbf{r}|$ by the component of $\mathbf{p}$ perpendicular to it, namely, mv_x. Designating clockwise as positive,

$$(3\text{m})(4\text{kg})(2\text{m/s}) = (3\text{m})(4\text{kg})(2\text{m/s})\cos\phi + (3\text{kg}\cdot\text{m}^2)\left(4/\sqrt{3}\text{s}^{-1}\right)$$

Solving for $\cos\phi$ and taking the arccosine gives

$$\phi = 44.7°$$

4. (a) This is an inelastic collision, so we solve by using conservation of momentum. Note that after the mass sticks to the rod, they move with a common speed.

$$mv_0 = (2m)v_f \Rightarrow v_f = \frac{v_0}{2}$$

(b) Because there are no external forces, the velocity of the center of mass remains constant throughout the collision. Therefore, the velocity of the center of mass after the collision is equal to the velocity of the center of mass before the collision, which is given by

$$v_{CM} = \frac{\sum m_i v_i}{\sum m_i} = \frac{m(v_0) + m(0)}{m + m} = \frac{v_0}{2}$$

(c) The center of mass moves with the velocity of the center of mass. Because the velocity of the center of mass is constant, this point moves in a straight line. Setting the origin at the puck (as specified in the problem), the distance between the center of mass and the puck is

$$x_{CM} = \frac{m(0) + m(l/2)}{m + m} = \frac{l}{4}$$

Therefore, a point along the rod a distance of $l/4$ from the puck moves in a straight line, and the system rotates about this point.

(d) The combined system rotates about its center of mass. Because there is no net external torque, the angular momentum about the center of mass is conserved.

- The parallel axis theorem gives us the rotational inertia of the rod about its new center of mass (a distance of $l/4$ from the center of the rod): $I = \frac{1}{12}ml^2 + m(l/4)^2 = \frac{7}{48}ml^2$.
- The rotational inertia of the puck about the new center of mass is $m(l/4)^2$.
- The initial angular momentum of the puck (about the center of mass of the combined system) is given by the equation

$$L = pr_\perp = (mv_0)(l/4)$$

Thus, conservation of angular momentum yields

$$mv_0 \frac{l}{4} = \left[\frac{7}{48}ml^2 + m\left(\frac{l}{4}\right)^2\right]\omega$$

Solving,

$$\omega = \frac{6v_0}{5l}$$

(e) Plot ω versus v_0, as shown in Figure 8.33. Fit the data to the straight line shown. Then calculate the slope of the line:

$$\frac{(7.2 - 4.75)}{(1.2 - 0.8)} = 6.125\ \text{m}^{-1}$$

Compare this to the theoretical result found for part (d):

$$\frac{6}{5\ell} = 6\ \text{m}^{-1}$$

This is reasonable agreement. You would have to know more about the uncertainties in the measurements of ω and v_0 to assess how good the agreement is between theory and experiment.

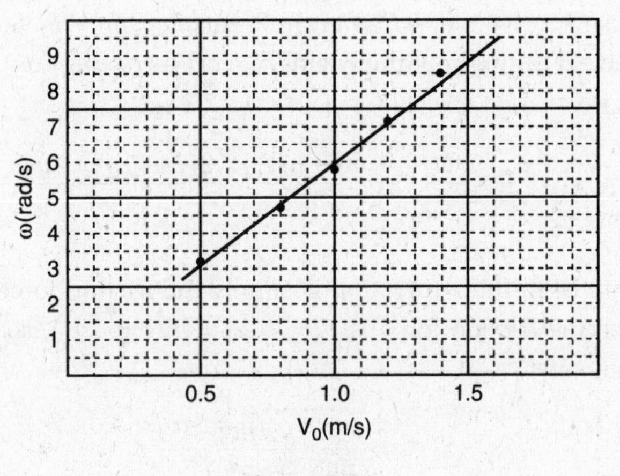

Figure 8.33

5. (a) The free-body diagram is shown in Figure 8.34. In the absence of friction, the spool rotates clockwise, slipping to the left at its point of contact with the table. The friction force opposes this slipping by pointing to the right.

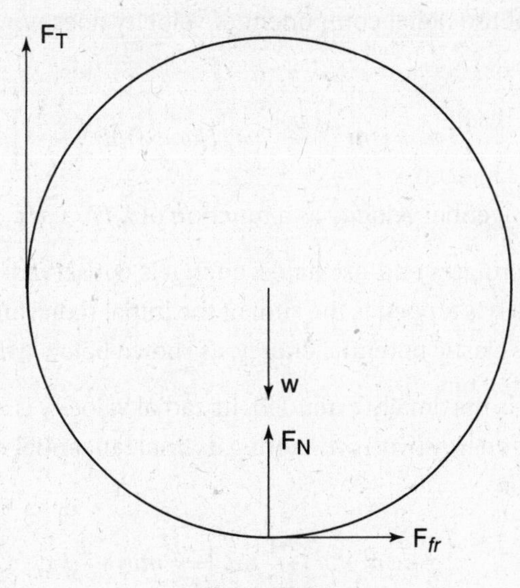

Figure 8.34

(b) Writing Newton's second-law equations (designating clockwise acceleration as positive so that its sign is compatible with the sign of the linear acceleration),

$$F_{\text{net},y} = 0 = F_T + F_N - mg$$

$$F_{\text{net},x} = ma = F_{\text{fr}}$$

$$\tau_{\text{net}} = I\alpha = \left(\frac{1}{2}mr^2\right)\alpha = F_T r - F_{\text{fr}} r$$

This gives us three equations with five variables, a, α, F_{fr}, F_N, and F_T. The last two equations required to solve the system are $F_{fr} = \mu_s F_N$ and $a = r\alpha$. (*Note*: Because we are looking for the maximum tension force that can be exerted without causing slipping, the static friction force will be at its maximum value. Also, $a = r\alpha$ because the spool rolls without slipping.)

(c) The magnitude of the torque is $\tau_{net} = rF_T - rF_{fr}$. Because the spool is slipping, the friction force is the kinetic friction force $F_{fr} = \mu_k F_N$. The net force in the y-direction must still be zero, so $F_{net,y} = 0 = F_T + F_N - mg$ and thus $F_N = mg - F_T$. Substituting these equations into the net torque equation yields $\tau_{net} = rF_T - r\mu_k(mg - F_T)$, such that the angular acceleration is

$$\alpha = \frac{\tau_{net}}{I} = \frac{r(F_T - \mu_k mg + \mu_k F_T)}{\frac{1}{2}mr^2} = \frac{2(F_T - \mu_k mg + \mu_k F_T)}{mr}$$

(d) The net force in the x-direction is simply the friction force. Based on the results from part (c), the friction force is $F_{net,x} = F_{fr} = \mu_k(mg - F_T)$, so the magnitude of the acceleration is

$$a = \frac{F_{net}}{m} = \frac{\mu_k(mg - F_T)}{m}$$

6. (a) The spring force must provide the centripetal force for the mass: $kx = mv_0^2/r$. Therefore, the extension of the spring is $x = mv_0^2/rk$, and the unstretched length of the spring is $r - x = r - mv_0^2/rk$.

(b) The radial tap imparts no torque to the mass, so angular momentum is conserved. (Note that the radial component of velocity does not contribute to the angular momentum.)

$$\left[I_0\omega_0 = (mr^2)\left(\frac{v_0}{r}\right)\right] = \left[I_f\omega_f = (ml^2)\left(\frac{v_f}{l}\right)\right]$$

Solving for the tangential velocity as a function of l, $v_f = v_0 r / l$.

(c) As the spring rotates and oscillates, energy is conserved. The initial energy the instant after the mass is tapped is the sum of the initial radial and tangential kinetic energies as well as the elastic potential energy, as shown below (where $x_0 = mv_0^2/rk$).

When the spring is maximally extended, its radial velocity is zero, so the mass has only tangential kinetic energy (with v_f denoting its final tangential velocity and x_f denoting its maximal extension).

$$\left[\frac{1}{2}m(v_0^2 + v_r^2) + \frac{1}{2}kx_0^2\right] = \frac{1}{2}mv_f^2 + \frac{1}{2}kx_f^2$$

(Note that m, v_0, v_r, k, and r are given quantities and $x_0 = mv_0^2/rk$.) This gives us one equation with two variables, x_f and v_f. These variables are related by the conservation of angular momentum as shown in part (b): $v_f = v_0 r / l$, where l, the final length of the spring, is equal to the unstretched length of the spring calculated in part (a) $(r - mv_0^2/rk)$ plus the extension x_f: $[r - (mv_0^2/rk) + x_f]$.

Simple Harmonic Motion 9

→ **DEFINITION OF SIMPLE HARMONIC MOTION (SHM)**
→ **PHYSICAL SYSTEMS EXHIBITING SHM**
→ **DEFINITIONS OF AMPLITUDE, FREQUENCY, AND PERIOD**
→ **USING NEWTON'S LAWS OR CONSERVATION OF ENERGY IN SHM PROBLEMS**

Simple harmonic motion (SHM) is any oscillation that is governed by a linear restoring force or restoring torque such that the displacement from equilibrium can be described using sine and/or cosine functions. SHM is a special case of *periodic motion*, which is any oscillatory repetitive motion due to a restoring force of any form.

What Is the General Mathematical Description of a Mass Connected to a Spring Oscillating on a Horizontal Frictionless Surface? We start with Hooke's law:

$$\mathbf{F}_{\text{net}} = m\mathbf{a} = m\frac{d^2x}{dt^2}\hat{i} = -kx\hat{i}$$

$$\frac{d^2x}{dt^2} = -\frac{k}{m}x$$

This is a second-order differential equation. Although you are not responsible for solving it, you should be able to see how the following two equations are solutions to this equation (you can do so by taking their derivatives and inserting them into the above equation).

Simple harmonic motion: a special kind of oscillatory motion

$$x(t) = A\cos\left(t\sqrt{\frac{k}{m}} + \phi\right) \quad \text{or} \quad x(t) = A\sin\left(t\sqrt{\frac{k}{m}} + \varphi\right)$$

You should be aware of the following properties of these equations:

- Based on the trigonometric identity $\sin\alpha = \cos(\alpha - \pi/2)$, these two solutions are equivalent if $\phi = \varphi - \pi/2$. Therefore, when solving problems, the choice of which equation to use is arbitrary.
- There are only two parameters that completely define the spring's motion: its position and velocity at time $t = 0$. These two parameters tell us everything there is to know about the system, and from them we can calculate the spring's velocity and position at any other time. The fact that there are two physical parameters is mirrored by the fact that there are two parameters (A and ϕ or φ) appearing in $x(t)$. This can also be explained mathematically in that two integrations are involved in moving from d^2x/dt^2 to $x(t)$, each of which introduces a constant of integration.

Important: The key equation in the last example was

$$F = -kx \Leftrightarrow \frac{d^2x}{dt^2} = -\frac{k}{m}x$$

Whenever the second time derivative of a position function (such as d^2x/dt^2 or $d^2\theta/dt^2$) is equal to a negative constant multiplied by the position (x or θ), the solutions of this differential equation involve sines or cosines and are examples of SHM. For example, if you were to encounter the following equation,

$$\tau = I\alpha = I\frac{d^2\theta}{dt^2} = -(\text{constant})\theta$$

you should immediately recognize this as simple harmonic motion.

What Causes SHM? In Which Situations Does SHM Arise? As discussed above, SHM occurs whenever there is a linear restoring force (a force that pushes the object back to its equilibrium position whose magnitude is proportional to the displacement from equilibrium). When do such linear restoring forces occur? We begin by exploring restoring forces in general.

Consider a particle trapped in a potential energy valley as in Figure 9.1. The total energy is constant (energy is conserved), so the kinetic energy equals the total energy minus the potential energy. Consequently, the particle has zero kinetic energy (and is temporarily at rest) at turning points a and b, whereas it has maximum kinetic energy (and maximum speed) at point c. It cannot go beyond the well, which would require it to have negative kinetic energy.

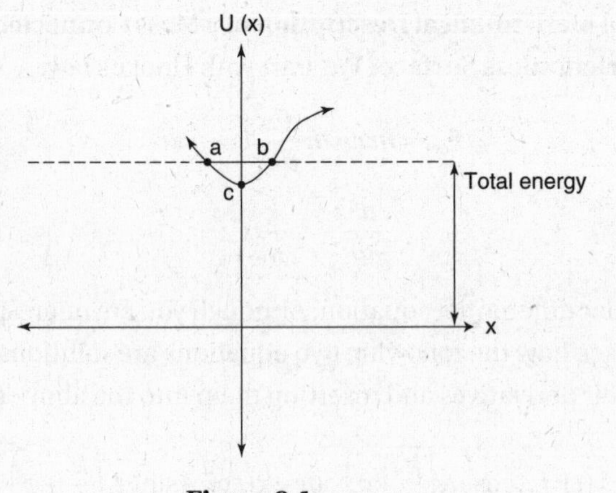

Figure 9.1

Recall the relationship between potential energy (U) and force:

$$F = -\frac{dU}{dx}$$

By taking the negative derivative of $U(x)$, we can obtain a force graph as shown in Figure 9.2. Examination of the slope of the $U(x)$ curve reveals that the force is zero at point c, positive between a and c, and negative between c and b, so that dF/dx near the equilibrium point is negative. This is what we call a *restoring force* because it always tries to push the particle toward the equilibrium point c (to restore the object's position to point c).

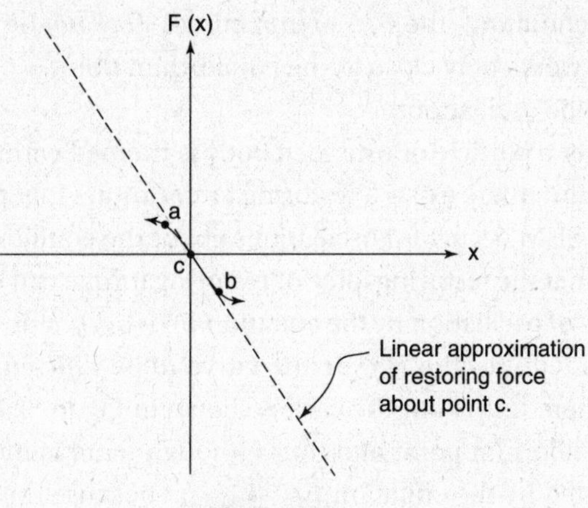

Figure 9.2

Note the way we defined the origin ($x = 0$) to lie at the equilibrium point; this is generally a good choice in SHM problems. Because we are interested in the motion close to the bottom of the well, it is useful to make a linear approximation of $F(x)$ about the equilibrium point ($x = 0$), as shown above. Since the force at this point must be zero and the slope is negative, the linear approximation is

$$F \approx \left(\frac{dF}{dx}\bigg|_{x=0}\right)x = \left(\frac{-d^2U}{dx^2}\bigg|_{x=0}\right)x = -k_{\text{effective}}x$$

(We call the slope $k_{\text{effective}}$ because near $x = 0$ the force is approximately equal that of a spring with a spring constant $k_{\text{effective}}$.) Applying Newton's second law,

$$F = ma = m\frac{d^2x}{dt^2} \approx -k_{\text{effective}}x$$

Rearranging, we find

$$\frac{d^2x}{dt^2} = -\frac{k_{\text{effective}}}{m}x$$

TIP

SHM occurs when the force on the object returns it to equilibrium and is proportional to the displacement from equilibrium.

We recognize the classic SHM equation (the second derivative being set equal to a negative constant multiplied by the position).

Rotational analog: If we are working in rotational coordinates, the above discussion corresponds to having an energy well in a $U(\theta)$ function and approximating the torque about the equilibrium point (where $\theta = 0$) using the following equation.

$$\tau \approx \left(\frac{d\tau}{d\theta}\bigg|_{\theta=0}\right)\theta = \left(\frac{-d^2U}{d\theta^2}\bigg|_{\theta=0}\right)\theta = -k_{\text{effective}}\theta$$

Applying Newton's second law (rotational form) yields

$$\tau = I\alpha = I\frac{d^2\theta}{dt^2} \approx -k_{\text{effective}}\theta$$

$$\frac{d^2\theta}{dt^2} \approx -\frac{k_{\text{effective}}}{I}\theta$$

Another SHM equation! But these equations only approximate the restoring force or torque. When are they valid? This depends on the particular $F(x)$ or $\tau(\theta)$ curve. The spring force, $F = -kx$, is already linear, so that the linear approximation always holds true. In general situations (e.g., the pendulum), the $F(x)$ or $\tau(\theta)$ curves may *not* be linear and therefore the linear approximation works only close to the equilibrium point. Consequently, SHM occurs only for small-amplitude oscillations.

Summary: Whenever a particle or extended body is trapped within a potential energy valley (a stable equilibrium), there exists a restoring force/torque that pushes the object toward its equilibrium point. SHM occurs for oscillations about the equilibrium point whose amplitude is small enough that the restoring force or restoring torque can be linearly approximated within the boundaries of oscillation by the equation $F \approx -k_{\text{effective}}x$ or $\tau = -k_{\text{effective}}\theta$.

Caution: SHM also requires the $F(x)$ or $\tau(\theta)$ curve to be *differentiable* at the equilibrium point. An example where this is not the case is shown in Figure 9.3. In this case, $F(x)$ is not continuous at the equilibrium point and thus although there is indeed a restoring force, it cannot be approximated by the equation $F \approx -k_{\text{effective}}x$ because the first derivative does not exist.

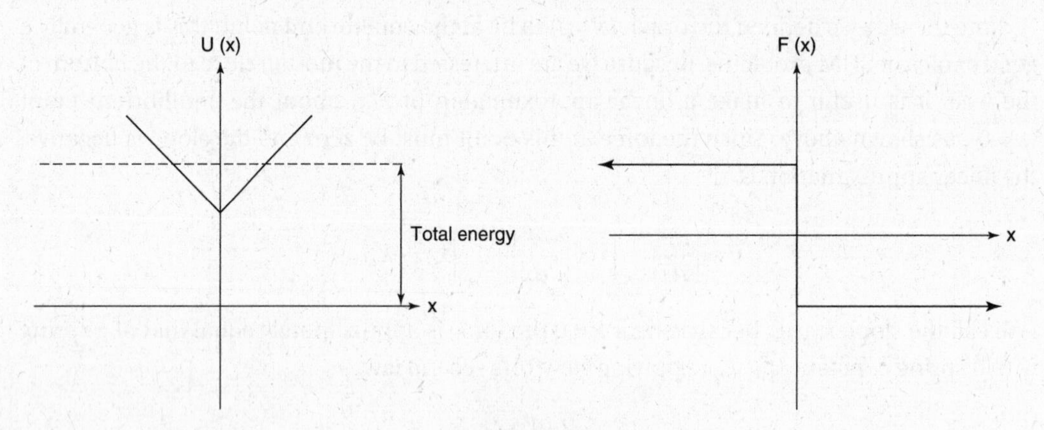

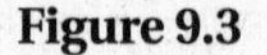

Figure 9.3

We Have Calculated the General Equation Describing the Oscillation of a Mass Connected to a Spring. What Is the General Equation for Simple Harmonic Motion? Starting with the SHM equations for linear and rotational motion,

$$\frac{d^2x}{dt^2} = -\frac{k_{\text{effective}}}{m}x$$

$$\frac{d^2\theta}{dt^2} = -\frac{k_{\text{effective}}}{I}\theta$$

The general position equation resulting from these differential equations is completely analogous to the general position equation for a mass oscillating on a spring:

$$x(t) \text{ or } \theta(t) = A\cos(\omega t + \phi) \quad \text{or} \quad x(t) \text{ or } \theta(t) = A\sin(\omega t + \varphi), \text{ given that}$$

$$\omega = \sqrt{\frac{k_{\text{effective}}}{m}} \quad \text{(for linear systems)}$$

$$\omega = \sqrt{\frac{k_{\text{effective}}}{I}} \quad \text{(for angular systems)}$$

Most general form of SHM

The argument of the sine or cosine function, $\omega t + \phi$, is called the *phase*. ϕ or φ, the *phase shift angle*, determines the horizontal shifting of the cosine or sine curve and thus influences the initial position and velocity. ω, the *angular velocity* or *angular frequency*, is the time derivative of the phase, with units radians/second. ω is a measure of how rapidly the position and velocity cycle back and forth. Note that in general the unmodified word *frequency* generally refers to linear frequency (discussed below) rather than angular frequency.

PROBLEM SOLVING: CALCULATING THE ANGULAR FREQUENCY OF OSCILLATION

STEP 1 Obtain an equation for the restoring force or restoring torque, $F(x)$ or $\tau(\theta)$.

STEP 2 Differentiate this equation at the equilibrium point to calculate the effective spring constant $k_{\text{effective}}$:

$$k_{\text{effective}} = -\left.\frac{dF}{dx}\right|_{\text{equilibrium distance}} \quad \text{or} \quad k_{\text{effective}} = -\left.\frac{d\tau}{d\theta}\right|_{\text{equilibrium distance}}$$

STEP 3

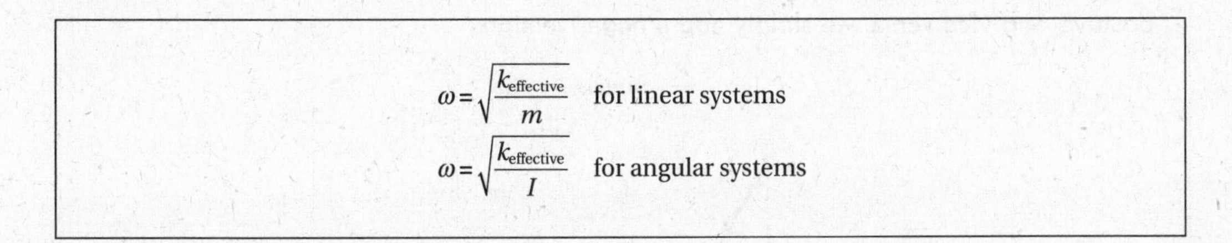

How Is the Angular Frequency from this General Position Equation Related to the Linear Frequency and Period of Oscillation?

A *cycle* is one complete oscillation (e.g., moving from $-A$ to $+A$ and back to $-A$). A cycle is like a radian in that it is dimensionless. *Linear frequency* (symbol f or sometimes υ, the Greek letter *nu*) is the number of cycles per second (having units of 1 hertz = 1 cycle/second). Linear frequency is related to angular frequency as follows:

$$f = \frac{\omega \text{ radians/second}}{2\pi \text{ radians/cycle}} = \frac{\omega}{2\pi}$$

The period, defined as the *time* per cycle, is the inverse of the frequency:

$$T \frac{\text{seconds}}{\text{cycle}} = \frac{1}{f \text{ cycles/second}}$$

One period of oscillation corresponds to the time required to change the phase angle, $\omega t + \phi$, by 2π (the period of the cosine and sine functions):

$$T = \frac{2\pi \text{ radians/cycle}}{\omega \text{ radians/second}} = \frac{2\pi}{\omega} \frac{\text{seconds}}{\text{cycle}}$$

These relationships can be summarized in one equation:

$$f = \frac{\omega}{2\pi} = \frac{1}{T}$$

General relationship among angular frequency, linear frequency, and period

EXAMPLE 9.1

Calculate the frequency of oscillation of a simple pendulum of length *l*, assuming small oscillations.

SOLUTION

STEP 1 **The pendulum undergoes rotational motion, so we use τ and θ. Referring to Figure 9.4,**

$$\tau = F_{\text{perpendicular}}\, l = mg\sin\theta l$$

It is clear that the force is a restoring force, so to make it negative when θ is positive, and vice versa, we simply add a negative sign:

$$\tau = -mgl\sin\theta$$

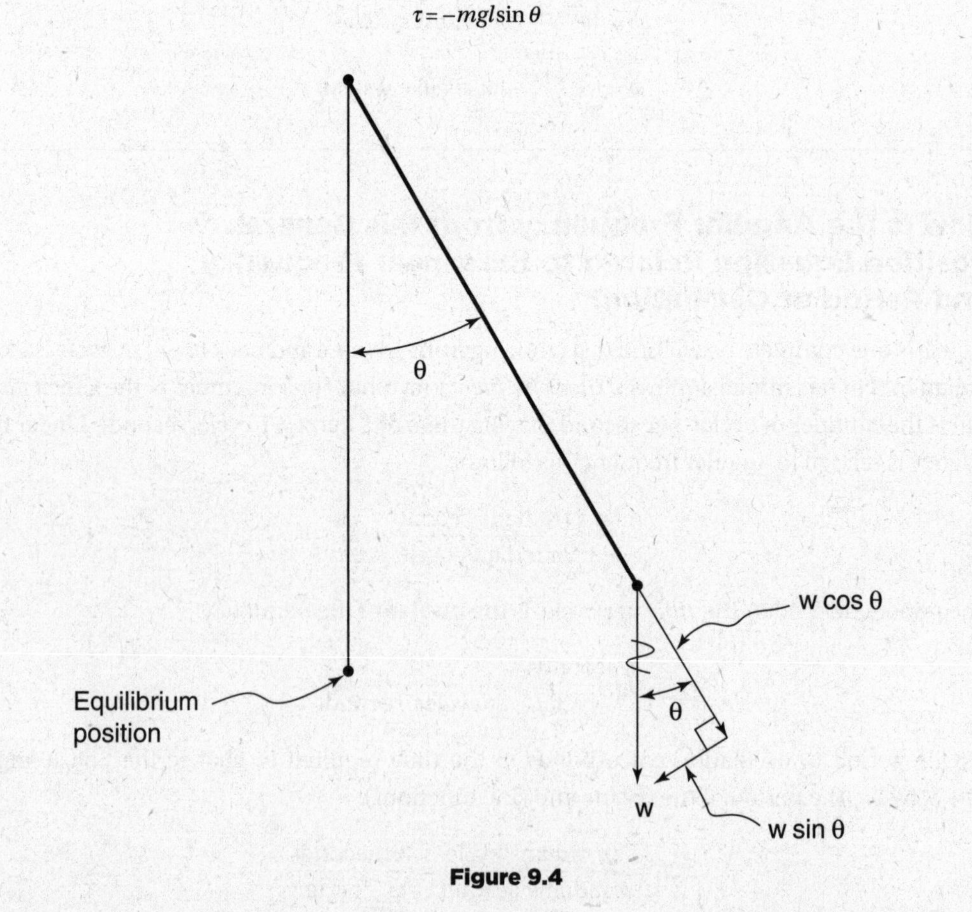

Figure 9.4

STEP 2

$$k_{\text{effective}} = -\left.\frac{d\tau}{d\theta}\right|_{\text{equilibrium angle}} = mgl\cos\theta\big|_{\theta=0} = mgl$$

STEP 3

$$\omega = \sqrt{\frac{k_{\text{effective}}}{I}} = \sqrt{\frac{mgl}{ml^2}} = \sqrt{\frac{g}{l}}$$

$$f = \frac{\omega}{2\pi} = \frac{1}{2\pi}\sqrt{\frac{g}{l}}$$

Angular and linear frequencies of a simple pendulum

Reality check: It might be a little surprising that the mass of the pendulum bob does not appear in the final equation. This makes sense though—increasing the mass increases the torque, but it also increases the rotational inertia, two effects that cancel each other out. It also makes sense that the frequency of the pendulum increases with increasing gravity and decreases with increasing length of the pendulum (the torque is proportional to l, but the rotational inertia is proportional to l^2, so the net effect of increasing l is to *decrease* the angular acceleration).

Alternative solution:

$$\tau = -mgl\sin\theta$$

For small θ, the approximation $\sin\theta \approx \theta$ is valid. (Why? This is a linear approximation of the sine curve about $x = 0$, given that $\sin 0 = 0$ and the slope of the sine curve is 1 at $\theta = 0$. The same is true of the tangent function, which can thus also be approximated as $\tan\theta \approx \theta$ for small angles.)

$$\tau = -mgl\theta$$

At this point, we extract the constant expression in front of θ:

$$k_{\text{effective}} = mgl$$

This yields the same answer as the previous method.

TIP

When angles are measured in radians (but not degrees!), you can approximate $\sin\theta \approx \theta$ for small θ.

EXAMPLE 9.2 LINEAR FREQUENCY AND THE PERIOD OF THE MASS-SPRING SYSTEM

What are the linear frequency and period of a horizontal mass-spring system as described at the beginning of this chapter?

SOLUTION

We have derived the angular frequency of the horizontal mass-spring system and can relate this result to the linear frequency and period as follows.

$$\omega = \sqrt{\frac{k}{m}}$$

$$f = \frac{\omega}{2\pi} = \frac{1}{2\pi}\sqrt{\frac{k}{m}}$$

$$T = \frac{1}{f} = \frac{2\pi}{\omega} = 2\pi\sqrt{\frac{m}{k}}$$

Angular frequency, linear frequency, and period of horizontal mass-spring system

Note that the angular frequency, linear frequency, and period do not depend on the amplitude of the oscillation. This is a general property of SHM that can be understood in that increasing the amplitude increases the forces involved (because force is always proportional to position) yet also increases the distance that must be traversed in a cycle.

EXAMPLE 9.3 A VERTICAL SPRING SYSTEM

A mass hanging from a spring (spring constant k) is at rest.

(a) What is the equilibrium extension of the spring?

(b) What is the angular frequency of oscillation about this equilibrium point?

SOLUTION

(a) At the equilibrium point, there is no net force on the spring. Therefore, with $y = 0$ denoting the end of the unstretched spring,

$$F_{net} = -ky_{eq} - mg = 0$$

$$y_{eq} = \frac{-mg}{k}$$

The extension of the spring is given by $|y_{eq}| = \frac{mg}{k}$.

(b) If the spring is now displaced by a distance d from the equilibrium point,

$$F_{net} = -k(y_{eq} + d) - mg = -kd$$

Gravity does not explicitly appear in the equation.

The rest of the solution is identical to that for the horizontal spring, with the same final results. This is an important result: The frequency of vertical spring systems (where masses are hanging from springs or sitting on springs) is the same as that of horizontal spring systems. Why? Hanging or sitting produces a certain extension or compression that exactly cancels out gravity. Displacements about the new equilibrium position cause the same net restoring force as displacements about the horizontal equilibrium. Vertical spring systems have different equilibrium points but frequencies and periods equal to their horizontal counterparts.

PROBLEM SOLVING: USING ENERGY CONSERVATION

Energy conservation is often required for SHM problems or can serve as a powerful shortcut. Using conservation of energy does not require any novel concepts, but we include it because it is a core technique involved in SHM problems.

TIP

As in Chapter 5, conservation of energy is a very helpful problem-solving tool when you are not asked a question involving time.

EXAMPLE 9.4

A 2.0-kg mass is oscillating on a spring with a spring constant of 10 N/m. When the spring is stretched 1.0 m, the velocity of the mass is 13 m/s. What is the amplitude of the oscillation?

SOLUTION

Energy conservation: (Remember that when the spring is maximally stretched, it has zero speed for an instant, so that the total mechanical energy is equal to the elastic potential energy.)

$$\frac{1}{2}mv_1^2+\frac{1}{2}kx_1^2=\frac{1}{2}m(0)^2+\frac{1}{2}kx_{\text{maximum}}^2$$

$$\frac{1}{2}(2\,\text{kg})\left(13\frac{\text{m}}{\text{s}}\right)^2+\frac{1}{2}\left(10\frac{\text{N}}{\text{m}}\right)(1\,\text{m})^2=\frac{1}{2}\left(10\frac{\text{N}}{\text{m}}\right)x_{\text{maximum}}^2 \Rightarrow x_{\text{maximum}}=5.9\,\text{m}$$

Conceptual Bridge: How Is Uniform Circular Motion (UCM) Related to Simple Harmonic Motion? We know that the force acting on an object in UCM is equal in magnitude to mv^2/r and points toward the center of the circle. Based on Figure 9.5, we can resolve this force into components (taking advantage of the fact that r is constant).

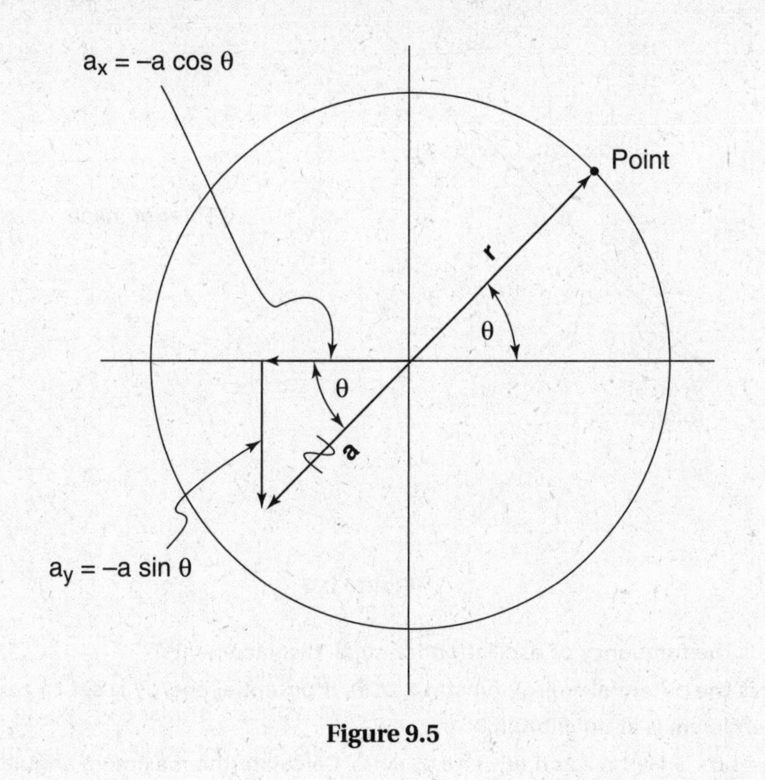

Figure 9.5

$$F_x = -\frac{mv^2}{r}\cos\theta = -\left(\frac{mv^2}{r^2}\right)(r\cos\theta) = -\left(\frac{mv^2}{r^2}\right)x$$

$$F_y = -\frac{mv^2}{r}\sin\theta = -\left(\frac{mv^2}{r^2}\right)(r\sin\theta) = -\left(\frac{mv^2}{r^2}\right)y$$

Note that each of these equations obeys SHM. Now recall the general form of uniform circular motion we used in Chapter 3:

$$\mathbf{r}(t) = r\left[\cos(\omega t + \phi)\hat{i} + \sin(\omega t + \phi)\hat{j}\right] \Leftrightarrow \begin{cases} x = r\cos(\omega t + \phi) \\ y = r\sin(\omega t + \phi) \end{cases}$$

Thus, each component of UCM obeys the equations for SHM. This link allows us to interpret the angular velocity $\omega = d\theta/dt$ and the phase shift angle ϕ as the rate at which the angle is swept out and the initial angle, respectively. Now we can more fully understand why the formulas relating angular frequency, linear frequency, and period, $f = \omega/2\pi = 1/T$, are exactly the same for UCM and SHM.

EXAMPLE 9.5

A *physical pendulum* is an extended object that oscillates like a pendulum as shown in Figure 9.6. A physical pendulum has a mass of 2.0 kg, a distance between the axle and the center of mass of 1.0 m, and a rotational inertia of 0.5 kg·m². (Take $g = 10\ \text{m/s}^2$.)

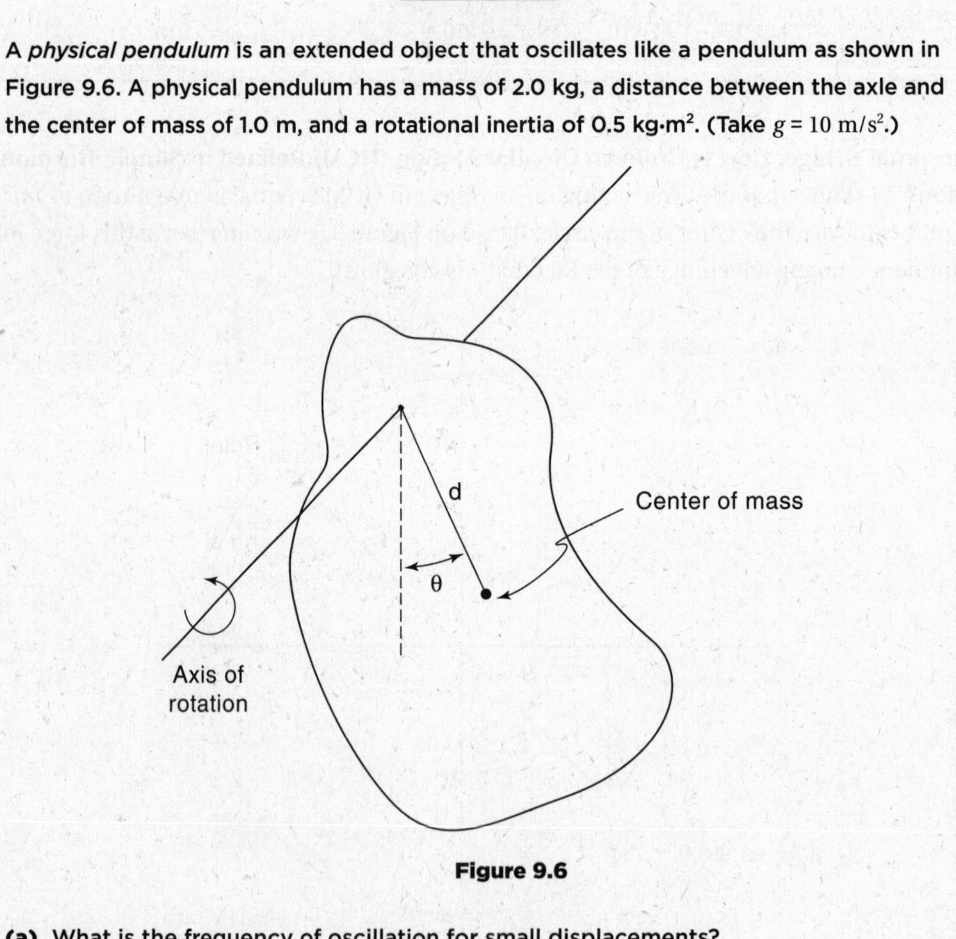

Figure 9.6

(a) **What is the frequency of oscillation for small displacements?**

(b) **What is the potential energy function, *U*(*θ*), if potential energy is set to zero when the pendulum is at equilibrium?**

(c) **If at $t = 1.0$ s, $\theta = 0.1$ rad and $d\theta/dt = 0.05$ rad/s. Calculate the maximum angular velocity $\omega = d\theta/dt$.**

SOLUTION

(a) The restoring torque (exactly like the restoring torque for the simple pendulum) is

$$\tau = -mgd\sin\theta$$

$$k_{\text{effective}} = -\left.\frac{d\tau}{d\theta}\right|_{\text{equilibrium angle}} = mgd\cos\theta\big|_{\theta=0} = mgd$$

$$\omega = \sqrt{\frac{k_{\text{effective}}}{I}} = \sqrt{\frac{mgd}{I}} = 6.32\text{s}^{-1}$$

Symbolic answer check: As in the case of the simple pendulum, the frequency increases with increasing gravity. It makes sense that increasing inertia decreases the angular frequency of the pendulum, while a larger restoring torque (produced by a larger *m*, *g*, and/or *d*) increases the angular frequency.

(b) (See Figure 9.7 for the geometry used here.)

$$U = mgh = mgd(1 - \cos\theta) = (20\text{ J})(1 - \cos\theta)$$

(c)

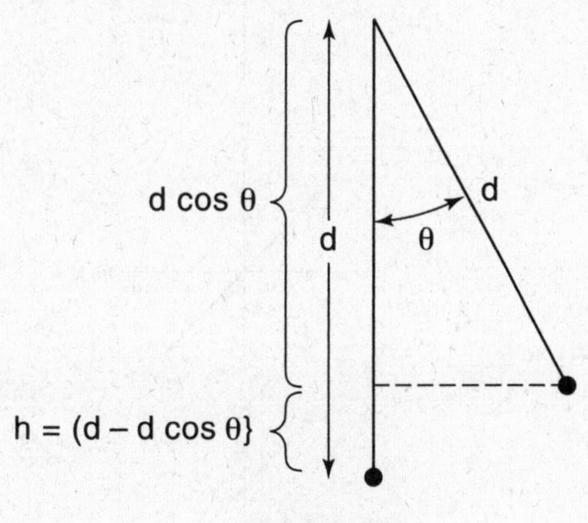

Figure 9.7

$$E_{\text{total}} = U + \text{KE} = mgd(1-\cos\theta) + \frac{1}{2}I\omega^2$$

When the velocity is maximized, the potential energy is zero and all the energy is kinetic:

$$(20\text{ J})(1-\cos 0.1) + \frac{1}{2}\left(\frac{1}{2}\text{ kg m}^2\right)(0.05\text{s}^{-1})^2 = 0 + \frac{1}{2}\left(\frac{1}{2}\text{ kg m}^2\right)\omega_{\text{max}}^2$$

$$\omega_{\text{max}} = 0.63\text{s}^{-1}$$

TIP

Note that in part (a), we cannot use conservation of energy.

CHAPTER SUMMARY

An oscillating object is said to exhibit simple harmonic motion (SHM) if its acceleration is proportional to its displacement from equilibrium. Equivalently, if the displacement obeys $x(t) = A\cos\left(t\sqrt{\frac{k}{m}} + \phi\right)$, then the object undergoes SHM. This motion arises when an object oscillates with small amplitude about its equilibrium position. The curvature of the potential energy function at the equilibrium position determines $k_{\text{effective}}$, which plays the role of a spring constant. Once this constant is determined from the potential energy (or force), the frequency or period of the oscillations can be determined. Energy conservation is a useful problem-solving tool in SHM situations when the time dependence of the motion is not of interest.

PRACTICE EXERCISES

Multiple-Choice Questions

1. Five graphs of a force as a function of position are shown in Figure 9.8. In how many cases can an object subject to this force undergo simple harmonic motion?

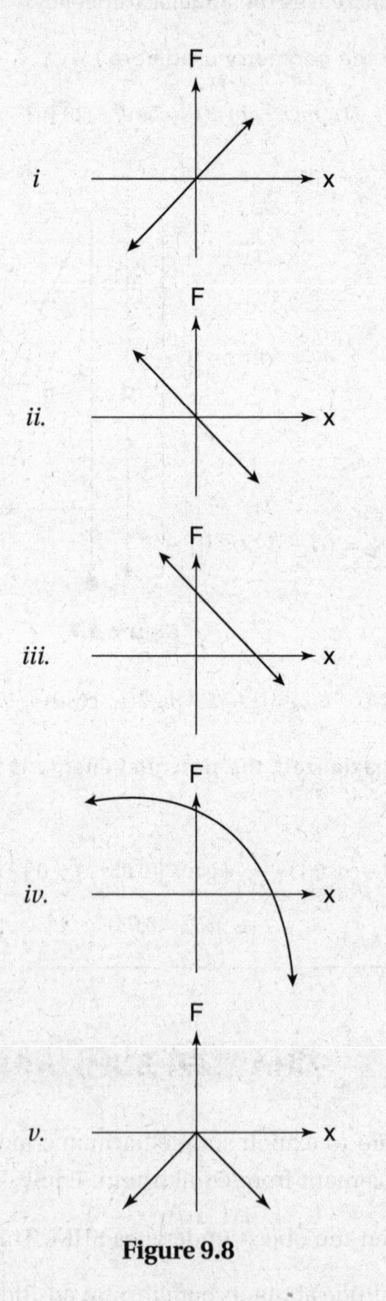

Figure 9.8

(A) none
(B) one
(C) two
(D) three
(E) four

2. A mass m hanging on a spring oscillates vertically. If the equilibrium point of the oscillation is a distance d below the relaxed length of the spring and if the amplitude of the oscillation is A, what is the maximum kinetic energy of the oscillation?

(A) $\frac{1}{2}(d/mg)A^2$
(B) $\frac{1}{2}(mg/d)A^2$
(C) $\frac{1}{2}(d/mg)A^2 + mgd$
(D) $\frac{1}{2}(d/mg)A^2 - mgd$
(E) $(d/mg)A^2 + mgd$

3. Shown in Figure 9.9 is the $v(t)$ curve for an object undergoing SHM. Which of the following is the $x(t)$ function for the object?

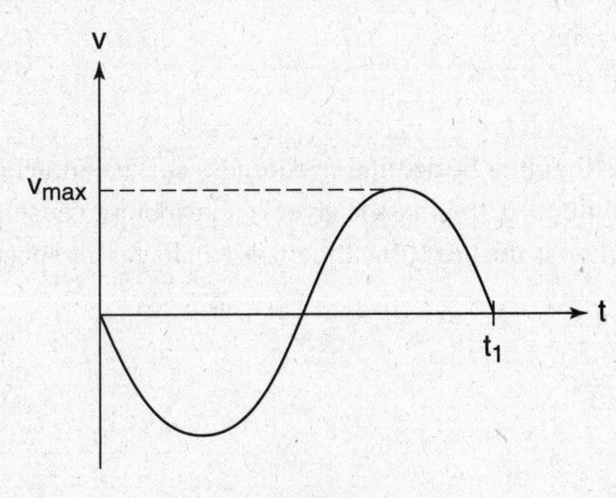

Figure 9.9

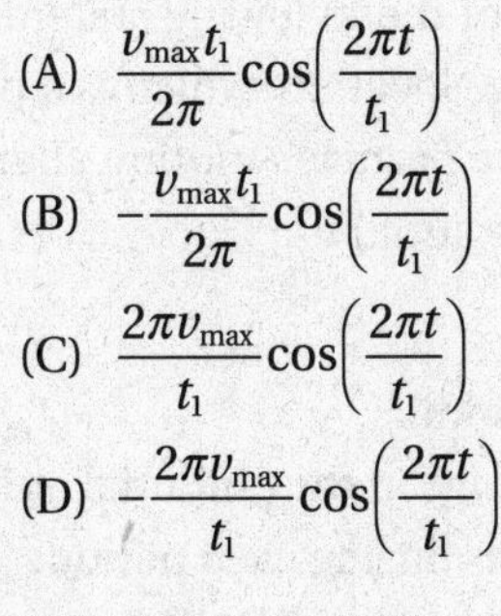
(A) $\frac{v_{max}t_1}{2\pi}\cos\left(\frac{2\pi t}{t_1}\right)$
(B) $-\frac{v_{max}t_1}{2\pi}\cos\left(\frac{2\pi t}{t_1}\right)$
(C) $\frac{2\pi v_{max}}{t_1}\cos\left(\frac{2\pi t}{t_1}\right)$
(D) $-\frac{2\pi v_{max}}{t_1}\cos\left(\frac{2\pi t}{t_1}\right)$

(E) $2\pi v_{max}t_1 \cos(t/2\pi t_1)$

4. The mass in Figure 9.10 undergoes simple harmonic motion as it slides back and forth along the frictionless incline. The angular frequency of the motion depends on which of the following variables?

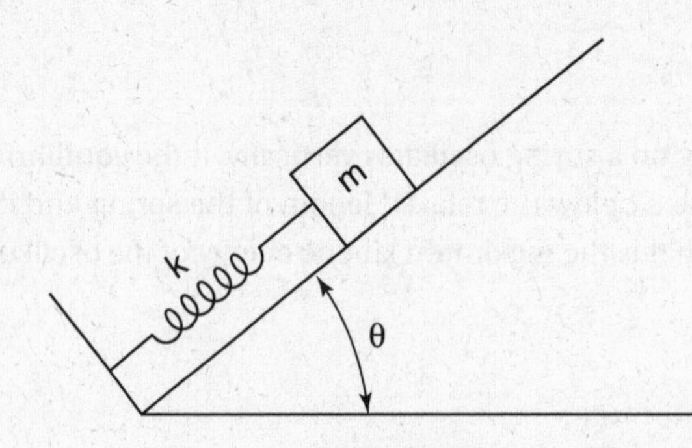

Figure 9.10

I. the spring constant k
II. the mass m
III. the angle of elevation of the incline, θ
IV. the acceleration due to gravity, g

(A) I only
(B) I and II only
(C) I, II, and III only
(D) I, II, and IV only
(E) I, II, III, and IV

5. A mass m is resting on a horizontal frictionless surface attached to a spring of spring constant k. At time $t = 0$, the mass is given a sharp blow, causing it to move to the right with speed v. When is the next time the mass will have this speed?

(A) $t = (1/4\pi)\sqrt{m/k}$
(B) $t = (1/2\pi)\sqrt{m/k}$
(C) $t = (1/\pi)\sqrt{m/k}$
(D) $t = \pi\sqrt{m/k}$
(E) $t = 2\pi\sqrt{m/k}$

6. For a system undergoing simple harmonic motion, which of the following statements is true? (The *equilibrium point* refers to the particle's average location, which is a point of static equilibrium; the turning points are points where the particle's motion changes direction. Assume that the potential energy is zero at equilibrium.)

(A) The kinetic energy is maximized at the turning points.
(B) The potential energy is maximized at the equilibrium points.
(C) The work done by the restoring force is equal to the change in the potential energy.
(D) The kinetic energy equals the potential energy four times during each period.
(E) none of the above

7. Which of the graphs in Figure 9.11 shows a possible relationship between the displacement and velocity of a mass oscillating on a spring?

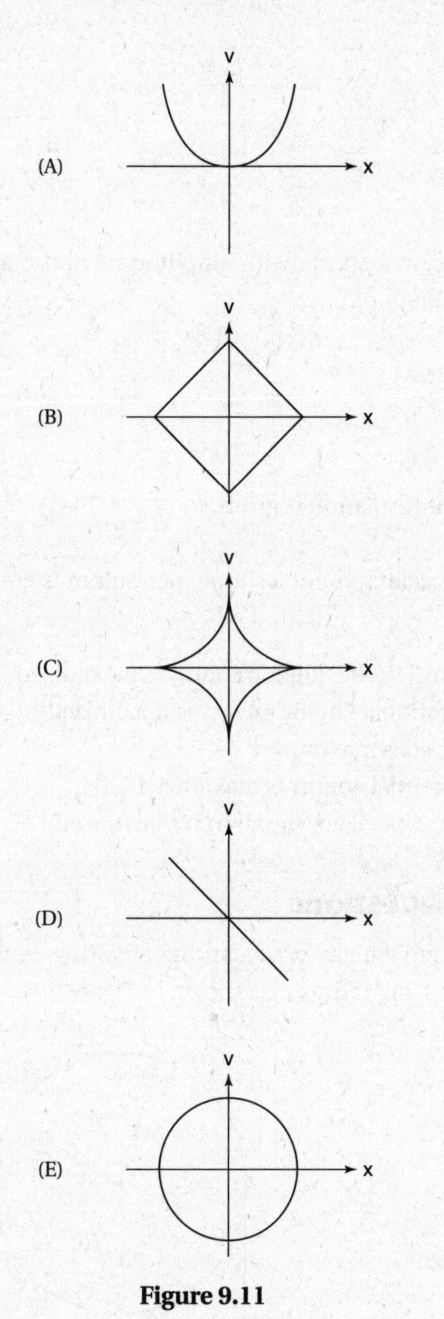

Figure 9.11

8. An object's angular acceleration is given by the equation $d^2\theta/dt^2 = -2\pi\theta$, where θ is in radians and t is in seconds. Which of the following is the *linear frequency* of the object?

   (A) $1/2\pi$Hz
   (B) $1/\sqrt{2\pi}$ Hz
   (C) 1 Hz
   (D) $\sqrt{2\pi}$ Hz
   (E) 2πHz

9. A mass m attached to a vertical spring (with spring constant k) is released from rest from the relaxed position of the spring. What is the amplitude of the resulting oscillation?

   (A) mg/k
   (B) $2\,mg/k$
   (C) $mg/2k$
   (D) $\sqrt{2}(mg/k)$
   (E) $4\,mg/k$

10. A mass oscillates on a spring with amplitude A and maximum velocity v. What is the period of the oscillation?

   (A) $2\pi\sqrt{A/v}$
   (B) $\pi\sqrt{2A/v}$
   (C) $2\pi(A/v)$
   (D) $\pi\sqrt{2}(A/v)$
   (E) Not enough information is given.

11. The tangential acceleration of a simple pendulum is greatest at which of the following times?

   (A) when the translational kinetic energy is maximized
   (B) when the rotational kinetic energy is maximized
   (C) when the speed is maximized
   (D) when the potential energy is maximized
   (E) when the centripetal acceleration is maximized

Free-Response Questions

1. A string connecting a mass m to a spring of spring constant k passes over a pulley of rotational inertia I and radius r, as shown in Figure 9.12.

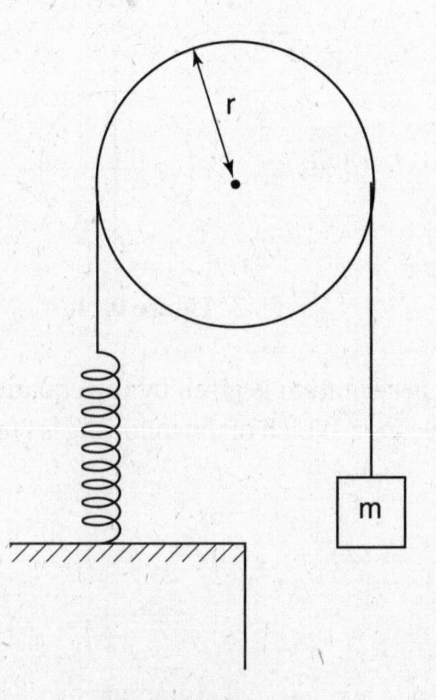

Figure 9.12

(a) At equilibrium, what is the extension of the spring?

(b) The mass is pulled downward a distance Δx. Write Newton's second-law equations for the pulley and the mass.

(c) Solve these equations for the linear acceleration of the system in terms of m, g, I, r, and Δx.

(d) For small displacements, what is the angular frequency of the system?

2. Two equal masses m lying on a frictionless horizontal surface are attached to each other by a spring of spring constant k, as in Figure 9.13. The spring is compressed a distance Δx, and then both masses are released from rest at the same instant.

Figure 9.13

(a) What is the maximum speed of the masses?

(b) What is the period of oscillation of each mass?

The same experiment is then performed with the difference that the two blocks attached to the spring have *different* masses (m and M). The spring constant is still k, and the initial compression distance is Δx.

(c) What is the maximum speed of mass M?

(d) What is the amplitude of oscillation of mass M?

3. Consider the simple pendulum discussed in Example 9.1. For small angles θ (in radians) such that $\sin\theta \approx \theta$, the frequency of oscillation is independent of the mass and is given by $f = \frac{1}{2\pi}\sqrt{\frac{g}{\ell}}$. An experimental study of this system in this limit can yield a very accurate measurement of g.

(a) Approximately, what is the largest value of θ (in degrees) that will make the difference between θ and $\sin\theta$ no greater than 0.2%?

(b) An experiment is carried out to check the validity of $f = \frac{1}{2\pi}\sqrt{\frac{g}{\ell}}$ and obtain a value of g. The period of oscillation is measured for a pendulum with strings of different lengths and with small angles of oscillation. The following data are obtained:

ℓ (meters)	0.25	0.5	0.75	1.0
***T* (seconds)**	1.0	1.4	1.7	2.0

What quantities should be plotted on the grid below to obtain a straight line?

(c) Plot the data, fit the points as well as possible to a straight line, and obtain a value of g.

ANSWER KEY

1. **D**	4. **B**	7. **E**	10. **C**
2. **B**	5. **D**	8. **B**	11. **D**
3. **A**	6. **D**	9. **A**	

ANSWERS EXPLAINED

Multiple-Choice

1. **(D)** Simple harmonic motion requires a linear restoring force: a force proportional to displacement that always points toward the equilibrium point [the point where there is no force on the object, which is the point where the $F(x)$ curve passes through the x-axis]. This corresponds to a point where the $F(x)$ curve passes across the x-axis with a differentiable negative slope. Equivalently, this corresponds to a point of static equilibrium in a potential energy function, $U(x)$. $U(x)$ curves corresponding to each of the $F(x)$ functions appear in Figure 9.14.

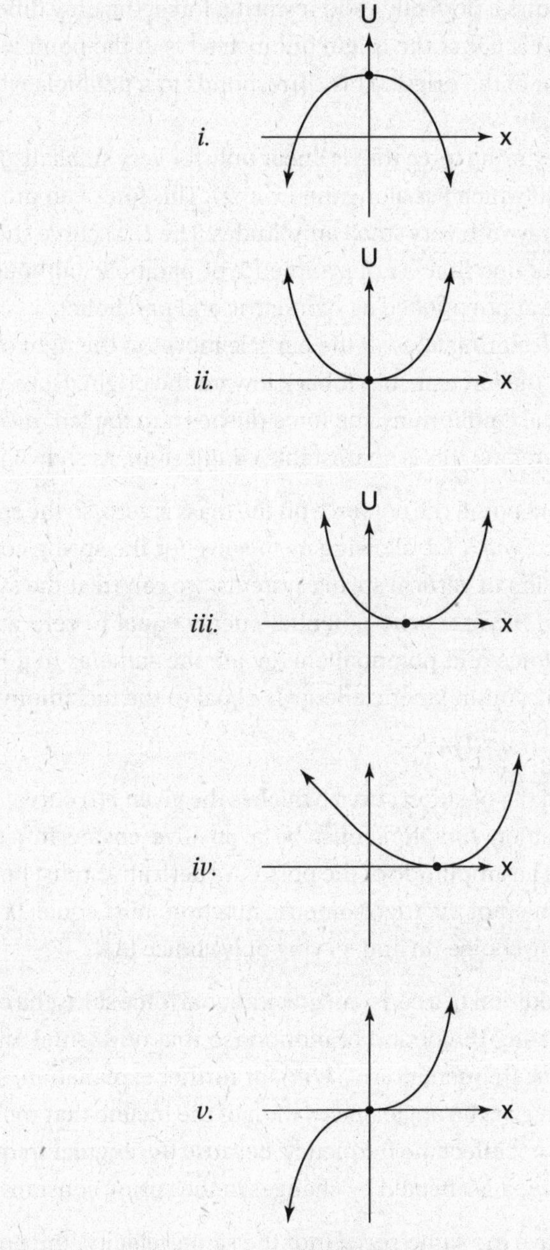

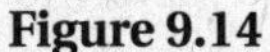

Figure 9.14

Graph (i) is not a restoring force: When the object moves to the right of the equilibrium point (the origin), the force pushes it farther to the right, farther away from the equilibrium point. Similarly, when the object moves to the left of the equilibrium, the force pushes it farther away from the equilibrium. This corresponds to unstable equilibrium: a downward-pointing parabola for $U(x)$ with a peak at $x = 0$.

Graph (ii) shows the quintessential linear restoring force. When the object is to the right of the equilibrium point (the origin), the force points to the left; when the object is to the left of the equilibrium point, the force points to the right. This corresponds to the simplest $U(x)$ curve that produces simple harmonic motion: an upward-pointing parabola with a minimum at $x = 0$.

Graph (iii) also shows a perfectly valid restoring force, the only difference being that the equilibrium point is not at the origin but instead is at the point where $F(x)$ crosses the x-axis, to the right of the origin. This corresponds to a parabola whose minimum is to the right of the origin.

Graph (iv) has a restoring force that is linear only for very small displacements about the equilibrium point (which lies along the $+x$-axis). This force can produce simple harmonic motion, but only with very small amplitudes. The $U(x)$ curve shows a situation of static equilibrium, but one that is not symmetric or parabolic (although, for small displacements, it can be approximated as symmetric and parabolic).

Graph (v) is not a restoring force: If the particle moves to the right of the equilibrium position (the origin), the force pushes it back toward the origin. However, if the particle moves to the left of the equilibrium, the force pushes it to the left, *away* from the equilibrium position. Therefore, this is an unstable equilibrium, as shown in the $U(x)$ curve.

2. **(B)** At the equilibrium point, the net force on the mass is zero, so the spring force equals the gravitational force, $mg = kd$, allowing us to solve for the spring constant $k = mg/d$. Based on our discussion of vertical spring systems, we can treat the system like a horizontal spring system. (If we set the potential energy equal to zero at the equilibrium point, the restoring force and potential energy are the same as in a horizontal spring system.) Thus, the maximum kinetic energy is equal to the maximum potential energy given by $U = \frac{1}{2}kA^2 = \frac{1}{2}(mg/d)A^2$.

3. **(A)** The derivative of the position curve [which is the given $v(t)$ curve] is a negative sine function, so the position function must be a positive cosine function, eliminating choices (B) and (D). The amplitude of the position's derivative must be v_{max}, eliminating choice (C). The argument of the trigonometric function must equal 2π when the time is equal to t_1, eliminating choice (E) and leaving only choice (A).

4. **(B)** Recall that the addition of a constant gravitational force shifts the equilibrium point of SHM without affecting the period of motion (so that horizontal and vertical spring systems have the same frequency, $\omega = \sqrt{k/m}$; for further explanation, see Example 9.3). Therefore, changes in g or the angle of elevation of the incline that modify the constant force on the mass do not affect the frequency. Because the angular frequency is given by the equation $\omega = \sqrt{k/m}$, it is affected by changes in the spring constant or the mass.

5. **(D)** The mass will have the same *speed* (not the same velocity) the next time the mass passes the equilibrium point, which will occur after half of a period ($T = 2\pi\sqrt{m/k}$). (One period is the time required for the mass to return to its original velocity and position, during which time it moves to the right and then back to the center and then to the left and finally back to the center. Thus, only moving to the right and back to the center is exactly half of a period given that the mass starts at the equilibrium position.)

6. **(D)** Choice (A): The kinetic energy is maximized at the equilibrium point. Choice (B): The potential energy is maximized at the turning points. Choice (C): The work done by the restoring force is equal to the change in kinetic energy (which is equal to the *negative* of the change in potential energy). Choice (D): This is correct. Every time the particle moves between either turning point and the equilibrium point, all the energy is transferred between kinetic and potential (such that, at some point during this transfer, the kinetic energy equals the potential energy). The object moves between a turning point and an equilibrium point four times during a complete period.

7. **(E)** According to conservation of energy, $\frac{1}{2}mv^2 + \frac{1}{2}kx^2 = \text{total energy} = \text{constant}$. Mathematically, this is the equation for an ellipse centered at the origin.

Alternative solution: The velocity is maximized at the equilibrium point (the point where the displacement is zero), eliminating choices (A) and (D). As the mass passes through the equilibrium position, the force on it is zero, so the rate at which the velocity changes is zero. Therefore, dv/dx is zero at this point; this condition is met only by graph (E); in graphs (B) and (C) this derivative is undefined.

8. **(B)** For an object like this one, executing angular simple harmonic motion, the angular frequency ω is the square root of the proportionality constant between the angular acceleration and the angular position: $\omega = \sqrt{2\pi}$. The linear frequency is then given by the equation $f = \omega/2\pi = \sqrt{2\pi}/2\pi = 1/\sqrt{2\pi}$. The units of frequency are usually written in terms of hertz, where one hertz is one cycle per second.

9. **(A)** The equilibrium position of the system is the point where the net force is zero: $0 = kx - mg$, so the equilibrium position is $x = mg/k$. The distance between the point at which the velocity is zero (when the mass is initially released at extension $x = 0$) and the equilibrium position, $x = mg/k$, is equal to the amplitude.

Alternative solution: Because the spring is initially at the highest point of its motion, it comes to rest after moving down a distance equal to *twice* the amplitude of the motion. Conservation of energy can be used to calculate this distance: $mg(2A) = \frac{1}{2}k(2A)^2$, yielding the same solution.

10. **(C)** Conservation of energy between the turning points and the equilibrium point yields the equation $\frac{1}{2}kx^2 = \frac{1}{2}mv^2$. Solving for the spring constant yields $k = mv^2/x^2$. Substitution into the equation $T = 2\pi\sqrt{m/k}$ yields choice (C).

Tip: What should you do if it appears that there aren't enough values given to solve the problem? Introduce as many variables as you need to solve the problem. If the extra variables cancel out in the final answer, they weren't needed in the first place (as in this problem). If the extra variables do not cancel out in the final answer, they are indeed needed to solve the problem and you should choose the "not enough information is given" option.

11. **(D)** Consider Figure 9.4. The tangential force is equal to $mg\sin\theta$, which is maximized when θ is greatest (at the turning points of the pendulum's trajectory). At these turning points, the pendulum bob is at its maximum height, so the potential energy is maximized.

At the turning points, the speed is equal to zero, so the kinetic energy is zero. The centripetal acceleration, v^2/r, is maximized when the speed is maximized (at the very bottom of the pendulum's trajectory).

Free-Response

1. (a) At equilibrium, the torque on the pulley is zero, requiring that $(mg)r = (kx)r$, so that the equilibrium extension is $x_{eq} = mg/k$.

(b) The Newton's second-law equation for the mass m (designating upward as positive so that a is positive) is

$$F_{net,y} = ma = F_T - mg$$

The Newton's second-law equation for the pulley (designating counterclockwise as positive so that α is positive) is

$$\tau_{net} = I\alpha = (kxr - F_T r) = [kr(x_{eq} + \Delta x) - F_T r]$$

(c) The relationship between the linear and angular acceleration is $\alpha = a/r$. (Because when the pulley rotates clockwise, the mass moves downward and the relative directions of a and α are consistent.)

$$[kr(x_{eq} + \Delta x) - F_T r] = \left[I\alpha = I\left(\frac{a}{r}\right)\right]$$

The equation for the net force acting on the hanging mass, $F_{net,y} = ma = F_T - mg$, can be solved for the tension force $F_T = mg + ma$ and substituted into the above equation. Also using the fact that $mg = kx_{eq}$, the above equation can be solved for the acceleration, yielding

$$|a| = \frac{k\Delta x}{m + I/r^2}$$

(d) Because the net force is a restoring force, the acceleration can be written as $a = -k\Delta x/(m + I/r^2)$. By comparison with the equation $a = -\omega^2 x$ (or alternatively, with $a = -(k/m)x$ and $\omega = \sqrt{k/m}$ for a mass attached to a spring), the angular frequency is the square root of the coefficient of the $-\Delta x$ term:

$$\omega = \sqrt{\frac{k}{m + I/r^2}}$$

2. (a) The maximum speed occurs when all the initial elastic potential energy has been converted to kinetic energy. Because of conservation of momentum, both equal masses must always have the same speed with opposite directions.

$$\frac{1}{2}k(\Delta x)^2 = 2\left(\frac{1}{2}mv^2\right)$$

Solving for the maximum speed yields $v = \sqrt{k(\Delta x)^2/2m}$.

(b) When the mass on the right is displaced a distance x to the right, by symmetry the mass on the left must be displaced the same distance to the left, making the net extension of the spring $2x$. (This can also be understood in terms of the fact that the center of mass of the system is stationary.) Therefore, the restoring force on the right mass is $F = -k(2x)$, and the effective spring constant acting on the right mass is $2k$. The period is then given by the equation $T = 2\pi\sqrt{m/k_{eff}} = 2\pi\sqrt{m/2k}$.

(c) Again, the maximum speed occurs when the spring is at its equilibrium position. However, because the masses are unequal, they do not have equal speeds; instead, their speeds must be related using conservation of momentum, which requires that $mv_m = Mv_M$. The second equation involves conservation of energy (the maximum speed occurs when all the potential energy has been converted to kinetic energy).

$$\frac{1}{2}k(\Delta x)^2 = \frac{1}{2}mv_m^2 + \frac{1}{2}Mv_M^2$$

Substituting for $v_m = Mv_M/m$ and solving for v_M yields

$$v_M = \sqrt{\frac{k(\Delta x)^2}{M(1+M/m)}}$$

(d) The amplitude of the oscillation is equal to the difference in position between the initial position (when the spring is maximally compressed) and the equilibrium position. During this time, the distance between the masses increases by an amount Δx. Again, because the masses are not equal, they do not move by the same amount. We must use the fact that the center of mass remains stationary to calculate the displacements of the two different masses. Based on the equation $x_{CM} = \Sigma x_i m_i / \Sigma m_i$, the displacement of the center of mass (which must be zero here because after the initial compression of the spring by an external force, the center of mass is motionless) is given by the equation $\Delta x_{CM} = \Sigma \Delta x_i m_i / \Sigma m_i$. Therefore,

$$0 = \frac{M\,\Delta x_M - m\Delta x_m}{m+M} \Rightarrow \Delta x_m = \frac{M\,\Delta x_M}{m}$$

Also, $\Delta x_m + \Delta x_M = \Delta x$. Solving these equations for Δx_M yields

$$\Delta x_M = \text{amplitude of } M = \frac{\Delta x}{1+M/m}$$

3. (a) Recall that 1 radian is approximately 60°, so try 10° first. After converting to radians and comparing $\theta = 0.174$ radians to $\sin\theta = 0.175$, you find a difference of 0.5%. For an angle of 5°, however, the difference is 0.1%.

(b) If you plot period-squared versus length, you should obtain a straight line: $T^2 = \frac{4\pi^2}{g}\ell$. See Figure 9.15.

(c) The slope of the line is $\frac{(4.0-1.0)}{(1.0-0.25)} = 4\ \text{s}^2/\text{m}$. Thus, $\frac{4\pi^2}{g} = 4\ \text{s}^2/\text{m}$, and $g = \pi^2 \approx 9.87\ \text{m/s}^2$.

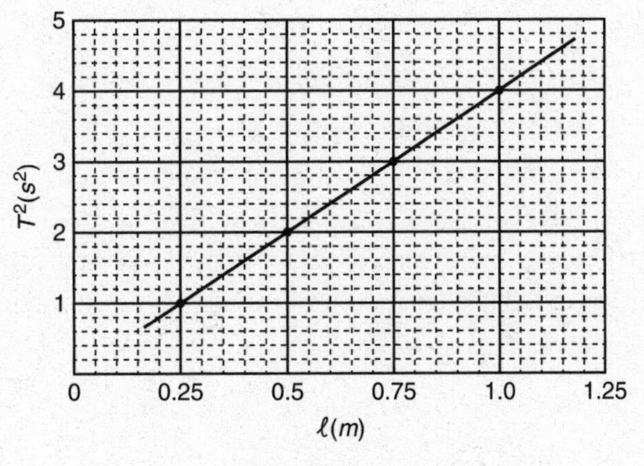

Figure 9.15

Universal Gravitation

10

- → LAW OF UNIVERSAL GRAVITATION
- → PRINCIPLE OF SUPERPOSITION
- → GRAVITATION DUE TO SPHERICALLY SYMMETRIC MASSES
- → RELATIONSHIP OF g AND *G*
- → KEPLER'S LAWS OF PLANETARY MOTION
- → GRAVITATIONAL POTENTIAL ENERGY

Newton's law of universal gravitation

Every particle in the universe attracts every other particle with a force whose magnitude is given by the equation

$$F = G\frac{m_1 m_2}{r^2}$$

where $G = 6.67 \times 10^{-11}\ \text{N} \cdot \text{m}^2/\text{kg}^2$.

This law makes intuitive sense in that the force is proportional to both masses and inversely proportional to the squared distance between them (having the same form as Coulomb's law).

Vector form: The law of universal gravitation can be written in full vector form as

$$\mathbf{F}_{\text{exerted on 1 by 2}} = -\frac{Gm_1 m_2}{r^2}\hat{r}_{2\text{ to }1} = +\frac{Gm_1 m_2}{r^2}\hat{r}_{1\text{ to }2}$$

(where $\hat{r}_{2\text{ to }1}$ is a unit vector pointing from m_2 to m_1 and $\hat{r}_{1\text{ to }2}$ is a unit vector pointing from m_1 to m_2). Thus, the two masses attract each other along a line joining them.

GRAVITATION DUE TO SPHERICALLY SYMMETRIC MASS DISTRIBUTIONS

What happens if there are more than two masses present? Then the net gravitational force on any mass is the vector sum of the gravitational forces exerted on that mass by each of the other masses (according to the superposition of forces). With the principle of superposition, it is possible to prove the following results:

- Result 1: The gravitational force exerted on an object *outside* any spherically symmetric mass distribution is the same as if all the mass in the distribution were concentrated at its center.
- Result 2: The gravitational force exerted on an object *inside* any spherically symmetric mass distribution a distance r from the center is the same as if all the mass located inside an imaginary sphere of radius r (the *enclosed mass*) were concentrated at its center and all the mass distributed outside this imaginary sphere did not exist.

You've probably used these observations many times without thinking about it [e.g., calculating the force of gravity on a mass lying on the surface of Earth using the formula $F = G(mM_{Earth}/r^2_{Earth})$.

We will not prove these results here. Rather when we discuss electrostatics (which is governed by Coulomb's law, a law that is mathematically similar to Newton's law of gravity), we will use Gauss's law to provide simple proofs of analogous results in the context of electrostatics.

Problem Solving: Gravity Inside Spherically Symmetric Mass Distributions

In calculating the force on an object within a spherically symmetric mass distribution, the only tricky part is determining the amount of mass located at distances closer to the center of the sphere than the given object (the mass *enclosed* in an imaginary spherical surface that passes through the object). There are generally two cases:

Case 1: If there is a uniform volume density, ρ = constant, you can take advantage of the fact that mass is proportional to volume (which can be calculated using solid geometry). Mathematically,

$$\rho = \frac{m_1}{V_1} = \frac{m_2}{V_2}$$

Case 2: If the density is a function of distance from the center, $\rho(r)$, you must integrate to obtain the enclosed mass. Your differential region (which must have a constant density, that is, a constant value of r) is a spherical shell of radius r and thickness dr with a differential volume $dV = 4\pi r^2 dr$. (You can understand this formula by noting that because the spherical shell is very thin, its volume is equal to its surface area, $4\pi r^2$, multiplied by its thickness, dr.) The differential mass is simply the volume density multiplied by this differential volume:

$$dm = \rho dV = [\rho(r)] 4\pi r^2 \, dr$$

The total mass enclosed within $r < d$ can then be calculated by integration:

$$m = \int dm = \int_0^d \rho(r) 4\pi r^2 \, dr$$

EXAMPLE 10.1

Calculate the gravitational force on an object mass *m* as a function of distance *r* from the center of a solid sphere of total mass *M*, radius *a*, and uniform mass density.

SOLUTION

The force has different forms inside and outside the sphere, so we must treat these situations separately. First, consider the force inside the sphere. To calculate the mass enclosed within a sphere of radius *r,* we take advantage of the fact that the mass density is constant (such that mass is proportional to volume).

$$\rho = \frac{m_1}{V_1} = \frac{m_2}{V_2} = \frac{M}{\frac{4}{3}\pi a^3} = \frac{m_{\text{enclosed}}}{\frac{4}{3}\pi r^3}$$

$$m_{\text{enclosed}} = \frac{r^3}{a^3} M$$

This mass acts as if it is all located at the center of the sphere:

$$F = \frac{Gm_1 m_2}{r^2} = \frac{Gm[(r^3/a^3)M]}{r^2} = \frac{GmM}{a^3} r \quad \text{for } r < a$$

Outside the sphere, all the mass acts as if it is at the center of the sphere:

$$F = \frac{Gm_1 m_2}{r^2} = \frac{GmM}{r^2} \quad \text{for } r > a$$

What happens at $r = a$? The gravitational force is continuous, with the two equations agreeing on the force.

Symbolic answer check: (1) Because of symmetry, the force at the center is zero. (2) Within the sphere, as *r* increases, the enclosed volume is proportional to r^3 (volume is proportional to linear dimensions cubed), while a factor of r^{-2} is introduced in the universal gravitation equation, so the net result is that force is proportional to *r*.

EXAMPLE 10.2

Calculate the gravitational force on an object of mass *m* as a function of distance *d* from the center of a sphere of radius *a* with a mass density given by the function $\rho = br$ (where *b* is a constant with units of kg/m^4).

SOLUTION

Again, the force is different inside and outside the sphere, so we must consider these situations separately. First, we calculate the force inside the sphere. To calculate the enclosed mass, we must integrate because the mass density is not constant.

$$m = \int dm = \int_0^d [\rho(r)] 4\pi r^2 \, dr = \int_0^d (br) 4\pi r^2 \, dr = \pi b d^4$$

This mass acts as if it is all located at the center of the sphere:

$$F = \frac{Gm_1 m_2}{r^2} = \frac{Gm(\pi b d^4)}{d^2} = (\pi b G m) d^2 \quad \text{for } d < a$$

For the force outside the sphere, we must calculate the enclosed mass:

$$M = \int dm = \int_0^a [\rho(r)] 4\pi r^2 \, dr = \int_0^a (br) 4\pi r^2 \, dr = \pi b a^4$$

This mass acts as if it is all located at the center of the sphere:

$$F = \frac{Gm_1 m_2}{r^2} = \frac{Gm(\pi b a^4)}{d^2} = \frac{\pi a^4 b G m}{d^2} \quad \text{for } d > a$$

As before, the two equations have the same limiting value as the position approaches $d = a$. It is useful to rewrite these equations in terms of the total mass $M = \pi b a^4$:

$$F = (\pi b G m) d^2 = \frac{GmMd^2}{a^4} \quad \text{for } d < a$$

$$F = \frac{\pi a^4 b G m}{d^2} = \frac{GmM}{d^2} \quad \text{for } d > a$$

As expected, outside the mass (no matter what the density function is, as long as it has spherical symmetry), we recover the equation for the gravitational force of a point mass.

Symbolic answer check: Force is zero at $d = 0$ as required by symmetry.

EXAMPLE 10.3

Consider two spheres, each of which has mass *M* and radius *a*. Sphere A has a constant mass density, while sphere B has a mass density proportional to the distance from the center of the sphere, $\rho = br$. Based on the results from Examples 10.1 and 10.2, sketch the gravitational force on a mass *m* as a function position for both these spheres on the same set of axes. Explain the differences between the graphs in terms of the enclosed mass at various radii. (Assume that the spheres are very far apart from each other, so that one sphere does not influence the gravitational force experienced by *m* when placed near the other sphere.)

SOLUTION

The two graphs are shown in Figure 10.1. Because sphere B has a fixed total mass, the net effect of making the mass density proportional to the radius is to shift mass away from the center of the sphere and toward the edges of the sphere. Thus, the gravitational force for $d < a$ is smaller for sphere B because the enclosed mass at a given distance from B's center is smaller than the enclosed mass at the same distance *d* from the center of sphere A. At $d = a$, measured from either center, the force is the same (because the enclosed mass is the same for both spheres, namely, *M*). Also, outside the sphere, the enclosed mass is the same for both spheres (*M*), so the forces are equal.

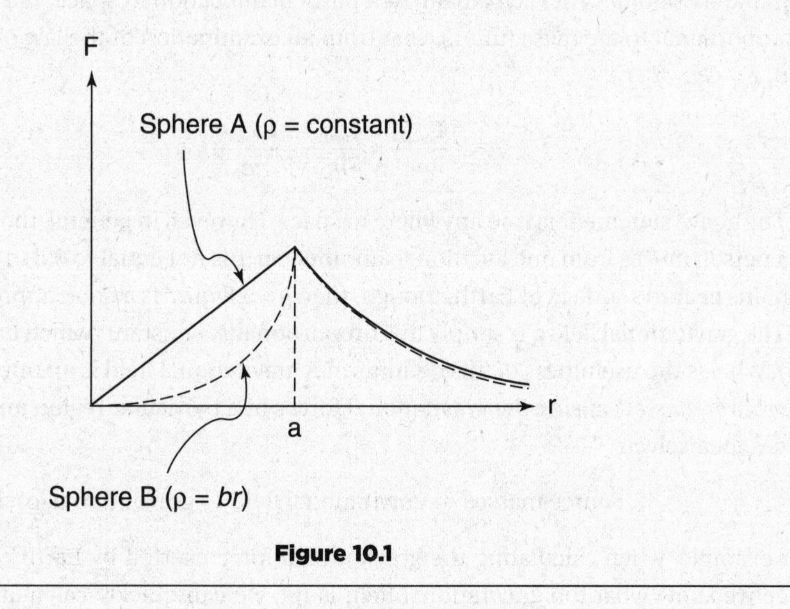

Figure 10.1

RELATING g TO *G*

For several chapters we have been calculating gravitational force as $\mathbf{F} = m\mathbf{g}$ with $g = 9.8\ \text{m/s}^2$. Is this incorrect? No, it is a valid approximation. As we just discussed, the gravitational force exerted by Earth on a mass a distance r from the center of Earth is

$$F = G\frac{mM_{\text{Earth}}}{r^2} = m\left(\frac{GM_{\text{Earth}}}{r^2}\right)$$

TIP

The weight of an object *mg* is the gravitational force exerted by Earth when the object is close to Earth's surface.

For an object at height h above Earth, the distance between the object and the center of Earth is equal to the radius of Earth plus h: $r = r_{\text{Earth}} + h$. If the object's height above Earth is much smaller than Earth's radius, $h << r_{\text{Earth}}$, it is valid to make the approximation $r = r_{\text{Earth}} + h \approx r_{\text{Earth}}$. The force is then

$$F = G\frac{mM_{\text{Earth}}}{r^2} \approx G\frac{mM_{\text{Earth}}}{r_{\text{Earth}}^2} = m\left(\frac{GM_{\text{Earth}}}{r_{\text{Earth}}^2}\right)$$

The term in parentheses above, $GM_{\text{Earth}}/r_{\text{Earth}}^2$, is defined to be g. Plugging in values for G, M_{Earth}, and r_{Earth} yields $g = 9.8\ \text{m/s}^2$.

GRAVITATIONAL FORCE FIELD

The gravitational acceleration **g** is an example of a gravitational force field and a demonstration of the usefulness of such a field. At a particular location in space, the gravitational force is proportional to the mass (this is clear from an examination of the law of universal gravitation, $F = Gm_1m_2/r^2$):

$$\frac{\mathbf{F}_{\text{mass 1}}}{m_1} = \frac{\mathbf{F}_{\text{mass 2}}}{m_2} = \frac{\mathbf{F}_{\text{mass } N}}{m_N} = g$$

The above statement is true anywhere in space. However, in general, the value of g will vary as a person moves from one location to another. So, g is not equal to $9.8\ \text{m/s}^2$. If an individual remains near the surface of Earth, though, then $g = 9.8\ \text{m/s}^2$ is a good approximation.

The gravitational field **g** is simply this proportionality constant (which happens to be a vector). What is the usefulness of this definition? A gravitational field is an intermediary between the *source masses* causing the gravitational forces in a particular region and the gravitational forces themselves:

Source masses $\leftrightarrow$ gravitational field $\leftrightarrow$ gravitational force

For example, when calculating the gravitational force exerted by Earth close to its surface, once we know what the gravitational field is (**g**), we can quickly calculate the gravitational forces using the equation $\mathbf{F} = m\mathbf{g}$ without worrying about what the mass or radius of Earth is. Alternatively, once we know what the force on a known mass (a *test mass*) is, we can calculate the field using the following equation:

$$\mathbf{g} = \frac{\mathbf{F}_g}{m_{\text{test particle}}}$$

Definition of a gravitational field

Gravitational fields are also useful because they immediately specify the acceleration of any object acting under the influence of gravity.

The gravitational field is a vector field: It assigns to every point in space a vector. A few examples of such fields are illustrated in Figure 10.2. Although you will learn more about electric fields than gravitational fields, keep in mind that the same mathematics underlies both and therefore many of the results obtained with electric fields can be carried over to gravitational fields.

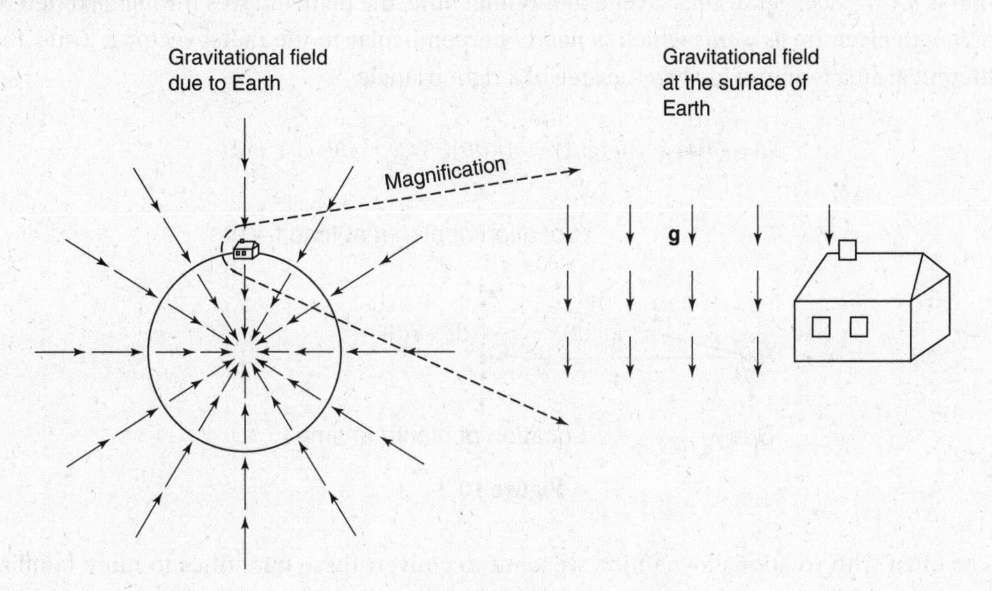

Figure 10.2

KEPLER'S LAWS

1. **The trajectories of all the planets are ellipses with the sun at one focal point.**
2. **The radial vector from the sun to a given planet sweeps out equal areas in equal time intervals.**
3. **For planets orbiting the same sun, the square of an orbit's period is proportional to the cube of its semimajor axis (or radius, for circular orbits):**

$$\frac{T_1^2}{r_1^3} = \frac{T_2^2}{r_2^3} = \text{constant}$$

Note that a circle is a special case of an ellipse (when both foci coincide with each other).

Kepler's first law can be derived using Newton's laws but involves solving differential equations at a level beyond that of Physics C.

Proof of Kepler's Second Law

What the second law really says is that the time derivative of the rate at which the radial vector sweeps out area is a constant:

$$\frac{dA}{dt} = \text{constant}$$

What is dA/dt? See Figure 10.3. Over a differential time, the planet moves through a differential length given by $ds = rd\theta$, which is nearly perpendicular to the radial vector **r**. Thus, the differential area is approximately the area of a right triangle:

$$dA = \frac{1}{2}(\text{base})(\text{height}) = \frac{1}{2}(r\,d\theta)(r) = \frac{1}{2}r^2\,d\theta = \frac{1}{2}r^2\omega\,dt$$

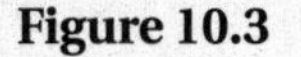

Figure 10.3

As often with rotational dynamics, we want to convert these quantities to more familiar expressions. This can be done by multiplying and dividing the right side by the mass of the planet M:

$$\frac{dA}{dt} = \frac{Mr^2\omega}{2M} = \frac{I\omega}{2M} = \frac{L}{2M}$$

Thus, if dA/dt is a constant, so is L. Now all we need to do to prove Kepler's second law is show that L is constant. This in turn is equivalent to the absence of a net torque on the planet because $\tau_{net} = dL/dt$. The net torque is zero because the only force acting on the planet, the gravitational force from the sun, cannot exert a torque; the force is parallel to **r**, making the cross-product $\boldsymbol{\tau} = \mathbf{r} \times \mathbf{F}$ equal to zero as shown in Figure 10.4. Therefore, L is constant, and thus dA/dt is constant.

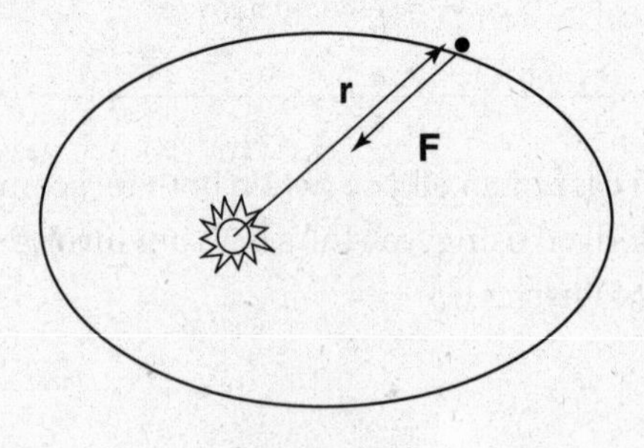

Figure 10.4

EXAMPLE 10.4

See Figure 10.5. If the speed of a satellite is v_1 at r_1, what is its speed at r_2?

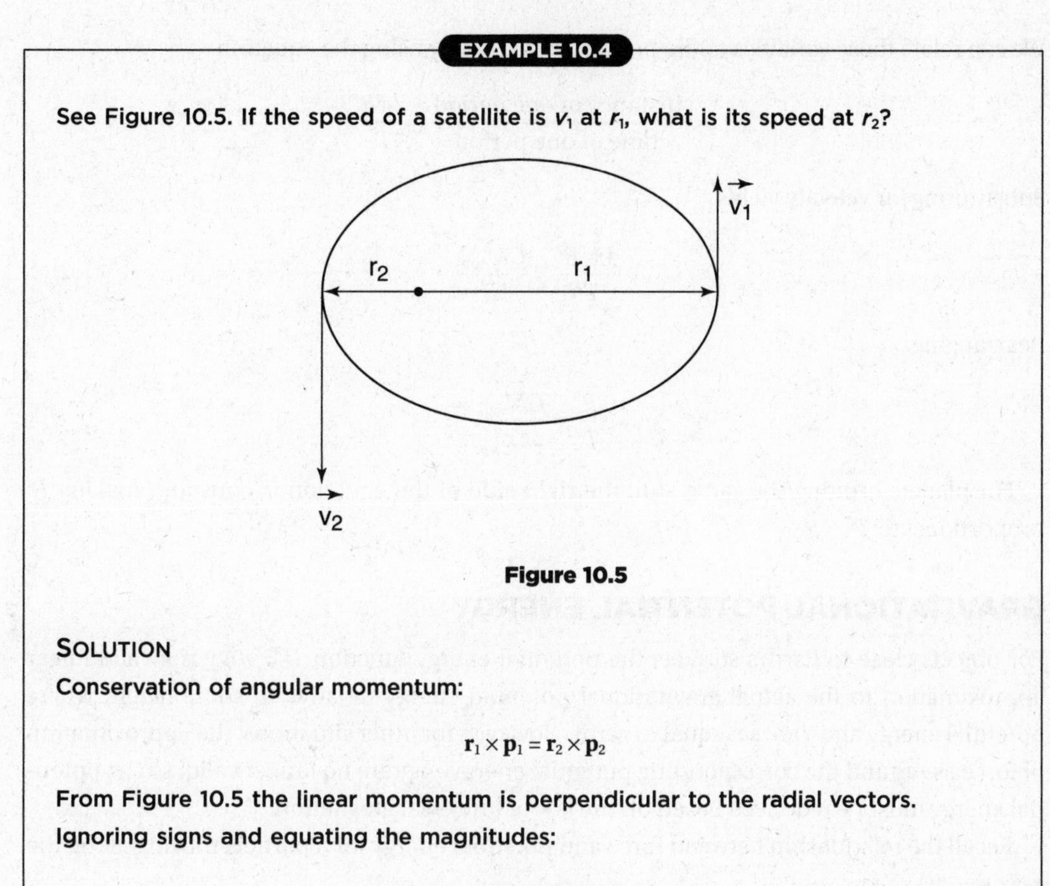

Figure 10.5

SOLUTION

Conservation of angular momentum:

$$\mathbf{r}_1 \times \mathbf{p}_1 = \mathbf{r}_2 \times \mathbf{p}_2$$

From Figure 10.5 the linear momentum is perpendicular to the radial vectors. Ignoring signs and equating the magnitudes:

$$mv_1r_1 = mv_2r_2$$

$$v_2 = \frac{v_1r_1}{r_2}$$

Proof of Kepler's Third Law for the Special Case of Circular Orbits

We start, as is often the case for problems of circular planetary motion, with the statement that the gravitational attraction to the sun provides the centripetal force:

$$F_{\text{centripetal}} = \frac{mv^2}{R} = \frac{GmM_{\text{sun}}}{R^2}$$

Simplifying,

$$v^2 = \frac{GM_{\text{sun}}}{R}$$

We can relate these variables to the period of rotation by using the equation

$$v = \frac{\text{distance of one period}}{\text{time of one period}} = \frac{2\pi R}{T}$$

Substituting for velocity yields

$$\frac{4\pi^2 R^2}{T^2} = \frac{GM_{\text{sun}}}{R}$$

Rearranging,

$$\frac{R^3}{T^2} = \frac{GM_{\text{sun}}}{4\pi^2}$$

For planets orbiting the same sun, the right side of the equation is constant, making R^3 proportional to T^2.

GRAVITATIONAL POTENTIAL ENERGY

TIP

The potential energy function *U = mgy* is a good approximation when the object is near Earth's surface.

For objects close to Earth's surface, the potential energy function $U = mgy$ is a valid linear approximation to the actual gravitational potential energy (relative to some height where potential energy and y are set equal to zero). However, for other situations, the approximation of force as mg and the corresponding potential energy mgy are no longer valid, so the potential energy must be rederived based on the law of universal gravitation.

Recall the relationship between force and potential energy for a particle moving along the x-axis:

$$F_x = -\frac{dU}{dx} \Leftrightarrow dU = -F_x dx$$

Now imagine that m_1 is fixed at the origin while m_2 is brought toward the origin from infinity $(\infty, 0)$, to a final location $(r, 0)$. The change in the potential energy while moving m_2 is equal to

$$\Delta U = \int dU = -\int_{x=\infty}^{x=r} F_x dx$$

The attractive force is along the $-x$-direction:

$$F_x = -\frac{Gm_1 m_2}{r^2}$$

Substituting,

$$\Delta U = \int dU = -\int_{x=\infty}^{x=r} F_x\,dx = -\int_{x=\infty}^{x=r}\left(-\frac{Gm_1m_2dx}{x^2}\right) = \left[-\frac{Gm_1m_2}{x}\right]_{x=\infty}^{x=r} = -\frac{Gm_1m_2}{r}$$

$$\Delta U = U_{x=r} - U_{x=\infty} = -\frac{Gm_1m_2}{r}$$

By convention, the potential energy is generally set equal to zero when objects are infinitely far apart from each other, so $U_{x=\infty} = 0$ and

$$U_{x=r} = -\frac{Gm_1m_2}{r}$$

What about the potential energy if m_2 does not lie along the x-axis? Based on the definition of work, no work is required to move m_2 perpendicular to the gravitational force exerted by m_1, in this case along any circular arc centered at the origin. Therefore, the potential energy is constant along a sphere centered at m_1, and the above equation $U_{x=r} = -Gm_1m_2/r$ is valid in three dimensions, where r is the distance between m_1 and m_2.

$$U_{\text{gravitational PE}} = -\frac{Gm_1m_2}{r}$$

Gravitational potential energy (energy set to zero at infinite separation)

TIP

Unlike kinetic energy, potential energy can be negative.

Gravitational potential energy is conservative. Changes in gravitational potential energy depend only on the initial and final positions (not the path), indicating that the universal gravitational force is conservative.

Problem Solving: Calculating Gravitational Potential Energy

Based on the above derivation, the most useful equation when calculating the gravitational potential energy due to various spherical mass distributions is

$$U(r) = U(x) = -\int_{x=\infty}^{x=r} F_x dx$$

Equation for calculating potential energy due to spherical mass distributions

EXAMPLE 10.5 DERIVING ANOTHER POTENTIAL ENERGY CURVE

In Example 10.1, we derived the gravitational force acting at a distance d from the center of a sphere of mass M, radius a, and uniform volume density:

$$F = \frac{GmM}{a^3} r \quad \text{for } r < a$$

$$F = \frac{GmM}{r^2} \quad \text{for } r > a$$

Calculate the $U(r)$ function for a mass m located a distance r from the center of this distribution.

SOLUTION

We will use the same general approach used in the above derivation. As in Example 10.1, the potential energy in the two regions must be calculated separately.

Beginning with the potential energy outside the sphere, consider the integral used to calculate the potential energy:

$$U(r) = U(x) = -\int_{x=\infty}^{x=r} F_x \, dx$$

Because this force outside the mass is *identical* to the force due to a point mass, the work integral is also identical, with the same final result:

$$U(r) = \int_{\infty}^{r} \frac{GmM \, dr}{r^2} = -\frac{GmM}{r} \quad \text{for } r > a$$

The tricky part is calculating the potential energy inside the sphere for $r < a$. We take the same basic approach, evaluating the negative work required to move the mass m from infinity to r:

$$U(r) = U(x) = -\int_{x=\infty}^{x=r} F_x \, dx$$

However, the mass m moves through two different regions governed by different force equations. We therefore need to split this integral into two parts that we can then evaluate (recall that the attractive force points in the negative direction)

$$U(r) = -\int_{\infty}^{a} F_x \, dx - \int_{a}^{r} F_x \, dx = -\int_{\infty}^{a} \left(-\frac{GmM}{x^2}\right) dx - \int_{a}^{r} \left(-\frac{GmM}{a^3}\right) x dx$$

$$U(r) = -\frac{3GmM}{2a} + \frac{GmMr^2}{2a^3}$$

Symbolic answer check: (1) Units check. (2) As r increases, $U(r)$ increases, which makes sense (the distribution of the mass changes the form of the force equation, but the force is still toward the origin, so $U(r)$ still increases with increasing r).

EXAMPLE 10.6 TOTAL ENERGY OF A SATELLITE

A satellite of mass m is orbiting Earth (mass M) at a distance of R in a circular orbit (R is measured from the center of Earth).

(a) Express the satellite's velocity as a function of R.

(b) Express the total energy of the satellite as a function of R.

(c) To increase the satellite's velocity (while keeping it in a circular orbit), should you increase or decrease the total energy?

SOLUTION

(a) We start with the fact that the universal gravitational force provides the centripetal force:

$$F_{\text{centripetal}} = \frac{mv^2}{R} = \frac{GmM_e}{R^2}$$

$$v = \sqrt{\frac{GM_e}{R}}$$

(b) Substituting the above velocity into the KE term,

$$E_{\text{tot}} = U + \text{KE} = -\frac{GmM}{R} + \frac{1}{2}mv^2 = -\frac{GmM}{2R}$$

(c) Based on the equation $v = \sqrt{GM_e/R}$, to increase v we must decrease R. Based on the result from part (b), $E_{\text{tot}} = -GmM/2R$, so to decrease R, we must *decrease* the total energy! This is counterintuitive: To increase the velocity of the satellite, we decrease the total energy of the system. Where does the extra energy come from to increase the satellite's kinetic energy? We can understand this using the equation $E_{\text{tot}} = U + KE$; the potential energy must decrease even more than the total energy decreases, allowing for an increase in KE.

Gravitational Potential Energy in Systems of Masses

What is the potential energy of a mass M if it is gravitationally attracted to several other masses ($m_1, m_2, m_3, \ldots, m_n$)? The gravitational potential energy is simply the sum of the potential energy of each interaction:

$$U = \sum_{i=1}^{i=n} -\frac{GMm_i}{r_i}$$

Why? Recall that superposition of forces says that one mass does not affect the force that another mass exerts on M. Therefore, no mass affects the potential energy between M and any other mass. Mathematically,

$$\Delta U = -\int \mathbf{F}_{\text{net}} \cdot d\mathbf{r} = -\int \left(\sum \mathbf{F}_i\right) \cdot d\mathbf{r} = \sum \left(-\int \mathbf{F}_i \cdot d\mathbf{r}\right) = \sum \Delta U_i$$

Solving Problems Involving Universal Gravitation

There is no simple paradigm for solving these problems. The following is a summary of useful formulas and facts from this chapter:

$$F = G\frac{m_1 m_2}{r^2}$$

$$U_{\text{gravitational PE}} = -\frac{Gm_1 m_2}{R}$$

Definition of a gravitational field:

$$\mathbf{g} = \frac{\mathbf{F}}{m_{\text{test particle}}} = \frac{\mathbf{F}_{\text{mass }1}}{m_1} = \frac{\mathbf{F}_{\text{mass }2}}{m_2} = \frac{\mathbf{F}_{\text{mass }N}}{m_N}$$

For circular planetary motion the centripetal force is provided by the gravitational force:

$$F_{\text{centripetal}} = \frac{mv^2}{R} = \frac{GmM_{\text{sun}}}{R^2}$$

The following relationship among the velocity, radius, and period of a circular orbit is often useful.

$$v = \frac{\text{distance of one period}}{\text{time of one period}} = \frac{2\pi R}{T}$$

The equation used to calculate gravitational potential energy for spherical mass distributions is

$$U(r) = U(x) = -\int_{x=\infty}^{x=r} F_x dx$$

When the only force acting on a planet is the gravitational force, the angular momentum is conserved (as discussed in the derivation of Kepler's second law).

Conservation of energy

CHAPTER SUMMARY

The gravitational attraction between any pair of masses in the universe is given by the law of universal gravitation. When this law is supplemented by the principle of superposition, the gravitational force due to an object with an arbitrary distribution of mass can be calculated. The weight mg of an object near Earth's surface is an approximation to the more accurate form predicted by the universal law for the gravitational force exerted by Earth on m. Kepler's laws of planetary motion correctly describe the motion of the planets in our solar system. These laws are consequences of Newton's laws of motion and the universal law of gravitation.

PRACTICE EXERCISES

Multiple-Choice Questions

1. An asteroid moves around the sun in a path as shown in Figure 10.6. The asteroid is an unbound particle (and is free to continue an infinite distance away from the sun). At the point indicated, which of the following statements is true of the asteroid's velocity?

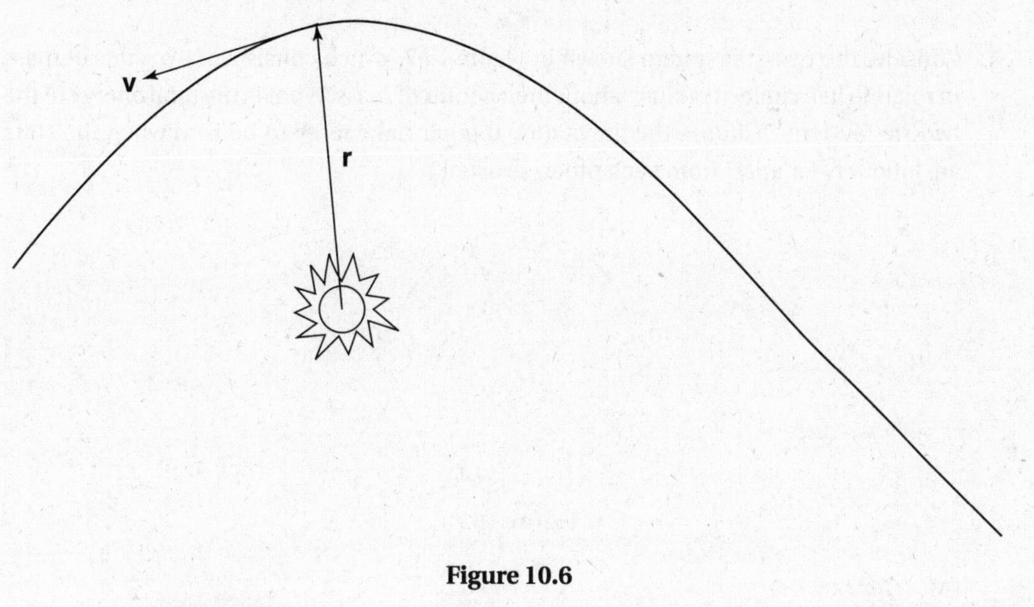

Figure 10.6

(A) $v < \sqrt{\frac{2Gm_{ast}m_{sun}}{r^2}}$

(B) $v > \sqrt{\frac{2Gm_{ast}m_{sun}}{r^2}}$

(C) $v < \sqrt{\frac{Gm_{sun}}{r}}$

(D) $v = \sqrt{\frac{Gm_{sun}}{r}}$

(E) $v > \sqrt{\frac{2Gm_{sun}}{r}}$

2. A planet revolves in a circular orbit around the sun. Which of the following relationships is true regarding the planet's kinetic and potential energies?

(A) $U = -2\text{KE}$
(B) $U = -\text{KE}$
(C) $U = \text{KE}/2$
(D) $U = -\text{KE}/2$
(E) Not enough information is given.

3. The gravitational acceleration on the surface of a planet (of mass M) with uniform volume density is g. What is the gravitational acceleration at a point halfway to the center of the planet (i.e., at $r = r_{planet}/2$)?

(A) $g/4$
(B) $g/2$
(C) g
(D) $2g$
(E) $4g$

4. Consider the two-star system shown in Figure 10.7, which consists of two stars of mass m rotating in a circle of radius r about their center of mass. What is the total energy of the two-star system? (Choose the gravitational potential energy to be zero when the stars are infinitely far apart from each other, as usual.)

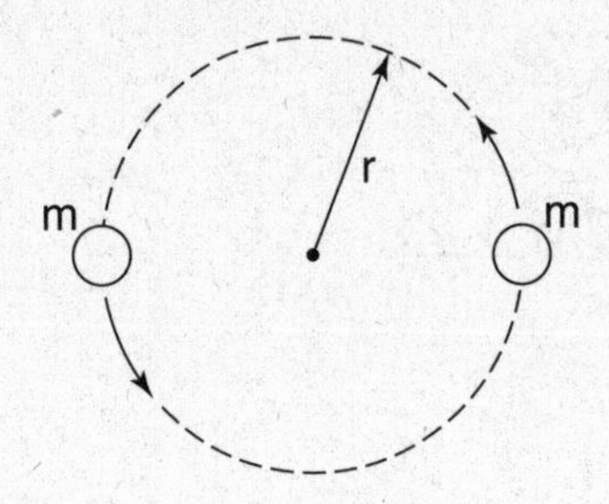

Figure 10.7

(A) $-Gm^2/2r$
(B) $Gm^2/2r$
(C) $Gm^2/4r$
(D) $3Gm^2/4r$
(E) $-Gm^2/4r$

5. A planet moves in an elliptical orbit around the sun. Which of the following statements is true?

I. The angular momentum of the planet around the sun is constant.
II. The speed of the planet is constant.
III. The total energy (potential plus kinetic) of the planet is constant.

(A) I only
(B) II only
(C) I and II only
(D) I and III only
(E) I, II, and III

6. On a planet of radius r and mass m, what is the escape velocity (i.e., the minimum velocity a particle must have to escape the gravitational field of the planet)?

   (A) $v = \sqrt{Gm/r}$
   (B) $v = \sqrt{2Gm/r}$
   (C) $v = \sqrt{Gm/r^2}$
   (D) $v = \sqrt{Gm/2r^2}$
   (E) The velocity depends on the mass of the escaping particle.

7. Consider a planet's orbit around the sun, as shown in Figure 10.8. Which of the following comparisons are valid concerning the planet's speed and angular momentum?

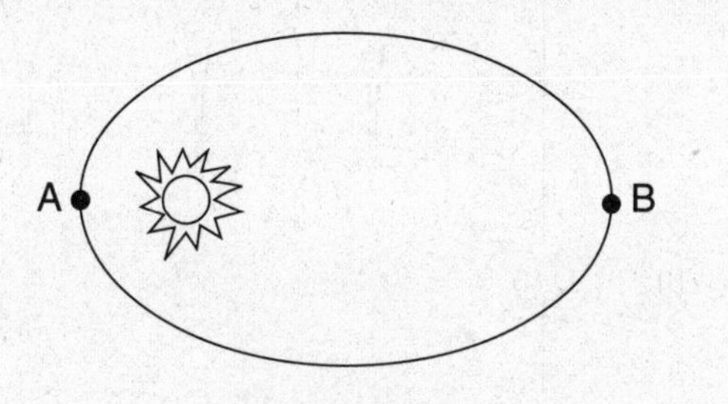

Figure 10.8

   (A) $L_A = L_B$; $v_A > v_B$
   (B) $L_A = L_B$; $v_A < v_B$
   (C) $L_A > L_B$; $v_A > v_B$
   (D) $L_A < L_B$; $v_A < v_B$
   (E) $L_A > L_B$; $v_A = v_B$

8. If a planet has twice the radius of Earth and half of Earth's density, what is the acceleration due to gravity on the surface of the planet (in terms of the gravitational acceleration g on the surface of Earth)?

   (A) $4g$
   (B) $2g$
   (C) g
   (D) $g/2$
   (E) $g/4$

9. A mass is moved by a nongravitational force (of unknown magnitude and direction) from point A to point B within a known gravitational field. Which of the following is true?

   (A) The change in the object's kinetic energy depends only on its mass and the locations of points A and B.
   (B) The change in the object's kinetic energy depends only on its mass and the path taken between points A and B.
   (C) The change in the object's kinetic energy is equal to the work done by the external force, $W = \int \mathbf{F} \cdot d\mathbf{r}$.
   (D) The total energy (kinetic plus potential) of the mass is constant.
   (E) The object's potential energy depends only on its mass and the locations of points A and B.

10. Consider a two-mass system where one mass is fixed and the second mass is initially moving directly away from the fixed mass. If this constitutes a *bound system*, such that the second mass does not have sufficient velocity to escape to an infinite distance from the fixed mass, which of the graphs in Figure 10.9 represents the gravitational potential energy of the system as a function of time? (Treat the two masses as points that essentially have no volume and so can "pass through" each other.)

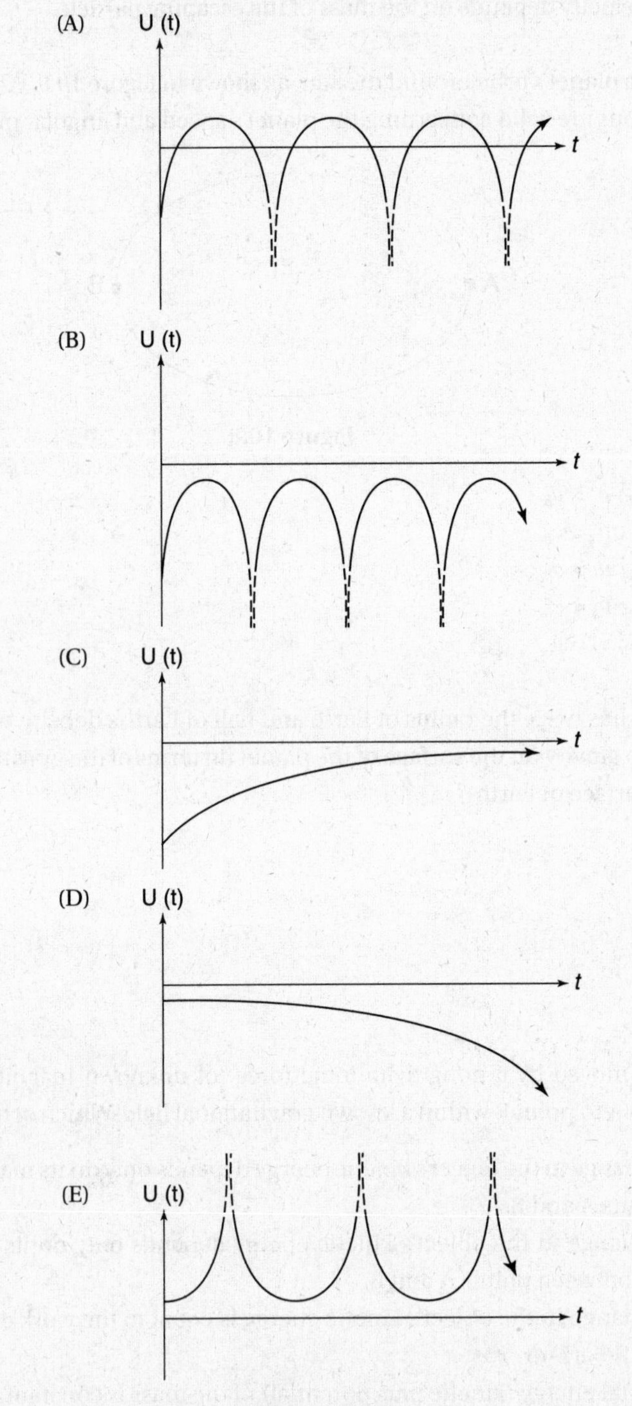

Figure 10.9

Free-Response Questions

1. A space shuttle moving with an initial velocity as shown in Figure 10.10 "slingshots" around the sun in order to reverse its direction. The sun's mass is m_{sun}, and you can make the assumption that the sun remains stationary.

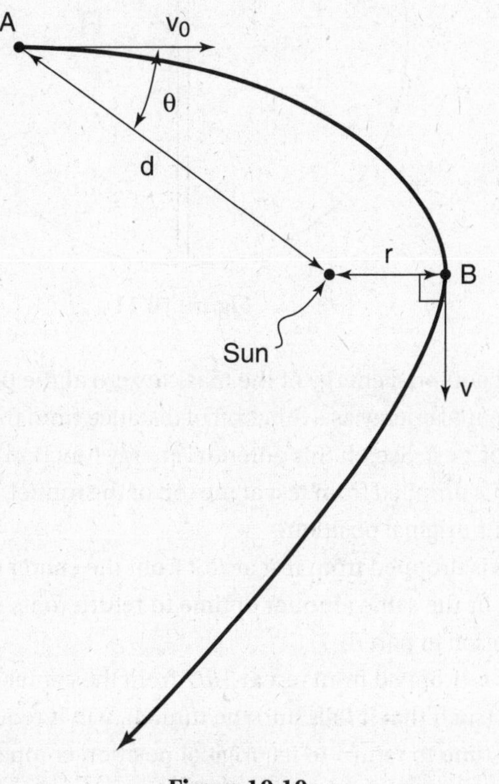

Figure 10.10

   (a) What is the minimum initial speed required by the shuttle to escape the sun's gravitational field and move in a linear trajectory toward infinity?

   (b) What is the minimum initial speed v_0 that the shuttle must have to avoid falling into the sun? (Treat the sun and the shuttle as points.)

   (c) Write down the equations required to calculate the initial angle θ in terms of v_0, d, m_{sun}, G, and r.

2. A planet of mass M, radius R, and uniform density has a small tunnel drilled through the center of the planet as shown in Figure 10.11. As discussed in Example 10.1, when the mass is within the tunnel, it experiences a force of $F = (GmM/R^3)r$, whereas when the mass is outside the planet, it experiences a gravitational force of $F = GmM/r^2$.

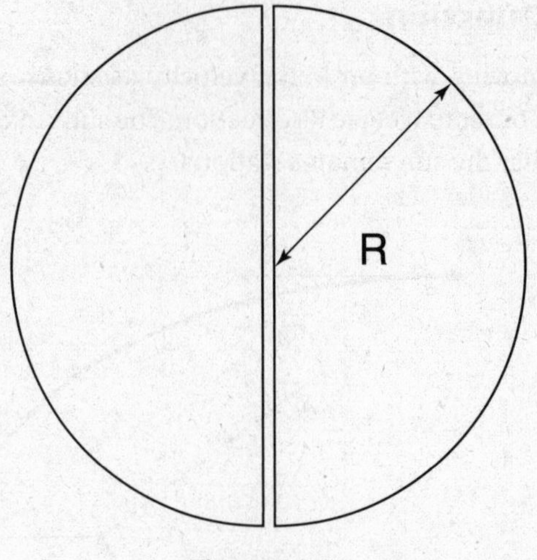

Figure 10.11

(a) Setting the potential energy of the mass to zero at the planet's center, calculate the mass's potential energy as a function of distance from the center of the planet, $U(r)$, for values of $r < R$. Sketch this potential energy function.

(b) If the mass is dropped from rest at the top of the tunnel, how long will it take until it returns to its original position?

(c) If the mass is dropped from rest at $R/2$ from the center of the planet, will it require more, less, or the same amount of time to return to its original position compared to the situation in part (b)?

(d) If the mass is dropped from rest at $3R/2$ from the center of the planet directly above the tunnel (such that it falls into the tunnel), will it require more, less, or the same amount of time to return to its original position compared to the situation in part (b)?

3. A satellite of mass m orbits the sun (of mass m_{sun}) in a circular orbit of radius R.

(a) Express the speed of the satellite in terms of G, R, and m_{sun}.

(b) An asteroid of mass $2m$ moving with the same speed collides elastically with the satellite. At the instant the asteroid hits the satellite, their velocities are oppositely directed and equal in magnitude. Calculate the final speed of the satellite if the collision is exactly head-on (i.e., it is not a glancing collision; the satellite's final velocity is along the line of its original velocity).

(c) After the collision, is the satellite still bound to the earth or is it free to move an infinite distance away from the earth?

4. Two planets of different masses rotate about each other in circular orbits as shown in Figure 10.12.

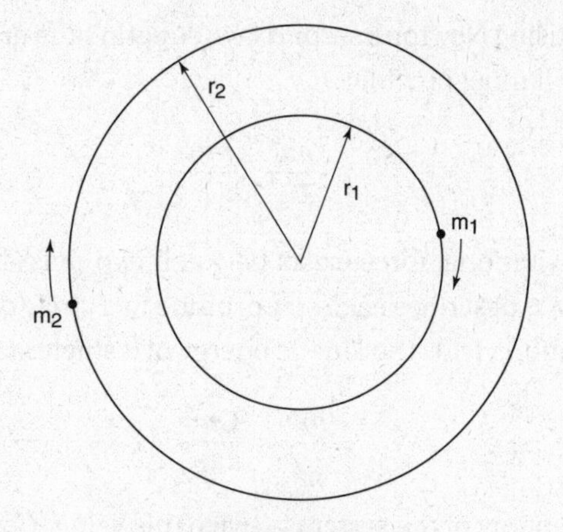

Figure 10.12

(a) What is the ratio of the periods of the two planets' orbits?
(b) What is the ratio of the two planets' speeds?
(c) Calculate m_2 in terms of m_1, r_1, and r_2.

ANSWER KEY

1. **E**	4. **E**	7. **A**	10. **B**
2. **A**	5. **D**	8. **C**	
3. **B**	6. **B**	9. **E**	

ANSWERS EXPLAINED

Multiple-Choice

1. **(E)** Using the condition that the asteroid's total energy must be greater than or equal to zero (allowing it to escape from Earth's gravitational field) we have $E = \frac{1}{2}mv^2 - \frac{Gm_{\text{sun}}m}{r} > 0$ where m is the mass of the asteroid, yielding choice (E). Note that the right-hand side of the inequality in the answer is the escape speed of an object a distance **r** from the sun.

2. **(A)** Because kinetic energy is always positive and gravitational potential energy, $U = -Gm_1m_2/r$, is always negative, choice (C) can be immediately eliminated.

   A good way to approach universal gravitation problems is to use Newton's second law, equating the gravitational force to the centripetal force that it must provide: $Gm_{\text{planet}}m_{\text{sun}}/r^2 = m_{\text{planet}}v^2/r$. Multiplying both sides by r and dividing by 2 yields: $Gm_{\text{planet}}m_{\text{sun}}/2r = m_{\text{planet}}v^2/2$. On the right, we recognize the kinetic energy of the planet, and on the left, $-U/2$. Therefore, $U = -2\text{KE}$.

   Note: If you don't immediately realize that Newton's second law can easily be manipulated to yield the kinetic energy, a more straightforward solution would be to solve for the velocity $v = \sqrt{Gm_{\text{sun}}/r}$ and then substitute the velocity into the kinetic energy equation $\text{KE} = \frac{1}{2}m_{\text{planet}}v^2$ in order to obtain the same final result.

3. **(B)** The acceleration due to gravity is given by the equation $g = GM/r^2$. At $r = r_{\text{planet}}/2$, the enclosed mass that gives rise to the net gravitational force is smaller by a factor of 8 (because mass is proportional to the cube of the radius) and the radius is smaller by a factor of 2 (making r^2 in the denominator smaller by a factor of 4). Therefore, the ratio $g = GM/r^2$ is smaller by a factor of 2.

4. **(E)** We begin by using Newton's second law, equating the gravitational force to the centripetal force that it must provide:

$$\frac{Gm^2}{(2r)^2}=\frac{mv^2}{r}$$

Note that the gravitational force exists between two masses a distance $2r$ apart, while the centripetal force describes each star orbiting in a circle of radius r. A little rearrangement of this equation yields the kinetic energy of a single star:

$$\frac{mv^2}{2}=\frac{Gm^2}{8r}$$

The total kinetic energy of the system is twice this value, $Gm^2/4r$, because there are two stars. The gravitational potential energy is $U = -Gm^2/2r$. Summing the kinetic and potential energies yields

$$\text{Total energy}=\frac{Gm^2}{4r}-\frac{Gm^2}{2r}=-\frac{Gm^2}{4r}$$

5. **(D)** Because the gravitational force is always parallel to the position vector from the sun to the planet (as shown in the derivation of Kepler's second law), the torque is zero and the angular momentum of the planet is constant. Since the orbit is an ellipse (rather than a circle), the planet moves closer and farther from the sun as it goes around the orbit. As this happens, the gravitational potential energy, $U = Gm_1m_2/r$, varies, causing the kinetic energy and the speed to vary along the orbit (alternatively, the speed would be constant for a circular orbit). However, the total energy, kinetic plus potential, is conserved (as long as we neglect any interstellar frictional drag).
6. **(B)** At its escape velocity, the particle must have a total energy of zero. (Why? The minimum energy required to escape the planet leaves the particle with zero kinetic energy when it reaches an infinite distance from the planet. At this position, the particle's total energy is zero (both potential and kinetic); thus, by energy conservation, the particle's energy must be zero as it leaves the planet.) Therefore, $0=\frac{1}{2}m_{\text{particle}}v^2-Gmm_{\text{particle}}/\text{r}$. Solving this equation for v yields choice (B).
7. **(A)** The gravitational force is always parallel to the radial position vector, so the gravitational force exerts no torque and the planet orbits with constant angular momentum, eliminating choices (C), (D), and (E). Conservation of angular momentum implies the following (note that the velocities are perpendicular to the radial position vectors at points A and B):

$$[L_a = |\mathbf{r}_a \times \mathbf{p}_a| = r_amv_a] = [L_B = |\mathbf{r}_b \times \mathbf{p}_b| = r_bmv_b]$$

Simplifying,

$$r_av_a = r_bv_b$$

Therefore, the smaller r is associated with the larger v, such that $v_a > v_b$.

8. **(C)** The acceleration on the surface of a planet is given by the equation $a = Gm_{\text{planet}}/r^2_{\text{planet}}$. The new planet has twice the radius of Earth, and *four times* the mass (doubling the radius increases the volume by a factor of 8, but halving the density decreases the mass by a factor of 2; the net result is an increase by a factor of 4). Because the radius is squared in the above formula, these effects cancel out exactly, leaving the planet's surface with the same acceleration due to gravity as on the surface of Earth.

9. **(E)** Because a nonconservative force (the applied force) is present, energy is not conserved, eliminating choice (D). Therefore, the change in the kinetic energy depends on *both* the change in the gravitational potential energy and the work done by the applied force, eliminating choices (A) and (B) (which do not allow for calculation of the work by the unknown applied force), and choice (C) (which does not allow for calculation of the change in the potential energy). Finally, choice (E) is correct: The gravitational potential energy depends only on the mass and the initial and final points of the path. (Because gravity is a conservative force, it is path independent.)

10. **(B)** The mobile mass will oscillate along a linear path that passes through the fixed mass. Because the system is bound, the potential energy must be negative, ruling out choices (A) and (E). The only other periodic graph is choice (B), which is correct.

    Initially, the mobile mass is moving away from the fixed mass, so its potential energy is increasing. However, eventually the mobile mass will pause, reverse its direction, and move through the fixed mass (causing the gravitational potential energy, $G = -m_1 m_2 / r$, to approach negative infinity). Because the mobile mass is bound, however, after passing through the fixed mass it will eventually come to rest, reverse its direction, and continue oscillating indefinitely.

Free-Response

1. (a) The shuttle must have an escape velocity, which corresponds to a total energy of zero:

$$E_{\text{total}} = 0 = \frac{1}{2} m_{\text{shuttle}} v_0^2 - \frac{Gm_{\text{sun}} m_{\text{shuttle}}}{d}$$

$$v_0 = \sqrt{\frac{2Gm_{\text{sun}}}{d}}$$

(b) When the shuttle collides with the sun, it will have a radial position vector of $\mathbf{r} = 0$ and therefore will have zero angular momentum. Thus, conservation of angular momentum requires that for the shuttle to reach the sun, it must initially have zero angular momentum, which requires that $v_0 = 0$.

This makes sense if we consider the possible trajectories the shuttle could take. If v_0 is greater than zero, the shuttle will slingshot around the sun (the smaller v_0 is, the closer the shuttle will come to the sun, but it will never make actual contact). Alternatively, if v_0 is zero, the shuttle will follow a linear path and be pulled directly into the sun. Remember that we have assumed that the sun is a point in this problem; if we include the finite radius of the sun, then the shuttle could collide with the sun even with nonzero initial speed.

(c) Angular momentum is conserved:

$$(d)\, m_{\text{shuttle}} v_0 \sin\theta = (r)\, m_{\text{shuttle}} v$$

Additionally, energy is conserved:

$$\frac{1}{2} m_{\text{shuttle}} v_0^2 - \frac{G m_{\text{shuttle}} m_{\text{sun}}}{d} = \frac{1}{2} m_{\text{shuttle}} v^2 - \frac{G m_{\text{shuttle}} m_{\text{sun}}}{r}$$

2. (a) Imagine moving the mass m from the center of the planet to a distance r from the center of the planet. The gravitational potential energy is equal to the negative of the work done by the gravitational force. (Note that the gravitational force points in the negative radial direction.)

$$U(r) = U(r) - U(0) = -W_{\text{gravity}} = -\int F_r dr = -\int_0^{r_{\text{final}}} \left(-\frac{GmMr}{R^3}\right) dr = \frac{GmMr^2}{2R^3}$$

The sketch of this potential energy curve is a parabola as shown in Figure 10.13.

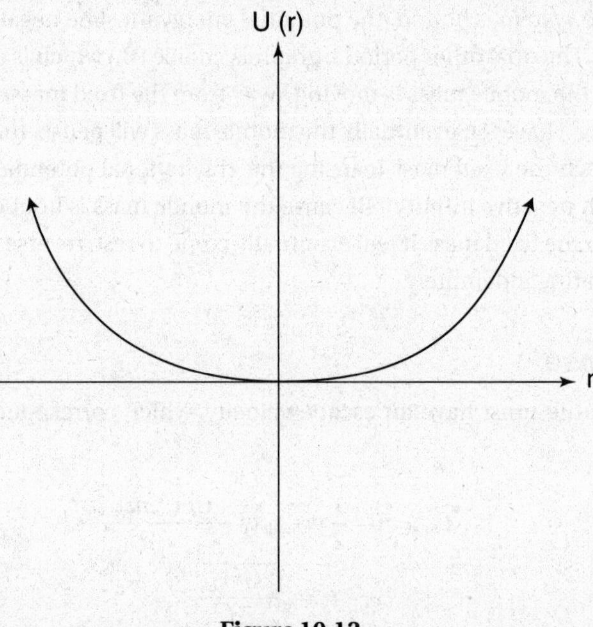

Figure 10.13

(b) By now, we are beginning to get the idea that this problem might involve simple harmonic motion. The force found in part (a) is proportional to r and directed toward $r = 0$, the center of the planet. We also see from part (a) the characteristic parabolic potential energy of simple harmonic motion.

The time required for the mass to return to its original position is given by the period of this simple harmonic motion. Examination of the equation $F(r) = -(GmM/R^3)r$ and comparison with the equation $F(x) = -kx$ indicates that the effective spring constant is GmM/R^3. Therefore, the period is

$$T = 2\pi\sqrt{\frac{m}{k_{\text{eff}}}} = 2\pi\sqrt{\frac{m}{GmM/R^3}} = 2\pi\sqrt{\frac{R^3}{GM}}$$

(c) A particle released at $R/2$ is influenced by the same linear restoring force, so it will have the same period (and thus the same amount of time will be required for it to return to its original position). Recall that the period does not depend on the amplitude as long as the particle is under the influence of the same linear restoring force (as in an ideal mass-spring system).

(d) Outside the planet, the restoring force *decreases* rather than continuing to increase linearly with distance. However, the mass will still oscillate, albeit not in simple harmonic motion. Because the restoring force is weaker than the linear restoring force $F(r) = -(GmM/R^3)r$, the period will be larger and the mass will require *more time* to return to its original position.

3. (a) The gravitational force provides the centripetal force for the satellite's orbit, and thus Newton's second law reads

$$\frac{Gm_{sat}m_{sun}}{R^2} = \frac{m_{sat}v^2}{R}$$

$$v = \sqrt{\frac{Gm_{sun}}{R}}$$

(b) This is a one-dimensional collision problem where both momentum and energy are conserved. (We designate the initial common speed of the asteroid and the satellite as $v_0 = \sqrt{Gm_{sun}/R}$.)

Setting the x-axis parallel to the initial velocity of the satellite (with the asteroid moving in the $-x$-direction), we have

Conservation of momentum:

$$mv_0 - 2mv_0 = mv_{sat} + 2mv_{ast}$$

Conservation of energy:

$$\frac{1}{2}mv_0^2 + \frac{1}{2}(2m)v_0^2 = \frac{1}{2}mv_{sat}^2 + \frac{1}{2}(2m)v_{ast}^2$$

Solving these simultaneous equations (by solving the first equation for v_{ast}, substituting into the second equation, and applying the quadratic formula) yields

$$v_{sat} = -\frac{5v_0}{3} = -\frac{5}{3}\sqrt{\frac{Gm_{sun}}{R}}$$

Given that the x-axis was designated to point parallel to the original velocity of the satellite, the negative sign here indicates that the satellite has reversed direction (as expected).

Alternative solution: Since this is an elastic collision in one dimension, the relative velocity of the two particles retains the same magnitude while switching direction (an example of special case 2 discussed in Chapter 6). Therefore, $v_{1,sat} - v_{1,ast} = -(v_{2,sat} - v_{2,ast})$. Using this equation together with conservation of momentum allows for a quick solution that involves solving a system of two linear equations rather than using the quadratic equation.

(c) If the total energy (gravitational potential plus kinetic) is greater than or equal to zero, the satellite is unbound. The total energy is

$$E = U + \text{KE} = -\frac{Gm_{sat}m_{sun}}{R} + \frac{1}{2}m_{sat}\left[\frac{5}{3}\sqrt{\frac{Gm_{sun}}{R}}\right]^2$$

$$E = \frac{7}{18}\frac{Gm_{sat}m_{sun}}{R}$$

Because the total energy is positive, the satellite is not bound.

4. (a) In order for the planets to provide each other with the necessary centripetal force, they must always be diametrically opposite each other. Therefore, they must move with the same angular velocity ω and they must share the same period, $T = 2\pi/\omega$.

(b) A planet's period, the amount of time required for it to make one complete revolution, is equal to the length of its orbit divided by the planet's speed. As discussed above, the two planets have the same period. Therefore,

$$\left[T_1 = \frac{2\pi r_1}{v_1}\right] = \left[T_2 = \frac{2\pi r_2}{v_2}\right]$$

Rearranging this equation to solve for the ratio of the two planets' speeds yields $v_1/v_2 = r_1/r_2$.

(c) The planets are held in orbit by the same centripetal force (which is equal to the gravitational attractive force that each planet exerts on the other). Therefore,

$$F_{centripetal} = \frac{Gm_1m_2}{(r_1+r_2)^2} = \frac{m_1v_1^2}{r_1} = \frac{m_2v_2^2}{r_2}$$

Using the last two terms in this equation and solving for m_2 yields $m_2 = m_1v_1^2r_2/r_1v_2^2$. Substituting for $v_1/v_2 = r_1/r_2$ gives the final answer:

$$m_2 = \frac{m_1r_2}{r_1}\left(\frac{v_1}{v_2}\right)^2 = \frac{m_1r_2}{r_1}\left(\frac{r_1}{r_2}\right)^2 = \frac{m_1r_1}{r_2}$$

Reality check: The larger mass orbits with a smaller radius. This makes sense in the extreme case that one mass is much larger than the other: The larger mass remains approximately stationary while the smaller mass rotates about the larger mass (much like Earth orbiting the sun).

Coulomb's Law and Electric Fields Due to Point Charges

11

→ COULOMB'S LAW
→ PRINCIPLE OF SUPERPOSITION
→ ELECTRIC FIELDS AND FIELD LINES
→ ELECTRIC POTENTIAL AND VOLTAGE

We begin our study of electric and magnetic phenomena by considering the interaction of stationary point electric charges. This is easy to do if we go back to Chapter 10 and replace the words *gravitational force* with *electric force*, and *mass* with *charge*. That's right—the physics and math behind stationary point electric charges and universal gravitation are just about the same, with the exception that in gravitation there are no repulsive forces.

COULOMB'S LAW

The *magnitude* of the force between two point charges Q_1 and Q_2 separated by a distance R is given by the equation

$$F = \frac{|Q_1 Q_2|}{4\pi\varepsilon_0 R^2}$$

Scalar part of Coulomb's law

TIP

Note the similarity with Newton's law of universal gravitation, Chapter 10.

The *direction* of the force is along the line between the two charges. Like charges repel, while unlike charges attract.

Coulomb's law, like Newton's law of universal gravitation, is an experimental result. However, it does make sense that the force increases with increasing charges and decreases with increasing distance between charges.

ε_0 is a constant called the permittivity of free space, and is given $\varepsilon_0 = 8.85 \times 10^{-12} \mathrm{C}^2/\mathrm{N} \cdot \mathrm{m}^2$. From your physics class, you may also be familiar with Coulomb's law written in the form $F = k|Q_1 Q_2|/R^2$, where k is called the electrostatic constant and has a value of $k = 1/4\pi\varepsilon_0 = 8.99 \times 10^9 \mathrm{N} \cdot \mathrm{m}^2/\mathrm{C}^2$. ε_0 is used exclusively in Physics C; in later chapters you will see how it simplifies formulas such as Gauss's law. The SI unit of charge is the coulomb (C).

The Principle of Superposition and Coulomb's Forces

Recall from universal gravitation that the net force of several masses on a body obeys the principle of superposition such that the net force is the vector sum of the forces due to each of the individual masses. Given the strong parallels between universal gravitation and Coulomb's law, it shouldn't surprise you that the same is true of electric forces. That is, the force on a charge Q' due to a group of charges is the vector sum of the force that each charge within the group exerts on Q'. In addition to being a key principle, this provides a basic problem-solving technique. If we ask you to find the net force on a charge due to a number of point charges, you must find the force vectors due to each of the individual charges and add them (vector addition).

EXAMPLE 11.1 COULOMB'S LAW AND SUPERPOSITION OF FORCES

Two 1-μC charges lie at (1, 0) and at (0, 1). A third negative charge, –*Q*, lies along the *y*-axis at (0, 10).

(a) Calculate the magnitude of the force that the 1-μC charge at (0, 1) exerts on the 1-μC charge at (1, 0).

(b) Calculate the components of this force in the *x*- and *y*-directions and express the force as a vector.

(c) Given that the *y*-component of the *net* force (due to the other 1-μC charge and the charge –*Q*) on the 1-μC charge at (1, 0) is zero, calculate the unknown charge –*Q*.

SOLUTION

(a) See Figure 11.1. Plugging into Coulomb's law (calculating the distance between the two points using the Pythagorean theorem),

$$F=\frac{|Q_1Q_2|}{4\pi\varepsilon_0R^2}=\frac{(1\times10^{-6}\text{ C})^2}{4\pi(8.85\times(10^{-12}\text{ C}^2)/\text{N}\cdot\text{m}^2)\left[(1\text{m})^2+(1\text{m})^2\right]}=4.5\times10^{-3}\text{ N}$$

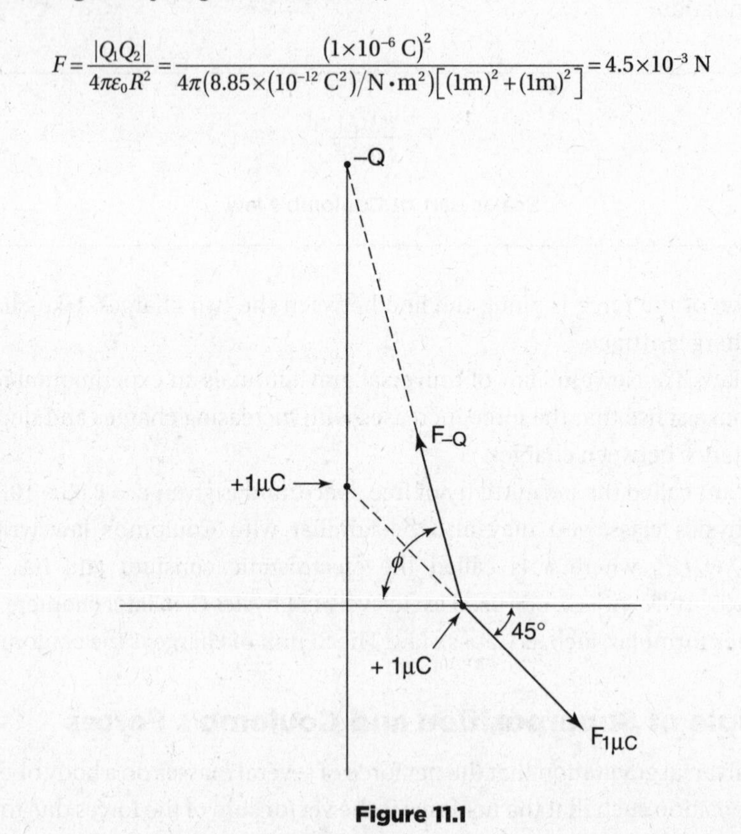

Figure 11.1

(b) **We resolve this vector into components using geometry**

$$\begin{cases} F_x = (4.5\times10^{-3}\ \text{N})\cos 45° = 3.18\times10^{-3}\ \text{N} \\ F_y = -(4.5\times10^{-3}\ \text{N})\sin 45° = -3.18\times10^{-3}\ \text{N} \end{cases}$$

Thus, the force vector can be written as

$$\mathbf{F} = (3.18\times10^{-3}\ \text{N})\hat{\imath} - (3.18\times10^{-3}\ \text{N})\hat{\jmath}$$

(c) **For the *y*-component of the net force to be zero, the unknown charge must exert a force that has a *y*-component of $3.18 \cdot 10^{-3}$ N. The magnitude of the unknown force (again using the Pythagorean theorem) is**

$$F = \frac{|Q_1Q_2|}{4\pi\varepsilon_0 R^2} = \frac{(1\times10^{-6}\ \text{C})Q}{4\pi(8.85\times(10^{-12}\ \text{C}^2)/\text{Nm}^2)\left[(1\text{m})^2 + (10\text{m})^2\right]} = (89Q)\ \text{N/C}$$

The *y*-component of the force (more vector geometry) is

$$F_y = (\sin\phi)(89Q)\text{N/C} = \frac{10\text{m}}{\left[(10\text{m})^2 + (1\text{m})^2\right]^{1/2}}(89Q)\text{N/C} = (88.6Q)\text{N/C}$$

Setting this force equal to 3.18×10^{-3} N = (88.6Q)N/C yields *Q* equal to –35.9μC.

THE CONCEPT OF FIELDS: GRAVITATIONAL FIELDS AND ELECTRIC FIELDS

We've already seen the concept of fields when dealing with gravity. To review: The gravitational force on a mass *M* at a point in space due to some specified mass distribution is proportional to *M*. Therefore, the ratio of the force to the mass *M* is independent of *M* and has the same value no matter what mass we place at that point (as long as it is not large enough to disturb the specified mass distribution). Thus, we can write

$$\frac{\mathbf{F}}{M} = \frac{\mathbf{F}_{\text{any mass}}}{M_{\text{any mass}}} = \mathbf{g}$$

By specifying the ratio $\mathbf{F}/M = \mathbf{g}$ (the gravitational field vector), we can quickly calculate the force on any other mass placed at that point by multiplying the gravitational field vector by the mass using the equation $\mathbf{F}/m = \mathbf{g}$. Note that the gravitational field is a *vector field*—it assigns a vector to every location in space. The electric field is completely analogous to the gravitational field:

$$\frac{\mathbf{F}}{Q} = \frac{\mathbf{F}_{\text{any charge}}}{Q_{\text{any charge}}} = \mathbf{E}$$

Definition of an electric field

Based on the above definition, because force is a vector quantity, the electric field must also be a vector quantity. What is its direction? If the charge Q is positive, **E** and **F** are parallel. Therefore,

The electric field points in the same direction as the force a positive charge would feel at that location.

How do you actually *use* this definition to calculate the electric field? The electric field can be calculated directly from Coulomb's law as follows (later on we will use faster methods).

1. Imagine placing a small positive charge (a test charge) at the location where you are trying to calculate the electric field.
2. Use Coulomb's law to calculate the force on the test charge, $\mathbf{F}_{\text{test charge}}$.
3. Calculate the electric field by dividing the force on the test charge by its charge according to the above definition: $\mathbf{E} = \mathbf{F}_{\text{test charge}} / Q_{\text{test charge}}$.

It immediately follows that the force on a point charge in an electric field is given by the equation

$$\mathbf{F} = q\mathbf{E}$$

Force on a point charge *q* in an electric field E
(analogous to the universal gravitation equation F = *mg*)

From these equations, it should be clear that the SI units of electric field are newtons/coulomb.

Electric Field Due to a Point Charge

TIP

The electric field is independent of the test charge.

What is the electric field due to a point charge $+Q$, located at the origin, at a point a distance R from it? Let's introduce a test charge of charge Q_{test} at our point a distance R from $+Q$. Coulomb's law states that the force on the test charge has magnitude

$$F = \frac{|Q_1 Q_{\text{test}}|}{4\pi\varepsilon_0 R^2}$$

As usual, the direction of the force is along the line between the charges, either attractive (unlike charges) or repulsive (like charges). According to the definition of electric field,

$$\mathbf{E} = \frac{\mathbf{F}_{\text{on test charge}}}{Q_{\text{test}}}$$

Therefore, the magnitude of the E field is given by

$$E = \frac{F_{\text{on test charge}}}{Q_{\text{test}}} = \frac{|Q|}{4\pi\varepsilon_0 R^2}$$

As usual, the electric field points in the direction of the force on a positive charge. In this case, this is directly away from the positive charge $+Q$. Of course, if the point charge at the origin is negative, the direction will be reversed (because a positive test charge would be attracted to it).

Magnitude of electric field due to a point charge (direction is away from positive point charges and toward negative point charges):

$$E = \frac{|Q|}{4\pi\varepsilon_0 R^2}$$

The Principle of Superposition and Electric Fields

We have already discussed the principle of superposition as applied to electric forces: The force on a charge Q_0 due to a number of charges $Q_1, \ldots, Q_N$ is equal to the vector sum of the individual forces of each charge on Q_0. Mathematically:

$$\mathbf{F}_{net} = \mathbf{F}_1 + \mathbf{F}_2 + \mathbf{F}_3 + \ldots + \mathbf{F}_N$$

Dividing by charge Q_0,

$$\frac{\mathbf{F}_{net}}{Q_0} = \frac{\mathbf{F}_1}{Q_0} + \frac{\mathbf{F}_2}{Q_0} + \frac{\mathbf{F}_3}{Q_0} + \ldots + \frac{\mathbf{F}_N}{Q_0}$$

Recalling the definition of electric field (here Q_0 acts like a test charge),

$$\mathbf{E}_{net} = \mathbf{E}_1 + \mathbf{E}_2 + \mathbf{E}_3 + \ldots + \mathbf{E}_N = \text{electric field at the location of } Q_0$$

This equation reveals that electric fields also obey the principle of superposition.

***Principle of superposition for the electric field*: The electric field at a location in space due to a number of point charges $Q_1, \ldots, Q_N$ is equal to the vector sum of the electric field produced by each of the individual point charges $Q_1, \ldots, Q_N$ at that location in space.**

EXAMPLE 11.2 ELECTRIC FIELDS DUE TO POINT CHARGES

A point charge +Q is located at the origin and a point charge −2Q is located at (1, 0). Where is the electric field equal to zero? (Distances are measured in meters, and charges in coulombs.)

SOLUTION

The first point to realize is that the only place the electric field can ever be zero is along the x-axis. Why? For two nonzero vectors to add to zero (cancel each other out), they must be *antiparallel*. As soon as you move off the x-axis, the electric field vectors due to the two charges can never be parallel.

So that's a big help; now, where on the x-axis should we look for solutions? Between (0, 0) and (1, 0) both charges produce electric field vectors pointing in the +x-direction, so they won't cancel and there won't be any solutions between (0, 0) and (1, 0). What about to the right of (1, 0)? In this region, we are always closer to the larger charge, so the field due to −2Q is always dominant, producing a nonzero field. However, this situation is reversed in the region to the left of the origin where we are *closer* to the *smaller* charge. Thus, very close to the origin, our proximity to +Q causes it to dominate, producing a net electric field to the *left*. However, as we move farther away from the origin, the difference in our distance from the +Q and the −2Q becomes less pronounced [e.g., at (−10, 0) the distances are 10 and 11, respectively], causing the −2Q charge to dominate and producing a net electric field to the *right*. Because the electric field is continuous (as long as we avoid point charges), it must go from positive to negative somewhere to the left of the origin; at this point it will be zero! Thus, with a little thinking we have figured out that there is only *one* place where the electric field will be zero, and that is along the x-axis to the left of the origin, at a point we will call (−x, 0). Now, time for the math. Using superposition we have

$$0=-\frac{Q}{4\pi\varepsilon_0 x^2}+\frac{2Q}{4\pi\varepsilon_0 (x+1)^2}$$

This equation simplifies nicely, yielding $(x+1)^2 = 2x^2$. Of the two solutions to this equation, $x = 1 \pm \sqrt{2}$, we choose the positive solution $x = 1+\sqrt{2}$ (because x was defined as positive above). The final answer is that the electric field is zero at $(-1-\sqrt{2}, 0)$.

If the two charges were equal, that is, if +Q were located at the origin and −Q were located at (1, 0), where would the electric field be zero? The answer is that it would never be zero! The same arguments hold as above, with the exception that to the right of (1, 0) or to the left of (0, 0), the charge closer to the point would *always* dominate and there would never be exact cancellation. (The key to a solution in the above example is the interplay between one charge being closer and the other charge being larger. Without this dynamic, the closer charge always produces a larger field, so the net field is always nonzero.)

Visualizing Electric Fields: Electric Field Lines

As we already discussed, the electric field is a vector field (i.e., it assigns a vector to every point in space). A vector field can be plotted as shown in Figure 11.2; at a variety of representative points, the field vector is calculated and drawn, with the tail of the vector lying at the point where the vector was calculated.

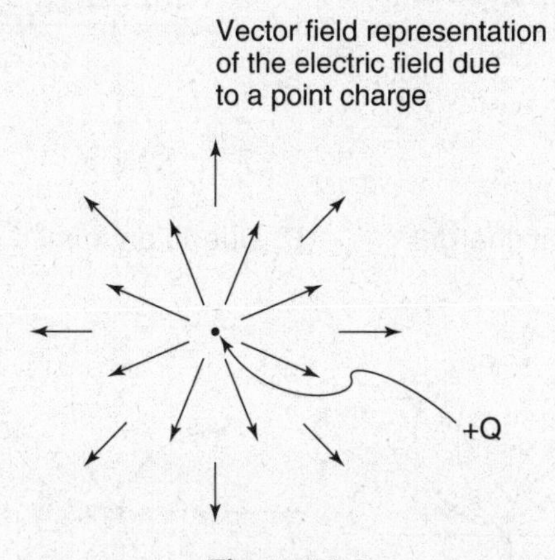

Figure 11.2

Another way to visualize the electric field is to draw electric *field lines* as in Figure 11.3. Electric field lines are lines (or curves) drawn such that the electric field is everywhere tangent to them. A few common diagrams of electric field lines are shown in the figure.

A few important properties of electric field lines are as follows.

1. The lines begin or end only on charges (this is a direct consequence of Gauss's law, as discussed in Chapter 13). As a result of this fact, the lines are spaced such that they are bunched more closely together in areas of strong field and spaced more widely apart in areas of weak field.
2. Because the electric field can have only one unique direction at any location in space, the lines may never intersect in a location where the direction of the electric field is well defined (this excludes locations where the volume charge density is infinite, that is, point charges and surface charge distributions, where the electric field is undefined and discontinuous).
3. The lines represent the *direction* of the force. In general, the lines do not represent the *trajectory* of a particle moving only under the influence of the electric field. Why? From mechanics we know that for a particle to move along a curved path, it must have a radial acceleration. If a particle were moving along a curved electric field line, because the force is parallel to the field line, there would be no force providing the radial acceleration—this would be impossible!

Examples of electric field lines

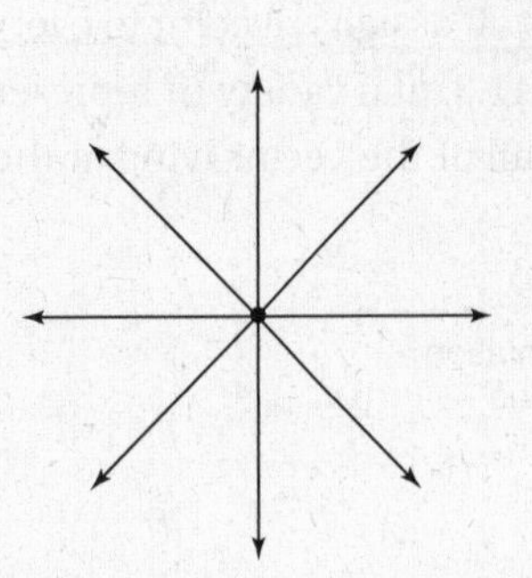

(A) Due to a positive point charge

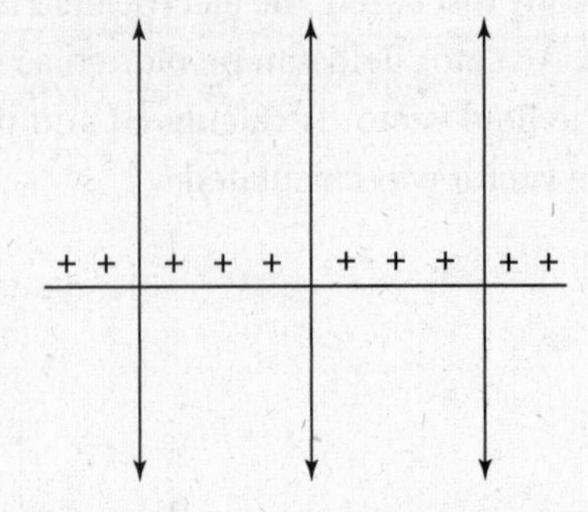

(B) Due to an infinite positively charged sheet

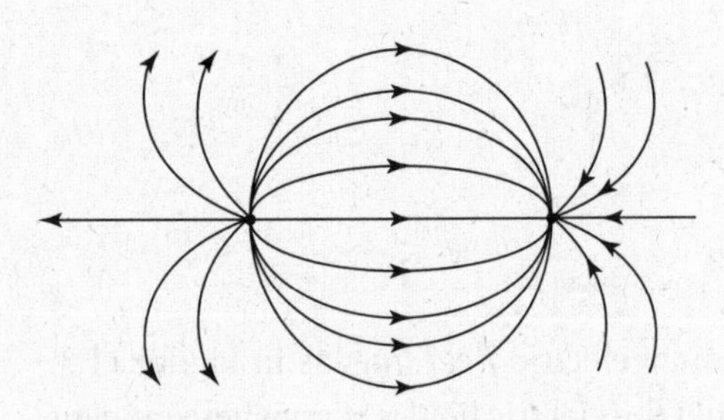

(C) Due to two unlike point charges

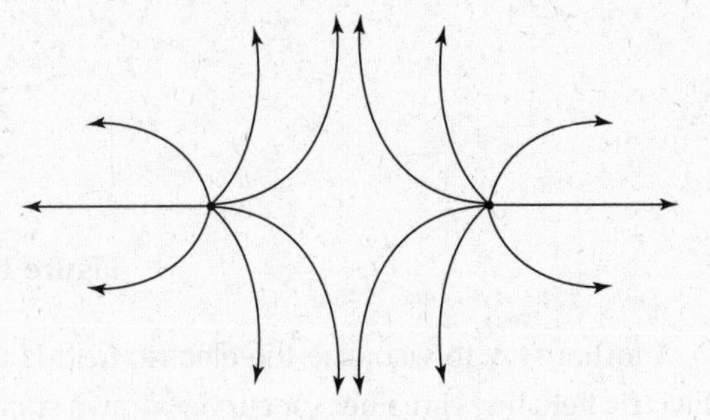

(D) Due to two positive point charges

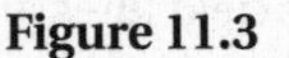

Figure 11.3

Electrostatic Induction

What happens if you place a point charge $+Q$ near an uncharged metal sphere where charges are free to move around? The $+Q$ attracts negative charge on the sphere and repels positive charge on the sphere. Therefore, the charges on the sphere rearrange so that the negative charge is closer to $+Q$ than the positive charge. Because the electrostatic force is inversely proportional to (distance)^2, this causes the attraction between $+Q$ and the negative charge on the sphere to be greater than the repulsion between $+Q$ and the positive charge on the sphere, giving rise to a *net attractive force.* The magnitude of this force depends on the magnitude of $+Q$ and the degree to which charges can move around on the sphere.

POTENTIAL AND VOLTAGE

The Definitions of Potential and Voltage

Recall from universal gravitation that we introduced a potential energy function $U(x)$ associated with a system of particles as a function of their relative position (e.g., $U(r) = -Gm_1m_2/r$). It would be perfectly valid to develop an analogous potential energy function associated with the electrostatic force. However, this $U(x)$ function would only tell us about the potential energy of a particular set of charges. It would be more useful to somehow generalize $U(x)$ so that it could tell us about the potential energy of a charge of *any* value in the presence of a set of source charges (the same way the electric field allows us to quickly calculate the force on a charge of any value using the equation $\mathbf{F} = Q\mathbf{E}$ and the gravitational field allows us to quickly calculate the force on a mass of any value).

Because $U(x)$ is obtained by integrating force, and because force is proportional to the magnitude of the test charge, $U(x)$ is also proportional to the charge of the test charge. Thus, the ratio between $U(x)$ and the charge is *independent* of the value of the charge and we can define electric potential V (a scalar quantity) as this ratio:

$$V = \frac{U_{\text{test charge}}}{Q_{\text{test charge}}} = \frac{U_{\text{any charge}}}{Q_{\text{any charge}}}$$

If we define potential as above, we can easily calculate the potential energy U of a charge of *any* value simply by multiplying the value of the charge by potential. For example, for charge 1, $U_{\text{charge 1}} = Q_1 V$.

Recall that the magnitude of any form of potential energy is arbitrary; only *differences* in potential energy have meaning. We then define voltage as the difference between the potentials at points 1 and 2 in space:

Voltage is the potential difference between two points.

$$\text{Voltage} = \Delta V = V_1 - V_2 = \frac{U_1}{Q} - \frac{U_2}{Q} = \frac{\Delta U}{Q}$$

Repeating the above derivation for very small ΔU and ΔV (such that these are differential differences), we can obtain a differential statement:

$$dV = \frac{dU}{Q}$$

$$V = \frac{U}{Q} \overset{\text{equivalent}}{\Leftrightarrow} \Delta V = \frac{\Delta U}{Q} \overset{\text{equivalent}}{\Leftrightarrow} dV = \frac{dU}{Q}$$

Definitions of potential and voltage

The units of potential and change in the potential energy of voltage are volts, given by 1 volt = 1 joule/coulomb. One *electronvolt* is the amount of energy stored by a charge of e when it moves through a potential difference of 1 volt: $1 \text{ eV} = (1.60 \times 10^{-19}\text{C})(1 \text{ V}) = 1.60 \times 10^{-19} \text{ J}$. As indicated by this equation, an electronvolt is a measure of energy, not voltage (it's simply an alternative unit for energy that is handy when working at the atomic scale).

Electrostatic Force as a Conservative Force

Recall from mechanics that not all forces have an associated potential energy function; only conservative forces do. Thus, by saying that a potential energy or potential function even exists, we are assuming that the electrostatic force is conservative. This is fairly clear from the fact that Coulomb's law has the same mathematical form as the universal gravitational force, which we know to be conservative. The fact that we will be able to determine potential functions later in this chapter confirms this assumption.

The Relationship Between Potential and Electric Field

We have created electrical counterparts of force (electrical field = force/charge) and the potential energy (potential = potential energy/charge) functions. Therefore, we might expect that mechanics equations that relate force and potential energy have electrical analogs that relate electrical field and potential. Starting with the relationship of potential energy, work, and force,

$$\Delta U = -\int \mathbf{F} \cdot d\mathbf{s}$$

Dividing through by charge yields

$$\frac{\Delta U}{q} = -\int \frac{\mathbf{F}}{q} \cdot d\mathbf{s}$$

Recognizing potential on the left and electric field on the right,

$$\Delta V = -\int \mathbf{E} \cdot d\mathbf{s}$$

Relationship between potential difference and electric field

In the case where motion is along the x-direction, this equation can be simplified as

$$\Delta V = -\int \mathbf{E} \cdot dx\hat{i} = -\int \left(E_x\hat{i} + E_y\hat{j} + E_z\hat{k}\right) \cdot dx\hat{i} = -\int E_x dx$$

Differentiating both sides with respect to x allows us to calculate the electric field from the potential function:

$$dV = -E_x dx \Rightarrow E_x = -\frac{dV}{dx}$$

Note that if the electric field is constant, we can write this equation as

$$E_x = -\frac{\Delta V}{\Delta x}$$

Relationship between potential and electric field with the integration path along the *x*-axis:

$$\Delta V = -\int E_x dx$$

$$E_x = -\frac{dV}{dx}$$

$$E_x = -\frac{\Delta V}{\Delta x}$$

(valid only if electric field is constant)

Happily, we will be using these one-dimensional equations much more than complicated line integrals. If you understand how to solve problems using the mechanics equation $F = -dU/dx$, you're 90% of the way to understanding this equation. (It really amounts to writing the same equation in a more convenient form.)

Calculation of Potential Due to a Point Charge by Integration

Unless otherwise stated, it is generally assumed that we take $U = 0$ (and thus $V = 0$) at an infinite distance from the source charges. Situations arise when this is not useful, and we let $U = 0$ wherever it is most convenient for us. Recall that the actual value of U is meaningless;

only differences in U have significance. Therefore, we can let $U = 0$ wherever we please, so long as we are consistent within a given problem.

When we have a charge $+Q$ located at the origin, we want to calculate the potential at a point $(0, a)$ on the x-axis. As usual, V is taken to be zero at $x \to \infty$.

> ***Mechanical analog*: It is helpful to remember how we calculated the potential energy of two masses a distance R from each other. We started with the masses infinitely far apart (one mass fixed at the origin, the other located at $x = \infty$) and then integrated the differential $dU = -F_x dx$ as we brought one of the masses in from $x = \infty$ to $x = a$. The integral we used was**
>
> $$\int_{\infty}^{a} dU = U(a) - U(\infty) = U(a) - 0 = U(a)$$
>
> **A gravitational "potential" could be defined if we divided $U(a)$ by the value of the mass moved from infinity.**

Our solution will be completely analogous; we will calculate potential by integrating dV as we go from $x = \infty$ to $x = a$. To find dV, we use the electrostatics analog of $dU = -F_x dx$ derived above, $dV = -E_x dx$ (for this equation to be valid, we choose an integration path along the x-axis). The electric field of a point charge is

$$E = \frac{|Q|}{4\pi\varepsilon_0 R^2} \quad \text{away from positive point charges and toward negative point charges}$$

Because a charge of $+Q$ is located at the origin, in this example the electric field points in the positive x-direction such that $E_x = Q/4\pi\varepsilon_0 x^2$. Substituting this into the equation for dV:

$$dV = -E_x dx = -\frac{Q}{4\pi\varepsilon_0 x^2} dx$$

Integrating,

$$\int_{x=\infty}^{x=a} dV = \int_{x=\infty}^{x=a} -\frac{Q}{4\pi\varepsilon_0 x^2} \cdot dx$$

$$V(a) - V(\infty) = \frac{Q}{4\pi\varepsilon_0 a} - \frac{Q}{4\pi\varepsilon_0 \infty}$$

Recall that we set $V(\infty) = 0$, so we have the following.

> $$V(a) = \frac{Q}{4\pi\varepsilon_0 a}$$ **Potential due to a point charge Q at distance a**

We derived this formula by calculating the potential at a point on the $+x$-axis of a positive charge at the origin, so is it valid to generalize it for negative charges and for points not lying on the x-axis? Yes, for the following two reasons.

1. If the charge is negative instead of positive, the electric field and thus the differential potential ($dV = -E_x dx$) change sign; the sign of the equation $V = Q/4\pi\varepsilon_0 R$ *also* changes sign so everything matches up.
2. What about the potential at points not on the x-axis? Well, we can just move the x-axis so they *do* lie on the x-axis! There's no predetermined x-axis, so we're free to do this. Note what a powerful trick this is: We just generalized a result from one dimension to three dimensions without any work because of the spherical symmetry of a point charge.

The Principle of Superposition and Potential

So now we know how to calculate the potential due to a point charge, but what about the potential due to a bunch of point charges? Do we have to set up and evaluate a messy integral? No. Once again superposition comes to the rescue. To see how this works, let's calculate the potential at some point $x = a$ due to a number of point charges Q_1, ..., Q_n that lie along the x-axis. Once again, we integrate dV from $x = \infty$ to $x = a$:

$$\int_{x=\infty}^{x=a} dV = V(a) - V(\infty) = V(a)$$

Substituting for dV,

$$V(a) = \int_{\infty}^{a} dV = -\int_{\infty}^{a} E_{x,\text{net}} dx$$

Because the net electric field is the superposition of the fields due to each of the point charges Q_1, ..., Q_n, it can be written as follows (of course, each of these terms is a function of x).

$$E_{\text{net}} = E_1 + E_2 + E_3 + \ldots + E_N$$

Plugging this into the potential equation,

$$V(a) = -\int_{\infty}^{a} (E_1 + E_2 + E_3 + \ldots + E_n) dx$$

The integral of a sum is equal to the sum of the integrals:

$$V(a) = -\int_{\infty}^{a} E_1 dx - \int_{\infty}^{a} E_2 dx - \int_{\infty}^{a} E_3 dx - \ldots - \int_{\infty}^{a} E_n dx$$

On the right side, we recognize integrals, each of which gives the potential at $x = a$ due to a given charge. For example, the integral $-\int_{\infty}^{a} E_2 dx$ is equal to the voltage at $x = a$ due to the charge Q_2. Therefore, we can rewrite this equation as

$$V_{\text{net}}(a) = V_{\text{due to } Q_1}(a) + V_{\text{due to } Q_2}(a) + V_{\text{due to } Q_3}(a) + \ldots + V_{\text{due to } Q_n}(a)$$

This equation reveals that potentials superpose or, in other words (because they are scalars and not vectors), add.

***Principle of superposition for potential*: The potential at a point due to a number of charges is equal to the sum of the potentials at that point due to each of the individual charges.**

We derived this formula in one dimension, but it also holds in three dimensions for the same reasons that it holds in one dimension: Superposition of electric fields leads to superposition of potentials. Thus,

$$V = \frac{Q_1}{4\pi\varepsilon_0 R_1} + \frac{Q_2}{4\pi\varepsilon_0 R_2} + \ldots + \frac{Q_n}{4\pi\varepsilon_0 R_n}$$

Superposition of potentials due to an array of point charges; R_i is the distance from charge Q_i to the point where we calculate V

Problem Solving and Potential: Energy Conservation

Because potential is a measure of potential energy per charge, it is easily used to calculate the potential energy of a specified charge and thus is useful in problems involving energy conservation.

Visualizing Potential: Equipotential Lines

Although potential exists in three dimensions, we will only try to describe the potential along a plane. Potential is a scalar, so it assigns a value to every point in the plane. Thus, one way to visualize potential along a plane is simply to make a three-dimensional plot with two position dimensions and a potential dimension. Such a plot might look a bit like a landscape, with conical "mountains" extending to infinite heights over positive point charges or equally deep conical "craters" extending below negative point charges.

Alternatively we can represent potential in two dimensions using equipotential lines, which are lines of constant potential in the two-dimensional plane. By providing a series of equipotential lines at equally spaced intervals (i.e., plotting when potential is 1V, 2V, 3V, and so on), a contour plot can be created (similar to topographical maps that display different altitudes in different colors). Some properties of equipotential lines are the following.

1. Because the potential is the same along an equipotential line, no force or work is required to move a charge along a given equipotential line.
2. Recall that work is defined as $W = \mathbf{F} \cdot d\mathbf{r}$. Thus, the fact that no work is required to move a charge along an equipotential line reveals that the line must be *perpendicular* to the electric force and thus perpendicular to the electric field.
3. As discussed above, along a line joining any two points in space, $E_{\text{parallel to line}} = -dV/d(\text{position})$. We can use this equation to estimate the electric field from equipotential lines: Because we know from property 2 that the field is perpendicular to the equipotential lines, for any line segment perpendicular to the equipotential lines, $E = E_{\text{parallel to line}} \approx -\Delta V/\Delta(\text{position})$. Qualitatively, this equation shows that the electric field is stronger (and thus the electric field lines are bunched closer together) in regions where the potential changes rapidly (which corresponds to regions where equipotential lines are spaced closely together). Some examples of electric field lines and equipotential curves are shown in Figure 11.4.

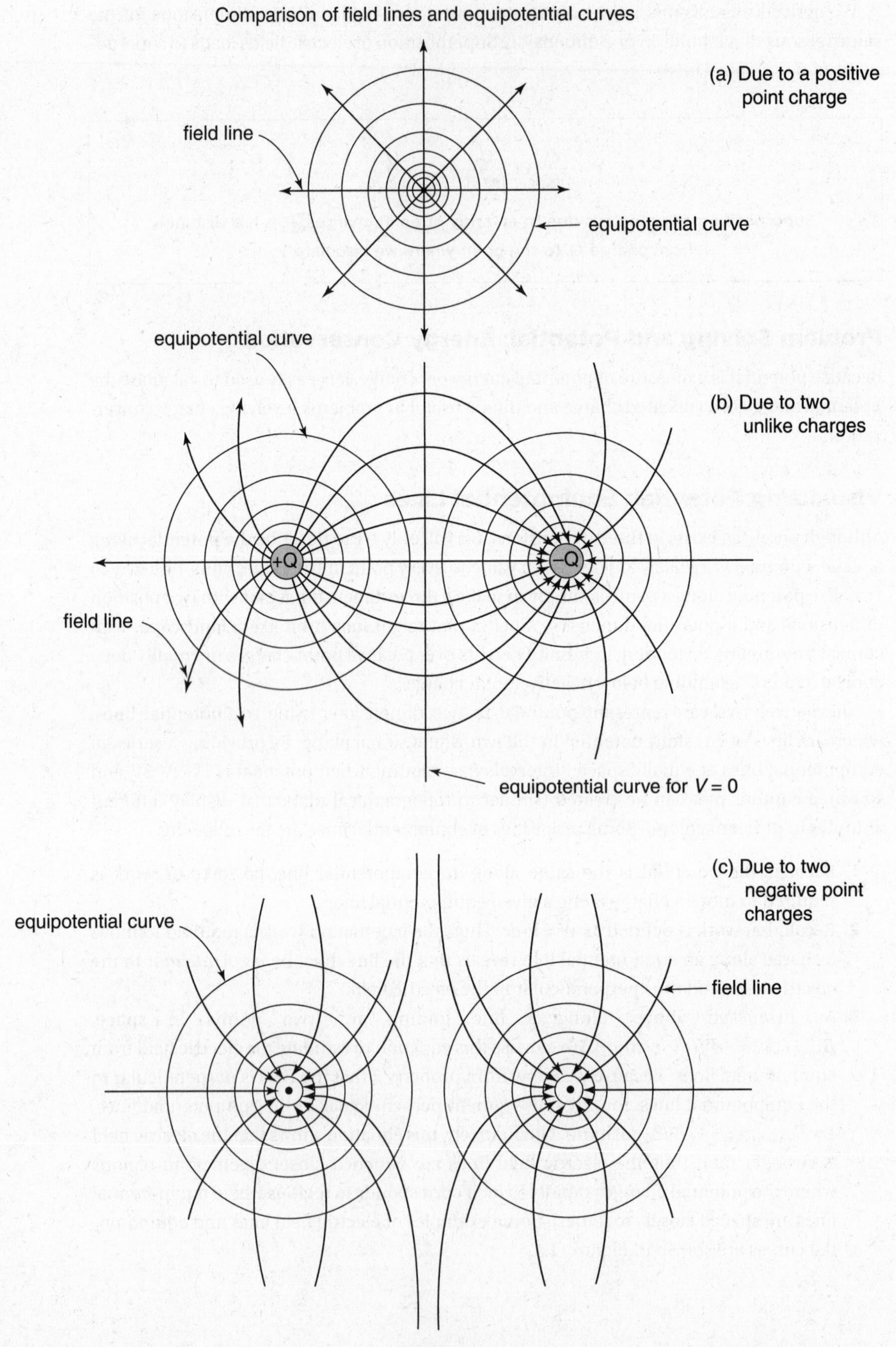
Comparison of field lines and equipotential curves
(a) Due to a positive point charge
field line
equipotential curve
equipotential curve
(b) Due to two unlike charges
+Q
−Q
field line
equipotential curve for V = 0
(c) Due to two negative point charges
equipotential curve
field line

Figure 11.4

CHAPTER SUMMARY

The force between two electric charges that are motionless is given by Coulomb's law. When more than two charges are present, the force on any one of them can be found using Coulomb's law and the principle of superposition. The electric field is defined as the force per unit charge. Electric field lines provide a convenient way to visualize the electric field produced by a charge distribution. The electric potential is the electric potential energy per unit charge, and the voltage is the difference in potential between two points in space.

PRACTICE EXERCISES

Multiple-Choice Questions

1. Two experiments are performed using a pair of solid metal spheres connected by a switch, as shown in Figure 11.5. Both spheres are grounded before each experiment so that they initially have no charge.

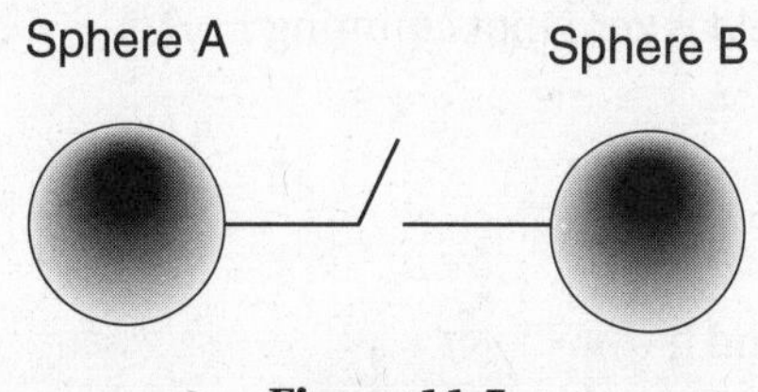

Figure 11.5

Experiment I: A positively charged rod is brought near sphere A while the switch is closed, the switch is opened, and the rod is then removed.

Experiment II: A positively charged rod is touched to sphere A while the switch is closed, the switch is opened, and the rod is then removed.

Which of the following describes the charge on the spheres after the two experiments?

(A) After experiment I both spheres have no charge; after experiment II both spheres are positively charged.
(B) After experiment I sphere A is negatively charged and sphere B is positively charged; after experiment II sphere A is positively charged and sphere B is negatively charged.
(C) After experiment I sphere A is negatively charged and sphere B is positively charged; after experiment II both spheres are positively charged.
(D) After both experiments, both spheres are positively charged.
(E) After both experiments, sphere A is positively charged and sphere B is negatively charged.

2. Which of the three distributions in Figure 11.6 has a point or points where the potential is zero (excluding $\pm\infty$, but assuming that the potential equals zero at these latter points)?

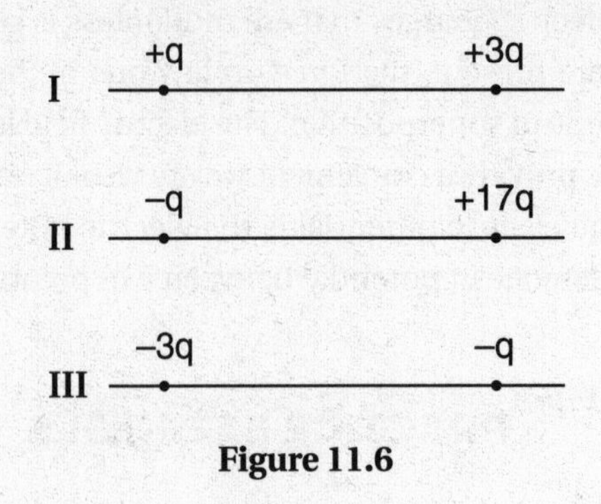

Figure 11.6

(A) distribution I
(B) distribution II
(C) distribution III
(D) distributions I and II
(E) distributions I, II, and III

3. Which of the three charge distributions introduced in question 2 has a point or points where the electric field is zero (not counting $\pm\infty$)?

(A) distribution I
(B) distribution II
(C) distribution III
(D) distributions I and II
(E) distributions I, II, and III

4. Consider the potential-vs-position function shown in Figure 11.7 (the potential asymptotically approaches zero as *x* goes to + infinity or – infinity). At which of the points—*A*, *B*, *C*, *D*, or *E*—does a positive charge feel the greatest force in the +*x*-direction?

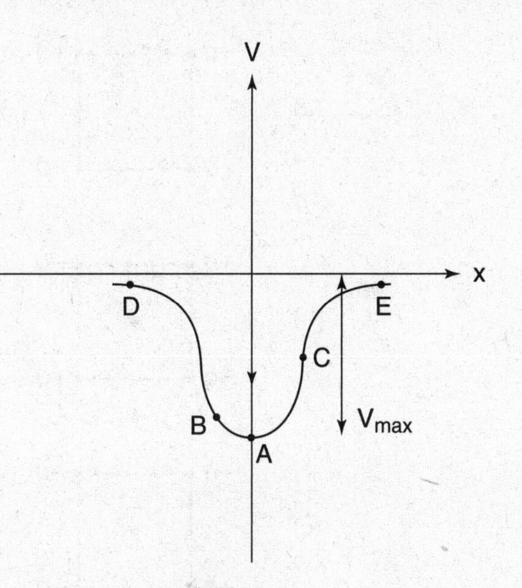

Figure 11.7

(A) point *A*
(B) point *B*
(C) point *C*
(D) point *D*
(E) point *E*

5. Considering the *V*(*x*) curve introduced in question 4, at what point does any charge feel no force?

(A) point *A*
(B) point *B*
(C) point *C*
(D) point *D*
(E) point *E*

6. In the five charge distributions shown in Figure 11.8, all have identical square geometry and all charges have identical magnitudes q. In how many of the charge distributions is the electric field zero at the center of the square?

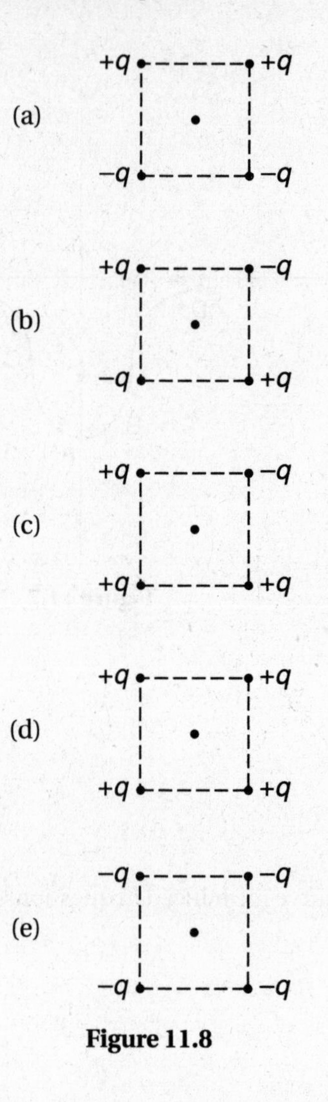

Figure 11.8

(A) one
(B) two
(C) three
(D) four
(E) five

7. In how many of the charge distributions introduced in question 6 is the potential zero at the center of the distribution?

(A) one
(B) two
(C) three
(D) four
(E) five

8. In which of the distributions shown in question 6 is the magnitude of the net electric field at the center of the distribution the greatest?

   (A) distribution A
   (B) distribution B
   (C) distribution C
   (D) distribution D
   (E) distribution E

Free-Response Questions

1. Two positive charges $+Q$ are located on the y-axis at $(0, a)$ and $(0, -a)$, and a third charge $-q$ is on the positive x-axis at $(x, 0)$.

   (a) Find the electric field along the x-axis as a function of position, $\mathbf{E}(x)$, due to the positive charges.
   (b) Find the force on $-q$ as a function of its position, $\mathbf{F}(x)$.
   (c) Using $V = \Sigma(Q_i/4\pi\varepsilon_0 r_i)$, calculate the potential energy function, $U(x)$.

2. Two point charges $+Q$ are located on the x-axis at $(-d, 0)$ and $(d, 0)$.

   (a) Sketch the potential along the x-axis as a function of position, $V(x)$.
   (b) Sketch the electric field along the x-axis as a function of position, $E(x)$.

   A charge of $+q$ and mass m is then placed at the origin.

   (c) What is the force on $+q$ due to the two $+Q$ charges?
   (d) If the charge $+q$ is given a nudge in the y-direction, it will move along the y-axis away from the origin out to infinity. What will its speed be when it reaches $y = \infty$?

3. Consider a set of equipotential lines arising from a static charge distribution, as shown in Figure 11.9,

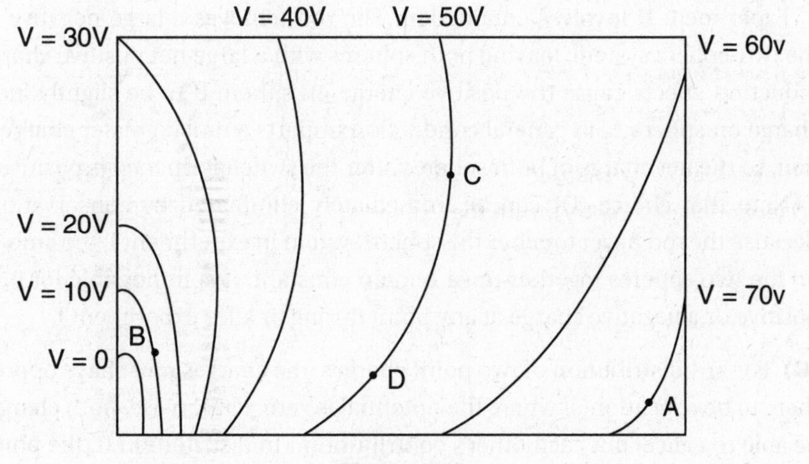

Figure 11.9

(a) Sketch an electric field line starting at point A.
(b) Could a charge ever move along this path due to the electrostatic force? Explain.
(c) Of the points A, B, and C, which has the greatest electric field magnitude?
(d) What is the work done by the electric force if a charge $+Q$ is moved from point C to point D?
(e) What is the work done by the electric force if a charge $+Q$ is moved from point C to point B and then to point D?

4. A charge of $-Q$ is located at the origin, and two charges of $+Q$ are located along the y-axis at $(0, -a)$ and $(0, a)$.

(a) Calculate the electric field along the x-axis, $E(x)$.
(b) Calculate the potential along the x-axis, $V(x)$.
(c) Calculate the approximate form of $E(x)$ for very small x. What physical situation approximately similar to the present one would give the same expression for $E(x)$?
(d) Calculate the approximate form of $E(x)$ for very large x. What physical situation approximately similar to the present one would give the same expression for $E(x)$?

ANSWER KEY

1. **C**	3. **E**	5. **A**	7. **B**
2. **B**	4. **B**	6. **C**	8. **A**

ANSWERS EXPLAINED

Multiple-Choice

1. **(C)** Experiment I involves electrostatic induction. When the positively charged rod is brought near sphere A, it attracts electrons from both spheres, causing sphere A to have a slight negative charge and sphere B to have a slight positive charge. Because the switch is opened before the rod is removed, the charge never has a chance to redistribute itself, so after the experiment is over, sphere A is left negatively charged and sphere B is left positively charged.

   Experiment II involves conduction. The rod removes a large negative charge from the two-sphere system, leaving both spheres with a large net positive charge. Although induction effects cause the positive charge on sphere B to be slightly larger than the charge on sphere A, in general conduction imparts a much greater charge than induction, so the net charge of both spheres after the switch is opened is positive.

   Note that choice (D) can be immediately eliminated by conservation of charge. Because the rod never touches the sphere system in experiment I, the amount of charge on the two spheres together must remain constant; it is impossible for both to have a positive or a negative charge at any point during or after experiment I.

2. **(B)** For any distribution of two point charges, the charges must have opposite signs for there to be a location(s) where the potential is zero. (Otherwise, both charges will never be able to cancel out each other's contribution.) In distribution II, the potential is zero in between the two charges, much closer to the $-q$ charge because this charge has a smaller magnitude (which can be compensated for by having a greater $1/r$ term in its

potential contribution), and also to the left of the $-q$ charge. This behavior is illustrated in Figure 11.10, which shows the potential due to point charges.

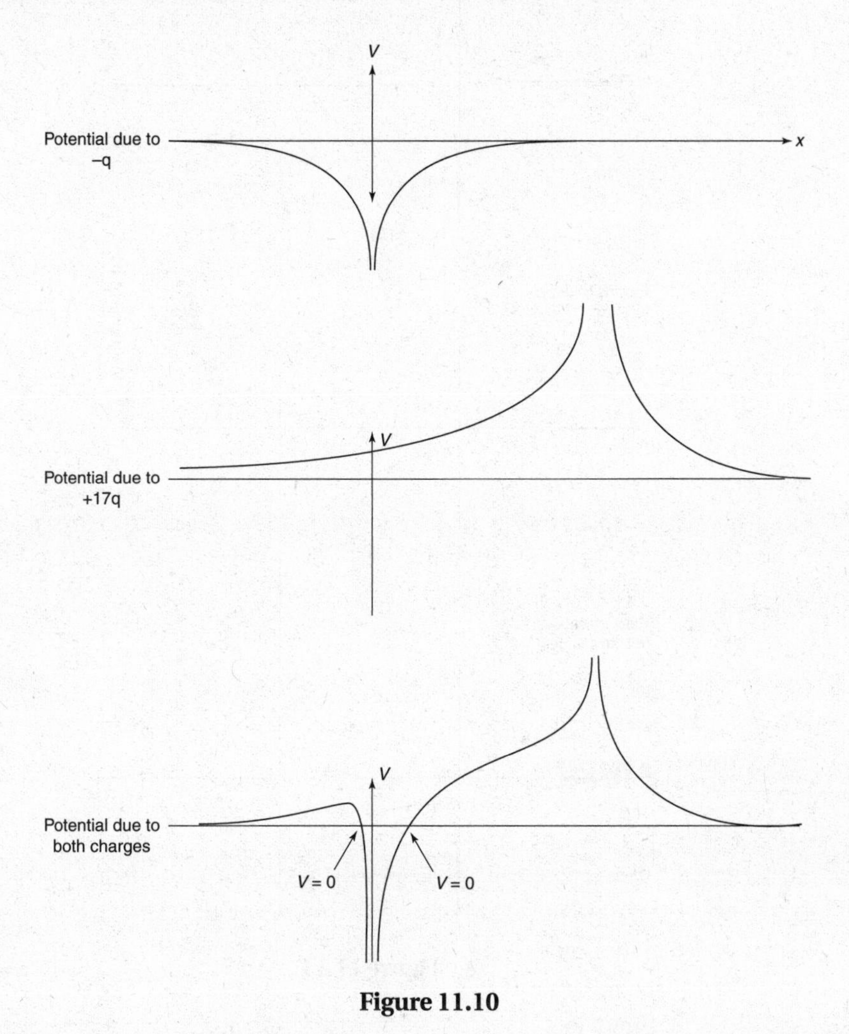

Figure 11.10

3. **(E)** Distributions I and III definitely have one point in between the two charges (closer to the charge of lower magnitude) where the two fields cancel. Distribution II is more tricky. In between the charges, both charges produce a field pointing to the left, so they cannot cancel and the net field is never zero. To the right, the $+17q$ charge creates a field to the right and the $-q$ charge produces a field to the left, but because the $+17q$ charge is always both closer and greater in magnitude at a point on the right, it is always greater than the field due to $-q$, and the net field is never zero. Finally, consider the region to the left of both charges. In this region, the $-q$ charge produces a field to the right, and the $+17q$ charge produces a field to the left. Points to the left of the charge are *closer* to the *smaller* charge, so the two fields can cancel. At some point, the smaller distance to $-q$ compensates for the smaller charge of $-q$, and the magnitudes of the fields contributed by these two charges are equal, producing a net zero field. Figure 11.11, showing the electric fields for distribution II, should explain this further.

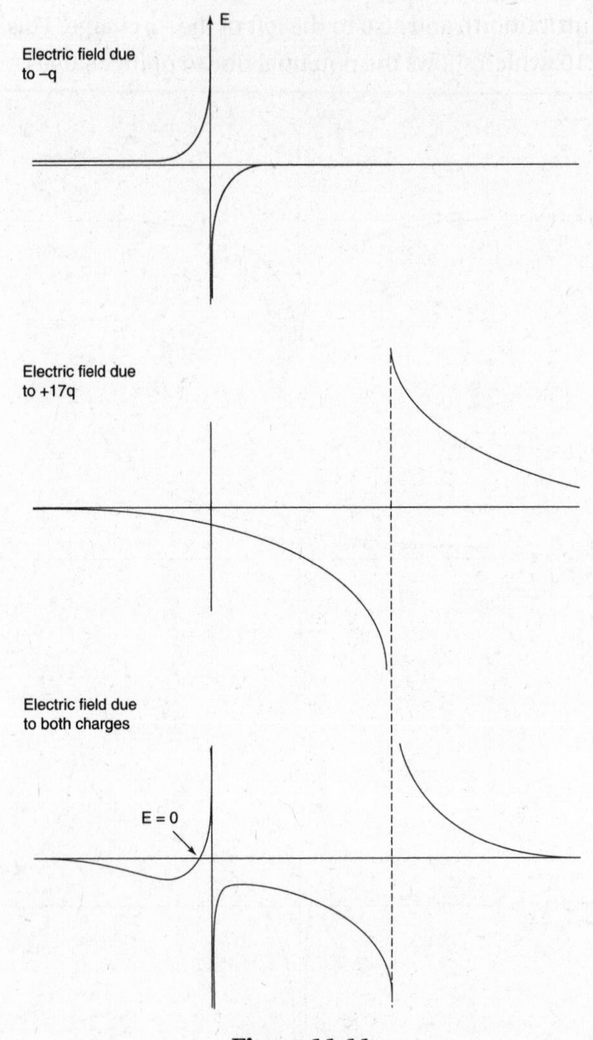

Figure 11.11

4. **(B)** The electric field is the negative slope of the potential curve, $E_x = -dV/dx$, so the place where the force to the right (the positive force) is maximized is the place where the slope of the potential curve has a negative value with the greatest magnitude, at point *B*.
5. **(A)** As discussed in question 4, the electric field is the negative derivative of the potential curve. Where the force is zero, $E_x = 0 = -dV/dx$, so the potential curve must have zero slope (which occurs at point *A*).
6. **(C)** Because of symmetry distributions (b), (d), and (e) all have an electric field of zero at the center of the square. Said another way: If we pair charges along the diagonals of the square, we will see that in distributions (b), (d), and (e), each pair consists of two charges of the same sign, and hence produce a net field of zero at the center of the square.
7. **(B)** Recall the formula for the potenital due to a set of point charges: $V = \Sigma Q_i/4\pi\varepsilon_0 r_i$. In this situation, because all the charges are an equal distance *r* from the center of the square,

$$V=\sum\frac{Q_i}{4\pi\varepsilon_0 r_i}=\frac{1}{4\pi\varepsilon_0 r}\sum Q_i$$

Therefore, for the potential to be zero, the sum of all the charges must equal zero. This is the case for only two of the distributions, distributions A and B.

8. **(A)** In distributions B, D, and E, the electric field is zero at the center of the distribution as discussed in question 6. Therefore, we must choose between only distributions A and C.

   Each charge produces an electric field of equal magnitude at the center of the distribution. It is useful to designate the magnitude of this field due to a single charge as E and then calculate the field in terms of E.

   In distribution C, the upper left and lower right charges cancel each other's electric field, while the electric field due to the upper right and lower left charges are parallel. Therefore, the magnitude of the net field is $2E$.

   In distribution A, because of symmetry, the net field points downward. Because of the vector geometry of the situation, each of the four charges contributes a downward magnitude of $E\cos 45°$. Therefore, the magnitude of the net field is $4E\cos 45° = (4/\sqrt{2})E = (2E)(2/\sqrt{2})$. Because $(2/\sqrt{2}) > 1$, the magnitude of the net field due to distribution A is greater than the magnitude due to distribution C.

Free-Response

1. (a) See Figure 11.12. Symmetry considerations dictate that the direction of the electric field for points on the x-axis is parallel to the x-axis. (You can also see this by noting that the y-components of the electric fields due to the two point charges exactly cancel each other at these points.) Both point charges $+Q$ make equal contributions to E_x:

$$\mathbf{E}=2\left[\frac{Q}{4\pi\varepsilon_0(x^2+a^2)}\cos\theta\right]\hat{i}=2\left[\frac{Q}{4\pi\varepsilon_0(x^2+a^2)}\frac{x}{(x^2+a^2)^{1/2}}\right]\hat{i}=\frac{Qx\hat{i}}{2\pi\varepsilon_0(x^2+a^2)^{3/2}}$$

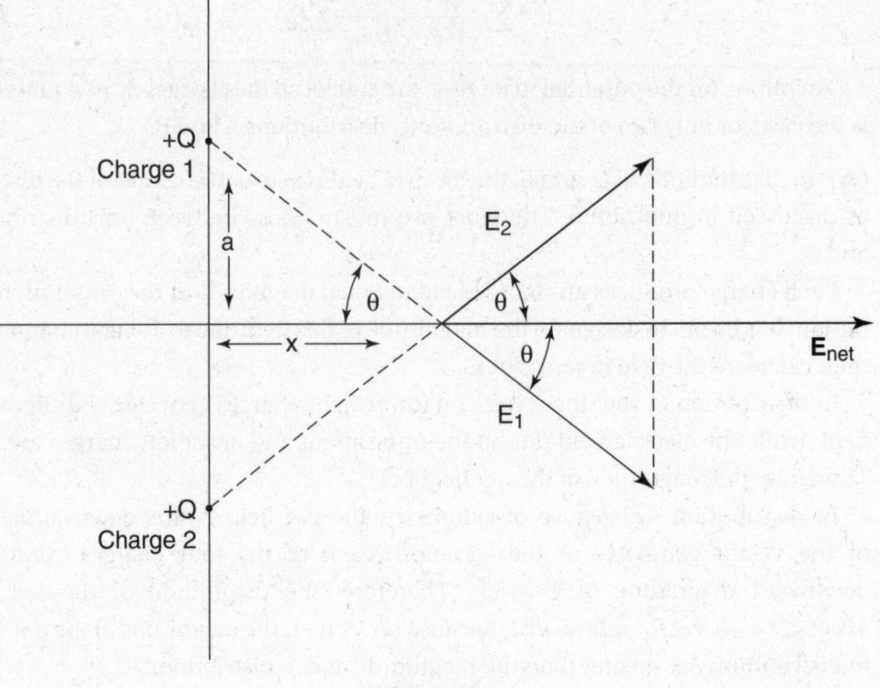

Figure 11.12

This result makes sense. At $x = 0$, the electric field is zero (because of symmetry). As $x \to \infty$, $x >> a$, so that $x^2 + a^2 \approx x^2$ and the electric field is approximately given by $\mathbf{E} \approx Q\hat{i}/2\pi\varepsilon_0 x^2$, the same as the electric field due to a charge of $+2Q$ at the origin (as you move farther and farther along the x-axis, the separation between the two charges becomes negligible and all you can "see" is one effective charge equal to $+2Q$).

(b) Using the definition of electric field, the force is obtained by multiplying the electric field by the charge:

$$\mathbf{F} = -q\mathbf{E} = \frac{-qQx\hat{i}}{2\pi\varepsilon_0 (x^2 + a^2)^{3/2}}$$

(c) Calculating potential using the equation $V = \Sigma(Q_i/4\pi\varepsilon_0 r_i)$ yields

$$V = \frac{Q}{4\pi\varepsilon_0 (a^2 + x^2)^{1/2}} + \frac{Q}{4\pi\varepsilon_0 (a^2 + x^2)^{1/2}} = \frac{Q}{2\pi\varepsilon_0 (x_0^2 + a^2)^{1/2}}$$

2. (a) and (b) Consider Figures 11.13 and 11.14. The key point here is to use superposition. Graph the field and potential due to one charge, graph the field and voltage due to the other charge, and then sum the graphs. Thus, superposition reduces the two-charge problem to two one-charge problems, allowing you to worry about only one charge at a time.

   (c) The force is zero; the forces due to each point charge cancel each other (from the sketch of the electric field, you can see that the field at the origin is zero).

(d) Use conservation of energy: Electrical potential energy is converted to kinetic energy:

$$U = qV = q \cdot 2\left(\frac{Q}{4\pi\varepsilon_0 d}\right) = \frac{1}{2}mv^2$$

$$v = \sqrt{\frac{qQ}{\pi\varepsilon_0 dm}}$$

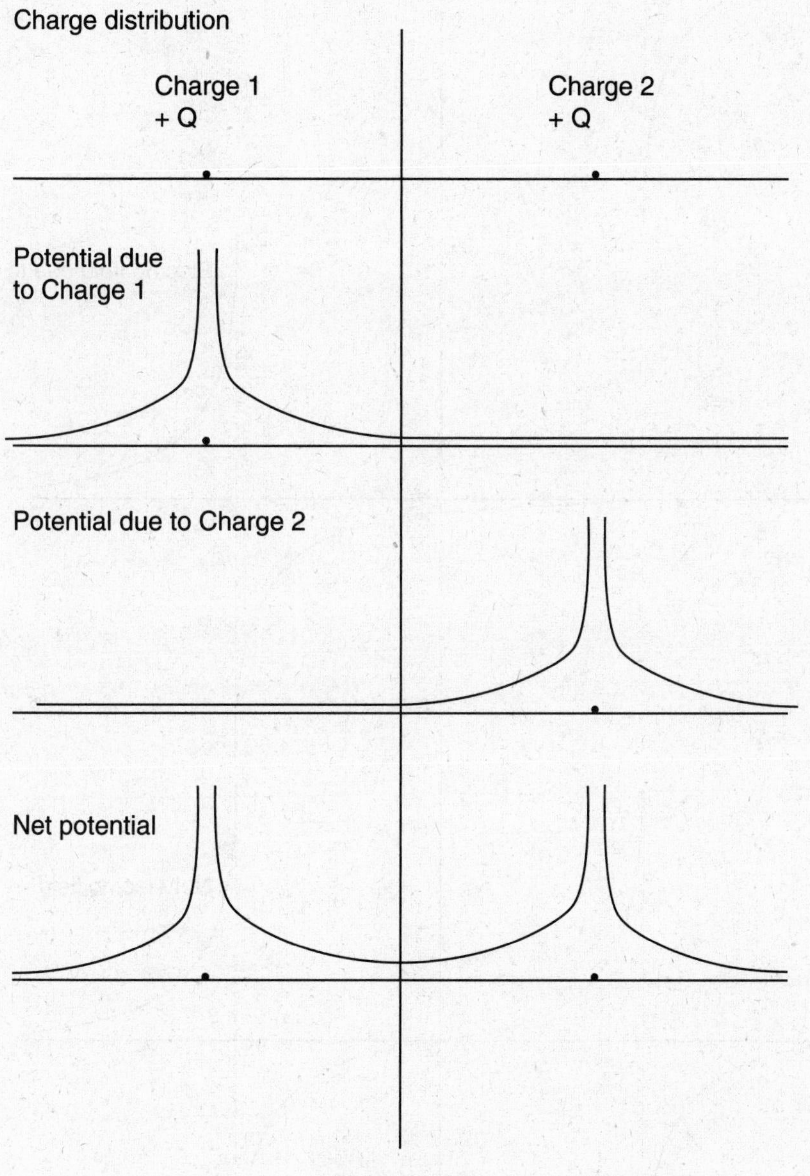

Figure 11.13

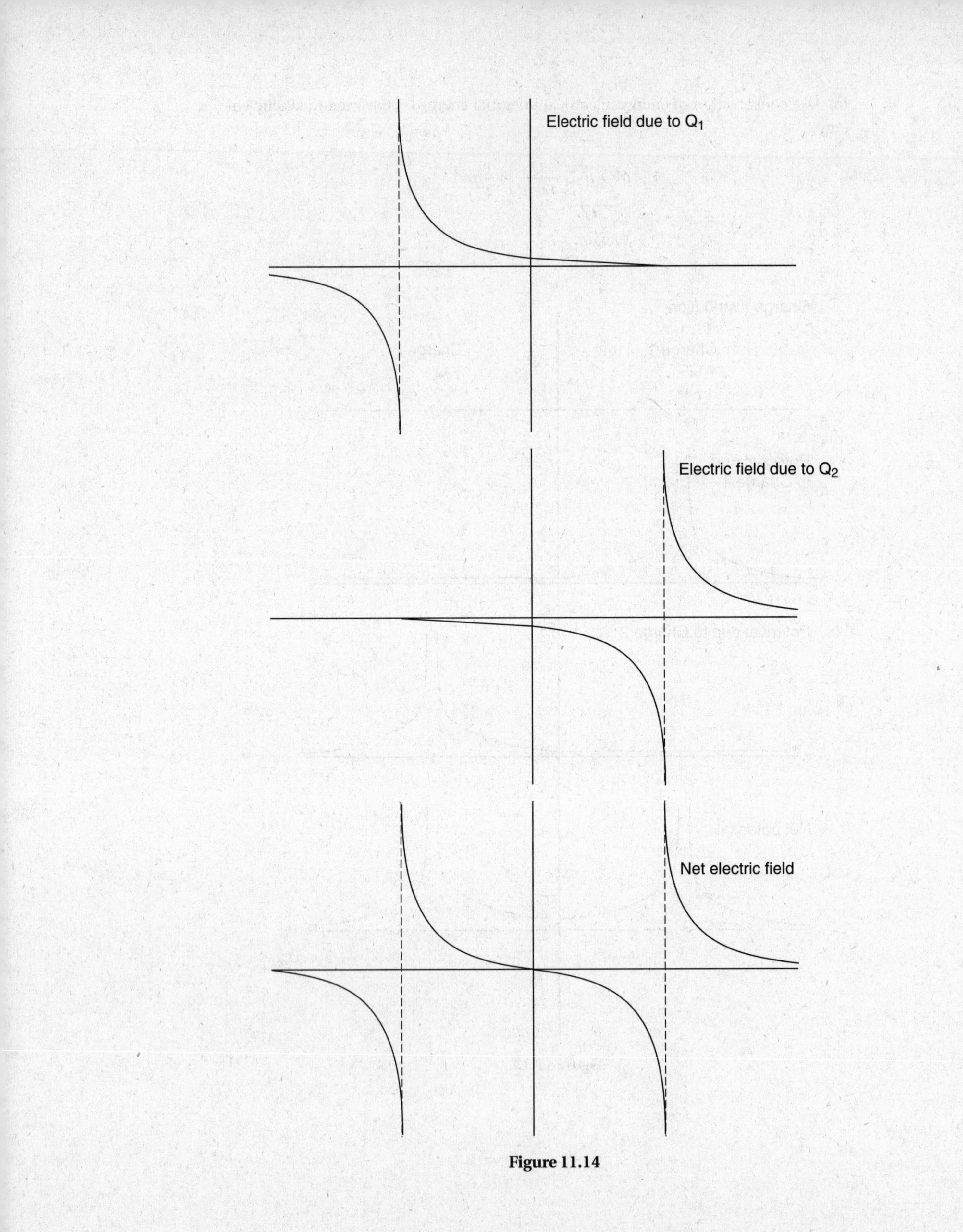
Electric field due to Q1
Electric field due to Q2
Net electric field

Figure 11.14

3. (a) The electric field line is shown in Figure 11.15, which was made using the facts that (i) the electric field is always perpendicular to equipotential lines, and (ii) the electric field points from higher to lower potential (as can be seen from the relationship $E_x = -dV/dx$).

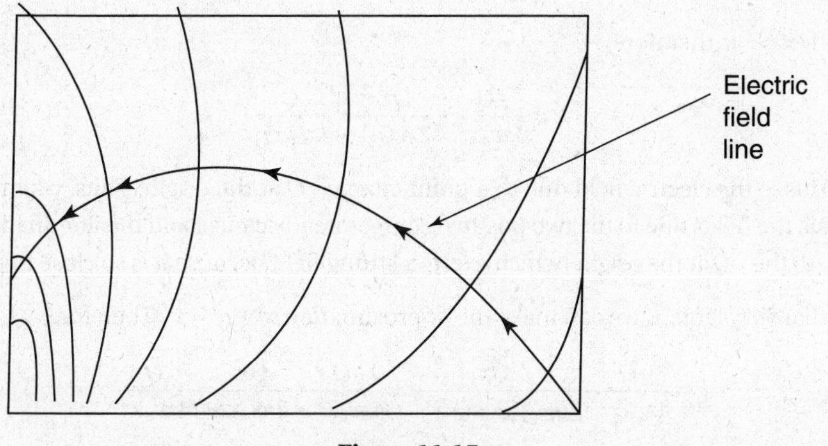

Figure 11.15

(b) No. The path is curved, and therefore a radial force would be required to provide the centripetal acceleration. Because the electric force along a field line is by definition tangential, it couldn't provide the required radial force.

(c) The field is strongest at point *B* and weakest at point *C*. The magnitude of the electric field is proportional to the change in potential with position ($E_{\text{parallel line}} = -dV/ds \approx -\Delta V/\Delta s$), which is greatest when the lines are spaced closest together (Δs small) and smallest when the lines are spaced farthest apart (Δs large).

(d) Points *C* and *D* lie on an equipotential line, so no work is done in moving a charge from point *C* to point *D*; the electric force is perpendicular to the displacement vector, making $dW = \mathbf{F} \cdot d\mathbf{r} = 0$.

(e) The work is again zero. Because the electric force due to any static charge distribution is conservative, the work done between any points (e.g., points *C* and *D*) is path independent. Therefore, if the answer is zero in part (D), it must be zero here as well.

4. (a) Along the $+x$ axis, symmetry dictates that the field must point in the x-direction. Summing the contributions of the x-components of the electric fields due to the three point charges yields the following (the vector geometry is the same as in question 1).

$$E = \frac{-Q}{4\pi\varepsilon_0 x^2} + 2\left[\frac{Q}{4\pi\varepsilon_0 (x^2 + a^2)} \frac{x}{(x^2 + a^2)^{1/2}}\right]$$

$$E = \frac{-Q}{4\pi\varepsilon_0 x^2} + \frac{Qx}{2\pi\varepsilon_0 (x^2 + a^2)^{3/2}}$$

(b) Applying the equation $V = \Sigma(Q_i/4\pi\varepsilon_0 r_i)$,

$$V = \frac{-Q}{4\pi\varepsilon_0 x} + 2\frac{Q}{4\pi\varepsilon_0 (a^2 + x^2)^{1/2}}$$

(c) For very small x, we can make the following two approximations:

(1) $x^2 + a^2 \approx a^2$; therefore,

$$E = \frac{-Q}{4\pi\varepsilon_0 x^2} + \frac{Qx}{2\pi\varepsilon_0 (x^2 + a^2)^{3/2}} \approx \frac{-Q}{4\pi\varepsilon_0 x^2} + \frac{Qx}{2\pi\varepsilon_0 a^3}$$

(2) $1/x^2 >> x$; therefore,

$$E \approx \frac{-Q}{4\pi\varepsilon_0 x^2} + \frac{Qx}{2\pi\varepsilon_0 a^3} \approx \frac{-Q}{4\pi\varepsilon_0 x^2}$$

This is the electric field due to a point charge $-Q$ at the origin. Thus, when x is very small, the fields due to the two positive charges nearly cancel and the dominant effect is due to the $-Q$ at the origin (which exerts a strong field because it is so close to point x).

(d) For very large x, we can make the approximation $x^2 + a^2 \approx x^2$. Therefore,

$$E = \frac{-Q}{4\pi\varepsilon_0 x^2} + \frac{Qx}{2\pi\varepsilon_0 (x^2 + a^2)^{3/2}} \approx \frac{-Q}{4\pi\varepsilon_0 x^2} + \frac{Qx}{2\pi\varepsilon_0 x^3} = \frac{Q}{4\pi\varepsilon_0 x^2}$$

This is the electric field due to a charge $+Q$ at the origin. When you are *very* far from the origin, it's impossible to see the separation of the three charges—it seems that they're all at the origin. The resulting field is due to the *sum* of the three charges, $Q_{\text{effective}} = Q + Q - Q = Q$, at the origin.

Calculating Electric Fields and Potentials Due to Continuous Charge Distributions

12

→ **CALCULATING THE POTENTIAL DUE TO A CONTINUOUS CHARGE DISTRIBUTION**
→ **FINDING THE ELECTRIC FIELD GIVEN THE POTENTIAL**
→ **CALCULATING THE ELECTRIC FIELD DUE TO A CONTINUOUS CHARGE DISTRIBUTION**

Now that we have learned how to deal with the electric fields and potentials due to point charges, we want to generalize to continuous charge distributions. How can we apply what we learned about point charges to continuous charges? The basis of all our work with continuous charges will be that a continuous charge can be broken up into tiny differential pieces that can then be treated as point charges. Because any continuous charge consists of an infinite number of these differential bits of charge, the superposition properties of potential and electric field will come in handy.

Conceptually this is straightforward. The problem is that many of these charge distributions are two- or three-dimensional and you probably know only single-variable calculus. How can single-variable calculus be used to sum over a two- or three-dimensional region? As usual, we can use *integral building*: If we're clever and take advantage of symmetry, we'll always be able to design single-variable integrals for the problems we encounter here.

This chapter is therefore all about integral building—dividing some region into differential regions that have a certain property in common (e.g., when we built integrals to calculate rotational inertia, we looked for differential volumes or differential areas that were a common distance from the axis of rotation).

CALCULATING THE POTENTIAL DUE TO A CONTINUOUS CHARGE DENSITY: A GENERAL APPROACH

STEP 1 Decide what type of differential regions you want to divide your charge distribution into. All points in a differential region must be a fixed distance away from the point whose potential you are calculating. There are only a few types of charge distributions you may be given, and they each generally call for the same type of differential region:

One-Dimensional: For a linear charge distribution or any one-dimensional distribution (e.g., a curved wire), use differential line segments. A special case is a ring of charge. If you're looking for the potential at a point on the axis of the ring, then all the charge is *already* the same distance away from the point of interest. You can simply apply $V = Q/4\pi\varepsilon_0 R$ (R being the common distance) to get an immediate answer.

Two-Dimensional: For a planar charge distribution, use concentric rings (washers) of charge located on the plane centered where the point of interest drops a perpendicular onto the plane, as shown in Figure 12.1.

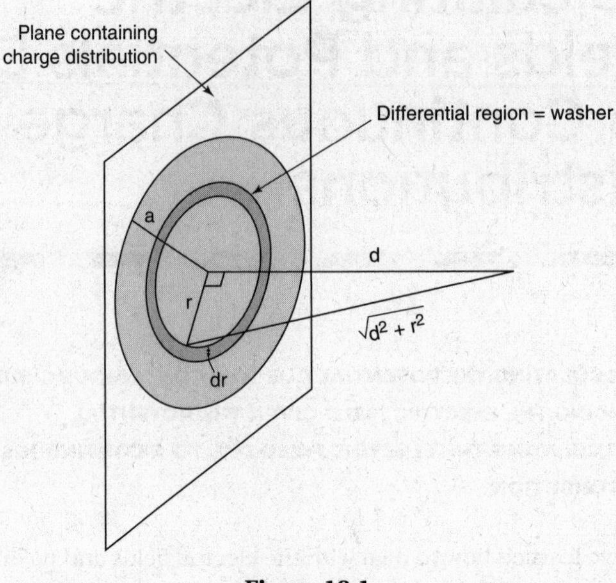

Figure 12.1

Three-Dimensional: For a spherical distribution, divide the region into spherical shells.

The following equations are useful in dealing with these differential regions.

A spherical shell is infinitesimally thin, so its volume is equal to its surface area $A = 4\pi R^2$ multiplied by its thickness, dR:

$$d(\text{volume}) = 4\pi R^2 dR$$

Volume of a spherical shell

A washer is infinitesimally thin, so its area is equal to its length, $l = 2\pi R$, multiplied by its thickness, dR:

$$dA = 2\pi R dR$$

Area of a washer

STEP 2 Determine the dimensions of the differential region. One of the dimensions should be a differential (e.g., dx, dr), which is the single variable that you integrate with respect to.

STEP 3 Determine the charge in the differential region, dQ. This generally involves one of the following situations.

- **a.** Linear charge distribution: $dQ = \lambda\, d(\text{length}) = \lambda dL$. ($\lambda$ is customarily the symbol for linear charge density.)
- **b.** Planar or surface charge distribution: $dQ = \sigma\, d(\text{area}) = \sigma dA$. ($\sigma$ is customarily the symbol for area charge density.)
- **c.** Volume charge distribution: $dQ = \rho\, d(\text{volume}) = \rho dV$. ($\rho$ is customarily the symbol for volume charge density.)

One of two situations may arise in terms of λ, σ, or ρ:

Case 1: The problem gives you a constant λ, σ, or ρ *or* λ, σ, or ρ as a continuous function of position. You're in good shape—keep on going.

Case 2: The problem doesn't give you λ, σ, or ρ but rather gives you the total charge Q and tells you that charge is evenly distributed (i.e., there is constant charge density). In this case, introduce your own constant λ, σ, or ρ and go ahead. (You will have to eliminate it later in step 6.)

STEP 4 Find the differential potential dV due to a given differential region. You can use the following equation (the entire differential region is at the same distance R from the point whose potential you are calculating, so this *point charge equation* is valid).

$$dV = \frac{dQ}{4\pi\varepsilon_0 R}$$

STEP 5 Integrate the expression for dV, using the geometry of the situation to find the limits of the integral. Always integrate from the lower limit to the upper limit (exactly why is explained at the end of Example 12.4). This results in the net voltage due to all the differential potentials.

STEP 6 If you have introduced a λ, σ, or ρ (step 3, case 2), you must get rid of it. Solve for λ, σ, or ρ by dividing the total charge by the total length, area, or volume and then substitute this expression for λ, σ, or ρ into the formula for potential from step 5.

STEP 7 Check the units in your final result.

EXAMPLE 12.1 POTENTIAL DUE TO A FINITE ROD ALONG ITS AXIS

A rod containing a uniformly distributed charge $+Q$ extends along the x-axis from $x = a$ to $x = b$. Calculate the potential at the origin.

SOLUTION

STEP 1 Linear charge distribution ⇒ use differential line segments.

STEP 2 Dimension of differential region = differential line segment length dx.

STEP 3 There is no given charge density, but the density is constant, so we introduce our own symbol λ for the charge density (situation 2): $dQ = \lambda dx$.

STEP 4

$$dV = \frac{dQ}{4\pi\varepsilon_0 x} = \frac{\lambda dx}{4\pi\varepsilon_0 x}$$

STEP 5

$$V = \int_{x=a}^{x=b} \frac{\lambda dx}{4\pi\varepsilon_0 x} = \frac{\lambda}{4\pi\varepsilon_0} \ln\left(\frac{b}{a}\right)$$

STEP 6 Calculating the charge density,

$$\lambda = \frac{Q}{\text{length}} = \frac{Q}{b-a}$$

Substituting into the expression for potential,

$$V = \frac{Q}{4\pi\varepsilon_0 (b-a)} \ln\left(\frac{b}{a}\right)$$

STEP 7 We have $C/4\pi\varepsilon_0 m$, the familiar form of potential, so the units are correct. Alternatively, you could use the fact that ε_0 has the units $C^2/N{\cdot}m^2$ to determine that the units of the answer are J/C, which is equal to volts.

EXAMPLE 12.2 POTENTIAL DUE TO A NONUNIFORMLY CHARGED SPHERICAL SHELL AT ITS CENTER

Consider a spherical charge distribution of inner radius *a* and outer radius *b*. The volume charge distribution is given by the formula $\rho(r) = kr$, where *k* is a proportionality constant with units $C \cdot m^{-4}$. Compute the potential at the center of the spherical shell.

SOLUTION

STEP 1 **Spherical distribution ⇒ use spherical shells.**

STEP 2 **The dimensions of a spherical shell of radius *r* and thickness *dr* are given by the equation**

$$d(\text{volume}) = 4\pi r^2 dr$$

STEP 3

$$dQ = \rho[d(\text{volume})] = (kr)(4\pi r^2\, dr) = 4\pi k r^3\, dr$$

STEP 4

$$dV = \frac{dQ}{4\pi\varepsilon_0 r} = \frac{kr^2 dr}{\varepsilon_0}$$

STEP 5

$$V = \int_{r=a}^{r=b} dV = \int \frac{kr^2 dr}{\varepsilon_0} = \left(\frac{k}{3\varepsilon_0}\right)(b^3 - a^3)$$

STEP 6 **Recalling from the above that the units of *k* are C/m^4, the answer has units of $C/\varepsilon_0 m$, the familiar form of potential (the same units as the standard expression for voltage $V = Q/4\pi\varepsilon_0 r$).**

EXAMPLE 12.3 POTENTIAL DUE TO A RING OF CHARGE ON ITS AXIS

Consider a ring of charge with radius *a* containing total charge *Q* evenly distributed over the ring. What is the potential a distance *b* from the center of the ring along the axis passing through the center of the ring?

SOLUTION

For a ring of charge, everything is already the same distance *R* from the point of interest, and so we can get an immediate answer:

$$V = \frac{Q}{4\pi\varepsilon_0 R}$$

We need to express this answer in terms of values given in the problem and fundamental constants, which requires getting rid of *R*. We can express it in terms of *a* and *b* using the Pythagorean theorem:

$$R = \sqrt{a^2 + b^2}$$

Substituting yields

$$V = \frac{Q}{4\pi\varepsilon_0 \left(a^2 + b^2\right)^{1/2}}$$

Units check: The answer is of the usual form for potential, $C/\varepsilon_0 m$.

EXAMPLE 12.4 POTENTIAL DUE TO A FINITE PLANAR CIRCULAR CHARGE DISTRIBUTION ALONG ITS AXIS

Consider a charge +*Q* distributed evenly along a flat circular surface of radius *a*. What is the potential a distance *d* from the surface along the perpendicular running through the center of the circle?

SOLUTION

STEP 1 Planar charge distribution ⇒ concentric washers of charge located on the plane centered where the point charge drops a perpendicular onto the plane (see Figure 12.1).

STEP 2 The area of the washer is given by the equation

$$dA = 2\pi r\, dr$$

STEP 3 Designating σ as the constant charge density,

$$dQ = \sigma dA = 2\pi\sigma r\, dr$$

STEP 4 Careful! The way we defined *r*, it is *not* the distance between the charge and the point of interest (but rather the distance from the center of the circular charge distribution).

$$dV = \frac{dQ}{4\pi\varepsilon_0 \left(r^2 + d^2\right)^{1/2}} = \frac{\sigma r\, dr}{2\varepsilon_0 \left(r^2 + d^2\right)^{1/2}}$$

STEP 5

$$V=\int_{r=0}^{r=a}\frac{\sigma r\,dr}{2\varepsilon_0(r^2+d^2)^{1/2}}$$

We can solve this with a *u*-substitution: $u = r^2 + d^2 \Rightarrow du = 2r\,dr$.

$$V=\frac{\sigma}{4\varepsilon_0}\int_{r=0}^{r=a}\frac{2r\,dr}{(r^2+d^2)^{1/2}}=\frac{\sigma}{4\varepsilon_0}\left[2\sqrt{r^2+d^2}\,\right]_{r=0}^{r=a}=\frac{\sigma}{2\varepsilon_0}\left(\sqrt{a^2+d^2}-d\right)$$

STEP 6

$$\sigma=\frac{\text{charge}}{\text{area}}=\frac{Q}{\pi a^2}$$

Substituting into the equation for *V*,

$$V=\frac{Q}{2\varepsilon_0\pi a^2}\left(\sqrt{a^2+d^2}-d\right)$$

Answer check: The answer is consistent with the following situations:

1. As *d* approaches infinity, the potential decreases to zero. (How does this work mathematically? As *d* grows large, $d^2 >> a^2$, so $\sqrt{a^2+d^2}\approx d$.)
2. As *a* approaches infinity, the charge (as a whole) is stretched farther and farther from the origin, and the potential approaches zero.
3. As *a* approaches zero, both the numerator and the denominator approach zero; we could calculate the limit using l'Hôpital's rule, but as an answer check it's probably not worth the effort. (*Note*: We expect the potential to approach that due to a point charge in this limiting case.)

DIRECT CONVERSION BETWEEN ELECTRIC FIELD AND POTENTIAL

Generally, when you are asked to calculate an electric field or potential, the easiest way is simply to evaluate it directly via an integral. However, if you already have an expression for the potential, and you need to calculate the electric field, one of the following equations (derived in the last chapter) provides a shortcut:

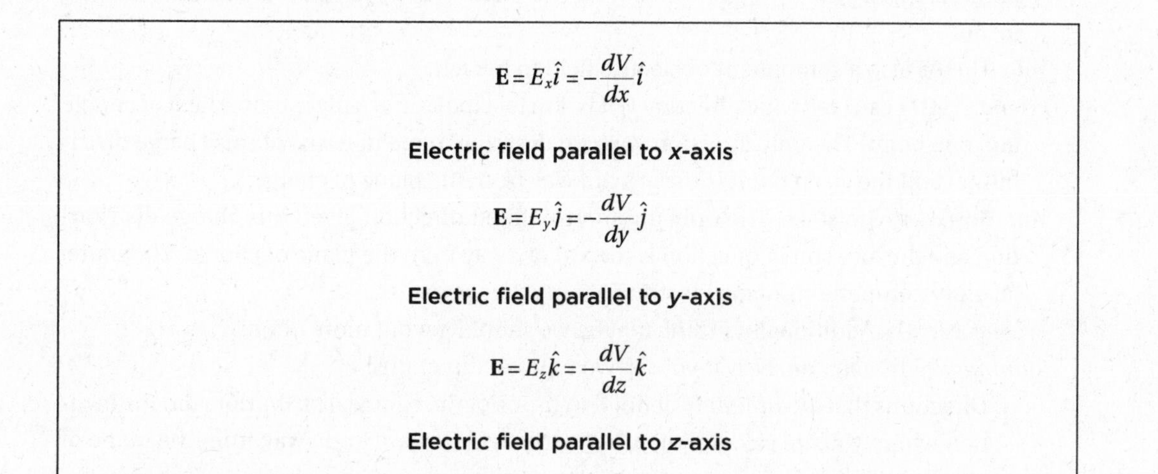

$$\mathbf{E} = E_x\hat{i} = -\frac{dV}{dx}\hat{i}$$

Electric field parallel to *x*-axis

$$\mathbf{E} = E_y\hat{j} = -\frac{dV}{dy}\hat{j}$$

Electric field parallel to *y*-axis

$$\mathbf{E} = E_z\hat{k} = -\frac{dV}{dz}\hat{k}$$

Electric field parallel to *z*-axis

As you will see in the remainder of this chapter, calculating electric fields via integration is tricky, so if you already have an expression for potential, it is generally easiest to simply differentiate this expression rather than start from scratch.

CALCULATING THE ELECTRIC FIELD DUE TO A CONTINUOUS CHARGE DENSITY: A GENERAL APPROACH

Calculating electric field via integration is very similar to calculating potential. However, electric field calculations are slightly more complicated because unlike potential (which is a scalar), electric field is a vector. Therefore, mathematically speaking, to integrate electric field (a three-dimensional vector) we need to carry out three scalar integrals:

$$\int d\mathbf{E} = \hat{i}\int dE_x + \hat{j}\int dE_y + \hat{k}\int dE_z$$

Don't worry—you almost never have to carry out three integrations. Symmetry usually causes one, two, or even all three of the above integrals to be zero, allowing you to ignore them.

STEP 1 Use symmetry to determine the nonzero components of the electric field. At this point, "using symmetry" probably sounds like so much physics voodoo, so here are two useful ways of determining symmetry.

Symmetry can make your life much easier!

a. The electric field cannot have a component that points in a direction that is not uniquely determined by the charge distribution itself. Imagine sending the charge distribution and the point where you are calculating the electric field in the mail to a friend. You should be able to describe to your friend the direction of the electric field over the telephone with reference to the point and the charge distribution. (You can't send your friend the artificial coordinate system that you imposed on the charge distribution.)

For example, suppose you have just finished calculating the electric field 1 cm above an infinite planar charge distribution. You send your friend the infinite charge distribution in the mail with a little point in space marked off where you calculated the electric field. The following is your phone conversation:

You: The electric field has two components. One points toward the plane of charge.

Friend: Great. That direction is uniquely determined by the charge distribution and the point in space where we are calculating the electric field—I can tell exactly what you mean.

You: There's also a component of electric field to the left.

Friend: Left? I can't tell what direction left is. I'm just looking at a big infinite sheet of charge and one point. The only direction you can describe to me in terms of this charge distribution and the given point is toward or away from the plane of charge.

You: Sorry—my mistake. The only uniquely defined direction given this charge distribution and the location in question is toward or away from the plane of charge. Therefore, all other components of the electric field must be zero.

Friend: Clearly. Additionally, I think maybe we should get out more often.

Summary: Symmetry analysis involves two types of directions:

i. Directions that are uniquely defined in terms of the charge distribution and the location where you are calculating the electric field (toward and away from the plane of charge in the above conversation).

ii. Directions that are not uniquely defined (left and right in the above conversation).

The electric field must equal zero in directions that are not uniquely defined. The electric field may or may not equal zero in directions that are uniquely defined.

Example 12.9 provides an argument as to exactly why the electric field cannot have components in directions that are not uniquely defined.

b. Here is an example of another way of seeing symmetry. Consider calculating the field due to a uniformly charged rod which extends on the x-axis from $x = -a$ to $x = a$ at a point on the y-axis $(0, b)$ as shown in Figure 12.2.

Imagine pairing every differential length dx along the positive x-axis with its symmetric partner, a differential length dx along the $-x$ axis that is an equal distance from the origin. The x-component of the electric field due to every one of these pairs is zero. (The two differential lengths, one along the $+x$ axis and one along the $-x$ axis, produce electric fields whose x-components exactly cancel each other.) Therefore, the *net* x-component of the electric field due to the entire rod must be zero. (This a two-dimensional analog of the situation you discussed with your friend on the telephone. The only uniquely defined directions are toward the charged rod and away from the charged rod.) Note that if we calculated the field at a point not on the y-axis, the net x-component of E would not necessarily be zero.

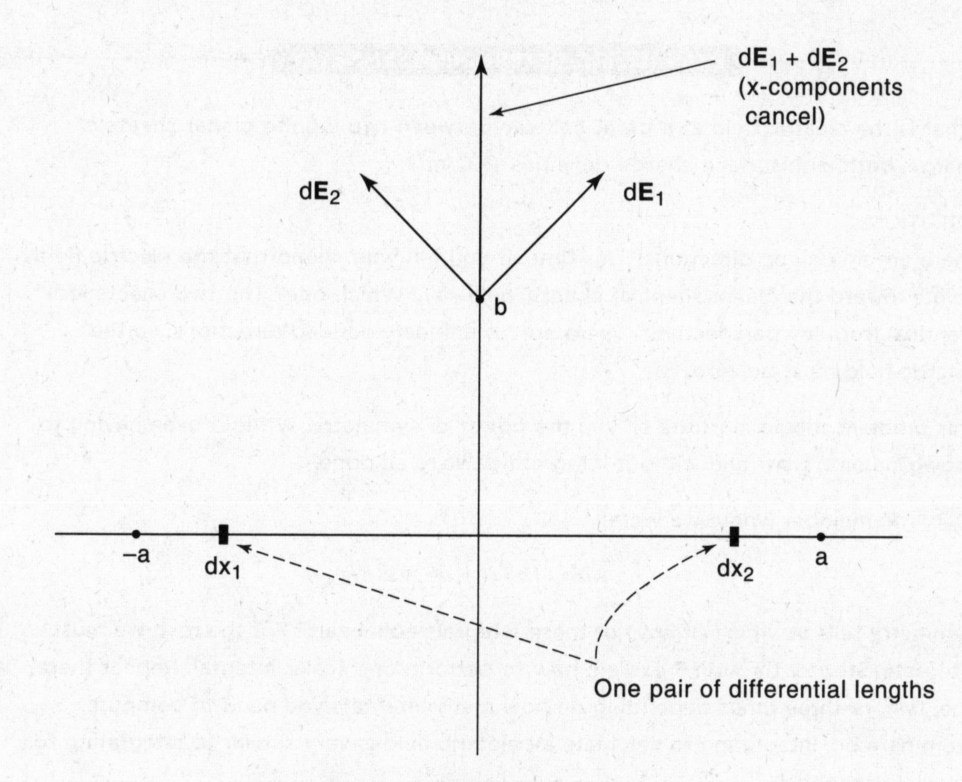

Figure 12.2

EXAMPLE 12.5 DETERMINING SYMMETRY

We are calculating the electric field at a point (1, 0, 0) sandwiched between infinite planar charge distributions with surface charge densities of 4, 3, and 1 C/m² located at $x = 0$, $x = 2$, and $x = 3$, respectively, as in Figure 12.3. What directions are uniquely defined by the charge distribution and the point where we are calculating electric field?

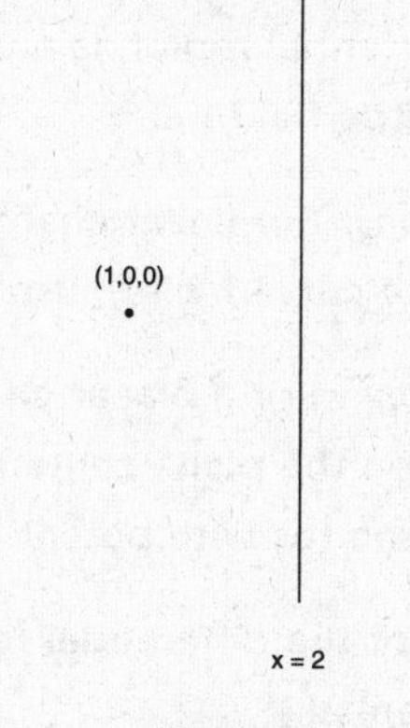

Figure 12.3

SOLUTION

This is very similar to the phone conversation. The only uniquely defined direction is perpendicular to the sheets, either toward the +4 –C/m² sheet or toward the +3 –C/m² sheet (the *x*-direction). By symmetry, we know that the electric field's components in the other directions (*y* and *z* according to our axes) must be zero. According to symmetry, the electric field may have a component in the *x*-direction. It turns out that it doesn't—the electric field is zero. (After further work in this chapter and in the following chapter on Gauss's law, you will see why.)

EXAMPLE 12.6 DETERMINING SYMMETRY

What is the electric field at a point halfway between two infinite planar sheets of charge, both with surface charge densities +1 C/m^2?

SOLUTION

There are no unique directions! This time, if you tell your friend that the electric field points toward the planar sheet of charge, he'll say "Which one? The two sheets look identical from my perspective." There are no uniquely defined directions, so the electric field must be zero.

This problem should illustrate to you the power of symmetry. Without even having to know Coulomb's law and without integrating, we're all done.

STEP 1 Remember where we were:

$$\int d\mathbf{E} = \hat{i}\int dE_x + \hat{j}\int dE_y + \hat{k}\int dE_z$$

Symmetry tells us which (if any) of these integrals equal zero. For the rest, we must integrate. Steps 2 through 5 explain how to perform one scalar integral. Repeat them one, two, or three times depending on how many integrals you need to compute. From here on, integrating to calculate an electric field is very similar to integrating to calculate potential.

Note: If you fail to detect all the symmetry present, don't worry. After crunching through an integral that you could have ignored, you'll simply calculate its value to be zero.

STEP 2 Decide what type of differential regions you want to divide your charge distribution into. You must choose the differential regions such that every bit of charge in a region makes an *equal contribution* to the component of the electric field being evaluated. (*Careful*: Electric field is a vector, unlike scalar potential, so it's no longer enough that the entire differential region is a uniform distance from the point where you are calculating the electric field.)

One-dimensional charge distributions: For linear charge distribution or any one-dimensional distribution (e.g., a curved wire), use differential line segments.

Two-dimensional charge distributions: For a planar charge distribution, use concentric rings (washers) of charge located on the plane centered where the point of interest drops a perpendicular onto the plane (as with potential).

STEP 3 Determine the dimensions of the differential region. (This step is exactly the same as step 1 for calculating potential.)

STEP 4 Determine the charge in the differential region. (This step is exactly the same as step 2 for calculating potential.)

STEP 5 Find the scalar magnitude of the electric field vector in the direction you are integrating due to a differential region. Often, the electric field due to each bit of charge dQ does *not* point in the direction of the component we are calculating, so we need to multiply by the fraction of the electric field in that direction. This fraction is given by the vector geometry of the problem and is generally either unity or a simple trigonometric function such as sine or cosine. What we are doing is vectorially adding all of the electric fields of each bit of charge dQ component by component.

STEP 6 Integrate your expression for $dE_{x,\,y,\,\text{or}\,z}$ using the geometry of the situation to find the limits of the integral. As with potential integrals, always integrate from the lower limit to the upper limit.

STEP 7 If you have introduced a constant charge density λ, σ, or ρ, to get rid of it as when calculating potential, solve for λ, σ, or ρ by dividing the total charge by the total (length, area, or volume) and substitute this expression for λ, σ, or ρ into the expression for electric field from step 6.

STEP 8 Returning to our main equation,

$$\mathbf{E} = \int d\mathbf{E} = \hat{i}\int dE_x + \hat{j}\int dE_y + \hat{k}\int dE_z = E_x\hat{i} + E_y\hat{j} + E_z\hat{k}$$

Once you know all the scalar magnitudes of the electric field, insert them into the above equation to assemble the electric field from its components.

Summary: Differences between calculating electric field and potential due to continuous charge distributions:

a. The electric field is a vector and may involve as many as three integrals in three directions.
b. Symmetry is used to determine which components of the electric field are possibly nonzero.
c. The proportion of the contribution of dQ to the electric field in each direction must be considered.
d. Different equations are used for dV and for dE ($dQ/4\pi\varepsilon_0 R$ vs $dQ/4\pi\varepsilon_0 R^2$ proportion in direction integrating, respectively).

EXAMPLE 12.7 ELECTRIC FIELD DUE TO A FINITE ROD ALONG ITS AXIS

A rod containing a uniformly distributed charge $+Q$ extends along the x-axis from $x = a$ to $x = b$ (same situation as in Example 12.1). Calculate the electric field at the origin.

SOLUTION

STEP 1 Symmetry should be considered. Send your friend the rod and point in the mail. The only direction that you can tell him about is toward or away from the rod (along the x-axis). The y- and z-directions are not uniquely defined, and thus the components of the electric field along these directions must be zero.

STEP 2

$$\int d\mathbf{E} = \hat{i}\int dE_x + \hat{j}\int dE_y + \hat{k}\int dE_z$$

From symmetry: $\int dE_y = \int dE_z = 0$. All we need to worry about is $\int dE_x$.

STEP 3 Linear charge distribution $\Rightarrow$ use differential line segments.

STEP 4 Dimension of a differential line segment = dx.

STEP 5 As in Example 12.1, designate λ as the constant linear charge density:

$$dQ = \lambda dx$$

STEP 6 In this case, the differential electric field $d\mathbf{E}$ (at the origin) due to a given dQ already points in the $-x$-direction, so the proportion of the $d\mathbf{E}$ that points along the x-direction is one.

$$dE_x = -\frac{dQ}{4\pi\varepsilon_0 x^2} = -\frac{\lambda dx}{4\pi\varepsilon_0 x^2}$$

STEP 7

$$E_x = \int_{x=a}^{x=b} dE_x = \int_{x=a}^{x=b} -\frac{\lambda dx}{4\pi\varepsilon_0 x^2} = -\frac{\lambda}{4\pi\varepsilon_0}\left(\frac{1}{a} - \frac{1}{b}\right)$$

STEP 8 Solving for λ and then eliminating λ by substitution,

$$\lambda = \frac{Q}{b-a}$$

$$E_x = -\frac{Q}{4\pi\varepsilon_0 (b-a)}\left(\frac{1}{a} - \frac{1}{b}\right)$$

STEP 9 From our scalars, we assemble a vector:

$$\mathbf{E} = E_x\hat{i} + E_y\hat{j} + E_z\hat{k} = -\frac{Q}{4\pi\varepsilon_0 (b-a)}\left(\frac{1}{a} - \frac{1}{b}\right)\hat{i}$$

(Recall that we already decided, based on symmetry, that $E_y = E_z = 0$.)

EXAMPLE 12.8 ELECTRIC FIELD DUE TO A SPHERICAL SHELL AT ITS CENTER

What is the electric field at the center of a spherical shell of inner radius a and outer radius b that contains a volume charge distribution given by the formula $\rho(r) = kr$? (This is the same situation as in Example 12.2.)

SOLUTION

At the center of the sphere, there are no uniquely defined directions. Therefore, the electric field must be zero!

We've already seen that the electric field must be zero if there are no uniquely defined directions, but we haven't really explained why yet. We'll try to make a proof by contradiction, using the above situation.

Suppose that there *were* a nonzero electric field. Because there is no way to define a unique direction, it could just as well be in any direction. Thus, if there is one solution, it's really equivalent to an infinite number of solutions. Now, let's remember the way we defined the electric field: the force per charge experienced by a test charge at the center of the sphere. But if we put a test charge at the center of the sphere, there's no way that it could experience a net force in an infinite number of directions. Therefore, our initial assumption, the existence of a nonzero electric field, is incorrect. The field must be zero.

EXAMPLE 12.9 ELECTRIC FIELD DUE TO A RING OF CHARGE ON ITS AXIS

A charge +Q is evenly distributed in a ring of radius a on the xy-plane centered at the origin. What is the magnitude of the electric field at a point along the z-axis (0, 0, b)?

SOLUTION

STEP 1 There is one uniquely defined direction: away from the ring along the z-axis (the charge is positive).

STEP 2 By symmetry, $E_x = E_y = 0$ and all we have to worry about is $\int dE_z$.

STEP 3 Because we are interested in the electric field at a point along the axis passing through the center of the ring, we don't need to divide the distribution into differential regions (we already have a situation where every dQ makes an equal contribution to the magnitude of the electric field in the z-direction). No integral building is required.

Every piece of charge causes an electric field that can be broken into two components: one that points in the z-direction and one that lies in the xy-plane. As required by symmetry, the components that lie in the xy-plane due to the entire ring of charge cancel each other and add to zero. We're interested only in the component in the z-direction. We can use geometry to convert from $|E|$ to E_z as shown in Figure 12.4.

$$E_z = E\cos\theta = \frac{Q}{4\pi\varepsilon_0 R^2}\cos\theta = \frac{Q}{4\pi\varepsilon_0 R^2}\frac{b}{\left(a^2+b^2\right)^{1/2}}$$

Note: You may be wondering why we don't have to use differentials here, given that the differential electric field $d\mathbf{E}$ due to each piece of charge dQ in the ring is indeed a different vector. The answer is that the z-component of each $d\mathbf{E}$ is the same. The z-component is all we care about, so we can ignore the fact that the x- and y-components of $d\mathbf{E}$ vary as we go around the ring.

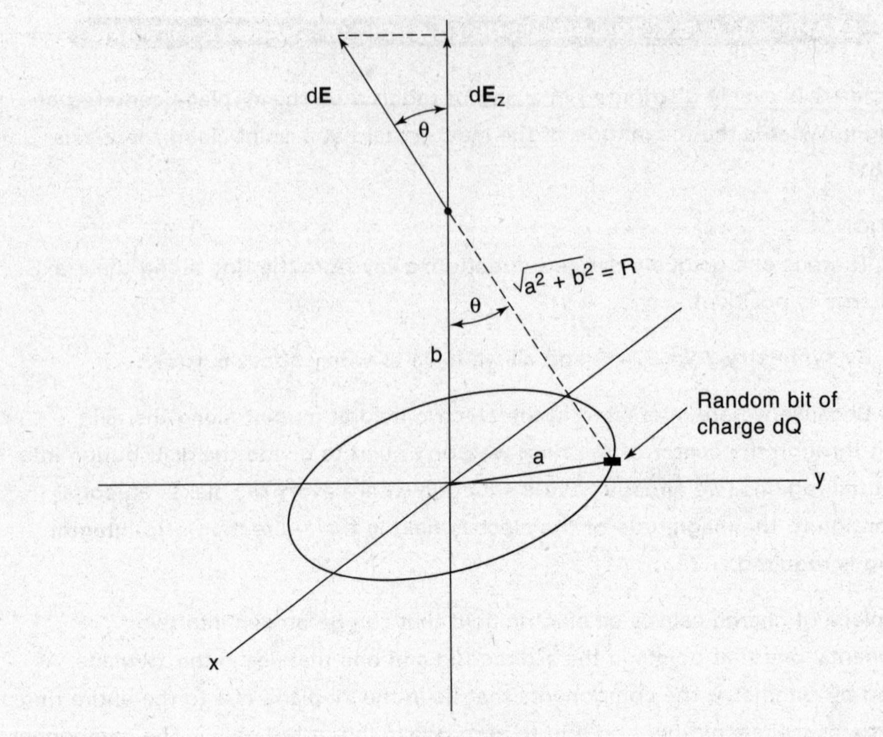

Figure 12.4

Noting that in this case $R = \sqrt{a^2 + b^2}$, the final result for E_z simplifies to

$$E_z = \frac{Qb}{4\pi\varepsilon_0 (a^2 + b^2)^{3/2}}$$

STEP 4 From scalars come vectors:

$$\mathbf{E} = E_x\hat{i} + E_y\hat{j} + E_z\hat{k} = \frac{Qb}{4\pi\varepsilon_0 (a^2 + b^2)^{3/2}}\hat{k}$$

Answer check: From Example 12.3, we know that

$$V = \frac{Q}{4\pi\varepsilon_0 (a^2 + b^2)^{1/2}}$$

We can differentiate this potential to obtain the electric field using the equation

$$E_z = -\frac{dV}{dz}$$

Now, we must realize that *b* is the distance in the *z*-direction that we are from the ring of charge: It is the *z*-coordinate of our position. Therefore, we need to differentiate the potential equation with respect to *b*:

$$E_z = -\frac{dV}{dz} = -\frac{dV}{db} = -\frac{d}{db}\left(\frac{Q}{4\pi\varepsilon_0 (a^2 + b^2)^{1/2}}\right)$$

We'll leave the final crunching out of the derivative to you.

Answer check: The answer is consistent with the following situations.

1. If $b = 0$, the field must equal zero (because of symmetry at the center of the ring).
2. As *a* or *b* approaches infinity, the field approaches zero (because the charge is infinitely far away from the point of interest).

EXAMPLE 12.10 ELECTRIC FIELD DUE TO A FINITE PLANAR CIRCULAR CHARGE DISTRIBUTION ALONG ITS AXIS

Consider a charge $+Q$ distributed evenly along a flat circular surface of radius a lying on the xy-plane. What is the electric field at a distance d away from the surface along the perpendicular passing through the center of the circle?

SOLUTION

STEP 1 As in Example 12.9, the only uniquely defined direction is away from the charge surface (the z-axis).

STEP 2 By symmetry, $E_x = E_y = 0$, and all we have to consider is $\int dE_z$.

STEP 3 Planar charge distribution ⇒ differential regions are concentric washers of charge centered at the center of the circular charge distribution (as shown at the beginning of this chapter).

STEP 4

$$dA = 2\pi r\, dr$$

STEP 5 Designating σ the constant area charge density:

$$dQ = \sigma dA = 2\pi\sigma r\, dr$$

STEP 6 (same geometry as in the last example):

$$dE_z = \frac{dQd}{4\pi\varepsilon_0 (d^2 + r^2)^{3/2}}$$

Substituting for dQ,

$$dE_z = \frac{\sigma(2\pi r\, dr)d}{4\pi\varepsilon_0 (d^2 + r^2)^{3/2}} = \frac{\sigma r d\, dr}{2\varepsilon_0 (d^2 + r^2)^{3/2}}$$

STEP 7

$$E_z = \int_{r=0}^{r=a} \frac{d\sigma r dr}{2\varepsilon_0 (d^2 + r^2)^{3/2}} = \frac{\sigma d}{2\varepsilon_0} \int_{r=0}^{r=a} \frac{r\, dr}{(d^2 + r^2)^{3/2}}$$

We can solve this integral with a U-substitution: $u = d^2 + r^2 \Rightarrow du = 2r dr$.

$$E_z = \frac{\sigma d}{2\varepsilon_0}\left[\frac{1}{d} - \frac{1}{(d^2 + a^2)^{1/2}}\right]$$

STEP 8 Solving for σ,

$$\sigma = \frac{Q}{\pi a^2}$$

Substituting,

$$E_z = \frac{Qd}{2\pi\varepsilon_0 a^2}\left[\frac{1}{d} - \frac{1}{(d^2 + a^2)^{1/2}}\right]$$

STEP 9

$$\mathbf{E}=\frac{Qd}{2\pi\varepsilon_0 a^2}\left[\frac{1}{d}-\frac{1}{(d^2+a^2)^{1/2}}\right]\hat{k}$$

Answer check 1: From Example 12.4 we know that

$$V=\frac{Q}{2\pi\varepsilon_0 a^2}\left(\sqrt{a^2+d^2}-d\right)$$

As in Example 12.9 we must realize that the *z*-coordinate of our position with respect to the ring of charge is *d*. Therefore,

$$E_z=-\frac{dV}{dz}=-\frac{dV}{d(d)}$$

Again, we'll leave the final crunching out of the derivative to you.

Answer check 2: The answer is consistent with the following situations.

1. As *d* approaches infinity, the electric field approaches zero (the point moves an infinite distance from the charge distribution).
2. As *d* approaches zero, the charge distribution acts like an infinite charged sheet (if you were at the point, you wouldn't be able to see the edge of the charge distribution; the edges are too far away to matter). In this situation, the electric field approaches:

$$\lim_{d\to 0}\mathbf{E}=\lim_{d\to 0}\frac{Q}{2\pi\varepsilon_0 a^2}\left[1-\frac{d}{(d^2+a^2)^{1/2}}\right]\hat{k}=\frac{Q}{2\pi\varepsilon_0 a^2}\hat{k}=\frac{\sigma}{2\varepsilon_0}\hat{k}$$

In later chapters, we will see that this is indeed the equation for the electric field of an infinite sheet of charge. Note that we didn't ask what happens at *d* = 0. As we will learn in the next chapter, because there is infinite volume charge density at that point, the field is discontinuous, so it's not a good place to confirm our results.

CHAPTER SUMMARY

Integral calculus is used to find the electric field or potential due to a continuous charge distribution. This method is based on Coulomb's law and the principle of superposition. The charge distribution is divided into differential regions whose shapes reflect the symmetry of the distribution. The electric potential is easier to calculate because it is a scalar quantity. Once the potential is determined, the electric field can be found by differentiation.

PRACTICE EXERCISES

Multiple-Choice Questions

1. The axis of the cone in Figure 12.5 lies on the y-axis. The cone has a uniform charge density. Based on symmetry considerations alone, in what direction(s) can the electric field at the origin point? (The origin lies within the cone along its axis of symmetry.)

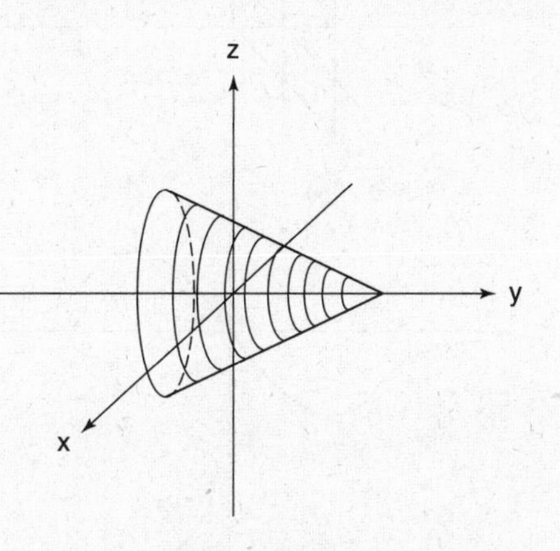

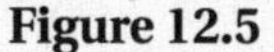
Figure 12.5

(A) the x-axis
(B) the y-axis
(C) the z-axis
(D) anywhere along the xy-plane
(E) anywhere along the yz-plane

2. Which of the following statements is true regarding the value of the electric field and the potential along a line passing perpendicularly through an infinite planar charge distribution (with uniform surface charge density)?

(A) The potential and the electric field are discontinuous.
(B) The potential and the electric field are continuous.
(C) The potential is continuous, and the electric field is discontinuous.
(D) The potential is discontinuous, and the electric field is continuous.
(E) The continuity of the potential and electric field depend on the sign of the charge.

3. A square of side length a with uniformly distributed positive charge lies on the yz-plane with its center at the origin as in Figure 12.6. Which of the graphs is of the electric field along the x-axis?

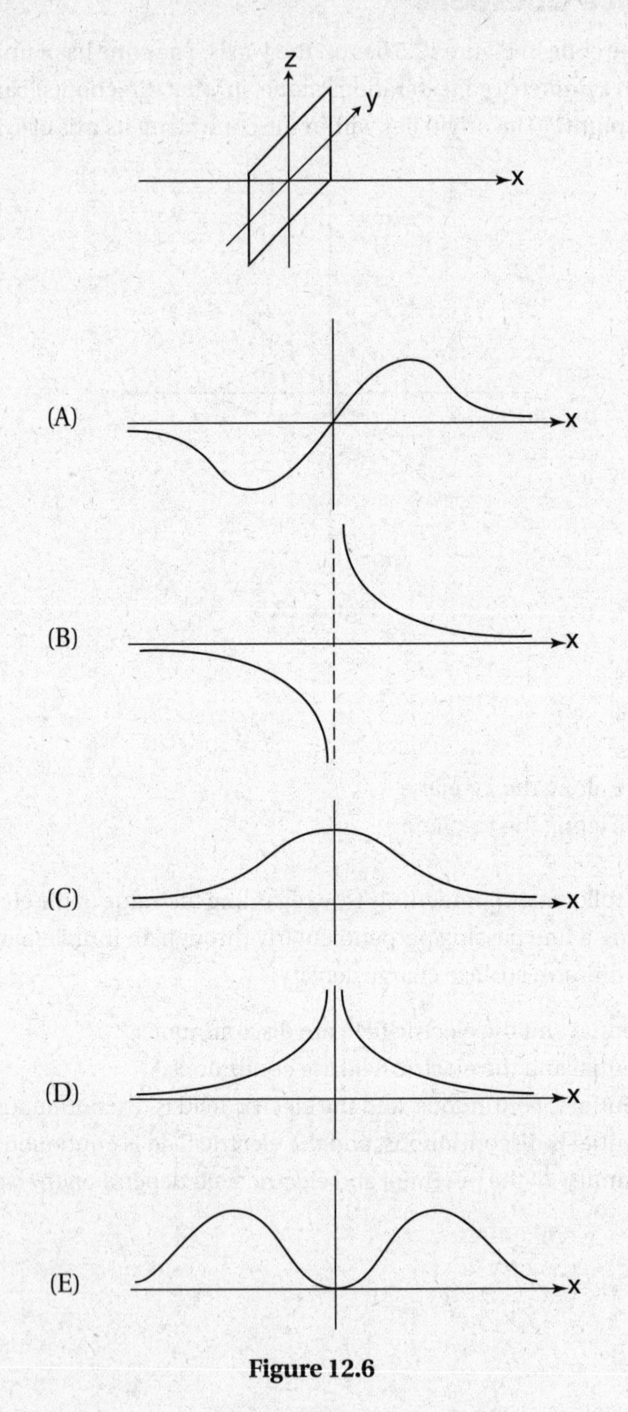

Figure 12.6

4. Which of the graphs in Figure 12.6 shows the potential along the x-axis due to the charge distribution shown in question 3?

5. Three nonconducting spheres all have the same net charge $+Q$ and radius R. The charge is uniformly distributed throughout the *volume* of sphere A, uniformly distributed on the *surface* of sphere B, and distributed with a density proportional to the radial variable, $\rho = cr$ throughout the volume of sphere C ($c = Q/\pi R^4$). Which of the following is a valid comparison of the potentials at the centers of each of the spheres?

(A) $V_A > V_B > V_C$
(B) $V_A > V_C > V_B$
(C) $V_C > V_A > V_B$
(D) $V_C > V_B > V_A$
(E) $V_B = V_C = V_A$

Free-Response Questions

1. A uniform ring of charge of radius a and total charge $+Q$ is centered at the origin and lying on the yz-plane as shown in Figure 12.7.

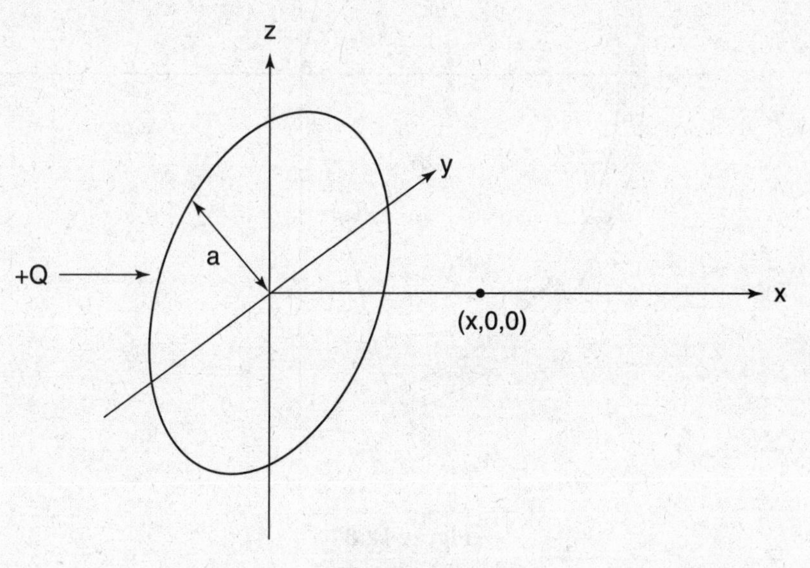

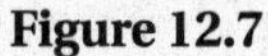

Figure 12.7

(a) Calculate the potential along the x-axis as a function of position, $V(x)$.
(b) Calculate the electric field along the x-axis as a function of position, $\mathbf{E}(x)$.
(c) Sketch these functions, $V(x)$ and $\mathbf{E}(x)$. Discuss the sketch in terms of:
   (i) maxima and minima
   (ii) parity symmetry (are these functions even, odd, or neither?)
   (iii) the equation $E_x = -dV/dx$
(d) A charge of $-Q$ and mass m passes through the origin with a speed v. How large must v be for the charge to continue forever without reversing direction (i.e., to achieve escape velocity)?

2. Charge is distributed along a linear semicircular rod with a linear charge density $+\lambda$ as in Figure 12.8.

   (a) Calculate the potential at the origin.
   (b) Calculate the electric field at the origin.
   (c) What value of θ (between 0 and π) maximizes the potential at the origin?
   (d) What value of θ (between 0 and π) maximizes the magnitude of the electric field at the origin?

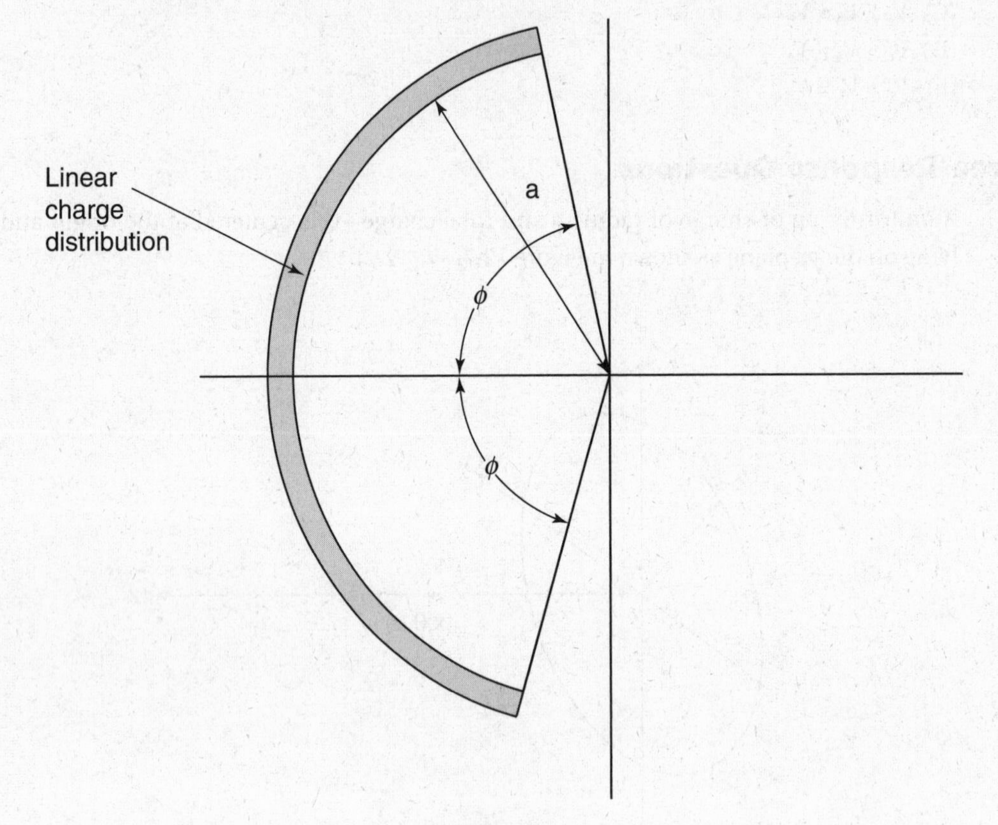

Figure 12.8

Now consider the situation in which a *fixed* charge of $+Q$ is distributed evenly in an identical circular arc.

   (e) Calculate the potential at the origin.
   (f) Calculate the electric field at the origin.
   (g) What value of θ (between 0 and π) maximizes the potential at the origin?
   (h) What value of θ (between 0 and π) maximizes the magnitude of the electric field at the origin?

ANSWER KEY

1. **B**
2. **C**
3. **A**
4. **C**
5. **B**

ANSWERS EXPLAINED

Multiple-Choice

1. **(B)** Imagine describing to a friend what direction the electric field points; the only description you could give would be "toward the tip of the cone" or "away from the tip of the cone" (i.e., along the *y*-axis).

2. **(C)** Figure 12.9 shows the behavior of the electric field and the potential along the *x*-axis due to a constant positive surface charge density along the *yz*-plane. The electric field due to an infinite planar charge distribution has magnitude $E = \sigma/2\varepsilon_0$ whose direction abruptly changes as the point passes through the planar charge distribution (making the electric field discontinuous at this point). Because the electric field is equal to the negative slope of the potential along the *x*-axis, this implies that the potential's *slope* is discontinuous at $x = 0$ but the potential itself is continuous. Changing the sign of the charge switches the sign of the electric field and potential, but does not affect whether they are continuous or discontinuous.

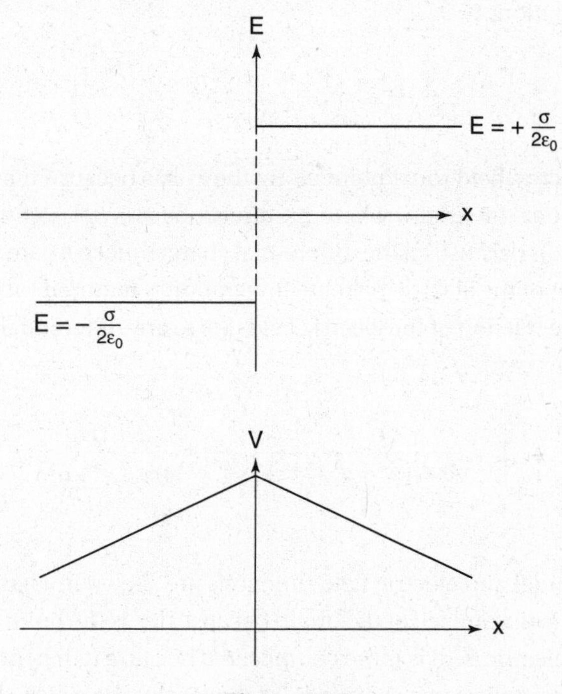

Figure 12.9

3. **(A)** Symmetry requires the field is zero at the origin, ruling out choices (B), (C), and (D). The charge is positive, so the field points in the +*x*-direction for $x > 0$ and in the −*x*-direction for $x < 0$, indicating that the correct answer is (A).

4. **(C)** Based on the formula $V = \int dQ/4\pi\varepsilon_0 r$, the potential can never be negative, ruling out choices (A) and (B). Furthermore, because the origin is the point closest to the positive charge, the potential must be maximum here, indicating that the correct answer is (C).

   Alternative approach: The electric field, shown in graph (A), is the negative of the slope of the potential graph. Therefore, the potential must have a negative slope to the right of the *y*-axis and a positive slope to the left of the *y*-axis. The only graph with these properties is choice (C).

5. **(B)** Although the field (and thus the potential) is identical outside the spheres, this is not the case inside the spheres. The spheres all have the same net charge. Therefore, based on the equation for potential, $V = \int dQ/4\pi r\varepsilon_0$, the closer the charge is to the center of the sphere, the greater the potential is at the center of the sphere. Sphere B, with all the charge on the surface of the sphere, has its charge located farthest away from the center of the sphere and therefore has the lowest potential [eliminating all choices except (B) and (C)]. Sphere C, with a charge density proportional to its distance from the sphere, tends to distribute more charge at a larger *r*, farther from the center of the sphere, compared to sphere A. Therefore, $V_C < V_A$ and the correct answer is choice (B).

Free-Response

1. (a) Because all the charge is at a constant distance from the point (*x*, 0), we can simply plug into the equation:

$$V = \frac{Q}{4\pi\varepsilon_0 r} = \frac{Q}{4\pi\varepsilon_0 (a^2 + x^2)^{1/2}}$$

   (b) First, the electric field must point along the *x*-axis because that is the only defined direction (except at the origin, where no directions are well defined, and the electric field must be zero). Because all the differential charges make a common contribution to the *x*-component of the electric field, no integration is required (of course, we still have to multiply by the fraction of the electric field due to any differential charge that is in the *x*-direction, in this case $x/\sqrt{x^2 + a^2}$):

$$E_x = \frac{Q}{4\pi\varepsilon_0 (x^2 + a^2)} \frac{x}{(x^2 + a^2)^{1/2}} = \frac{Qx}{4\pi\varepsilon_0 (x^2 + a^2)^{3/2}}$$

   (c)

   (i) The potential and electric field functions are shown in Figure 12.10. The potential function is maximized at the origin because this is the point closest to the ring of charge. The electric field is more complicated because it depends on both proximity to the charges *and* vector geometry (at the origin, the point closest to the charge, vector geometry causes the electric field to be zero). The result, as shown, is a compromise (with extrema lying close to the origin but not at the origin).

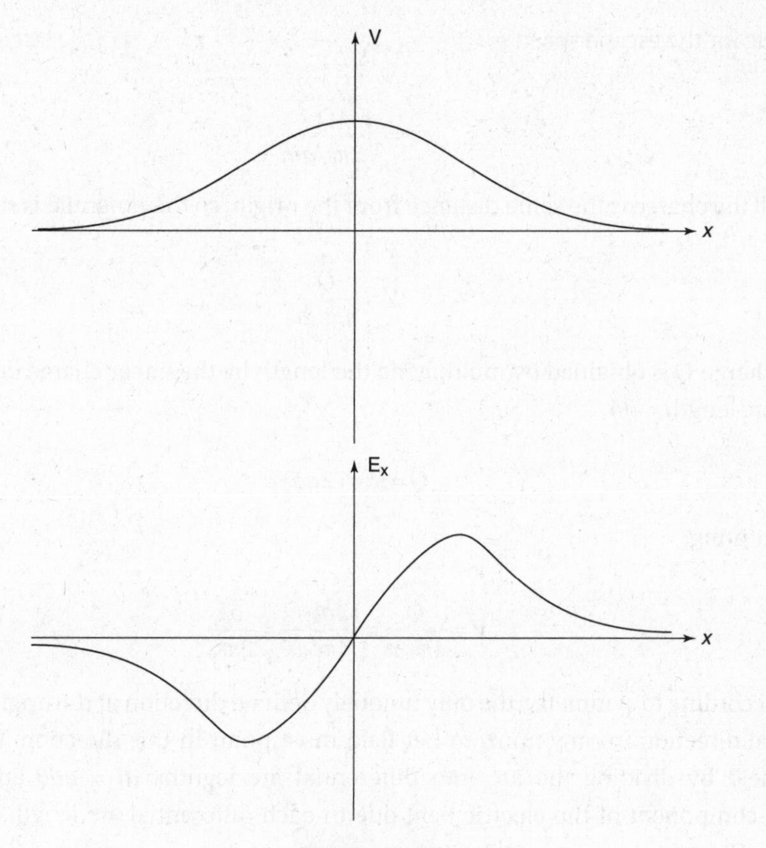

Figure 12.10

(ii) The potential is an even function of x (because it depends only on distance from the ring), whereas the electric field is an odd function of x (because it is a vector, it *does* depend on what side of the ring the point is on).

(iii) The electric field is equal to the negative slope of the potential. This can be observed in several ways. At the origin, the potential curve is flat (zero slope) and the electric field is zero. For positive x, the potential has a negative slope and the electric field is positive (the reverse is true for negative x). The electric field is maximized and minimized (it has a first derivative equal to zero) at the inflection point of the potential curve (the point where the second derivative of the potential curve is zero).

(d) As with gravitation, for the particle to escape to infinity (where the minimum energy is zero potential energy [by definition] and zero kinetic energy), the particle must have at least zero total energy (potential plus kinetic). At the origin, the particle's potential energy is equal to

$$U = qV = (-Q)\frac{Q}{4\pi\varepsilon_0 a} = -\frac{Q^2}{4\pi\varepsilon_0 a}$$

Therefore, to have a total energy of zero it must have a kinetic energy of

$$\text{KE} = \frac{Q^2}{4\pi\varepsilon_0 a} = \frac{1}{2}mv^2$$

Solving for the escape speed v,

$$v = \sqrt{\frac{Q^2}{2\pi\varepsilon_0 am}}$$

2. (a) All the charge is the same distance from the origin, so the potential is simply:

$$V = \frac{Q}{4\pi\varepsilon_0 a}$$

The charge Q is obtained by multiplying the length by the linear charge density (recall that arc length = $r\theta$).

$$Q = l\lambda = (2a\phi)\lambda$$

Substituting,

$$V = \frac{Q}{4\pi\varepsilon_0 a} = \frac{(2a\phi)\lambda}{4\pi\varepsilon_0 a} = \frac{\phi\lambda}{2\pi\varepsilon_0}$$

(b) According to symmetry, the only uniquely defined direction at the origin is the horizontal direction, so any nonzero net field must point in this direction. We calculate the field by dividing the arc into differential arc legnths dl = $ad\theta$ and summing the x-component of the electric field due to each differential arc length as shown in Figure 12.11.

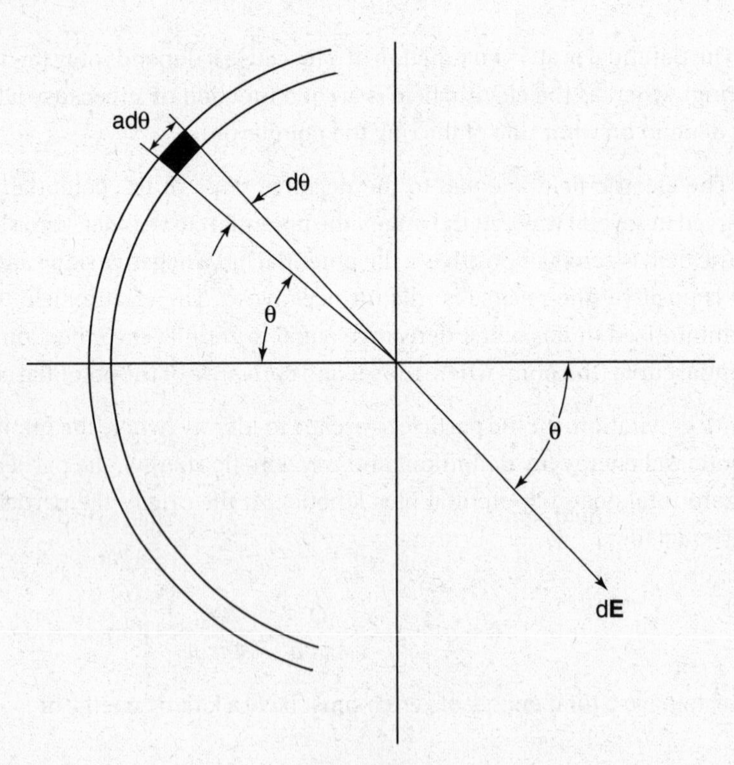

Figure 12.11

$$dE_x = \frac{dQ}{4\pi\varepsilon_0 a} = \cos\theta$$

As in part (a), the differential charge is obtained by multiplying the arc length by the linear charge density:

$$dQ=(dl)\lambda=(ad\theta)\lambda$$

Substituting for dQ,

$$dE_x=\frac{ad\theta\lambda}{4\pi\varepsilon_0 a^2}\cos\theta=\frac{\lambda\cos\theta d\theta}{4\pi\varepsilon_0 a}$$

Integrating,

$$E_x=\int_{-\phi}^{\phi}\frac{\lambda\cos\theta d\theta}{4\pi\varepsilon_0 a}=\frac{\lambda\sin\phi}{2\pi\varepsilon_0 a}$$

and

$$\mathbf{E}=\frac{\lambda\sin\phi}{2\pi\varepsilon_0 a}\hat{i}$$

(c) From the potential calculated in part (a),

$$V=\frac{\phi\lambda}{2\pi\varepsilon_0}$$

Because the potential is proportional to ϕ, it is maximized when ϕ is maximized, at $\phi=\pi$ (creating a complete circular ring of charge).

(d) From the electric field calculated in part (b),

$$\mathbf{E}=\frac{\lambda\sin\phi}{2\pi\varepsilon_0 a}\hat{i}$$

The potential is maximized when $\sin\phi$ is maximized, at $\phi=\pi/2$ (creating an exact seimicircle of charge).

Answer check for parts (c) and (d): Although the math immediately gave us the answers, we could have figured out the answers simply based on physics. As for the potential, the more charge there is nearby, the higher the potential. Therefore, the potential is maximized when the charge is maximum, which occurs when the entire ring contains charge.

As for the electric field, symmetry always requires the net field to be in the x-direction. Therefore, the magnitude of the field is equal to the x-component. Now let's consider a few extreme situations: (i) $\phi=0$, no charge, no electric field; (ii) $\phi=\pi/2$, semicircle of charge, electric field in the $+x$ direction; (iii) $\phi=\pi$, circle of charge, symmetry requires that electric field is zero.

Clearly, the maximal charge lies somewhere between $\phi=0$ and $\phi=\pi$. Now let's think about how the field changes as ϕ increases from zero to π. As ϕ increases from 0 to $\pi/2$, matching differential regions above and below the x-axis produces a net field in the positive x-direction, causing the electric field to increase. However, as ϕ increases beyond $\pi/2$, matching differential regions above and below the x-axis causes a field in the negative x-direction, causing the total field to shrink back to zero. Thus, calculating extrema from the first derivative, the maximum must lie where the field is about to stop growing and start decreasing, at $\pi/2$.

(e) Again, all the charge is the same distance from the origin:

$$V = \frac{Q}{4\pi\varepsilon_0 a}$$

(f) Same as in part (b), except that we must use the substitution $\lambda = Q/l = Q/2a\phi$:

$$\mathbf{E} = \frac{\lambda \sin\phi}{2\pi\varepsilon_0 a}\hat{i} = \frac{Q \sin\phi}{4\phi\pi\varepsilon_0 a^2}\hat{i}$$

(g) Potential doesn't depend on ϕ, so it is maximized for any value of ϕ.

(h) When ϕ is very small ($\phi \approx 0$), all the charge is concentrated at $(-a, 0)$, and it acts like a point charge. In this situation, all the differential electric fields due to differential bits of charge point along the x-axis, and there is no cancellation. As ϕ increases, the charge is spread around the circle, causing the differential charges to produce differential electric fields with components in the y-direction that cancel each other. This cancellation "wastes" electric field, decreasing the field to the point where the field equals zero when $\phi = \pi$. Therefore, the field is maximized when $\phi = 0$.

Gauss's Law

13

→ ELECTRIC FLUX
→ GAUSS'S LAW
→ APPLICATION OF GAUSS'S LAW TO SYSTEMS WITH PLANAR, CYLINDRICAL, AND SPHERICAL SYMMETRY
→ GAUSS'S LAW AND CONDUCTORS

THE CONCEPT OF FLUX

The flux Φ of a constant vector field **F** through a flat surface is defined as

$$\Phi = \mathbf{A} \cdot \mathbf{F} = AF\cos\theta$$

where **A**, the *area vector*, is defined as a vector with magnitude equal to the surface's area and direction perpendicular to the area. (Actually, there are two possible directions perpendicular to any flat surface. Later in the chapter we will define a unique direction.)

So what is flux? If the vector field is the constant-velocity field of a liquid, and the surface is a square wire frame, the flux has a simple physical interpretation—the volume rate of flow of the liquid through the wire frame. This makes sense: The flow is maximized when the wire frame is perpendicular to the velocity (corresponding to **A** parallel to **v**, as shown in Figure 13.1, such that the cosine function is maximized; $\cos 0 = 1$), and the flow is zero when the wire frame is parallel to the velocity [corresponding to **A** perpendicular to **v**, as shown in Figure 13.1, such that the cosine function is minimized; $\cos(\pi/2) = 0$].

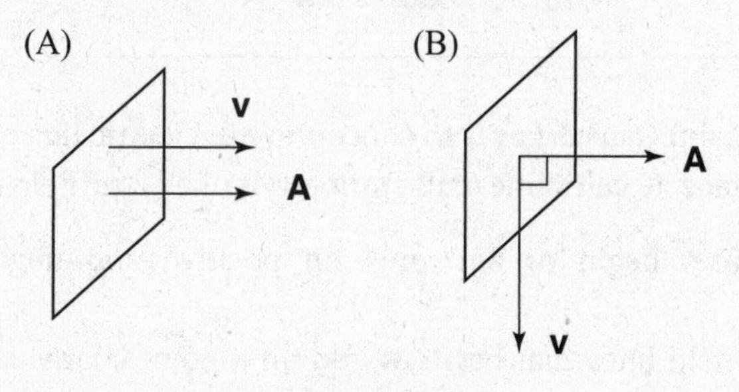

Figure 13.1

What about electric flux (the flux when the vector field is an electric field)? Electric flux is more difficult to conceptualize because the electric field is more abstract than fluid velocity; furthermore, we generally deal with electric flux through *imaginary* surfaces (surfaces that are not associated with physical objects but rather are imagined to suit our mathematical needs). One way of conceptualizing electric flux is to imagine the number of electric field lines that pierce a surface (the stronger the field, the greater the *density* of the field lines; the more parallel the field is to **A**, the greater the number of lines piercing the surface).

What if the vector field is not constant, or if we're dealing with surfaces that are not flat? Our definition of flux still applies on a microscopic level. If we look at small enough pieces of area, the surface is approximately flat and the field is approximately constant, allowing us to use the equation $d\Phi = d(\mathbf{A} \cdot \mathbf{F})$.

To calculate the macroscopic flux through a complicated surface, we integrate this differential flux over the entire surface:

$$\Phi_{electric} = \oint \mathbf{E} \cdot d\mathbf{A}$$

Electric flux

The circle in the integral sign indicates that we are integrating over a closed surface, which is always the case for Gauss's law.

GAUSS'S LAW

TIP

Gauss's law is a restatement of Coulomb's law, but it's valid even if the charges are in motion.

By now you should be familiar with calculating the electric field of a charge distribution via Coulomb's law. However, calculating the electric field due to complicated continuous-charge distributions can be difficult. In this chapter, we introduce another way to calculate electric fields—using Gauss's law. Coulomb's law is equivalent to Gauss's law[1]; which approach we use depends on the problem (in some cases, both can be used).[2]

Gauss's law, as written below, states that the electric flux *out* of a closed surface equals $1/\varepsilon_0$ times the charge enclosed within the surface. Note that here we are defining the area vector **A** to point *out* of the closed surface (such that outward flux is positive and inward flux is negative).

$$\text{Electric flux} = \Phi_{electric} = \underset{\text{surface integral}}{\oint} \mathbf{E} \cdot d\mathbf{A} = \frac{Q_{enclosed}}{\varepsilon_0}$$

Gauss's law

One way to understand Gauss's law is to conceptualize electric flux as the number of field lines that pierce a surface. Recall some of the properties of electric field lines:

1. Electric field lines begin or end only on positive and negative point charges, respectively.
2. The number of field lines that begin or end on a point charge is proportional to the charge's magnitude.

[1]After reading this chapter, you should be able to easily derive Coulomb's law from Gauss's law. Alternatively, deriving Gauss's law from Coulomb's law requires math beyond BC calculus.
[2]However, Gauss's law has a greater range of validity, as it correctly describes the electric fields of moving charges, whereas Coulomb's law is valid only for static charge distributions.

Within this framework, it is easy to understand why charges outside a surface contribute no net flux: The electric field lines due to these charges pass completely through the surface; the *net* flux due to their entering and exiting the surface is zero. Additionally, it makes sense that the net flux depends only on the algebraic sum of the enclosed charge if you think about the electric field lines produced by a collection of charges.

Solving Gauss's Law Problems

In a typical Gauss's law problem, you are given a charge distribution (e.g., an infinite line of charge with linear charge density λ) and your goal is to calculate the electric field created by this charge. While Gauss's law, as a statement of a physical principle, is always valid, it is useful only as a calculational tool for charge distributions with the following types of symmetry: planar, cylindrical, or spherical. To find the electric field at a point in space, you must create a gaussian surface—an imaginary mathematical surface that passes through the point in question. A useful gaussian surface has symmetry that mirrors the symmetry of the charge distribution. For spherical symmetry we use a gaussian sphere, for cylindrical symmetry a gaussian cylinder, and for planar symmetry a gaussian box. With these choices of surfaces, there are regions of constant electric flux (due in part to constant electric field) and possibly regions of zero flux (because the component of the electric field parallel to $d\mathbf{A}$ is zero because the field is perpendicular to the surface or because it equals zero), making it easy to calculate the electric flux integral as we will now illustrate.

TIP

When using Gauss's law to solve problems, we always exploit symmetry.

EXAMPLE 13.1 PLANAR SYMMETRY

Calculate the electric field produced by an infinitely large sheet of charge with uniform charge density $+\sigma$ C/m^2.

SOLUTION

First we choose a gaussian surface. Because the charge distribution has planar symmetry, a reasonable gaussian surface is a box divided evenly by the charged sheet as shown in Figure 13.2.

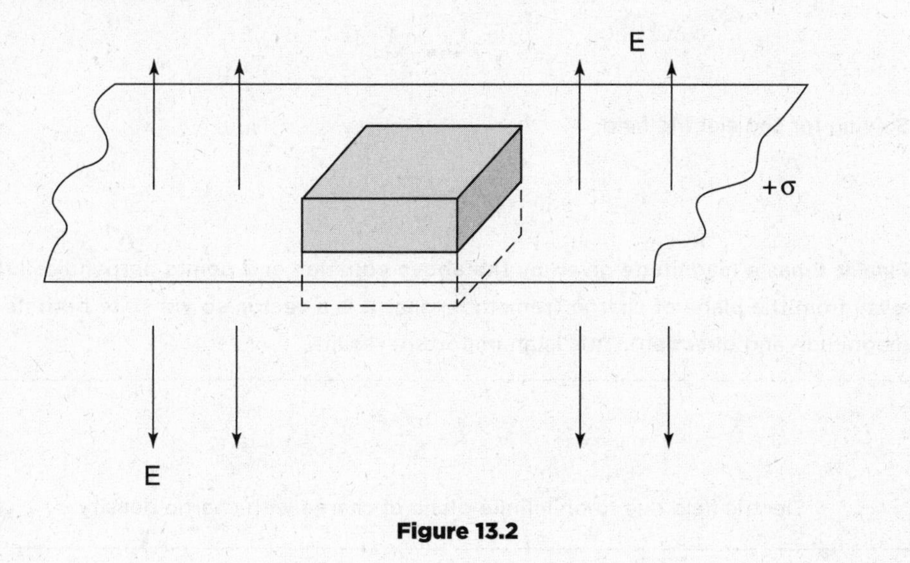

Figure 13.2

(Why should the box be divided evenly by the plane? In Gauss's law problems it is generally best to choose gaussian surfaces with as much symmetry as possible.) To calculate the electric flux, recall from calculus that the integral over a region is equal to the sum of the integrals of its component regions:

$$\Phi_{\text{electric}} = \oint_{\text{wholebox}} \mathbf{E} \cdot d\mathbf{A} = \oint_{\text{sides}} \mathbf{E} \cdot d\mathbf{A} + \oint_{\text{top}} \mathbf{E} \cdot d\mathbf{A} + \oint_{\text{bottom}} \mathbf{E} \cdot d\mathbf{A}$$

We must separately consider the integral of $\mathbf{E} \cdot d\mathbf{A}$ over the sides, top, and bottom. Symmetry requires the electric field to be perpendicular to the plane (no other direction is uniquely defined). Therefore, on the sides of the box, **E** is perpendicular to $d\mathbf{A}$ and $\mathbf{E} \cdot d\mathbf{A} = 0$. What about the top and bottom of the box? In these regions, **E** is parallel to $d\mathbf{A}$ (defined to point away from the center of the box), so that $\mathbf{E} \cdot d\mathbf{A} = E dA$. Therefore,

$$\Phi_{\text{electric}} = 0 + \oint_{\text{top}} E_{\text{top}} dA + \oint_{\text{bottom}} E_{\text{bottom}} dA$$

Symmetry requires the magnitude of **E** to be the same on the top and bottom of the box (because we chose the box to be cut evenly by the charged surface), so $E_{\text{top}} = E_{\text{bottom}} = E$. Because E is a constant, we can move it in front of the integrals. We are left with the integral of dA, which is the surface area of the top or bottom of the box (we will call it A).

$$\Phi_{\text{electric}} = 2AE$$

Applying Gauss's law,

$$[\Phi_{\text{electric}} = 2AE] = \frac{Q_{\text{enclosed}}}{\varepsilon_0}$$

What is Q_{enclosed}? The box contains an area A of the charged sheet of surface density $+\sigma$. Therefore, the enclosed charge is

$$Q_{\text{enclosed}} = \left(\frac{\text{charge}}{\text{area}}\right)\text{area} = \sigma A$$

Substituting this in Gauss's law yields

$$2AE = \frac{\sigma A}{\varepsilon_0}$$

Solving for the electric field,

$$E = \frac{\sigma}{2\varepsilon_0}$$

TIP

You need to understand three types of symmetry: planar, cylindrical, and spherical, to solve Gauss's law problems.

Finally, **E** has a magnitude given by the above equation and points perpendicularly away from the plane of charge (remember that **E** is a vector, so we state both its magnitude and direction). This is an important result:

$$E = \frac{\sigma}{2\varepsilon_0}$$

Electric field due to an infinite plane of charge with charge density σ

EXAMPLE 13.2 CYLINDRICAL SYMMETRY

Calculate the electric field produced by an infinitely long line of charge of constant linear charge density λC/m.

SOLUTION

First, we choose a gaussian surface. Because the charge distribution has cylindrical symmetry, a reasonable gaussian surface is a cylinder of radius *r* and length *L* whose axis is along the line of charge, as in Figure 13.3.

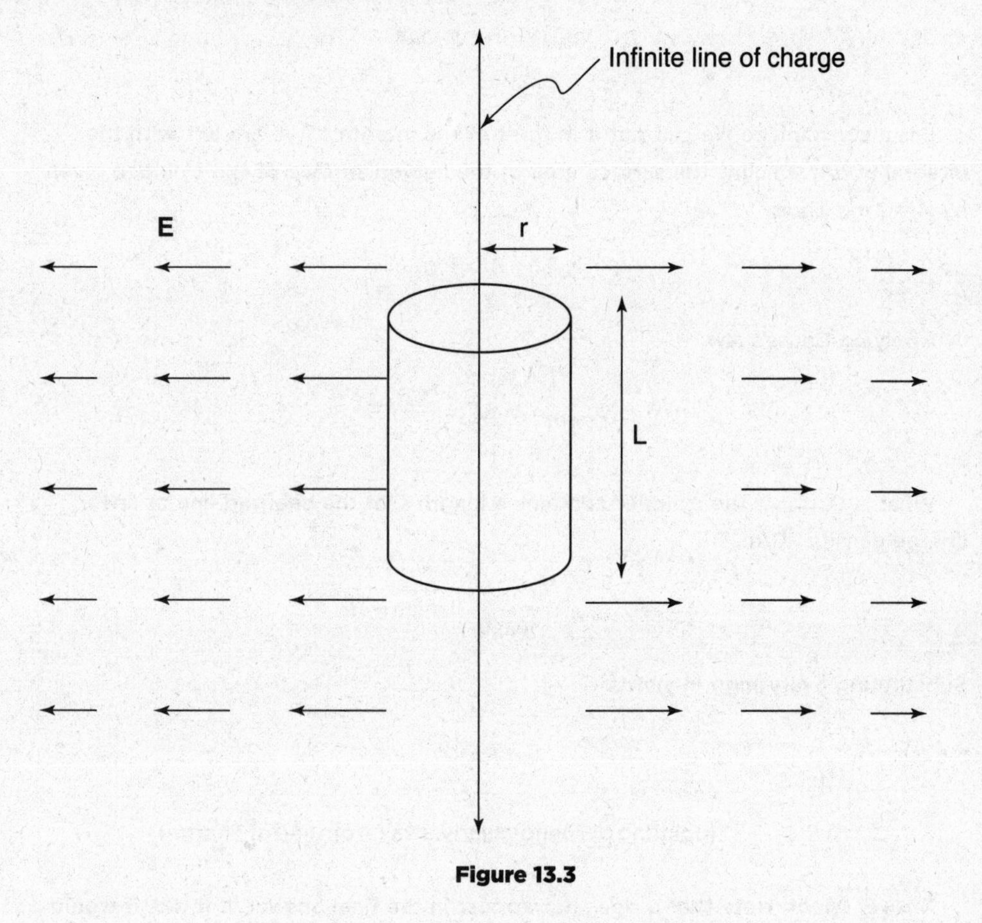

Figure 13.3

$$\Phi_{\text{electric}} = \oint_{\text{cylinder}} \mathbf{E} \cdot d\mathbf{A} = \oint_{\text{top}} \mathbf{E} \cdot d\mathbf{A} + \oint_{\text{bottom}} \mathbf{E} \cdot d\mathbf{A} + \oint_{\text{curved}} \mathbf{E} \cdot d\mathbf{A}$$

Symmetry requires the direction of **E** to be perpendicular to the line of charge (no other direction is unique). Therefore, on the top and bottom of the cylinder, **E** is perpendicular to d**A** and **E**·d**A** = 0. On the curved portion of the surface of the cylinder, **E** is parallel to d**A** (defined to point away from the center of the cylinder), so **E**·d**A** = EdA. Symmetry requires the magnitude of **E** to be constant on this surface. (Why? If not for our imposed coordinate system and gaussian surface, you wouldn't be able to tell the difference between any two points on the cylinder; they are all equivalent.)

$$\Phi_{\text{electric}} = 0 + \oint_{\text{curved}} EdA$$

E is a constant, so we can put it in front of the integrals. We are left with the integral of dA, which is the surface area of the curved surface of the cylinder, given by $A = 2\pi rL$. Thus,

$$\Phi_{\text{electric}} = AE = 2\pi rLE$$

Applying Gauss's law,

$$[\Phi_{\text{electric}} = 2\pi rLE] = \frac{Q_{\text{enclosed}}}{\varepsilon_0}$$

What is Q_{enclosed}? The cylinder contains a length L of the charged line of linear charge density λC/m.

$$Q_{\text{enclosed}} = \left(\frac{\text{charge}}{\text{length}}\right)\text{length} = \lambda L$$

Substituting everything in yields

$$2\pi rLE = \frac{\lambda L}{\varepsilon_0}$$

$$E = \frac{\lambda}{2\pi r\varepsilon_0} \quad \text{(pointing perpendicularly away from line of charge)}$$

Answer check: Note that L does not appear in the final answer. If it did, it would indicate that our answer is incorrect. L is an arbitrary constant that we invented. It has no basis in the actual physical situation and should not appear in the final physical answer.

EXAMPLE 13.3 SPHERICAL SYMMETRY

Given a thin spherical shell of surface charge density σ C/m^2 and radius R, calculate both (a) the electric field outside the shell and (b) the electric field inside the shell as functions of r, the distance from the center of the sphere.

SOLUTION

(a) The charge distribution has spherical symmetry, so we use a spherical gaussian surface with radius r concentric with the charged shell. At every point on the surface of our gaussian surface the electric field is constant in magnitude and parallel to $d\mathbf{A}$ (because of symmetry, the only allowed direction is toward or away from the center of the sphere; because the sphere is positively charged, the electric field points away from it, parallel to the directed area vector). Therefore,

$$\Phi_{\text{electric}} = \oint_{\text{sphere}} \mathbf{E} \cdot d\mathbf{A} = E \oint_{\text{sphere}} dA = 4\pi r^2 E$$

What is Q_{enclosed}? The gaussian surface contains a sphere of radius R and surface charge density σ C/m^2:

$$Q_{\text{enclosed}} = \left(\frac{\text{charge}}{\text{area}}\right)\text{area} = 4\pi\sigma R^2$$

Applying Gauss's law:

$$\Phi_{\text{electric}} = 4\pi r^2 E = \frac{4\pi\sigma R^2}{\varepsilon_0}$$

$$E = \frac{\sigma R^2}{\varepsilon_0 r^2}\begin{pmatrix}\text{pointing away from the center of}\\ \text{the positively charged shell}\end{pmatrix}$$

(b) As before, the spherical symmetry of the situation establishes that

$$\Phi_{\text{electric}} = \oint_{\text{sphere}} \mathbf{E} \cdot d\mathbf{A} = E \oint_{\text{sphere}} dA = 4\pi r^2 E$$

However, there is no charge enclosed! Therefore, according to Gauss's law, $E = 0$. *Note*: If the charge outside the gaussian surface didn't have a simple type of symmetry, the net flux through the gaussian surface would still be zero (Gauss's law always holds). However, the electric field would not be constant, so we wouldn't be able to put it outside the integral and conclude that $E = 0$.

Shell Theorems: Gauss's Law and Analogies with Gravity

Recall the following theorems that were introduced in Chapter 10.

Case 1: The gravitational force exerted on any object *outside* any spherically symmetric mass distribution is the same as if all the mass in the distribution were concentrated at its center.

Case 2: The gravitational force exerted on an object *inside* any spherically symmetric solid mass distribution a distance r from the center is the same as if all the mass located inside an imaginary sphere of radius r (the enclosed mass) were concentrated at its center and all the mass distributed outside this imaginary sphere didn't exist.

Now you should finally be able to understand the basis of these statements! Because Coulomb's law has the same form as the law of universal gravitation, an analog of Gauss's law applies to universal gravitation. In Gauss's law and its gravitational analog, as long as the charge (or mass) distribution is spherically symmetric, the exact mathematical form of the distribution doesn't matter. The only thing that does matter is the amount of charge (or mass) that is enclosed. Therefore, the field (or force) is equal to the field (or force) produced as if all the enclosed charge (or mass) were concentrated at the center of the sphere. Specifically, if we are outside any kind of spherically symmetric object, we can treat it as a point charge of charge Q_{enclosed} at the sphere's center and simply apply Coulomb's law to calculate the electric field. You should be able to understand how electrostatic analogs of these *shell theorems* can be written (the result would read the same and differ only by replacing *mass* by *charge* and *gravitational force* by *electrostatic force*).

Applying this shortcut to calculate the electric field in Example 13.3(a) immediately yields the answer

$$E = \frac{Q}{4\pi\varepsilon_0 r^2} = \frac{\left(4\pi\sigma R^2\right)}{4\pi\varepsilon_0 r^2} = \frac{\sigma R^2}{\varepsilon_0 r^2}$$

Variations on Gauss's Law Problems: Calculation of Q_{enclosed}

The above cases deal with charges that are spread out on surfaces and lines. When charges are distributed in volumes, the calculation of Q_{enclosed} is more complicated. There are generally two situations that you will encounter: (1) constant volume charge density and (2) charge density as a function of position. Each case dictates a different approach to calculating Q_{enclosed}:

Case 1: In the case of constant volume charge density, the charge is proportional to the volume. Thus, simply using the equation

$$\rho = \frac{\text{charge}}{\text{volume}} = \frac{Q_1}{V_1} = \frac{Q_2}{V_2}$$

is generally the best way to deal with these situations.

Case 2: When the charge density is *not* constant, integration is required. Partition Q_{enclosed} into convenient regions of constant charge density and integrate over the enclosed region (another case of integral building). As you might expect, the differential volumes you choose will mirror the symmetry of the charge distribution (spherical shells, cylindrical shells, and rectangular plates for systems with spherical, cylindrical, and planar symmetries, respectively).

EXAMPLE 13.4 VARYING VOLUME CHARGE DENSITY

Consider an insulating sphere of radius *a*. Within the sphere the charge density is given by the equation $\rho(r) = br^3$. Calculate the electric field at a point r_0 with (a) $r_0 < a$, (b) $r_0 > a$, and (c) $r_0 = a$.

SOLUTION

(a) This problem has spherical symmetry, so we use a spherical gaussian surface with radius $r_0 < a$. Because the charge density is not constant, we must calculate $Q_{enclosed}$ via integration. To mirror the spherical symmetry of the charge distribution, we select spherical shells as our differential regions. Recall from the last chapter that a spherical shell has radius *r*, thickness *dr*, and volume

$$dV = 4\pi r^2 dr$$

The macroscopic statement $Q = \rho V$ over the entire sphere is no longer true because the charge density is not constant over this region. However, over an infinitely thin shell, ρ is approximately constant, and we can correctly state that $dQ = \rho(r)dV$.

$$dQ = \rho(r)dV = br^3(4\pi r^2 dr)$$

Integrating *dQ* to obtain *Q*,

$$Q = \int_{r=0}^{r=r_0} dQ = \int_{r=0}^{r=r_0} 4\pi br^5\, dr = \frac{2\pi br_0^6}{3}$$

By symmetry the electric field is constant over the gaussian surface. Because the enclosed charge is positive, the field points away from the center of the sphere, parallel to *d*A:

$$\Phi_{electric} = \oint \mathbf{E} \cdot d\mathbf{A} = E\oint dA = 4\pi r_0^2 E$$

Applying Gauss's law,

$$\Phi_{electric} = 4\pi r_0^2 E = \frac{Q_{enclosed}}{\varepsilon_0} = \frac{2\pi br_0^6}{3\varepsilon_0}$$

$$E = \frac{br_0^4}{6\varepsilon_0}$$

(b) Outside the sphere, we perform the same integral and integrate over the entire volume in calculating $Q_{enclosed}$:

$$Q = \int_{r=0}^{r=a} dQ = \int_{r=0}^{r=a} 4\pi br^5\, dr = \frac{2\pi ba^6}{3}$$

Applying Gauss's law, the final result is

$$E = \frac{Q}{4\pi\varepsilon_0 r_0^2} = \frac{ba^6}{6\varepsilon_0 r_0^2} \quad \text{(away from the center of the sphere)}$$

(c) On the surface of the sphere, the enclosed charge is the same as in part (b) and the radius is $r_0 = a$. Applying Gauss's law yields

$$E = \frac{Q}{4\pi\varepsilon_0 r_0^2} = \frac{ba^4}{6\varepsilon_0}$$

You can verify that this is the same answer as would be obtained by plugging $r_0 = a$ into the results from part (a) or (b). That is, the electric field is continuous at the surface of the sphere, which makes sense: Neither the enclosed charge nor the area is discontinuous. This illustrates a useful principle:

As long as the volume charge density is finite (which is *not* true of surface charge distributions, line charge distributions, or point charges), the electric field is continuous.

When the volume charge density is infinite, the enclosed charge is discontinuous, causing the electric field to be discontinuous as well.

EXAMPLE 13.5 CONSTANT CHARGE DENSITY

An infinitely long nonconducting cylindrical rod of radius a has a linear charge density of λ C/m distributed evenly through its volume. Calculate the electric field (a) outside and (b) inside the rod at a distance R from its center.

SOLUTION

(a) Recall that outside the rod we don't care how the charge is distributed within the gaussian surface (so long as it is cylindrically symmetric). Therefore, outside the rod, the electric field is the same as if the charge were concentrated on a line through the center of the rod. This situation has already been worked out in Example 13.2, so we reproduce the answer here without derivation:

$$E = \frac{\lambda}{2\pi R \varepsilon_0} \quad \text{for } R > a \text{ (pointing perpendicularly away from line of positive charge)}$$

(b) Inside the rod there is cylindrical symmetry, so we use a cylindrical gaussian surface of radius R and length L. As in Example 13.2, the electric flux is expressed as

$$\Phi_{\text{electric}} = AE = 2\pi RLE$$

What is Q_{enclosed}? We have a constant charge density, so we can use the equation ρ = charge/volume. The volume of a cylinder is given by the equation $V = \pi r^2 L$. Let's compare two cylinders: one of length L and radius a, and one of length L and radius R. The cylinder of radius a contains the entire rod, so we know that it contains charge:

$$Q_{\text{cyclinder}R=a} = \frac{\text{charge}}{\text{length}} \cdot \text{length} = \lambda L$$

Additionally, we know that the volume charge density is constant; therefore,

$$\rho = \frac{\text{charge}}{\text{volume}} = \frac{\lambda L}{\pi a^2 L} = \frac{Q_{\text{enclosed}}}{\pi R^2 L}$$

Rearranging this equation yields

$$Q_{\text{enclosed}} = (\lambda L)\frac{R^2}{a^2}$$

Now we have expressions for flux and Q_{enclosed}, so it's time to apply Gauss's law:

$$2\pi RLE = \frac{\lambda LR^2}{\varepsilon_0 a^2}$$

$$E = \frac{\lambda R}{2\pi\varepsilon_0 a^2}$$

Answer check: As discussed at the end of Example 13.2, L is an arbitrary dimension not present in the physical statement of the problem, so it shouldn't appear in the final answer. As another check, both expressions agree on the value for electric field at the surface of the sphere (for $R = a$), $E = \lambda/2\pi\varepsilon_0 a$, as they must because the volume charge density is never infinite.

Further Variations on Gauss's Law Problems: Electric Fields and Charge Distributions of Conductors

So far, all the charge distributions discussed here have been on insulators (also called nonconductors). Insulators are materials that have no "free" electrons, so charge is not free to move within them. Consequently, all sorts of charge distributions are possible on nonconductors: You can put charge wherever you want to and it stays there. The opposite is true of metals. Conductors (which, as far as Physics C is concerned, are equivalent to metals) have free electrons, so charge is free to move within them. This gives rise to the following rule:

The electric field inside a metal is zero in static equilibrium.

Why? If the electric field were not zero (e.g., if an external field were suddenly applied to a metal), electrons would feel a force and would move until they reached new positions such that their fields canceled out the external electric field, resulting in zero net field and thus static equilibrium. Therefore, unlike insulators, only certain charge distributions are possible in metals (because of the requirement that the electric field inside the metal be zero). (When we apply an external field to an insulator, there are opposing forces from the material itself that prevent the charges from moving.)

Gauss's law yields much useful info about conductors of any shape.

Gauss's law problems involving metals often require that you figure out what the charge distribution is on a metal. The following guidelines will help you solve these problems.

Rule 1: Any gaussian surface that is entirely inside a metal must enclose a net charge of zero. We know that the electric field in a metal is zero, so the electric flux through the surface is zero. According to Gauss's law, this means that Q_{enclosed} must also be zero.

Rule 2: Charge can exist only on the surface of a metal. If charge were located in the interior of a metal, this would violate *Rule 1* because a gaussian surface drawn within the metal would enclose nonzero charge.

Rule 3: The net charge on a metal is constant and is equal to the sum of all the charges present on all the surfaces of the metal (because of conservation of charge).

EXAMPLE 13.6 GAUSS'S LAW AND METALS

Suppose we have a hollow metal spherical shell of inner radius *a* and outer radius *b*. A point charge +Q is located in the center of the shell. The shell itself has a net charge of +2Q.

(a) Find the charge distribution on the spherical shell.

(b) Find the *E* field for *r* < *a*.

(c) Find the *E* field for *b* > *r* > *a*.

(d) Find the *E* field for *r* > *b*.

SOLUTION

(a) *Rule 1*: If we draw a spherical gaussian surface of radius *r* with *a* < *r* < *b*, it will be completely within the metal and therefore must contain zero net charge. This means there must be a charge of −Q present on the spherical shell within this gaussian surface. *Rule 2*: Because charge cannot exist within the metal, this −Q charge must be distributed on the inner surface of the metal. *Rule 3*: The charge on the inside of the shell plus the charge elsewhere on the shell must equal the net charge (+2Q). We know the charge on the inner surface is −Q, so subtraction yields a charge of +3Q elsewhere on the shell. Again, because *Rule 2* requires that there be no charge within the metal, the only place that +3Q can be is on the outside surface of the metal.

$$\text{Charge on inside surface} = -Q$$

$$\text{Charge on outside surface} = +3Q$$

(b) Inside the cavity, we draw a gaussian sphere of radius *r* and conclude that the spherically symmetric charge on the metal shell has no impact on $Q_{enclosed}$ or on the *E* field. Therefore, the electric field in this region is simply the electric field due to a point charge, given by Coulomb's law:

$$E = \frac{Q}{4\pi\varepsilon_0 r^2} \quad \text{(pointing away from the positive point charge)}$$

Of course, we could also have obtained this answer using Gauss's law, with an enclosed charge of Q.

(c) Inside the metal, the electric field is always zero. We don't have to know anything about the charge distribution or Gauss's law to find the answer to this part.

(d) For *r* > *b*, consider a gaussian sphere of radius *r* that encloses a net charge $Q_{enclosed} = Q + 2Q = 3Q$. We can apply Gauss's law as usual (spherical symmetry), with the result

$$E = \frac{3Q}{4\pi\varepsilon_0 r^2} \quad \text{(pointing away from the point charge)}$$

IMPLICATIONS OF ELECTRIC FIELDS IN METALS: METALS ARE AT CONSTANT POTENTIAL

Recall that the potential difference between any two points is equal to the line integral $V = -\int \mathbf{E} \cdot d\mathbf{s}$. Therefore, the fact that electric fields are zero within a metal guarantees that the metal is at a constant potential (this holds only in the electrostatic case—where charges have reached equilibrium—as is assumed by our arguments that the electric field in a metal is zero).

A Final Variation on Gauss's Law Problems: Calculating Potential by Integrating Electric Field

Recall the relationship between potential and electric field for integration paths along the x-axis:

$$\Delta V = -\int \mathbf{E} \cdot d\mathbf{s} \Rightarrow \text{when } d\mathbf{s} = dx\hat{i}, \quad \Delta V = -\int E_x dx$$

Using this equation and the expression for the electric field of a point charge, we calculated the potential due to a point charge in Chapter 11. We can do the same thing with the electric fields of more complicated charge distributions calculated with Gauss's law. There are a few points you should be aware of:

1. Unless otherwise stated, zero potential is generally taken, by convention, to lie an infinite distance away from the charge distribution. Therefore, to find the potential at a certain point, you must evaluate an integral from infinity to the point:

$$-\int_{\infty}^{p} E_x \, dx = V(p) - V(\infty) = V(p)$$

Integrating to calculate potential

2. What should you do if your integral passes through multiple regions where the electric field has different functional forms? Recall the following property from calculus:

$$\int_{a}^{c} f(x)dx = \int_{a}^{b} f(x)dx + \int_{b}^{c} f(x)dx$$

We can do the same thing when integrating over multiple regions: Sum the integrals over each region using the appropriate form of $f(x)$ in each region.

EXAMPLE 13.7 CALCULATING POTENTIAL FROM ELECTRIC FIELDS (AND A REVIEW OF GAUSS'S LAW)

An insulating sphere of radius *a* contains a spherically symmetric charge distribution specified by the volume charge density $\rho(r) = br$, where *b* is a positive constant of units C/m^4 and *r* is measured from the center of the sphere. The charge density outside the sphere is zero.

(a) Calculate the total charge *Q* in the sphere.

(b) Determine the electric field for $R > a$.

(c) Determine the electric field for $R < a$.

(d) Do your results from parts (b) and (c) agree at $R = a$? Explain.

(e) Determine the potential at $R = a$.

(f) Determine the potential at $R = 0$.

SOLUTION

(a) The charge density is not uniform, so we need to integrate. Spherical symmetry dictates the use of spherical shells in the integral for the charge. The volume of a spherical shell is

$$dV = 4\pi r^2 dr$$

The charge of a differential volume is equal to the differential volume multiplied by the volume charge density function:

$$dQ = \rho(r)dV = 4\pi br^3 dr$$

To calculate the total charge, we integrate *dQ* (from the lower to the upper limit):

$$Q = \int_{r=0}^{r=a} dQ = \int_0^a 4\pi br^3\, dr = \pi ba^4$$

(b) For $R > a$, $Q_{enclosed}$ is still the total charge on the sphere, $Q = \pi ba^4$. Applying Gauss's law as usual in the case of spherical symmetry (or the shortcut discussed above that the field outside a spherically symmetric charge distribution is the same as would be produced by having all the enclosed charge concentrated at the center of the sphere, allowing the use of Coulomb's law), we obtain

$$E = \frac{Q_{enclosed}}{4\pi\varepsilon_0 R^2} = \frac{ba^4}{4\varepsilon_0 R^2} \quad \text{(pointing away from the center of the sphere)}$$

(c) For $R < a$, $Q_{enclosed}$ must be calculated by integration. The same integral is used as in part (a) except that as the upper limit we use *R* rather than *a*:

$$Q = \int_{r=0}^{r=R} 4\pi br^3\, dr = \pi bR^4$$

Again, we apply Gauss's law with spherical symmetry, with the final result

$$E = \frac{Q_{enclosed}}{4\pi\varepsilon_0 R^2} = \frac{bR^2}{4\varepsilon_0} \quad \begin{pmatrix}\text{pointing away from the center} \\ \text{of a positively charged sphere}\end{pmatrix}$$

(d) Plugging $R = a$ into the results from parts (c) and (b) we obtain the same result for the electric field at this point: $E = ba^2/4\varepsilon_0$. This is to be expected: There are no surface or point charge distributions (i.e., no infinite volume charge densities), so the electric field must be continuous.

(e) The potential at $R = a$ depends on an integral of the electric field evaluated from infinity to $R = a$. Noting that the values for the electric field at the limits of this integral (infinity to $r = a$) are the same as the corresponding values for the field of a point charge $Q = \pi ba^4$, as discussed in part (b), the voltage at $R = a$ is the same as the potential due to a point charge:

$$V = \frac{Q}{4\pi\varepsilon_0 a} = \frac{ba^3}{4\varepsilon_0}$$

Of course, if you didn't see this shortcut, integration of the electric field is a perfectly valid solution that would also yield the correct answer.

(f) Potential at $R = 0$. The electric field inside the sphere is *not* the electric field due to a point charge of value $Q = \pi ba^4$ (because the enclosed charge varies from this value), so we must integrate breaking the integral up into two regions:

$$V = -\int_{r=\infty}^{r=0} E\,dr = -\int_{r=\infty}^{r=a} E\,dr - \int_{r=a}^{r=0} E\,dr$$

We already know the potential at $r = a$:

$$V_{r=a} = -\int_{r=\infty}^{r=a} E\,dx = \frac{ba^3}{4\varepsilon_0}$$

So we have only one new integral to evaluate. Using the expression from part (c) for the electric field,

$$-\int_{r=a}^{r=0} E\,dr = -\int_{r=a}^{r=0} \frac{br^2 dr}{4\varepsilon_0} = \frac{ba^3}{12\varepsilon_0}$$

Plugging these two integrals into the equation for the potential at $x = 0$,

$$V = -\int_{r=\infty}^{r=a} E\,dx - \int_{r=a}^{r=0} E\,dx = \frac{ba^3}{4\varepsilon_0} + \frac{ba^3}{12\varepsilon_0} = \frac{ba^3}{3\varepsilon_0}$$

This is one situation where, because of symmetry, we can also calculate potential by integrating $dV = dQ/4\pi\varepsilon_0 r$ (as in Chapter 12) to confirm our answer. We have spherical symmetry, so we choose spherical shells. Each shell has a charge

$$dQ = \rho(r)\,d(\text{volume}) = 4\pi br^3 dr$$

and is located a distance r from the center of the sphere. Therefore,

$$dV = \frac{dQ}{4\pi\varepsilon_0 r} = \frac{br^2\,dr}{\varepsilon_0}$$

The charge distribution extends from $r = 0$ to $r = a$:

$$V = \int_{r=0}^{r=a} dV = \int_{r=0}^{r=a} \frac{br^2\,dr}{\varepsilon_0} = \frac{ba^3}{3\varepsilon_0}$$

Happily, this agrees with the answer obtained from integrating electric fields.

CHAPTER SUMMARY

Coulomb's law can be recast into a different form called Gauss's law, which provides a very powerful way to find the electric field when the charge distribution exhibits a high degree of symmetry, such as a sphere, cylinder, or plane of charge. Gauss's law is also of greater validity than Coulomb's law as it applies even when charges move. Gauss's law also provides much insight into the electrical properties of conductors.

PRACTICE EXERCISES

Multiple-Choice Questions

1. Which of the following statements is correct?
   (A) If a gaussian surface contains zero net charge, the electric field at every location on the surface must be zero.
   (B) If net flux through a gaussian surface is zero, the surface must enclose no charge.
   (C) If the electric field at every location on a gaussian surface is zero, the surface must contain no net charge.
   (D) The electric field on a gaussian surface is generally not influenced by charge that is not enclosed by the surface.
   (E) Charges outside the gaussian surface affect the electric field on the surface and therefore affect the net flux through the surface.

2. A spherical metal shell with zero *net charge* encloses a point charge. The magnitude of the surface charge density
   (A) is greater on the outside surface of the shell than on the inside surface of the shell
   (B) is greater on the inside surface of the shell than on the outside surface of the shell
   (C) is zero because the net charge is zero
   (D) is zero because the surface charge density of a metal must be zero
   (E) Not enough information is given.

3. Two metal spherical shells, with radii $r_1 > r_2$, each contain a charge $+Q$ distributed on their metal surfaces. The two shells are located far apart from each other compared to their radii. When a wire connects the two shells, what happens?
   (A) Negative charge flows from sphere 1 to sphere 2.
   (B) Negative charge flows from sphere 2 to sphere 1.
   (C) No charge flows between the spheres.
   (D) An electric field builds with time within the wire.
   (E) No magnetic fields are produced.

4. Referring to the situation described in question 3 after the two spheres reach equilibrium, what is the ratio of their charges (assuming negligible charge accumulates on the wire)?
   (A) $Q_1/Q_2 = r_1/r_2$
   (B) $Q_1/Q_2 = r_2/r_1$
   (C) $Q_1/Q_2 = (r_1/r_2)^2$
   (D) $Q_1/Q_2 = (r_2/r_1)^2$
   (E) none of the above

5. A point charge $-Q$ is inserted inside a cubic uncharged metal box. Which of the following statements is true?

   (A) A net charge of $+Q$ is distributed on the inner surface of the box.
   (B) The electric field outside the box is unaffected by the box and is determined solely by the point charge.
   (C) The potential within the walls of the metal box is zero.
   (D) There is a net charge of $+Q$ on the box.
   (E) The electric field inside the region bounded by the metal box is zero.

6. Suppose that two concentric spherical shells with radii $r_1 < r_2$ contain net charges $Q_2 > 0$ and $Q_1 = -Q_2$. In Figure 13.4 which graph is a plot of the electric field as a function of radial position? (Defining away from the center as positive.)

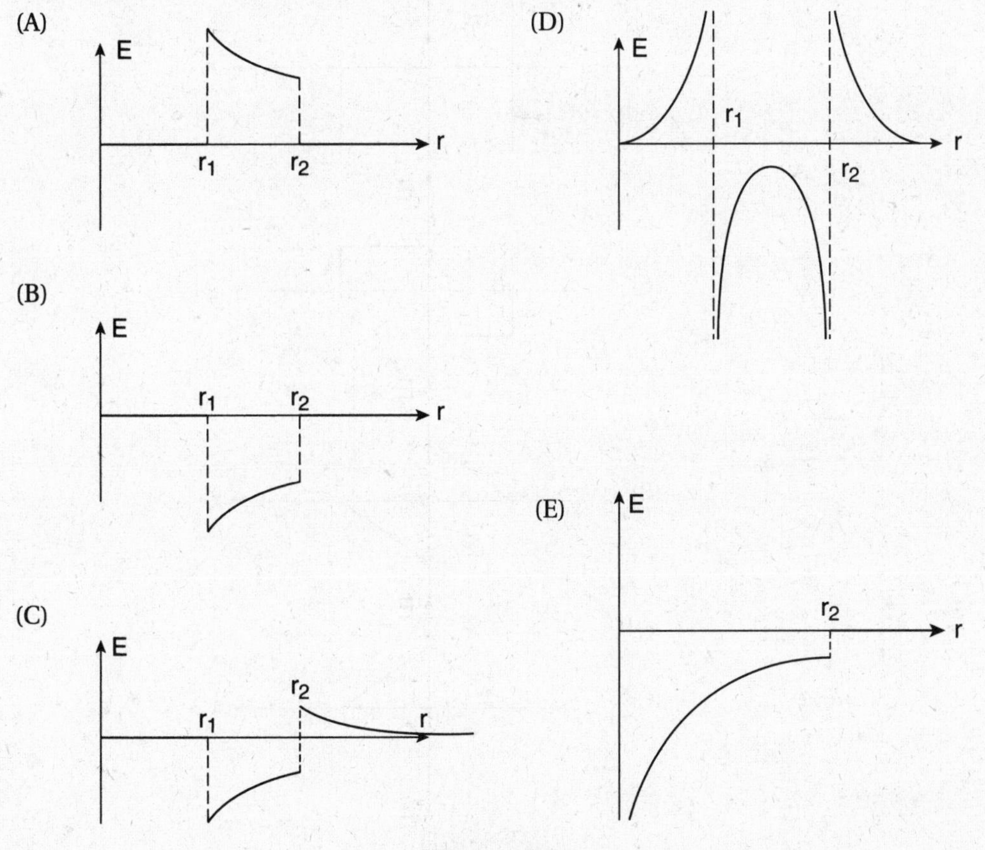

Figure 13.4

7. Three nonconducting spheres all have the same net charge $+Q$ and radius R. The charge is uniformly distributed throughout the volume of sphere A, uniformly distributed on the surface of sphere B, and distributed with a density proportional to the radial variable $\rho = cr$ throughout the volume of sphere C ($c = Q/\pi R^4$). Which is a valid comparison of the electric fields at a distance r from the center of these spheres, assuming that $r > R$?

   (A) $E_A < E_B < E_C$
   (B) $E_A < E_C < E_B$
   (C) $E_C < E_A < E_B$
   (D) $E_C < E_B < E_A$
   (E) $E_B = E_C = E_A$

8. Consider the three spheres introduced in question 7. Which of the following is a valid comparison of the potentials at the surface of each of the spheres?

(A) $V_A < V_B < V_C$
(B) $V_A < V_C < V_B$
(C) $V_C < V_A < V_B$
(D) $V_C < V_B < V_A$
(E) $V_B = V_C = V_A$

9. Three charged sheets lie parallel to the *yz*-plane as shown in Figure 13.5. Which graph shows the electric field as a function of position?

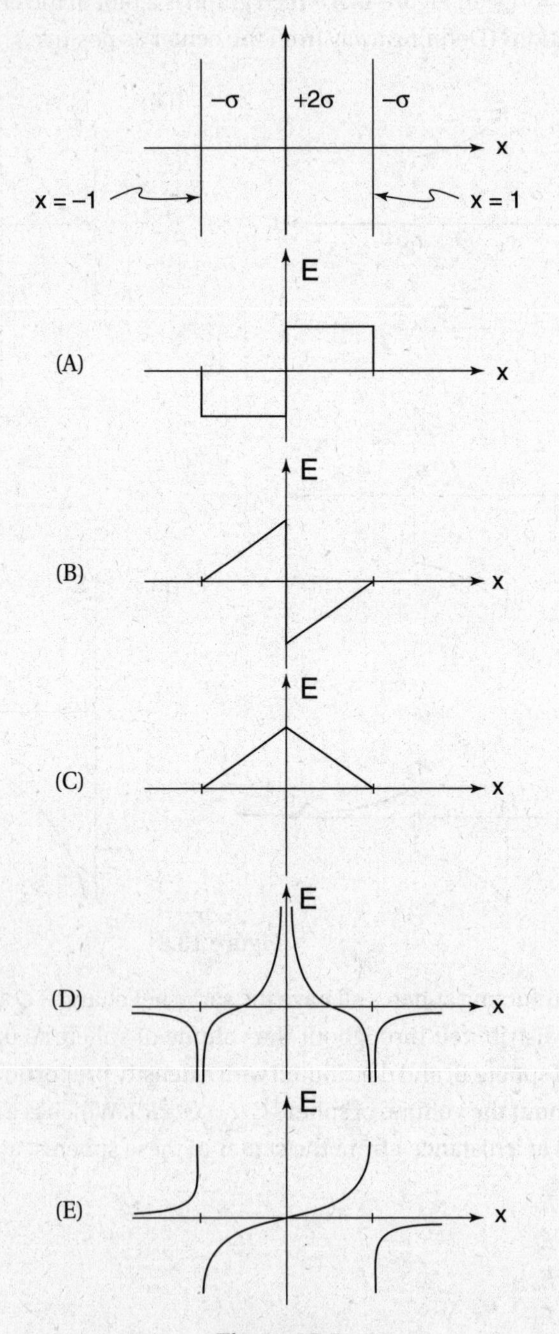

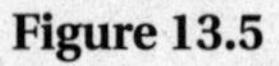
Figure 13.5

(A) graph (A)
(B) graph (B)
(C) graph (C)
(D) graph (D)
(E) graph (E)

Free-Response Questions

1. Consider the two concentric imaginary spherical surfaces shown in Figure 13.6 that surround a spherically symmetric charge distribution. There is no charge between surface a and surface b. Express all answers in terms of a, b, and the electric flux through surface a, Φ_a.

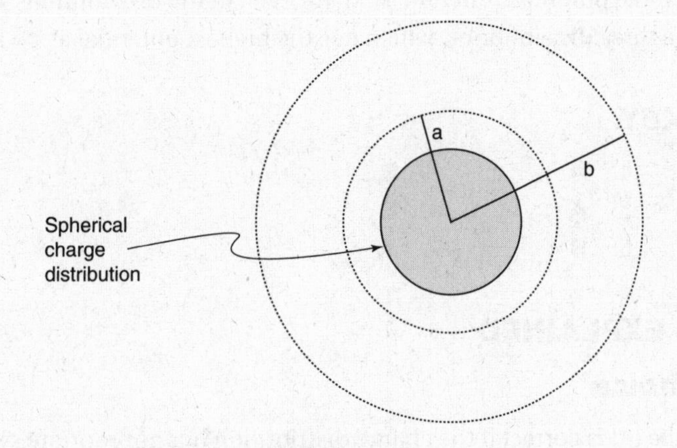

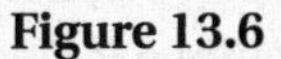
Figure 13.6

(a) Calculate the electric field at $r = a$.
(b) What is the charge enclosed by surface a?
(c) Compare the flux through the two surfaces.
(d) Calculate the electric field at $r = b$.
(e) A metallic shell containing net charge $+Q$ is now introduced, as shown in Figure 13.7. Surface b passes through the thickness of the shell. How does this shell affect the charges enclosed within the two surfaces and the electric fields at $r = a$ and $r = b$? How is charge distributed on the metal shell?

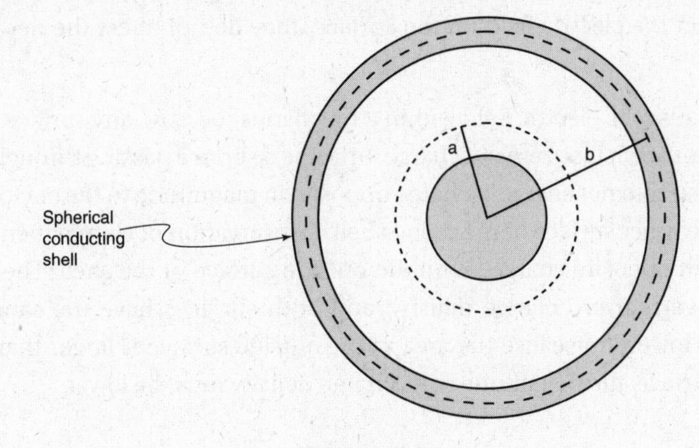

Figure 13.7

2. Consider three charge distributions:

   Distribution A: a point charge $+Q$
   Distribution B: a solid metal sphere (radius a) with net charge $+Q$
   Distribution C: a solid metal sphere (radius a) with net charge $+Q$, surrounded by a metal shell (inner radius b, outer radius c; b and c both larger than a) with a net charge of zero

   (a) Describe the charge distribution on the surfaces of the metals in distributions B and C.
   (b) Using Gauss's law, calculate the electric field in all regions of space for each of the charge distributions.
   (c) Sketch the electric field for each of the distributions.
   (d) Sketch the potential function, $V(x)$, for each of the distributions.
   (e) Of the three distributions, which has the highest potential at $x = a/2$? The lowest?

ANSWER KEY

1. **C**	4. **A**	7. **E**
2. **B**	5. **A**	8. **E**
3. **B**	6. **B**	9. **A**

ANSWERS EXPLAINED

Multiple-Choice

1. **(C)** Choice (A) is correct if the charge distribution has appropriate symmetry but is not true in general. If the enclosed charge is zero, the *net* flux must be zero. However, this does not require the field at every point along the surface to be zero: It is possible for the flux to be positive in some locations along the gaussian surface and negative along others, such that the net flux is zero (even if the field is *not* zero throughout the surface). Choice (B) is almost true. If the electric flux is zero, the *net* charge must be zero. However, it is possible that the surface contains charge, as long as it contains equal amounts of positive and negative charge. Choice (C) is true. [The addition of the word *net* fixes the flaw in choice (B).] Choice (D) is completely false. Charge outside the gaussian surface does not affect the net electric flux through the surface, but it does affect the field along the gaussian surface. Choice (E) is also false. Although charges outside the gaussian surface affect the electric field on the surface, they do not affect the net flux through the surface.

2. **(B)** Because the electric field within a metal must be zero, any surface passing through a metal must enclose zero net charge. Imagine a surface passing through the metal shell. To enclose zero net charge, a charge opposite in magnitude to the enclosed charge must lie on the inner surface of the metal shell. Conservation of charge then requires that an equal and opposite charge lie on the outside surface of the shell. Therefore, both surfaces have nonzero charge density, and both surfaces have the same magnitude of charge. However, because the area of the outside surface is larger than the area of the inside surface, the magnitude of its charge density must be lower.

3. **(B)** The potentials of spheres 1 and 2 are approximately $Q/4\pi\varepsilon_0 r_1$ and $Q/4\pi\varepsilon_0 r_2$, respectively, because the spheres are far apart and thus essentially isolated. Because $r_1 > r_2$, $V_1 < V_2$ and charge flows from the higher to the lower potential. The fact that a sphere of charge $+Q$ with radius r has the same potential as a point a distance r from a point charge $+Q$ is discussed in Example 13.9, part (E).

4. **(A)** In equilibrium a conductor is an equipotential. The two spheres connected by a wire form a single conductor. Therefore,

$$\left[V_1 = \frac{Q_1}{4\pi\varepsilon_0 r_1}\right] = \left[V_2 = \frac{Q_2}{4\pi\varepsilon_0 r_2}\right]$$

This equation can be rearranged into choice (A).

5. **(A)** Choice (A): Since the electric field within the metal is zero, a gaussian surface lying within the metal of the box experiences zero flux and thus must enclose zero net charge. Therefore, the charge on the inner surface of the metal must be the opposite of the point charge, $+Q$. Choice (B): Based on the above discussion, the net electric field outside the box arises from the charge $-Q$ induced on the outer surface of the box. The distribution of this charge is not spherically symmetric (it has the symmetry of the box), and thus the field outside the box is not that of a point charge except at very large distances. Choice (C): The electric field is zero within the metal, so that the potential is constant but not equal to zero. The electric field outside the box arises from the induced negative charge on the surface of the box, so the potential is negative. Choice (D): Conservation of charge requires the net charge on the box remain zero. Choice (E): Within the box, electric field lines begin on the inner surface of the box at the positive charge $+Q$ and end on the point charge $-Q$. The electric field is not zero.

6. **(B)** Gaussian surfaces for $r > r_2$ and $r < r_1$ enclose zero net charge, so the electric field is zero in these regions [eliminating choices (C), (D), and (E)]. In the region $r_1 > r > r_2$, the enclosed charge is negative, so the electric field points toward the center of the sphere (in the $-\hat{r}$ direction); therefore, the correct answer is (B). Note that the magnitude of the E field in this region is proportional to $1/r^2$.

7. **(E)** Because the enclosed charge is the same in every case (for a gaussian sphere passing through the point of interest), the electric fields are the same.

8. **(E)** Because the electric field is the same outside each of the spheres, the potential at the surface of each of the spheres, which is an integral of this electric field,

$$V = -\int_{\infty}^{R} E_r \, dr$$

is also the same.

9. **(A)** Recall that the formula for the magnitude of an electric field due to a sheet of charge is $E = \sigma/2\varepsilon_0$. The magnitude of the electric field does not depend on the distance from the charged sheets (only on which side of the sheets the point is), so the field must remain constant within any region not divided by the plates (eliminating all the choices except (A). To find the electric field, compute the vector sum of the fields caused by each of the three sheets.

Free-Response

1. (a) Applying Gauss's law,

$$\Phi_a = \oint \mathbf{E} \cdot d\mathbf{A} = E\oint dA = 4\pi a^2 E$$

$$E = \frac{\Phi_a}{4\pi a^2}$$

As for the direction of the field, if the flux is positive, the electric field must be parallel to the directed area vector (pointing away from the surface), whereas if the flux is negative, the electric field will point toward the center of the surface.

(b) The enclosed charge is related to the flux:

$$\Phi_a = \frac{Q_{\text{enclosed}}}{\varepsilon_0} \Rightarrow Q_{\text{enclosed}} = \varepsilon_0 \Phi_a$$

(c) The flux through the two surfaces is equal (because they both contain the same enclosed charge).

(d) Making a straightforward application of Gauss's law:

$$\oint \mathbf{E} \cdot d\mathbf{A} = \pm 4\pi b^2 E = \frac{Q_{\text{enclosed}}}{\varepsilon_0} = \Phi$$

$$E = \frac{\Phi}{4\pi b^2}$$

Again, if the flux is positive, the electric field will point away from the surface, whereas if the flux is negative, the electric field will point toward the center of the surface.

(e) At $r = b$, the electric field is zero because the point is located inside a metal. Therefore, the net flux must be zero and the enclosed charge must also be zero. This implies that a charge $-\varepsilon_0\Phi$ must exist on the inside surface of the metal shell. Charge conservation then requires a charge of $Q + \varepsilon_0\Phi$ to reside on the outside surface of the metal shell. The flux through surface a and the field at $r = a$ is unchanged because the surrounding charge is spherically symmetric and the enclosed charge is unchanged.

2. (a) In distribution B, the charge on the solid metal sphere distributes itself evenly on the outside surface of the sphere. This can be understood very simply in terms of charge repulsion. Another way to think about it is that, because the electric field inside a metal is zero, any gaussian surface drawn within the sphere must contain no charge. In distribution C, the charge of the solid sphere is again distributed on its surface. There is a charge of $-Q$ on the inside surface of the metal shell (so that the electric field inside the shell will be zero) and a charge of $+Q$ on the outside surface of the shell (required by conservation of charge).

(b) A possible shortcut here is to realize that because Gauss's law depends only on the magnitude of the enclosed charge and *not* its distribution (as long as the distribution is spherically symmetric), an enclosed sphere (or sphere and shell) of net charge $+Q$ produces the same field as a point charge of the same magnitude located at the center of the sphere ($E = Q/4\pi\varepsilon_0 r^2$). Bearing this in mind, the answers are:

In distribution A, the electric field points away from the point charge and has magnitude $E = Q/4\pi\varepsilon_0 r^2$. In distribution B, for $r < a$, the field is zero (inside a metal). For $r > a$, the enclosed charge is Q, so the field is the same as that due to a point charge of $+Q$, namely, $E = Q/4\pi\varepsilon_0 r^2$. In distribution C, as in distribution B, the field is zero for $r < a$ and $E = Q/4\pi\varepsilon_0 r^2$ for $a < r < b$. For $b < r < c$, the field is zero inside a metal. For $r > c$, the enclosed charge is once again $+Q$, so the field is again equal to $E = Q/4\pi\varepsilon_0 r^2$.

(c) The functions are shown in Figure 13.8.

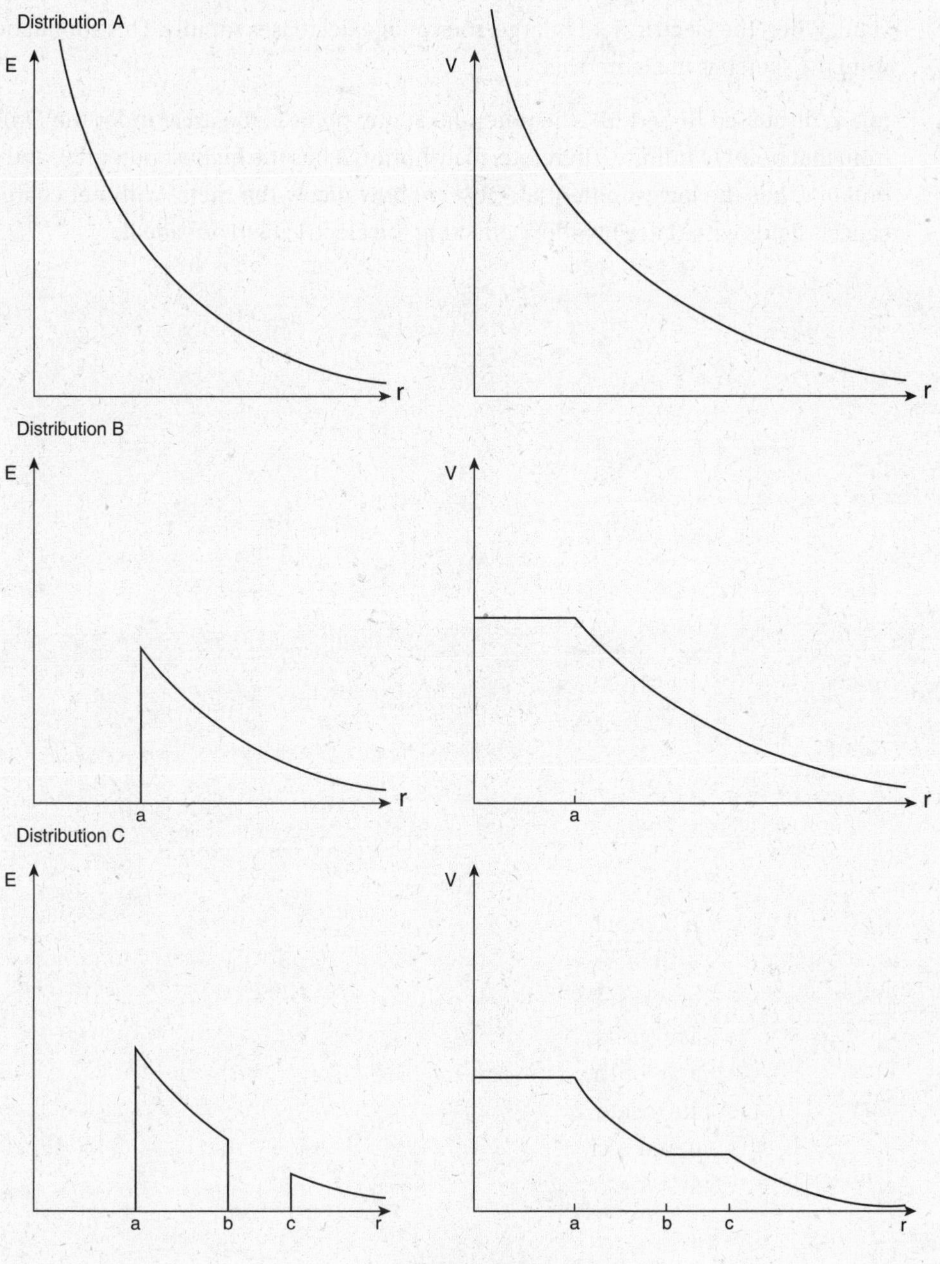

Figure 13.8

(d) How do we proceed from the sketched electric field to a sketched potential? Recall the equation

$$V(x_0) = -\int_{\infty}^{x_0} E_x \, dx = \int_{x_0}^{\infty} E_x \, dx$$

One consequence of this equation is that the potential at any point is equal to the area under the $E(x)$ curve from x_0 to infinity. Therefore, the electric field is the negative derivative of the potential: When the electric field is zero, the potential remains constant. When the electric field is large, the voltage decreases rapidly. This should be clear from the sketches in Figure 13.8.

(e) As discussed in part (d), the potential at any point is the area under the $E(x)$ curve from that point to infinity. Therefore, distribution A has the highest potential, and distribution C has the lowest potential. Observe how nicely the metal with net charge zero cancels fields within itself without affecting the electric field outside it.

Analysis of Circuits Containing Batteries and Resistors

14

- → **DEFINITIONS OF CURRENT, DRIFT VELOCITY, AND RESISTANCE**
- → **OHM'S LAW**
- → **RESISTORS IN SERIES AND PARALLEL**
- → **CIRCUIT ANALYSIS; KIRCHHOFF'S LAWS**
- → **INTERNAL RESISTANCE**

Current is defined as the rate of charge flow per unit time through some surface (generally the cross-sectional area of the wire) with units of amps (A = C/s). Mathematically,

$$I = \frac{dQ}{dt}$$

Definition of the magnitude of current

In addition to magnitude, current also has direction (e.g., current might be flowing in either direction through a wire), but it is not a vector; it does not obey the law of vector addition.

EXAMPLE 14.1

The current moving through a wire is given by the equation $I(t) = ce^{-at}$. How much charge passes through a cross-sectional area of the wire between time $t = 0$ and $t = \infty$?

SOLUTION

$$Q = \int dQ = \int I\,dt = \int_0^\infty ce^{-at}\,dt = \frac{c}{a}$$

We perform unit analysis to verify the answer. Because e^{-at} is a dimensionless number and ce^{-at} has the units of current, c must have the units of current (C/s). Since exponents must be dimensionless numbers, the presence of e^{-at} requires at to be a dimensionless number, which in turn requires a to have units of s^{-1}. Therefore, c/a has units of $(C/s)/s^{-1} = C$.

CONVENTIONAL CURRENT

Conventional current flows from high to low potential. The electrons are actually flowing from low to high potential.

A metal is composed of stationary positive ions and mobile negative electrons that are free to move and produce a current. However, by convention, when analyzing circuits, we generally pretend that current is caused by the motion of positively charged particles that move from high to low potential (*conventional current*). Why? Well, that's the way people started analyzing circuits before they knew that the negative charge was actually the charge that moved, and the convention stuck. Therefore, whenever we talk about conventional current as a stream of positive charges moving in one direction, what is actually happening is that a stream of negative charges are moving in the opposite direction.

Fortunately, consistent use of the conventional current description is completely valid because the two "mistakes" the conventional current description makes (the sign and the direction of the charge) cancel each other out, much like a double negative. For example, a stream of negatively charged electrons moving in one direction causes the same net charge movement as a stream of positive charges flowing in the opposite direction (from our macroscopic point of view in which we can't see the electrons).

A more involved example is the behavior of charges in an electric field. As discussed in previous chapters, positive charges tend to move from high to low potential, while negative charges (which experience an opposite electric force) tend to move from low to high potential difference. Because both the direction in which the charge carriers move and their signs are opposite, we calculate the same net current whether we regard the charge carriers as positive or negative. Throughout this text and the AP exam, when we analyze circuits we will always use conventional currents.

CURRENT DENSITY AND DRIFT VELOCITY

TIP

Current has a direction (describing the motion of charges through an extended object), but it is not a vector quantity. The current density *J* is a vector that describes the motion of the charges at a particular point.

To examine current within a wire more closely, it is useful to define and familiarize ourselves with current density and drift velocity.

Current density, **J**, is a vector field within a wire. The vector at each point in the wire points in the direction of conventional current with magnitude equal to the current transported in this direction *per area*. The current is equal to the flux of the current density through a given area:

$$I = \int \mathbf{J} \cdot d\mathbf{A}$$

Relationship between current and current density

In the case of uniform current flow straight through a wire, **J** is constant and parallel to *d***A**, so that

$$I = \int \mathbf{J} \cdot d\mathbf{A} = \int J dA = J \int dA = JA$$

$$J = \frac{I}{A}$$

Relationship between current and current density for the case of a straight wire where the current density is uniform and parallel to the wire

Drift velocity: Even when there is no current through a wire, the electrons are in constant motion with very high speeds (on the order of 10^6 m/s). However, because the velocity of the electrons is random in direction, the net charge transport (equal to the vector sum of the velocities of the electrons) is zero. While current is flowing through a wire, a slight bias is imposed on the random motion of the electrons such that, on average, they move with a *drift velocity* defined as

$$\mathbf{v}_D = \text{average electron velocity} = \frac{\sum_{i=1}^{N} (\text{velocity of electron}_i)}{n}$$

Drift speeds are generally very small (10^{-4} to 10^{-5} m/s), but because of the very large number of electrons in the metal, they may give rise to considerable currents.

Charge carrier density, n, is generally defined as the number of mobile electrons per volume (n is a characteristic property of the metal or conducting material; insulators, which have no free electrons, have $n \approx 0$).

Relationship between $\mathbf{J}$, $\mathbf{v}_D$, *and n:* Consider a wire of cross-sectional area A that is carrying current of uniform current density straight down the wire. We said above that in this situation $|J| = I/A$. How can we relate current to drift velocity and n?

The velocity of the electrons in the wire can be conceptualized as the drift velocity $\mathbf{v}_D$ superimposed on random motion of the electrons. Because the random motion causes no net current, we can ignore it here and imagine that electrons are flowing through the conductor in nice straight paths with velocity $\mathbf{v}_D$ (like water flowing straight through a pipe with uniform speed). Therefore, in a time increment dt, a volume of electrons equal to $Adx = Av_Ddt$ passes through a cross-section of the wire (illustrated in Figure 14.1b). This volume of electrons contains a number of electrons $dN = n(dV) = nAv_Ddt$ and thus a charge of $dQ = (dN)e = neAv_Ddt$. Therefore, the rate of charge transport through the wire is

$$I = \frac{dQ}{dt} = \frac{neAv_Ddt}{dt} = neAv_D$$

(Note that because we are using the conventional-current description, the charge of an electron, e, is given a positive value.) Combining this with results from above,

$$I = JA = neAv_D$$

And thus,

$$J = nev_D$$

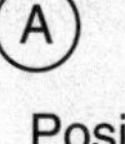

Positive charge-carriers moving through a wire with constant velocity

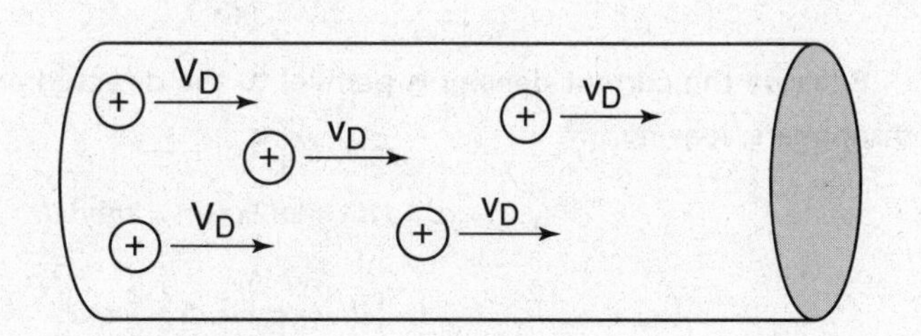

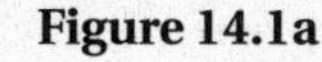
Figure 14.1a

B

Volume of charge carriers that pass through a cross-section of the wire during dt

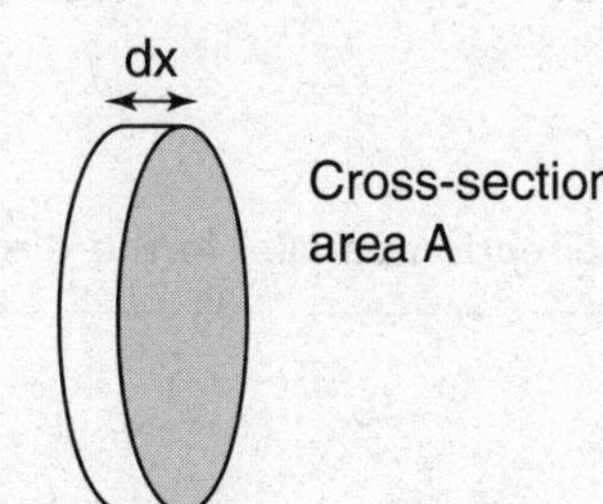

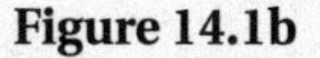
Figure 14.1b

It shouldn't surprise you that this is a general result that can be written as a vector, as shown below. (In the case of nonuniform current, you can simply imagine differential wirelike cylinders parallel to the local current density that experience constant current density parallel to the drift velocity. These differential wirelike cylinders behave exactly like the macroscopic wire used in this example, yielding the same result.)

$$\mathbf{J} = ne\mathbf{v}_D$$

Relationship between current density and drift velocity

EXAMPLE 14.2

The current density through a wire with a circular cross-section of radius *a* is given by the equation *J* = *br*, where *r* is the distance from the center of the wire (the direction of J is parallel to the wire).

(a) What is the total current flowing through the wire?

(b) What is the magnitude of the drift velocity as a function of *r* (given that *n* is the volume density of the electrons)?

SOLUTION

(a) $I = \int \mathbf{J} \cdot d\mathbf{A}$. How do we perform this integral of the two-dimensional circular cross-sectional area of the wire? As always when integral building, we need to select differential regions along which the function we are integrating (current density) is constant. In this case, because current density depends only on *r*, we divide the area into differential washers, each of which has a constant value of *r* (and thus, constant current density).

$$dI = \mathbf{J} \cdot d\mathbf{A}$$

Because the current density is parallel to the directed area vector, the dot product disappears, leaving

$$dI = J\,dA = br(2\pi r\,dr) = 2\pi br^2\,dr$$

$$I = \int dI = \int_0^a 2\pi br^2\,dr = \frac{2}{3}\pi ba^3$$

(b) Using the equation $\mathbf{J} = ne\mathbf{v}_D$ yields the magnitude of the drift velocity to be

$$v_D = \frac{J}{ne} = \frac{br}{ne}$$

with its direction parallel to the direction of the current.

RESISTANCE

The potential difference (i.e., voltage) across a resistor (the amount of potential energy per unit charge that electrons lose when they flow across it) is proportional to the current through the resistor, with a ratio defined as the resistance of the resistor:

$$V = IR \Leftrightarrow R = \frac{V}{I} \Leftrightarrow I = \frac{V}{R}$$

Ohm's law

Ohm's law is an experimental observation that V is independent of R and applies to many, but not all, materials. It does make intuitive sense though. By increasing the voltage at constant resistance, we can increase the current through a resistor; by increasing the resistance at constant voltage, we can decrease the current.

The SI units for resistance are ohms (Ω), defined as Ω = V/A. Different materials provide more or less resistance to current, a property that is called the resistivity of the material and defined as the ratio of the electric field to the current density:

Resistivity

$$\mathbf{E} = \rho \mathbf{J} \Rightarrow \rho = \frac{E}{J}$$

Definition of resistivity

Thus, materials with high resistivity produce less current density in response to the same electric field. (Note that it makes sense that the electric field is always parallel to the current density **J** because we are pretending that charge carriers are positive and thus feel a force parallel to the electric field.)

To confuse you with more new terms, conductivity is defined as the reciprocal of resistivity:

$$\sigma = \frac{1}{\rho} = \frac{J}{E}$$

Definition of conductivity

Calculating Resistance from Resistivity

What is the difference between resistance and resistivity? Resistivity is the property of a certain material, whereas resistance is the property of a particular object that depends on both the resistivity and the geometry of the object (called a resistor).

What is the resistance of a cylindrical resistor of length L, cross-sectional area A, and resistivity ρ? We can answer this question, and similar questions to come, by (1) imagining that we apply a potential difference (or voltage) V across the resistor, (2) calculating the current in response to this voltage, and (3) computing the resistance based on its definition, $R = I/V$.

Resistivity and conductivity are properties of materials. Resistance depends on the material and the geometry of the object.

Suppose we apply a voltage across the resistor. Assuming that the resistor is a perfectly symmetric cylinder, the electric field should be constant throughout the resistor, parallel to the resistor, and equal to $E = V/L$. According to the relationship between current density and electric field, the current density should also be parallel to the resistor, of magnitude $J = E/\rho = V/\rho L$. Because the current density is uniform and perpendicular to the cross-sectional area, we can use the equation $I = JA$ to relate current density to current and cross-sectional area: $I = JA = VA/\rho L$. Finally,

$$R = \frac{V}{I} = V\left(\frac{\rho L}{VA}\right) = \frac{\rho L}{A}$$

In fact, this is true of *any* resistor of constant cross-sectional area:

$$R = \frac{\rho L}{A}$$

Resistance of a resistor with length L, cross-sectional area A, and resistivity ρ

Resistors in Parallel: Equivalent Resistance

Consider two resistors in parallel. How do they behave? One way to answer this question is to calculate their combined resistance, the resistance of an *equivalent resistor* that would act exactly like both resistors combined. To calculate the equivalent resistance, we (1) imagine that a voltage V is placed across the resistor group, (2) calculate the resulting current through the resistor array, and (3) apply the definition of resistance, $R_{eq} = I/V$.

Suppose a voltage V is applied across both resistors. As will become evident as we analyze circuits further, elements in parallel experience the *same* voltage drop.

In this case, a current of $I_1 = V/R_1$ will flow through the first resistor, and a current of $I_2 = V/R_2$ will flow through the second resistor. Thus, the total current through both resistors will be

$$I_{net} = I_1 + I_2 = \frac{V}{R_1} + \frac{V}{R_2}$$

The equivalent resistance of the array is then

$$R = \frac{V}{I_{net}} = \frac{V}{(V/R_1) + (V/R_2)} = \frac{1}{(1/R_1) + (1/R_2)}$$

Inverting this equation gives it a simpler form and yields the result shown below. (Although we illustrated this derivation in the case of two resistors, it isn't hard to generalize it to n resistors in parallel.)

$$\frac{1}{R_{equivalent}} = \frac{1}{R_1} + \frac{1}{R_2} + \frac{1}{R_3} + \cdots + \frac{1}{R_n}$$

Resistors in parallel

Thus, resistors in parallel behave such that the equivalent resistance is *less* than the resistance of any of the individual resistors.

Resistors in Series: Equivalent Resistance

The same approach can be used to understand resistors in series. However, unlike resistors in parallel, which share a common voltage drop, elements in series share a common current.

(This makes sense: Whatever current flows through one resistor must flow through the adjacent resistor because of conservation of charge.) Therefore, it is easier to begin by (1) supposing that a certain current I flows through the array, then (2) calculating the voltage across the array, and only then (3) calculating the equivalent resistance using $R_{eq} = V/I$.

If a current I flows through the resistors, Ohm's law indicates that the voltage across the first resistor is $V_1 = IR_1$ and the voltage across the second resistor is $V_2 = IR_2$. Thus, the net voltage is

$$V_{net} = V_1 + V_2 = IR_1 + IR_2$$

The equivalent resistance is then

$$R_{eq} = \frac{V_{net}}{I} = \frac{IR_1 + IR_2}{I} = R_1 + R_2$$

As before, this is easily generalized to the case of n resistors:

$$R_{equivalent} = R_1 + R_2 + R_3 + \cdots + R_n$$

Resistors in series

Thus, resistors in series add, so that the equivalent resistance is *greater* than the resistance of any of the individual resistors.

More Complicated Resistor Arrays: Equivalent Resistance

How can we calculate the equivalent resistance of more complicated networks of resistors? This can be done by repetitively applying the above two formulas. More specifically, this involves three steps:

STEP 1 Identify a small, isolated group of resistors in series or in parallel.

STEP 2 Calculate their equivalent resistance using one of the above equations.

STEP 3 Rewrite the resistor network, replacing the isolated group by a single resistor whose resistance is equal to the equivalent resistance of the isolated group.

Repeat steps 1–3 until only one equivalent resistor is left (whose resistance is the equivalent resistance of the entire network).

EXAMPLE 14.3 EQUIVALENT RESISTANCE

Calculate the equivalent resistance of the resistor network shown in Figure 14.2.

SOLUTION

STEP 1 First we choose the group containing resistors with resistances of 5 and 20 Ω.

STEP 2 They're in series, so the equivalent resistance is the sum of the two individual resistances, $R_{eq} = 5\ \Omega + 20\ \Omega = 25\ \Omega$.

STEP 3 Rewrite the circuit as shown in Figure 14.2.

STEP 1 Next we choose the group containing 2- and 3-Ω resistors (*Note*: We could have chosen this group first, before the 5- and 20-Ω resistors).

STEP 2 They're in parallel (look carefully at the circuit; they share a common terminal on the left), so the equivalent resistance is

$$R_{\text{equivalent}} = \frac{1}{(1/2\,\Omega)+(1/3\,\Omega)} = 1.2\,\Omega$$

STEP 3 Rewrite the circuit as shown in Figure 14.2.

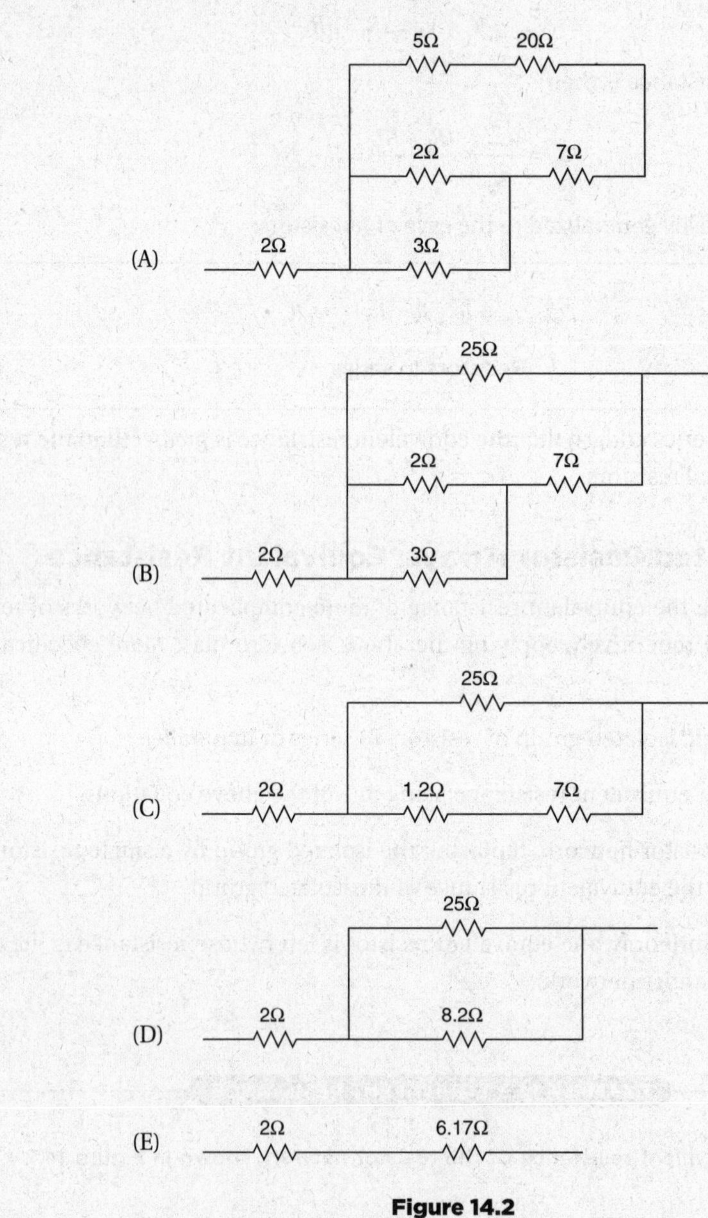

Figure 14.2

STEP 1 Then we choose the group containing the 1.2- and 7-Ω resistors.

STEP 2 They're in series, so the equivalent resistance is $R_{eq} = 1.2\ \Omega + 7\ \Omega = 8.2\ \Omega$.

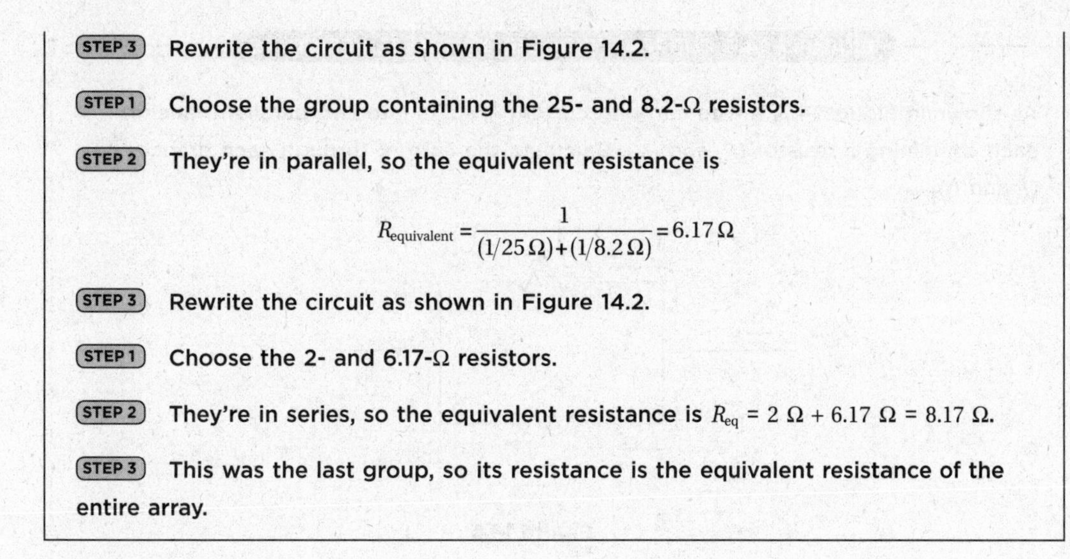

STEP 3 Rewrite the circuit as shown in Figure 14.2.

STEP 1 Choose the group containing the 25- and 8.2-Ω resistors.

STEP 2 They're in parallel, so the equivalent resistance is

$$R_{\text{equivalent}} = \frac{1}{(1/25\,\Omega)+(1/8.2\,\Omega)} = 6.17\,\Omega$$

STEP 3 Rewrite the circuit as shown in Figure 14.2.

STEP 1 Choose the 2- and 6.17-Ω resistors.

STEP 2 They're in series, so the equivalent resistance is $R_{eq} = 2\ \Omega + 6.17\ \Omega = 8.17\ \Omega$.

STEP 3 This was the last group, so its resistance is the equivalent resistance of the entire array.

Analysis of Circuits Containing a Single Battery

When analyzing circuits with only one battery attached to a network of resistors, the only equation you need is Ohm's law and the equations for equivalent resistance. The following rules are also useful.

The potential at any point along a continuous stretch of wire (assumed to have negligible resistance) is constant.

Why? Ohm's law says that $V = IR$. Unless otherwise indicated, we generally assume that the resistance of the wire is negligible compared to that of the resistors in the circuit. Thus, to a reasonable approximation, $R = 0$ in the wires, and this makes the voltage $V = IR$ also equal to zero. Because the change in potential is zero, the potential is constant throughout the wire. (If there are no resistors in the circuit, that is, if a battery is simply connected to a piece of wire, the resistance is no longer negligible and the wire is not at a single potential.)

At any intersection of wires, the current entering the intersection must equal the current leaving the intersection (as required by conservation of charge).

TIP

Kirchhoff's first law: node rule

When analyzing circuits with a single battery, there are generally many ways to arrive at the correct answer. If you get stuck, one useful strategy is often to calculate the equivalent resistance of a network of resistors and apply Ohm's law across the entire array to calculate the potential drop or current of the entire array.

The battery symbol is two parallel lines, one longer than the other. (The longer line is the side of the battery with the higher potential.) Recall that the magnitude of potential, like that of energy, is meaningless—only *differences* in potential are important. For that reason, we are free to assign zero potential wherever we please. In a circuit with only one battery, we generally set the side of the battery at lower potential (the shorter line) equal to zero potential, making the other side of the battery have a positive potential equal to the potential difference across the battery (e.g., +9 V).

EXAMPLE 14.4 CURRENT THROUGH PARALLEL RESISTORS

As shown in Figure 14.3, a wire carrying current I_0 splits into two parallel branches each containing a resistor (R_1 and R_2). Calculate the current through each branch (I_1 and I_2).

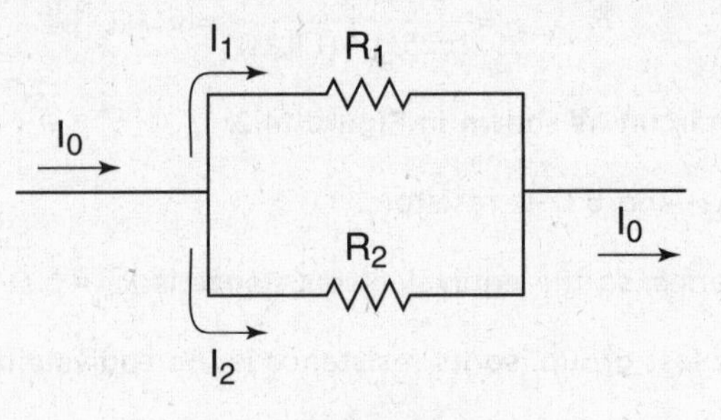

Figure 14.3

SOLUTION

Elements in parallel have the same voltage:

$$V = I_1 R_1 = I_2 R_2$$

TIP

Remember: voltage is the potential difference between two points, in this case the two sides of the resistor.

Additionally, because of rule 2 (the current entering the intersection equals the current leaving the intersection),

$$I_1 + I_2 = I_0$$

Now we have two linear equations with two variables. It is straightforward to solve for the currents, which yields the same answer as the following alternative solution.

***Alternative solution*: We can calculate the equivalent resistance of the two resistors as usual:**

$$R_{\text{equivalent}} = \frac{1}{(1/R_1)+(1/R_2)} = \frac{R_1 R_2}{R_1 + R_2}$$

The voltage across the array is then

$$V = IR = I_0\left(\frac{R_1 R_2}{R_1 + R_2}\right)$$

Because this is a parallel arrangement, the voltage across each resistor is equal to the voltage across the network. Therefore, we can use this voltage to apply Ohm's law across each resistor:

$$I_1 = \frac{V}{R_1} = \frac{I_0 R_2}{R_1 + R_2}$$

$$I_2 = \frac{V}{R_2} = \frac{I_0 R_1}{R_1 + R_2}$$

Symmetry check: The two branches of this equivalent resistor are *mathematically symmetric*—there is nothing different about them other than the way we name them. Therefore, the answer should be symmetric—that is, by switching R_1 and R_2, we can convert between the expressions for I_1 and I_2. You can verify that this is indeed the case. This type of symmetry is subtle and should not be relied on. However, if our answers are grossly asymmetric, it might be an indication that something is amiss.

We might expect that current tends to take the path of less resistance. As a check, we can consider some limiting cases. From the above equations, as $R_1 \to \infty$, $I_1 \to 0$ and $I_2 \to I_0$. What this means is that as R_1 increases to infinity, all the current passes through R_2, which makes sense (R_1 has such a large resistance that no current can get through it. This is equivalent to cutting open the wire attached to resistor 1, introducing an *open circuit*. Alternatively, as $R_1 \to 0$, $I_1 \to I_0$ and $I_2 \to 0$. That is, if R_1 is zero, all the current flows through R_1 (this is the situation where we short out a resistance by connecting a wire to either side of it; all the current flows through the wire rather than the resistor).

EXAMPLE 14.5 POTENTIAL DROP ACROSS SERIES RESISTORS

A battery of voltage V is connected to two resistors R_1 and R_2 in series. Calculate the potential drop across each of the resistors (V_1 and V_2).

SOLUTION

The equivalent resistance of the circuit is $R_{\text{equivalent}} = R_1 + R_2$. Therefore, the current flowing through the circuit is

$$I = \frac{V}{R_{\text{equivalent}}} = \frac{V}{R_1 + R_2}$$

We can apply Ohm's law across each resistor separately: $V_1 = IR_1 = VR_1/(R_1 + R_2)$ and similarly $V_2 = IR_2 = VR_2/(R_1 + R_2)$.

Symmetry check: The symmetry of the circuit is similar to the symmetry in Example 14.4 (in that it doesn't matter which order the resistors are in; the current has to go through both resistors either way) and is mirrored in the symmetry of these solutions (by switching the 1s and 2s, you can convert between the equations for V_1 and V_2).

The voltage is greater across the larger resistor. If one resistor is much greater than the other, the larger resistor will experience the vast majority of the voltage drop. This is what happens in real life when we connect a wire to a resistor (if we think of the wire as a "resistor" with very low resistance). The resistance of the resistor is so much greater than the resistance of the wire that the vast majority of the potential difference occurs over the resistor (this is another way to explain rule 1, the approximation that wires generally have a constant potential).

VOLTMETERS AND AMMETERS

Examples 14.4 and 14.5 explain how voltmeters (devices that measure voltage) and ammeters (devices that measure current) are designed.

TIP

Voltmeters are connected in parallel with circuit elements; ammeters are connected in series.

To measure voltage, a voltmeter is attached in *parallel* across the resistor (or other circuit element) whose voltage it is measuring. Because elements in parallel exhibit the same voltage, the voltmeter exhibits (and thus is able to measure) the voltage of the element it is connected across. In order that the addition of the voltmeter not perturb the circuit too much, voltmeters are designed to have very *high* resistance so that little current flows through them (i.e., nearly the same amount of current continues flowing through the circuit after connecting the voltmeter).

To measure current, an ammeter is inserted in *series* into the wire whose current it is measuring. Because elements in series have the same current, the current flowing through the ammeter is the same as that flowing through the circuit elements. In order that the ammeter not perturb the circuit, it is designed to have very *low* resistance so that it doesn't cause a noticeable change in the voltage (i.e., it acts like a piece of wire).

ANALYSIS OF CIRCUITS CONTAINING MULTIPLE BATTERIES: KIRCHHOFF'S LAWS

More complicated circuits that contain more than one battery cannot always be solved using Ohm's law and equivalent resistance formulas. Kirchhoff's laws provide a methodical way to solve such circuits. These laws always hold and can be used to solve simpler circuits (although it is generally not the easiest method in these cases).

KIRCHHOFF'S LAWS

1. *The node rule*: At any intersection of wires, the total current entering the intersection must equal the total current leaving the intersection.
2. *The loop rule*: The sum of the potential differences around any closed loop must be zero.

We introduced the node rule in the last section. As for the loop rule, if the sum of the potential differences around a loop were not zero, there would be a net gain or loss of energy every time an electron went around the circuit. This would violate conservation of energy because an electron could continue to go around the circuit, continually generating energy.

Applying Kirchhoff's Laws: A General Approach

STEP 1 Assign a current to each wire segment (both a magnitude and a direction). It is generally easiest to assign the directions by guessing which way the current will flow. If you happen to be wrong, don't worry—the value of the current will simply end up being negative, indicating that you've chosen the wrong direction.

STEP 2 Using loop equations and node equations, create as many unique equations as there are currents.

HOW TO CREATE LOOP EQUATIONS

Pick a closed loop in a circuit. Go completely around the loop in one direction, summing potential differences as you go, and set this sum equal to zero. The next two paragraphs explain how to evaluate the potential differences across batteries and resistors.

Calculating potential changes across batteries: The positive terminal (the longer line in the battery symbol) is at a higher potential than the negative terminal. Therefore, going from the negative to the positive terminal involves an increase in potential (a positive voltage), whereas going from the positive terminal to the negative terminal involves a decrease in potential (a negative voltage).

Calculating potential drops across resistors: When current passes through a resistor, its potential energy decreases as it expends energy to transverse the resistor (this energy is converted to heat). Therefore, the side of the resistor where current enters is at higher potential than the side where current exits by an amount given by Ohm's law, $V = IR$. Thus, when constructing loop rule equations, when you go across a resistor in the direction of the current, there is a decrease in potential (a negative voltage), whereas when you go across a resistor in the direction opposite the current, there is an increase in potential (a positive voltage).

Other applications: In addition to helping you calculate various currents, these rules can also be used to calculate the potential difference between any two points in a circuit. The potential differences are "real" (you can measure them with a voltmeter); they're not just a mathematical trick we use to calculate currents, and they are involved in other problems besides Kirchhoff's laws problems.

HOW TO CREATE NODE EQUATIONS

Simply pick a node and equate the current entering the node to the current leaving the node. You have already seen this in Example 14.5 (the equation $I_0 = I_1 + I_2$ is a node equation).

HOW TO CREATE THE RIGHT NUMBER OF NODE AND LOOP EQUATIONS

Consider the circuit shown in Figure 14.4. There are three current variables. To find these values, you need three independent equations. However, if you try to create three loop equations, you will find that you are unable to solve your system of equations. Why? The fact that loop 1 and loop 2 sum to zero guarantees that loop 3 sums to zero! To prove this, suppose we let V_{AD} be the voltage difference when traveling directly from point A to point D. Then,

$$\text{Loop 1 loop equation} \rightarrow V_{ABCD} = V_{AD}$$

$$\text{Loop 2 loop equation} \rightarrow V_{AFED} = V_{AD}$$

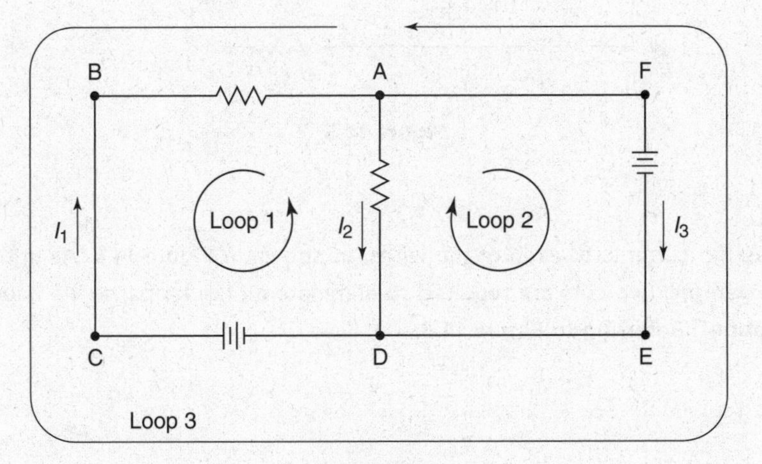

Figure 14.4

We travel along loop 3 as follows: A → B → C → D → E → F → A. This is equal to $V_{ABCD} - V_{AFED}$ (when we go in the opposite direction, all the potential jumps switch signs). From above, we already know that $V_{ABCD} = V_{AFED} = V_{AD}$. Therefore, the fact that the loop 3 potential differences sum to zero is guaranteed by the loop equations from loops 1 and 2. Because the loop 3 equation doesn't give us any extra information, it cannot serve as a third independent equation.

How do we know if we are creating too many or too few loop equations? There is a simple rule that governs the maximum number of independent loop equations: The number of loop equations needed is equal to the minimal number of cuts in the wires needed to eliminate all the loops. To obtain any additional equations required to solve the system of linear equations, use node equations. (Recall that you must have as many independent equations as you have variables.) Another rule you might find useful is that each unique loop equation should involve a new circuit or new current that has not been used in a previous equation.

How do you know if you have enough equations to solve the problem?

For example, in the circuit shown in Figure 14.4, only two cuts are needed to eliminate all the loops; therefore, only two independent loop equations can be written. Because there are three variables (I_1, I_2, and I_3), we need three independent equations. The last required equation is a node equation ($I_1 = I_2 + I_3$).

STEP 3 Now you simply have to solve the system of linear equations. The most straightforward way to do this is to systematically pick variables one by one and eliminate them by substitution. Eventually you will be able to solve for one variable. You can then work backward and solve for all the others.

EXAMPLE 14.6 KIRCHHOFF'S LAWS

For the circuit in Figure 14.5, calculate the magnitude and direction of current through each wire.

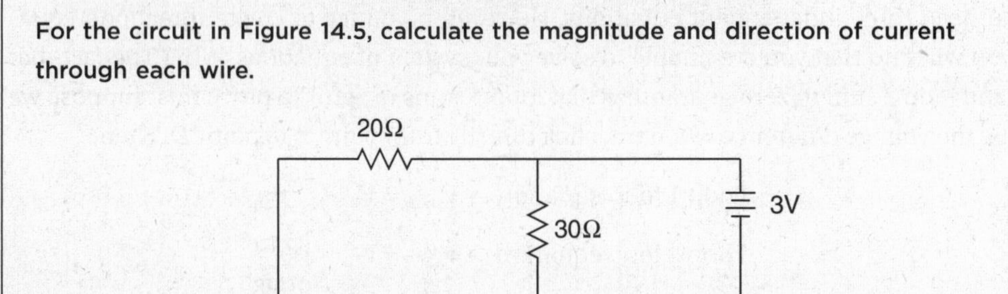

Figure 14.5

SOLUTION

First we assign currents to each of the wires, as shown in Figure 14.6. As in the previous example, two cuts are required to eliminate all the loops, so we need two loop equations. Referring to Figure 14.6,

$$\text{Loop 1: } 10 - 20I_1 - 30I_2 = 0$$

$$\text{Loop 2: } 30I_2 - 3 - 40I_3 = 0$$

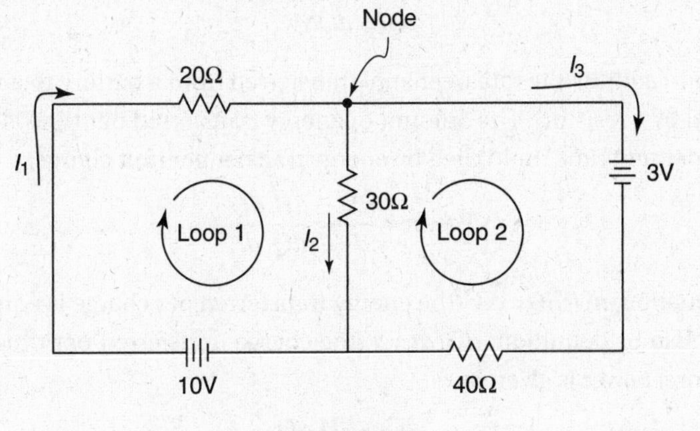

Figure 14.6

We are solving for three variables (I_1, I_2, and I_3), so we need three independent equations. We have two loop equations, so we need one node equation:

$$\text{Node equation: } I_1 = I_2 + I_3$$

So now we just have to solve a system of linear equations:

$$\begin{cases} 10 - 20I_1 - 30I_2 = 0 \\ 30I_2 - 3 - 40I_3 = 0 \\ I_1 = I_2 + I_3 \end{cases}$$

Solving these equations is straightforward: We eliminate I_1 by substitution using the last equation, leaving two simultaneous equations that can then be solved by substitution. The result is $I_1 = 0.23\text{ A}$, $I_2 = 0.18\text{ A}$, and $I_3 = 0.058\text{ A}$.

Power Dissipated Across Resistors A battery is a chemical device that boosts the potential energy of electrons flowing through it. In a simple circuit, these electrons then flow through various resistors that impede their progress. As electrons flow through a resistor, they lose their electrical potential energy to various forms of energy (e.g., heat dissipated by resistors, light given off by lightbulbs, or mechanical work produced by a motor). What is the rate at which energy stored in the battery is transferred to the resistor; that is, what is dU/dt?

Recall that, by definition, potential is the potential energy per charge:

$$V = \frac{U}{Q}$$

Now consider moving a charge from point A to point B. The change in potential is related to the change in potential energy as follows:

$$V_B - V_A = \frac{U_B}{Q} - \frac{U_A}{Q}$$

More generally,

$$\Delta V = \frac{\Delta U}{Q} \Leftrightarrow \Delta U = Q\Delta V$$

What, then, is the potential energy lost by a very small amount of charge, dQ, as it passes through some fixed potential difference ΔV? The energy and charge become infinitesimal, but the potential difference ΔV is fixed:

$$dU = dQ\Delta V$$

Now we're ready to address the rate of energy transferred from a battery to a resistor (i.e., the power dissipated by a resistor). The amount of energy transferred per time is simply equal to charge transfer per unit time multiplied by energy transfer per unit charge:

$$\text{Power} = \frac{dU}{dt} = \frac{dU}{dQ}\frac{dQ}{dt}$$

From the last equation, $dU/dQ = \Delta V$ (the energy transferred per charge is equal to the potential difference). Also by definition, $dQ/dt = I$ (the charge transferred per time is equal to the current). Therefore, power is given by

$$\text{Power} = \frac{dU}{dt} = \frac{dU}{dQ}\frac{dQ}{dt} = (\Delta V)I$$

(This is more often written as $P = IV$, where V is implied to be the potential jump across the circuit element.) Using Ohm's law ($V = IR$), power can be expressed in terms of any two of the three quantities V, I, and R:

$$P = IV = I^2R = \frac{V^2}{R}$$

Rate of energy transfer to a resistor

The units of power are watts, defined as W = V·A.

Internal Resistance In real life, batteries and wire have some resistance. Although we generally ignore these resistances because they are low compared to the resistance of resistors, it is sometimes useful to discuss and quantify the *internal resistance* of a battery. Despite the fact that all the components of a battery have resistance, it is useful to think of a real battery as a series arrangement of an ideal resistance-free battery and an internal resistor (this is valid because all the charge that flows through the battery encounters the internal resistance) as shown in Figure 14.7(A).

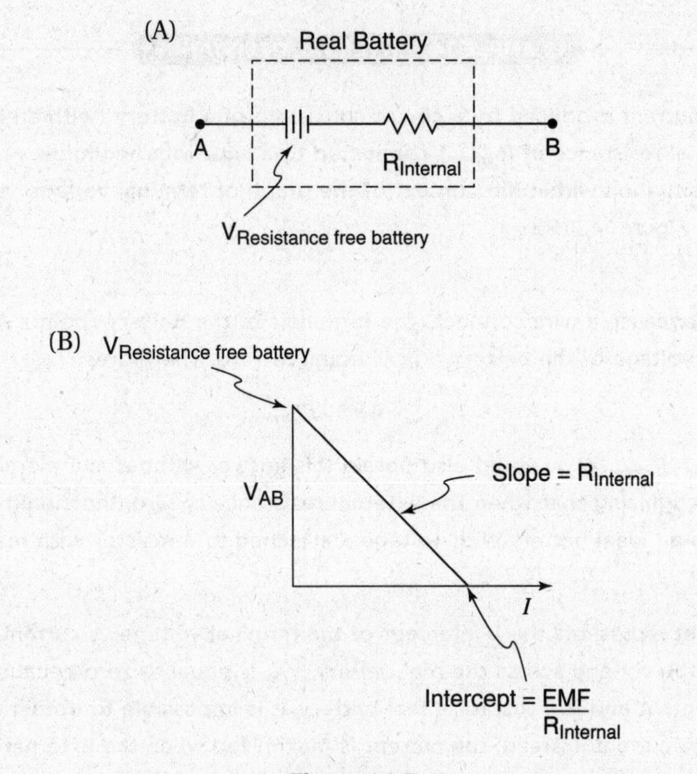

Figure 14.7

Therefore, the voltage across real batteries (the *terminal voltage*) is actually a function of the current flowing through them (a linear function with a negative slope), as shown

$$V_{\text{terminal}} = V_{\text{resistance-free battery}} - IR_{\text{internal}} = \text{EMF} - IR_{\text{internal}}$$

In general, the term EMF (electromotive force) is used interchangeably with the term *voltage*. In this situation, however, EMF generally refers to the voltage of the resistance-free battery, that is, to the total electromotive force of the battery. The terminal voltage refers to the voltage of the terminals of the real battery (the difference in potential between points *A* and *B* in Figure 14.7(A)).

As the current increases, the effect of the internal resistance increases and the actual voltage across the battery decreases. Therefore, the voltage of the EMF is the *y*-intercept of this function, and the internal resistance is the slope of the voltage vs current plot, $R_{\text{internal}} = dV_{\text{terminal}}/dI$.

EXAMPLE 14.7 INTERNAL RESISTANCE

What is the current produced by a circuit consisting of a battery (with an EMF of $\mathcal{E}$ and an internal resistance of R_{internal}) connected to a wire with negligible resistance? Explain this situation within the context of the graph of terminal voltage vs internal resistance in Figure 14.7(B).

SOLUTION

In this case, because a wire connects the terminals of the battery (points A and B), the terminal voltage of the battery, V_{AB}, is equal to zero. Therefore,

$$V_{\text{terminal}} = 0 = \mathcal{E} - IR_{\text{internal}}$$

And thus, $I = \mathcal{E}/R_{\text{internal}}$. (You could also obtain this answer without any memorization by simply recognizing that when the external resistance is zero, the circuit is equivalent to an ideal battery with voltage $\mathcal{E}$ attached to a resistor with resistance R_{internal}).

This current represents the *x*-intercept of the terminal voltage vs current graph, the point where the voltage across the real battery, V_{AB}, is equal to zero because a wire connects points A and B. Thus, for a real battery, it is impossible to obtain an infinitely large current; instead, the current is maximized when the external circuit has zero resistance (assuming, of course, that the circuit consists only of resistors; if another battery is added, it might be possible to increase the current beyond $I = \mathcal{E}/R_{\text{internal}}$).

EXAMPLE 14.8 CONSIDER THE TWO CIRCUITS SHOWN IN FIGURE 14.8.

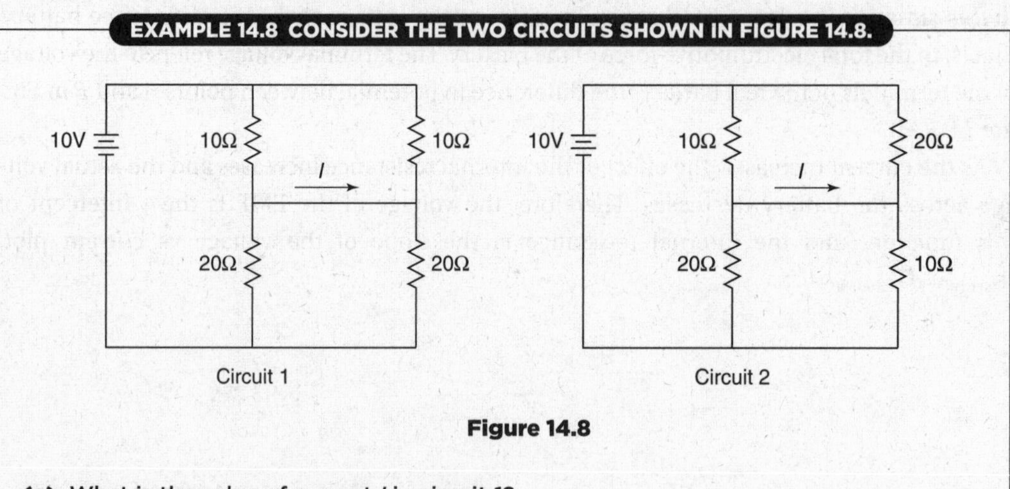

Figure 14.8

(a) What is the value of current *I* in circuit 1?

(b) What is the value of current *I* in circuit 2?

SOLUTION

(a) The current is zero. Because of symmetry, the two branches of the circuit are identical. Therefore, there is no uniquely defined direction for *I*. (If you didn't immediately see this, correct application of Kirchhoff's laws will provide the correct answer.)

(b) The most straightforward way to solve this problem is to use Kirchhoff's laws. Consider the currents shown in Figure 14.9.

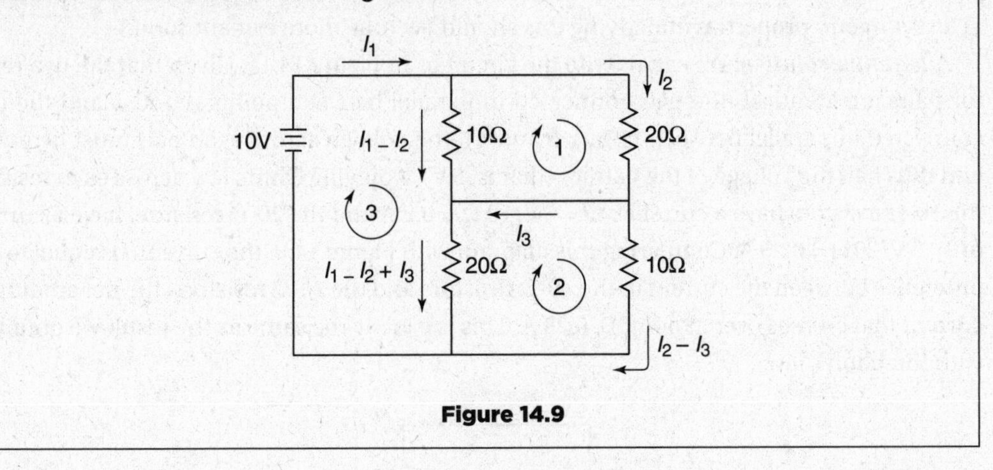

Figure 14.9

Instead of labeling each wire with a different current, we saved some time by drawing as few independent currents as possible. In so doing, we avoided having to write any node equations—all we need are loop equations. (If we introduced more currents, we would still get the correct answer; the only difference would be that we would end up with more node equations and the solution would take longer.)

Because three cuts are needed to eliminate all the loops, there are three independent loop equations:

$$\text{Loop 1: } 10\ \Omega(I_1 - I_2) - 20\ \Omega I_2 = 0$$

$$\text{Loop 2: } 20\ \Omega(I_1 - I_2 + I_3) - 10\ \Omega(I_2 - I_3) = 0$$

$$\text{Loop 3: } 10\ \text{V} - 10\ \Omega(I_1 - I_2) - 20\ \Omega(I_1 - I_2 + I_3) = 0$$

Solving this by successive substitution to eliminate I_1 and I_2 yields

$$I_1 = 0.75\ \text{A}$$

$$I_2 = 0.25\ \text{A}$$

$$I_3 = -0.25\ \text{A}$$

Therefore, the current the question asks for is $I = 0.25$ A (the fact that we calculated a negative value for I_3 reveals that our current points in the wrong direction; the current in Figure 14.10 points in the correct direction).

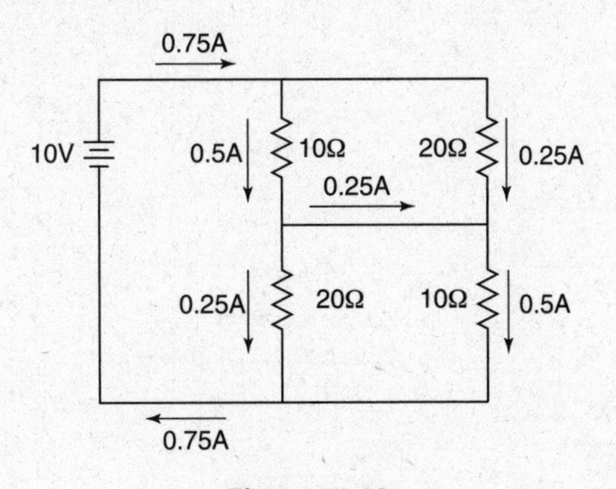

Figure 14.10

Answer check: Consider the currents in the circuit shown in Figure 14.10. It makes sense that the same amount of current flows through both 10-Ω resistors and both 20-Ω resistors. (The symmetry properties underlying this should become more evident soon.)

Alternative solution: We can rewrite the circuit as in Figure 14.11. Given that the two resistor pairs are identical (the pair connected in parallel between points *A* and *B* and the pair connected in parallel between points *B* and *C*), the voltage across each pair must be equal, and thus half the voltage of the battery (that is, 5 V). Applying Ohm's law across each resistor, the 10-Ω resistors have a current of $I = 5\text{ V}/10\ \Omega = 0.5\text{ A}$ and the 20-Ω resistors have a current of $I = 5\text{ V}/20\ \Omega = 0.25\text{ A}$. Comparing this diagram with Figure 14.9, the current I is equal to the difference between the current in the 10-Ω resistors and the 20-Ω resistors, the net amount of current that crosses over at point B, 0.25 A. This is exactly the same as the result we obtained with Kirchhoff's law.

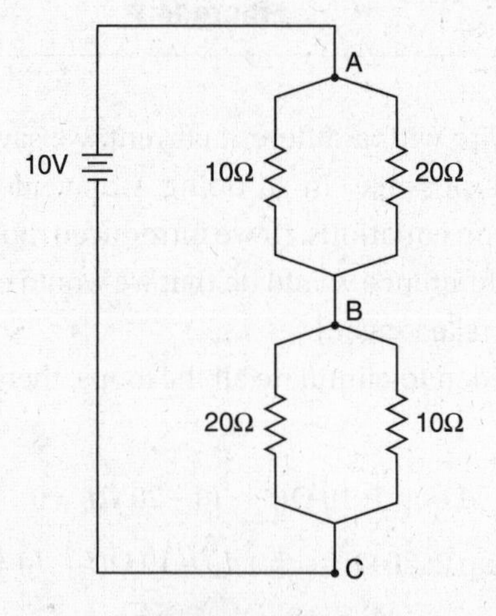

Figure 14.11

CHAPTER SUMMARY

The behavior of electrical circuits can be understood in terms of quantities including current, resistance, and potential differences. For materials that obey Ohm's law, the potential difference across the material is proportional to the resistance. Kirchhoff's laws, which are based on the fundamental principles of charge conservation and the conservative nature of the electrostatic force, provide the basis for analyzing the flow of charge through a circuit.

PRACTICE EXERCISES

Multiple-Choice Questions

1. The equivalent resistance of the circuit in Figure 14.12 is

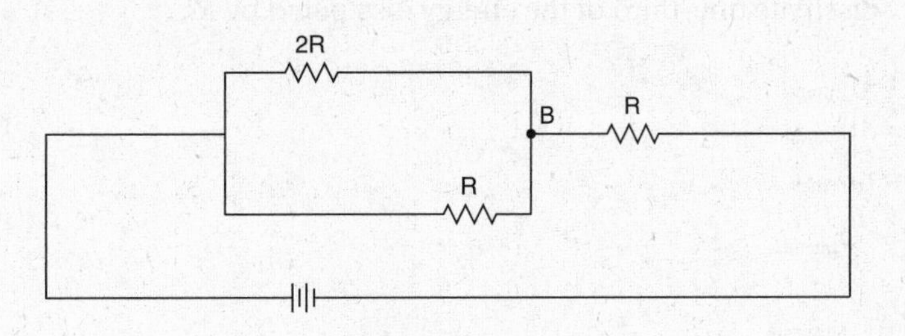

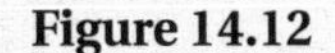

Figure 14.12

(A) $4R$

(B) $\frac{4}{3}R$

(C) $\frac{3}{4}R$

(D) $\frac{5}{3}R$

(E) $2R$

2. Referring to the circuit in question 1, if the voltage across the battery is 10V, what is the potential at point B? (Choose the potential of the negative terminal of the battery to be zero.)

(A) 5 V
(B) 2.5 V
(C) 7.5 V
(D) 6 V
(E) 8 V

3. If a battery with EMF V_0 and internal resistance r_{internal} is connected to a resistor with resistance R, what is the terminal voltage of the battery (the voltage across the two terminals of the actual battery)?

(A) $V_0\left(1-\frac{R}{R+r_{\text{internal}}}\right)$

(B) $V_0\left(1-\frac{r_{\text{internal}}}{R+r_{\text{internal}}}\right)$

(C) $V_0\left(1-\frac{r_{\text{internal}}^2}{R+r_{\text{internal}}}\right)$

(D) $V_0\left(1-\frac{R+r_{\text{internal}}}{r_{\text{internal}}}\right)$

(E) $V_0\left(1+\frac{R}{R+r_{\text{internal}}}\right)$

4. If a battery has no internal resistance, the power dissipated in a resistor attached to the battery is maximized for very small resistances according to the equation $P = IV = V^2/R$. However, this is not the case if the battery has internal resistance. Consider a circuit in which a battery with EMF V_0 and internal resistance $r_{internal}$ is connected to an external resistor whose resistance is R. For what value of R does the internal resistance of the battery dissipate one-third of the energy dissipated by R?

(A) $R = 4r_{internal}$
(B) $R = 3r_{internal}$
(C) $R = r_{internal}$
(D) $R = \frac{1}{3}r_{internal}$
(E) $R = \frac{1}{4}r_{internal}$

5. For the circuit in Figure 14.13, which of the following is true?

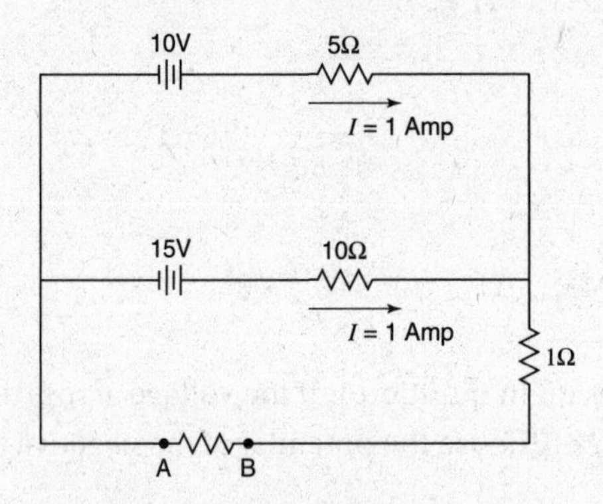

Figure 14.13

(A) $V_B - V_A = -2\text{ V}$
(B) $V_B - V_A = +3\text{ V}$
(C) $V_B - V_A = -3\text{ V}$
(D) $V_B - V_A = -5\text{ V}$
(E) $V_B - V_A = +5\text{ V}$

6. An ammeter and a voltmeter connected in series

(A) could correctly measure current through a circuit element if inserted in series with the element
(B) could correctly measure current through a circuit element if inserted in parallel with the element
(C) could correctly measure the voltage across a circuit element if inserted in parallel with the element
(D) could correctly measure the current across a circuit element if inserted in parallel with the element
(E) none of the above

7. What is the equivalent voltage of the two batteries arranged in series in Figure 14.14?

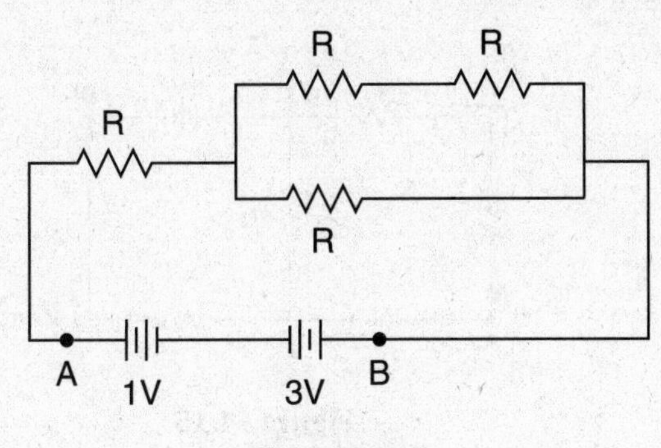

Figure 14.14

(A) 2 V, with point *A* at higher potential
(B) 2 V, with point *B* at higher potential
(C) 4 V, with point *A* at higher potential
(D) 4 V, with point *B* at higher potential
(E) It is not valid to treat the two batteries as a single equivalent battery.

8. What is the equivalent resistance of the circuit in question 7?

(A) R
(B) $\frac{3}{2}R$
(C) $\frac{2}{3}R$
(D) $\frac{3}{5}R$
(E) $\frac{5}{3}R$

9. How much power is dissipated in the circuit discussed in questions 7 and 8?

(A) $P = 3/5R$
(B) $P = 12/5R$
(C) $P = 6/5R$
(D) $P = 20/3R$
(E) none of the above

10. In the circuit shown in Figure 14.15, if the potential at point A is chosen to be zero, what is the potential at point B?

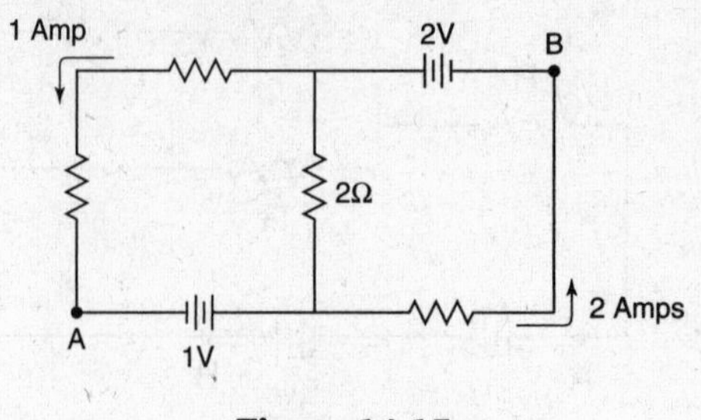

Figure 14.15

(A) +1 V
(B) −1 V
(C) +2 V
(D) −2 V
(E) +3 V

Free-Response Questions

1. (a) In the circuit shown in Figure 14.16, calculate the current through each resistor.
   (b) Indicate the direction of current in each branch of the circuit.
   (c) Calculate the potential difference $V_a - V_b$.

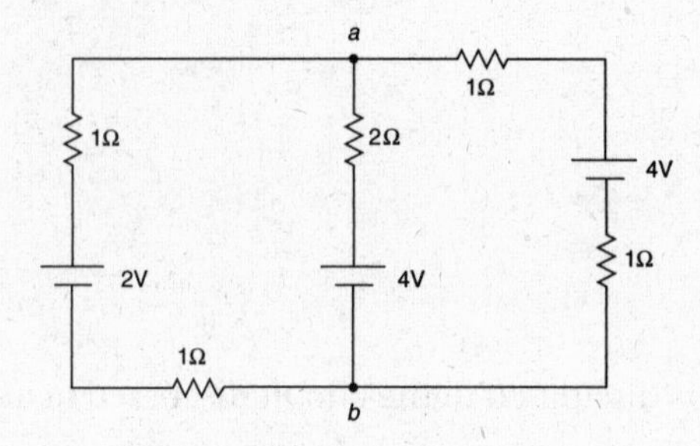

Figure 14.16

2. (a) A battery is connected to the two ends of the resistor network shown in Figure 14.17. What is the current flowing through the 5-Ω resistor?
   (b) Find the equivalent resistance of the network.
   (c) If the battery has a voltage of 9 V, how much current will flow through the network?

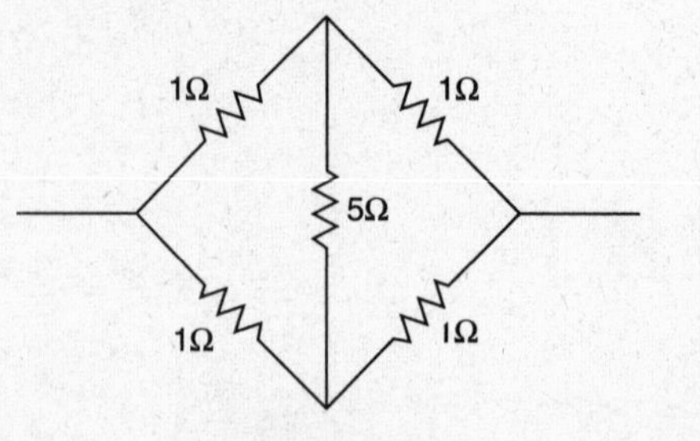

Figure 14.17

ANSWER KEY

1. **D**	4. **B**	7. **B**	10. **A**
2. **D**	5. **B**	8. **E**	
3. **B**	6. **C**	9. **B**	

ANSWERS EXPLAINED

Multiple-Choice

1. **(D)** The parallel combination of the two resistors with resistances R and $2R$ has an equivalent resistance of

$$R_{eq} = \frac{1}{(1/R)+(1/2R)} = \frac{2R}{3}$$

Adding the resistor R, which is in series with this parallel combination, yields an equivalent resistance of $\frac{5}{3}R$.

2. **(D)** In order to calculate the potential at point B, we start at the negative terminal of the battery (where the potential is zero) and trace our way to point B, keeping track of the potential as we go using the same rules we used when writing Kirchhoff's loop rule equations. In fact, this only involves going across one resistor (of resistance R). The equivalent resistance of the circuit is $\frac{5}{3}R$, so the current passing through this resistor, which equals the current flowing through the battery, is $I = V/R = 10V\Big/\frac{5}{3}R = 6V/R$. Because we are going in the direction opposite the direction of the current, this is a *positive* potential jump, with magnitude $V = IR = (6\text{ V}/R)R = 6\text{ V}$. Thus, the potential at point B is 6 V.

3. **(B)** The current flowing through the circuit is the EMF of the battery divided by the total resistance of the circuit (including the resistance within the battery itself):

$$I = \frac{V_0}{R_{circuit}} = \frac{V_0}{R + r_{internal}}$$

The terminal voltage across the battery is then

$$V_{terminal} = V_0 - Ir_{internal} = V_0 - \left(\frac{V_0}{R + r_{internal}}\right)r_{internal} = V_0\left(1 - \frac{r_{internal}}{R + r_{internal}}\right)$$

Alternative solution: If you understand the principle of internal resistance, you can get the correct answer without doing any algebra. The terminal voltage is never greater than the EMF, so choice (E) is incorrect. Also, the terminal voltage cannot be negative in this situation (although the internal resistance can decrease the battery's terminal voltage, it can never change the polarity of the battery), so choice (D) is incorrect. The expression inside the brackets in choice (C) has mixed units (the 1 has no units, and the ratio has units of resistance); the complete expression also has mixed units, and so choice (C) is incorrect. To distinguish between the remaining choices (A) and (B), consider what happens when the external resistance R approaches zero. When this happens, the only resistance in the circuit is internal to the battery. The two terminals of the battery are connected by a resistanceless wire, and thus the two terminals must be at the same potential; that is, the terminal voltage is zero. This eliminates choice (A), which approaches V_0 as R approaches zero. Therefore, choice (B) is the correct answer.

4. **(B)** The most useful formula for power here is $P = I^2R$. Because the internal and external resistors are in series, the same current flows through each, so the ratio of the power dissipated in each resistor is equal to the ratio of their resistances. For the internal resistance to dissipate one-third of the energy dissipated by R, its resistance must be one-third of the resistance of R.

5. **(B)** This problem involves Kirchhoff's loop rule. First, Kirchhoff's node equation indicates that the current flowing down the 1-Ω resistor is 2 A. Then, writing a Kirchhoff's loop equation containing points A and B in either possible loop yields the correct answer.

   For example, we will use the outer loop and move clockwise. Starting at point A and summing potential jumps as we move from point A to point B yields

$$V_B = V_A + 10\text{V} - (1\text{A})(5\Omega) - (2\text{A})(1\Omega) = 3\text{V}$$

   Therefore,

$$V_B - V_A = +3\text{V}$$

6. **(C)** Recall the way voltmeters and ammeters work.

   - Voltmeters are attached in parallel to the element they are measuring. They have a very high resistance so that they do not draw too much current from the circuit and thus do not perturb the circuit they are measuring.
   - Ammeters are attached in series to the element they are measuring. They have a very small resistance so that they produce a negligible voltage drop in the circuit and thus do not perturb the circuit they are measuring.

   A voltmeter attached in series to an ammeter has a large resistance (the small resistance of the ammeter plus the large resistance of the voltmeter). Therefore, inserting a voltmeter-ammeter combination *in series* into the circuit inserts a large resistance into the circuit, significantly disturbing the circuit (in particular, it reduces the current dramatically) and making any sort of simple measurement impossible. Alternatively, what will happen if the voltmeter-ammeter is connected to the circuit *in parallel*? The combination has a large equivalent resistance like a voltmeter, so it does not draw significant current from the circuit and thus doesn't disturb the circuit. Since the ammeter has a very small resistance, it acts like a piece of wire and doesn't affect the voltmeter's ability to measure the voltage, making choice (C) correct. (The ammeter measures the very tiny current running in series through the ammeter and the voltmeter.)

7. **(B)** Summing the potential jumps indicates that point B always has a potential 2 V higher than point A. Therefore, the two batteries together have the same effect as one 2-V battery with its positive terminal at point B and its negative terminal at point A.

8. **(E)** The two resistors in series on the top leg of the circuit are equivalent to a single resistor of resistance $2R$. The parallel combination of this resistance with the resistor R below it has an equivalent resistance of

$$R_{\text{eq, parallel resistors}} = \frac{1}{(1/R)+(1/2R)} = \frac{1}{(3/2R)} = \frac{2}{3}R$$

   Because this parallel array is in series with the resistor R, the equivalent resistance of the entire circuit is $R_{\text{eq}} = R + \frac{2}{3}R = \frac{5}{3}R$.

9. **(B)** $P = V_{eq}^2/R_{eq} = 2^2\Big/\left(\frac{5}{3}R\right) = 12/5R$. This problem illustrates how simplifying a circuit and calculating equivalent voltages and resistances simplifies calculating quantities such as power.

10. **(A)** We start at point A and work our way to point B, summing potential jumps as we go. The only path to take from A to B about which we have enough information to calculate the potential jumps is from A to C to D to B, as in Figure 14.18. The potential jump from A to C is +1 V (moving from the negative terminal to the positive terminal of a battery is a positive potential jump). Using Kirchhoff's node equation at point C or D indicates that a current of 1 A moves downward through the central resistor, so that the potential jump from C to D is +2 V (moving across a resistor in the direction opposite the current is a positive potential jump). Finally, moving from D to B yields a potential jump of −2.0 V (moving from the positive terminal of a battery to the negative terminal is a negative potential jump). Summing these potential changes, we find

$$V_{AB} = V_{ACDB} = V_{AC} + V_{CD} + V_{DB} = (+1\text{V}) + (+2\text{V}) + (-2\text{V}) = +1\text{V}$$

Figure 14.18

Free-Response

1. (a) We label the currents in the branches as shown in Figure 14.19. Note that our choice of the arrow directions is arbitrary. The application of Kirchhoff's laws will give us the correct directions.

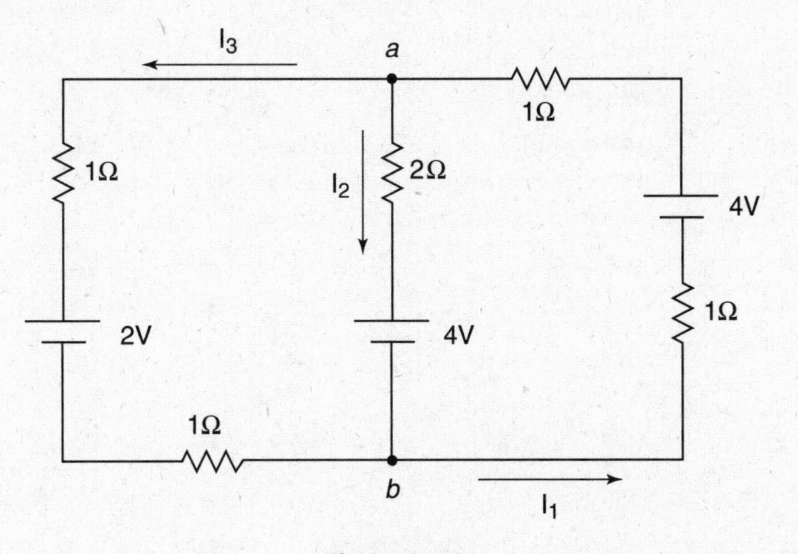

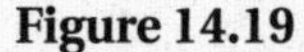

Figure 14.19

The node equation applied at either point a or point b yields $I_1 = I_2 + I_3$. The loop rule applied to the left loop yields $4 + 2I_2 - 2I_3 - 2 = 0$ if we traverse the loop counterclockwise. The loop rule applied to the right hand loop yields $-4 - 2I_2 - 2I_1 + 4 = 0$. From these equations we determine that

$$I_1 = \frac{1}{3}\text{A}, I_2 = -\frac{1}{3}\text{A}, I_3 = \frac{2}{3}\text{A}$$

(b) The signs of the currents we found in part (a) yield the directions of the current. Thus, I_1 and I_3 flow as shown in Figure 14.19, while I_2 flows in the opposite direction to that shown in the figure.

(c) Starting at point b, the potential increases by 4 V as we traverse the battery and then drops by $2|I_2|$ as we traverse the resistor. Thus, $V_a - V_b = 4 - 2|I_2| = 3.33$ V.

2. (a) Given the symmetry of the network, there will be no current flowing through the 5-Ω resistor. The current splits equally at the first node, which means that there will be no potential difference across the 5-Ω resistor.

(b) From the answer to part (a) we see that the branch with the 5-Ω resistor has no current flowing through it and no potential difference. It is as if we collapsed the two middle nodes together, leaving us with a parallel combination of two 1-Ω resistors on the left and a similar parallel combination on the right. Each of these parallel combinations has an equivalent resistance of $\frac{(1)(1)}{1+1} = 0.5\,\Omega$. These parallel combinations are in series with each other; thus, the equivalent resistance of the entire network is 2(0.5) = 1 Ω.

(c) With an equivalent resistance of 1 Ω, the current delivered by a 9-V battery will be 9 A.

15 Capacitors

→ DEFINITION OF CAPACITORS AND CAPACITANCE
→ CALCULATION OF CAPACITANCE
→ CAPACITORS IN SERIES AND PARALLEL
→ ENERGY STORED IN CAPACITORS
→ CAPACITORS WITH DIELECTRICS

QUALITATIVE INTRODUCTION TO CAPACITORS

Suppose two metal plates are connected to a battery as shown in Figure 15.1. At first, the battery deposits positive charge on the left plate of the capacitor.[1] Because like charges repel, the positive charge on the left plate repels positive charge on the right plate of the capacitor, causing positive charge on the right plate to flow back into the battery. Thus, a current is established that results in the accumulation of positive charge on the left plate of the capacitor and the removal of positive charge from the right plate of the capacitor (leaving it negatively charged). Will this continue forever? No. Eventually, enough positive charge builds up on the left plate so that the battery does not have enough voltage to deposit any more charge (which requires energy because like charges repel), and the current ceases. In slightly more sophisticated terms, as charge is deposited on the capacitor, a potential difference builds up across the capacitor. When the potential difference across the capacitor equals the potential difference across the battery, the battery is unable to force any more positive charge onto the capacitor and the current stops.

Capacitors store electrical energy.

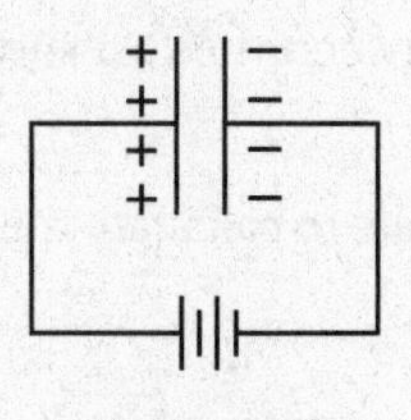

Figure 15.1

Suppose you disconnect the battery. You now have a charged capacitor—the plates have opposite charges, which creates an electric field and a potential difference (voltage) across the capacitor. If a conductor is connected between the two plates (e.g., a wire or an unlucky person), the capacitor will push charge through the conductor, demonstrating that charged capacitors store energy.

Capacitors can store energy (deposited slowly while being charged up by a battery) and then quickly give up that energy (e.g., to power a camera flash or a portable defibrillator shock). The usefulness of a capacitor is that it can achieve a much greater power than the

[1]Note that we're using the conventional-current description here (in reality, the electrons are actually moving).

battery used to charge it (if the resistance present during discharging is less than the resistance present during charging).

CAPACITANCE

As discussed above, imposing a potential difference (voltage) across a capacitor causes charge separation. As we will show many times in this chapter, voltage is proportional to charge accumulation; thus, a capacitor's ability to store charge can be quantified by the ratio of charge accumulation to voltage, defined as capacitance.

TIP

Capacitance is a property of the device. It does not depend on the charge placed on the device or the voltage across the device.

$$C = \frac{Q}{V}$$

Definition of capacitance

Like resistance, capacitance is a physical property of the capacitor (and thus is independent to the capacitor's particular charge or voltage at any given time).

Capacitance is measured in farads (F), defined as 1 farad = 1 coulomb/volt. Like coulombs, farads are very large, so most real-life capacitors are in the range of microfarads (10^{-6}F), nanofarads (10^{-9}F), or picofarads (10^{-12}F), though capacitors with values of several farads are available.

Note that when a capacitor is charged, its two plates have opposite charges. When dealing with capacitors, Q is generally taken to represent the magnitude of the charge on either plate of the capacitor. (The net charge of both plates of a capacitor is generally zero.) Capacitance is always positive.

Calculating Capacitance Based on Geometry

When given the dimensions of a capacitor, you should be able to calculate its capacitance.

STEP 1 Assume a charge Q is on your capacitor (this generally means that there is $+Q$ on one surface [or plate] of the capacitor and $-Q$ on the other surface of the capacitor).

STEP 2 Calculate the electric field between the two surfaces of the capacitor. Gauss's law is generally the best way to do this.

STEP 3 Given this electric field, integrate to calculate the voltage between the plates using the integral

$$V_{\text{capacitor}} = -\int_{\substack{\text{path between}\\ \text{the plates}}} \mathbf{E} \cdot d\mathbf{R}$$

Because the plates are metal, they each are at a constant potential. Therefore, any integration path from any point on one plate to any point on the other plate yields the correct voltage. The key, then, is finding an integration path that is easy to evaluate, which generally involves choosing a path that lies parallel to the electric field (so that the dot product simplifies nicely). Another factor that makes our lives easier is that we don't have to worry about the sign of the voltage because capacitance is the ratio of the magnitudes of charge and voltage. Therefore, what we're actually interested in is

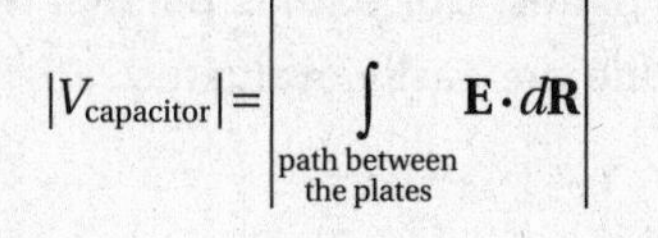

$$|V_{\text{capacitor}}| = \left| \int_{\substack{\text{path between}\\ \text{the plates}}} \mathbf{E} \cdot d\mathbf{R} \right|$$

STEP 4 Now that we know Q and V, we can compute the capacitance by applying the definition $C = Q/V$.

EXAMPLE 15.1 CALCULATING THE CAPACITANCE OF A PARALLEL PLATE CAPACITOR

Calculate the capacitance of a parallel plate capacitor with plates of area A separated by a distance d. Assume that the dimensions of the plates are much greater than the separation d so that "edge effects" can be ignored.

Note: This solution assumes that the plates are separated by a vacuum. Later in the chapter, we will learn how the solution to this problem is modified if the plates are separated by a dielectric.

SOLUTION

STEP 1 Let's suppose there's $+Q$ on one plate and $-Q$ on the other plate.

Capacitors always have plates with equal and opposite charges.

STEP 2 We must calculate the electric field between the plates. Recall that the electric field due to an infinitely large charged plate with charge density σ is

$$E = \frac{\sigma}{2\varepsilon_0}$$

Can we apply this equation here? As discussed above, all we need is one line integral where **E** is well defined and parallel to $d\mathbf{R}$. In this case, the best choice of integration path is a line between the two plates at the center of the capacitor (because the electric field is easily calculated along the path and parallel to the path). Because this is far from the edges, it is a reasonable approximation to treat the plates of the capacitor as infinitely large.

What is the field along this integration path? Each plate has a charge magnitude Q and an area A. Therefore, the magnitude of the charge density on each plate is as follows (of course, the charges on the different plates have opposite signs):

$$\sigma = \frac{Q}{A}$$

Now, inside the capacitor, points along the integration path are in between two oppositely charged plates. Because of the geometry, the fields add, so that the net field is

$$E = 2\left(\frac{\sigma}{2\varepsilon_0}\right) = \frac{Q}{\varepsilon_0 A}$$

STEP 3 We must evaluate the line integral

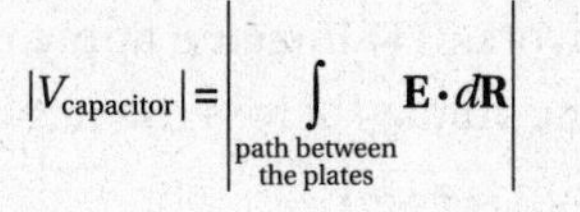

$$|V_{\text{capacitor}}| = \left| \int_{\substack{\text{path between}\\ \text{the plates}}} \mathbf{E} \cdot d\mathbf{R} \right|$$

along a path connecting the plates. Our path is parallel to the electric field, so the dot product and absolute value are easily evaluated:

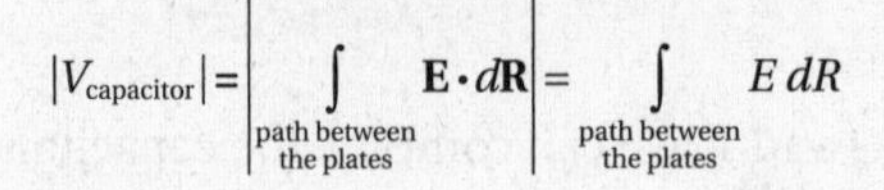

$$|V_{\text{capacitor}}| = \left| \int_{\substack{\text{path between}\\ \text{the plates}}} \mathbf{E} \cdot d\mathbf{R} \right| = \int_{\substack{\text{path between}\\ \text{the plates}}} E\,dR$$

Along this path, E is constant, so that we can put it in front of the integral. This leaves only dR, and the integral of dR is simply the distance between the plates, d:

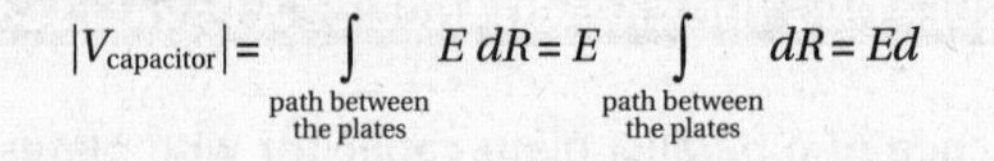

$$|V_{\text{capacitor}}| = \int_{\substack{\text{path between}\\ \text{the plates}}} E\,dR = E \int_{\substack{\text{path between}\\ \text{the plates}}} dR = Ed$$

$$|V| = Ed \quad \text{(valid only for parallel plate capacitors)}$$

This is an important result: The electric field inside a parallel plate capacitor is constant, so the voltage of a parallel plate capacitor is simply $|V| = |E|d$ (the positively charged plate being at higher potential).

STEP 4 Plugging our values of Q and V into the equation $C = Q/V$, we obtain

$$C = \frac{Q}{V} = \frac{\sigma A}{(\sigma/\varepsilon_0)d} = \frac{\varepsilon_0 A}{d} \quad \text{Capacitance of parallel plate vacuum capacitor}$$

1. *Why capacitance increases with increasing plate area:* The voltage of a parallel plate capacitor is determined only by the distance between the plates and the plate charge density (as evident from the equation $V = Ed$, where E is proportional to the charge density). Therefore, if we fix voltage and distance, the charge density will also be fixed. If we then increase the plate area, charge ($Q = \sigma A$) will also increase without affecting the voltage. Therefore, according to the definition $C = Q/V$, increasing area increases capacitance.
2. *Why plate separation distance* d *is inversely related to capacitance:* Consider charging a capacitor and disconnecting it from its battery, so that the charge on the plates remains fixed. Moving the plates closer together does not change the electric field strength (which depends only on the charge density of the two plates and thus remains constant). However, decreasing d causes a decrease in voltage (according to the equation $V = Ed$). Thus, according to the definition $C = Q/V$, a decrease in V with no change in Q causes an increase in capacitance.

Combinations of Capacitors in Series

Consider two capacitors in series as in Figure 15.2. How do they behave? As in our treatment of networks of resistors, we can understand the behavior of combinations of capacitors by calculating the capacitance of the entire group (called the *equivalent capacitance*). This can be done using a method similar to the method we used in Chapter 14 to calculate the resistance of parallel and series resistor networks: (1) Imagine applying a charge Q across the capacitor network, (2) calculate the resulting voltage across the network, and (3) calculate the equivalent capacitance using the equation $C = Q/V$.

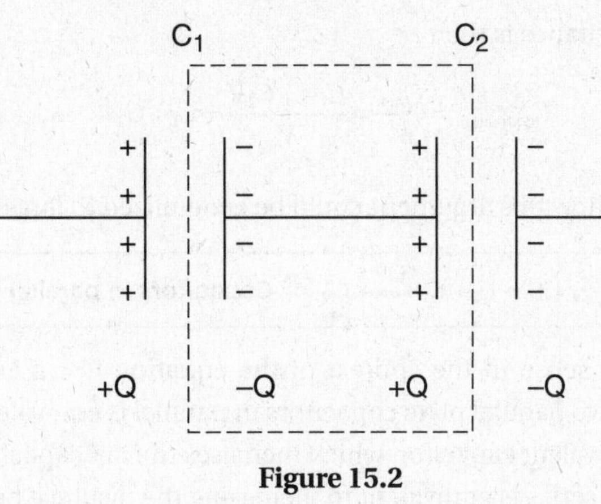

Figure 15.2

The only trick here is to realize that each capacitor in a series network stores exactly the *same* charge. Why? Consider the piece of metal within the dotted box in the figure. Because of conservation of charge, the net charge on this piece of metal is zero; so if one plate contains $+Q$, the other must contain $-Q$. Thus, if the array of capacitors in series is charged to a charge Q, each individual capacitor also contains Q.

Given this insight, calculation of voltage is straightforward. The voltage across the first capacitor is $V_1 = Q/C_1$, and the voltage across the second capacitor is $V_2 = Q/C_2$. The voltage across the two capacitors is then

$$V_{\text{net}} = V_1 + V_2 = \frac{Q}{C_1} + \frac{Q}{C_2}$$

Applying the definition of capacitance,

$$C_{\text{net}} = \frac{Q}{V_{\text{net}}} = \frac{Q}{(Q/C_1)+(Q/C_2)} = \frac{1}{(1/C_1)+(1/C_2)}$$

Similar to calculating parallel resistance, the result is easier to write after it is inverted. (As with resistors, the result is easily generalized to n capacitors in series.)

$$\frac{1}{C_{\text{equivalent}}} = \frac{1}{C_1} + \frac{1}{C_2} + \frac{1}{C_3} + \cdots + \frac{1}{C_N}$$

Capacitors in series

Combination of Capacitors in Parallel

Unlike capacitors in series, capacitors in parallel (like any elements in parallel) are characterized by the same potential differences. Thus, unlike previous derivations, it is easier to start by first assigning a voltage to the network and then calculating the charge on each of the capacitors before finding the ratio of Q/V to calculate capacitance.

When we apply a voltage V, the plates of the first capacitor have charge $Q_1 = C_1V$ and the plates of the second capacitor have charge $Q_2 = C_2V$. Thus, the net charge is

$$Q_{\text{net}} = Q_1 + Q_2 = C_1V + C_2V$$

And the equivalent capacitance is then

$$C_{\text{equivalent}} = \frac{Q_{\text{net}}}{V} = \frac{C_1V + C_2V}{V} = C_1 + C_2$$

It's not difficult to see how this argument could be generalized to larger parallel networks:

$$C_{\text{equivalent}} = C_1 + C_2 + C_3 + \cdots + C_N \qquad \textbf{Capacitors in parallel}$$

These formulas make sense in the context of the equation $C = \varepsilon_0 A/d$ for parallel plate capacitors. Connecting two parallel plate capacitors in parallel is equivalent to increasing the plate area of a larger equivalent capacitor, which increases the net capacitance. Alternatively, connecting capacitors in series is equivalent to increasing the distance between the plates of a larger equivalent capacitor.

COMPARISON WITH RESISTORS IN SERIES AND IN PARALLEL

These results are the reverse of those for combining resistors in series (which add) and in parallel (whose reciprocals add). This makes sense because unlike capacitance, which is proportional to area/d, resistance is *inversely* proportional to this ratio. For example, as with capacitors, combining resistors in series is analogous to increasing d, but because resistance is proportional to d, this causes an increase in the equivalent resistance. Alternatively, combining resistors in parallel is analogous to increasing the area, but because resistance is inversely proportional to area, this causes a decrease in the equivalent resistance.

Calculating the Equivalent Capacitance of Complicated Capacitor Arrays

The basic strategy used here is similar to the one we used to calculate the equivalent resistance of complex arrays of resistors:

STEP 1 Identify a small, isolated group of capacitors in series or parallel.

STEP 2 Calculate their equivalent capacitance using one of the above equations.

STEP 3 Replace the selected group by a single capacitor whose capacitance is equal to the equivalent capacitance just calculated.

Repeat steps 1 through 3 until only one capacitor is left whose capacitance is equal to the equivalent capacitance of the entire array.

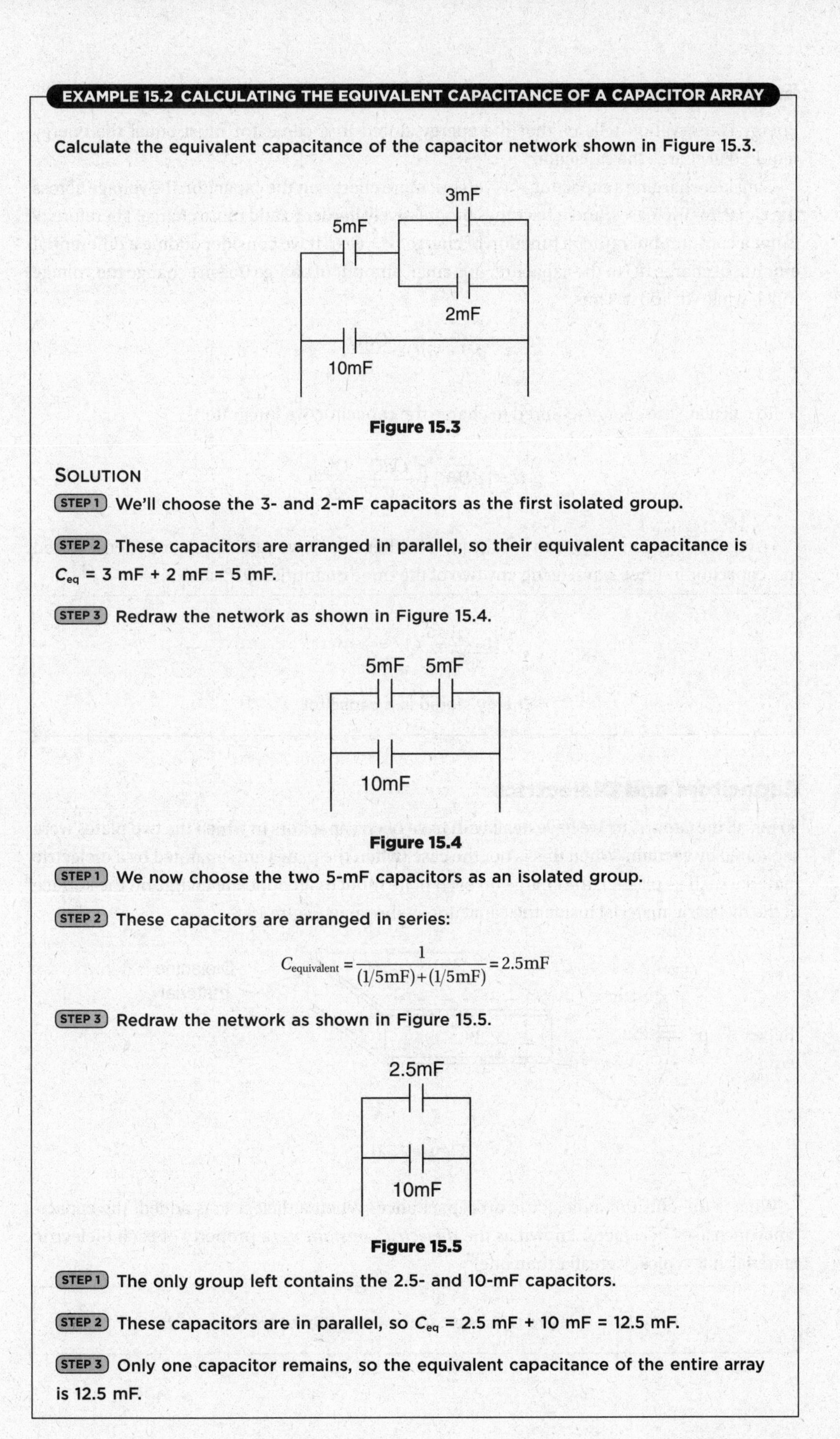

EXAMPLE 15.2 CALCULATING THE EQUIVALENT CAPACITANCE OF A CAPACITOR ARRAY

Calculate the equivalent capacitance of the capacitor network shown in Figure 15.3.

3mF
5mF
2mF
10mF

Figure 15.3

SOLUTION

STEP 1 **We'll choose the 3- and 2-mF capacitors as the first isolated group.**

STEP 2 **These capacitors are arranged in parallel, so their equivalent capacitance is C_{eq} = 3 mF + 2 mF = 5 mF.**

STEP 3 **Redraw the network as shown in Figure 15.4.**

5mF 5mF
10mF

Figure 15.4

STEP 1 **We now choose the two 5-mF capacitors as an isolated group.**

STEP 2 **These capacitors are arranged in series:**

$$C_{\text{equivalent}} = \frac{1}{(1/5\text{mF}) + (1/5\text{mF})} = 2.5\text{mF}$$

STEP 3 **Redraw the network as shown in Figure 15.5.**

2.5mF
10mF

Figure 15.5

STEP 1 **The only group left contains the 2.5- and 10-mF capacitors.**

STEP 2 **These capacitors are in parallel, so C_{eq} = 2.5 mF + 10 mF = 12.5 mF.**

STEP 3 **Only one capacitor remains, so the equivalent capacitance of the entire array is 12.5 mF.**

Energy Stored in Capacitors

Energy conservation tells us that the energy stored in a capacitor must equal the energy required to charge the capacitor.

Consider charging a capacitor—as you put more charge on the capacitor, the voltage across the capacitor increases, and it becomes progressively harder to add more charge. Therefore, V is not a constant but rather a function of charge, $V = Q/C$. If we consider adding a differential amount of charge dQ to the capacitor, this small amount of charge doesn't change the voltage much while we add it. Thus,

$$dU = VdQ = \frac{QdQ}{C}$$

To calculate the energy required to charge the capacitor, we integrate:

$$U = \int dU = \int_0^{Q_{\text{final}}} \frac{QdQ}{C} = \frac{Q_{\text{final}}^2}{2C}$$

Using the equation $C = Q/V$ to substitute for either C or Q, we can express the energy stored in a capacitor in three ways, using any two of the three quantities Q, C, and V.

$$U = \frac{Q^2}{2C} = \frac{1}{2}CV^2 = \frac{1}{2}QV$$

Energy stored in a capacitor

Capacitors and Dielectrics

So far, all the capacitors we have dealt with have been capacitors in which the two plates were separated by vacuum. When this is not the case (when the plates are separated by a dielectric material such as plastic), the charge on each plate induces an opposite charge on the surface of the dielectric material inside the capacitor as shown in Figure 15.6.

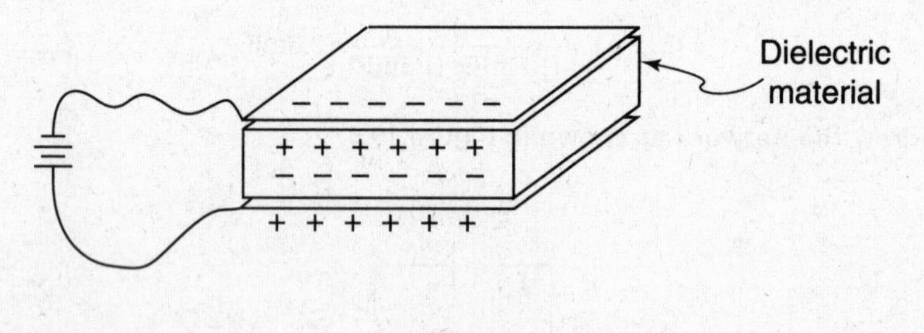

Figure 15.6

What is the effect of a dielectric on capacitance? When a dielectric is added, the capacitance increases by a factor known as the *dielectric constant* κ_D (a property of each dielectric material; it is typically greater than one):

TIP

Adding a dielectric material increases the capacitance.

$C = \kappa_D C_0$ **(where C_0 is the capacitance without the dielectric)**

It makes sense that adding a dielectric increases the capacitance, as illustrated by the following thought experiment. Suppose we charge a parallel plate vacuum capacitor and disconnect it from its power source. When we add a dielectric, charge is induced on the dielectric as shown in Figure 15.6. This charge on the dielectric *reduces* the electric field inside the capacitor, which in turn *decreases* the voltage of the capacitor (recall that for a parallel plate capacitor, $V = Ed$). Because addition of the dielectric decreases the voltage without affecting the amount of charge on the plates, the equation $C = Q/V$ indicates that it *increases* the capacitance. In general, you can imagine the charges on the dielectric partially "canceling" the charges on the plates, decreasing the field within the capacitor.

Calculating the capacitance of capacitors containing a dielectric is straightforward: First calculate the capacitance without the dielectric and then multiply the result by the dielectric constant, κ_D.

Important: Gauss's law (in the form that we have been using here) does not hold in the case of dielectrics. Therefore, it is imperative that we first calculate the capacitance for a vacuum case using Gauss's law and then modify the result to account for the dielectric by multiplying by the dielectric constant.

EXAMPLE 15.3 CALCULATING THE CAPACITANCE OF CAPACITORS WITH DIELECTRICS

Calculate the capacitance of a parallel plate capacitor with plates of area A separated by a distance d. Assume that the dimensions of the plates are much greater than the separation d. The volume between the plates is filled with a dielectric material of dielectric constant κ_D.

SOLUTION

We already derived the capacitance of a parallel plate vacuum capacitor in Example 15.1:

$$C = \frac{Q}{V} = \frac{\varepsilon_0 A}{d}$$

Capacitance of parallel plate vacuum capacitor

Therefore, the capacitance of a parallel plate capacitor with a dielectric is simply

$$C = \kappa_D C_0 = \frac{\kappa_D \varepsilon_0 A}{d}$$

Parallel plate capacitor with a dielectric

Note that, although we generally do not think of vacuum as a dielectric material, it has a dielectric constant of 1 (by definition). Therefore, the equation we wrote previously for a parallel plate vacuum capacitor can be thought of as a special case of the equation above.

CHAPTER SUMMARY

Capacitors are devices for storing electrical energy and consist of two conductors separated by an insulator or vacuum. A capacitor stores energy when the two conductors are charged with equal but opposite charges. The capacitance of a capacitor is the ratio of the charge to the potential difference and depends only on the geometry of the device. Arrays of capacitors can be analyzed with a strategy similar to the one used for resistors, bearing in mind that the equivalent capacitance of capacitors in series is analogous to resistors in parallel and vice versa. When the vacuum between the two conductors in a capacitor is replaced by an insulating material, the capacitance will increase by a factor called the dielectric constant of the material.

PRACTICE EXERCISES

Multiple-Choice Questions

1. The dielectric slab filling the space between the plates of a charged, isolated capacitor is removed. The energy stored in the capacitor

   (A) remains constant
   (B) increases by a factor of κ_D
   (C) decreases by a factor of κ_D
   (D) increases by a factor of κ_D^2
   (E) decreases by a factor of κ_D^2

2. Which one of the networks of four identical capacitors in Figure 15.7 can store the greatest amount of energy when connected to a given battery?

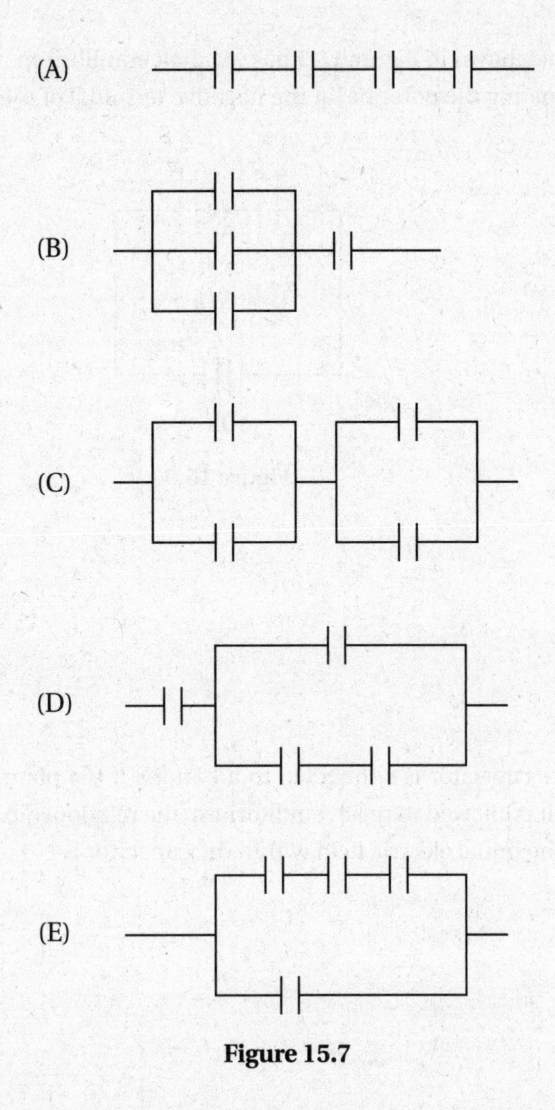

Figure 15.7

3. A sphere is suspended by an insulating cable between the plates of a charged capacitor, resulting in the electric field lines as shown in Figure 15.8. Which of the following is true?

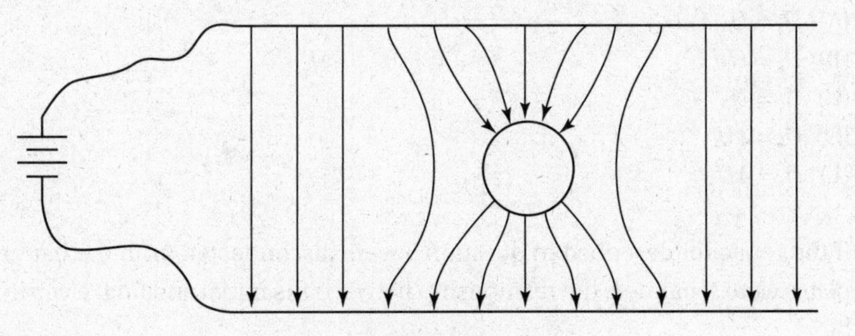

Figure 15.8

(A) The sphere is an uncharged insulator.
(B) The sphere is a conductor with zero net charge.
(C) The sphere is a positively charged insulator.
(D) The sphere is a negatively charged insulator.
(E) The sphere is a positively charged conductor.

4. After the circuit shown in Figure 15.9 has reached equilibrium, what is the potential at point A, designating the potential at the negative terminal of the battery to be zero?

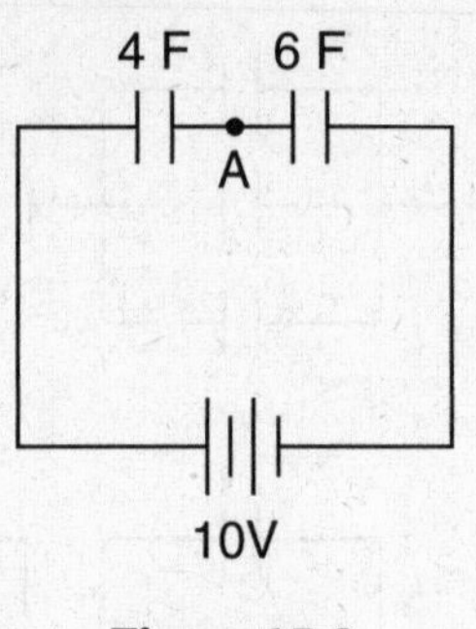

Figure 15.9

(A) 4 V
(B) 6 V
(C) 6.66 V
(D) 3.33 V
(E) 5 V

5. A parallel plate capacitor is connected to a battery. If the plate separation is doubled, after the circuit is allowed to reach equilibrium, the relationship between the final electric field and the initial electric field within the capacitor is

(A) $E_F = E_0/4$
(B) $E_F = E_0/2$
(C) $E_F = E_0$
(D) $E_F = 2E_0$
(E) $E_F = 4E_0$

6. In the situation described in question 5, which of the following is a valid comparison of the energy stored in the capacitor before and after the plate separation is doubled?

(A) $U_F = U_0/4$
(B) $U_F = U_0/2$
(C) $U_F = U_0$
(D) $U_F = 2U_0$
(E) $U_F = 4U_0$

7. If the capacitor described in question 5 were disconnected from the battery before the plates were separated, the relationship between the initial and final electric fields would be

(A) $E_F = E_0/4$
(B) $E_F = E_0/2$
(C) $E_F = E_0$
(D) $E_F = 2E_0$
(E) $E_F = 4E_0$

8. Again, supposing that the capacitor described in question 5 were disconnected from the battery before the plates were separated, the relationship between the initial and final energies would be

(A) $U_F = U_0/4$
(B) $U_F = U_0/2$
(C) $U_F = U_0$
(D) $U_F = 2U_0$
(E) $U_F = 4U_0$

9. When the circuit shown in Figure 15.10 has reached a steady state, what is the charge on capacitor 1 (which has a capacitance C_1, as shown)?

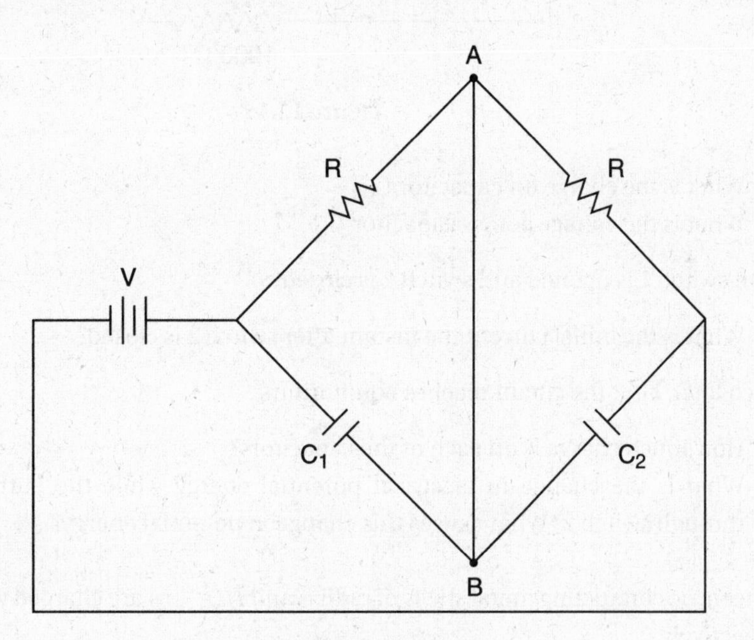

Figure 15.10

(A) $C_1^2V/(C_1+C_2)$
(B) $C_1C_2V/(C_1 + C_2)$
(C) $C_2^2V/(C_1+C_2)$
(D) C_1V
(E) $\frac{1}{2}C_1V$

10. For a parallel plate vacuum capacitor, the electric field within the capacitor can be calculated from each of the following combinations of variables *except*

(A) the magnitude of the charge density on the plates, σ
(B) the voltage of the capacitor V and the plate separation distance d
(C) the charge on the plates Q, the capacitance C, and the plate separation distance d
(D) the magnitude of charge on the plates Q and the area of the plates A
(E) the energy stored in the capacitor U, the voltage across the capacitor V, and capacitance C

Free-Response Questions

1. Consider the circuit shown in Figure 15.11. A long time after switch 1 is closed (and switch 2 is opened),

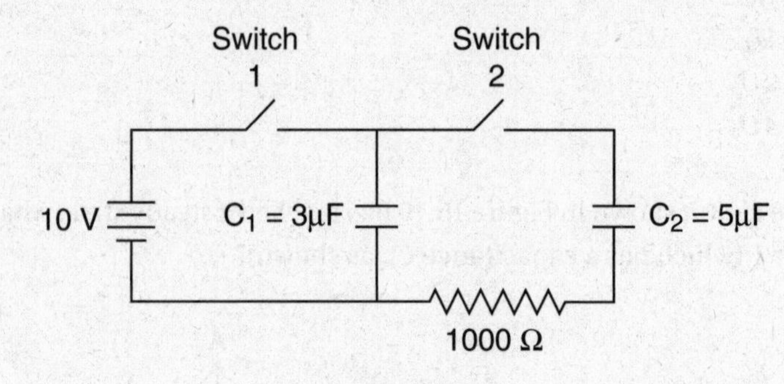

Figure 15.11

(a) What is the charge on capacitor C_1?
(b) What is the voltage across capacitor C_1?

Then switch 1 is opened and switch 2 is closed.

(c) What is the initial current the instant after switch 2 is closed?

Much later, after the circuit reaches equilibrium,

(d) How much charge is on each of the capacitors?
(e) What is the change in electrical potential energy while the current is running through switch 2? What causes this change in potential energy?

2. Concentric conducting metal shells of radii r_1 and r_2 ($r_1 < r_2$) are charged with $-Q$ and $+Q$, respectively.

(a) Calculate the electric field in the following regions:
    (i) $r > r_2$
    (ii) $r_2 > r > r_1$
    (iii) $r < r_1$
(b) Calculate the potential at
    (i) $r = r_2$
    (ii) $r = r_1$
(c) What is the capacitance of this spherical capacitor?

3. A parallel plate capacitor with thick metal plates is shown in Figure 15.12. The plates have opposite charge density, which is distributed entirely on the inside faces of the plates. Express all answers in terms of σ and κ_D.

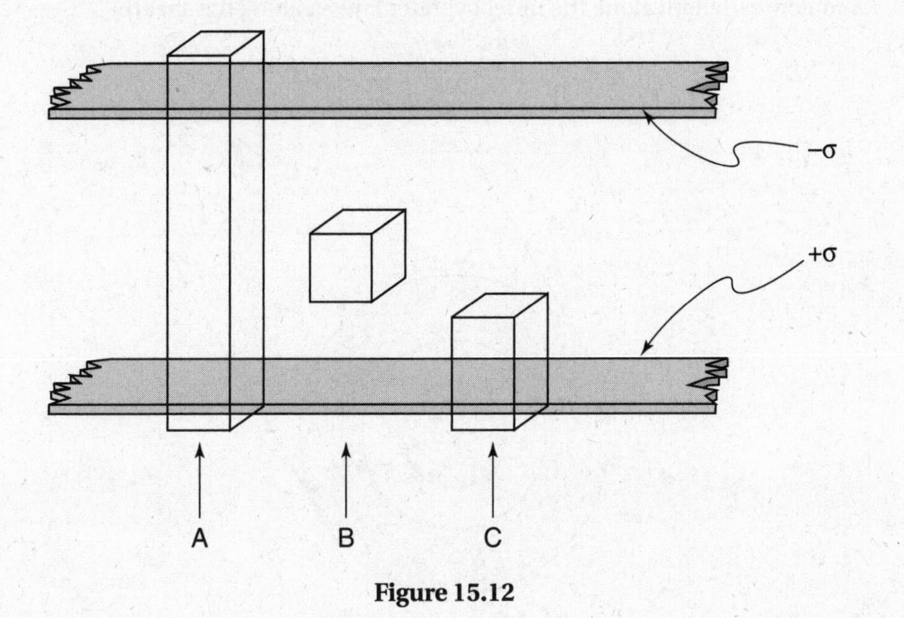

Figure 15.12

(a) Sketch the electric field lines and equipotential curves in the region between the plates of the capacitor.

(b) Three gaussian surfaces are shown in Figure 15.12 (all are boxes whose faces parallel to the charged sheets have area *A*). In terms of the electric field *E*, what is the flux through each surface? Which surface can be used to calculate *E* from σ?

(c) Use this gaussian surface to calculate *E*.

Then a dielectric material (of dielectric constant κ_D) is added, completely filling the space between the plates. The charge on the plates induces opposite charge on the adjacent dielectric material as previously discussed.

(d) Given that the capacitor is isolated (i.e., not connected to a battery), what is the new electric field strength?

4. Consider the coaxial cable in Figure 15.13, which contains a central cylindrical core of metal (radius a) surrounded by a cylindrical sheath of metal (inner radius b, outer radius c). Charge is moved from the sheath to the inner cylinder, so that charge density is then λ coulombs/length along the inner cylinder and $-\lambda$ along the sheath.

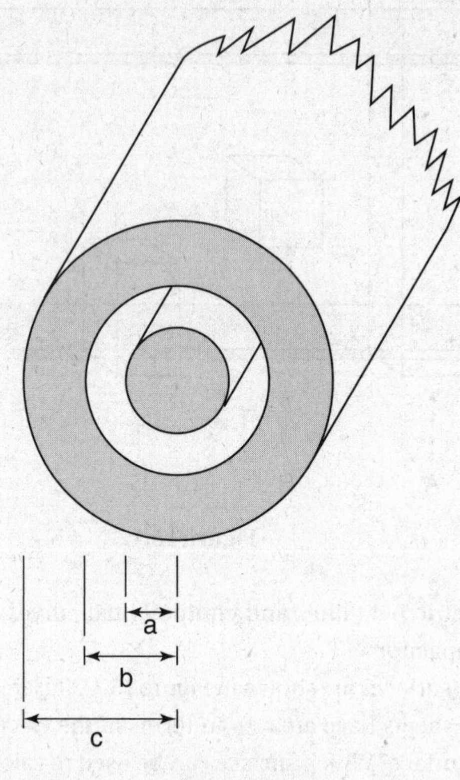

Figure 15.13

(a) Sketch electric field lines and equipotential curves in the cross-section of coaxial cable shown in Figure 15.14.

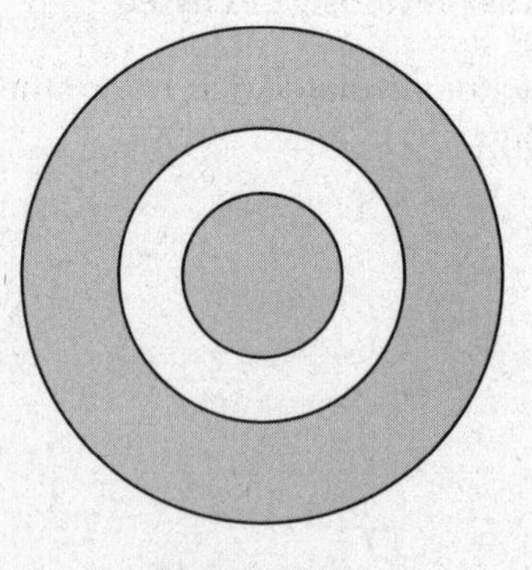

Figure 15.14

(b) Find an expression for the electric field in the region $a < r < b$ away from the edges of the capacitor.

(c) Calculate the potential difference between $r = a$ and $r = b$. Which surface is at higher potential?

(d) What is the capacitance of a coaxial cable of length l (ignore edge effects, as in our treatment of a parallel plate capacitor).

5. A parallel plate capacitor with plate separation d is charged to a potential difference V_0. Using the infinite-sheet approximation,

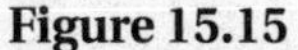
Figure 15.15

(a) What is the electric field inside and outside the capacitor?
(b) What is the charge density on each of the plates?

Then, a metal sheet of thickness t is slipped inside the capacitor parallel to the two plates, as shown in Figure 15.15.

(c) What is the charge density on the upper and lower surfaces of the metal sheet?
(d) What is the electric field above, within, and below the metal sheet?
(e) What is the voltage between the two plates of the capacitor?
(f) By what factor has the energy increased or decreased? Where did the energy come from or go to?

ANSWER KEY

1. **B**	4. **A**	7. **C**	10. **E**
2. **E**	5. **B**	8. **D**	
3. **B**	6. **B**	9. **E**	

ANSWERS EXPLAINED

Multiple-Choice

1. **(B)** Recall that charge is induced on the surfaces of the dielectric so that the induced charge attracts the adjacent, oppositely charged plates of the capacitor. Thus, removing the dielectric requires work, increasing the potential energy of the capacitor, immediately eliminating choices (A), (C), and (E).

   The energy stored on a capacitor can be expressed in terms of any two of the three variables Q, C, and V. Now, when we add or remove a dielectric, Q remains the same (the capacitor is isolated), and we know that C decreases by a factor of κ_D. Therefore, the most convenient equation to use here expresses the potential energy in terms of Q and C: $U = Q^2/2C$. Because Q remains constant and C decreases by a factor of κ_D, the energy increases by a factor of κ_D.

2. **(E)** Considering the equation $U = \frac{1}{2}CV^2$ and the fact that the voltage V of the battery is fixed, the network with the greatest capacitance stores the most energy. Which network has the greatest equivalent capacitance? The most straightforward way to solve the problem is to calculate the equivalent capacitance of each network (the equivalent capacitances are $0.25C$, $0.75C$, C, $0.6C$, and $1.33C$, respectively).

   However, the quicker way to solve the problem is to compare the various networks. Recall the following semiqualitative rules that govern equivalent capacitances:

   - The equivalent capacitance of capacitors in *parallel* is *greater* than that of each of the individual capacitors.
   - The equivalent capacitance of capacitors in *series* is *smaller* than that of each of the individual capacitors.

   Because choices (A), (B), and (D) all contain a single capacitor in series with other elements, the equivalent capacitances of these arrays are all smaller than $1C$. Because choice (E) contains a single capacitor in parallel with other elements, its capacitance is greater than $1C$. Therefore, we have immediately eliminated choices (A), (B), and (D). At this point, it is probably best to simply calculate the equivalent capacitances of choices (C) and (E) and choose the larger equivalent capacitance.

3. **(B)** Recall that electric field lines begin on positive charge and end on negative charge. Because an equal number of lines begin and end at the sphere, its net charge must be zero. (Imagine drawing an imaginary sphere around the suspended sphere and applying Gauss's law: The net electric flux would be zero.) However, there is definitely a nonzero charge density on the sphere: negative charge on the top of the sphere and positive charge on the bottom of the sphere. This is the situation we would expect because of electrostatic induction if the sphere were a conductor with zero net charge. (The net charge remains zero, but the external electric field causes the charge to redistribute itself.)

   Note that the induced charge is on the surface of the sphere, distributed in such a way that the electric field within the metal is zero (as displayed by the fact that no field lines penetrate the sphere).

4. **(A)** Because the two capacitors are in series, they each have the same charge in equilibrium. Using the equation $Q = CV$, the voltage across the 4-F capacitor is $V_4 = Q/4\text{F}$, and the voltage across the 6-F capacitor is $V_6 = Q/6\text{F}$. A Kirchhoff's loop rule equation then yields $10\ V = Q/4\text{F} + Q/6\text{F}$; solving for Q shows that $Q = 24C$. (*Note*: What we just did is equivalent to simply applying the equation for the equivalent capacitance of two capacitors in series.) Finally, to calculate the potential at point A, we start at the negative terminal of the battery (where $V = 0$) and trace the circuit counterclockwise to point A, keeping track of potential jumps. The only jump is across the 6-F capacitor, which is a positive jump (because we are going from the negatively charged plate to the positively charged plate) whose value is $V = 24C/6\text{F} = 4\ \text{V}$. Therefore, the potential at point A is 4 V.

5. **(B)** Recall that for a parallel plate capacitor, the relationship among voltage, electric field, and plate separation is $|V| = |E|d$. Because the capacitor is equilibrated with a battery of the same voltage before and after doubling the plate separation, the voltage is the same in both cases. Therefore, doubling the plate separation causes the electric field to decrease by a factor of 2.

6. **(B)** The energy of a capacitor can be expressed using any two of the three variables Q, C, and V; in this case, the most convenient formula to use is $U = \frac{1}{2}CV^2$ because it is easiest to figure out what happens to the voltage and capacitance before and after plate separation. As in question 5, because the capacitor is allowed to equilibrate with the same battery before and after increasing the plate separation, the voltage is the same in both cases. Since the capacitance of a parallel plate capacitor is given by the equation $C = \varepsilon_0 \kappa_D A/d$, doubling the plate separation decreases the capacitance by a factor of 2. Thus, using the equation $U = \frac{1}{2}CV^2$, we see that the potential energy decreases by a factor of 2.

7. **(C)** Recall that the electric field due to an infinite sheet of charge does not depend on how close you are to the infinite sheet of charge. Because the capacitor is isolated, the charge has nowhere to go, so the charge density on the plates remains unchanged. Since the plates act approximately like infinite sheets of charge, the field between them does not depend on the distance between the sheets, so it remains unchanged.

   Perhaps a more straightforward, mathematical solution would involve the equations $Q = CV$ and $V = Ed$. Because Q remains constant, CV must remain constant. As discussed in the last section, C decreases by a factor of 2 when the plate separation is doubled, so V must then increase by a factor of 2 for CV to remain constant. Now consider the equation $V = Ed$. Because both V and d increase by a factor of 2, E must remain constant.

8. **(D)** In this case, the most convenient formula to use for the energy stored on the capacitor is $U = Q^2/2C$. Because the capacitor is disconnected before the plates are separated, the charge has nowhere to go, so it remains constant. As before, because the distance is doubled, the capacitance is halved. Combining these observations with the formula $U = Q^2/2C$ indicates that the amount of energy is doubled. This makes sense: Physically pulling apart the plates (which attract each other) requires energy, so we expect the potential energy of the capacitor to increase. (Pulling the plates apart in question 7 also required work, but this work was more than compensated for by charge that moved across the battery, increasing its potential energy.) From this reasoning, we could have immediately eliminated choices (A), (B), and (C).

9. **(E)** At equilibrium, there is zero current through each of the capacitors and thus no current across the wire connecting points A and B. The voltage across each resistor is equal to $V/2$ (as required by a Kirchhoff's loop equation for the outer loop and the fact that the same current runs through both resistors), so that the voltage across each capacitor is $V/2$. [The voltage across R and C_1 (and across R and C_2) must be equal because of the parallel connection.] The charge on capacitor C_1 is then $Q = CV = (C_1)(V/2)$.

10. **(E)** Calculation of the electric field from the parameters given in choices (A)–(D) is shown below, demonstrating that they are not the correct answer. For choice (A), approximating the plates as infinite sheets of charge, the electric field due to an infinite sheet of charge is $E = \sigma/2\varepsilon_0$ (which does not depend on the distance from the sheet of charge). Within the capacitor, the electric fields due to the two plates add, so that the net electric field is $E = \sigma/\varepsilon_0$. For choice (B), for a parallel plate capacitor, $V = Ed$, so $E = V/d$. For choice (C), from Q and C, the voltage can be calculated using the equation $Q = CV$. Then, given the voltage and the separation distance, the electric field can be calculated using the formula $E = V/d$ as in choice (B). For choice (D), from the magnitude of charge per plate and the plate area, the charge density can be calculated as $\sigma = Q/A$. Then, the

electric field can be calculated from the charge density as in choice (A). The variables in choice (E) do not allow calculation of the electric field. From U, V, and C, one could calculate the charge $Q = CV$. However, without knowing the area of the plates or the plate separation distance, it is impossible to calculate the electric field.

Free-Response

1. (a) $Q = C_1V = 3 \times 10^{-5}$C

(b) The battery stops charging the capacitor when the capacitor's voltage equals the battery's voltage. Therefore, the voltage across the capacitor when it is fully charged is equal to the voltage across the battery, in this case 10 V.

(c) Initially, capacitor 2 has no charge and thus no potential difference, so it acts like a short circuit (like any circuit element with no potential drop). Alternatively, capacitor 1 has a voltage of 10 V, so it acts like a 10-V battery. Therefore, the current is $I = V/R = \frac{10\text{ V}}{1000\Omega} = 0.01$ A.

(d) This is the trickiest part of this problem. We must make use of two facts: (1) After equilibrium is reached, no current flows through the circuit, so the voltage across the resistor is zero. Kirchhoff's loop rule then requires that the capacitors have the same voltage. (If they did not have the same voltage, a current would flow, and the circuit would not be at equilibrium.) (2) Charge must be conserved. Therefore, designating the total charge on both capacitors as $Q = C_1V$ [the answer found in part (a)], we have two simultaneous equations in two variables:

$$Q_1 + Q_2 = 3 \times 10^{-5}\text{C}$$

$$\text{Common voltage} = \frac{Q_1}{C_1} = \frac{Q_2}{C_2}$$

Solving these equations yields

$$Q_2 = 1.88 \times 10^{-5}\text{C}$$

$$Q_1 = 1.13 \times 10^{-5}\text{C}$$

(e) The initial potential energy when switch 1 is opened and switch 2 is closed is given by

$$U_0 = \frac{Q^2}{2C} = \frac{(3\times10^{-5}\text{C})^2}{2(3\times10^{-6}\text{F})} = 1.5\times10^{-4}\text{J}$$

Much later after the circuit reaches equilibrium, the potential energy is given by

$$U_f = \frac{Q_1^2}{2C_1} + \frac{Q_2^2}{2C_2} = \frac{(1.13\times10^{-5}\text{C})^2}{2(3\times10^{-6}\text{F})} + \frac{(1.88\times10^{-5}\text{C})^2}{2(5\times10^{-6}\text{F})} = 5.66\times10^{-5}\text{J}$$

Thus, the difference in potential energy is

$$\Delta U = U_f - U_0 = 5.66 \times 10^{-5}\text{ J} - 1.5 \times 10^{-4}\text{ J} = -9.34 \times 10^{-5}\text{ J}$$

Thus, the potential energy decreases. This decrease in energy is due to conversion of electrical potential energy to thermal energy as charge passes through the resistor.

2. (a) This section is a straightforward application of Gauss's law (in the case of spherical symmetry and thus spherical gaussian surfaces).

   (i) Because there is zero net charge enclosed, the field is zero for $r > r_2$.

   (ii) In this region, the enclosed charge is $-Q$, so the electric field points toward the center of the sphere. Because the directed area vector $d\mathbf{A}$ points away from the surface (by definition),

$$\oint \mathbf{E} \cdot d\mathbf{A} = -E \oint dA = -4\pi r^2 E = \frac{-Q}{\varepsilon_0}$$

$$E = \frac{Q}{4\pi\varepsilon_0 r^2}$$

   (iii) In this region the enclosed charge is again zero, so the field is zero.

   (b) This section involves applying the equation

$$V(r) = -\int_{\infty}^{r} E_x dx$$

   (i) In the region $x > r_2$, the electric field is zero. Therefore, the potential integral above is zero, and the potential is zero within this region.

   (ii) Here's where things get a little tricky:

$$V(r_1) = -\int_{\infty}^{r_1} E_x dx = -\int_{\infty}^{r_2} E_x dx - \int_{r_2}^{r_1} E_x dx$$

As just discussed in part (b)(i), the first of these integrals is equal to zero because the electric field is zero. In the second integral, we use the electric field calculated in part (a)(ii), noting that the direction of the field is in the negative x-direction:

$$V(r_1) = 0 - \int_{r_2}^{r_1} -\frac{Q}{4\pi\varepsilon_0 r^2} dx = \frac{Q}{4\pi\varepsilon_0}\left(\frac{1}{r_2} - \frac{1}{r_1}\right) = \frac{Q}{4\pi\varepsilon_0}\left(\frac{r_1 - r_2}{r_1 r_2}\right)$$

Therefore, the calculated potential at r_1 is negative. This makes sense: A positive charge is attracted toward the center of the sphere and tends to move downhill from r_2 to r_1; thus, r_2 is at a higher potential than r_1.

   (c) The definition of capacitance is the charge stored per potential difference:

$$C = \frac{Q}{|V|} = \frac{Q}{\frac{Q}{4\pi\varepsilon_0}\left(\frac{r_2 - r_1}{r_1 r_2}\right)} = 4\pi\varepsilon_0\left(\frac{r_1 r_2}{r_2 - r_1}\right)$$

3. (a) See Figure 15.16. The electric field is uniform and perpendicular to the plates. The equipotential curves are perpendicular to the electric field lines and equally spaced (because potential is a linear function that depends only on the relative distance between the two plates).

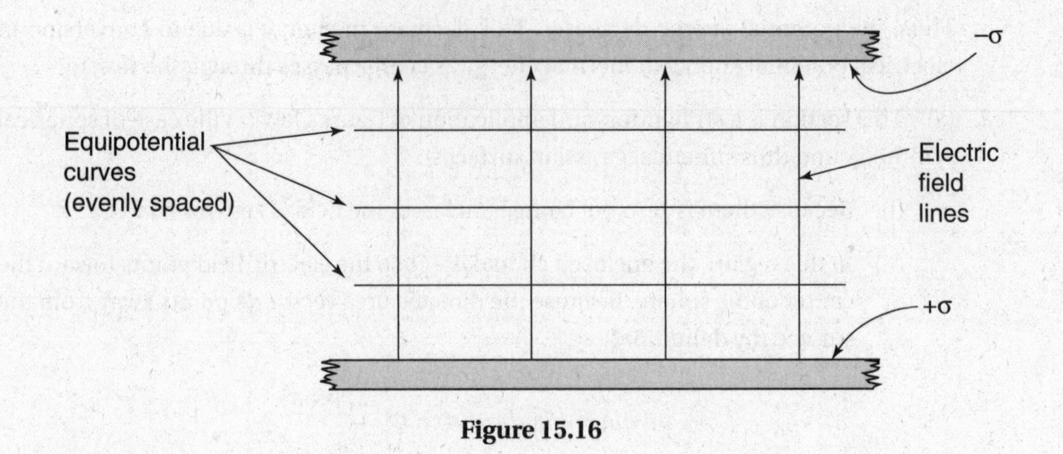

Figure 15.16

(b) The flux through all faces parallel to the field or immersed in the metal is zero (because the field is parallel to the surface or zero, respectively). Therefore, the net flux through box A is zero. The net flux through box B is zero because the inward flux through the bottom face cancels the outward flux through the top face (recall that the electric field within a parallel plate capacitor is constant, so these fluxes are equal in magnitude). Another way to realize that the flux through boxes A and B is zero is because they enclose zero net charge (box A contains equal amounts of positive and negative charge; box B contains no charge at all). Box C encloses net charge, so its flux is nonzero (there is nonzero flux through the top face and no flux through any other faces). Because box C is the only box with nonzero net flux, it is the only box where the application of Gauss's law is nontrivial.

(c) The flux through box C is

$$\Phi = \oint \mathbf{E} \cdot d\mathbf{A} = EA$$

Applying Gauss's law,

$$\left[\Phi = \oint \mathbf{E} \cdot d\mathbf{A} = EA\right] = \left[\frac{Q_{\text{enclosed}}}{\varepsilon_0} = \frac{\sigma A}{\varepsilon_0}\right]$$

$$E = \frac{\sigma}{\varepsilon_0}$$

Note that the field does not depend on the arbitrary area of our imaginary gaussian surface (that would be a problem). Also note that comparison with the equation for the electric field due to a sheet of charge, $E = \sigma/2\varepsilon_0$, immediately reveals that this equation has correct units.

(d) Consider the equations $C = Q/V$ and $V = Ed$ (the relationship between voltage and electric field for a parallel plate capacitor). We know that adding a dielectric increases the capacitance by a factor of κ_D. Because charge is constant (the capacitor is isolated), this means that the voltage decreases by a factor of κ_D. Now, considering the second equation, because d is constant, E also decreases by a factor of κ_D. Therefore, the new value of E is

$$E = \frac{E_0}{\kappa_D} = \frac{\sigma}{\kappa_D \varepsilon_0}$$

This answer makes intuitive sense: Induced charge on the dielectric partially cancels charge on the plates, reducing the effective charge on the plates and thus reducing the electric field.

4. (a) See Figure 15.17 (the only electric field is in the region $a < r < b$). The electric field points radially away from the positive charge on the inner cylinder. The circular equipotential curves are perpendicular to the field lines and thus more closely spaced near the inner cylinder (where the magnitude of the electric field is greater). Note that charge is arranged on the outer surface of the inner cylinder and the inner surface of the sheath (as required for the electric field within the metal to be zero).

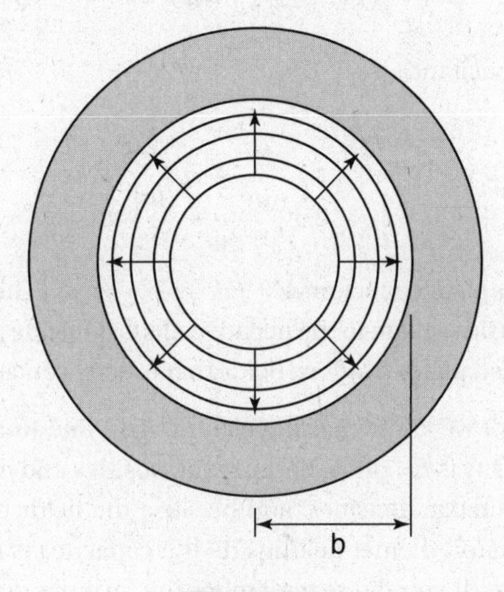

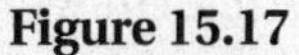

Figure 15.17

(b) Away from the edges of the cylinder, we can treat the cylinder as if it were infinitely long. Applying Gauss's law, we have cylindrical symmetry and so we use a cylindrical Gaussian surface of length ℓ and radius $a < r < b$. The flux through the ends of the Gaussian cylinder is zero because the field is parallel to the surface of the caps. The flux through the sides is equal to

$$\Phi = \oint \mathbf{E} \cdot d\mathbf{A} = E \oint dA = 2\pi \ell E r$$

Applying Gauss's law,

$$[\Phi = 2\pi r \ell E] = \left[\frac{Q_{\text{enclosed}}}{\varepsilon_0} = \frac{\lambda \ell}{\varepsilon_0}\right]$$

$$E = \frac{\lambda}{2\pi r \varepsilon_0} \quad \text{(direction is away from the axis of the cable)}$$

(c) To calculate this potential difference, we must find a path where the electric field is easily calculated (and, ideally, parallel to the path of integration). A suitable path is one perpendicular to the axis of the cylinder (so that it is parallel to the field) and at the center of the capacitor (so that edge effects can be ignored and the above expression for the electric field can be used). Integrating,

$$V(b)-V(a)=\int_a^b dV=-\int_a^b E dr=-\int_a^b \frac{\lambda dr}{2\pi r\varepsilon_0}=-\frac{\lambda}{2\pi\varepsilon_0}\ln\left(\frac{b}{a}\right)=\frac{\lambda}{2\pi\varepsilon_0}\ln\left(\frac{a}{b}\right)$$

Because $V(b) - V(a)$ is negative, the potential is greater at a (as always, the positive plate of the capacitor is at a higher potential).

(d) The charge on the capacitor is $Q = \lambda l$, and the magnitude of the potential difference across the capacitor is

$$|V|=\frac{\lambda}{2\pi\varepsilon_0}\ln\left(\frac{b}{a}\right)$$

Therefore, the capacitance is

$$C=\frac{Q}{|V|}=\frac{\lambda l}{\frac{\lambda}{2\pi\varepsilon_0}\ln\left(\frac{b}{a}\right)}=\frac{2\pi\varepsilon_0 l}{\ln\left(\frac{b}{a}\right)}$$

5. (a) Inside a parallel plate capacitor, $V = Ed$, so $E = V_0/d$. (The direction of the field is pointing from the positive plate to the negative plate.) Outside a parallel plate capacitor, the oppositely charged plates produce fields that exactly cancel each other.

(b) From Chapter 13 we know that the electric field due to a single infinite plane of constant charge density is $E = \sigma/2\varepsilon_0$. Because the positive and negative plates must have equal and opposite charge densities, and because the fields caused by the two plates add inside the capacitor, the net field inside the capacitor is equal to $E = \sigma/\varepsilon_0 = V_0/d$. Therefore, $\sigma = \varepsilon_0 V_0/d$. (If you didn't remember the formula for the electric field due to an infinite sheet of charge, you could quickly obtain this result by constructing a cubic gaussian surface half of which is immersed in the metal, as shown in Chapter 13.)

(c) In order to conserve charge, the two faces must have opposite charges. In order for the field inside the metal to be zero, the upper face of the sheet must have a surface charge density of $-\sigma$ and the lower surface must have a surface charge density of $+\sigma$. The field produced by these charge densities then exactly cancels the field due to the capacitor plates.

(d) Within the metal, the electric field is zero. Above and below the metal, the fields due to the two planes of charge on the metal sheet cancel each other exactly (just as the field outside the capacitor is zero), and the field is unchanged ($E = V_0/d$) from what we found in part (b).

(e) The potential difference between the negatively charged plate and the positively charged plate is equal to $V = -\int \mathbf{E}\cdot d\mathbf{r}$. This line integral can be broken into three segments. The middle segment of the line integral in the metal sheet yields zero because the electric field is zero in the metal. The two segments outside the metal yield the following (noting that the electric field is antiparallel to $d\mathbf{r}$).

$$v=-\int \mathbf{E}\cdot d\mathbf{r}=E\int dr=E(d-t)=\left(\frac{V_0}{d}\right)(d-t)=V_0\left(\frac{d-t}{d}\right)$$

(f) From the above equation, the voltage decreases by a factor of $(d - t)/d$ because of insertion of the metal sheet. Because the charge does not change (the capacitor is isolated), the energy (given by the equation $U = \frac{1}{2}QV$) decreases by a factor of $(d - t)/d$. The energy is lost to mechanical energy; the capacitor draws the metal sheet because of the attraction between the induced charges on the sheet and the charged capacitor plates. Note that the metal sheet acts like a dielectric material with an infinite dielectric constant; thus, this example should give you new insight into how induced charge on a dielectric can reduce the electric field.

RC Circuits

16

→ ANALYSIS OF CIRCUITS CONTAINING RESISTORS AND CAPACITORS
→ CHARGING A CAPACITOR
→ DISCHARGING A CAPACITOR
→ *RC* TIME CONSTANT

In the last chapter when we discussed the behavior of capacitors, we always assumed that enough time had elapsed for the battery to completely charge or discharge the capacitor. In this chapter, we explore what happens *while* a capacitor is charging and discharging.

GENERAL APPROACH TO DERIVING EQUATIONS FOR *RC* CIRCUITS

The following is an outline of the derivations that are performed in this chapter, which will give you some sense of direction. Don't worry about understanding all of this just yet.

STEP 1 Create a Kirchhoff's loop equation for the circuit.

STEP 2 Before we can solve this equation, all the variables must be expressed in terms of time and charge. This can be done by relating the voltage across the capacitor to its charge using the equation $Q = CV$ and relating the current through the circuit with the charge on the capacitor using the relationship $|I| = |dQ/dt|$. (We have to be careful about the signs in both these equations.)

STEP 3 Now we should have a differential equation involving various constants, Q, and t. After separating variables, indefinite integration yields an equation relating charge to time with a constant of integration.

STEP 4 Use the initial conditions to eliminate the constant of integration. (We can check our answer by making sure that the final condition, at $t = \infty$, is consistent with the results from the previous chapter.)

STEP 5 Now we should have charge as a function of time, $Q(t)$. If we want to calculate current as a function of time, we can plug $Q(t)$ into our previous formula relating current and time, $|I| = |dQ/dt|$.

STEP 6 We should be able to check that our new equation gives the right current values at $t = 0$ and $t = \infty$.

DISCHARGING A CAPACITOR THROUGH A RESISTOR

A capacitor (of capacitance C) is charged by a battery of voltage V and then attached to a resistor at time $t = 0$. We are interested in determining the charge on the capacitor and the current through the circuit as functions of time.

Qualitative solution: As the capacitor discharges, the charge across the capacitor decreases, resulting in a decrease in the voltage as well (according to the equation $Q = CV$). As voltage decreases, so does the current through the resistor. Thus, we expect charge, voltage, and current to all decrease with time, reaching zero at equilibrium (mathematically, at $t = \infty$).

Additionally, because the rate of change in charge (equal to the current) is proportional to charge, we expect that the decrease in charge will be exponential. (Whenever a quantity's derivative is proportional to its own value, as in $y' = cy$, exponential behavior occurs of the form $y = ke^{cx}$.)

Quantitative solution: We start with Kirchhoff's loop rule, which states that, referring to Figure 16.1,

$$V_C - IR = 0$$

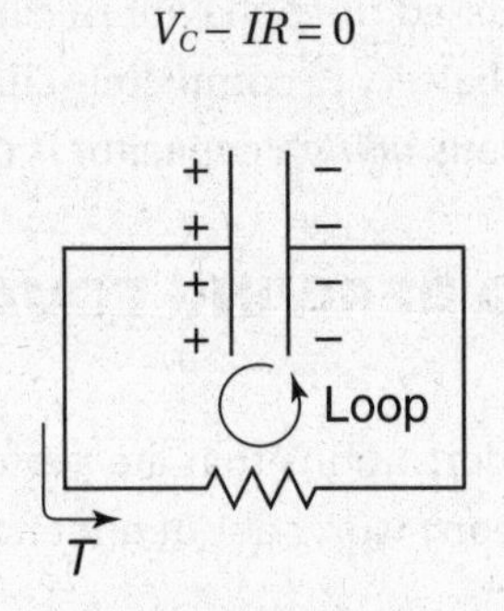

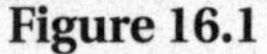

Figure 16.1

Note the positive sign before V_C. We never discussed calculating potential drops across capacitors in the sections on Kirchhoff's laws, but the same principles apply. As discussed in the last chapter, the positive plate of the capacitor is always at greater potential than the negative plate. Therefore, when we go around our loop, we go from lower to higher potential (positive voltage). Note that this equation is consistent with the fact that the current is zero at $t = \infty$ because at equilibrium $V_C = 0$.

The voltage of the capacitor is not constant; it is given by $V_C = Q/C$. Substituting this in Kirchhoff's loop rule yields

$$\frac{Q}{C} = IR$$

How can we use this equation to relate charge to time? By expressing the magnitude of the current as $|I| = |dQ/dt|$, we obtain a differential equation relating charge and time. However, we must be careful about the sign of the current. In the above voltage equation, Q refers to the amount of charge stored on the capacitor. When current flows, it causes a decrease in the amount of charge on the capacitor (Q). Therefore, $I = -dQ/dt$. Substituting,

$$\frac{Q}{C} = -\frac{dQ}{dt}R$$

Before we can integrate we must separate the variables:

$$\frac{dQ}{Q} = \frac{-dt}{RC}$$

Integrating,

$$\int \frac{dQ}{Q} = \int \frac{-dt}{RC}$$

$$\ln(Q) = \frac{-t}{RC} + K$$

where K is a constant of integration. Exponentiating, we find

$$Q = e^{(-t/RC)+K} = e^{-t/RC} e^{K}$$

Now we must use the initial conditions to calculate the constant of integration. We call the initial charge on the capacitor Q_0.

$$Q(0) = e^{K} e^{0} = e^{K} = Q_0$$

Therefore,

$$Q = e^{-t/RC} e^{K} = Q_0 e^{-t/RC}$$

Note that we didn't have to solve for the actual constant of integration, K; instead we only had to solve for e^K.

$$Q(t) = Q_0 e^{-t/RC}$$

Charge on a discharging capacitor

Answer check: We know that the final charge on the capacitor is zero; mathematically $Q(t = \infty) = 0$. The above equation for $Q(t)$ indeed approaches zero as time approaches infinity.

What about current as a function of time? We can calculate current from time using the equation $I = -dQ/dt$ (from above). Taking this derivative yields

$$I = \frac{Q_0}{RC} e^{-t/RC}$$

Answer check: We should make sure this equation is consistent with what we already know from the previous chapter:

1. At $t = \infty$, the capacitor is completely drained and no current flows through the circuit.
2. At the instant the capacitor is attached to the resistor, the voltage across the capacitor is equal to $V = Q/C$. Therefore, Ohm's law gives the initial current $I(t = 0) = V/R = Q/RC$. Designating this initial current as I_0,

$$I(t) = I_0 e^{-t/RC}$$

Current driven by a discharging capacitor

Therefore, both current and charge decrease exponentially. This confirms our earlier qualitative predictions.

CHARGING A CAPACITOR

At time $t = 0$ a battery (voltage V) is attached to an uncharged capacitor (capacitance C). Again, we are interested in the charge and current as a function of time.

We already know the following information:

1. Initial condition: at $t = 0$, there is no charge (and thus no voltage) across the capacitor. Thus, the entire potential drop due to the battery occurs across the resistor, and the current through the circuit is $I = V/R$. We observe that initially the uncharged capacitor behaves like a short circuit (a piece of wire).[1]
2. Final condition: At $t = \infty$, the capacitor is completely charged and the battery is unable to force any more charge onto it, causing the current to be zero. Thus, at equilibrium, a fully charged capacitor behaves like an open circuit (a break in the circuit). The final charge on the capacitor is $Q = CV$.
3. In between: As charge builds up, the potential across the capacitor increases, making it harder to push more charge onto the capacitor. Therefore, we expect the current to decrease with time.

Quantitative solution: Writing a Kirchhoff's loop rule equation using Figure 16.2,

$$V - IR - V_{capacitor} = 0$$

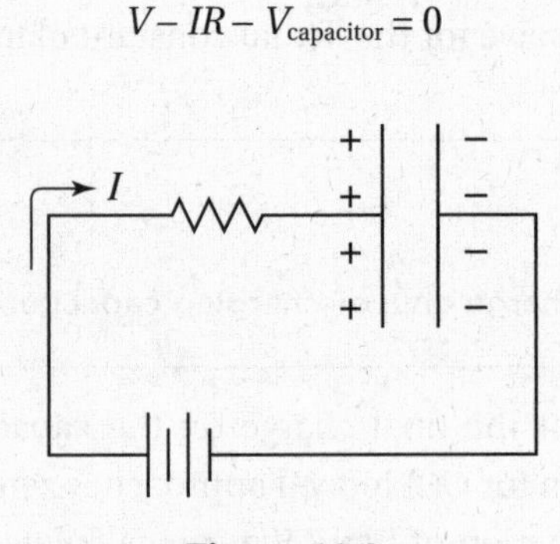

Figure 16.2

As in the first derivation, we express all the variables in terms of charge and time to obtain a differential equation with only two variables. We begin by substituting for $V_{capacitor} = Q/C$:

$$V - IR - \frac{Q}{C} = 0$$

Again, we can express current in terms of charge and time using the equation $|I| = |dQ/dt|$. When the current is positive (i.e., it flows in the direction indicated in Figure 16.2), it increases the charge stored on the capacitor (Q). Therefore, $I = +dQ/dt$. Substituting this in yields

$$V - R\frac{dQ}{dt} - \frac{Q}{C} = 0$$

[1]Of course, charge doesn't actually flow *through* the capacitor. Rather, a charge imbalance between the two capacitor plates develops with a buildup of positive charge on one plate and negative charge on the opposite plate.

Separating variables,

$$VC - Q = RC\frac{dQ}{dt}$$

$$\frac{dQ}{CV - Q} = \frac{dt}{RC}$$

Does this seem familiar? It should—it's similar to the way we separated variables when calculating the velocity of a falling object subject to gravity and air resistance.

$$\int\frac{dQ}{CV - Q} = \int\frac{dt}{RC}$$

Integrating,

$$\ln(CV - Q) = \frac{-t}{RC} + K$$

Exponentiating,

$$CV - Q = e^{(-t/RC)+K} = e^{K}e^{-t/RC}$$

$$Q = CV - e^{K}e^{-t/RC}$$

We can eliminate the constant of integration using the fact that the initial charge on the capacitor is zero:

$$Q(t = 0) = CV - e^{K}e^{0} = CV - e^{K} = 0$$

Therefore, $e^{K} = CV$

$$Q = CV - CVe^{-t/RC} = CV(1 - e^{-t/RC})$$

Answer check: This answer is consistent with our expectation that the charge on the fully charged capacitor should be $Q(t = \infty) = CV$. Designating this final charge as Q_{final}, the answer can be rewritten as follows.

$$Q(t) = Q_{\text{final}}(1 - e^{-t/RC})$$

Charge on a charging capacitor

As before, we can take the time derivative of the charge to obtain the current as a function of time. Recall that our equation relating current and charge is $I = +dQ/dt$. Substituting our expression for $Q(t)$ into this equation yields

$$I = \frac{Q_{\text{final}}}{RC}e^{-t/RC}$$

Answer check: We expect that $I(0) = V/R$ and that $I(\infty) = 0$. Recalling that by definition $Q_{\text{final}} = CV$, the above equation satisfies both these conditions.

Generally, we call $Q_{\text{final}}/RC = V/R = I_0$ (the initial current) and write the final result more neatly in the following form.

$$I(t) = I_0 e^{-t/RC}$$

Current through a charging capacitor

The following generalizations about charging capacitors are worth reiterating.

1. **At the instant charging begins, the charge and potential on the capacitor are zero, so the capacitor acts like a wire (a short circuit) in the sense that there is no voltage across this circuit element.**
2. **After the circuit reaches equilibrium (at $t = \infty$), the voltage of the capacitor is equal to the voltage of the power source, so no current flows through the capacitor and it acts like an open circuit.**

TIME CONSTANTS

Let's compare our final results:

$$Q(t) = Q_0 e^{-t/RC}$$

Charge on a discharging capacitor

$$I(t) = I_0 e^{-t/RC}$$

Current through a discharging capacitor

$$Q(t) = Q_{\text{final}}(1 - e^{-t/RC})$$

Charge on a charging capacitor

$$I(t) = I_0 e^{-t/RC}$$

Current through a charging capacitor

In all these equations, time enters in the form of the exponential $e^{-t/RC}$. Note that RC must have units of time (seconds) such that the factor $-t/RC$ is a dimensionless number, as shown below. (Recall that exponents must be dimensionless numbers.)

$$\begin{aligned} &\text{(ohms)(farads)} \\ RC = \Omega F &= \left(\frac{\text{volts}}{\text{amperes}}\right)\left(\frac{\text{coulombs}}{\text{volt}}\right) \\ &= \frac{\text{coulombs}}{\text{ampere}} = \left(\frac{\text{coulombs}}{\text{coulombs/second}}\right) = \text{second} \end{aligned}$$

This quantity, RC, with units of time, is defined as the time constant, τ.

$$\tau = RC$$

With this definition, the time constant is the amount of time it takes for the exponential term $e^{-t/RC}$ to decrease from its maximum value of 1 at $t = 0$ to $1/e \approx \frac{1}{3}$. If you haven't taken the

SAT yet, here's an analogy for you: The time constant τ is to base e as the half-life is to base 2. If we're discharging a capacitor, the time constant is the amount of time required for the charge on the capacitor (and the current) to decrease by a factor of e. If we're charging a capacitor, the time constant is the amount of time required for the *difference* between the charge and the final charge, $Q(t) - Q_{\text{final}}$, to decrease by a factor of e (and the time required for the current to decrease by a factor of e).

Note the following parallels between charging and discharging capacitors.

- Both when charging and discharging a capacitor, the current starts at some initial value and decreases exponentially with time constant τ.
- Both when charging and discharging a capacitor, the difference between the charge and its final value decreases exponentially with time constant τ.

One time constant describes charging and discharging a capacitor.

The following statement is a useful generalization of these results, which is valid even under different initial conditions: The difference between currents, energies, voltages, and their final values always decreases exponentially.

CHAPTER SUMMARY

The analysis of circuits including resistors and capacitors uses Kirchhoff's laws and the definitions of capacitance and current. Kirchhoff's loop rule yields a differential equation for the charge on the capacitor. The time dependence of the charge is an exponential function with a time constant given by the product of the resistance and capacitance.

PRACTICE EXERCISES

Multiple-Choice Questions

Questions 1–4 refer to the circuit shown in Figure 16.3.

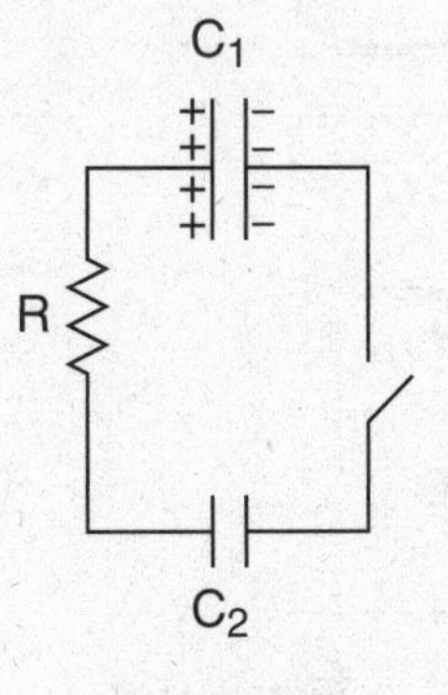

Figure 16.3

1. In the circuit shown, capacitor C_1 is initially charged and capacitor C_2 is initially uncharged when the switch is closed at $t = 0$. Which of the graphs in Figure 16.4 represents the current through the circuit as a function of time?

   To solve this problem, you will be required to apply the general principle of *RC* circuits that all quantities exponentially approach their final values.

2. Which of the graphs in Figure 16.4 represents the charge on C_2 as a function of time?

3. Which of the graphs in Figure 16.4 represents the charge on C_1 as a function of time?

4. Which of the following is Kirchhoff's loop equation for the circuit? (*I* is designated as the counterclockwise current.)

   (A) $-Q_1/C_1 - IR + Q_2/C_2 = 0$
   (B) $Q_1/C_1 - IR + Q_2/C_2 = 0$
   (C) $-Q_1/C_1 + IR - Q_2/C_2 = 0$
   (D) $Q_1/C_1 - IR - Q_2/C_2 = 0$
   (E) $Q_1/C_1 + IR + Q_2/C_2 = 0$

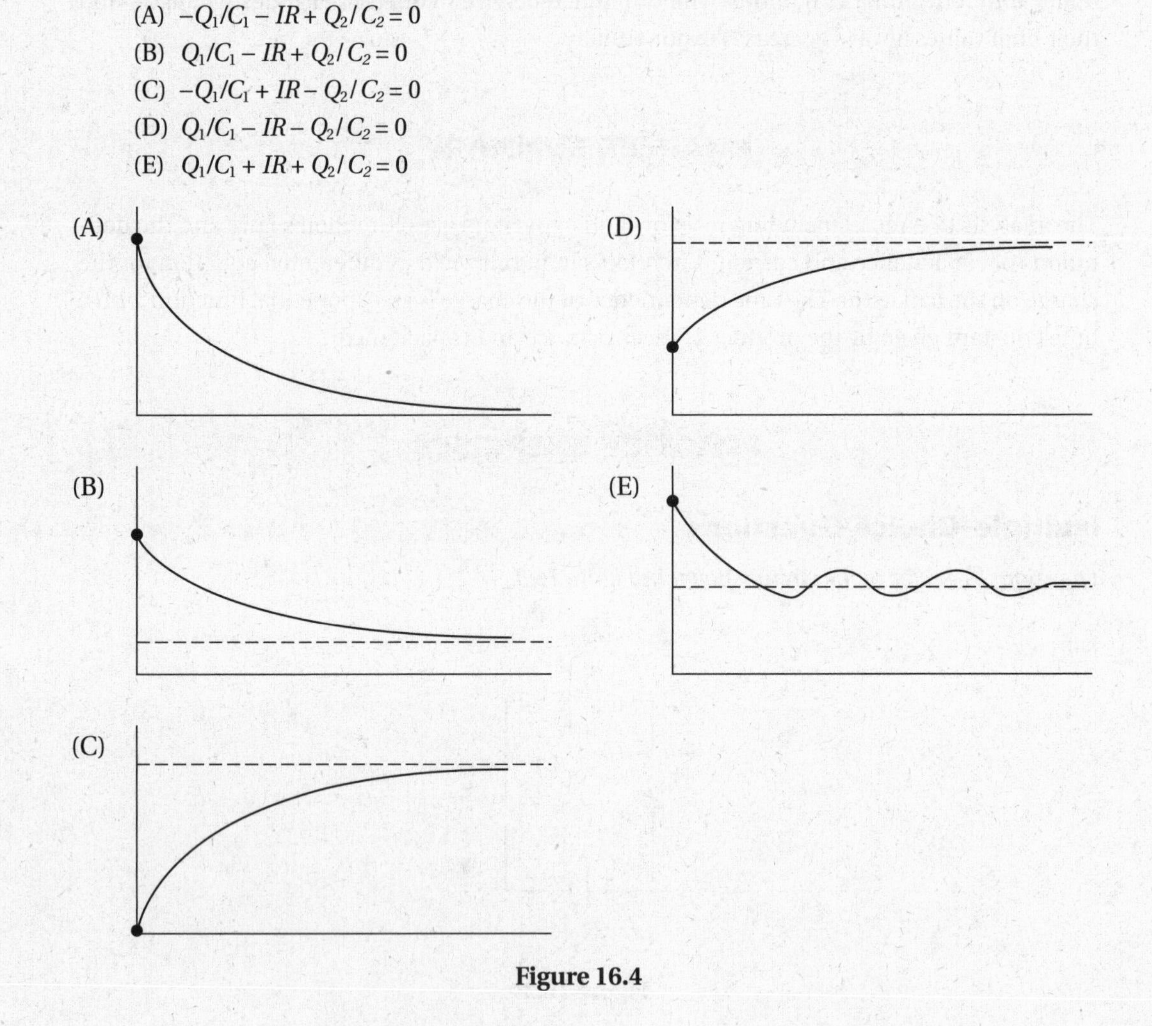

Figure 16.4

5. Which of the following functions of time have the same general form when both charging and discharging a capacitor?

I. $U(t)$, the energy stored in the capacitor
II. $I(t)$, the current through the circuit
III. $Q(t)$, the charge on the capacitor

(A) I
(B) II
(C) III
(D) I and II
(E) I, II, and III

6. A capacitor that stores energy for a camera flash is charged up slowly and then discharged very rapidly. Which of the graphs in Figure 16.5 represents the current through the capacitor as a function of time? (Current is defined here as the rate of change in the amount of charge on the positive terminal of the capacitor.)

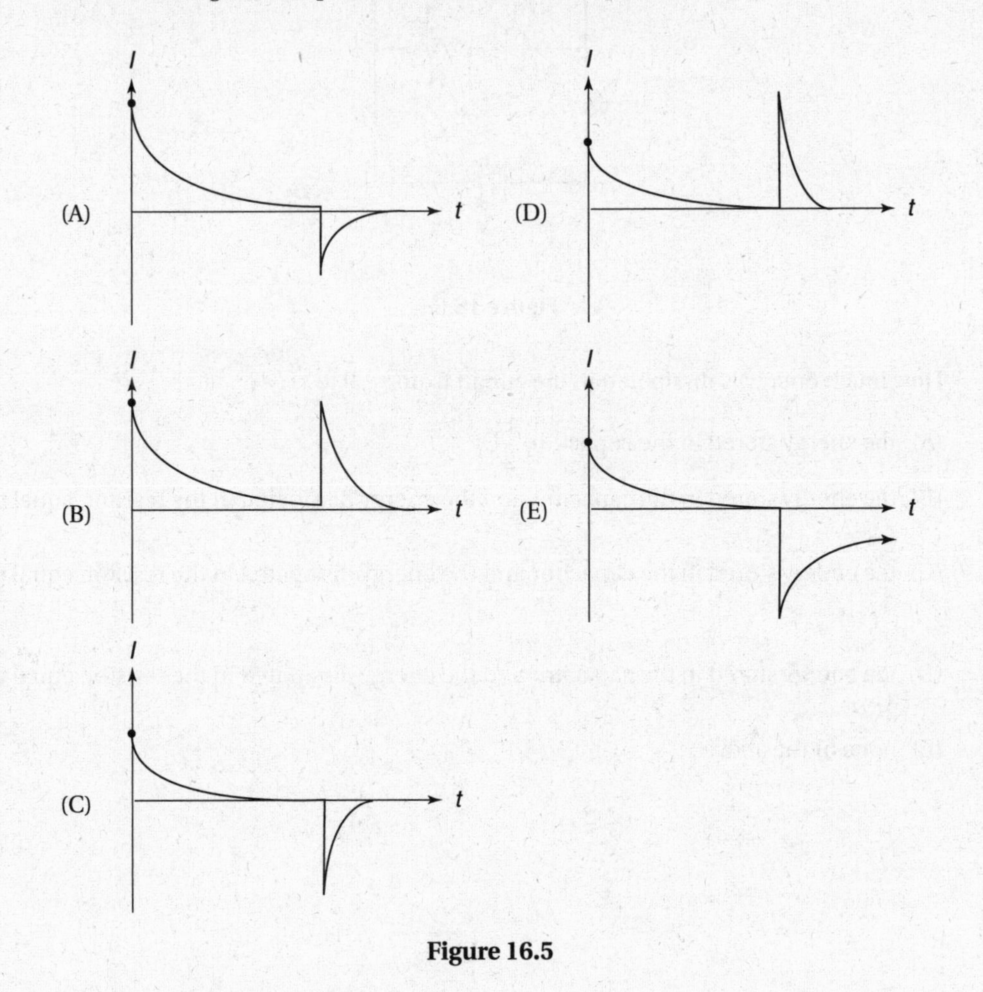

Figure 16.5

7. Which of the following could increase the time constant of a circuit containing a resistor of resistance R and a capacitor of capacitance C?

(A) adding another resistor of resistance R in parallel with the resistor
(B) adding a resistor of resistance $0.5R$ in parallel with the resistor and a capacitor of capacitance C in parallel with the capacitor
(C) adding another capacitor of capacitance C in series with the capacitor
(D) adding a resistor of resistance $2R$ in series with the resistor and a capacitance of capacitance C in series with the capacitor
(E) adding a resistor of resistance R in series with the resistor and a capacitor of capacitance $0.2C$ in series with the capacitor

8. The capacitor in the circuit in Figure 16.6 is uncharged when the switch is closed at $t = 0$.

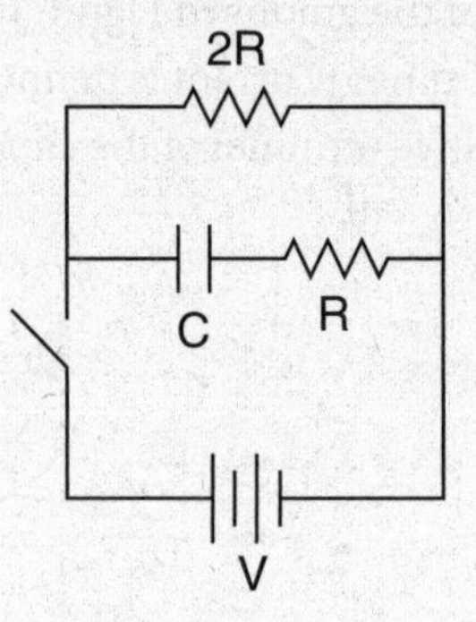

Figure 16.6

How much energy is dissipated in the circuit from $t = 0$ to $t = \infty$?

(A) the energy stored in the capacitor, $\frac{1}{2}CV^2$
(B) the energy stored in the capacitor and the energy dissipated in the resistor, equal to CV^2
(C) the energy stored in the capacitor and the energy dissipated in the resistor, equal to $\frac{3}{2}CV^2$
(D) the energy stored in the capacitor and the energy dissipated in the resistor, equal to $2CV^2$
(E) none of the above

9. For the circuit in question 8, which of the graphs in Figure 16.7 shows the current through the battery as a function of time?

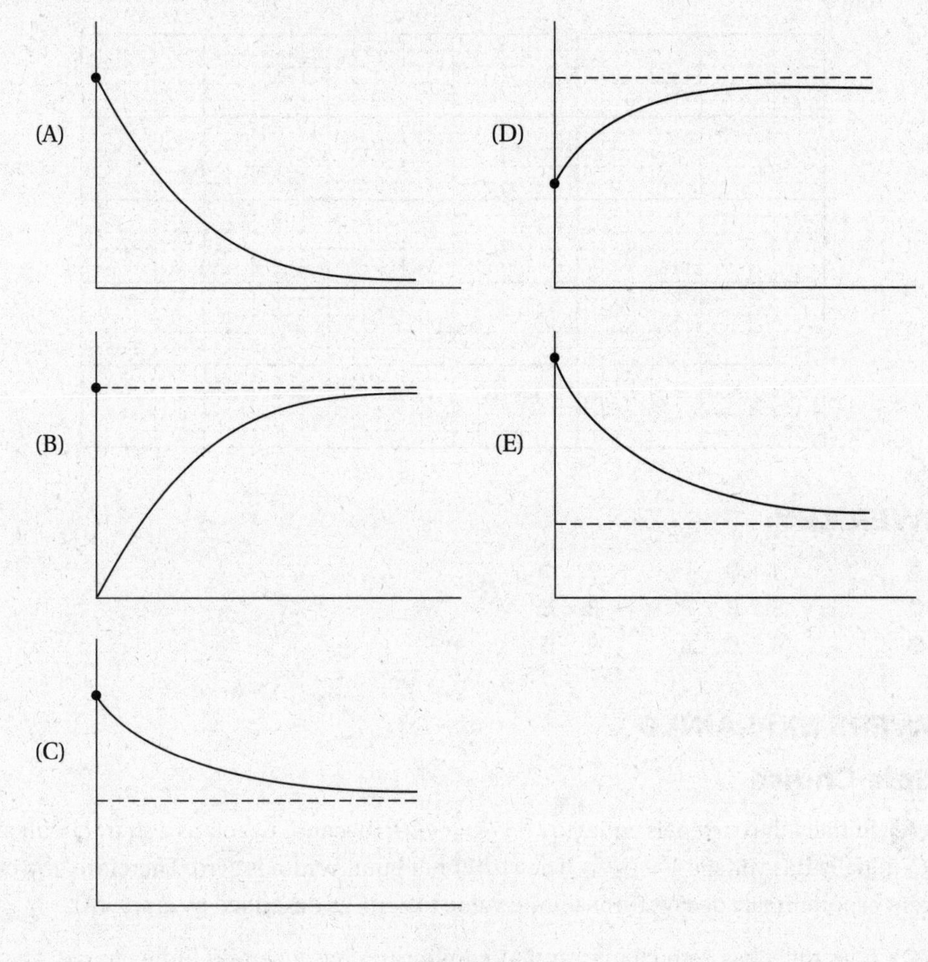

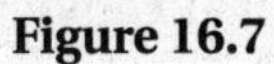

Figure 16.7

Free-Response Questions

1. A battery of voltage V is attached to a resistor and an uncharged capacitor at $t = 0$. The current is given by the function $I(t) = I_0 e^{-t/\tau}$. Express all answers in terms of I_0, V, and τ.

   (a) What is the resistance of the resistor?
   (b) What is the capacitance of the capacitor?
   (c) Calculate the magnitude of charge on the capacitor as a function of time, $Q(t)$, by integration.
   (d) Which quantities should be plotted on the grid below to yield a straight line whose slope equals $-1/\tau$?

(e) Assuming that $V = 10$ V, $R = 50{,}000\ \Omega$, and $C = 200\ \mu F$, plot the straight line whose slope equals $-1/\tau$ for the time range $0 < t < 3\tau$. Scale and label the axes, including units.

ANSWER KEY

1. **A**	4. **D**	7. **D**
2. **C**	5. **B**	8. **E**
3. **B**	6. **C**	9. **E**

ANSWERS EXPLAINED

Multiple-Choice

1. **(A)** Initially the current is equal to $I = V/R = Q/C_1R$ (because C_2 acts as a short circuit and C_1 initially has voltage $V = Q/C_1$). The current at equilibrium is zero. Therefore, the current exponentially decays from a finite value to zero, as described by graph (A).

2. **(C)** C_2 initially has zero charge, and at equilibrium has a certain finite charge. Therefore, the charge increases from zero and exponentially approaches a positive value, as described by graph (C).

3. **(B)** C_1 is initially charged. At equilibrium, it has a finite, lower charge. The only graph that begins at a positive value and exponentially approaches a lower positive value is graph (B).

4. **(D)** Beginning at the top right corner of the circuit and moving counterclockwise:

   1. The potential jump across C_1 is positive because we are going from the negative plate to the positive plate (which is at a higher potential).
   2. The potential jump across the resistor is negative because we are going in the same direction as the current (which flows counterclockwise).
   3. The potential jump across C_2 is negative because we are going from the positive plate to the negative plate.

   Note: Of course, if you choose to go in the opposite direction, all the signs will be reversed and an equivalent equation will be obtained (which can be converted to the equation in choice (D) by multiplying both sides by -1).

5. **(B)** The current function, $I(t)$, exponentially decays toward zero when both charging and discharging a capacitor. $Q(t)$ and $U(t)$ behave similarly because they are related via the equation $U = Q^2/2C$. Both quantities exponentially decay toward zero when discharging a capacitor but exponentially approach their final values when charging a capacitor.

6. **(C)** The charge on the positive plate of the capacitor increases while the capacitor is being charged and decreases while the capacitor is being discharged; therefore, the current is first positive and then negative, eliminating choices (B) and (D). Choice (E) can be eliminated because the discharging current decays very slowly, which indicates that the capacitor is discharging slowly rather than rapidly.

   Because charging is slower than discharging, the time constant ($\tau = RC$) must be greater while charging. The capacitor is the same in both cases, so this requires that the resistance experienced while charging must be greater than the resistance experienced while discharging ($R_{\text{charging}} > R_{\text{discharging}}$). Now, if V_0 is the voltage of the battery used to charge the capacitor, the initial charging current is V_0/R_{charging} and the initial discharging current is $V_0/R_{\text{discharging}}$. Because $R_{\text{charging}} > R_{\text{discharging}}$, the initial current magnitude while charging is smaller than the initial current magnitude while discharging, eliminating choice (A).

7. **(D)** The time constant of an RC circuit is $\tau = RC$. In choice (A), adding a resistor R in parallel will decrease the equivalent resistance by a factor of 2, decreasing the time constant by a factor of 2. In choice (B), the new circuit will have an equivalent resistance of $\frac{1}{3}R$ and an equivalent capacitance of $2C$, so overall the new time constant will be $\tau' = \frac{2}{3}RC$. In choice (C), adding a capacitor in series with the capacitor decreases the equivalent capacitance by a factor of 2, decreasing the time constant by a factor of 2. In choice (D), the new circuit will have an equivalent resistance of $3R$ and an equivalent capacitance of $\frac{1}{2}C$, so overall the new time constant will be $\tau' = \frac{3}{2}RC$. In choice (E), the new circuit will have an equivalent resistance of $2R$ and an equivalent capacitance of $\frac{1}{6}C$, so overall the new time constant will be $\tau' = \frac{1}{3}RC$.

8. **(E)** Although the rate at which energy is dissipated by the RC loop of the circuit is finite, energy continues to be dissipated by the $2R$ resistor forever, so the energy dissipated from $t = 0$ to $t = \infty$ is undefined (infinite).

9. **(E)** Because the battery imposes a voltage of V across each branch of the circuit, the current through the branch consisting of the resistor and capacitor is independent of the current through the $2R$ resistor. Therefore, the current through the RC branch decreases exponentially while the current through the $2R$ resistor remains constant.

   The current through the battery is equal to the current through the $2R$ resistor plus the current through the RC segment of the circuit (because of Kirchhoff's node rule). Therefore, the graphs of the currents through the two wires add together (superimpose) to yield a net current that exponentially decreases to approach a positive value equal to the current through the $2R$ resistor, as shown in Figure 16.8.

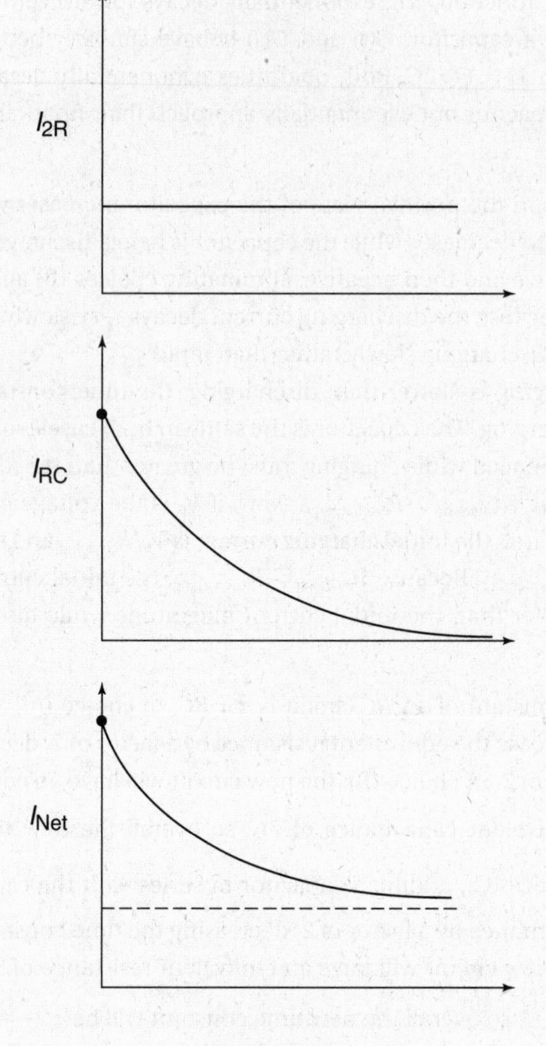

Figure 16.8

Initial current values allow selection between choices (C) and (E). The initial current through the battery is $I_0 = (V/R) + (V/2R) = \frac{3}{2}(V/R)$. The final current is equal to the current flowing through the $2R$ resistor, $V/2R$. Therefore, the initial current is three times the final current, which is true only of choice (E).

Free-Response

1. (a) At time $t = 0$, the voltage across the capacitor is zero, so the capacitor acts like a piece of wire (a short circuit). Therefore, at time $t = 0$, current is equal to I_0 and we can use Ohm's law to calculate the resistance of the resistor: $R = V/I_0$.

   (b) The time constant $\tau = RC$. Therefore, $C = \tau/R = I_0\tau/V$.

   (c) The magnitude of the current is equal to the time derivative of the charge on the capacitor: $|I(t)| = |dQ/dt|$. Because the capacitor is being charged, current increases the charge on the capacitor, so the sign of this relationship is positive: $I(t) = +dQ/dt$. Therefore, we can integrate charge as follows.

$$Q = \int_{t=0}^{t=t_f} dQ = \int_0^{t_f} I(t)dt = \int_0^{t_f} I_0 e^{-t/\tau} dt = I_0\tau\left(1 - e^{-t_f/\tau}\right)$$

Therefore,

$$Q(t) = I_0\tau(1 - e^{-t/\tau})$$

Answer check: Given the relationships $R = V/I_0$ and $\tau = RC$, it is not difficult to see that the final charge will be $Q = CV$, as expected. We also observe that, as always, the difference between the charge at any instant and the final charge decreases exponentially with a time constant τ.

(d) Given that $I(t) = I_0 e^{-t/\tau}$, plot ln $(I(t)/I_0)$ versus t.

(e) You have $I_0 = \dfrac{V}{R} = \dfrac{10\text{ V}}{50{,}000\ \Omega} = 2.0\times10^{-4}$ A and $\tau = RC = (50{,}000\ \Omega)(200\ \mu\text{F}) = 10$ s. The plot is shown in Figure 16.9.

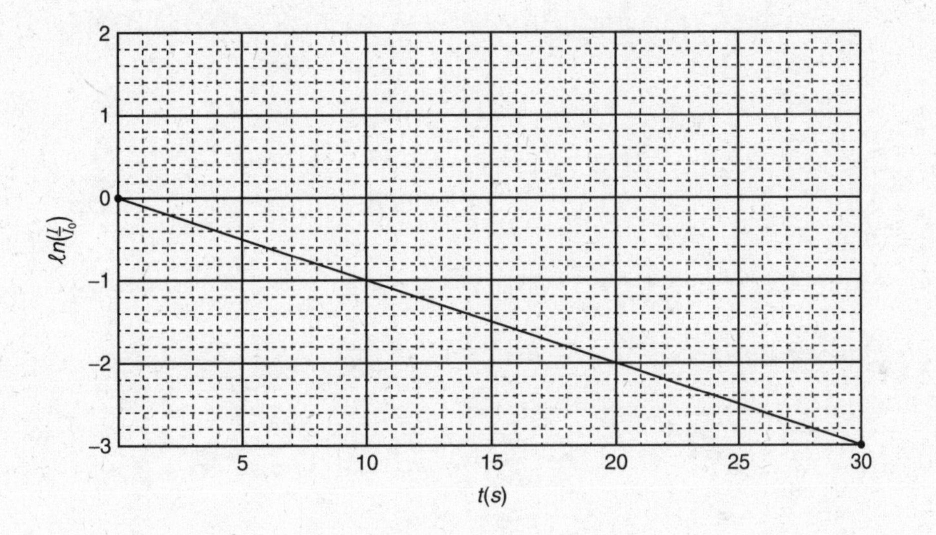

Figure 16.9

Magnetic Fields

17

- → DEFINITION OF MAGNETIC FIELD AND MAGNETIC FORCE ON A MOVING CHARGE
- → MAGNETIC FORCE ON A CURRENT-CARRYING WIRE
- → MAGNETIC FIELD DUE TO A MOVING CHARGE
- → BIOT-SAVART LAW: MAGNETIC FIELD DUE TO A PORTION OF WIRE
- → AMPERE'S LAW

WHAT IS A MAGNETIC FIELD?

A magnetic field is in some ways similar to an electric field; it is a vector field caused by charges that determines the force felt by other charges. The rest of this chapter will clarify this vague definition in detail, first exploring the effects of magnetic fields and then the sources of magnetic fields.

Magnetic field is generally designated with the vector **B** and has units of teslas,

$$T = \frac{N}{C \cdot m/s} = \frac{N}{A \cdot m}$$

Sources of Magnetic Field Magnetic fields are caused by moving charge (moving point charges or current moving through a wire). On a microscopic level, the intrinsic *spin* of electrons (which at a very crude level can be thought of as the electron spinning about its axis) can cause magnetic fields in "permanent magnets" like the ones on your refrigerator. (Don't worry if you don't understand this; the point is that even in permanent magnets where there are no obvious moving charges, if you look closely enough, the magnetic field is still due to moving charges.)

Force on a Moving Charge in a Magnetic Field The force on a point charge moving through a magnetic field is given by the following equation.

A charge must be in motion to feel a magnetic force.

$$\mathbf{F}_{\text{magnetic}} = q\mathbf{v} \times \mathbf{B}$$

Force exerted on point charge by magnetic field

There are a few observations that we can immediately make from this equation:

1. The magnetic force on a charge at rest is zero (i.e., magnetic force is exerted only on moving charges).
2. The magnetic force on a charge moving parallel or antiparallel to the magnetic field is zero.

3. Magnetic force, when nonzero, is perpendicular to both velocity and magnetic field. Thus, it contributes only to radial acceleration and never performs work on an object or changes an object's speed (it can only change an object's direction).
4. Magnetic force is proportional to the charge of a point charge and thus exerts forces only on charged objects.
5. The direction of the magnetic force can be obtained using the right-hand rule (when the fingers of the right hand sweep from **v** to **B**, the thumb of the right hand points in the direction of the force, as discussed in Chapter 1. When the charge is negative, the direction is reversed).

We can combine this equation with the force due to an electric field to write the net force on a point charge due to both electric and magnetic fields:

$$\mathbf{F}_{\text{electric}} = q\mathbf{E}$$

$$\mathbf{F}_{\text{net}} = \mathbf{F}_{\text{electric}} + \mathbf{F}_{\text{magnetic}} = q(\mathbf{E} + \mathbf{v} \times \mathbf{B})$$

Net force on a point charge due to electric and magnetic fields

This equation states that the electric and magnetic fields superpose; the net force is the vector sum of the two independent forces.

Magnetic Field Lines Like electric fields, magnetic fields can be illustrated using field lines. The general rules are the same: (1) the magnetic field is tangent to the lines, (2) the density of lines is proportional to the strength of the field, and (3) because the direction of the magnetic field is unique, the field lines never intersect. By definition, the lines point from the north pole to the south pole outside a magnet. (Note that magnetic poles are not equivalent to geographic poles. Because the needle of a compass points to the geographic *north* pole, the geographic north pole of Earth is actually a magnetic *south* pole.) Unlike electric field lines, there are no magnetic monopoles (discussed further in Chapter 20), so magnetic field lines always form closed loops. Figure 17.1 shows the magnetic field lines produced by a bar magnet and by a loop of wire.

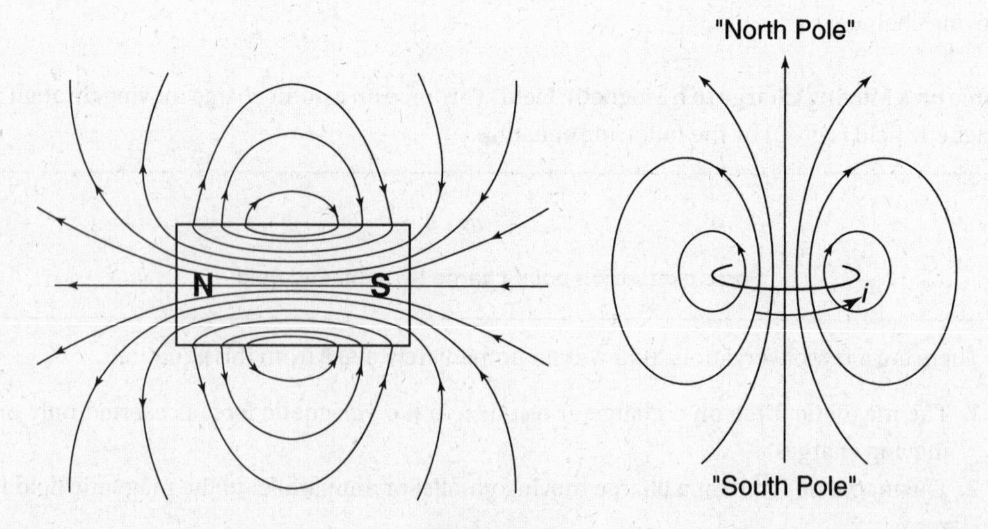

Figure 17.1

General Description of the Motion of Charged Particles in a Constant Magnetic Field There are three types of motion that can occur when a charged particle moves in a constant magnetic field.

1. Charged particles moving *parallel* to the magnetic field experience no force and therefore move in a straight line.
2. Charged particles moving *perpendicular* to the magnetic field experience a force perpendicular to their velocity. This force supplies a centripetal force, causing these particles to move in uniform circular motion. The strength of the field determines the radius of this motion according to the following equation that sets the magnetic force equal to the centripetal force:

$$\frac{mv^2}{r} = q|\mathbf{v} \times \mathbf{B}|$$

3. The motion of charged particles whose velocity has components both parallel and perpendicular to the magnetic field is a superposition of case #1 and case #2. These particles experience a constant velocity parallel to the magnetic field while moving in uniform circular motion in the plane perpendicular to the magnetic field. The net effect is a helical (corkscrew) path through space, with the axis of the corkscrew parallel to the magnetic field.

TIP

If a charge moves parallel to the magnetic field, it feels no force.

Force on a Current-Carrying Wire What is the force on a length of wire l carrying a current I in a magnetic field **B**?

The force of a magnetic field on a wire is equal to the sum of the forces of the magnetic field on all the electrons moving within the wire (because a magnetic field exerts forces only on moving charges). The net force on a wire carrying no current is zero: The random motion of electrons gives rise to no *net force* for the same reason that the random motion of electrons gives no rise to *net current*: The forces on the individual electrons are vectors pointing in random directions that sum to zero.

Now consider a current-carrying wire. As discussed in Chapter 14, we can consider the motion of the electrons a superposition of drift velocity on random motion. As above, the random motion component of the electrons gives rise to no net force; it is the drift velocity of the electrons that produces a net force. Recall from Chapter 14 that the drift velocity is equal to the average velocity of the electrons, mathematically given by the equation

$$\mathbf{v}_D = \text{average electron velocity} = \frac{\sum_{i=1}^{N} \text{velocity of electron}_i}{N}$$

As a first step toward calculating the net force on the wire, let's consider the force on a single electron moving in a magnetic field:

$$\mathbf{F}_{\text{magnetic}} = e\mathbf{v} \times \mathbf{B}$$

Therefore, the net force on two electrons in the same field can be simplified using the property of the cross-product that $\mathbf{A} \times \mathbf{B} + \mathbf{C} \times \mathbf{B} = (\mathbf{A} + \mathbf{C}) \times \mathbf{B}$.

$$\mathbf{F}_{\text{net}} = e\mathbf{v}_1 \times \mathbf{B} + e\mathbf{v}_2 \times \mathbf{B} = e(\mathbf{v}_1 \times \mathbf{B} + \mathbf{v}_2 \times \mathbf{B}) = e(\mathbf{v}_1 + \mathbf{v}_2) \times \mathbf{B}$$

The force on a current-carrying wire is the sum of the forces on *all* the electrons in the wire. Generalizing the above formula to N electrons in a wire,

$$\mathbf{F}_{\text{magnetic}} = e\left(\sum_{i=1}^{N} \mathbf{v}_i\right) \times \mathbf{B}$$

Based on the definition of the drift velocity, $\mathbf{v}_D = \frac{\sum_{i=1}^{N} \mathbf{v}_i}{N}$, we can express this as

$$\mathbf{F}_{\text{magnetic}} = e(N\mathbf{v}_D) \times \mathbf{B}$$

Now we have an answer, but we need to express it in more convenient terms (generally the number of electrons in the wire N and the drift velocity are not readily measurable). This involves a few substitution steps based on definitions involving drift velocity and current density from Chapter 14.

In terms of n, the volume density of electrons in the conductor, the number of electrons N in a section of wire of length l and cross-sectional area A, is

$$N = n(\text{volume}) = nlA$$

Therefore,

$$\mathbf{F}_{\text{magnetic}} = (nlA)\, e\mathbf{v}_D \times \mathbf{B}$$

Again, from Chapter 14, we know that for the special case of uniform current flow (which applies to the current-carrying wire we are considering here),

$$I = JA$$

and

$$\mathbf{J} = ne\mathbf{v}_D$$

Thus, the magnitude of the current is given by

$$I = nev_DA$$

Using these equations, we can express the force in terms of current:

$$\mathbf{F}_{\text{magnetic}} = I\mathbf{l} \times \mathbf{B}$$

Force on a straight current-carrying wire in a constant magnetic field

where we have defined the vector **l** to point in the direction of the current, with a magnitude equal to the length of the wire.

What about a curved wire? The above equation no longer applies because the vector **l** is no longer uniform. However, if we break the wire into tiny differential lengths, each differential length is approximately straight, so the above equation is valid on the differential level:

$$d\mathbf{F}_{\text{magnetic}} = Id\mathbf{l} \times \mathbf{B}$$

Force on a differential length *dl* of current-carrying wire in a magnetic field

To find the net force on the curved wire, we then have to sum the differential forces via integration. The problem-solving approach to this equation is essentially identical to the approach used with Biot-Savart's law, so we defer problem-solving for this equation until later in the chapter.

Note: This is a conventional-current derivation: We have pretended that electrons have positive charge e and that they move in the same direction as conventional current. Of course, because you are a smart Physics C student, you know that electrons have negative charge and move in the direction opposite conventional current. However, this derivation illustrates why the conventional-current convention works. Consider the equation for the force on a point charge: $\mathbf{F}_{\text{magnetic}} = q\mathbf{v} \times \mathbf{B}$. The force on a particle with charge $+e$ moving in the same direction as conventional current ($\mathbf{v}_D$) is exactly the same as the force on a particle with charge $-e$ moving in the direction opposite conventional current ($-\mathbf{v}_D$):

$$(-e)(-\mathbf{v} \times \mathbf{B}) = (+e)(\mathbf{v} \times \mathbf{B})$$

Therefore, the conventional-current approach gives the correct answer.

Be careful when you calculate a magnetic force. Just like the electric force, its direction depends on the sign of the charge.

Forces on Wires in Uniform Magnetic Fields In the case of a uniform magnetic field, the magnetic field comes out of the integral, so the net force on a wire of arbitrary shape is

$$\mathbf{F}_{\text{magnetic}} = \int I\, d\mathbf{l} \times \mathbf{B} = I\left(\int d\mathbf{l}\right) \times \mathbf{B}$$

The term $\int d\mathbf{l}$ is simply the net displacement of the wire. Therefore, when the wire is a closed loop, the net displacement is zero and the force is zero. Alternatively, when the loop is a curve that does not close on itself, the force depends only on the endpoints and is equal to the force on a straight wire with the same endpoints, as shown in Figure 17.2.

Of course, in the case of a straight wire, this integral reduces to the previous result, $\mathbf{F}_{\text{magnetic}} = I\mathbf{l} \times \mathbf{B}$.

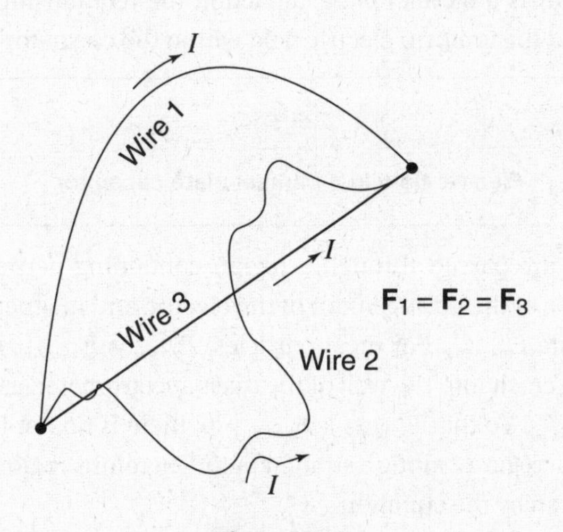

Figure 17.2

PROBLEM SOLVING: MASS SPECTROMETER PROBLEMS

A mass spectrometer is a device that measures the mass/charge ratio of charged atoms and molecules. Figure 17.3 is a schematic diagram of a mass spectrometer.

In region 1 charged particles are accelerated by passing them through a parallel plate capacitor with holes in the plate to allow for the charges to be inserted into the capacitor and then for the charges to fly out of the capacitor. The particles gain an amount of energy equal to qV in the form of kinetic energy.

$$qV = \frac{mv^2}{2}$$

Energy gained by passing through a capacitor converted to KE

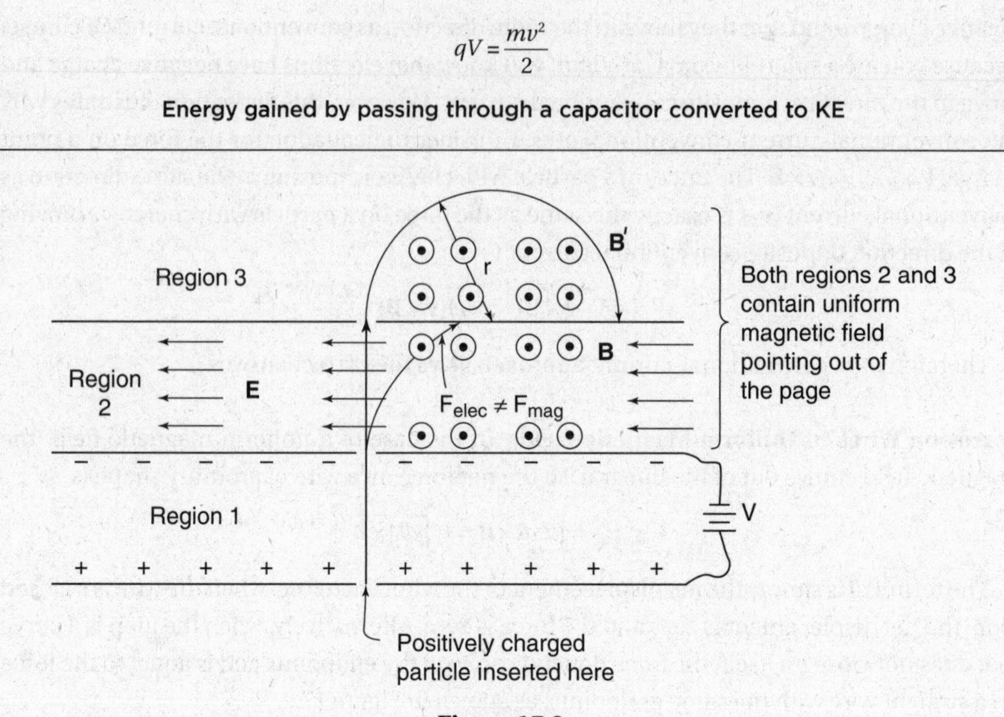

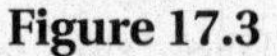

Figure 17.3

Because the capacitor is a parallel plate capacitor, the relationship between the voltage across the capacitor and the uniform electric field within the capacitor is

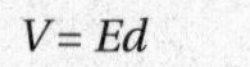

$$V = Ed$$

Electric field in a parallel plate capacitor

In region 2 charges are passed through a region containing perpendicular electric and magnetic fields. Because of the arrangement of the electric and magnetic fields, $\mathbf{F}_{\text{electric}}$ points in the direction opposite $\mathbf{F}_{\text{magnetic}}$. For most particles, $F_{\text{electric}} \neq F_{\text{magnetic}}$, so there is a net force causing the particles to crash into the wall of the mass spectrometer, as shown in Figure 17.3. However, if $F_{\text{electric}} = F_{\text{magnetic}}$ so that $\mathbf{F}_{\text{magnetic}} + \mathbf{F}_{\text{electric}} = 0$, there is no net force and the particles pass effortlessly through region 2 along a straight path. Therefore, region 2 selects for particles of a certain velocity, given by the equation

$$\mathbf{F}_{\text{net}} = \mathbf{F}_{\text{electric}} + \mathbf{F}_{\text{magnetic}} = q(\mathbf{E} + \mathbf{v} \times \mathbf{B}) = 0$$

$$q\mathbf{E} = -q\mathbf{v} \times \mathbf{B}$$

$$\mathbf{E} = -\mathbf{v} \times \mathbf{B}$$

Thus, only particles with a known velocity given by the ratio E/B exit from this region.

Region 3 contains only a magnetic field $\mathbf{B}'$ perpendicular to the diagram. Therefore, the moving charges experience a force perpendicular to their velocity; this force provides a centripetal force, causing the charges to move in circles of various radii according to the equation

$$\frac{mv^2}{r} = q|\mathbf{v} \times \mathbf{B}'|$$

Overall design and function of mass spectrometers. The term "mass spectrometer" is a misnomer; mass spectrometers do not allow us to calculate the mass of a particle but instead its mass-to-charge ratio (m/q). Additionally, any combination of two of the above regions is sufficient to calculate the mass-to-charge ratio from experimental parameters of the mass spectrometer, such as the strengths of the magnetic and electric fields. (Thus, the combination of all three regions is redundant and is generally not used in real-life mass spectrometers.)

SOURCES OF MAGNETIC FIELDS

Magnetic Field Due to a Point Charge

As mentioned at the beginning of this chapter, magnetic fields are caused by moving charge. Specifically, the magnetic field due to a point charge moving with constant velocity is

TIP

Moving charges create magnetic fields.

$$\mathbf{B} = \frac{\mu_0}{4\pi} \frac{q\mathbf{v} \times \hat{r}}{r^2}$$

Magnetic field due to a point charge

In this equation, $\mathbf{r}$ is defined to be a position vector pointing from the instantaneous location of the point charge q to the point in space where we are calculating the magnetic field ($\hat{r}$ is a unit vector parallel to $\mathbf{r}$). μ_0 is a new constant, called the *permeability of free space*, whose value is defined to be $\mu_0 = 4\pi \cdot 10^{-7} \text{T} \cdot \text{m/A}$.

Getting a feel for the way magnetic fields behave is much more difficult than for electric fields because the cross-product in the equation above makes magnetic fields behave in decidedly unusual ways. The following descriptions should help you understand the direction and magnitude of magnetic fields. Additionally, Figure 17.4 shows the magnetic field along the yz-plane produced by a positive point charge at the origin moving in the $+x$-direction. (The field along other planes parallel to the yz-plane has the same general shape but smaller magnitudes.)

1. Consider another right-hand rule. The magnetic field lines caused by a moving point charge (or a number of charges moving in a wire) form concentric circles around the path of the charges. (The magnetic field at any point is tangent to these circles.) The direction of these concentric circles is given by the right-hand rule: If you point your right thumb in the direction of conventional current or the velocity vector of a positive point charge, your fingers curl in the direction of the magnetic field. (If you are dealing with a negative point charge, the direction of the field will be reversed.)
2. Along any line that passes through the point charge (such that the sin θ component of the cross-product is constant), the magnitude of the magnetic field is proportional to the inverse square of the distance to the point charge.
3. On the surface of an imaginary sphere centered at the moving point charge, the magnetic field is maximized at the equator of the sphere where $\hat{r} \perp \mathbf{v}$ and is zero at the poles of the sphere where $\hat{r}$ is parallel to $\mathbf{v}$, as in Figure 17.5.

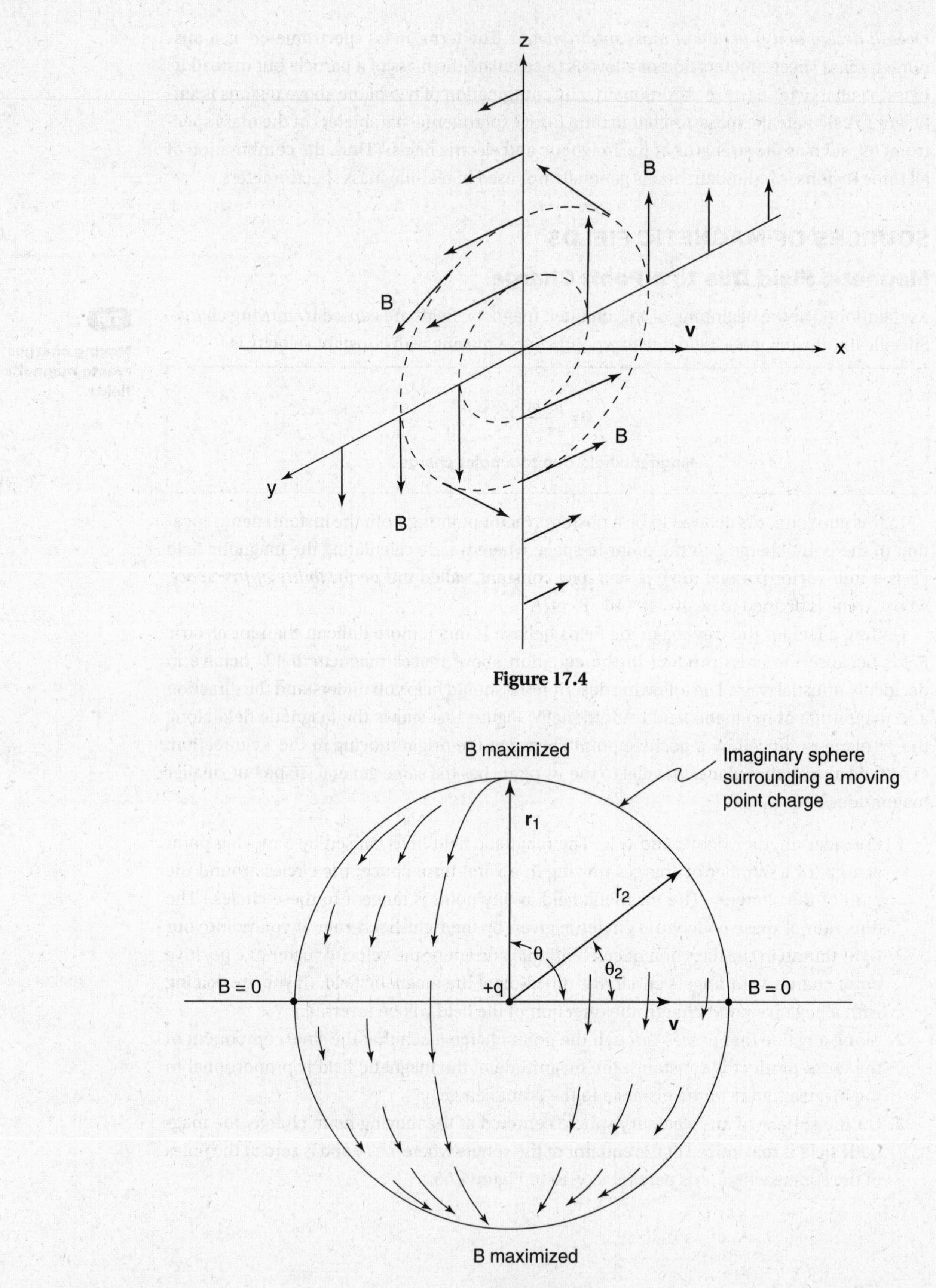

Figure 17.4

Figure 17.5

Magnetic Field Due to a Differential Segment of Current-Carrying Wire: Biot-Savart Law

Now we want to convert our equation for a magnetic field created by a single point charge to an equation for the magnetic field created by a current-carrying wire. This is very similar to our conversion of the formula for a force on a single moving charge to the formula for a force acting on a current-carrying wire (in the first part of this chapter).

Consider a magnetic field caused by an infinitely long wire at a certain point in space. Because of the form of the magnetic field created by a point charge, we can see that the contribution to the magnetic field by different segments of wire is different because the vector **r** from the segment to the point P varies. For example, wire segments closer to point P contribute more to the magnetic field than wire segments farther away. Therefore, to find the magnetic field due to the entire wire, we need to divide the wire into differential segments of length dl. For each segment, all the electrons have the same **r**, so we can treat them together and find the $d\mathbf{B}$ due to the differential length of wire. (One approximation we must make here is that the dimensions of the wire segment, such as its cross-sectional area, are negligible compared to the **r** vector, so that all the electrons have approximately the same **r**.) We can then integrate the differential magnetic fields due to each differential wire segment to find the net magnetic field due to the entire wire.

Note that this derivation relies on the *principle of superposition of magnetic fields*: The magnetic field due to a number of point charges is equal to the vector sum of the magnetic fields due to each point charge.

So, what is the differential magnetic field due to a differential wire segment? If we designate N as the number of electrons in a segment of wire,

$$d\mathbf{B}=\sum_{i=1}^{N}\left(\frac{\mu_0}{4\pi}\frac{q\mathbf{v}_i\times\hat{r}}{r^2}\right)=\frac{\mu_0}{4\pi}\frac{q\left(\sum_{i=1}^{N}\mathbf{v}_i\right)\times\hat{r}}{r^2}$$

As before in this chapter, based on the definition of drift velocity,

$$\mathbf{v}_D=\frac{\sum_{i=1}^{N}(\text{velocity of electron})_i}{N}$$

We can substitute:

$$d\mathbf{B}=\frac{\mu_0}{4\pi}\frac{q(N\mathbf{v}_D)\times\hat{r}}{r^2}$$

Again, we would like to express this answer in terms of more useful quantities, specifically current. We begin by introducing n, the volume density of electrons, so that

$$N=n(\text{volume})=n(A\,dl)$$

where A is the cross-sectional area of the wire. Substituting this into the expression for the magnetic field yields

$$d\mathbf{B}=\frac{\mu_0}{4\pi}\frac{q(nA\,dl)}{r^2}\mathbf{v}_D\times\hat{r}$$

As in the previous derivation, $I = qnAv_D$. Substitution then yields

$$d\mathbf{B} = \frac{\mu_0}{4\pi}\frac{I}{r^2}d\mathbf{l} \times \hat{r}$$

Biot-Savart law: magnetic field due to a differential length of wire

Note that calculating the magnetic field due to a current reflects many of the properties of the magnetic field due to a point charge. The magnetic field increases if the wire carries more current, if it is closer to the point of interest, or if $\hat{r}$ is more nearly perpendicular to $d\mathbf{l}$.

USING THE BIOT-SAVART LAW TO CALCULATE THE MAGNETIC FIELD DUE TO A WIRE

This is very similar to calculating the electric field due to a continuous charge distribution by direct integration.

STEP 1 Use symmetry to determine the allowed directions of the magnetic field. In Chapter 12, we discussed the technique of determining which directions are well-defined. When dealing with magnetic fields, because of the presence of directions defined by the direction of the current and the right-hand rule, this is generally not very helpful. Instead, it is best to consider the differential magnetic fields produced by each segment of wire and whether (1) symmetric pairs of differential fields might cancel each other in a certain direction, or (2) if the differential magnetic field never has a component in a certain direction. Additional symmetry considerations of magnetic fields will be addressed in Chapter 20.

STEP 2 Like the electric field, the magnetic field is a vector. Therefore, mathematically,

$$\mathbf{B} = \int d\mathbf{B} = \hat{i}\int dB_x + \hat{j}\int dB_y + \hat{k}\int dB_z$$

It is possible that one or more of these integrals will be zero if, using symmetry, we determine that the field has no component along that direction.

For the remaining components, we must integrate. Steps 3 and 4 explain how to perform an integration for one component of the magnetic field; repeat these steps for all remaining nonzero components.

STEP 3 Unlike in the case of electric fields, we do not have to worry about what type of differential regions are needed; we always use differential line segments. Therefore, performing a scalar integral isn't so difficult. We already know what the differential magnetic field due to a wire segment is:

$$d\mathbf{B} = \frac{\mu_0}{4\pi}\frac{I}{r^2}d\mathbf{l} \times \hat{r}$$

All we have to do is to find the component of $d\mathbf{B}$ in the direction we are integrating. This may involve multiplying by some sine or cosine factor dictated by the geometry of the situation. For example,

$$B_x = \left|\frac{\mu_0}{4\pi}\frac{I}{r^2}d\mathbf{l} \times \hat{r}\right| \times (\text{Proportion of } \mathbf{B} \text{ pointing in } x\text{-direction})$$

Note: In cases where the wire is straight, expressing dl as dx, dy, or dz works just fine. However, in the case of circular segments of wire, it is generally more convenient to work with differential arc lengths of wire whose lengths are given by the equation

$$dl = rd\theta$$

(which is simply the differential form of the equation arc length = $r\theta$).

When using this equation, express everything in terms of θ and integrate with respect to θ.

STEP 4 You then need to integrate your scalar differential magnetic field (e.g., dB_x) over the limits dictated by the problem. As usual, always integrate from the lower limit to the upper limit.

STEP 5 Once you have obtained the components of the magnetic field in all three directions (either by integration or symmetry), reassemble the magnetic field vector:

$$\mathbf{B} = B_x\hat{i} + B_y\hat{j} + B_z\hat{k}$$

EXAMPLE 17.1 MAGNETIC FIELD DUE TO AN INFINITELY LONG CURRENT-CARRYING WIRE

What is the magnetic field at the point $(-r, 0)$ caused by an infinitely long wire carrying a current I along the y-axis in the $+\hat{j}$-direction, as shown in Figure 17.6?

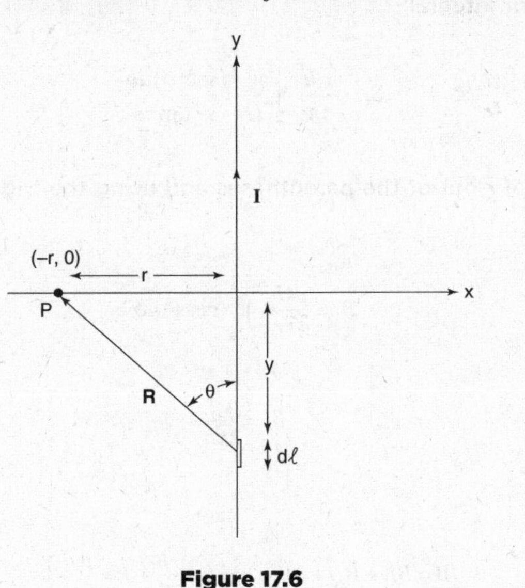

Figure 17.6

SOLUTION

STEP 1 The differential magnetic field $d\mathbf{B}$ due to every differential line segment points in the z-direction. Therefore, the net magnetic field must be in the z-direction (i.e., $B_x = B_y = 0$).

STEP 2 Based on symmetry, we must integrate only the magnetic field in the z-direction.

STEP 3

$$dB_Z = \left|\frac{\mu_0}{4\pi}\frac{I}{R^2}d\mathbf{l}\times\hat{R}\right| \times (\text{Proportion of } d\mathbf{B} \text{ pointing in } z\text{-direction})$$

We just said that $d\mathbf{B}$ points in the z-direction, so the proportion of $d\mathbf{B}$ pointing in the z-direction is 1. Also from Figure 17.6 and the definition of a cross-product,

$$dB_z = \frac{\mu_0}{4\pi}\frac{I}{(r^2+y^2)}dl\sin\theta$$

$$= \frac{\mu_0}{4\pi}\frac{I}{(r^2+y^2)}dy\frac{r}{(r^2+y^2)^{1/2}}$$

STEP 4

$$B_z = \int_{y=-\infty}^{y=+\infty} dB_z = \frac{\mu_0 Ir}{4\pi}\int_{-\infty}^{+\infty}\frac{dy}{(r^2+y^2)^{3/2}}$$

This integral requires a trigonometric substitution:

$$y = r\tan\phi \Leftrightarrow \phi = \tan^{-1}\left(\frac{y}{r}\right)$$

$$dy = r(\sec^2\phi)d\phi$$

Substituting into our integral,

$$B_z = \frac{\mu_0 Ir}{4\pi}\int_{-\pi/2}^{\pi/2}\frac{r(\sec^2\phi)d\phi}{(r^2+r^2\tan^2\phi)^{3/2}}$$

Taking the factors of r out of the parentheses and using the trigonometric identity $1 + \tan^2\phi = \sec^2\phi$,

$$B_z = \frac{\mu_0 I}{4\pi r}\int_{\phi=-\pi/2}^{\phi=\pi/2}(\cos\phi)d\phi$$

$$B_z = \frac{\mu_0 I}{2\pi r}$$

STEP 5

$$\mathbf{B} = B_x\hat{i} + B_y\hat{j} + B_z\hat{k} = 0\hat{i} + 0\hat{j} + \frac{\mu_0 I}{2\pi r}\hat{k} = \frac{\mu_0 I}{2\pi r}\hat{k}$$

$$B = \frac{\mu_0 I}{2\pi r}$$

Magnetic field due to an infinitely long current-carrying wire (direction given by the right-hand rule)

Because of the cylindrical symmetry of the wire, this is the magnetic field at any point a distance r from the wire. Thus, an infinitely long current-carrying wire produces magnetic field lines as shown in Figure 17.7. The magnetic field lines form concentric circles around the wire whose direction is indicated by the right-hand rule (when you point your right thumb in the direction of the current, your fingers curl in the direction of the field).

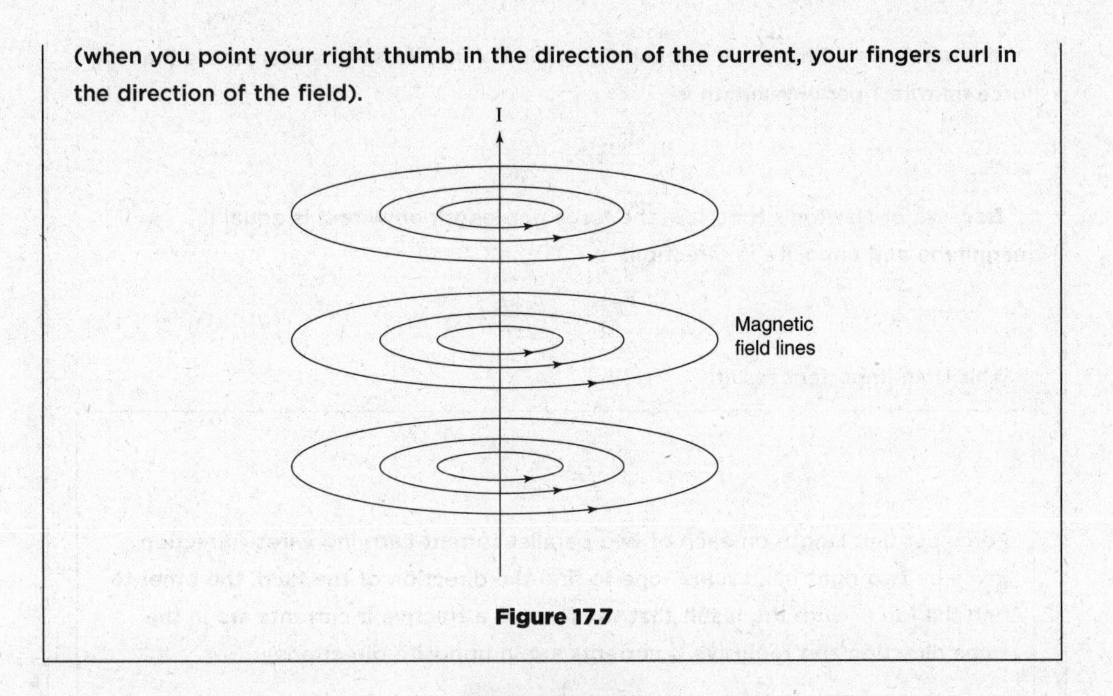

Figure 17.7

EXAMPLE 17.2

The force between parallel current-carrying wires: Given two infinitely long parallel wires lying on the *xy*-plane as shown in Figure 17.8, what is the force per length exerted on wire 1 by wire 2?

SOLUTION

Our general approach is first to calculate the magnetic field that wire 2 produces along the *y*-axis and then to find the force this field exerts on wire 1.

Along wire 1, the magnetic field due to wire 2 (as we just derived) is

$$\mathbf{B} = \frac{\mu_0 I_2}{2\pi d}\hat{k}$$

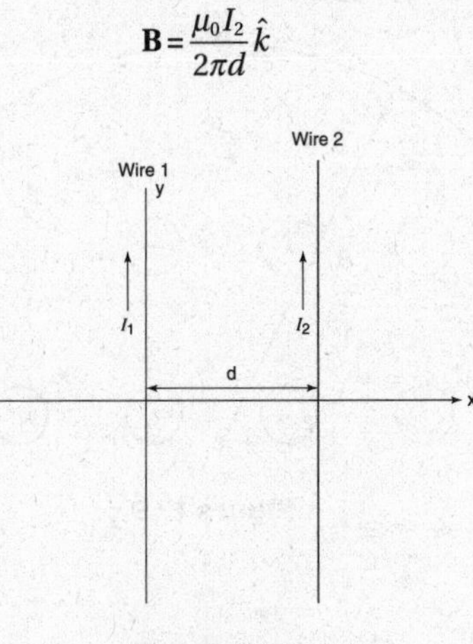

Figure 17.8

To find the force on wire 1, we use the equation for the force on a current-carrying wire in a constant magnetic field:

$$\mathbf{F}_{\text{on wire 1}} = I_1\mathbf{l} \times \mathbf{B} = I_1 l \frac{\mu_0 I_2}{2\pi d}\hat{i}$$

The wires are infinite, so $l = \infty$ and $\mathbf{F}_{\text{on wire 1}} = \infty\hat{i}$. A more useful statement is that the force on wire 1 per unit length is

$$\frac{\mathbf{F}_{\text{on wire 1}}}{l} = \frac{\mu_0 I_1 I_2}{2\pi d}\hat{i}$$

Because of Newton's third law, the force per length on wire 2 is equal in magnitude and opposite in direction:

$$\frac{\mathbf{F}_{\text{on wire 2}}}{l} = \frac{\mu_0 I_1 I_2}{2\pi d}\hat{i}$$

This is an important result:

$$\frac{F}{l} = \frac{\mu_0 I_1 I_2}{2\pi d}$$

Force per unit length on each of two parallel current-carrying wires (direction given by two right-hand rules—one to find the direction of the field, the other to find the force—with the result that the force is attractive if currents are in the same direction and repulsive if currents are in opposite directions)

EXAMPLE 17.3

Using the equation $d\mathbf{F}_{\text{magnetic}} = Id\mathbf{l} \times \mathbf{B}$, calculate the magnetic force on a semicircle of wire due to a uniform magnetic field as shown in Figure 17.9.

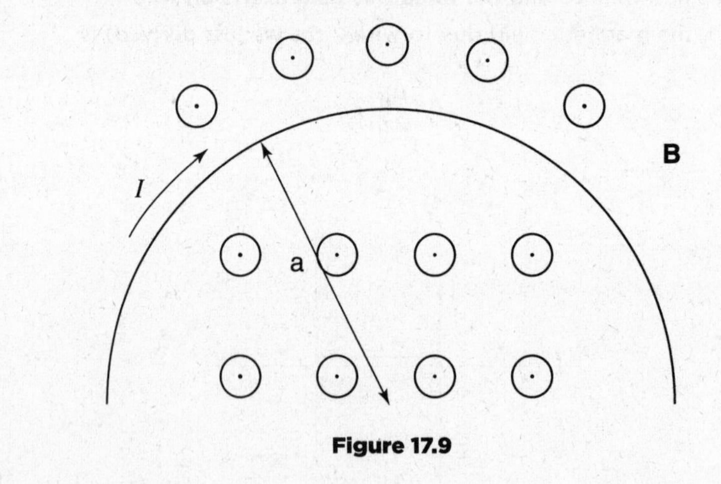

Figure 17.9

SOLUTION

Our approach is very similar to that used with the Biot-Savart law. By pairing matching differential line segments we see that the net magnetic force points in the y-direction, as shown in Figure 17.10 (because the net force on every pair of differential segments points in the y-direction, the sum of all the pairs must as well).

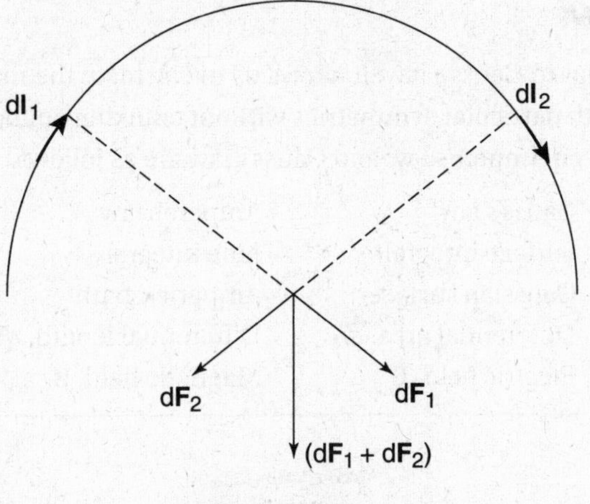

Figure 17.10

Therefore, our approach is to divide the semicircle into differential wire segments and sum the *y*-components of the magnetic forces on the segments. Because the wire is circular, the polar coordinate θ is preferable to dl. The geometry of the differential wire segments is shown in Figure 17.11.

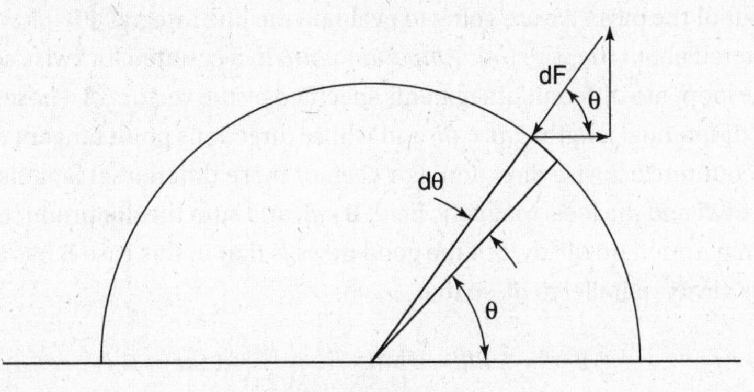

Figure 17.11

Now we can apply the equation $d\mathbf{F}_{\text{magnetic}} = Id\mathbf{l} \times \mathbf{B}$ (because $d\mathbf{l}$ is always perpendicular to the magnetic field, $d\mathbf{l} \times \mathbf{B} = dlB$). The *y*-component of the force on the line segment is given by

$$dF_y = -I(dl)B\sin\theta$$

To express everything in terms of the same variable, we can use the formula $dl = ad\theta$:

$$dF_y = -I(dl)B\sin\theta = -aIB(\sin\theta)d\theta$$

Finally, integrating,

$$F_y = 0 - \int_0^{\pi} aIB(\sin\theta)d\theta = -2aIB$$

Unit check: Comparison with the equation $\mathbf{F} = I\mathbf{l} \times \mathbf{B}$ reveals that the answer has the correct units (i.e., current × length × magnetic field).

Reality check: Recall that the magnetic force on a curved line segment due to a uniform magnetic field is equal to the force on a straight length of wire whose length is given by the distance between the endpoints of the curved wire. In the present problem, this distance is 2*a*, and thus our answer is consistent with this general result.

AMPERE'S LAW

Ampere's law is similar to Gauss's law: It allows us to calculate the magnetic field due to current distributions with particular symmetries without cranking out ugly Biot-Savart integrals. Some parallels between Ampere's law and Gauss's law are as follows.

Gauss's law	Ampere's law
Surface integral	Line integral
Gaussian surface	Amperian path
Differential area, $d\mathbf{A}$	Differential length, $d\mathbf{l}$
Electric field, **E**	Magnetic field, **B**

TIP

Ampere's law is a restatement of the Biot-Savart law.

$$\oint \mathbf{B} \cdot d\mathbf{l} = \mu_0 I_{\text{enclosed}}$$

Ampere's law

Note: This equation is valid only in cases where if an electric field is present as well, it is static. Later on we will generalize this equation to deal with situations where this is not the case.

What is a line integral? Consider Figure 17.12, which shows an infinitely long wire carrying a current I out of the page. We are going to evaluate the line integral $\oint \mathbf{B} \cdot d\mathbf{l}$ around a circle of radius r centered about the wire (our *Amperian path*) in a counterclockwise sense. To do this, we divide the loop into differential segments specified by the vectors $d\mathbf{l}$ whose magnitudes are equal to the differential lengths, $|d\mathbf{l}| = dl$, and whose directions point tangent to the Amperian path in the counterclockwise direction. For each of these differential lengths $d\mathbf{l}$, we take the dot product of $d\mathbf{l}$ and the local magnetic field, $\mathbf{B} \cdot d\mathbf{l}$, and sum the dot products over the entire circle. This may sound unwieldy, but the good news is that in this case **B** has a constant magnitude and is always parallel to $d\mathbf{l}$, so that

$$\oint \mathbf{B} \cdot d\mathbf{l} = \oint B dl = B \oint dl = Bl = \left(\frac{\mu_0 I}{2\pi r}\right)(2\pi r) = \mu_0 I$$

(using the magnetic field that we calculated using the Biot-Savart law in the last chapter).

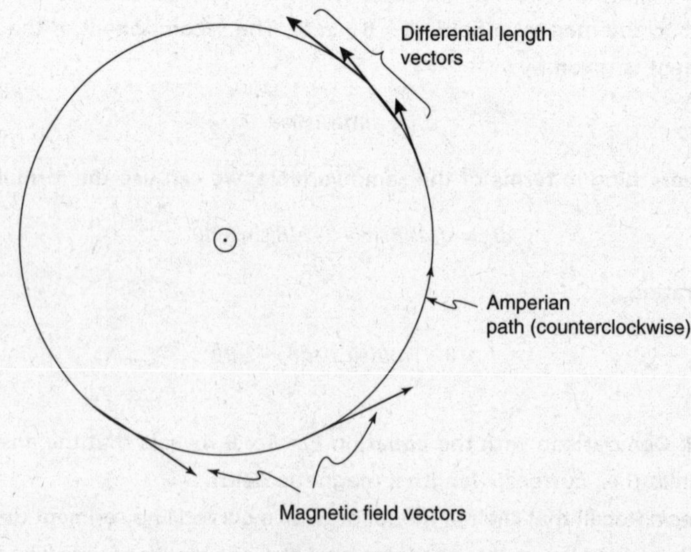

Figure 17.12

We have just verified Ampere's law, in this case using our result from the Biot-Savart law.

PROBLEM SOLVING: HOW TO APPLY AMPERE'S LAW

STEP 1 Choose an Amperian path based on the following guidelines.

a. The Amperian path must pass through the point where you wish to calculate the magnetic field.

b. Eventually, you will have to evaluate the integral $\oint \mathbf{B} \cdot d\mathbf{l}$ around your Amperian path. To make this possible, you should design this path so that it contains segments where $\mathbf{B} \cdot d\mathbf{l}$ is constant and possibly segments where $\mathbf{B} \cdot d\mathbf{l}$ is zero (either because $\mathbf{B} \perp d\mathbf{l}$ or because $\mathbf{B} = 0$). This generally requires a qualitative understanding of the magnetic field before you calculate its magnitude. Additionally, as with gaussian surfaces, it is generally best to pick paths with as much symmetry as possible.

c. To calculate a nonzero magnetic field, the path must enclose nonzero current.

Happily, there are only a few path types that you should be familiar with, all of which are featured in the examples below.

STEP 2 Calculate the magnitude of the enclosed current. There are a few general possibilities that you should be able to deal with. Read this section now but don't worry if it's not entirely clear; the examples at the end of this chapter should provide further explanation.

a. Constant current density:

   i. You are given a constant current density (current per area or current per length). To calculate the I_{enclosed}, just multiply this density by the enclosed area (if given current per area) or by the enclosed length (if given current per length).

   ii. You are given the current that passes through the entire wire and told that the current density is constant. In this case, you can use the fact that current is proportional to area to calculate the enclosed current.

b. You are given the current density as a function of position. In this case you must integrate to find I_{enclosed}. For differential regions, use differential areas across which the current density function is constant (typical integral building). Therefore, the geometry of the differential areas must mirror the current density function: If the current density is a function of radius (cylindrical symmetry), use washers; if the current density is a function of distance from the plane (planar symmetry), use rectangular strips.

c. In the case of a solenoid or a toroid, you may be given the number of coils per length or the total number of coils and the current passing through the solenoid or toroid. To calculate I_{enclosed}, multiply the number of wires passing through the Amperian path by the current passing through each wire, which is equal to the current passing through the entire solenoid or toroid. (Assume that these currents all pierce the Amperian path in the same direction; otherwise you will have to subtract some currents from others to obtain the net enclosed current.)

STEP 3 At this point, you're ready to apply Ampere's law:

$$\oint \mathbf{B} \cdot d\mathbf{l} = \mu_0 I_{\text{enclosed}}$$

Ampere's law

If you choose a good Amperian path, evaluating $\oint \mathbf{B} \cdot d\mathbf{l}$ won't be too difficult:

In some regions, **B** is parallel or antiparallel to $d\mathbf{l}$ and constant, so that the dot product is easily evaluated and we can put $|B|$ in front of the integral. In other regions, **B** may be perpendicular to $d\mathbf{l}$ or zero, so that $\mathbf{B} \cdot d\mathbf{l}$ is zero.

In which direction should you evaluate the line integral (clockwise or counterclockwise)? Should positive current be designated as passing into the page or out of the page? There are several conventions for setting all these signs correctly. The conventions that we will use here are the following.

- Always take the line integral in the counterclockwise direction.
- Currents coming *out* of the page are positive; currents going into the page are negative.

If you examine these conventions carefully, you will see that they have the right-hand rule built in. Therefore, if you figure out the direction of the magnetic fields correctly using the right-hand rule, the signs of Ampere's law will always match up correctly (which is a good way to check your work). Alternatively, it is possible to determine the direction of the magnetic field by carefully examining the signs in the Ampere's law equation (if a negative sign is missing, the magnetic field must be antiparallel to $d\mathbf{l}$), but we don't recommend this.

EXAMPLE 17.4 MAGNETIC FIELD DUE TO AN INFINITELY LONG CURRENT-CARRYING WIRE

Use Ampere's law to calculate the magnetic field due to an infinitely long wire carrying current out of the page a distance *r* from the wire.

SOLUTION

We are already quite familiar with the magnetic fields produced by infinitely long current-carrying wires; their magnitudes are dependent only on the distance from the wire (as required by symmetry), and the direction is given by the right-hand rule. Therefore, as covered in the introduction to Ampere's law, we can use a circular Amperian path that lies in a plane perpendicular to the wire. Along this path, $\mathbf{B}$ is always parallel to $d\mathbf{l}$ (both point in the counterclockwise direction) and the magnitude of $\mathbf{B} \cdot d\mathbf{l}$ is constant. Because the current is coming out of the page, it is designated as positive: $I_{enclosed} = I$. Applying Ampere's law,

$$\oint \mathbf{B} \cdot d\mathbf{l} = B\oint dl = B(2\pi r) = \mu_0 I_{enclosed}$$

$$B = \frac{\mu_0 I}{2\pi r}$$

Note how easy it is to calculate this field using Ampere's law compared to calculating it via a Biot-Savart integral.

EXAMPLE 17.5

The current density J(*r*) of a wire of radius *b* is parallel to the wire going into the page with a magnitude given by the function *J*(*r*) = *ar* (where *a* is a constant with units of A/m^3). Calculate the magnitude of the magnetic field in the following regions:

(a) $R < b$

(b) $R > b$

(c) $R = b$

(a) Qualitatively, the field is the same as the magnetic field due to a current-carrying wire (concentric circles around a wire of uniform magnitude). Therefore, our Amperian path will be circles of radius *R*, exactly as in Example 17.4. The magnetic field along the Amperian path is constant and antiparallel to the direction of integration (the magnetic field is in the clockwise direction), so the line integral is

$$\oint \mathbf{B} \cdot d\mathbf{l} = -2\pi RB$$

The tricky part here is calculating the magnitude of the enclosed current. As shown in Figure 17.13, we need to divide the cross-section of the wire into washers (each of which has a constant current density) and sum the current passing through each washer (given by the differential below).

$$dI = -[J(r)]dA = -[J(r)](2\pi r\, dr) = -(ar)(2\pi r\, dr) = -2\pi ar^2\, dr$$

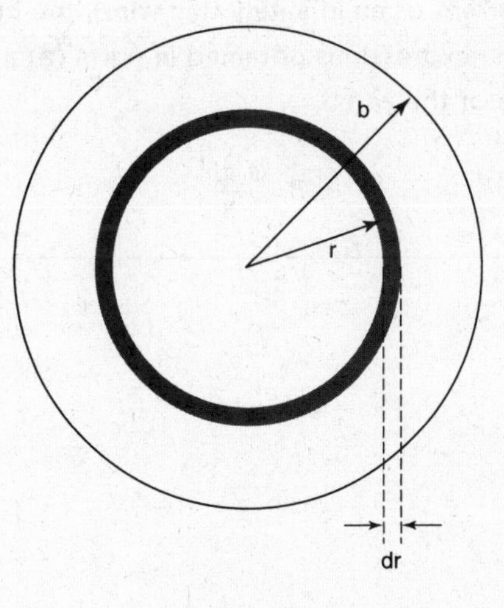

Figure 17.13

The current enclosed in our Amperian path of radius *R* is then as follows. (Note that because the current density J is parallel to the wire, we can use the equation $I = JA$; this is usually the case for the problems we will encounter.)

$$I_{\text{enclosed}} = \int dI = \int_{r=0}^{r=R} (-2\pi ar^2)dr = -\frac{2}{3}\pi aR^3$$

Applying Ampere's law,

$$\oint \mathbf{B} \cdot d\mathbf{l} = -2\pi RB = \left[\mu_0 I_{\text{enclosed}} = -\frac{2}{3}\pi\mu_0 aR^3\right]$$

$$B = \frac{\mu_0 aR^2}{3}$$

Answer check: The magnetic field is zero at the center of the wire, which makes sense (either because of symmetry arguments or because there is no enclosed current). This is analogous to the way the electric field is always zero at the center of a spherically symmetric charge distribution.

(b) Exactly the same as part (a), except that the Amperian path (of radius *R*) encloses the entire wire, so the enclosed current is

$$I_{\text{enclosed}} = \int dI = \int_{r=0}^{r=b} (-2\pi ar^2)\,dr = -\frac{2}{3}\pi ab^3$$

Therefore,

$$\oint \mathbf{B} \cdot d\mathbf{l} = -2\pi RB = \left[\mu_0 I_{\text{enclosed}} = -\frac{2}{3}\pi\mu_0 ab^3\right]$$

$$B = \frac{\mu_0 ab^3}{3R}$$

(c) As with Gauss's law, as long as there is no infinite current volume density (as in the case of a sheet of current or an infinitely thin wire), the magnetic field is continuous. Therefore, the expressions obtained in parts (a) and (b) yield the same value of *B* on the surface of the wire:

$$B = \frac{\mu_0 ab^2}{3}$$

CHAPTER SUMMARY

Electric charges in motion produce magnetic as well as electric fields. The magnetic field gives rise to a magnetic force on other charges, provided they are in motion, too. The magnetic field produced by a small segment of a current-carrying wire is given by the Biot-Savart law. The field due to the entire wire can be found using the principle of superposition. The magnetic field due to current distributions with particular symmetries such as cylindrical and planar can be easily calculated using Ampere's law, which is equivalent to the Biot-Savart law.

PRACTICE EXERCISES

Multiple-Choice Questions

1. Two loops of wire centered at the origin with equal radii carry current of equal magnitude, as shown in Figure 17.14. One loop lies on the *xz*-plane, while the other lies on the *xy*-plane. At the origin, which of the following vectors is parallel to the magnetic field?

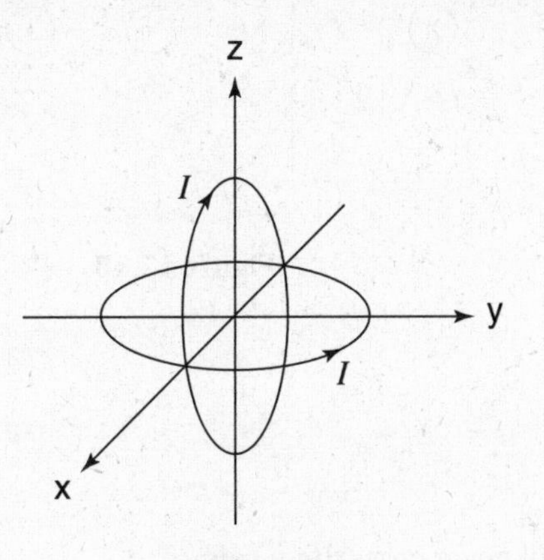

Figure 17.14

(A) $\hat{j}+\hat{k}$
(B) $\hat{j}-\hat{k}$
(C) $\hat{k}-\hat{j}$
(D) $-\hat{j}-\hat{k}$
(E) none of the above

2. A uniform magnetic field of magnitude B points out of the page in the region $x > 0$, and a uniform magnetic field of equal magnitude points into the page in the region $x < 0$, as shown in Figure 17.15. A particle with charge $+q$ and mass m initially at the origin moves to the right with speed v. At which of the following points does the particle cross the y-axis?

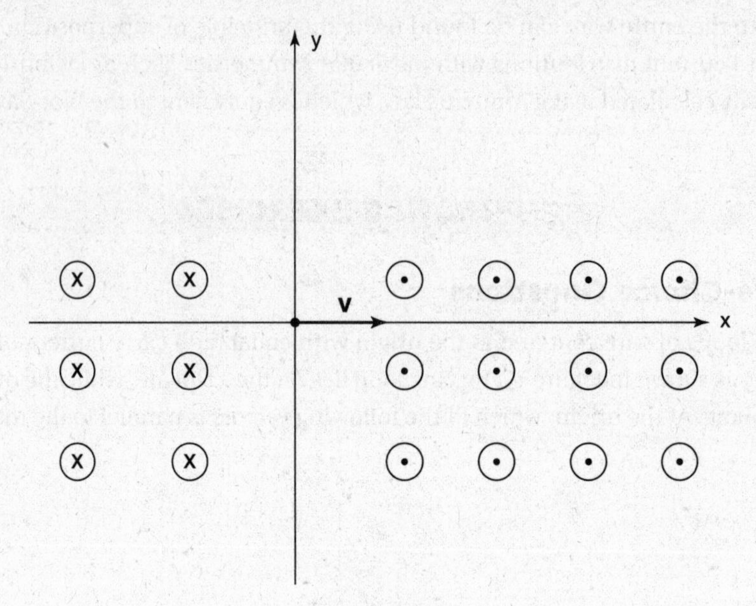

Figure 17.15

(A) $3\ mv/qB$
(B) $6\ mv/qB$
(C) $-3\ mv/qB$
(D) $-6\ mv/qB$
(E) The particle will never cross the y-axis.

3. An electron moves in a magnetic field along a certain path at a constant speed. The magnetic field does no work on the electron if

(A) the path is parallel to the magnetic field, so that the magnetic force is zero
(B) the path is perpendicular to the field, so that the electron moves in uniform circular motion
(C) the magnetic field is constant in space, so that there is no induced electric field
(D) the electron must move along a magnetic field line
(E) None of the above conditions are necessary.

4. Figure 17.16 shows a cross-section of a coaxial cable. A current of 1.0 A passes through the center of the cable, while a uniform current density of 1.0 A/m^2 comes out of the page through the outer sheath of the cable. At what distance from the center of the cable is the magnetic field equal to zero?

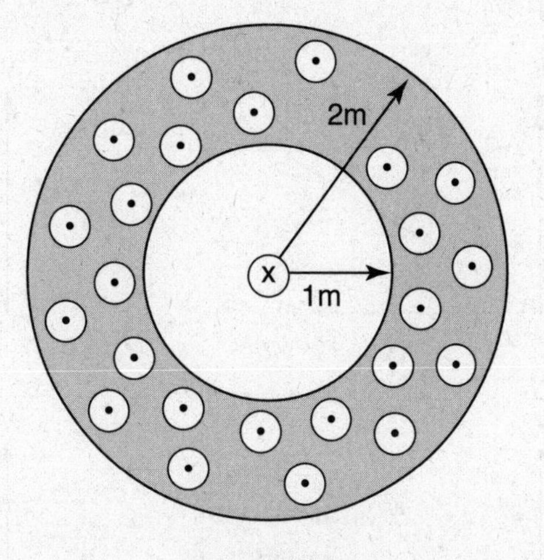

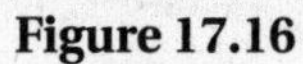
Figure 17.16

(A) $r = \left(1 + \frac{3}{4\pi}\right)^{1/3}$ m

(B) $r = \frac{1}{2\pi} + 1\text{m}$

(C) $r = \sqrt{\frac{1}{\pi} + 1\text{m}}$

(D) The magnetic field is zero, but not at the radii listed above.

(E) The magnetic field is never zero except at infinity.

5. A charged particle moves in uniform circular motion due to a magnetic field perpendicular to the plane of its motion. Which of the graphs in Figure 17.17 illustrates the relationship between the charge's velocity and the radius of its motion? (Assume that other variables, such as the magnetic field strength and the mass of the particle, are held constant.)

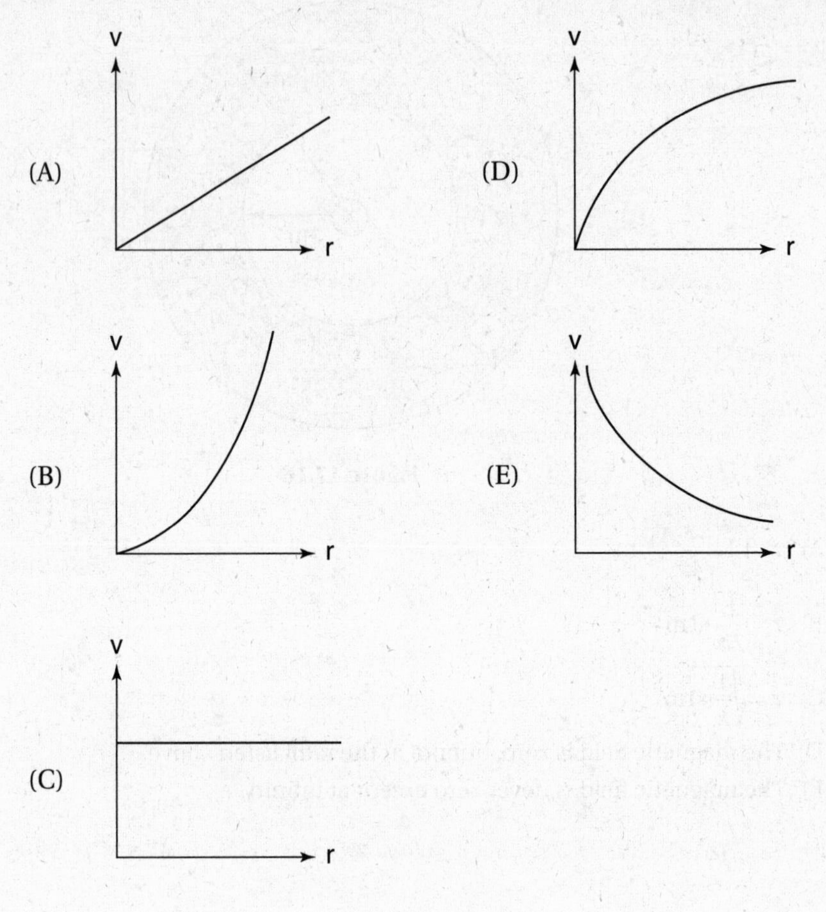

Figure 17.17

6. A particle with a charge $-q$ and mass m is moving with speed v through a mass spectrometer that contains a uniform outward magnetic field as shown in Figure 17.18. What is the magnitude and direction of the electric field required so that the net force on this particle is zero?

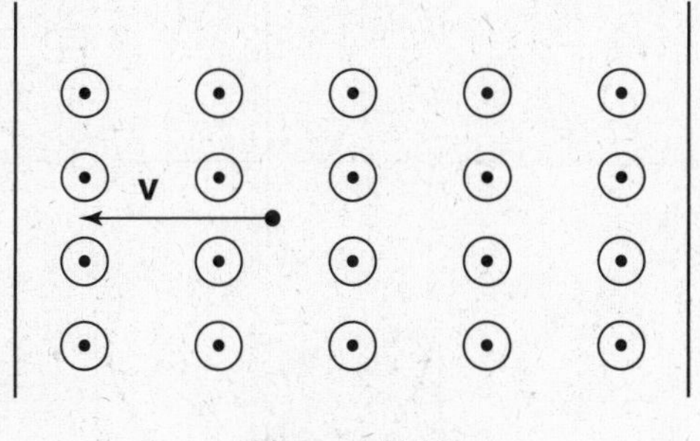

Figure 17.18

(A) $E = vB$, with a direction upward
(B) $E = vB$, with a direction downward
(C) $E = B/v$, with a direction upward
(D) $E = B/v$, with a direction downward
(E) $E = B/v$, with a direction into the page

7. Consider current passing through a short segment of a circular loop of wire as shown in Figure 17.19. The current passing through the rest of the circular wire produces a magnetic field at the location of the short segment, resulting in a magnetic force on the current-carrying segment of wire. Which of the following statements about this force is true?

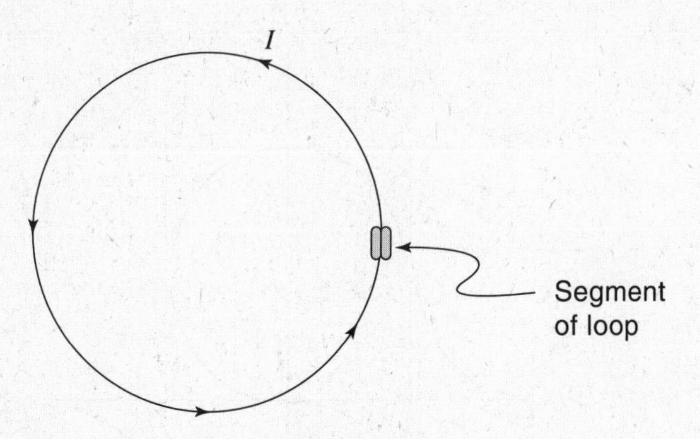

Figure 17.19

(A) The force is to the right.
(B) The force is to the left.
(C) The magnetic field is zero because of symmetry, so there is no force.
(D) The force is pointing out of the page.
(E) The force is pointing into the page.

8. Two wires located along the x-axis carry equal currents into the page as shown in Figure 17.20. Which of the graphs shows the y-component of the magnetic field along the x-axis?

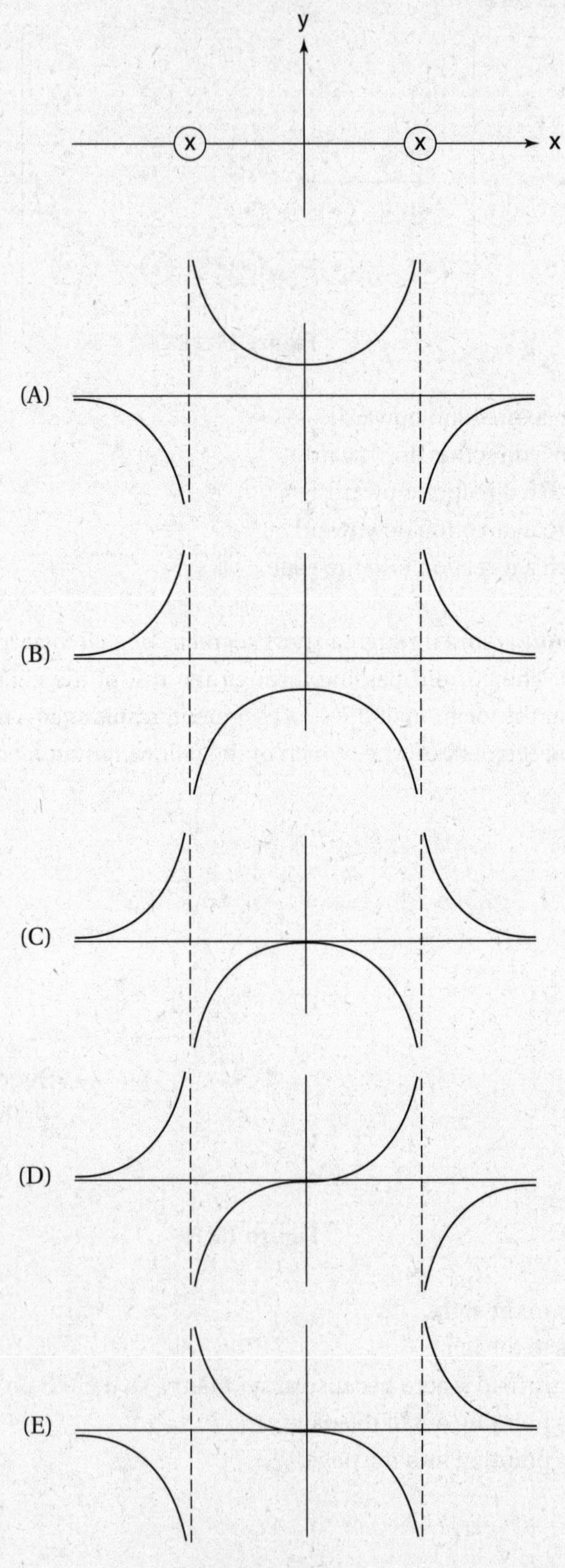

Figure 17.20

9. A particle moves through an electric field as shown in Figure 17.21. What is the direction of the magnetic field needed to ensure that the particle experiences zero net force?

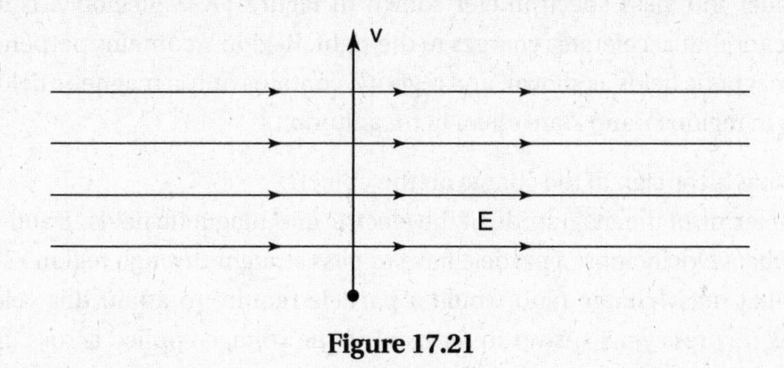

Figure 17.21

(A) into the page if the particle is positively charged, out of the page if the particle is negatively charged
(B) into the page if the particle is negatively charged, out of the page if the particle is positively charged
(C) into the page, irrespective of the particle's charge
(D) out of the page, irrespective of the particle's charge
(E) to the left, irrespective of the particle's charge

10. Three loops of equal radii carrying equal currents lie on planes parallel to the *yz*-plane as shown in Figure 17.22. Two of the loops are located equal distances *d* from the *yz*-plane as shown, while the third loop lies on the *yz*-plane. In what direction is the net magnetic field at the origin?

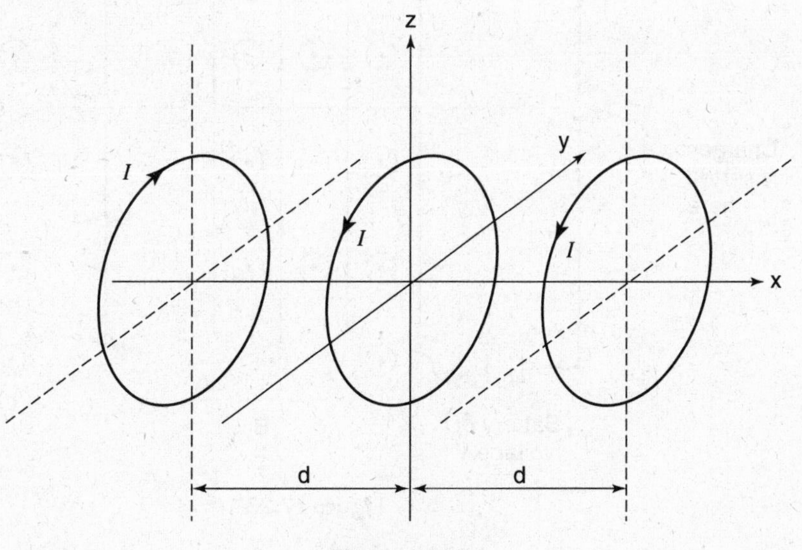

Figure 17.22

(A) The magnetic field is zero because of symmetry.
(B) The magnetic field is in the $+x$-direction.
(C) The magnetic field is in the $-x$-direction.
(D) The magnetic field lies in the xz-plane.
(E) The magnetic field lies in the xy-plane.

Free-Response Questions

1. Consider the mass spectrometer shown in Figure 17.23. Region X is a parallel plate capacitor that accelerates charges to the right. Region Y contains perpendicular electric and magnetic fields as shown, and region Z contains only a magnetic field (the magnetic fields in regions Y and Z are equal in magnitude).

   (a) What is the sign of the charge on the object?

   (b) In terms of the magnitude of the electric and magnetic fields, *E* and *B*, respectively, what velocity must a particle have to pass straight through region Y?

   (c) What mass/charge ratio would a particle require to attain this velocity in region X? (Express your answer in terms of *V*, the voltage applied across the plates of the accelerating capacitor, *B*, and *E*.)

   (d) What is the mass/charge ratio of a particle that moves straight through region Y and then in a circle of radius *r* in region Z as shown? (Express your answer in terms of *B*, *E*, and *r*.)

   (e) Is it possible to determine the mass of the particle given the experimental parameters *V*, *B*, *E*, and *r*?

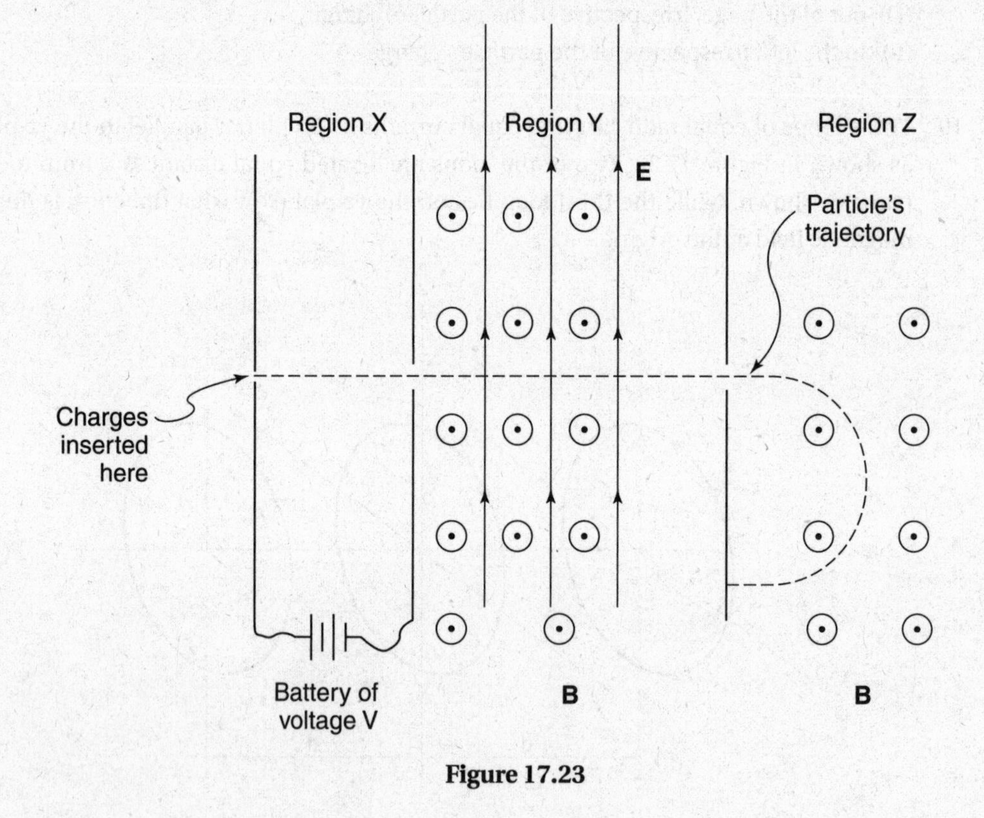

Figure 17.23

2. Two infinitely long wires carry equal currents of I into the page, as shown in Figure 17.24.

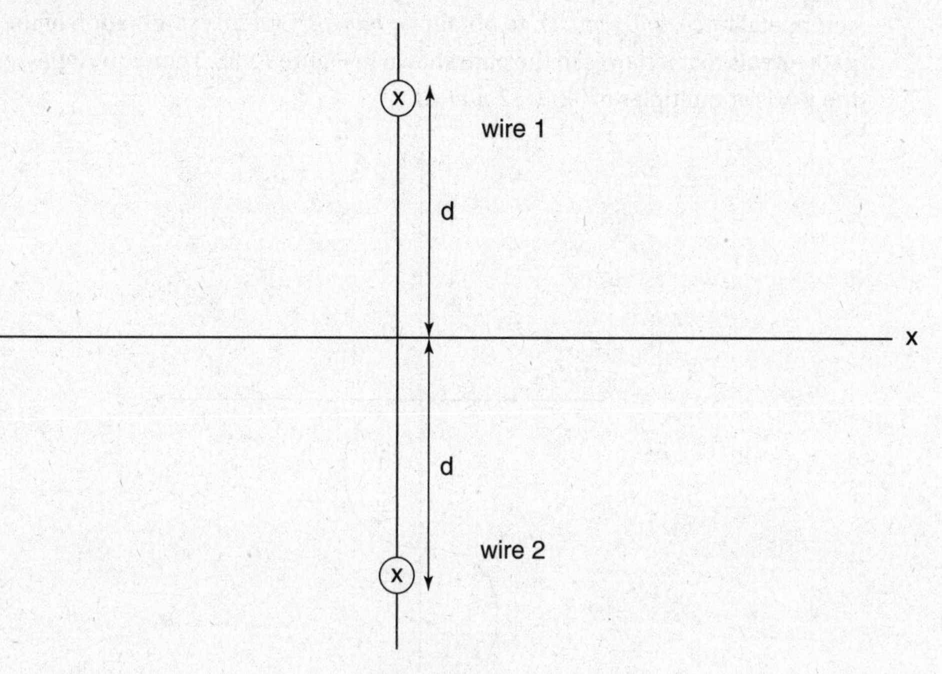

Figure 17.24

(a) Calculate the magnitude and direction of the magnetic field along the x-axis as a function of position, $B(x)$.

(b) Sketch the magnetic vector field along the x-axis.

(c) Calculate the work required to move a point charge of value $+q$ from $x = \infty$ to the origin. What does this result reveal about magnetic fields in general?

ANSWER KEY

1. **C**	4. **C**	7. **A**	10. **B**
2. **D**	5. **A**	8. **D**	
3. **E**	6. **B**	9. **C**	

ANSWERS EXPLAINED

Multiple-Choice

1. **(C)** Using the right-hand rule, the loop in the xy-plane produces a magnetic field in the $+z$-direction and the loop in the xz-plane produces a magnetic field in the $-y$-direction. Because these wires are carrying identical amounts of current, they produce magnetic fields of equal magnitudes so that the net magnetic field consists of two components of equal magnitude (one pointing in the $+z$-direction and one in the $-y$-direction). Therefore, the net magnetic field is parallel to the vector $\hat{k} - \hat{j}$. (Note that this problem relies on the superposition of magnetic fields.)

2. **(D)** The particle's velocity is perpendicular to the magnetic field, so it moves in a circular path. The radius of the path can be calculated by equating the magnetic force to the centripetal force, $qvB = mv^2/r$, to obtain $r = mv/qB$. Careful examination of the particle's path reveals that it moves in the path shown in Figure 17.25. Therefore, it passes through the y-axis at multiples of $-2r = -2\ mv/qB$.

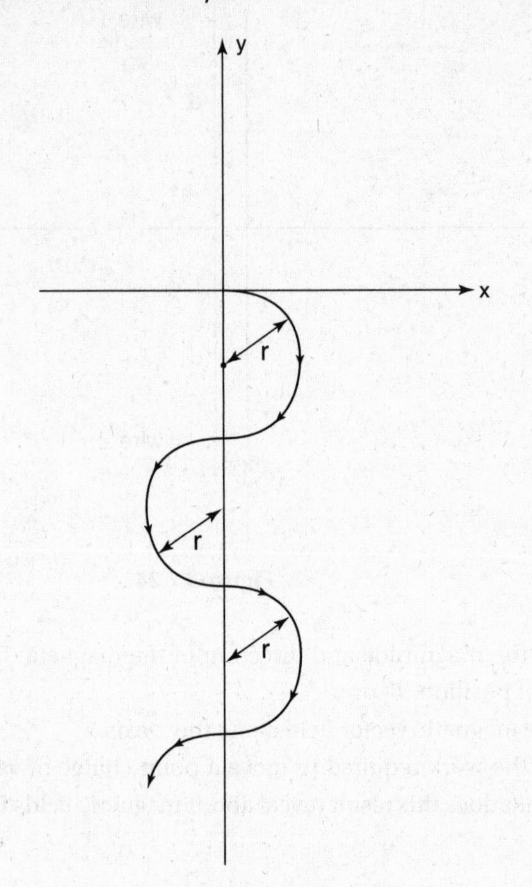

Figure 17.25

3. **(E)** The magnetic force is always perpendicular to velocity, so it can never do work (recall that $dW = \mathbf{F} \cdot d\mathbf{r}$).

4. **(C)** According to Ampere's law, the magnetic field is zero when the enclosed current is zero. For the inward current to balance the outward current, the outward current enclosed must equal 1.0 A. The enclosed outward current is 1.0 A when the Amperian path encloses an area of 1.0 m² of the sheath (because the current density is 1.0 A/m²). Geometrically, $1 = \pi r^2 - \pi(1.0)^2$. Solving this equation for r yields choice (C).

5. **(A)** The magnetic force provides the centripetal force:

$$|q\mathbf{v} \times \mathbf{B}| = qvB = \frac{mv^2}{r}$$

$$qB = m\frac{v}{r}$$

$$v = \left(\frac{qB}{m}\right)r$$

Thus, v is a linear function of r that passes through the origin.

6. **(B)** For the magnetic force to cancel the electric force, $|q\mathbf{v} \times \mathbf{B}| = qvB = qE$, which simplifies to $E = vB$. The direction of the magnetic force on the charge is downward (opposite the direction indicated by the right-hand rule because the charge is negative). The electric force must point in the opposite direction, upward. However, since the charge is negative, in order to produce an upward electric force the field must point downward.

7. **(A)** First, imagine dividing the loop into little segments and considering the magnetic field due to each segment (using the right-hand rule for the direction of a magnetic field produced by a current-carrying wire, or using the right-hand rule embedded in Biot-Savart's law). Because each segment produces a magnetic field pointing out of the page at the segment of wire, the net magnetic field points out of the page. Using the right-hand rule again, the force this field exerts on the segment of wire is to the right. In general, each segment of the loop feels a force directed radially outward due to the rest of the loop.

8. **(D)** At the origin, the fields due to the two wires cancel and the net field is zero, eliminating choices (A) and (B). As each wire is approached from the right, the field decreases to negative infinity, and as each wire is approached from the left, the field increases to positive infinity. This is consistent only with choice (D).

9. **(C)** Changing the sign of the charge of the particle changes the direction of both the electric and the magnetic forces, two changes that cancel each other out, so that in either case the magnetic field must point into the page in order to cancel out the electric field. (If the magnetic field is into the page, a positive charge will feel an electric force to the right and a magnetic force to the left. Alternatively, a negative charge will feel forces in the opposite directions.)

10. **(B)** The two loops lying off the *yz*-plane produce equal and opposite fields which cancel each other. (The loop at $x = -d$ produces a magnetic field in the $-x$-direction, while the loop at $x = d$ produces a magnetic field in the $+x$-direction.) The net field is therefore equal to the field produced by the middle loop, which produces a field in the $+x$-direction.

Free-Response

1. (a) Consider region Z: Using the right-hand rule for the magnetic charge on a moving point charge and realizing that the centripetal force must point toward the center of the circular trajectory reveals that the charge must be positive. We could also determine this by the fact that the charge accelerates parallel to the electric field in region X.

   (b) The electric and magnetic forces in region Y are in opposite directions; for the net force to be zero, they must cancel out exactly (the magnetic field is perpendicular to the velocity, so the sine in the cross-product is 1):

$$qE = qvB \Rightarrow v = \frac{E}{B}$$

   (c) Applying conservation of energy across region X,

$$qV = \frac{1}{2}mv^2 = \frac{1}{2}m\left(\frac{E}{B}\right)^2$$

Solving for m/q yields

$$\frac{m}{q} = \frac{2VB^2}{E^2}$$

(d) In region Z, the magnetic force provides the centripetal force:

$$\frac{mv^2}{r} = qvB$$

Substituting for $v = E/B$ and solving for m/q yields

$$\frac{m}{q} = \frac{rB^2}{E}$$

(e) No. As shown in this problem, experimental parameters only allow you to calculate the mass/charge ratio [the calculations in parts (c) and (d) are thus redundant, as discussed earlier in the chapter]. To calculate the mass, you must know the charge independently.

2. (a) The magnetic field due to an infinitely long wire is equal in magnitude to $B = \mu_0 I/2\pi r$ and has a direction indicated by the right-hand rule. The magnetic field along the x-axis is a superposition of the magnetic fields due to both of the wires, as shown in Figure 17.26.

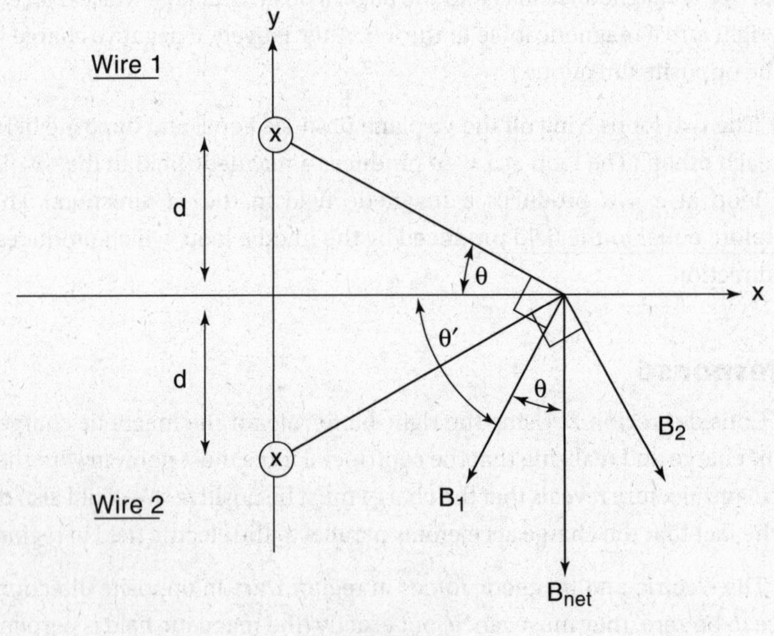

Figure 17.26

Adding the two vectors yields

$$\mathbf{B} = -2\left(\frac{\mu_0 I}{2\pi(d^2+x^2)^{1/2}}\cos\theta\right)\hat{j} = \frac{-\mu_0 I x \hat{j}}{\pi(d^2+x^2)}$$

(b) This expression is given in Figure 17.27. In sketching the curve, first we note that the function has odd symmetry; mathematically, $B(-x) = -B(x)$. Because x is raised to a greater power in the denominator than in the numerator, the function asymptotically approaches zero as x approaches positive or negative infinity. Finally, we note that the function is zero at the origin and that it passes through the origin with a negative slope.

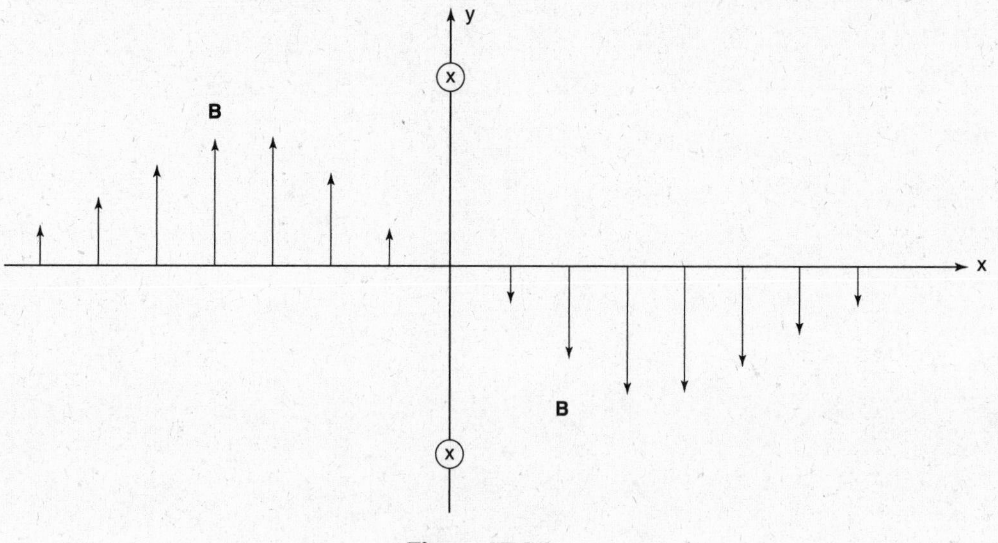

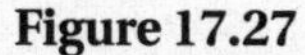

Figure 17.27

Note: A graphing calculator can also be used to sketch this curve: Simply graphing the function $f(x) = -x/(1 + x^2)$ is sufficient to get an idea of how this curve looks.

(c) The magnetic force is always perpendicular to the velocity, so it never does any work. (Thus, the magnetic force can change only an object's direction, not its speed.) Therefore, no work need be done by an external force to move the charge from infinity to the origin.

Faraday's Law and Lenz's Law 18

→ MAGNETIC FLUX
→ FARADAY'S LAW
→ INDUCED EMF
→ LENZ'S LAW
→ MOTIONAL EMF

CALCULATING THE MAGNETIC FLUX Φ_B

Before we introduce Faraday's law, we should discuss magnetic flux. You should already be familiar with the concept of flux and the mathematical definition of flux (phrased here in terms of magnetic flux).

$$\text{Magnetic flux} = \Phi_B = \int \mathbf{B} \cdot d\mathbf{A}$$

As before, $d\mathbf{A}$, the differential area vector, is a vector pointing perpendicularly to a differential slab of surface area with magnitude equal to the differential area $d\mathbf{A}$. In contrast to Gauss's law, where we defined $d\mathbf{A}$ to point away from the center of the gaussian surface, we do not make a universal choice for the direction of $d\mathbf{A}$ when using Faraday's law. However, we have to be consistent in the direction that we choose, as explained below.

Flux Through a Single Loop

If the magnetic field is uniform in space (at any particular time), the equation for flux simplifies to

$$\text{Magnetic flux} = \Phi_B = \int \mathbf{B} \cdot d\mathbf{A} = \mathbf{B} \cdot \int d\mathbf{A} = \mathbf{B} \cdot \mathbf{A}$$

If the magnetic field is not constant throughout the surface, we will need integration to calculate the magnetic flux. Here is a stepwise guide to carrying out the integration:

1. Determine the geometry of the differential areas needed. The magnetic field must be uniform across each differential area so that it has constant flux density. Therefore, the shape of these differential regions must mirror the symmetry of the magnetic field.
2. Determine the dimensions of the differential areas. You should arrive at an expression for dA in terms of various constants and one differential length (e.g., dx).
3. Calculate the differential flux through a differential area according to the equation $d\Phi_B = \mathbf{B} \cdot d\mathbf{A}$. This involves arbitrarily setting $d\mathbf{A}$ to point in one of the two directions perpendicular to the area (recall that you must be consistent about this choice).

4. To find the entire flux, integrate the expression for $d\Phi_B$ over limits determined by the geometry of the situation. Always integrate from the lower limit to the upper limit. Use the same choice for the direction of $d\mathbf{A}$ across the entire surface.

Flux Through a Coil

The flux through a coil is the sum of the fluxes through every loop of wire, so that the flux through the entire coil is equal to the number of loops multiplied by the flux through each loop. Mathematically,

$$\Phi_{B,\text{coil}} = (\text{number of loops})(\text{flux per loop}) = n \int_{\text{one loop}} \mathbf{B} \cdot d\mathbf{A}$$

The SI unit for flux is the Weber, defined as $1\ \text{Wb} = 1\ \text{T} \cdot \text{m}^2$.

FARADAY'S LAW

TIP

Note that the magnetic flux must change with time to induce an EMF.

$$|\mathcal{E}| = \left|\frac{d\Phi_B}{dt}\right|$$

Faraday's law (scalar form)

In words, a changing magnetic flux through a loop of wire induces an electromotive force (EMF, denoted by $\mathcal{E}$). We will use this equation to calculate the magnitude of induced EMFs, and later in the chapter we will use Lenz's law to figure out the direction of induced EMFs. (Recall that the term "EMF" means essentially the same thing as "voltage.")

Before we jump into applying Faraday's law, let's take a moment to gain some further insight by rewriting Faraday's law for a uniform B field as

$$|\mathcal{E}| = \left|\frac{d\Phi_B}{dt}\right| = \left|\frac{d}{dt}(AB\cos\theta)\right|$$

where θ is the angle between the field and the area vector.

What this reveals is that an EMF can be induced by changing the area, the magnetic field, the angle θ, or some combination of these variables. The AP exam very often asks about one of the first three of these situations, all of which are illustrated in examples below.

Using Faraday's Law

The situations where we use Faraday's law are much more varied than the situations where we use Gauss's or Ampere's law, so there isn't much of a general approach other than the guidelines presented for calculating flux. However, we will attempt to familiarize you with all of the scenarios that are likely to appear on the exam.

EXAMPLE 18.1 EMF IN A FIXED COIL NEAR A WIRE OF VARIABLE CURRENT

An infinitely long wire lies on the *y*-axis and carries a current given by $I(t) = ct^2$, where *c* is a constant with units of A/s^2, as shown in Figure 18.1. Calculate the EMF experienced by the loop of wire shown as a function of time.

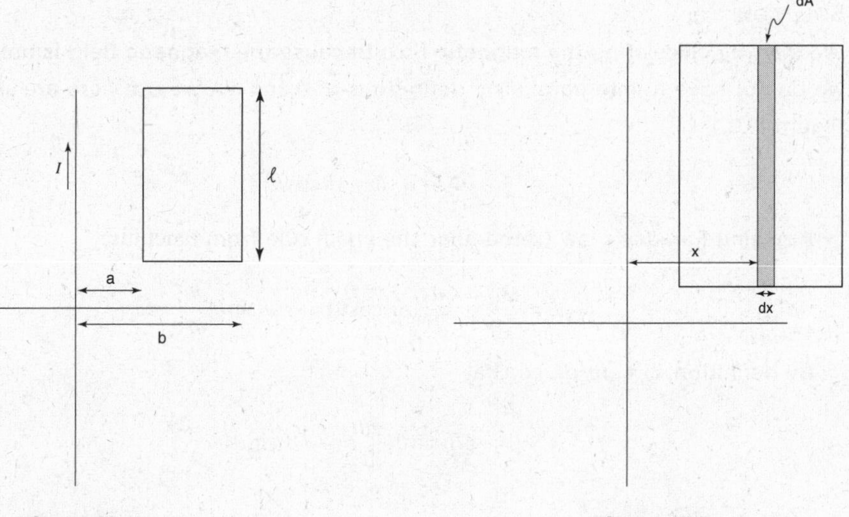

Figure 18.1 **Figure 18.2**

Solution

We begin by calculating the flux through the loop. The magnetic field at any particular time is not uniform across the loop, so we integrate. In order for the magnetic field to be constant over each differential area, we select rectangular strips as shown in Figure 18.2.

From the last chapter, we know that the magnetic field due to an infinite wire is given by the expression

$$B = \frac{\mu_0 I}{2\pi x}$$

According to the right-hand rule, the magnetic field points perpendicularly into the page. Arbitrarily setting *d***A** to point into the page (so that the dot product is positive), the differential flux through each differential area is

$$d\Phi_B = \mathbf{B} \cdot d\mathbf{A} = BdA = \left(\frac{\mu_0 I}{2\pi x}\right)(l\,dx) = \frac{\mu_0 Il\,dx}{2\pi x}$$

Integrating to calculate the net flux,

$$\Phi_B = \int_{x=a}^{x=b} d\Phi_B = \int_{x=a}^{x=b} \frac{\mu_0 Il\,dx}{2\pi x} = \frac{\mu_0 Il}{2\pi}\int_{x=a}^{x=b}\frac{dx}{x} = \frac{\mu_0 Il}{2\pi}\ln\left(\frac{b}{a}\right)$$

Now we can apply Faraday's law to calculate the magnitude of the EMF by taking the time derivative:

$$|\mathcal{E}| = \left|\frac{d\Phi_B}{dt}\right| = \left|\frac{d}{dt}\left[\frac{\mu_0 Il}{2\pi}\ln\left(\frac{b}{a}\right)\right]\right| = \frac{\mu_0 l}{2\pi}\ln\left(\frac{b}{a}\right)\left|\frac{dI}{dt}\right|$$

From the above, we know that $I(t) = ct^2$. Therefore, $|dI/dt| = 2ct$ and the answer is

$$|\mathcal{E}| = \frac{\mu_0 clt}{\pi}\ln\left(\frac{b}{a}\right)$$

EXAMPLE 18.2 ROTATING A LOOP OF WIRE IN A CONSTANT MAGNETIC FIELD

Consider a flat loop of wire area A rotating at a constant angular velocity ω within a constant magnetic field of magnitude B. What are the maximum and minimum magnitudes of the induced EMF in the loop?

SOLUTION

We start by calculating the magnetic flux. Because the magnetic field is uniform, we do not have to integrate. (The definitions of θ and $d\mathbf{A}$ we use here are shown in Figure 18.3.)

$$\Phi_B = \mathbf{B} \cdot \mathbf{A} = AB\cos\theta$$

Applying Faraday's law (remember the chain rule from calculus),

$$|\mathcal{E}| = \left|\frac{d\Phi_B}{dt}\right| = \left|\frac{d}{dt}(AB\cos\theta)\right| = \left|-AB\sin\theta\frac{d\theta}{dt}\right|$$

By definition, $\omega = d\theta/dt$, so that

$$|\mathcal{E}| = \left|-AB(\sin\theta)\frac{d\theta}{dt}\right| = |-AB(\sin\theta)\omega|$$

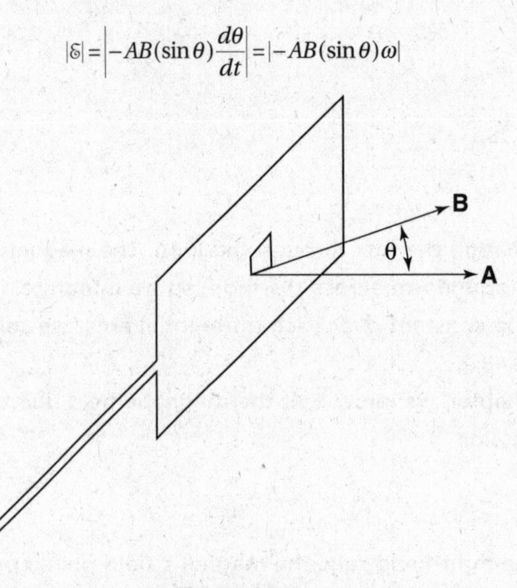

Figure 18.3

Now remember what the question is asking for: the maximum and minimum EMF. The |sin| function has a maximum of one and a minimum of zero, so the minimum magnitude of the induced EMF is zero and the maximum magnitude of the induced EMF is

$$|\mathcal{E}|_{\text{maximum}} = \omega AB(1) = \omega AB$$

The physical significance of this is that when the loop is perpendicular to the magnetic field (so that the area vector, which is perpendicular to the loop, is parallel to the magnetic field and $\theta = 0$), there is no instantaneous change in magnetic flux (the flux is maximized at the crest of a cosine curve, where the derivative is zero). When the loop is parallel to the magnetic field (so that the area vector, which is perpendicular to the loop, is perpendicular to the magnetic field and $\theta = \pi/2$), the change in magnetic flux is maximized (the flux is at zero, where the slope of the cosine curve is maximum).

EXAMPLE 18.3 SLIDING A RECTANGULAR LOOP OF WIRE THROUGH A REGION OF NONUNIFORM MAGNETIC FIELD

A rectangular loop of wire is moving at constant velocity v between two regions, each of which has a uniform magnetic field, as shown in Figure 18.4. The magnetic field in region 1 is pointing out of the page with magnitude B_1, and the magnetic field in region 2 has the same direction with magnitude B_2. What is the magnitude of the induced EMF?

SOLUTION

In this situation, instead of trying to first calculate the magnetic flux and then take its derivative, we begin by calculating the change in flux during an increment of time, dt. During a time dt, the loop slides to the right by an amount $dx = vdt$ as shown in Figure 18.5. This shift causes an increase in the area exposed to the field in region 2 (dA_2) and a decrease in the area exposed to the field in region 1 (dA_1). When we choose $d\mathbf{A}$ pointing out of the page, the net change in magnetic flux due to the "appearance" of dA_2 and the "disappearance" of dA_1 is

$$d\Phi_B = \Phi_{dA2} - \Phi_{dA1} = B_2 l\,dx - B_1 l\,dx$$

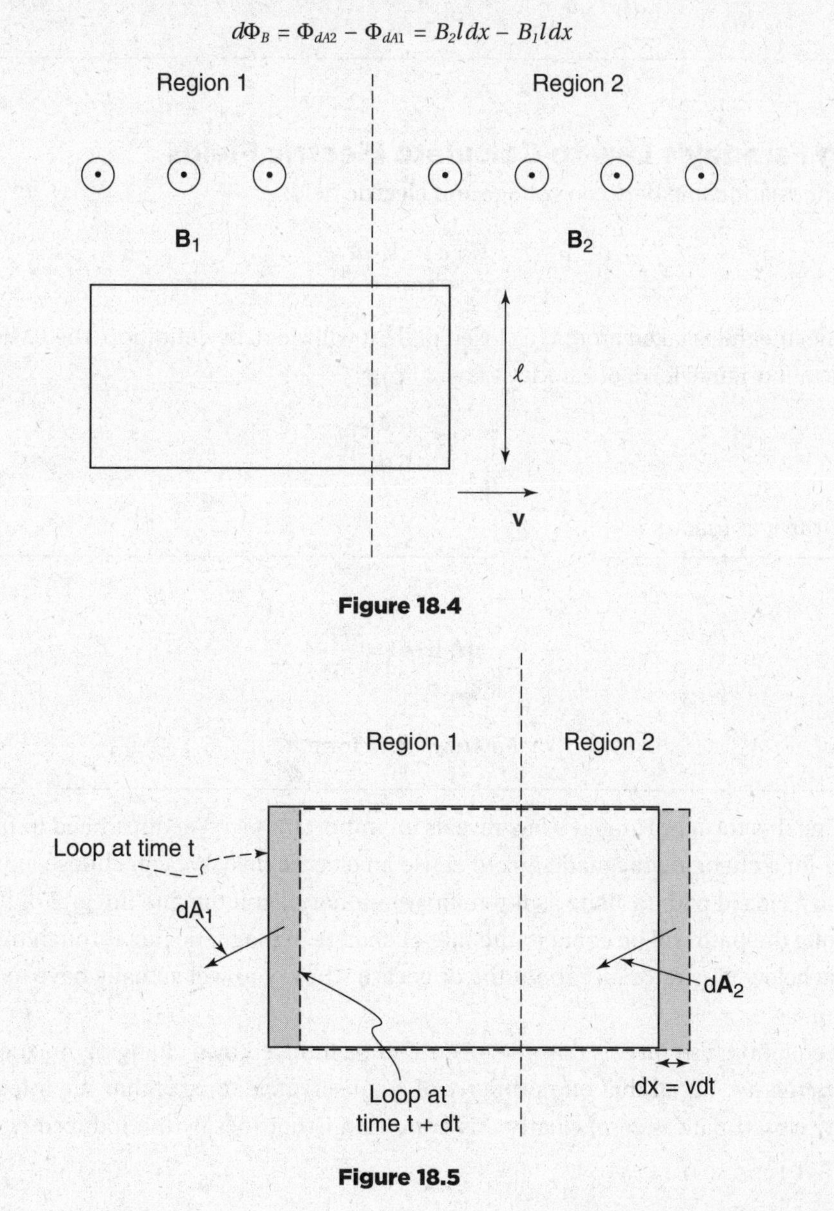

Figure 18.4

Figure 18.5

Applying Faraday's law,

$$|\mathcal{E}| = \left|\frac{d\Phi_B}{dt}\right| = \left|\frac{B_2 l\,dx - B_1 l\,dx}{dt}\right| = \left|l(B_2 - B_1)\frac{dx}{dt}\right| = |l(B_2 - B_1)v|$$

Of course, once the left end of the loop passes over the border between region 1 and region 2, there will no longer be a changing flux through the loop and there will be no induced EMF. The above equation applies only while the loop is passing over the border between region 1 and region 2.

Alternative solution: Perhaps a more straightforward way to solve the problem would be to define the width of the loop to be d and the horizontal length extending into region 2 to be x (leaving a horizontal length of $d - x$ extending into region 1). In this case, the flux would be

$$\Phi_B = xlB_2 + (d - x)lB_1$$

Taking the derivative of the flux to calculate the EMF yields

$$|\mathcal{E}| = \left|\frac{d\Phi_B}{dt}\right| = \left|\frac{d}{dt}(xlB_2 + (d-x)lB_1)\right| = \left|l(B_2 - B_1)\frac{dx}{dt}\right| = |l(B_2 - B_1)v|$$

Using Faraday's Law to Calculate Electric Fields

Recall the relationship between voltage and electric field:

$$V = \int_{\text{line integral}} \mathbf{E} \cdot d\mathbf{l}$$

If the line integral is taken around a closed path, it will yield, by definition, the EMF. Thus, we can take our original form of Faraday's law:

$$|\mathcal{E}| = \left|\frac{d\Phi_B}{dt}\right|$$

and rewrite it as follows

$$\left|\oint_{\substack{\text{line}\\\text{integral}}} \mathbf{E} \cdot d\mathbf{l}\right| = \left|\frac{d\Phi_B}{dt}\right|$$

Faraday's law, form II

Writing this form of Faraday's law reveals an important fact: We don't need to have a wire in order for a changing magnetic flux to cause an electric field. We can choose any arbitrary, imaginary, closed path (as long as it remains stationary), and the line integral of the electric field along the path will be equal to the rate of change in magnetic flux through the path. An example below illustrates just about the only case where you will actually have to apply this equation.

This equation also reveals that the electric force induced by a changing magnetic field is *not conservative.* Recall that one property of a conservative force is that the integral $\oint \mathbf{F} \cdot d\mathbf{l}$ over any closed path is zero; clearly this condition is not met by the induced electric field because

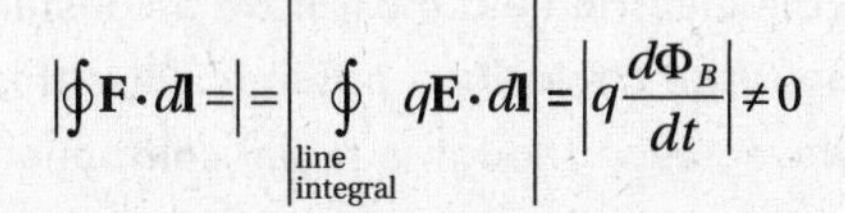

$$\left|\oint \mathbf{F}\cdot d\mathbf{l}\right| = \left|\oint_{\text{line integral}} q\mathbf{E}\cdot d\mathbf{l}\right| = \left|q\frac{d\Phi_B}{dt}\right| \neq 0$$

TIP

Faraday's law applies even when the flux is computed through an imaginary loop in space.

EXAMPLE 18.4 USING FARADAY'S LAW TO CALCULATE ELECTRIC FIELD

The current passing through an infinitely long solenoid of radius *R* with *n* turns per length is given as a function of time, *I*(*t*). What is the magnitude of the electric field a distance *a* > *R* from the center of the solenoid?

SOLUTION

Consider a circular path of radius *a* centered at the axis of the solenoid, as shown in Figure 18.6. What is the flux through the surface that this path defines? The magnetic field inside the solenoid is a constant and is given by

$$B = \mu_0 n I$$

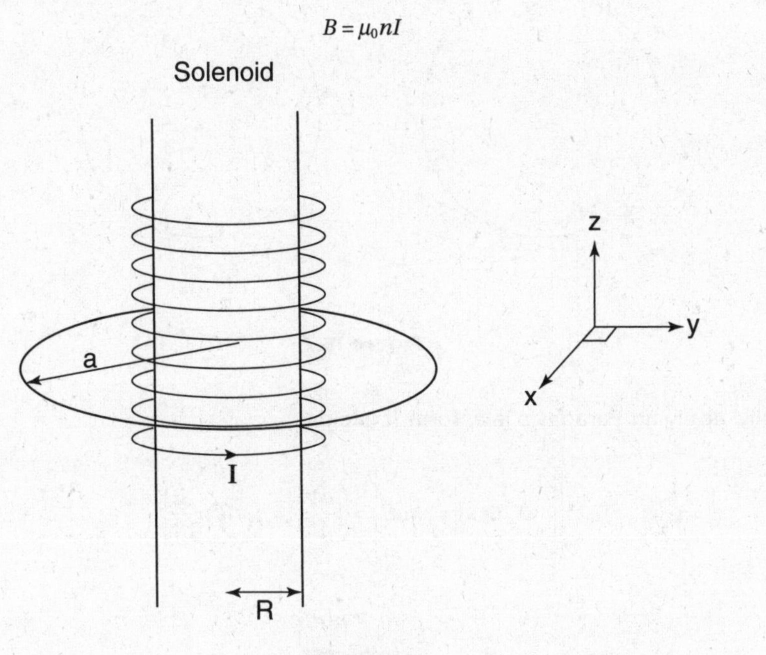

Figure 18.6

The magnetic field outside an infinite ideal solenoid is zero. Therefore (designating *d*A to point in the same direction as the magnetic field of the solenoid),

$$\Phi_B = \int_{\text{entire circle}} \mathbf{B}\cdot d\mathbf{A} = \int_{0<r<R} \mathbf{B}\cdot d\mathbf{A} + \int_{R<r<a} \mathbf{B}\cdot d\mathbf{A} = (\mu_0 n I)(\pi R^2) + 0$$

Differentiating the flux yields

$$\frac{d\Phi_B}{dt} = \mu_0 \pi R^2 n \frac{dI}{dt}$$

Now we are all set with the right side of the equation, so let's work out the line integral:

$$\oint_{\text{line integral}} \mathbf{E}\cdot d\mathbf{l}$$

Because of symmetry, the electric field must have a constant magnitude and relative direction throughout the circle. Two possible ways this might be achieved are shown in Figure 18.7. However, even though a radial component of the electric field is allowed to be nonzero by symmetry, Gauss's law requires that it be zero. Imagine drawing a cylindrical gaussian surface with radius *a*: Because the gaussian surface encloses no net charge, it must experience zero electric flux. This requires the radial component of the field to be zero. Therefore, the electric field must be tangent to the direction of integration. Therefore,

$$\left|\oint_{\substack{\text{line}\\ \text{integral}}} \mathbf{E}\cdot d\mathbf{l}\right| = 2\pi a E$$

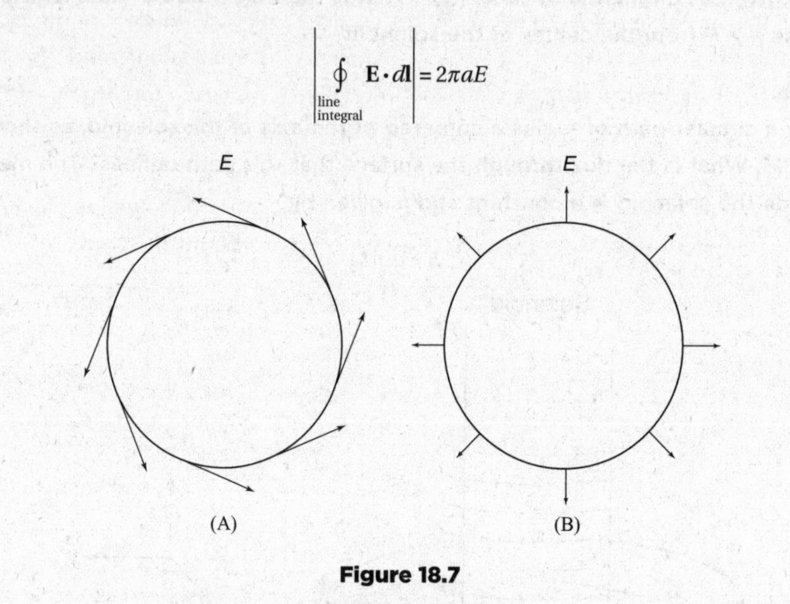

Figure 18.7

Finally, applying Faraday's law, form II, yields

$$\left\{\left|\oint_{\substack{\text{line}\\ \text{integral}}} \mathbf{E}\cdot d\mathbf{l}\right| = 2\pi a E\right\} = \left\{\left|\frac{d\Phi_B}{dt}\right| = \mu_0 \pi R^2 n \frac{dI}{dt}\right\}$$

$$E = \frac{\mu_0 R^2 n}{2a}\left|\frac{dI}{dt}\right|$$

LENZ'S LAW AND DETERMINING THE DIRECTION OF THE EMF

As mentioned in the introduction, it is generally easiest to use Faraday's law in the form we stated (without reference to signs) to find the magnitude of the induced EMF, and to use Lenz's law to find the direction of the induced EMF.

Lenz's law: The induced current produces a magnetic field, which in turn produces a flux that *opposes* the change in flux in the loop caused by the changing external magnetic field.

The best way to explain this is by presenting an example:

EXAMPLE 18.5 APPLYING LENZ'S LAW

A circular loop of wire is located in a region of space where an external magnetic field points out of the page. If the external magnetic field is decreasing, what is the direction of the induced current?

SOLUTION

The external magnetic field is decreasing, which causes a decreasing flux through the loop. According to Lenz's law, the induced current should flow in the direction that opposes this decrease in flux. Therefore, inside the loop, the induced magnetic field should point out of the page (to bolster the decreasing external magnetic field). Using the right-hand rule, you can see that a counterclockwise current causes a magnetic field that comes out of the page inside the loop of wire. Therefore, the induced current must flow in a counterclockwise direction.

***Note*: We don't care what the direction of the induced field is outside the loop because the field in this region doesn't affect the flux through the loop.**

Motional EMF

So far, we have learned that closed loops of wire with changing magnetic flux experience an induced EMF. The following example illustrates that wires moving within magnetic fields also experience an induced EMF.

EXAMPLE 18.6

Consider Figure 18.8. Both situations feature a bar of length ℓ, moving to the right with speed *v* in a uniform magnetic field *B* going into the page. The only difference is that in Figure 18.8(A) the bar is connected to a circuit, whereas in Figure 18.8(B) it is not part of a closed circuit.

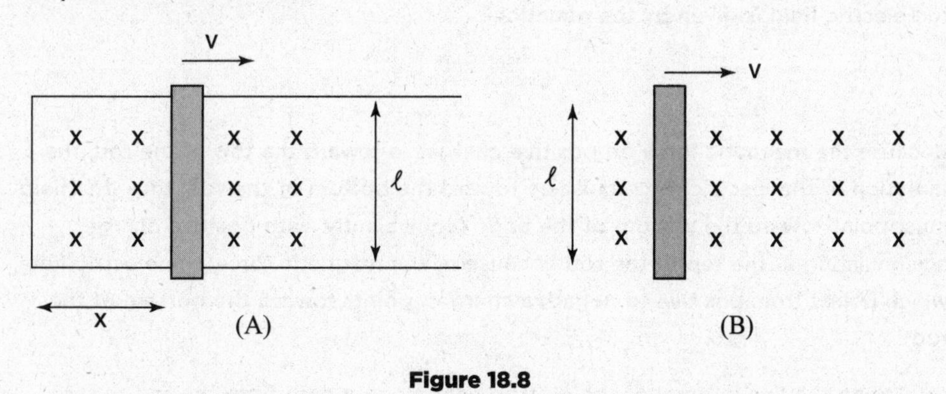

Figure 18.8

(a) Calculate the EMF induced in the loop to the left of the bar in Figure 18.8(A). In what direction does the current flow?

(b) Positive charges within the bar feel a magnetic force toward the top of the page, due to their motion to the right. Therefore, in Figure 18.8(B) positive charge accumulates at the top of the rod, producing a downward electric field. After the rod has been moving with constant velocity for a long time, the charge reaches

equilibrium so that the downward electric force on charges in the rod exactly cancels the upward magnetic force and charges do not move up or down the rod. What are the magnitude and direction of this electric field?

(c) What is the potential difference between the two ends of the rod? Which end is at a higher potential?

(d) Discuss the relationship between the two situations explored in this problem.

SOLUTION

(a) During a time interval *dt*, the area of the loop increases by a differential rectangular strip of area:

$$dA = lv\,dt$$

This change in area causes a differential change in flux given by

$$|d\Phi_B| = |\mathbf{B} \cdot d\mathbf{A}| = B\,dA = Bl\,dx$$

According to Faraday's law, the induced EMF is then

$$|\mathcal{E}| = \left|\frac{d\Phi_B}{dt}\right| = Bl\frac{dx}{dt} = Blv$$

Alternative solution: Defining *x* to be the horizontal dimension of the circuit (see Figure 18.8(A)),

$$|\Phi_B| = Blx$$

$$|\mathcal{E}| = \left|\frac{d\Phi_B}{dt}\right| = Bl\frac{dx}{dt} = Blv$$

Note: This equation, $|\mathcal{E}| = Blv$, often appears on AP examinations but is valid only if the wire is perpendicular to the velocity vector.

To oppose the increase in inward flux, the current must flow counterclockwise.

(b) After a long time, the magnitude of the electric force on a charge q must equal the magnitude of the magnetic force, $F = |q\mathbf{v} \times \mathbf{B}| = qvB$. Therefore, the magnitude of the electric field is given by the equation

$$E = \frac{F}{q} = \frac{qvB}{q} = vB$$

Because the magnetic force on positive charges is toward the top of the rod, the direction of the electric force must be toward the bottom of the rod; thus, the field must point toward the bottom of the page (equivalently, with positive charge accumulating at the top of the rod because of the magnetic force, the electric field, which points from positive to negative charges, points toward the bottom of the rod).

(c) Taking the line integral of the electric field along a path from the top to the bottom of the rod,

$$V_{\text{top}} - V_{\text{bottom}} = -\int \mathbf{E} \cdot d\mathbf{l}$$

Because the electric field points downward, parallel to the direction of integration, $\mathbf{E} \cdot d\mathbf{l} = E\,dl$:

$$V_{\text{top}} - V_{\text{bottom}} = -\int \mathbf{E} \cdot d\mathbf{l} = \int vB\,dl = vB\int dl = vBl$$

Therefore, the magnitude of the potential difference between the two ends of the bar is Blv, with the top end at higher potential (as expected, because the field must point from higher potential to lower potential).

(d) In the case of Figure 18.8(A), the magnetic force on the charges in the sliding bar causes a current to flow through the circuit. In the case of Figure 18.8(B), the magnetic force on the charges causes charge to redistribute itself until the net force on charges in the rod is zero (much like the redistribution of charges we examined in Chapter 13 such that the net electric field within metals in stationary charge distributions is zero).

In both cases, the induced EMF is Blv and the voltage at the top end of the bar is greater than that at the bottom end. We showed this explicitly for Figure 18.8(B). In Figure 18.8(A) we showed that the current flowed counterclockwise. If you think of the bar as a battery with a positive terminal at its top, this direction of current is consistent with this potential difference across the bar.

What is the link between these situations? The sliding rod may be thought of as a special case of Figure 18.8(A) in which the rest of the circuit has an infinite resistance (so that no current flows through the circuit).

Justification for Lenz's Law

Lenz's law can be justified in terms of energy considerations. Consider, for example, the sliding rod introduced in Example 18.6. Lenz's law requires the current to flow counterclockwise, which produces a magnetic force on the rod pointing to the left and retarding the bar's motion. What would happen if the current instead flowed clockwise? In that case, the current would produce a magnetic force on the rod pointing to the right, causing the rod to speed up. As the rod sped up, the current would also increase, causing further increases in the rod's speed, which would increase exponentially. This situation would be a violation of conservation of energy. This result is general: In any situation, Lenz's law is required by conservation of energy.

CHAPTER SUMMARY

Faraday's law states that whenever there is a changing magnetic flux through a loop of wire or even an imaginary loop in space, an electric field will be induced along the loop. In the case of a wire, this induced electric field gives rise to an electromotive force (EMF), which means that current will flow in the wire. The change in magnetic flux can be created by physically moving the source of the magnetic field (whether it is a magnet or current-carrying wire) or changing the strength of the field itself. Only relative motion matters; the loop of wire where the EMF is induced can be moved while the source of the magnetic field remains stationary. Lenz's law indicates the direction of the induced current in the loop.

PRACTICE EXERCISES

Multiple-Choice Questions

1. An external force pushes a square loop of wire through a region of constant magnetic field as shown in Figure 18.9. Which of the following is a plot of the magnetic flux as a function of the position of the loop?

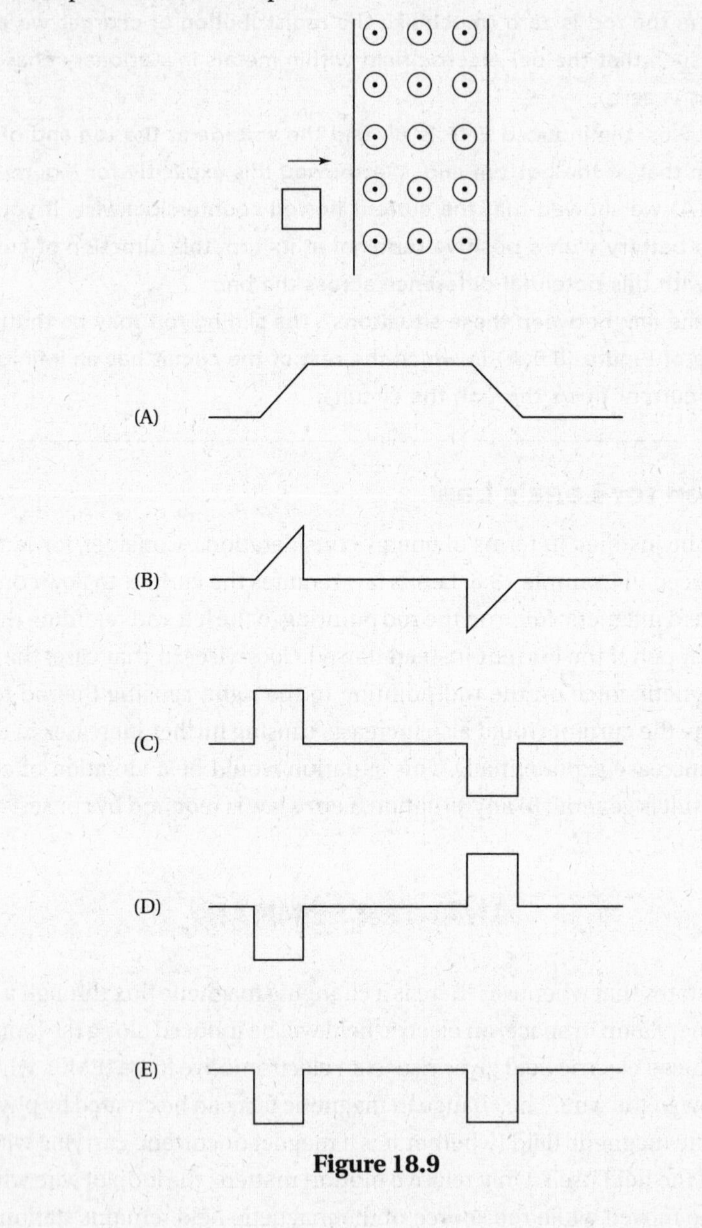

Figure 18.9

2. Referring to the situation introduced in question 1, which of the plots in Figure 18.10 shows the induced current in the loop (with clockwise current designated as positive)?

3. Referring to the situation in question 1, which of the plots in Figure 18.9 shows the force due to the field on the wire, now carrying induced current (designating rightward force as positive)?

4. Consider Figure 18.10. The current in the straight wire varies with time, inducing a current in the loop. Which of the following statements is true about the magnetic force exerted on the loop of wire due to the straight wire?

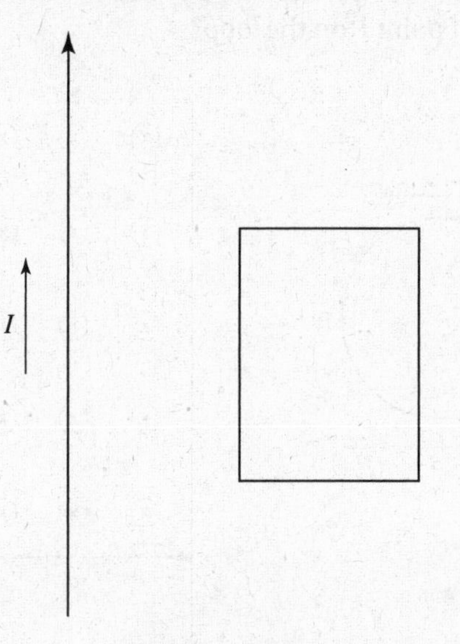

Figure 18.10

(A) The force is to the right if I is increasing, and to the left if I is decreasing.
(B) The force is to the left if I is increasing, and to the right if I is decreasing.
(C) The force is to the left regardless of whether I is increasing or decreasing.
(D) The force is to the right regardless of whether I is increasing or decreasing.
(E) The force is zero because the force on each side of the rectangular loop cancels the force on the opposite side of the loop.

5. A uniform magnetic field points out of the page through a circular area as shown in Figure 18.11. If the magnetic field strength is decreasing, the electric field at point P

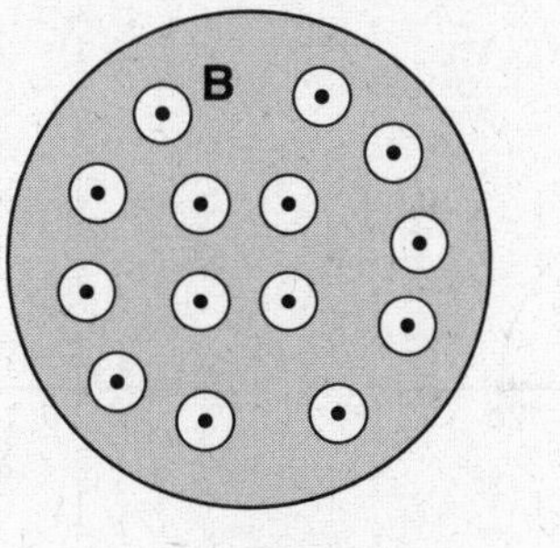

Figure 18.11

(A) points upward
(B) points downward
(C) points to the left
(D) points to the right
(E) is zero because of symmetry

6. A circular loop of wire of diameter d moves with constant velocity across a region with a constant magnetic field as in Figure 18.12. Choosing counterclockwise current as positive, which of the graphs in Figure 18.13 is a plot of the induced current as a function of the x-coordinate of point P on the loop?

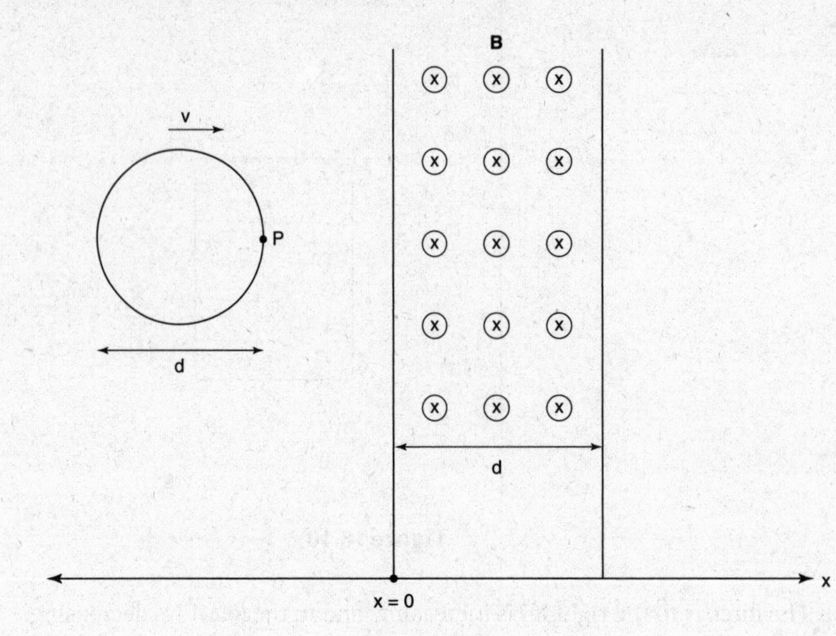

Figure 18.12

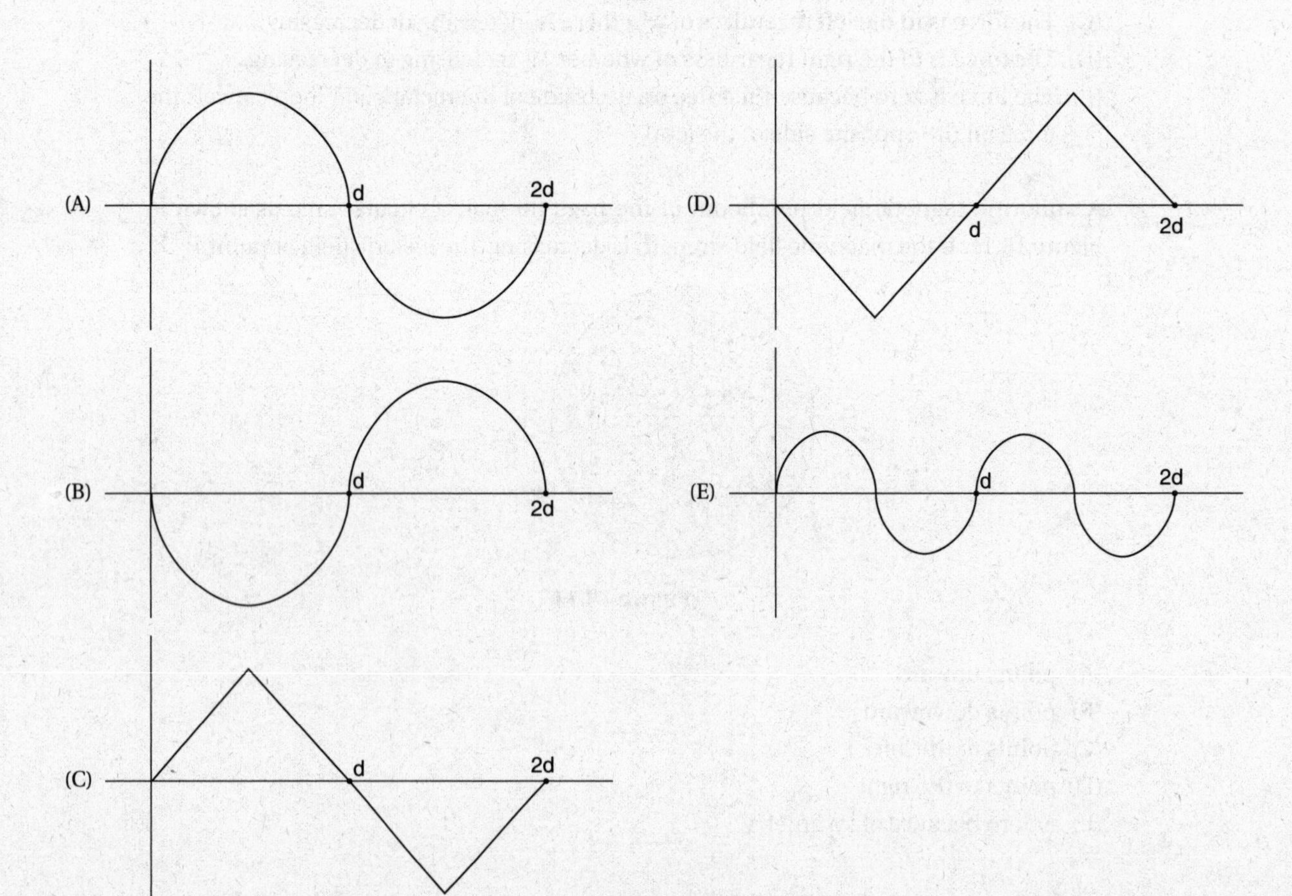

Figure 18.13

7. A loop of wire is moved toward a pole of a bar magnet as shown in Figure 18.14. The force on the loop is

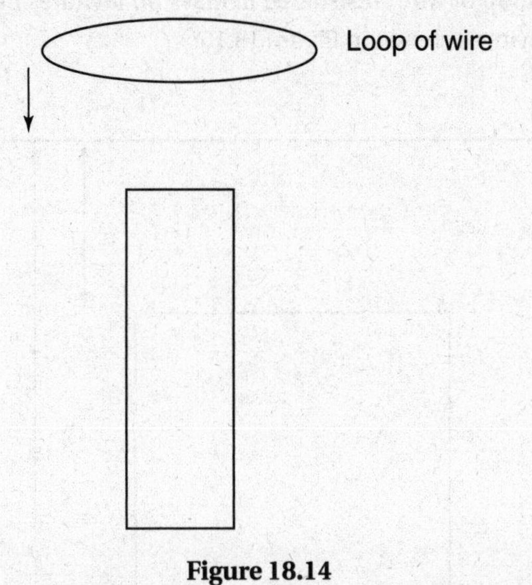

Figure 18.14

(A) toward the magnet if the loop is closer to the north pole of the magnet, and away from the magnet if the loop is closer to the south pole of the magnet
(B) toward the magnet if the loop is closer to the south pole of the magnet, and away from the magnet if the loop is closer to the north pole of the magnet
(C) toward the magnet regardless of which pole of the magnet is closer to the loop
(D) away from the magnet regardless of which pole of the magnet is closer to the loop
(E) zero because of vector cancellation and symmetry considerations

Free-Response Questions

1.[1] A rectangular loop of wire (resistance R, mass m) levitates below an infinitely long current-carrying wire as shown in Figure 18.15.

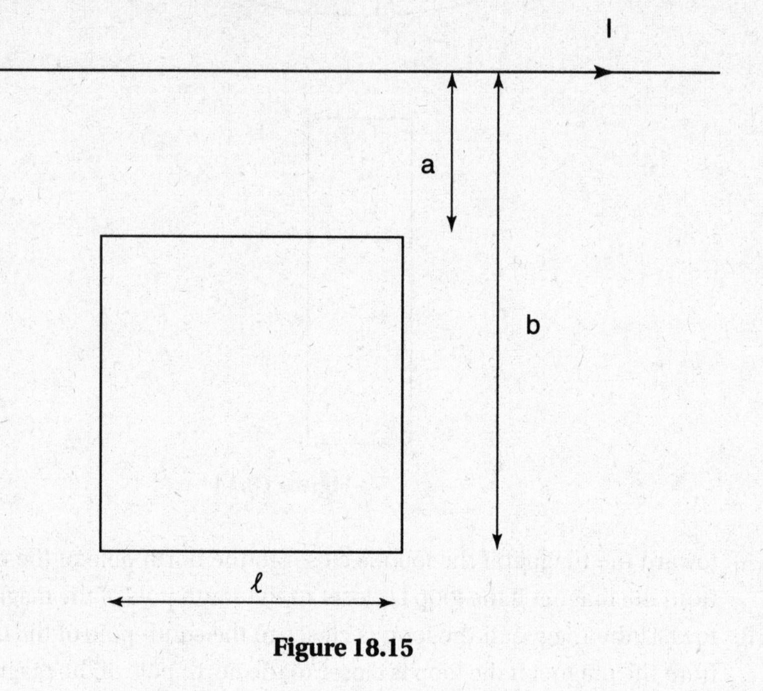

Figure 18.15

(a) If a current I_{wire} flows through the current-carrying wire, what is the magnitude of the magnetic flux in the loop?

(b) If the current in the wire is decreasing at a rate of dI_{wire}/dt, what is the magnitude and direction of the induced EMF in the loop?

(c) What is the induced current in the loop?

(d) What is the magnetic force on the loop? (Give magnitude and direction.)

[1]It may seem that there are a lot of Faraday's law problems. There are good reasons for this: (1) Faraday's law problems are very common on AP exams, rivaling Gauss's law problems for popularity with about one per exam, and (2) although these problems are centered around Faraday's law, they review other topics as well.

2. A square metal loop (sides of length l, mass m, resistance R) moves into a region of constant magnetic field B (directed out of the page) as in Figure 18.16. In terms of the loop's speed v and the above constants,

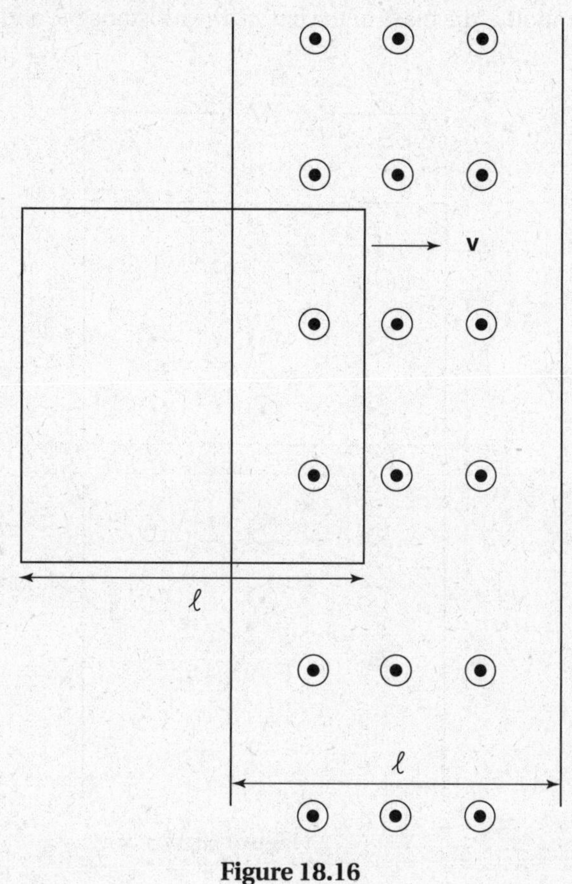

Figure 18.16

(a) What is the magnitude of the EMF of the loop as it enters the magnetic field? As it exits?

(b) What is the magnitude of the current through the loop as it enters the magnetic field? As it exits?

(c) What is the direction of the current as it enters the magnetic field? As it exits?

(d) What is the force on the loop as it enters the field? As it exits?

3. A metal rod of length l slides without friction down two wires as in Figure 18.17. A uniform magnetic field of magnitude B points out of the page. Express your answers to the following questions in terms of the velocity of the bar v, the magnetic field B, gravitational acceleration g, the mass of the bar m, the resistance R, and the length of the rod l.

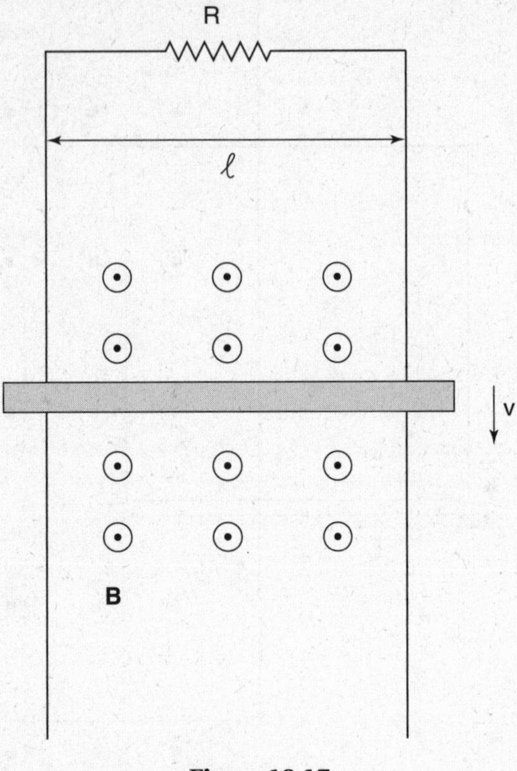

Figure 18.17

(a) What is the magnitude of the EMF induced in the circuit?
(b) What is the magnitude of the current in the circuit?
(c) What is the net acceleration on the bar?
(d) What is the terminal velocity of the bar?

4. The magnitude of the uniform magnetic field inside an infinite solenoid with radius a is given by the equation $\mathbf{B} = -f(t)\hat{k}$, with a direction pointing into the page. The magnetic field is increasing at a constant rate denoted as $f'(t) > 0$.

(a) Calculate the EMF on a circular loop of wire placed in the plane of the figure concentric with the axis of the solenoid. Consider the loop to have three possible radii r:
   (i) $r < a$
   (ii) $r = a$
   (iii) $r > a$

(b) Calculate the magnitude of the electric field along the circular loop of wire in each of the three cases defined in part (a).
(c) What is the direction of the field along the circular loop?
(d) Suppose the loop is removed. Would it be possible for an electron placed at a point outside the solenoid to move in a circular path due to the induced electric field? Explain.

5. A bar of length l and resistance R lies in a uniform magnetic field B as shown in Figure 18.18. At time $t = 0$, it abruptly begins to move to the right with a constant speed v as indicated.

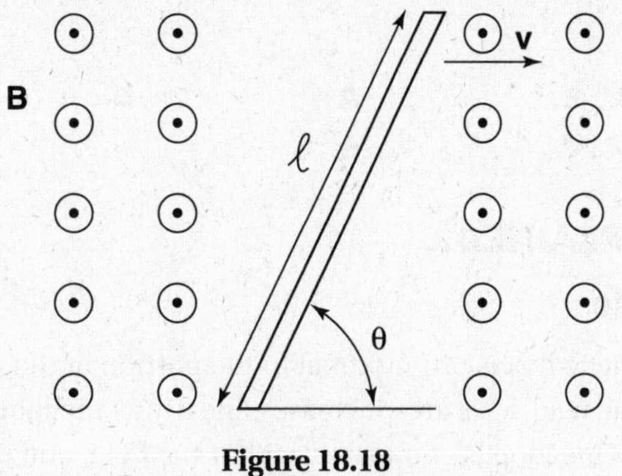

Figure 18.18

(a) Qualitatively discuss the current through the bar as a function of time.
(b) Calculate the electric field within the bar once equilibrium has been reached.
(c) What is the voltage across the bar after equilibrium has been reached? Which end of the bar is at higher potential?
(d) What is the initial current in the bar?
(e) After the charges in the bar have reached equilibrium, it is abruptly stopped. At the instant the bar is stopped, what is the current in the bar? This current produces a force in what direction?

6. Consider the circuit shown in Figure 18.19, which has a mass m and is free to move up and down.

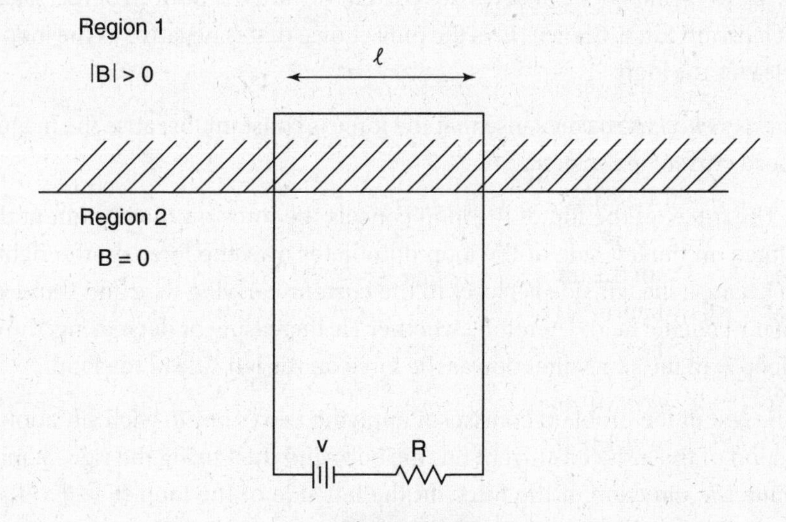

Figure 18.19

(a) Assuming that the loop starts from rest and initially accelerates upward, what is the magnitude of the induced EMF as a function of the loop's speed?
(b) What is the current as a function of the loop's speed?
(c) What is the net force on the loop as a function of its speed? (Designate upward as positive.)
(d) What is the loop's terminal speed?

(e) Considering the fact that the bottom of the loop eventually passes into region 1, qualitatively describe the motion of the loop when it is entirely in region 1.

ANSWER KEY

1. **A**	3. **E**	5. **A**	7. **D**
2. **C**	4. **A**	6. **A**	

ANSWERS EXPLAINED

Multiple-Choice

1. **(A)** The magnetic flux is proportional to the portion of the loop's area that is exposed to the magnetic field. This area increases linearly as the loop enters the field, remains constant while the loop is completely within the field, and decreases as the loop exits the field.

2. **(C)** According to Lenz's law, as the loop enters the field, the induced magnetic current flows in the clockwise direction to create flux into the page (opposing the increase in outward flux); this immediately eliminates choices (D) and (E). While the loop is completely within the magnetic field, the flux through the loop is constant, so the induced current is zero, eliminating choice (A). Finally, as the loop exits the field, the flux decreases at a constant rate, so the induced current is constant, eliminating choice (B).

3. **(E)** As the loop both enters and leaves the magnetic field, the forces produced by the field on the top and bottom of the current-carrying loop cancel each other. The only vertical wire (on the right segment as the loop enters and on the left segment as it exits) exposed to the magnetic field carries current toward the bottom of the page, producing a net leftward force. Choice (E) is the only choice that is negative as the loop both enters and leaves the loop.

   Answer check: It makes sense that the force is constant, because the magnitude of the induced current is constant.

4. **(A)** The force on the top of the loop cancels the force on the bottom of the loop, but the force on the left side of the loop dominates over the force on the right side of the loop because the left side is closer to the current-carrying wire and thus experiences a greater magnetic field. Therefore, whether I is increasing or decreasing, the net force on the loop is in the same direction as the force on the left side of the loop.

   The rest of the problem consists of applying Lenz's law in each situation (to find the direction of the induced current on the loop) and then using the right-hand rule to determine the direction of the force on the left side of the loop ($\mathbf{F} = I\mathbf{l} \times \mathbf{B}$). You should convince yourself that the following are true:

   I increasing → counterclockwise current → net force to the right

   I decreasing → clockwise current → net force to the left

5. **(A)** Imagine a loop of wire concentric with the circular region of magnetic field but passing through point *P*. From Lenz's law, current will flow counterclockwise through the loop. Because the flow of current is parallel to the electric field along the loop, and because of the symmetry considerations discussed in Example 18.4 that establish that the radial component of the field is zero, the field will point up, so choice (A) is correct.

6. **(A)** From $x = 0$ to $x = d$, the flux into the page across the loop is increasing, so a counterclockwise current is produced to oppose this increase in flux. This eliminates choices (B), (D), and (E). The magnitude of the current depends on the change in flux per unit time, which depends on the additional amount of area that enters the magnetic field region per unit time. The rate at which area enters the magnetic field is proportional to the height of the portion of the loop in the field region. This height is a continuous nonlinear function of the location of *P*. Thus, of the two remaining choices, (A) is correct.

7. **(D)** First, let's consider the situation where the north pole is closer to the loop, as shown in Figure 18.20. Recall that, outside the magnet, the field lines point away from the north pole and the magnitude of the field increases as we approach the poles. Therefore, as you move the loop closer to the north pole, the upward flux through the loop increases, inducing a current in the indicated direction (clockwise as viewed from above). Using the right-hand rule, the force on any differential length of wire points upward and toward the center of the loop. Because of symmetry, the horizontal components of these differential forces cancel, leaving a net force away from the magnet. Therefore, if the loop is closer to the north pole, the force is away from the magnet, eliminating choices (E), (A), and (C).

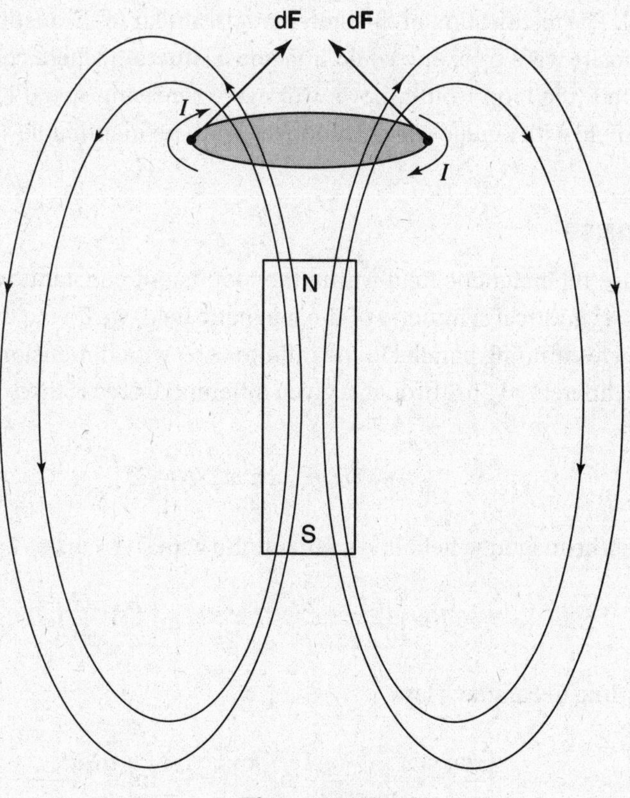

Figure 18.20

Another way to realize that the loop repels the magnet is to sketch the magnetic field lines produced by the current induced in the loop, as shown in Figure 18.21. The current through the loop creates a magnetic dipole much like that created by a bar magnet, with the north end of the dipole on the bottom, closer to the north pole of the magnet. Because like magnetic poles repel, the loop is expected to repel the magnet.

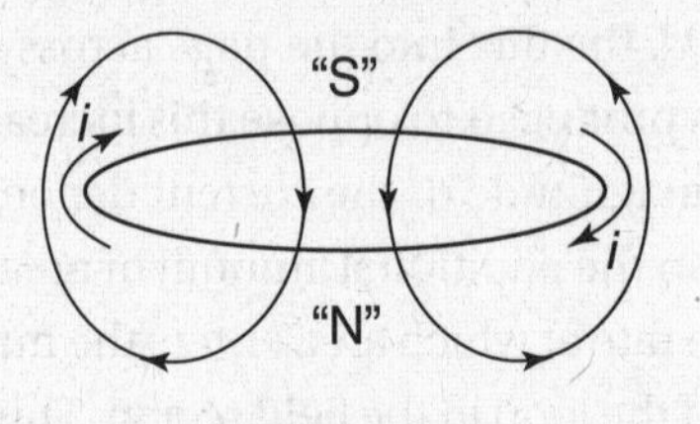

Figure 18.21

What if the loop is closer to the south pole of the magnet? The field lines, and therefore the direction of the current, will be reversed. However, because both the magnetic field and the current have reversed their directions, these opposite signs will cancel in the equation $d\mathbf{F} = Id\mathbf{l} \times \mathbf{B}$ and the same force will be produced. Therefore, the correct answer is (D).

The most powerful and simple way to solve this problem, however, is to realize that currents are induced such that the *force* on the current-carrying wire caused by the external magnetic field always tries to oppose the change in magnetic flux. For example, when a loop is being pushed into a region of nonzero magnetic field, the magnetic force on the loop (now carrying induced current) always tries to oppose the motion of the loop into the region of nonzero magnetic field. This is a restatement of Lenz's law at the force level. The justification of this statement is similar to the justification of Lenz's law: If the opposite were true, one could imagine a situation where conservation of energy was violated (the loop would move with ever-increasing speed into the region of the magnetic field with kinetic energy produced from no identifiable source).

Free-Response

1. (a) Because the magnetic field within the loop is not constant, we must integrate. To mirror the cylindrical symmetry of the magnetic field, we divide the loop into differential rectangles running parallel to the infinite wire, with dimensions dr by l and area $dA = ldr$. The differential flux through a given differential area is then

$$d\Phi = \mathbf{B} \cdot d\mathbf{A} = \frac{\mu_0 I_{\text{wire}}}{2\pi r}(l\, dr)$$

(given that the magnetic field due to an infinite wire is $B = \mu_0 I_{\text{wire}}/2\pi r$). Integrating,

$$\Phi = \int_a^b \frac{\mu_0 I_{\text{wire}} l\, dr}{2\pi r} = \frac{\mu_0 I_{\text{wire}} l}{2\pi} \ln\left(\frac{b}{a}\right)$$

(b) According to Faraday's law,

$$|\mathcal{E}| = \left|\frac{d\Phi}{dt}\right| = \left|\frac{d}{dt}\left[\frac{\mu_0 I_{\text{wire}} l}{2\pi r} \ln\left(\frac{b}{a}\right)\right]\right| = \frac{\mu_0 l}{2\pi} \ln\left(\frac{b}{a}\right) \frac{dI_{\text{wire}}}{dt}$$

Decreasing current causes a decrease of flux into the page, and the current flows clockwise to counter this decrease in inward flux.

(c) $I_{\text{loop}} = \frac{V}{R} = \frac{\mu_0 l}{2\pi R} \ln\left(\frac{b}{a}\right) \frac{dI_{\text{wire}}}{dt}$

(d) Given the direction of the current, the direction of the magnetic field, and the equation $\mathbf{F} = Il \times \mathbf{B}$, the following are the directions of the force on all the wire segments: top segment, upward force; bottom segment, downward force; left segment, leftward force; right segment, rightward force.

As a result of symmetry, the rightward and leftward forces have the same magnitude and cancel. However, since the top segment is closer to the wire than the bottom segment, it experiences a larger force, producing a net upward force on the loop. Quantitatively, this force (designating positive as upward) is

$$F = I_{\text{loop}} l_{\text{top}} \times \mathbf{B}_{\text{top}} + I_{\text{loop}} l_{\text{bottom}} \times \mathbf{B}_{\text{bottom}}$$

$$F = I_{\text{loop}} l \frac{\mu_0 I_{\text{wire}}}{2\pi a} - I_{\text{loop}} l \frac{\mu_0 I_{\text{wire}}}{2\pi b} = \frac{\mu_0 I_{\text{loop}} I_{\text{wire}} l}{2\pi} \left(\frac{1}{a} - \frac{1}{b}\right)$$

$$= \frac{\mu_0^2 l^2 I_{\text{wire}}}{4\pi^2 R} \ln\left(\frac{b}{a}\right)\left(\frac{1}{a} - \frac{1}{b}\right) \frac{dI_{\text{wire}}}{dt}$$

2. (a) As the loop enters the field, during a time interval dt, the area of the loop exposed to the field changes by a differential strip with height l and thickness dx (and thus, area $dA = ldx$). The change in flux during dt is thus (choosing flux out of the page as positive and the direction of dA as out of the page)

$$d\Phi = \mathbf{B} \cdot \mathbf{A} = Bl\,dx$$

Applying Faraday's law,

$$|\mathcal{E}| = \left|\frac{d\Phi}{dt}\right| = \left|\frac{Bldx}{dt}\right| = \left|Bl\frac{dx}{dt}\right| = Blv$$

Alternative solution: Letting x be the horizontal length of the loop exposed to the magnetic field,

$$\Phi = Blx$$

$$|\mathcal{E}| = \left|\frac{d\Phi}{dt}\right| = \left|\frac{d}{dt}(Blx)\right| = Bl\frac{dx}{dt} = Blv$$

As the loop exits the field, the rate at which the flux changes is the same, and thus the EMF has the same magnitude (the only difference is that the outward flux is decreasing rather than increasing).

(b) $I = \frac{V}{R} = \frac{Blv}{R}$

(c) As the loop enters the field, the current is clockwise; as the loop exits the field, the current is counterclockwise (Lenz's law).

(d) The magnetic field exerts a force on current-carrying wires equal to $\mathbf{F} = Il \times \mathbf{B}$. The forces on the top and bottom segments of the loop always cancel, and the net force is due to the vertical segment within the magnetic field, with magnitude

$$F = |Il \times B| = IlB = \left(\frac{Blv}{R}\right)lB = \frac{B^2l^2v}{R}$$

As the loop enters the region of the field, the induced current flows clockwise, and thus current flows downward through the segment exposed to the magnetic field (the right side of the square). As the loop exits the field, current flows counterclockwise, yet it also flows downward through the segment exposed to the magnetic field (the left side of the square). Thus, current is always flowing downward through the vertical segment exposed to the magnetic field, and the direction is thus always to the left (opposite the velocity).

3. (a) During a time increment dt, the area of the circuit increases by a differential strip of area, $dA = A\,dx$. The increase in flux during this time dt is equal to the flux through the new strip, which equals $d\Phi = B\,dA = Bl\,dx$ (setting $d\mathbf{A}$ pointing out of the page). Therefore, applying Faraday's law:

$$|\mathcal{E}| = \left|\frac{d\Phi_B}{dt}\right| = Bl\frac{dx}{dt} = Blv$$

(b) According to Ohm's law,

$$I = \frac{V}{R} = \frac{Blv}{R}$$

(c) The bar experiences two forces: a gravitational force equal to mg, and the force exerted by a magnetic field on a current-carrying wire. Designating the downward direction as positive,

$$F_{\text{gravity}} = +mg$$

What about the force exerted by the magnetic field? Lenz's law requires the current to flow in a clockwise direction, making this force point upward.

$$\mathbf{F}_{\text{magnetic}} = Il \times \mathbf{B} = -lB\left(\frac{Blv}{R}\right) = -\frac{B^2l^2v}{R}$$

$$F_{\text{net}} = F_{\text{gravity}} + F_{\text{magnetic}} = mg - \frac{B^2l^2v}{R}$$

The acceleration is then simply the net force divided by the mass:

$$a_{\text{net}} = \frac{F_{\text{net}}}{m} = g - \frac{B^2l^2v}{mR}$$

(d) At the terminal velocity the acceleration is zero. Therefore, we can solve for the terminal velocity by setting the acceleration equal to zero:

$$a_{\text{net}} = g - \frac{B^2l^2v}{mR} = 0 \Rightarrow v_{\text{term}} = \frac{mgR}{B^2l^2}$$

Answer check: Larger B and larger l or smaller R produces a greater magnetic force and thus a smaller terminal velocity. Greater mass and greater g produce a larger gravitational force and thus a larger terminal velocity.

4. (a)

(i) Making a straightforward application of Faraday's law,

$$|\mathbf{\Phi}_B| = \pi r^2 f(t)$$

$$|\mathcal{E}| = \left|\frac{d\Phi_B}{dt}\right| = \left|\frac{d}{dt}[\pi r^2 f(t)]\right| = \pi r^2 f'(t)$$

(ii)

$$|\mathbf{\Phi}_B| = \pi a^2 f(t)$$

$$|\mathcal{E}| = \left|\frac{d\Phi_B}{dt}\right| = \left|\frac{d}{dt}[\pi a^2 f(t)]\right| = \pi a^2 f'(t)$$

(iii) Because the magnetic field outside an infinite solenoid is zero, the flux and EMF are the same as in part (a)(ii).

(b)

(i) (For further explanation, see Example 18.4.)

$$|\mathcal{E}| = |V| = \left|\oint \mathbf{E} \cdot d\mathbf{l}\right| = E\oint dl = 2\pi r E$$

$$E = \frac{|\mathcal{E}|}{2\pi r} = \frac{\pi r^2 f'(t)}{2\pi r} = \frac{r f'(t)}{2}$$

(ii) Similarly,

$$E = \frac{|\mathcal{E}|}{2\pi a} = \frac{\pi a^2 f'(t)}{2\pi a} = \frac{a f'(t)}{2}$$

(iii) Again,

$$E = \frac{|\mathcal{E}|}{2\pi r} = \frac{\pi a^2 f'(t)}{2\pi r} = \frac{a^2 f'(t)}{2r}$$

(c) If $f(t)$ is increasing, flux into the page is increasing. To oppose this increase in flux, the induced current flows counterclockwise and the electric field (parallel to the current) points tangentially to the path in the counterclockwise direction (since the current flows in the direction of the electric field).

(d) No. The induced electric field lines are circles, so the electric field has no radial component. Since there is no force capable of providing centripetal force, the charge cannot move in a circular path.

5. (a) As soon as the bar moves, the positive current-carrying charges feel a magnetic force, $F = q\mathbf{v} \times \mathbf{B}$, that causes them to move down the bar. However, the current has nowhere to go, so positive charge quickly accumulates at the bottom of the bar and negative charge quickly accumulates at the top of the bar. This uneven charge distribution creates an electric force that exactly cancels out the magnetic force, eventually causing the current to stop.

(b) The magnetic force on a test charge $+q$ is downward with a magnitude of $|F| = qvB$. When equilibrium has been established, the electric force balances this magnetic force. Therefore, the electric field has a magnitude of $|E| = |F|/q = vB$.

(c) In order for the electric force to cancel the magnetic force, the electric field must point directly upward toward the top of the page (not parallel to the bar). Since the electric field points in the direction of decreasing potential, the bottom of the bar must be at a higher potential than the top. We can evaluate the magnitude of this potential difference by integrating the electric field:

$$V_{top} - V_{bottom} = -\int_{bottom}^{top} \mathbf{E} \cdot d\mathbf{l}$$

The geometry of the dot product in this integral is shown in Figure 18.22.

$$V_{top} - V_{bottom} = -\int_{bottom}^{top} \mathbf{E} \cdot d\mathbf{l} = -\int_{bottom}^{top} Edl\cos\theta' = -\int_{bottom}^{top} Edl\sin\theta = -El\sin\theta$$

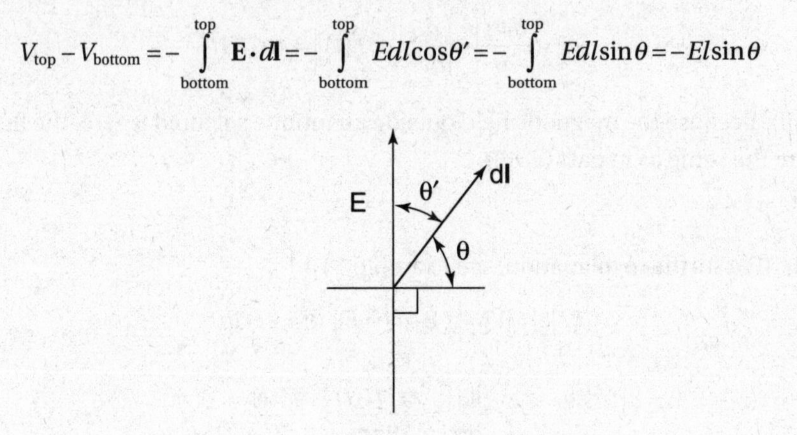

Figure 18.22

Given that the magnitude of the electric field is $E = vB$, this potential difference is then $V = -Blv\sin\theta$. (Consistent with our discussion above, the potential is greater at the bottom of the rod.)

(d) At the instant that the bar begins to move, the charge has not had time to redistribute itself, so the electric field and the electric force are both zero. However, the charges experience a magnetic force of magnitude $F = qvB$ directly downward. How can we calculate the current due to this force? The key is to realize that this magnetic force has the same exact effect on the charges as an electric field of magnitude $E = vB$ pointing directly downward would. Because this effective electric field produces identical forces on the mobile charges, it produces the same current as well.

Calculating the effective voltage by integrating the electric field [as in choice (D)], we find that the magnitude of the potential is $V = El\sin\theta = vBl\sin\theta$. Since the initial effective potential difference between the two ends of the bar is $V = vBl\sin\theta$ and the resistance of the bar is R, the initial current is $I = V/R = vBl\sin\theta / R$.

(e) The instant after the bar is stopped, the magnetic force acting on the charges is zero. However, the charge has not yet had time to move, so it is unevenly distributed, producing the electric field and voltage calculated in parts (c) and (d) (i.e., a voltage of magnitude $V = blv\sin\theta$ with the lower end of the rod at higher potential). This voltage causes a current of $I = (blv\sin\theta)/R$ to flow up the rod. Using the right-hand rule, this current produces a force perpendicular to the rod pointing toward the bottom right corner of the page.

6. (a) During a time dt, the area exposed to the magnetic field increases by a differential strip of area $dA = lvdt$, where v is the speed of the loop and the magnetic flux increases by an amount $d\Phi_B = \mathbf{B} \cdot d\mathbf{A} = Blv\,dt$. Therefore, the induced EMF is

$$|\mathcal{E}| = \left|\frac{d\Phi_B}{dt}\right| = \frac{Blv\,dt}{dt} = Blv$$

(b) From Lenz's law, the EMF opposes the battery (and pushes current counterclockwise). Therefore, the voltage across the resistor is

$$V_{\text{resistor}} = IR = V - Blv$$

The current is then

$$I = \frac{V - Blv}{R}$$

(c) The magnitude of the upward magnetic force is

$$F = I\mathbf{l} \times \mathbf{B} = IlB = \left(\frac{V - Blv}{R}\right) lB$$

The gravitational force is $-mg$, so the net force is

$$F_{\text{net}} = \left(\frac{V - Blv}{R}\right) lB - mg$$

(d) At the terminal speed, the net force is zero. Therefore,

$$F_{\text{net}} = \left(\frac{V - Blv_{\text{terminal}}}{R}\right) lB - mg = 0$$

$$v_{\text{terminal}} = \frac{V}{lB} - \frac{mgR}{l^2B^2}$$

(e) When the loop is entirely within region 1, there is no net magnetic force on the loop (because the force on each segment is canceled by the opposite segment). With no magnetic force, the net force is due to gravity alone, so the loop behaves as if in free fall (with an initial upward velocity given by its velocity as the bottom of the loop enters region 1) and the $y(t)$ function is parabolic.

Inductors

19

→ **SELF-INDUCTANCE**
→ **ENERGY STORED IN AN INDUCTOR**
→ **CIRCUITS CONTAINING RESISTORS AND INDUCTORS**
→ **CIRCUITS CONTAINING CAPACITORS AND INDUCTORS**

QUALITATIVE UNDERSTANDING OF SELF-INDUCTANCE

Consider increasing the current that passes through a lone solenoid. This causes an increasing magnetic field to pass through the center of the solenoid, which in turn produces an increasing magnetic flux through the solenoid. This changing flux, according to Faraday's law, then causes an electromotive force. What direction does this EMF push the current? According to Lenz's law, this electromotive force opposes the increasing magnetic flux. We just discussed how the increasing magnetic flux is due to the increasing current; therefore, to oppose the increase in magnetic flux, the induced EMF must oppose the increasing current. In summary, increasing the current passing through a solenoid induces an EMF that opposes the increase in current.

The opposite argument can be made in the case of decreasing the current that passes through a solenoid. The decreasing current causes a decreasing magnetic field and thus a decreasing magnetic flux. This decreasing magnetic flux, according to Faraday's and Lenz's laws, produces an electromotive force that opposes the decrease in magnetic flux. Again, because the decrease in flux is due to decreasing the current, for the induced EMF to oppose the decrease in flux it must oppose the decrease in current. In summary, decreasing the current passing through a solenoid induces an EMF that opposes the decrease in current.

Therefore, combining these two results, we see that whenever the current through a solenoid changes, this change induces an EMF that opposes the change in current. This phenomenon is common to any circuit element that produces magnetic flux through itself, and such objects are called *inductors.* A useful way to think of inductors is to imagine that they are trying to stabilize the amount of current flowing through a circuit by creating an EMF that opposes changes in current. Now that we have a qualitative understanding of inductors, we're ready for a more rigorous quantitative exploration of inductors.

Induction occurs even when there is only one circuit.

EXAMPLE 19.1 THE SELF-INDUCTANCE OF A SOLENOID

A solenoid of length l with n turns per length and radius R experiences a changing current dI/dt. Calculate the magnitude of the induced EMF in the solenoid.

SOLUTION

The magnetic field within the solenoid is given by

$$B = \mu_0 nI$$

Calculating the magnetic flux,

$$\Phi_B = (\text{number of loops})(\text{flux per loop}) = (nl)(B\pi R^2) = \mu_0 \pi n^2 l R^2 I$$

$$|\mathcal{E}| = \left|\frac{d\Phi_B}{dt}\right| = \mu_0 \pi n^2 l R^2 \left|\frac{dI}{dt}\right|$$

QUANTITATIVE DEFINITION OF SELF-INDUCTANCE

As we might expect, the induced EMF is proportional to the rate of change in the current. This proportionality constant is called *self-inductance.*

$$L = \frac{\Phi_B}{I}$$

Definition of self-inductance

It immediately follows that

$$|\mathcal{E}| = L\left|\frac{dI}{dt}\right|$$

Equation for the induced EMF

The *inductance* of an inductor, L, is a property of the geometry of the inductor itself. For example, from Example 19.1, you can verify that the inductance of a solenoid is given by the following expression that involves only physical constants and the dimensions of the inductor: $L = \mu_0 \pi n^2 l R^2$.

The SI unit for inductance is the henry (H), given by the equation $1\text{H} = 1\ (\text{V}\cdot\text{s})/\text{A}$.

ENERGY STORED IN AN INDUCTOR

As we increase the current through an inductor, an EMF is induced in the inductor; that is, a potential difference (voltage) develops across the two terminals of the inductor that opposes increases in current. Therefore, increasing the current in an inductor involves driving current against this potential difference, which requires energy. Where does this energy go? It is stored in the inductor. Later on, if the battery wears out and the current begins to decrease, a potential difference develops across the inductor that pushes the current in the same direction as the battery, in an effort to oppose the decrease in current. At this point, the energy stored in the inductor is being depleted to maintain the current.

Quantitatively, how much energy is stored in an inductor? As with capacitors, because it is difficult to directly calculate the energy stored, we instead calculate the energy required to produce a certain current through the inductor. (Conservation of energy then requires that this energy equal the energy stored in the current-carrying inductor.)

Whenever calculating the energy expended in a circuit element, we generally integrate power as a function of time. (Recall that power is defined as the rate at which the circuit exerts work on the circuit element. From the work-energy theorem, the rate at which the circuit does work on the inductor is equal to the rate at which the inductor's potential energy increases.)

$$U = \int dU = \int P\,dt = \int IV\,dt$$

Recall from above that the voltage across an inductor is given by Faraday's law (in its self-induction form):

$$V = L\frac{dI}{dt}$$

Therefore, the energy required to increase the current through an inductor from zero current to a final current I_F is

$$U = \int I\left(L\frac{dI}{dt}\right)dt = \int_{I=0}^{I=I_F} LI\,dI = \frac{LI_F^2}{2}$$

The energy in an inductor depends on the final current like the energy in a capacitor depends on the final charge and the gravitational potential energy of a climber depends only on her final height, not on how quickly she climbed or on whether or not she stopped for a break. Therefore, this is a general result.

$$U = \frac{LI^2}{2}$$

Energy stored in an inductor

Circuit Problems with Inductors

Later in this chapter we will discuss how current and voltage change with time in circuits containing an inductor and a resistor. However, before we launch into the complicated dynamics of such circuits, it is essential to make a few observations about the final and initial conditions.

1. The current through an inductor cannot change abruptly. If the current through an inductor were not continuous, the derivative dI/dt would be infinite and the induced current would be infinite (violating conservation of energy).

   This rule is very useful in figuring out the initial current through an inductor the instant after some change, because this initial current must be exactly the same as the current before the change was made. For example, the instant before you connect a battery to an inductor and a resistor, the current is zero. Therefore, the instant after you connect a battery to an inductor and a resistor, the current is still zero.
2. At equilibrium (e.g., steady state with constant current), the voltage across an inductor is zero, so the inductor acts like a piece of wire (a short circuit). When the change in current is zero, the voltage across the inductor is zero according to the equation $|\mathcal{E}| = L|dI/dt|$, so the inductor acts like a short circuit.

   Caution: In *LC* circuits, the circuit never reaches a steady state, so this rule doesn't apply.
3. Don't forget Kirchhoff's loop equation. As you will see in some of the following supplemental problems, solving circuit problems with inductors often involves the use of a simple Kirchhoff's loop equation.

RL CIRCUITS (CIRCUITS CONTAINING INDUCTORS AND RESISTORS)

We have just discussed the behavior of circuits containing inductors immediately after changes are made (at $t = 0$) and much, much later after the circuit has reached equilibrium (at $t = \infty$). In this section, we will quantitatively explore what happens between $t = 0$ and $t = \infty$.

LR circuits are analogous to *RC* circuits, and it is useful to keep in mind the following parallels:

RC circuit	*LR* circuit analog
Time	Time
Charge	Current
Current	Voltage
Time constant $\tau = RC$	Time constant $\tau = L/R$

General Approach to Deriving Equations for *RL* Circuits

The following is the general approach used to derive all the formulas in this chapter. (As you will see, it is quite similar to the approach we used to solve *RC* circuits.)

STEP 1 Write a Kirchhoff's loop equation for your circuit. Here's how to treat an inductor in your Kirchhoff's loop equation:

a. The magnitude of the voltage is given by Faraday's law:

$$|\text{voltage}| = L\left|\frac{dI}{dt}\right|$$

b. The sign of the voltage is given by Lenz's law. The sign of the voltage across an inductor always serves to oppose the change in current in the circuit (thereby opposing the change in magnetic flux).

STEP 2 Now you should have a differential equation involving various constants, *I*, and *t*. You can separate variables and perform an indefinite integral.

STEP 3 Use the initial conditions to eliminate the constant of integration. (The value of the current at very long times can serve as verification that your answer is correct.)

EXAMPLE 19.2 CURRENT GROWTH THROUGH AN INDUCTOR

At time $t = 0$, you attach a battery of voltage V to a circuit containing a resistor (of resistance R) and an inductor (of inductance L). What is the current as a function of time?

STEP 1 **Start with Kirchhoff's law equation (Figure 19.1)**

$$V - IR - L\frac{dI}{dt} = 0$$

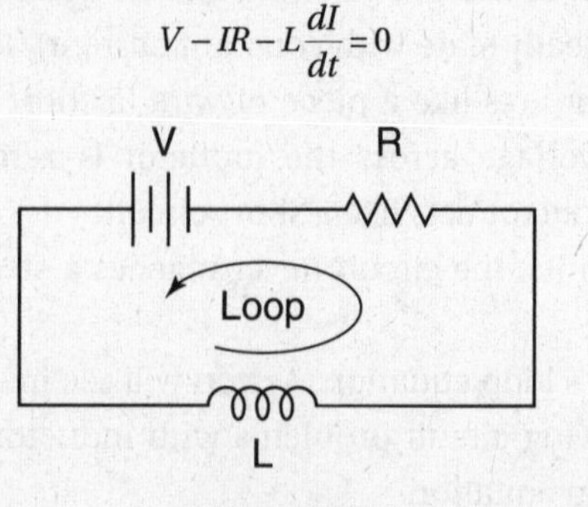

Figure 19.1

Note the sign of the voltage across the inductor. Because the inductor is pushing the current in the opposite direction compared to the battery (in order to oppose the increase in current), the voltage across the inductor must have a sign opposite that of the battery. The quantity $L(dI/dt)$ is positive because current is increasing, so we add a negative sign to make the expression negative, $-L(dI/dt)$.

STEP 2 Separate variables and integrate:

$$V - IR = L\frac{dI}{dt}$$

$$dt = \frac{LdI}{V - IR}$$

$$\frac{R}{L}dt = \frac{dI}{\frac{V}{R} - I}$$

$$\int \frac{R}{L}dt = \int \frac{dI}{(V/R) - I}$$

Indefinite integration introduces a constant of integration, K:

$$K - \frac{R}{L}t = \ln\left(\frac{V}{R} - I\right)$$

Exponentiating,

$$\frac{V}{R} - I = e^{K-(R/L)t} = e^{K} \cdot e^{-(R/L)t}$$

$$I = \frac{V}{R} - e^{K}e^{-(R/L)t}$$

STEP 3 We know that the current through an inductor is a continuous function of time, so at $t = 0$ it must have the same current that it had an instant before connecting the battery (zero current). Therefore, the current is initially zero (the inductor acting like an open circuit).

$$I(t=0) = \frac{V}{R} - e^{K}e^{0} = \frac{V}{R} - e^{K} = 0$$

$$e^{K} = \frac{V}{R}$$

Substituting this result back into the equation for $I(t)$, we find

$$I(t) = \frac{V}{R} - \frac{V}{R}e^{-(R/L)t} = \frac{V}{R}(1 - e^{-(R/L)t})$$

We can verify this equation by observing that it satisfies what we expect at very long times after the battery has been hooked up: At $t = \infty$, the inductor acts like a closed circuit, so the current is given by Ohm's law: $I(t = \infty) = V/R$. Noting that the final current is $I_F = V/R$, we often write the final result as

$$I(t) = I_F(1 - e^{-(R/L)t})$$

Growth of current in an *RL* circuit

EXAMPLE 19.3 CURRENT DECAY IN AN *RL* CIRCUIT

Consider the circuit shown in Figure 19.2. First, the switch is connected to position A, causing current to flow in the inductor, ultimately reaching the equilibrium value of $I_0 = V_A/R_A$. At time $t = 0$, the switch is instantaneously flipped from position A to position B, causing the inductor to be connected to a resistor of resistance R. What is the current through the resistor R as a function of time after the switch is flipped to position B?

SOLUTION

STEP 1 The Kirchhoff loop equation is

$$-IR - L\frac{dI}{dt} = 0$$

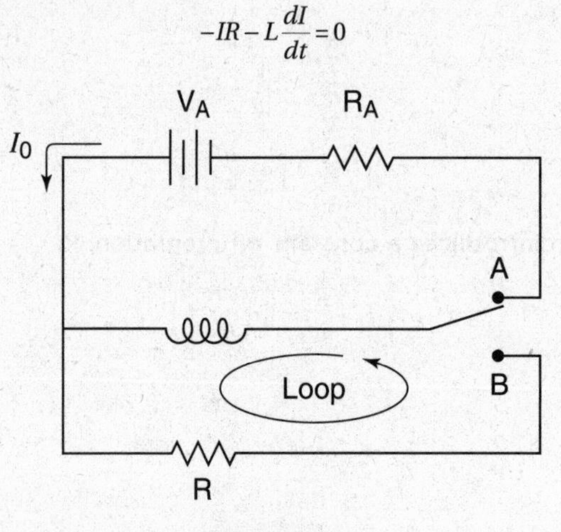

Figure 19.2

Note the sign of the voltage across the inductor. We know that the inductor acts as a battery that pushes the current in the direction it was flowing at $t = 0$ (in order to oppose the decrease in current). Therefore, the voltage across the inductor when the loop is traversed as indicated in Figure 19.2 must be positive. Because $L(dI/dt)$ is negative (because the current is decreasing), we need to add a negative sign as shown above to make the voltage across the inductor positive.

STEP 2 Separate variables and integrate:

$$-L\frac{dI}{dt} = IR$$

$$\frac{dI}{I} = -\frac{R}{L}dt$$

$$\int\frac{dI}{I} = \int -\frac{R}{L}dt$$

Performing indefinite integration,

$$\ln(I) = -\frac{R}{L}t + K$$

Exponentiating,

$$I = e^{-(R/L)t+K} = e^K e^{-(R/L)t}$$

STEP 3 From the problem we know that the initial current is $I_0 = V_A/R_A$ (because the current through an inductor is continuous). Thus,

$$I(t = 0) = e^K e^0 = e^K = I_0$$

Substituting into $I(t)$ yields the following.

$$I(t) = I_0 e^{-(R/L)t}$$

Decay of current in an *RL* circuit

As a check, we verify that the final condition (current approaches zero as time goes to infinity) is satisfied by the equation above.

Voltage Across an Inductor in an *RL* Circuit

Once we've calculated $I(t)$, the magnitude of the voltage across an inductor can easily be found by applying the equation $|\text{voltage}| = L|dI/dt|$ to the current function, as shown below.

Before we begin number crunching, let's figure out the initial and final voltages across an inductor with growing current.

- Initially, $I = 0$. Therefore, Kirchhoff's loop rule requires that the voltage across the inductor equals the voltage across the battery (with opposite polarity).
- At $t = \infty$, the current is constant, so $dI/dt = 0$ and the voltage across the inductor is zero.

Now, taking the derivative of the current function derived above to calculate voltage,

$$|V(t)| = L\left|\frac{dI}{dt}\right| = L\left|\frac{d}{dt}\left[I_F\left(1 - e^{-(R/L)t}\right)\right]\right| = I_F R e^{-(R/L)t}$$

Noting that $I_F R = V_{\text{battery}}$ (by Ohm's law) and that the voltage of the battery is the initial voltage across the inductor, we can rewrite this equation as follows.

$$|V(t)| = V_{\text{battery}} e^{-(R/L)t}$$

Voltage across an inductor while the current grows

Alternative solution: Another way to obtain $V(t)$ is to calculate the voltage across the resistor using Ohm's law, $V = I(t)R$, and then use a Kirchhoff's loop rule to obtain the voltage across the inductor.

Now let's determine the initial and final voltages across an inductor with decaying current.

- Initially, a current of I_0 flows though the resistor, so the resistor must have a voltage of $V_0 = I_0 R$. A Kirchhoff's loop equation then requires the inductor to have the same voltage (with the opposite polarity).
- At $t = \infty$, the current in the circuit is not changing, so $dI/dt = 0$ and the voltage across the inductor is zero.

Again plugging our expression for $I(t)$ into the equation $V = L(dI/dt)$ to calculate the voltage as a function of time,

$$|V(t)| = L\left|\frac{d}{dt}\left(I_0 e^{-(R/L)t}\right)\right| = I_0 R e^{-(R/L)t} = V_0 e^{-(R/L)t}$$

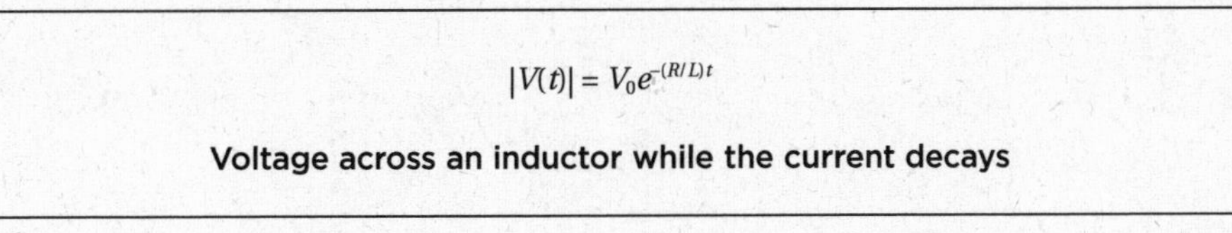

$$|V(t)| = V_0 e^{-(R/L)t}$$

Voltage across an inductor while the current decays

Time Constants

Let us rewrite the final results:

$$I(t) = I_F(1 - e^{-(R/L)t})$$

Current growth in an inductor

$$I(t) = I_0 e^{-(R/L)t}$$

Current decay in an inductor

$$|V(t)| = V_{\text{battery}} e^{-(R/L)t}$$

Voltage across an inductor during current growth

$$|V(t)| = V_0 e^{-(R/L)t}$$

Voltage across an inductor during current decay

You might notice that, as in *RC* circuits, time always enters these equations in the same factor, $e^{-(R/L)t}$. As in *RC* circuits, we define the time constant τ as the amount of time required to make the exponent of this expression change by 1, causing the factor $e^{-(R/L)t}$ to change by a factor of *e*.

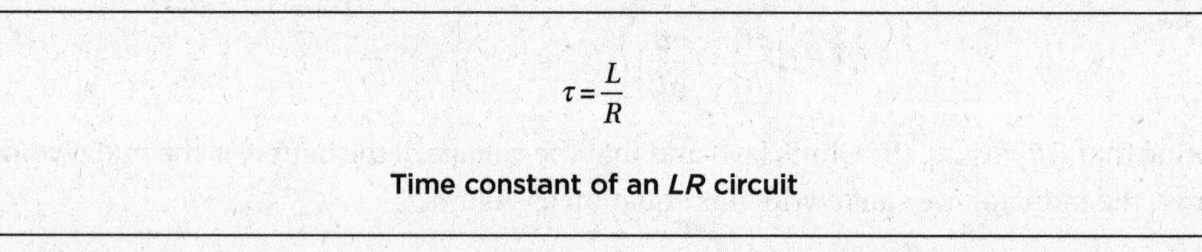

$$\tau = \frac{L}{R}$$

Time constant of an *LR* circuit

Comparing these results then yields a very compact way to summarize what we know about *RL* circuits: Everything exponentially approaches its final value with a time constant of $\tau = L/R$.

- Because the voltage across an inductor at equilibrium always equals zero, the voltage always approaches zero as an exponentially decreasing function, similar to its analog, current, in *RC* circuits.
- Current approaches a constant value (when current is growing) or zero (when current is decaying), similar to its analog, charge, in *RC* circuits.

CURVE SKETCHING

What do these functions look like? Given the strong parallels between *RL* and *RC* circuits, it shouldn't surprise you that they look exactly like their *RC* circuit counterparts (an exponential decay or an exponential approach to a finite value). See Figure 19.3.

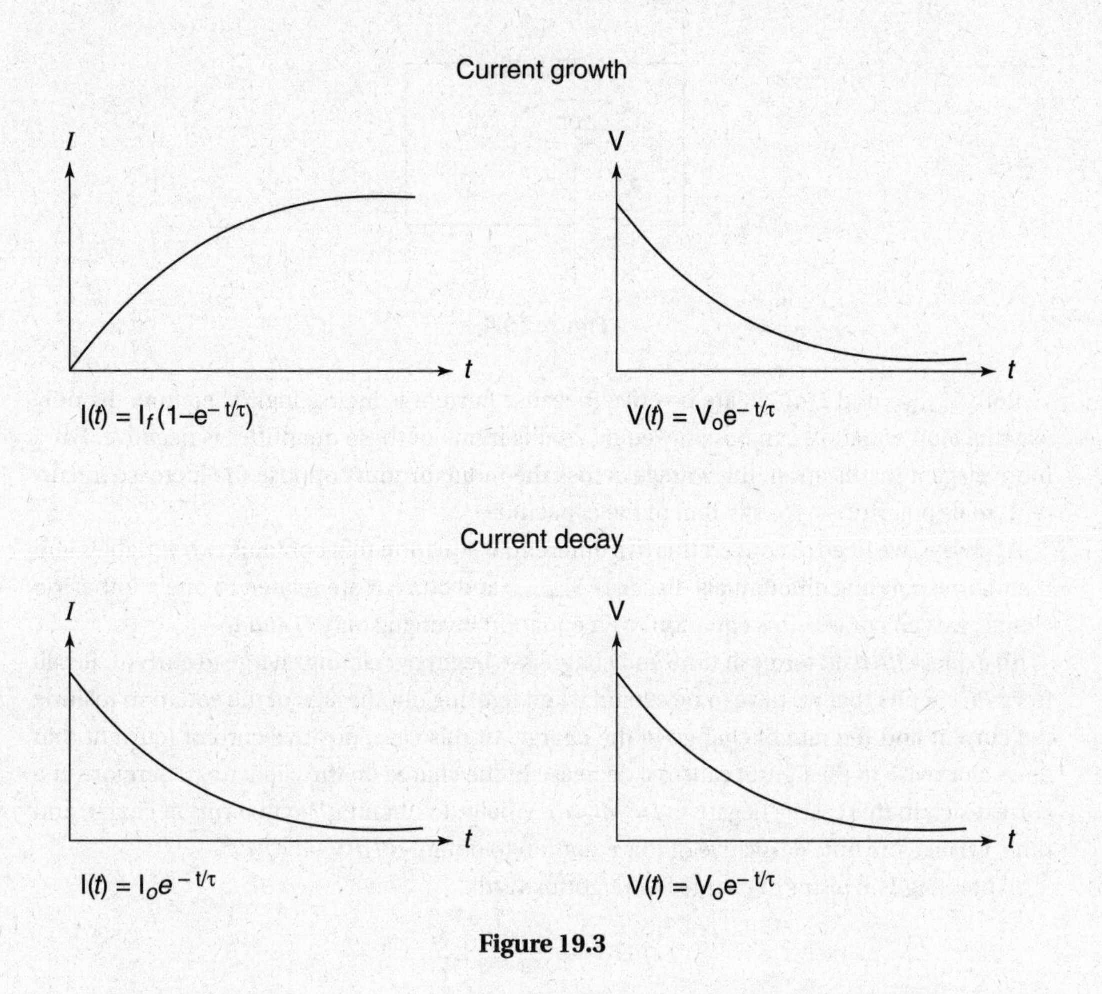

Figure 19.3

LC CIRCUITS (CIRCUITS CONTAINING A CAPACITOR AND AN INDUCTOR)

To understand the behavior of *LC* circuits we will follow the general pattern we have been using with *RC* and *RL* circuits; we start with a Kirchhoff loop equation, which is a differential equation. After solving the differential equation, the initial conditions allow us to calculate the constant(s) of integration.

General Behavior of *LC* Circuits

A capacitor of capacitance C is charged to a voltage V and then attached to an inductor of inductance L (which carries no current at the instant we attach the capacitor). What is the current through the circuit as a function of time?

We start with a Kirchhoff's loop equation (Figure 19.4):

$$V_{\text{capacitor}} - L\frac{dI}{dt} = 0$$

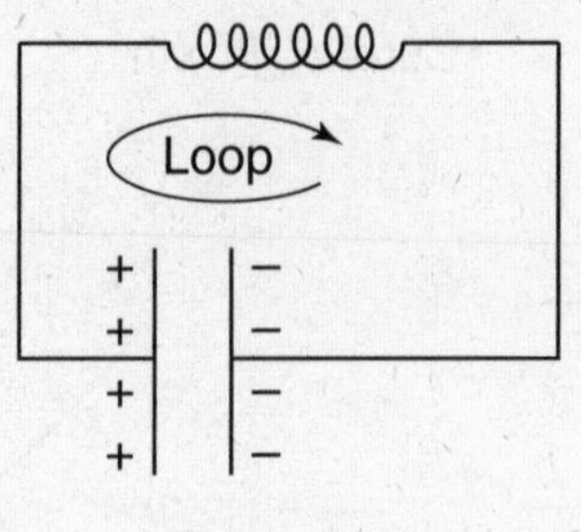

Figure 19.4

Both $V_{\text{capacitor}}$ and $L(dI/dt)$ are positive (because current is increasing). Therefore, the only way the loop equation can possibly equal zero is if one of these quantities is negative. For a more elegant justification, the voltage across the inductor must oppose the increase in current, so its polarity is opposite that of the capacitor.

As always, we need to convert this to a differential equation that contains two variables and their corresponding differentials. Because $V_{\text{capacitor}}$ and current are related to one another via charge, we can convert this equation to an equation involving only Q and t.

To express dI/dt in terms of time and charge, we begin by relating charge to current. Recall from *RC* circuits that we have to be careful when figuring out the sign of the equation relating the current and the rate of change in the charge. In this case, positive current (current that flows clockwise in the figure) causes a decrease in the charge on the capacitor. Therefore, the correct sign in this case is negative: $I = -dQ/dt$. Finally, to obtain dI/dt in terms of charge and time, we take the time derivative of this equation to obtain $dI/dt = -d^2Q/dt^2$.

Expressing V in terms of charge is straightforward:

$$Q = CV \Rightarrow V_{\text{capacitor}} = \frac{Q}{C}$$

Substituting all this into the loop equation yields

$$\frac{Q}{C} + L\frac{d^2Q}{dt^2} = 0$$

A more useful form of this equation is

$$Q = -LC\frac{d^2Q}{dt^2}$$

Hopefully you recognize this as the equation of simple harmonic motion. Even if you don't, if you are familiar with the differentiation of trigonometric functions, you should be able to see how the following is a solution to this equation (K and ϕ being constants of integration):

$$Q = K\sin\left(\frac{t}{(LC)^{1/2}} + \phi\right)$$

Note: As in simple harmonic motion, the following is also a solution:

$$Q = K\cos\left(\frac{t}{(LC)^{1/2}} + \phi\right)$$

The two solutions are equivalent (the only difference is that, depending on which you choose, the constant ϕ differs by $\pi/2$). For the rest of the chapter, we will use the sine solution for consistency.

$$Q = K\sin\left(\frac{t}{\sqrt{LC}} + \phi\right)$$

Capacitor's charge in an *LC* circuit (general solution)

Remember our original goal was to calculate the current as a function of time. We can calculate current by plugging the $Q(t)$ function into the equation for current from above, $I = -dQ/dt$, and differentiating to obtain the following.

$$I = -\frac{dQ}{dt} = -\frac{K}{\sqrt{LC}}\cos\left(\frac{t}{\sqrt{LC}} + \phi\right)$$

Current in an *LC* circuit (general solution)

These equations are completely general for *LC* circuits. Although we started with a specific problem, the only equation we have used so far is the Kirchhoff loop rule, which applies to any *LC* circuit with any set of initial conditions.

Specific solution for the given problem: As in *RC* and *RL* circuits and SHM problems, we can use initial conditions to determine the constants of integration. Unlike problems with *RC* and *RL* circuits, we don't already know what is going on at $t = \infty$; that's one of the things we are trying to find out.

We already have had lots of experience dealing with this type of differential equation, so it is worth pointing out the parallels between simple harmonic motion (mass-on-a-spring) problems and *LC* circuits:

Mass on a spring	***LC* circuit analog**
Position	Charge
Velocity, $v = dx/dt$	Current, $I = dQ/dt$
Angular frequency, $\omega = \sqrt{k/m}$	Angular frequency, $\omega = 1/\sqrt{LC}$
Linear frequency, $f = \frac{\omega}{2\pi} = \frac{1}{2\pi}\sqrt{k/m}$	Linear frequency, $f = \frac{\omega}{2\pi} = \frac{1}{2\pi}\sqrt{\frac{1}{LC}}$
Period, $T = \frac{1}{f} = \frac{2\pi}{\omega} = 2\pi\sqrt{m/k}$	Period, $T = \frac{1}{f} = \frac{2\pi}{\omega} = 2\pi\sqrt{LC}$
Kinetic energy $= \frac{1}{2}mv^2$	Energy due to moving charge stored in magnetic field of inductor $= \frac{1}{2}LI^2$
Potential energy $= \frac{1}{2}kx^2$	Energy due to static charge stored in capacitor $= Q^2/2C$

A helpful mathematical analogy between a mechanical system and an electrical system

EXAMPLE 19.4

Calculate the angular frequency of an *LC* circuit using only parallels between the mass-spring system and *LC* circuits (i.e., without solving any differential equations).

SOLUTION

Comparing the loop equation for the *LC* circuit,

$$\frac{Q}{C}+L\frac{d^2Q}{dt^2}=0$$

to the equation for a mass on a spring,

$$kx+m\frac{d^2x}{dt^2}=0$$

we see that L is analogous to m, the mass on the spring, and $1/C$ is analogous to k, the spring constant. Therefore, to obtain the angular frequency of the *LC* circuit, we simply start with the equation for angular frequency of the mass-spring system, $\omega=\sqrt{k/m}$, and replace k and m by their analogs ($1/C$ and L, respectively), to obtain $\omega=1/\sqrt{LC}$. This should illustrate how powerful these parallels are: We have put a lot of effort into mass-spring systems, and if we are clever we can put it to work here.

Problem Solving in *LC* Circuits

Problem solving for *LC* circuits is analogous to problem solving in simple harmonic motion situations. There are a few basic problem-solving skills that are covered more extensively in Chapter 9.

1. Given the initial conditions, you should be able to go from the general solution to the specific solution. From the specific solution, you should be able to calculate various quantities (such as current, charge, energy in the capacitor, energy in the inductor) at various times on demand. (As in simple harmonic motion, if you are clever, you can often avoid calculating the entire specific solution.)
2. Remember how we used conservation of energy in spring-mass and similar SHM problems? We can also use conservation of energy to solve *LC* circuit problems. Let us restate the general solution of *LC* circuits with the following substitution.

$$\theta=\frac{t}{\sqrt{LC}}+\phi$$

$$Q=K\sin\theta$$

$$I=\frac{K}{\sqrt{LC}}\cos\theta$$

Also recall the formulas for energy in a capacitor and in an inductor:

$$U_{\text{capacitor}}=\frac{Q^2}{2C}$$

$$U_{\text{inductor}}=\frac{LI^2}{2}$$

Therefore, when $\theta = 0 + n\pi$, the charge ($Q = 0$) and the current squared (I^2) are maximized so that $U_{capacitor} = 0$ and $U_{inductor}$ is maximized. All the energy is stored in the inductor. Alternatively, when $\theta = (\pi/2) + n\pi$, the current (I) is zero and the charge squared (Q^2) is maximized, so that $U_{inductor} = 0$ and $U_{capacitor}$ is maximized and all the energy is stored in the capacitor.

By the law of energy conservation, the energy stored in the inductor when $\theta = 0 + n\pi$ must be equal to the energy stored in the capacitor when $\theta = (\pi/2) + n\pi$, which must be equal to the total potential energy of the system. If we designate $|I_{max}|$ as the maximum magnitude of current experienced when $\theta = 0 + n\pi$ and $|Q_{max}|$ as the maximum magnitude of charge observed when $\theta = (\pi/2) + n\pi$, conservation of energy requires that

$$\frac{|Q_{max}|^2}{2C} = \frac{L|I_{max}|^2}{2}$$

Relationship between Q_{max} and I_{max}

Additionally, energy is conserved between any two times, so that

$$U_{cap,t1} + U_{ind,t1} = U_{cap,t2} + U_{ind,t2}$$

General statement of conservation of energy

This statement of conservation of energy can easily be proven by calculating the energy stored in the capacitor and the inductor as a function of time:

$$U_{cap}(t) = \frac{[Q(t)]^2}{2C} = \frac{K^2}{2C}\sin^2\theta$$

$$U_{ind}(t) = \frac{L[I(t)]^2}{2} = \frac{L}{2}\frac{K^2}{LC}\cos^2\theta = \frac{K^2}{2C}\cos^2\theta$$

Therefore, the total energy stored in the inductor and capacitor as a function of time is

$$U_{tot}(t) = U_{cap}(t) + U_{ind}(t) = \frac{K^2}{2C}(\sin^2\theta + \cos^2\theta) = \frac{K^2}{2C}$$

As expected, the total energy is constant.

CHAPTER SUMMARY

Faraday's law is applicable even when there is only one circuit (with a time-varying current) present. In that case, the induced EMF can be expressed in terms of the self-inductance of the circuit, a quantity that depends only on the geometry of the circuit. The self-inductance is a natural quantity to use when analyzing circuits containing inductors and other circuit elements, such as resistors and capacitors. Kirchhoff's loop rule yields differential equations for the current or charge in the circuit.

PRACTICE EXERCISES

Multiple-Choice Questions

1. Consider the two concentric solenoids shown in Figure 19.5, one of which is attached to a battery, a resistor, and a switch and the other of which is attached to a lightbulb. (No current passes directly between the two circuits.) When the switch is turned on, what is expected to happen?

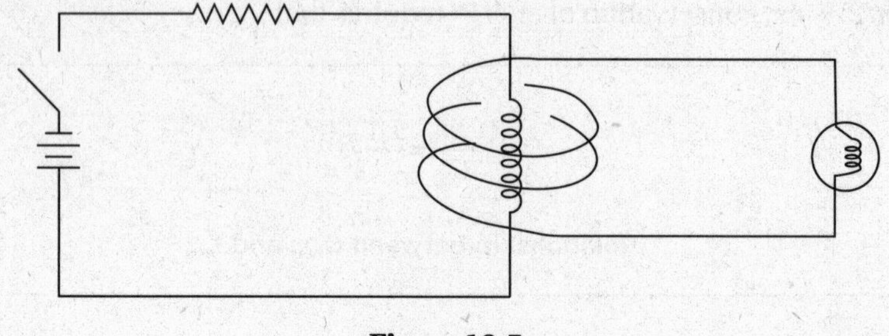

Figure 19.5

(A) No current passes through the lightbulb.
(B) The intensity of the lightbulb increases and quickly reaches its maximum, final intensity.
(C) The lightbulb flickers on and off with a regular frequency and constant intensity.
(D) The lightbulb flickers on and off with a regular frequency and decreasing intensity.
(E) The lightbulb flashes on once, and then no current passes through it.

2. An inductor is connected in series to a battery of voltage V and a resistor at $t = 0$. Which of the graphs in Figure 19.6 shows the voltage across the inductor as a function of time?

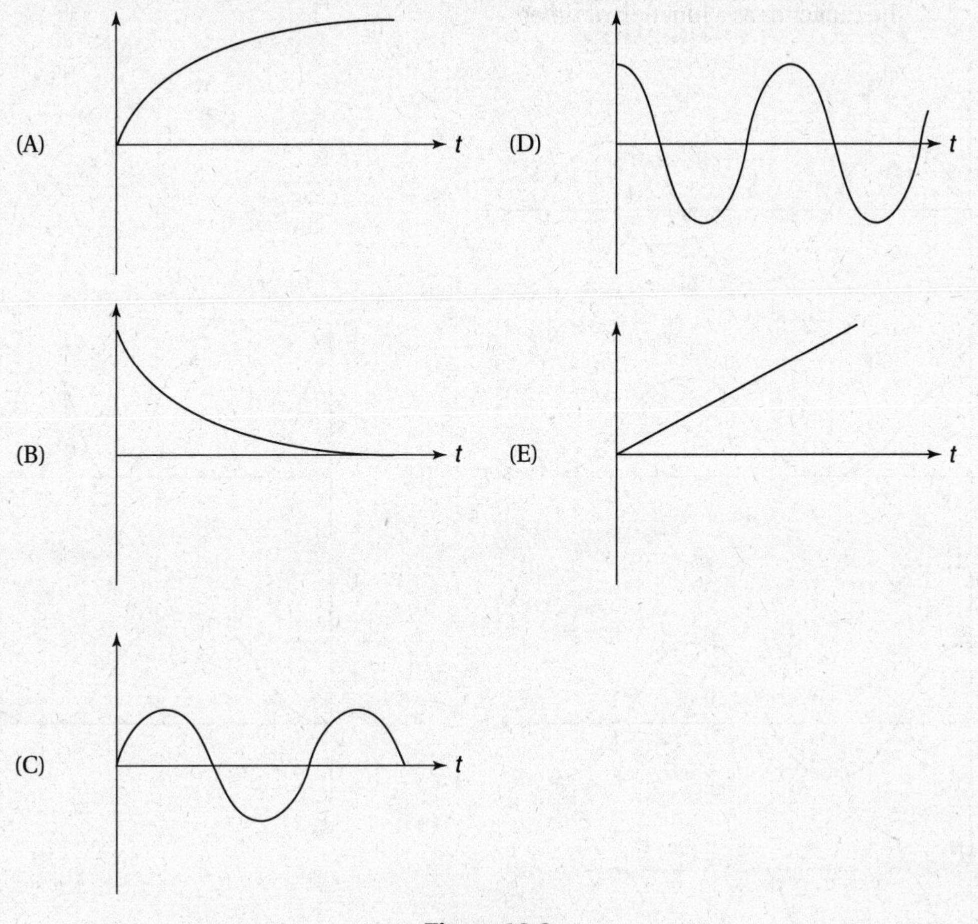

Figure 19.6

3. What is the period of the circuit shown in Figure 19.7?

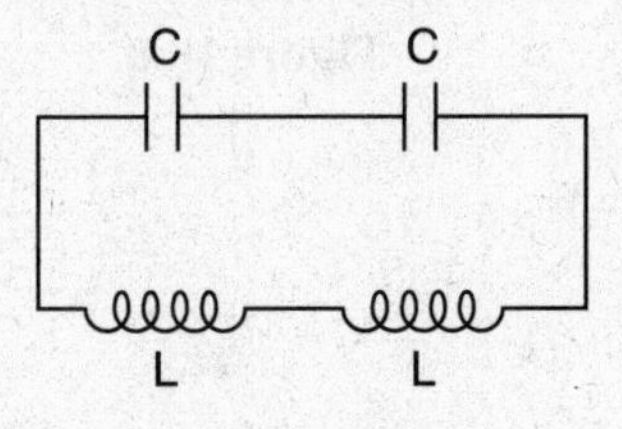

Figure 19.7

(A) $T = \sqrt{LC}$
(B) $T = \sqrt{LC/2}$
(C) $T = \sqrt{2LC}$
(D) $T = \pi\sqrt{2LC}$
(E) $T = 2\pi\sqrt{LC}$

4. A fully charged capacitor is put in series with an inductor at time $t = 0$. The current through this *LC* circuit is shown in Figure 19.8. Which graph shows the energy stored in the capacitor as a function of time?

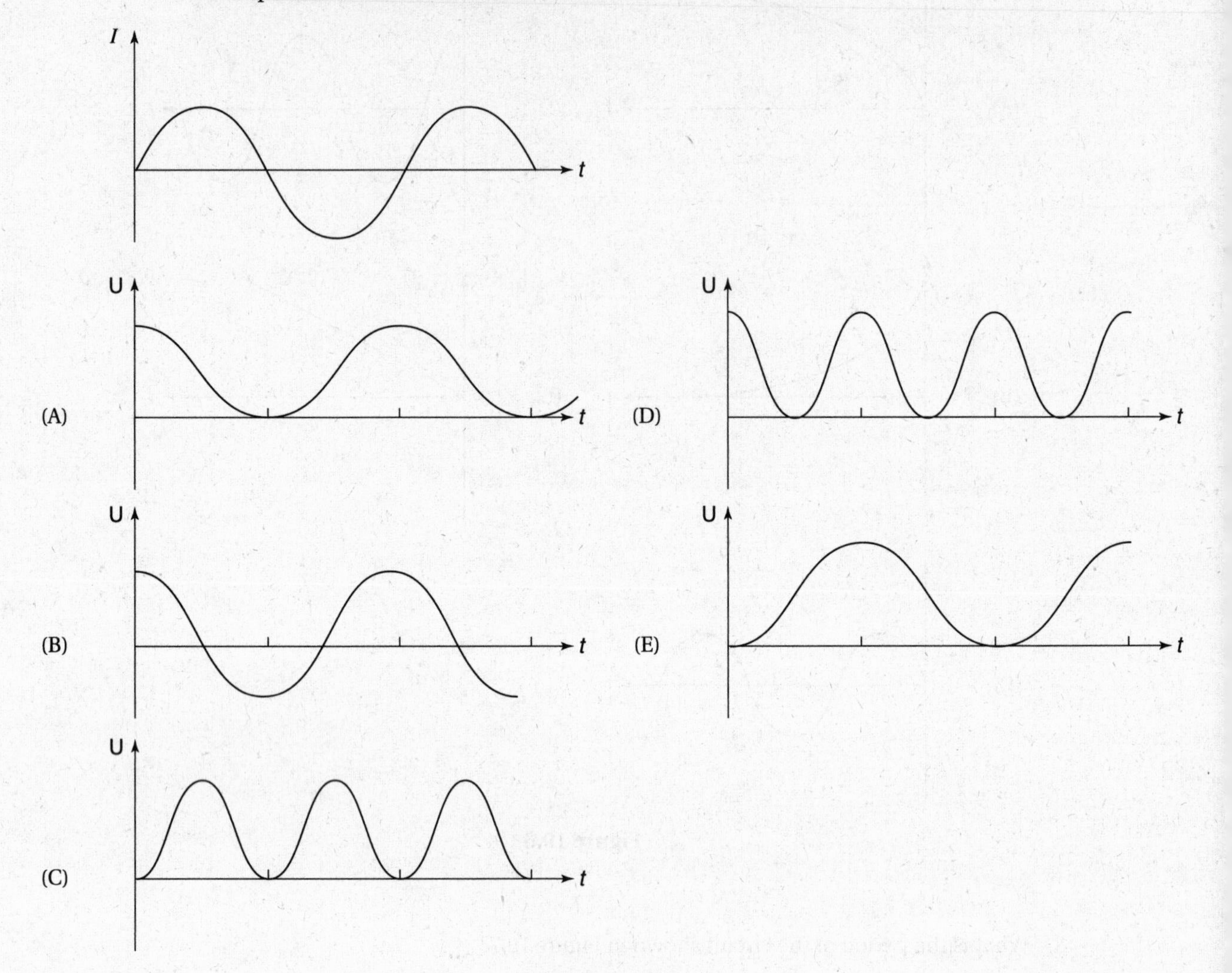

Figure 19.8

5. Consider the circuit in Figure 19.9. The switch is closed at $t = 0$, when the capacitor is uncharged and no current is flowing through the inductor. What is the initial current that passes through the battery the instant after the switch is closed?

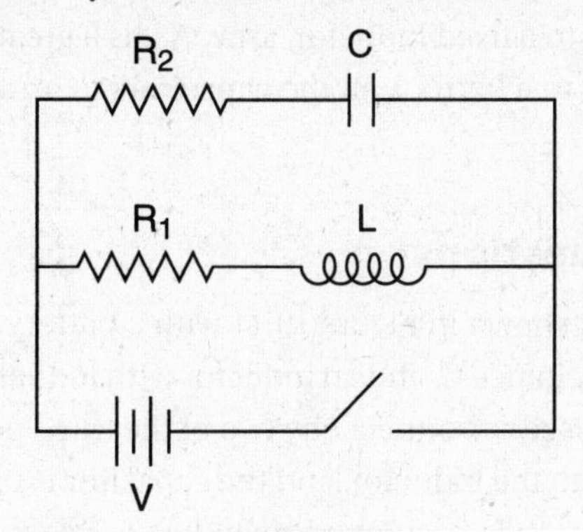

Figure 19.9

(A) $I = 0$
(B) $I = V/R_1$
(C) $I = V/R_2$
(D) $I = V/(R_1 + R_2)$
(E) $I = V(R_1 + R_2)/R_1R_2$

6. What is the final current in the circuit shown in question 5?

(A) $I = 0$
(B) $I = V/R_1$
(C) $I = V/R_2$
(D) $I = V/(R_1 + R_2)$
(E) $I = V(R_1 + R_2)/R_1R_2$

7. All the capacitors in each of the networks shown in Figure 19.10 have a capacitance of C. Which statement is correct?

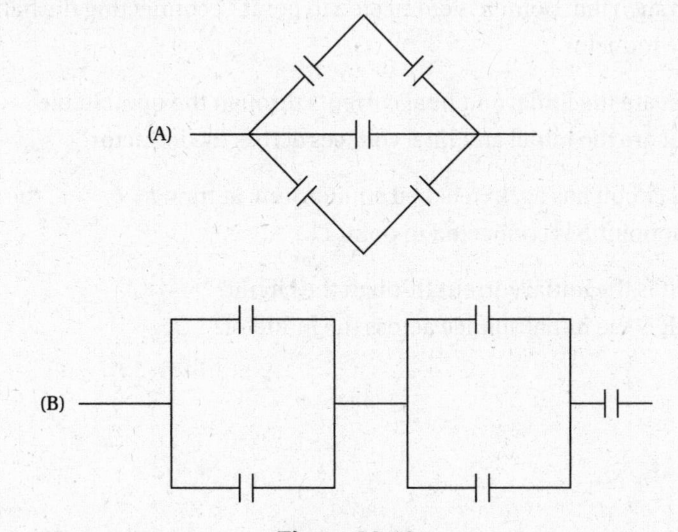

Figure 19.10

(A) When attached to a given battery, array A stores more energy.
(B) Equal amounts of charge are required to charge each array to the same final voltage.
(C) When attached to a fixed battery and resistor, array A has a lower time constant.
(D) When attached to a fixed inductor, array A has a greater angular frequency.
(E) When attached to a battery, all the capacitors in array A store the same amount of charge.

Free-Response Questions

1. Consider the circuit shown in Figure 19.11 with a battery with voltage *V*, resistance *R*, a capacitor with capacitance *C*, and an inductor with inductance *L*. The switch is designed such that it can be used to connect any two of the three points (*A*, *B*, and *C*). Initially, no current flows through the inductor, and the capacitor is uncharged (the switch is open). At time $t = 0$, the switch is positioned such that *A* is connected to *B*, causing current to flow to the capacitor.

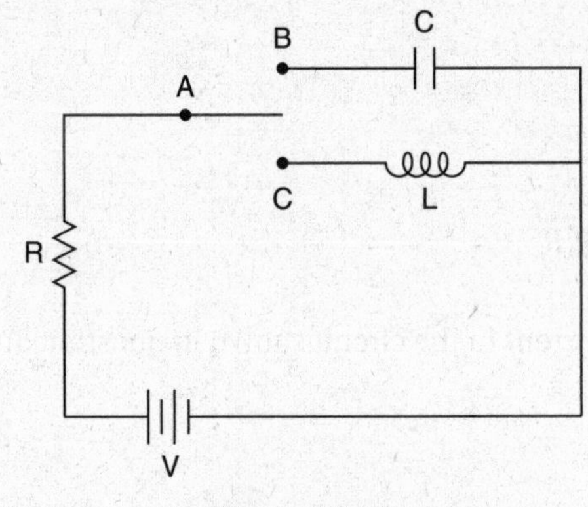

Figure 19.11

(a) What are the initial and final currents through the circuit?
(b) What are the initial and final voltages across the capacitor?

After the circuit has reached equilibrium, at time $t = t_1 >> 0$, the switch is instantaneously adjusted such that point *A* is connected to point *C*, connecting the battery and the resistor to the inductor.

(c) What are the initial and final currents through the new circuit?
(d) What are the initial and final voltages across the inductor?

After the circuit has again reached equilibrium, at time $t = t_2 >> t_1$, the switch is adjusted such that point *B* is connected to point *C*.

(e) What is the initial current through the circuit?
(f) What is the initial voltage across the inductor?

2. Consider the coaxial cable shown in Figure 19.12. At one end of the cable, the inner wire has been connected to the outer sheath, so that current coming out of the page through the inner wire equals the current going into the page through the outer sheath. If this current I has a uniform current density,

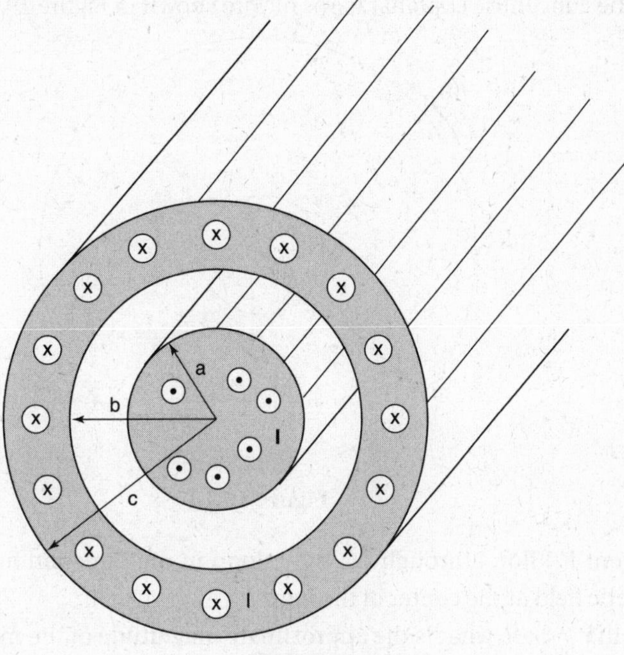

Figure 19.12

(a) Calculate the magnitude of the magnetic field in the region $a < r < b$.

Figure 19.13 shows a lengthwise cross section of the coaxial cable.

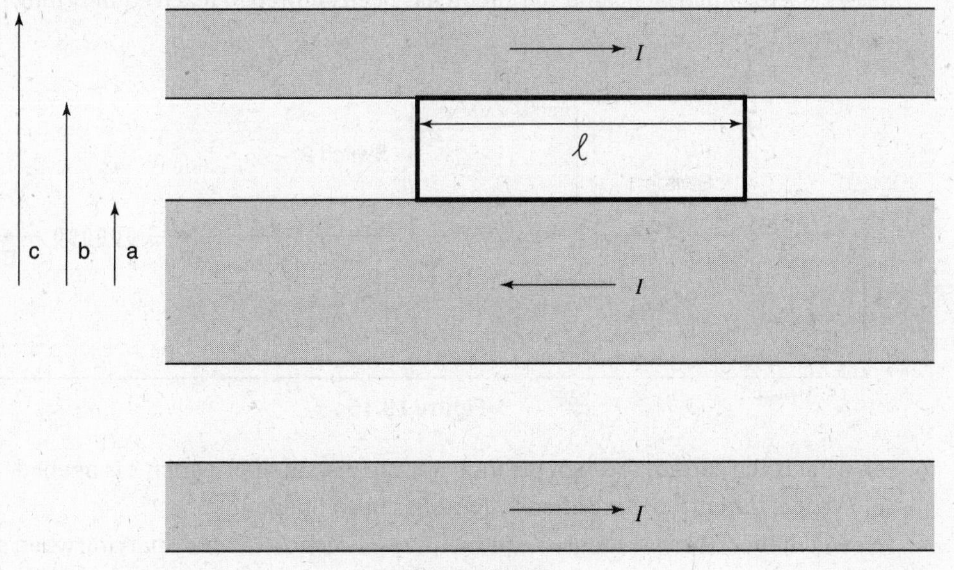

Figure 19.13

(b) If the current is increasing at a rate of dI/dt, what is the induced EMF through the area shown in the figure?

(c) How does this EMF affect the change in current?

(d) What is the inductance of a coaxial cable of length l?

3. (a) What is the equivalent inductance of two inductors, L_1 and L_2, in series?
   (b) What is the equivalent inductance of two inductors, L_1 and L_2, in parallel? (Assume that the mutual inductance of the two inductors involved is negligible.)

4. Consider the concentric coplanar loops of wire shown in Figure 19.14.

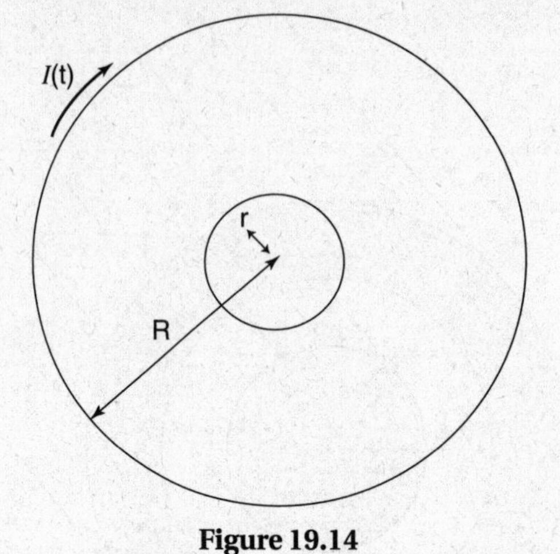

Figure 19.14

   (a) If current $I(t)$ flows through the outer loop in the direction indicated, what is the magnetic field at the center of the loop?
   (b) Given that $r << R$, what is the approximate magnitude of the magnetic flux through the small loop due to the current in the larger loop?
   (c) If $I(t) = I_0 e^{-t/\tau}$, calculate the EMF through the smaller loop as a function of time.

5. Consider the circuit in Figure 19.15. At time $t = 0$, switch 1 is closed (with switch 2 previously closed). Much later, after the circuit has been allowed to reach equilibrium, switch 2 is opened.

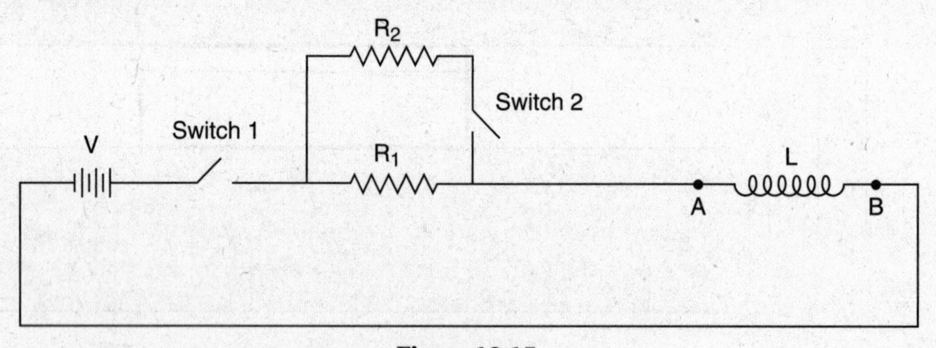

Figure 19.15

   (a) What is the current through the inductor the instant after switch 2 is opened?
   (b) What is the current long after switch 2 has been opened?
   (c) Sketch the current through the inductor as a function of time, marking when switch 1 is closed and when switch 2 is opened. Indicate relevant current values.
   (d) Designating the potential on the negative side of the battery to be zero, what is the potential at point *A* the instant after switch 1 is closed? How does this potential behave as time increases?
   (e) What is the potential at point *A* the instant after switch 2 is opened? How does this potential behave as time increases?

6. Consider the circuit shown in Figure 19.16. At time $t = 0$, the switch is closed. (Before the switch is closed, the capacitor is uncharged and there is no current flowing through the inductor.)

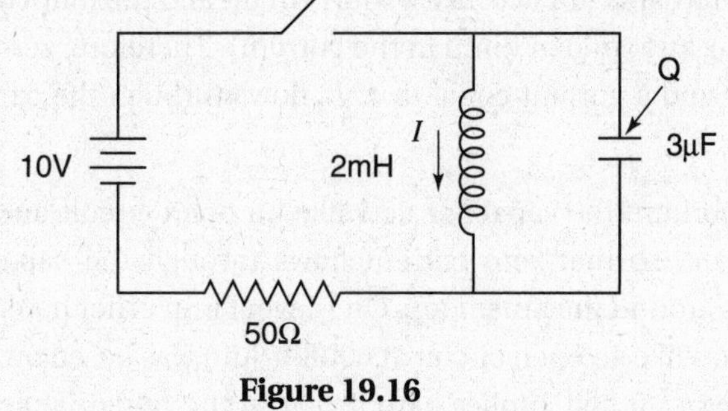

Figure 19.16

(a) What is the current through the battery the instant after the switch is closed, at $t = 0$?

(b) What is the current through the battery after the circuit has reached equilibrium, at $t = \infty$?

After the circuit has reached equilibrium, the switch is opened (for simplicity we will designate a new frame of reference so that this occurs at $t' = 0$).

(c) What is the period of oscillation of the *LC* circuit formed?

(d) What is the maximum current?

ANSWER KEY

1. **E**	3. **E**	5. **C**	7. **A**
2. **B**	4. **D**	6. **B**	

ANSWERS EXPLAINED

Multiple-Choice

1. **(E)** The instant after the switch is turned on, the flux through the inner solenoid increases sharply. This changing flux induces a current in the outer solenoid, which briefly causes the lightbulb to flash on. Very soon, the current in the left circuit reaches its final value, the flux through the inner solenoid ceases changing, and the induced current through the lightbulb drops to zero and remains zero.

2. **(B)** As the current exponentially approaches its final value, the change in current (which is proportional to the voltage) decreases exponentially, as discussed in the beginning of this chapter.

3. **(E)** In general, the angular frequency of an *LC* circuit is $\omega = 1/\sqrt{LC}$, so the period is $T = 2\pi/\omega = 2\pi\sqrt{LC}$. (For a review of period and angular frequency, see Chapter 9.) In this case, the equivalent capacitance is $C/2$ and the equivalent inductance is $2L$ (because the voltage jumps across the inductors add), so the product $L_{\text{effective}}C_{\text{effective}} = (2L)(C/2) = LC$, and the period is $T = 2\pi\sqrt{LC}$.

4. **(D)** Every time the current is zero, the energy in the inductor is zero and the energy in the capacitor is maximized. The only graph that has a peak every time the current of the inductor passes through zero is choice (D). Note that with the initial conditions given, $I(0) = 0$.

Note: Without any knowledge of *LC* circuits, choice (B) can be eliminated because the energy stored in a capacitor, $Q^2/2C$, is never negative.

5. **(C)** At $t = 0$, the capacitor acts like a short circuit and the inductor acts like an open circuit (opposing any sudden jump in the current). Therefore, zero current flows through the inductor, and a current equal to V/R_2 flows through the capacitor and around the outer loop.

6. **(B)** At equilibrium, the capacitor acts like an open circuit and the inductor acts like a closed circuit, so that zero current flows through the capacitor and a current of $I = V/R_1$ flows around the inner loop. One caveat is in order here, however: The rule that a capacitor acts like an open circuit at equilibrium was based on *RC* circuits, so we must consider whether it still applies here. Because the potential across the capacitor and the resistor R_2 is always V, the inductor branch of the circuit actually has absolutely no impact on the current flowing through the capacitor, so any generalizations we made for *RC* circuits apply here. Similarly, the voltage across the inductor and the resistor R_1 is always V, so the inductor acts independently of the capacitor. The only place where we actually see the loops in the circuit interacting is where the current is flowing through the battery, which is a sum of the currents flowing through each loop.

7. **(A)** By applying the formulas for parallel and series capacitance, it is not hard to determine that networks A and B have equivalent capacitance of $2C$ and $0.5C$, respectively. Choice (A): $U = \frac{1}{2}CV^2$, so network A, which has a greater equivalent capacitance, indeed stores more energy. Choice (B): The amount of charge required to charge a network to a given voltage is $Q = CV$. Because the two networks have different equivalent capacitances, they require different amounts of charge, making this choice false. Choice (C): The formula for the time constant is $\tau = RC$. Therefore, network A, which has a larger equivalent capacitance, produces a circuit with a greater time constant. Choice (D): The angular frequency of an *LC* circuit is given by the formula $\omega = \sqrt{1/LC}$. Because network A has a greater equivalent capacitance, it has a smaller angular frequency. Choice (E): The voltage across the central capacitor in network A is twice the voltage across the other capacitors, so the capacitors do not store the same amount of charge.

Free-Response

1. This problem drills you on determining initial and final conditions of circuits containing inductors and capacitors. The challenge here is figuring out which rule or equation to apply; once you have figured out how to correctly approach each section, the math is very simple. Some of the most useful equations and generalizations are listed here.

Capacitors:

1. Initially, capacitors act as short circuits.
2. At equilibrium, capacitors act as open circuits.

Inductors:

1. The initial condition of an inductor can generally be figured out using the rule that the current through an inductor is a continuous function of time.
2. At equilibrium, inductors act like short circuits.

Kirchhoff's loop rule: Once you have the voltage across one component of the circuit, you can use a loop rule to figure out the voltage across other parts of the circuit.

(a) Initially, the capacitor acts like a short circuit so that the current is $I = V/R$. At equilibrium when the capacitor is fully charged, it acts like an open circuit so that the current is zero.

(b) Initially, the capacitor is uncharged, so that its voltage is $V = Q/C = 0$. At equilibrium, the voltage across the resistor is zero (the current is zero, so $V = IR = 0$), so according to Kirchhoff's loop rule the voltage across the capacitor must be V (in the opposite polarity as the battery).

(c) Initially, the inductor acts like an open circuit so that the current is zero (as required by the fact that the current through an inductor must be continuous). At equilibrium, the inductor acts like a closed circuit so that the current is $I = V/R$.

(d) Initially, the voltage across the resistor is zero because the current is zero (according to Ohm's law, $V = IR = 0$). Therefore, Kirchhoff's loop rule requires the voltage across the inductor to equal V (in the polarity opposite that of the battery). At equilibrium, the change in current, dI/dt, is zero, so the voltage across the inductor is zero.

(e) The presence of an inductor requires the current to be continuous in time. (If the current were discontinuous, dI/dt would be infinite, generating an infinite voltage across the inductor and an infinite current and violating conservation of energy.) Therefore, the initial current through the inductor is the same as the current through the inductor before the switch is flipped, $I = V/R$. Because this current is to the right, it generates a counterclockwise current in the circuit containing the inductor and capacitor. This initial state of the circuit is shown in Figure 19.17.

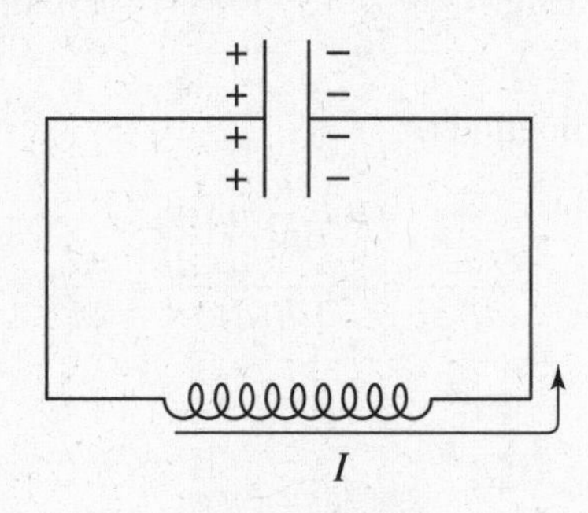

Figure 19.17

(f) The initial voltage across the capacitor is V. [The capacitor was left fully charged from its previous state, as calculated in part (b).] Kirchhoff's loop rule then requires the voltage across the inductor to have magnitude V with opposite polarity.

2. (a) This is a classic Ampere's law problem with cylindrical symmetry. The enclosed current is simply I, so the magnetic field is equal to the field due to a wire (recall that the field doesn't depend on the distribution of the current as long as it is cylindrically symmetric):

$$B = \frac{\mu_0 I}{2\pi r}$$

(b) We can use a standard Faraday's law application. Because the magnetic field varies, we must integrate to calculate the magnetic flux. To mirror the cylindrical symmetry of the magnetic field, we divide the area into differential areas that are thin rectangular strips running parallel to the axis of the cable with differential area:

$$dA = l dr$$

The differential flux through each differential area is

$$d\Phi_B = \mathbf{B} \cdot d\mathbf{A} = B dA = \left(\frac{\mu_0 I}{2\pi r}\right)(l dr)$$

Integrating,

$$\Phi_B = \int_a^b \left(\frac{\mu_0 I}{2\pi r}\right)(l dr) = \frac{\mu_0 I l}{2\pi} \int_a^b \frac{dr}{r} = \frac{\mu_0 I l}{2\pi} \ln\left(\frac{b}{a}\right)$$

Applying Faraday's law,

$$|\mathcal{E}| = \left|\frac{d\Phi_B}{dt}\right| = \left|\frac{d}{dt}\left[\frac{\mu_0 I l}{2\pi} \ln\left(\frac{b}{a}\right)\right]\right| = \frac{\mu_0 l}{2\pi} \ln\left(\frac{b}{a}\right)\frac{dI}{dt}$$

(c) We can apply Lenz's law directly: Increasing current causes an increase in inward flux through the loop, which induces a counterclockwise EMF, which decreases the current.

Or we can apply Lenz's law indirectly: Because the change in current causes the change in flux, in order to oppose the change in flux, the induced EMF must oppose the change in current.

(d) The inductance is defined as

$$L = \frac{\mathcal{E}}{dI/dt} = \frac{\frac{\mu_0 l}{2\pi} \ln\left(\frac{b}{a}\right)\frac{dI}{dt}}{dI/dt} = \frac{\mu_0 l}{2\pi} \ln\left(\frac{b}{a}\right)$$

3. Our general approach is the same as when we calculated equivalent resistance or equivalent capacitance. The goal is to find values of voltage and dI/dt to plug into the definition of inductance, $L_{eq} = V/(dI/dt)$. We do this by assigning either a voltage (V) or a change in current (dI/dt), then calculating the other value, and finally finding the ratio. Whether we begin by assigning V or dI/dt is simply a matter of convenience (parallel elements share a common voltage, so it's easier to start with the voltage V and calculate dI/dt; series elements share a common current, so it's easier to start with dI/dt and calculate the voltage).

(a) Elements in series share a common current, so let's begin by considering what will happen if the current through both inductors is changing at a rate dI/dt. Each inductor produces a voltage of $L(dI/dt)$ such that the net voltage is

$$V_{eq} = L_1 \frac{dI}{dt} + L_2 \frac{dI}{dt}$$

Then, according to the definition of inductance, the net inductance is

$$L_{eq} = \frac{V_{eq}}{dI/dt} = \frac{L_1(dI/dt) + L_2(dI/dt)}{dI/dt} = L_1 + L_2$$

Therefore, the equivalent inductance of inductors in series is the sum of their individual inductances. This makes sense: Adding two coils in series is quite similar to increasing the length of a single coil, and both cause increases in the net inductance.

(b) Elements in parallel share a common voltage, so let's begin by considering the situation where two inductors in parallel have a voltage V across them. The change in current through each inductor is then

$$\frac{dI_1}{dt} = \frac{V}{L_1} \quad \text{and} \quad \frac{dI_2}{dt} = \frac{V}{L_2}$$

The change in the total current throughout the inductor array, $I_{net} = I_1 + I_2$, is then

$$\frac{dI_{net}}{dt} = \frac{dI_1}{dt} + \frac{dI_2}{dt} = \frac{V}{L_1} + \frac{V}{L_2}$$

Therefore, applying the definition of inductance,

$$L_{eq} = \frac{V}{dI_{net}/dt} = \frac{V}{(V/L_1) + (V/L_2)} = \frac{1}{(1/L_1) + (1/L_2)}$$

Therefore, the net inductance of inductors in parallel is smaller than the inductance of either of the inductors alone. This makes sense if we consider an inductor as an element that resists changes in current. When more inductors are added in parallel, each allows a certain change in current for a given voltage; thus, the net effect of adding more inductors is to produce a greater change in current for the same voltage. (This is similar to the way adding more resistors in parallel produces a greater current for the same voltage, such that the equivalent resistance is smaller than the resistance of any of the individual resistors.)

4. (a) First apply Biot-Savart's law (note that the magnitude of **r** is constant and that it is always perpendicular to $d\mathbf{l}$).

$$d\mathbf{B}=\frac{\mu_0 I d\mathbf{l}\times\hat{r}}{4\pi R^2}=\frac{-\mu_0 I(t)\,dl}{4\pi R^2}\hat{k}$$

$$\mathbf{B}=\int\frac{-\mu_0 I(t)\,dl}{4\pi R^2}\hat{k}=\frac{-\mu_0 I(t)}{4\pi R^2}\hat{k}\int dl=\frac{-\mu_0 I(t)}{4\pi R^2}\hat{k}(2\pi R)=\frac{-\mu_0 I(t)}{2R}\hat{k}$$

(b) Because $r << R$, we can assume that the field is approximately uniform within the smaller loop and equal to the value at the center, calculated above in part (a). Therefore, the flux (choosing **A** to point into the page) is

$$\Phi_B=\mathbf{B}\cdot\mathbf{A}=\left[\frac{\mu_0 I(t)}{2R}\right](\pi r^2)=\frac{\pi\mu_0 r^2 I(t)}{2R}$$

(c) Using Faraday's law,

$$\mathcal{E}=\left|\frac{d\Phi_B}{dt}\right|=\left|\frac{d}{dt}\left[\frac{\pi\mu_0 r^2 I(t)}{2R}\right]\right|=\frac{\pi\mu_0 r^2}{2R}\left|\frac{dI(t)}{dt}\right|=\frac{\pi\mu_0 r^2 I_0}{2\tau R}e^{-t/\tau}$$

5. (a) Because the current through an inductor is a continuous function of time, the current through the inductor the instant after switch 2 is opened is the same as the current the instant before the switch is opened. This current, with switches 1 and 2 closed, is easily calculated as follows.

After switch 1 has been closed for a very long time, the inductor acts like a piece of wire (a short circuit). Because switch 2 is closed, the equivalent resistance of the circuit is

$$R_{eq}=\frac{1}{(1/R_1)+(1/R_2)}=\frac{R_1R_2}{R_1+R_2}$$

Based on Ohm's law, the current through the inductor before and right after switch 2 is opened is

$$I=\frac{V}{R}=\frac{V(R_1+R_2)}{R_1R_2}$$

(b) After reaching equilibrium, the inductor again acts like a short circuit. The resistance of the circuit is simply R_1, so the current is

$$I=\frac{V}{R_1}$$

(c) See Figure 19.18. When switch 1 is first closed, the current exponentially approaches the value calculated in part (a). Opening switch 2 increases the resistance, so it decreases the final current. Again, the difference between the current and its final value decreases exponentially.

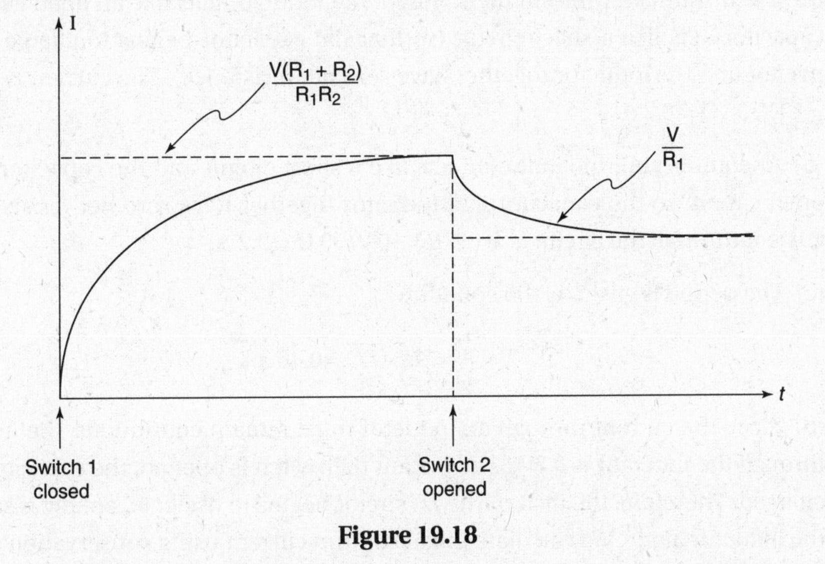

Figure 19.18

(d) What happens to the potential when switch 1 is first closed and later on as the circuit reaches equilibrium? At $t = 0$, the current is zero (since the current must be continuous). Summing potential drops from the positive terminal of the battery to point A (noting that the voltage across the resistors is zero because the current is zero) indicates that point A must have the same potential as the positive terminal of the battery, $+V$.

Since point A is separated from the negative terminal of the battery by the inductor, the magnitude of the potential at point A is equal to the magnitude of the voltage across the inductor. As time increases, the voltage across the inductor exponentially approaches zero (as is always the case with voltages across inductors in RL circuits); thus, the potential at point A also exponentially approaches zero.

(e) What happens to the potential when switch 2 is first opened and later on as the circuit reaches equilibrium? At the instant the switch is opened, the current through the circuit is $I = V(R_1 + R_2)/R_1R_2$ (because of the requirement that the current be continuous). As in part (d), we can calculate the potential at point A by starting at the positive terminal of the battery (where the potential is $+V$) and sum potential drops as we work our way to point A. Moving from the positive terminal of the battery to point A requires going across the resistor R_1, moving in the direction of current (hence a negative potential drop of magnitude IR_1). Plugging in our result for I we find

$$V_A = V - \left[\frac{V(R_1 + R_2)}{R_1R_2}\right]R_1 = -\frac{VR_1}{R_2}$$

As explained in part (e), as time increases, the potential at A exponentially decreases to zero as the voltage across the inductor exponentially approaches zero, causing point A to be at the same potential as the negative terminal of the battery.

Answer check: If R_1 is very small, all the current flows through it even before switch 2 is opened, opening switch 2 makes no difference, and the voltage across the inductor when switch 2 is opened is virtually zero.

6. (a) The instant after the circuit is closed, the inductor acts like an open circuit and the capacitor acts like a short circuit (so that the capacitor begins to charge and the capacitor and the inductor together have zero net resistance). The current is then $I = V/R = 10\text{ V}/50\ \Omega = 0.2\text{ A}$.

   (b) At equilibrium, the inductor acts like a short circuit and the capacitor acts like an open circuit, so the capacitor and inductor together have zero net resistance and the current through the circuit is $I = V/R = 10\text{ V}/50\ \Omega = 0.2\text{ A}$.

   (c) The period is given by the equation

$$T = \frac{2\pi}{\omega} = 2\pi\sqrt{LC} = 0.49\text{ ms}$$

   (d) Since the current through an inductor must remain continuous, the initial current through the inductor is 0.2 A. The instant the switch is opened, the capacitor is still uncharged. Therefore, the instant the *LC* circuit begins to oscillate, energy is stored within the inductor alone. We calculate the maximum current using conservation of energy:

$$\frac{LI_{max}^2}{2} = \frac{LI_0^2}{2}$$

   Thus the maximum current equals the initial current, 0.2 A.

20 Maxwell's Equations

→ GAUSS'S LAW FOR MAGNETISM
→ MAXWELL'S LAW OF INDUCTION
→ DISPLACEMENT CURRENT
→ AMPERE-MAXWELL LAW

Maxwell's equations are the four fundamental equations of electricity and magnetism:

$$\oint E \cdot d\mathbf{A} = \frac{Q_{\text{enclosed}}}{\varepsilon_0}$$

Gauss's law for electricity

$$\oint B \cdot d\mathbf{A} = 0$$

Gauss's law for magnetism

$$\oint E \cdot d\mathbf{s} = -\frac{d\Phi_B}{dt}$$

Faraday's law

$$\oint B \cdot d\mathbf{s} = \mu_0 I_{\text{enclosed}} + \mu_0 \varepsilon_0 \frac{d\Phi_E}{dt}$$

Ampere-Maxwell law

TIP

Maxwell's equations: the Holy Grail of electricity and magnetism

You are already familiar with the first and third equations. The other equations are explained below.

GAUSS'S LAW FOR MAGNETISM

It might help to first introduce some vocabulary. An *electric monopole* is an elementary particle having a net electric charge (e.g., an electron). An *electric dipole* is a pair of charges with equal magnitudes and opposite signs. Although together the two charges have zero net charge, they produce an electric field, as shown in Figure 20.1. Note how similar this field is to the magnetic field lines due to a bar magnet (a magnetic dipole).

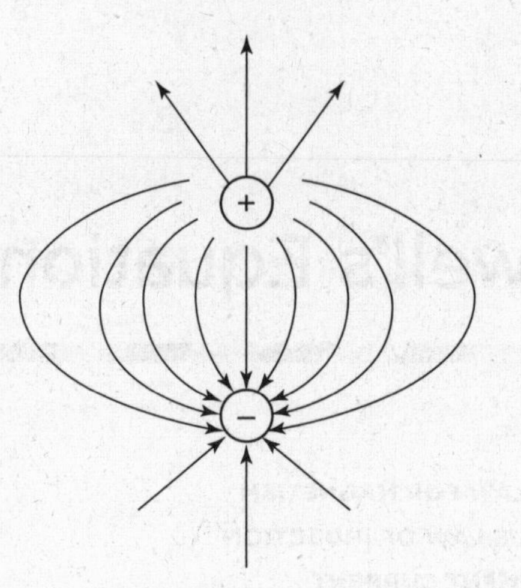

Figure 20.1

Recall Gauss's law:

$$\underset{\text{surface integral}}{\oint} \mathbf{E}\cdot d\mathbf{A} = \frac{Q_{\text{enclosed}}}{\varepsilon_0}$$

An electric monopole has net charge, so a gaussian surface enclosing an electric monopole has a nonzero net electric flux. Alternatively, an electric dipole has zero net charge, so a gaussian surface enclosing an electric dipole also has a zero net electric flux.

Now consider a bar magnet with north and south poles. If you cut it in half, could you isolate a piece of north pole and a piece of south pole (magnetic monopoles)? No. Remember what we said in Chapter 17 about how bar magnets are formed by summing the magnetic fields due to aligning the spin of electrons in iron atoms? What this means is that when we cut the bar magnet in half, we do not isolate magnetic monopoles. Instead, we are left with two chunks of metal containing polarized atoms, each piece with its own north and south poles, as shown in Figure 20.2. This reveals that there is nothing intrinsically different about the atoms in the north and south poles other than their relative positions in the magnet. Because the magnetic field is caused by the circulation of electrons, it is impossible to isolate north and south poles. This is a general result: Experimentally, no pieces of matter have been found with net magnetic "charge."

By analogy with the electric field case, Gauss's equation for magnetic fields would read

$$\oint \mathbf{B}\cdot d\mathbf{A} = (\text{enclosed magnetic charge})(\text{physical constant})$$

Given that there are no magnetic monopoles, the net magnetic charge enclosed by any surface must be zero, so the net magnetic flux through any surface is zero. Thus, the equation above becomes

$$\underset{\text{surface integral}}{\oint} \mathbf{B}\cdot d\mathbf{A} = 0$$

Absence of magnetic monopoles

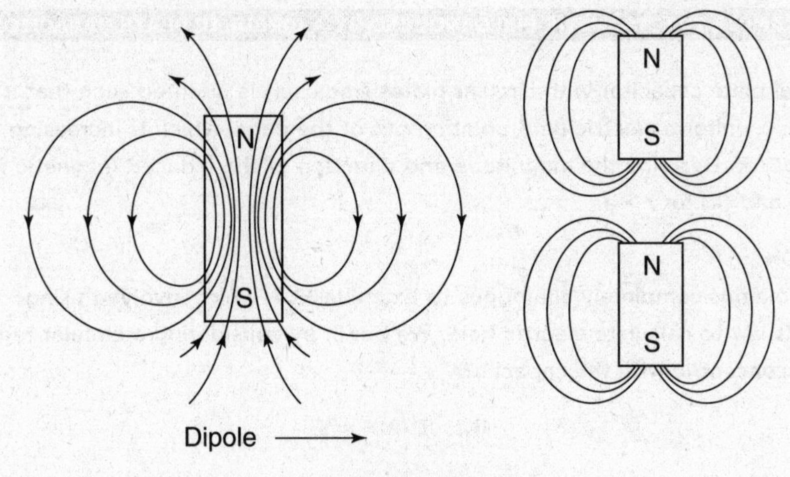

Figure 20.2

MAXWELL'S LAW OF INDUCTION

According to Faraday's law,

$$\oint E \cdot d\mathbf{s} = -\frac{d\Phi_B}{dt}$$

Faraday's law

Thus, a changing magnetic field produces an electric field. Given the considerable amount of symmetry between electric and magnetic fields, it shouldn't be too surprising that the reverse is true as well: A changing electric field produces a magnetic field. The equation governing this fact is very similar to Faraday's law.

$$\oint B \cdot d\mathbf{s} = \mu_0 \varepsilon_0 \frac{d\Phi_E}{dt}$$

Maxwell's law of induction, in the absence of a current

Applying this equation is essentially the same as applying Faraday's law; the only difference in form is that there's no negative sign.

EXAMPLE 20.1 MAGNETIC FIELD INDUCED BY A CIRCULAR PLATE CAPACITOR

A parallel plate capacitor with circular plates (radius a) is oriented such that it produces a uniform electric field pointing out of the page, which is increasing at a rate of dE/dt. Describe the magnitude and direction of the induced magnetic field (i) for $r < a$ and (ii) for $r > a$.

SOLUTION

This problem is completely analogous to Example 18.4, which involved using Faraday's law to calculate electric field. We begin by considering a circular region of radius r concentric with the capacitor.

$$|\Phi_E| = |\mathbf{E} \cdot \mathbf{A}| = \pi r^2 E$$

The direction of the magnetic field is tangent to the circular path (for the same symmetry reasons discussed in Example 18.4) and constant in magnitude at a particular r, so that

$$\left|\oint \mathbf{B} \cdot d\mathbf{s}\right| = 2\pi r B$$

Applying Maxwell's law of induction,

$$\left[\left|\oint \mathbf{B} \cdot d\mathbf{s}\right| = 2\pi r B\right] = \left[\mu_0 \varepsilon_0 \left|\frac{d\Phi_E}{dt}\right| = \mu_0 \varepsilon_0 \left|\frac{d}{dt}(\pi r^2 E)\right| = \mu_0 \varepsilon_0 \pi r^2 \frac{dE}{dt}\right]$$

$$B = \frac{\mu_0 \varepsilon_0 r}{2} \frac{dE}{dt}$$

What about the direction? If we were applying Faraday's law, then an increasing flux coming out of the page would produce clockwise current (and thus, clockwise field), according to Lenz's law. However, Maxwell's law of induction does not have a negative sign, and so the direction of the field is reversed, so that the induced magnetic field points counterclockwise tangent to the circular path.

Outside the capacitor, the direction is the same. Calculating the magnitude is very similar:

$$|\Phi_E| = |\mathbf{E} \cdot \mathbf{A}| = \pi a^2 E$$

$$\left[\left|\oint \mathbf{B} \cdot d\mathbf{s}\right| = 2\pi r B\right] = \left[\mu_0 \varepsilon_0 \left|\frac{d\Phi_E}{dt}\right| = \mu_0 \varepsilon_0 \left|\frac{d}{dt}(\pi a^2 E)\right| = \mu_0 \varepsilon_0 \pi a^2 \frac{dE}{dt}\right]$$

$$B = \frac{\mu_0 \varepsilon_0 a^2}{2r} \frac{dE}{dt}$$

EXAMPLE 20.2 AMPERE'S LAW AND CAPACITORS

Consider the form of Ampere's law that we are familiar with, $\oint \mathbf{B} \cdot d\mathbf{s} = \mu_0 I_{\text{enclosed}}$. In words, the line integral of the magnetic field around any path is equal to μ_0 multiplied by the flow of current through any surface bounded by the path. Now, examine Ampere's law in terms of charging a capacitor, as in Figure 20.3.

What is the current enclosed in the given path (equal to the flow of current through any surface bounded by the path)? If we consider surface A, the enclosed current is *I*. However, if we consider surface B, the enclosed current is zero. What is the solution to this contradiction?

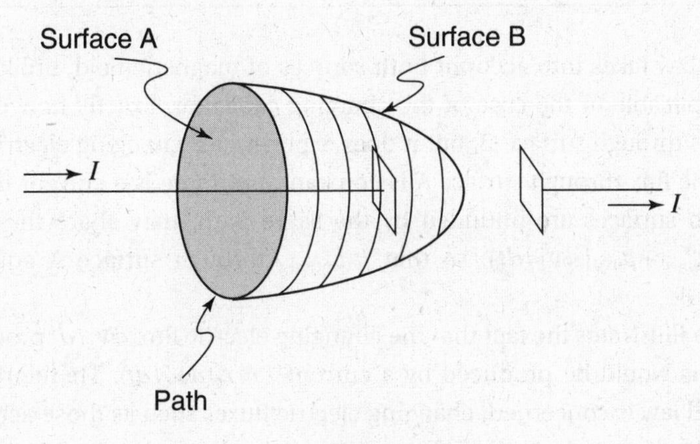

Figure 20.3

SOLUTION

Ampere's law, in the above form, is not applicable here because it does not take into account the magnetic fields induced by changing electric fields. As we have seen, magnetic fields are produced by two sources: moving charges and changing electric fields. Ampere's law as written above does not take into account the magnetic fields produced by changing electric fields, so it is valid only in the case of constant electric fields. We will soon amend Ampere's law so that it can handle this situation.

TIP

Ampere's law is not complete.

DISPLACEMENT CURRENT AND THE AMPERE-MAXWELL LAW

At this point, we can calculate magnetic fields due to moving charges using Ampere's law and magnetic fields due to changing electric fields using Maxwell's law of induction:

$$\oint \mathbf{B} \cdot d\mathbf{s} = \mu_0 \varepsilon_0 \frac{d\Phi_E}{dt}$$

Maxwell's law of induction

$$\oint \mathbf{B} \cdot d\mathbf{s} = \mu_0 I_{\text{enclosed}}$$

Ampere's law

How can we calculate the magnetic field due to both currents and changing electric fields? Because magnetic fields obey superposition, the fields due to current and due to changing electric fields simply add together. Therefore,

$$\oint \mathbf{B}_{\text{net}} \cdot d\mathbf{s} = \oint (\mathbf{B}_{\text{moving charge}} + \mathbf{B}_{\text{induced}}) \cdot d\mathbf{s} = \left[\left(\oint \mathbf{B}_{\text{moving charge}} \cdot d\mathbf{s}\right) + \left(\oint \mathbf{B}_{\text{induced}} \cdot d\mathbf{s}\right)\right]$$

TIP

Maxwell's amendment of Ampere's law

$$\oint \mathbf{B}_{\text{net}} \cdot d\mathbf{s} = \mu_0 I_{\text{enclosed}} + \mu_0 \varepsilon_0 \frac{d\Phi_E}{dt}$$

Ampere-Maxwell law

Because this law takes into account both sources of magnetic field, unlike Ampere's law it should be applicable to the case of the charging capacitor. Exactly how does this work? No current flows through surface B, but it does experience a changing electric flux. Alternatively, the electric flux through surface A is constant, but there is a current flowing through it. Because both surfaces are bounded by the same path, they share the same value of $\oint \mathbf{B}_{\text{net}} \cdot d\mathbf{s} = \mu_0 I_{\text{enclosed}} + \mu_0 \varepsilon_0 (d\Phi_E/dt)$, so that $\mu_0 I_{\text{enclosed}}$ through surface A equals $\mu_0 \varepsilon_0 d\Phi_E/dt$ through surface B.

This situation illustrates the fact that the changing electric flux $d\Phi_E/dt$ produces the same magnetic field as would be produced by a current $I = \varepsilon_0(d\Phi_E/dt)$. Therefore, as far as the Ampere-Maxwell law is concerned, changing electric fluxes such as those across a capacitor can be imagined to be carrying a *displacement current* (as opposed to a real current composed of moving charges) given by the following equation.

$$I_{\text{displacement}} = \varepsilon_0 \frac{d\Phi_E}{dt}$$

Displacement current

The displacement current points parallel to the increase in electric field (which is in the same direction as the real current through the wire in the capacitor example). Additionally, the displacement current across the entire area of the capacitor has the same magnitude as the real current through the wire. Therefore, this displacement current nicely resolves the seeming contradiction with Ampere's law encountered in Example 20.2. The Ampere-Maxwell law can be compactly written in terms of displacement current as follows.

$$\oint \mathbf{B} \cdot d\mathbf{s} = \mu_0 I_{\text{enclosed}} + \mu_0 I_{\text{displacement}}$$

Ampere-Maxwell law, compact form

This equation reveals that the displacement current gives rise to the same magnetic field that would be produced by a real current with the same direction, magnitude, and current density.

CHAPTER SUMMARY

Maxwell's equations comprise the four fundamental statements describing all electric and magnetic phenomena. Two of the equations, Gauss's law for electricity and Faraday's law of induction, were encountered in earlier chapters. Gauss's law for magnetism states that no magnetic charges have been found in the universe. Magnetic fields are created by electric charges in motion. The Ampere-Maxwell law includes the original Ampere law, which describes how currents create magnetic fields, and Maxwell's displacement current, which describes how a changing electric flux produces a magnetic field.

PRACTICE EXERCISES

Multiple-Choice Questions

1. Which of the following statements is true in the absence of moving charges?

   (A) The potential within a metal is zero.
   (B) The potential within a metal is constant.
   (C) The electric field as a function of position is continuous.
   (D) The potential is always well-defined.
   (E) The potential is nonzero only in the presence of an electric field.

2. A circular parallel plate capacitor of radius r is being charged. At the instant that a current I is flowing into the capacitor, what is the magnitude of the integral $\oint \mathbf{B} \cdot d\mathbf{s}$ for a circular path of radius $r/2$ centered within the capacitor?

   (A) $\mu_0 I$
   (B) $\frac{1}{2}\mu_0 I$
   (C) $2\mu_0 I$
   (D) $\frac{1}{4}\mu_0 I$
   (E) $4\mu_0 I$

ANSWER KEY

1. **B**
2. **D**

ANSWERS EXPLAINED

Multiple-Choice

1. **(B)** Choice (A): the potential within a metal can be nonzero, and the electric field within a metal must be zero. Choice (B): because the electric field within a metal must be zero and because the electric field is equal to the negative slope of the potential, the potential within a metal must indeed be constant. Choice (C): the electric field is continuous only in the absence of charge distributions with infinite volume density. (For example, the electric field is discontinuous at point charges, lines of charge, and sheets of charge.) Choice (D): Similarly, the potential at the location of a point charge is undefined. Choice (E): The slope of the potential graph is related to electric field; there is no necessary relationship between the magnitude of potential and the electric field. For example, it is possible for the potential to be constant and nonzero within a metal, where the electric field is zero.

2. **(D)** The total displacement current across the capacitor is equal to the real current flowing into the capacitor, *I*. Because the electric field (and thus the change in the electric field) is uniform between the plates, the displacement current density also is uniform. Therefore, the enclosed displacement current is proportional to the area enclosed by the path:

$$\text{Enclosed displacement current} = I \cdot \frac{\text{enclosed area}}{\text{total area}} = I \cdot \frac{\pi (r/2)^2}{\pi r^2} = \frac{I}{4}$$

Applying the compact form of the Ampere-Maxwell law then reveals that

$$\oint \mathbf{B} \cdot d\mathbf{s} = \mu_0 \left(I_{\text{enclosed}} + I_{\text{displacement}} \right) = \mu_0 \frac{I}{4}$$

ANSWER SHEET
Practice Test 1

Mechanics

1.	Ⓐ Ⓑ Ⓒ Ⓓ Ⓔ	10.	Ⓐ Ⓑ Ⓒ Ⓓ Ⓔ	19.	Ⓐ Ⓑ Ⓒ Ⓓ Ⓔ	28.	Ⓐ Ⓑ Ⓒ Ⓓ Ⓔ
2.	Ⓐ Ⓑ Ⓒ Ⓓ Ⓔ	11.	Ⓐ Ⓑ Ⓒ Ⓓ Ⓔ	20.	Ⓐ Ⓑ Ⓒ Ⓓ Ⓔ	29.	Ⓐ Ⓑ Ⓒ Ⓓ Ⓔ
3.	Ⓐ Ⓑ Ⓒ Ⓓ Ⓔ	12.	Ⓐ Ⓑ Ⓒ Ⓓ Ⓔ	21.	Ⓐ Ⓑ Ⓒ Ⓓ Ⓔ	30.	Ⓐ Ⓑ Ⓒ Ⓓ Ⓔ
4.	Ⓐ Ⓑ Ⓒ Ⓓ Ⓔ	13.	Ⓐ Ⓑ Ⓒ Ⓓ Ⓔ	22.	Ⓐ Ⓑ Ⓒ Ⓓ Ⓔ	31.	Ⓐ Ⓑ Ⓒ Ⓓ Ⓔ
5.	Ⓐ Ⓑ Ⓒ Ⓓ Ⓔ	14.	Ⓐ Ⓑ Ⓒ Ⓓ Ⓔ	23.	Ⓐ Ⓑ Ⓒ Ⓓ Ⓔ	32.	Ⓐ Ⓑ Ⓒ Ⓓ Ⓔ
6.	Ⓐ Ⓑ Ⓒ Ⓓ Ⓔ	15.	Ⓐ Ⓑ Ⓒ Ⓓ Ⓔ	24.	Ⓐ Ⓑ Ⓒ Ⓓ Ⓔ	33.	Ⓐ Ⓑ Ⓒ Ⓓ Ⓔ
7.	Ⓐ Ⓑ Ⓒ Ⓓ Ⓔ	16.	Ⓐ Ⓑ Ⓒ Ⓓ Ⓔ	25.	Ⓐ Ⓑ Ⓒ Ⓓ Ⓔ	34.	Ⓐ Ⓑ Ⓒ Ⓓ Ⓔ
8.	Ⓐ Ⓑ Ⓒ Ⓓ Ⓔ	17.	Ⓐ Ⓑ Ⓒ Ⓓ Ⓔ	26.	Ⓐ Ⓑ Ⓒ Ⓓ Ⓔ	35.	Ⓐ Ⓑ Ⓒ Ⓓ Ⓔ
9.	Ⓐ Ⓑ Ⓒ Ⓓ Ⓔ	18.	Ⓐ Ⓑ Ⓒ Ⓓ Ⓔ	27.	Ⓐ Ⓑ Ⓒ Ⓓ Ⓔ		

Electricity and Magnetism

1.	Ⓐ Ⓑ Ⓒ Ⓓ Ⓔ	10.	Ⓐ Ⓑ Ⓒ Ⓓ Ⓔ	19.	Ⓐ Ⓑ Ⓒ Ⓓ Ⓔ	28.	Ⓐ Ⓑ Ⓒ Ⓓ Ⓔ
2.	Ⓐ Ⓑ Ⓒ Ⓓ Ⓔ	11.	Ⓐ Ⓑ Ⓒ Ⓓ Ⓔ	20.	Ⓐ Ⓑ Ⓒ Ⓓ Ⓔ	29.	Ⓐ Ⓑ Ⓒ Ⓓ Ⓔ
3.	Ⓐ Ⓑ Ⓒ Ⓓ Ⓔ	12.	Ⓐ Ⓑ Ⓒ Ⓓ Ⓔ	21.	Ⓐ Ⓑ Ⓒ Ⓓ Ⓔ	30.	Ⓐ Ⓑ Ⓒ Ⓓ Ⓔ
4.	Ⓐ Ⓑ Ⓒ Ⓓ Ⓔ	13.	Ⓐ Ⓑ Ⓒ Ⓓ Ⓔ	22.	Ⓐ Ⓑ Ⓒ Ⓓ Ⓔ	31.	Ⓐ Ⓑ Ⓒ Ⓓ Ⓔ
5.	Ⓐ Ⓑ Ⓒ Ⓓ Ⓔ	14.	Ⓐ Ⓑ Ⓒ Ⓓ Ⓔ	23.	Ⓐ Ⓑ Ⓒ Ⓓ Ⓔ	32.	Ⓐ Ⓑ Ⓒ Ⓓ Ⓔ
6.	Ⓐ Ⓑ Ⓒ Ⓓ Ⓔ	15.	Ⓐ Ⓑ Ⓒ Ⓓ Ⓔ	24.	Ⓐ Ⓑ Ⓒ Ⓓ Ⓔ	33.	Ⓐ Ⓑ Ⓒ Ⓓ Ⓔ
7.	Ⓐ Ⓑ Ⓒ Ⓓ Ⓔ	16.	Ⓐ Ⓑ Ⓒ Ⓓ Ⓔ	25.	Ⓐ Ⓑ Ⓒ Ⓓ Ⓔ	34.	Ⓐ Ⓑ Ⓒ Ⓓ Ⓔ
8.	Ⓐ Ⓑ Ⓒ Ⓓ Ⓔ	17.	Ⓐ Ⓑ Ⓒ Ⓓ Ⓔ	26.	Ⓐ Ⓑ Ⓒ Ⓓ Ⓔ	35.	Ⓐ Ⓑ Ⓒ Ⓓ Ⓔ
9.	Ⓐ Ⓑ Ⓒ Ⓓ Ⓔ	18.	Ⓐ Ⓑ Ⓒ Ⓓ Ⓔ	27.	Ⓐ Ⓑ Ⓒ Ⓓ Ⓔ		

Practice Test 1

MULTIPLE-CHOICE QUESTIONS

Directions: Each multiple-choice question is followed by five answer choices. For each question, choose the best answer and fill in the corresponding circle on the answer sheet. You may refer to the formula sheet in the Appendix (pages 641–644).

Mechanics

1. A constant horizontal force F applied to a block on an incline causes the block to move a distance d along the incline, as shown. If the coefficient of friction between the block and the incline is μ_k, what is the work done by the applied force? (Do not assume that the block moves at constant velocity.)

   (A) $W = Fd\sin\theta$
   (B) $W = Fd\cos\theta$
   (C) $W = mgd\sin\theta$
   (D) $W = mgd\sin\theta - \mu_k mgd\cos\theta$
   (E) $W = mgd\sin\theta + \mu_k d(mg\cos\theta + F\sin\theta)$

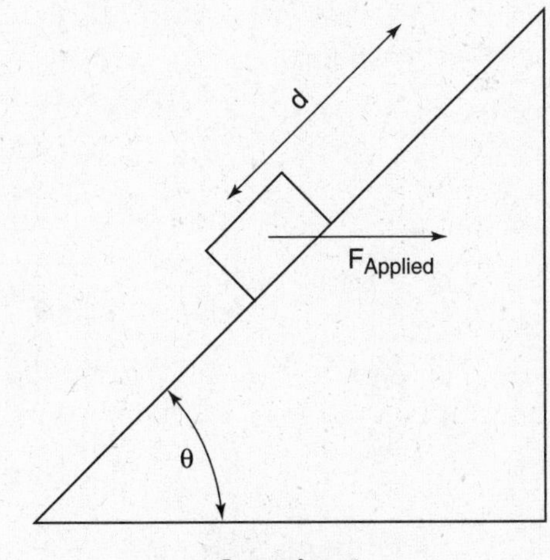

Question 1

2. The restoring force that acts on a mass is shown. For small amplitudes, the mass moves in simple harmonic motion with a period T. Which of the following is true of oscillations with larger amplitudes?

   (A) The motion is SHM with a period of T.
   (B) The motion is SHM with a period smaller than T.
   (C) The motion is SHM with a period larger than T.
   (D) The motion is no longer SHM, and its period is smaller than T.
   (E) The motion is no longer SHM, and its period is larger than T.

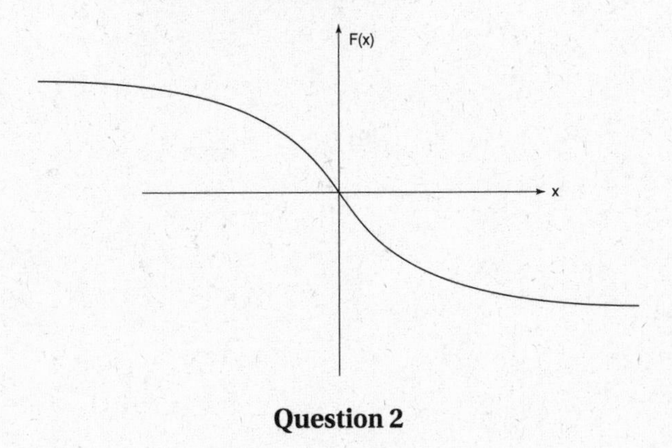

Question 2

GO ON TO THE NEXT PAGE

3. A simple pendulum of length ℓ is moved to Mars (where $g = 3.71\ \text{m/s}^2$). Its period on Earth is 2.0 s. Its period on Mars is

(A) 0.02 s
(B) 2.0 s
(C) 3.25 s
(D) 600 s
(E) 6,000 s

4. At the instant that a mass in simple harmonic motion passes through its equilibrium position its acceleration is

(A) maximum
(B) zero
(C) oppositely directed to its velocity
(D) parallel to its velocity
(E) cannot be determined without additional information

5. A solid sphere has a rotational inertia of $\frac{7}{5}Mr^2$ when rotated about an axis tangent to its surface as shown. What is the rotational inertia of the sphere when it is rotated about a line passing through the center of the sphere?

(A) $\frac{2}{5}Mr^2$
(B) $\frac{3}{5}Mr^2$
(C) $\frac{7}{5}Mr^2$
(D) $\frac{12}{5}Mr^2$
(E) $\frac{5}{14}Mr^2$

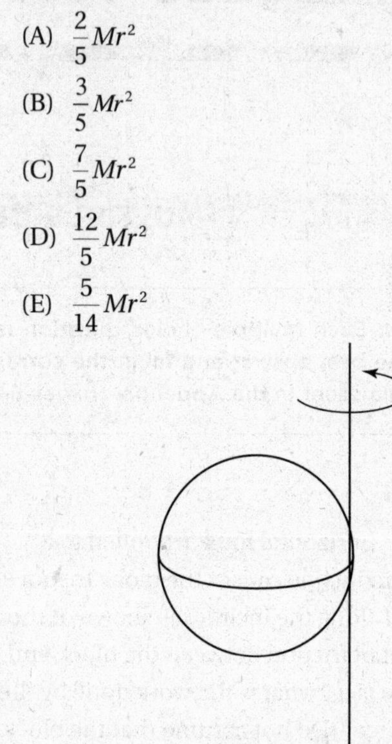

Question 5

GO ON TO THE NEXT PAGE

6. As shown in the figure, a pendulum is released with the string parallel to the ground at point I, resulting in its swinging through a 180° angle between points I and III. When the pendulum is at point II and moving downward, what is the direction of its acceleration?

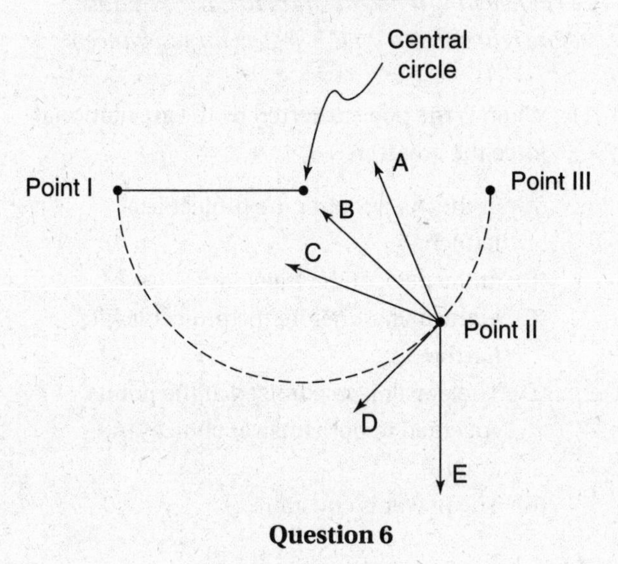

Question 6

7. A mass m moving to the right with speed v collides and sticks to a mass of $3m$ moving upward with a speed $2v$ as shown. What is the final velocity of the two masses?

(A) $\frac{3}{2}v\hat{i} + \frac{1}{4}v\hat{j}$

(B) $\frac{1}{4}v\hat{i} + \frac{3}{2}v\hat{j}$

(C) $\frac{1}{3}v\hat{i} + \frac{2}{3}v\hat{j}$

(D) $\frac{2}{3}v\hat{i} + \frac{1}{3}v\hat{j}$

(E) none of the above

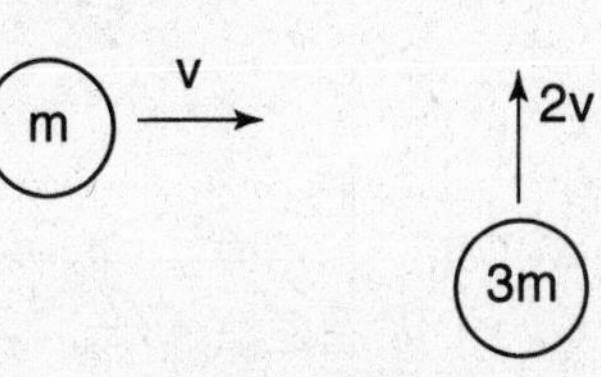

Question 7

8. If a force of 2 N does work at the rate of $-2\sqrt{2}$ W on an object moving with a speed of 2 m/s, the angle between the force and the velocity vector must be

(A) 45°
(B) 120°
(C) 135°
(D) 150°
(E) It is impossible to deliver a negative power.

PRACTICE TEST 1

GO ON TO THE NEXT PAGE

9. A mass m is released from rest at $t = 0$ in a parabolic depression as shown. If there is no friction between the mass and the surface, which of the graphs shows the mass's x-position as a function of time, assuming that the mass's displacement from the origin is small?

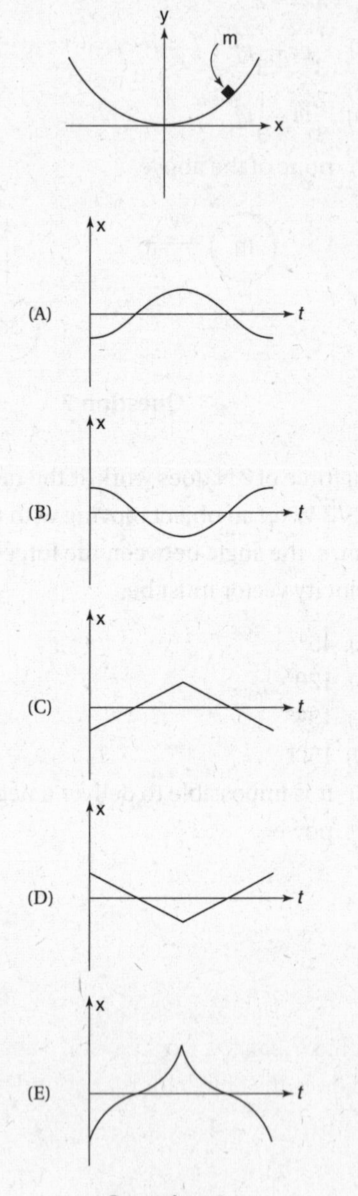

Question 9

10. Referring to the situation introduced in question 9, which of the graphs shows the x-component of the object's acceleration as a function of time?

Questions 11–13 refer to a projectile launched from Earth reaching a height equal to Earth's radius before returning to land. (Neglect air resistance.)

11. When is the power exerted by the gravitational force the greatest?

(A) at the instant after the projectile is launched
(B) at the peak of the trajectory
(C) at the instant before the projectile hits Earth
(D) The power is maximized at the points specified in both answer choices (A) and (B).
(E) The power is constant.

12. At which of the following points is the acceleration the greatest in magnitude?

(A) at the instant after the projectile is launched
(B) at the peak of the trajectory
(C) at the instant before the projectile hits Earth
(D) The acceleration is maximized at the points specified in both answer choices (A) and (C).
(E) The acceleration is constant.

13. Compare the gravitational potential energy when the particle is launched to the potential energy when the particle is at the peak of its trajectory.

(A) $U_{peak} = U_{launch}$ because of conservation of energy.
(B) $|U_{peak}| > |U_{launch}|$ and $U_{peak} > 0$
(C) $|U_{peak}| < |U_{launch}|$ and $U_{peak} < 0$
(D) $|U_{peak}| < |U_{launch}|$ and $U_{peak} > 0$
(E) $|U_{peak}| > |U_{launch}|$ and $U_{peak} < 0$

GO ON TO THE NEXT PAGE

14. A wheel accelerates at uniform angular acceleration α from rest. If its speed after one full rotation is ω, what was its speed after half a rotation?

(A) $\frac{1}{2}\omega$
(B) ω
(C) $\frac{3}{4}\omega$
(D) $(1/\sqrt{2})\omega$
(E) $\frac{1}{4}\omega$

15. A mass m is suspended in an elevator. In terms of the tension T of the string and the weight of the mass, what is the upward acceleration of the elevator?

(A) $T - mg$
(B) $mg - T$
(C) $(T/m) + g$
(D) $(T/m) - g$
(E) $g - T/m$

16. A mass initially at rest explodes into two fragments with masses m_1 and m_2. If m_1 subsequently moves with a speed v_1, how much kinetic energy was liberated in the explosion?

(A) $\frac{1}{2}m_1v_1^2$
(B) $\frac{1}{2}m_1v_1^2(1+m_1/m_2)$
(C) $\frac{1}{2}m_1v_1^2(1+m_2/m_1)$
(D) $\frac{1}{2}m_1v_1^2(1+m_2^2/m_1^2)$
(E) Not enough information is given.

17. A projectile is launched on level ground with an initial speed v_0 and angle of elevation θ. All of the following statements concerning the projectile right before it lands are correct *except*

(A) The acceleration is equal to g downward.
(B) The power exerted by the gravitational force is maximized.
(C) The time the projectile was in the air is equal to $2v_0\sin\theta/g$.
(D) The projectile has the same kinetic energy as when it was launched.
(E) The projectile has the same velocity as when it was launched.

18. For a system in simple harmonic motion, which of the following statements is true? (The equilibrium point refers to the point in the middle of the particle's oscillation, which could potentially be a static equilibrium point; the turning points are points where the particle reverses direction.)

(A) The acceleration is maximized at the equilibrium point.
(B) The velocity is maximized at the turning points.
(C) The restoring force is maximized at the equilibrium point.
(D) The displacement is maximized at the turning points.
(E) None of the above are true; it depends on the particular situation.

19. An object slides with horizontal velocity v_0 off a cliff of height h. How far from the base of the cliff does the object hit the ground?

(A) v_0g/h
(B) $v_0h/2g$
(C) $2gv_0/h$
(D) $v_0\sqrt{2h/g}$
(E) $v_0\sqrt{h/2g}$

GO ON TO THE NEXT PAGE

20. The angular position of a wheel is given by the equation $\theta(t) = 2t^3 - 6t^2$, where θ is measured in radians and t in seconds. When is the torque on the wheel equal to zero?

(A) $t = 0.33$ s
(B) $t = 0.5$ s
(C) $t = 1$ s
(D) $t = 2$ s
(E) $t = 3$ s

21. A small mass attached to a string rotates on a frictionless tabletop as shown. If the tension in the string is increased, causing the radius of the circular motion to decrease by a factor of 2, what effect will this have on the kinetic energy of the mass?

(A) It will decrease by a factor of 2.
(B) It will remain constant.
(C) It will increase by a factor of 2.
(D) It will increase by a factor of 4.
(E) It will increase by a factor of 8.

Question 21

22. Three point masses of equal mass are located as shown in the figure. What is the x-coordinate of the center of mass of this system?

(A) $\frac{a}{2}$

(B) $\frac{2a}{3}$

(C) $\frac{a}{3}$

(D) $\frac{b}{2}$

(E) $\frac{3a}{4}$

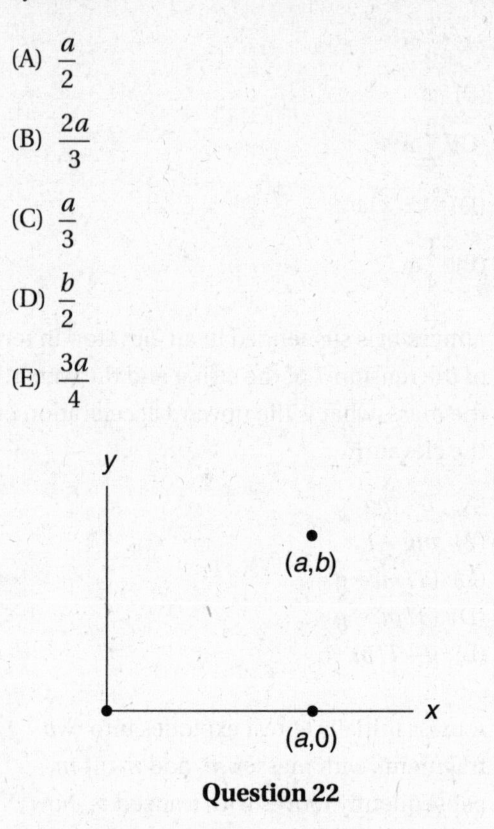

Question 22

GO ON TO THE NEXT PAGE

23. When two springs are attached in parallel to a mass m as shown, what is the equivalent spring constant?

(A) $k_{eq} = k_1 + k_2$

(B) $k_{eq} = \dfrac{1}{1/k_1 + 1/k_2}$

(C) $k_{eq} = \dfrac{k_1 k_2}{k_1 + k_2}$

(D) $k_{eq} = \sqrt{k_1^2 + k_2^2}$

(E) none of the above

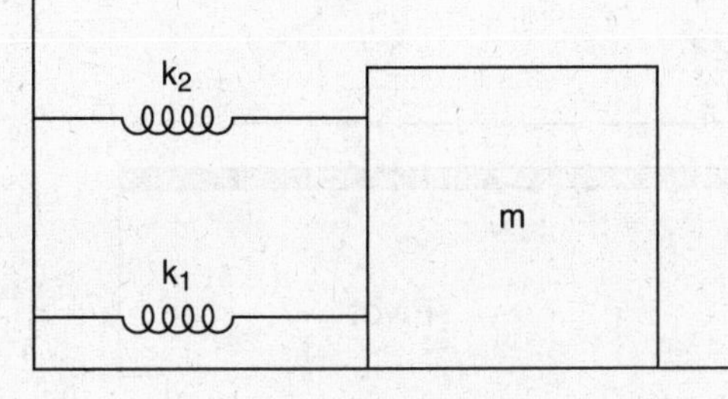

Question 23

24. A mass m is initially moving to the right with speed v toward a series of hills and valleys on a frictionless surface, as shown. What range of speeds must the mass have to become trapped in the valley between points A and B?

(A) $v > \sqrt{2gh_1}$

(B) $\sqrt{2g(h_1 + h_2)} > v > \sqrt{2gh_1}$

(C) $\sqrt{2gh_1h_2/(h_1 + h_2)} > v > \sqrt{2gh_1}$

(D) $\sqrt{2g(h_1^2 + h_2^2)^{1/2}} > v > \sqrt{2gh_1}$

(E) It is impossible for the mass to become trapped.

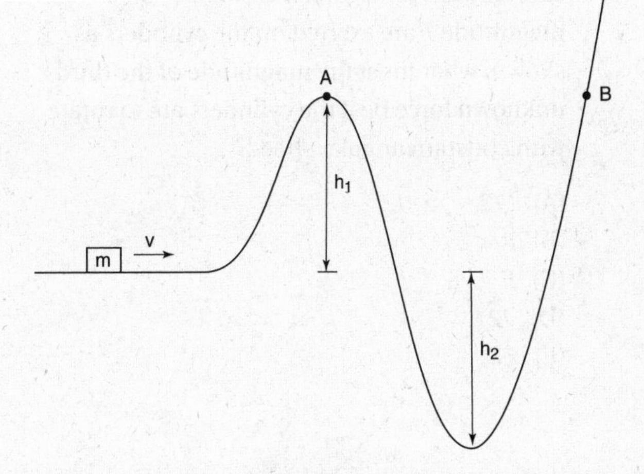

Question 24

GO ON TO THE NEXT PAGE

25. If a constant net force F causes an object to accelerate from rest to a final velocity of v while moving a distance Δx, what is the object's mass?

(A) $2\Delta xF/v_f^2$
(B) $2\Delta xF/v_f$
(C) $\Delta xF/v_f^2$
(D) $\Delta xF/v_f$
(E) Not enough information is given.

26. Two concentric cylinders of radii R and $2R$ are fastened together and can rotate about their centers as in the figure. If two forces of magnitude F are exerted on the cylinders as shown, what must the magnitude of the third unknown force be if the cylinders are to rotate with constant angular speed?

(A) $F/2$
(B) $F/\sqrt{2}$
(C) F
(D) $\sqrt{2}F$
(E) $2F$

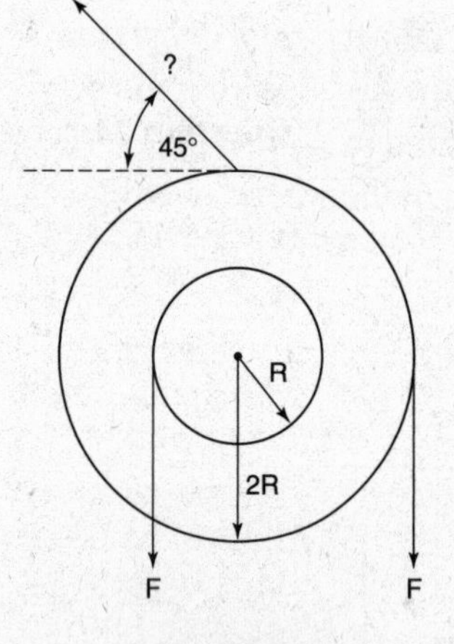

Question 26

27. A rod of uniform mass density and total mass M with a mass m attached to one end is pivoted as shown. What is the tension force F_T needed to prevent the rod from moving?

(A) $F_T = 2mg$
(B) $F_T = \frac{1}{2}mg - \frac{1}{4}Mg$
(C) $F_T = 2mg + \frac{1}{4}Mg$
(D) $F_T = \frac{1}{4}Mg - 2mg$
(E) $F_T = 2mg - \frac{1}{6}Mg$

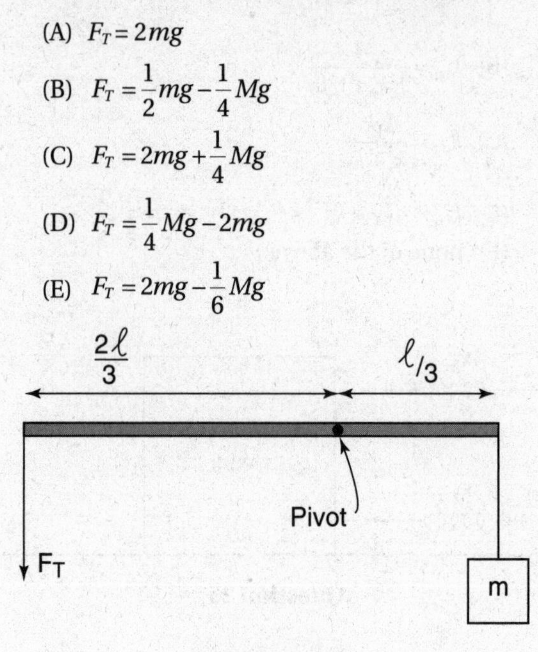

Question 27

28. A child riding in a car along a flat road tosses a ball straight up with a speed v. To an observer standing on the ground outside the car, the ball appears to be thrown with an angle of elevation of θ. What is the speed of the car?

(A) $v\sin\theta$
(B) $v\cos\theta$
(C) $v\sec\theta$
(D) $v\tan\theta$
(E) $v\cot\theta$

GO ON TO THE NEXT PAGE

29. A car of mass 400 kg is rounding a flat curve of radius 80 m. If the coefficient of static friction between the car's tires and the ground is 0.5, the maximum speed at which the car can round the curve is closest to which of the following?

(A) 10 m/s
(B) 20 m/s
(C) 30 m/s
(D) 40 m/s
(E) 80 m/s

30. If the acceleration of an object of mass m starting from rest is given as a function of time by $a(t) = A\sqrt{t}$, then the kinetic energy of this object is proportional to

(A) $\frac{mA^2}{t}$
(B) mA^2t
(C) mA^2t^3
(D) mA^2t^{-3}
(E) $mA^2t^{3/2}$

31. A mass m is thrown off the roof of a building of height h with an initial speed v_0 and angle of elevation θ *below* the horizontal. What is the object's speed when it reaches the ground? (Do not ignore air friction.)

(A) $v \geq \sqrt{gh + v_0^2/2}$
(B) $v < \sqrt{gh + v_0^2/2}$
(C) $v < \sqrt{2gh + v_0^2}$
(D) $v < \sqrt{gh + v_0^2 \sin^2\theta/2}$
(E) $v < \sqrt{2gh + v_0^2 \sin^2\theta}$

32. A force **F** is exerted on the rightmost of three masses tied together by strings as shown, causing the masses to accelerate to the right over a frictionless horizontal surface. What is the magnitude of the indicated tension force?

(A) $\frac{(m_2 + m_3)F}{m_1 + m_2 + m_3}$
(B) $\frac{m_1F}{m_1 + m_2 + m_3}$
(C) $F/3$
(D) $\frac{F(m_1 - m_2 - m_3)}{m_1 + m_2 + m_3}$
(E) $\frac{F(m_2 + m_3 - m_1)}{m_1 + m_2 + m_3}$

Question 32

33. A simple pendulum of length l has period T. A second pendulum of length $l/2$ is located at twice the distance from the center of Earth compared to the first pendulum. What is the period of the second pendulum?

(A) $T/2$
(B) $T/\sqrt{2}$
(C) T
(D) $T\sqrt{2}$
(E) $2T$

PRACTICE TEST 1

GO ON TO THE NEXT PAGE

34. A mass m is attached to a rod with uniform mass density at point C as shown. The rotational inertia of this system is greatest when it is rotated about an axis perpendicular to the rod that passes through

(A) point A
(B) point B
(C) point C
(D) a point between points A and B
(E) a point between points B and C

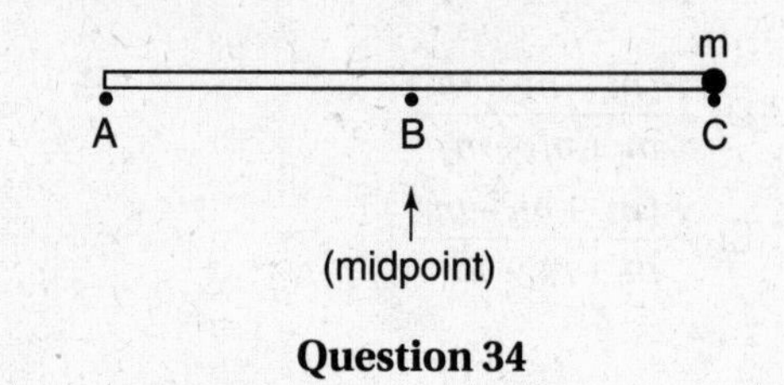

Question 34

35. A particle's position and velocity vectors at two different times are shown. In what direction does the average acceleration vector point?

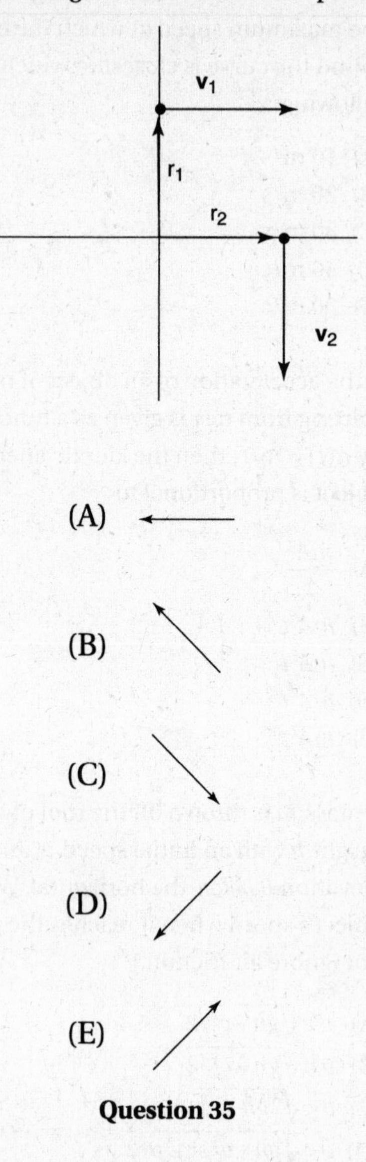

Question 35

GO ON TO THE NEXT PAGE

FREE-RESPONSE QUESTIONS

Directions: Use separate sheets of paper to write down your answers to the free-response questions. You may refer to the formula sheet in the Appendix (pages 641–644).

Mechanics

MECHANICS I

A person (mass m) is flying to the left at a speed v_0 when she manages to grab onto the rim of a cylinder of radius R, length l, mass m, and uniform mass density that can rotate without friction about its axis (see figure).

(a) Using integration, calculate the rotational inertia of the cylinder.
(b) What is the person's linear speed the instant after she grabs the cylinder?
(c) If the collision between the person and the cylinder lasts a time Δt, what is the minimum force the person's hands must be capable of sustaining? (Assume Δt is very small, so that the person's height does not change significantly during this time interval.)

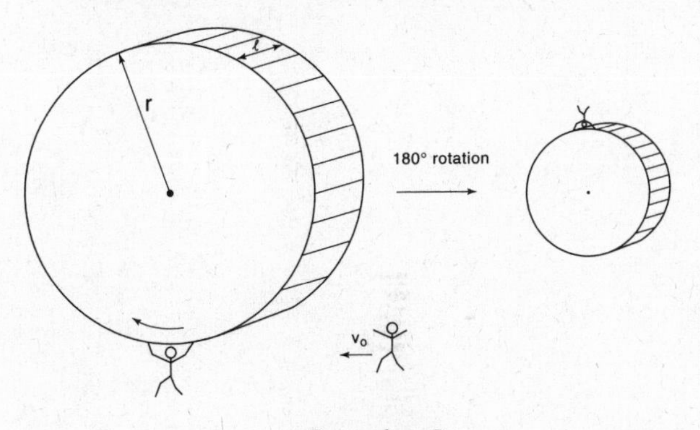

Question I

Then the cylinder rotates 180°, so that the person is at the top of the cylinder.

(d) What is the angular velocity at that instant?
(e) What force must the person's hands sustain now? (Assume that the wheel exerts a downward force on the person in order to keep her from flying off the wheel.)

MECHANICS II

A mass m slides down a frictionless incline and sticks to a spring (spring constant k), initiating simple harmonic motion (see figure).

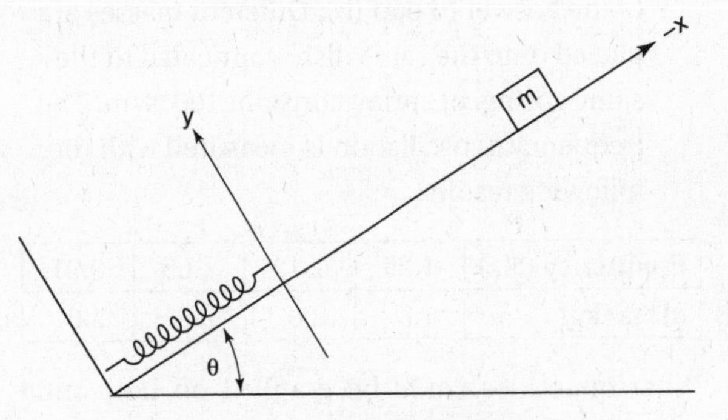

Question II

(a) Express the potential energy U of the oscillating mass and spring system in terms of x (given the axis as shown, where the origin lies at the spring's initial unstretched length).
(b) Graph this $U(x)$ function, indicating the points where $U(x) = 0$.
(c) Based solely on this $U(x)$ curve, where do you expect the equilibrium position of the simple harmonic motion to lie?
(d) Based solely on Newton's second law, where does the equilibrium point lie?
(e) What is the angular frequency of the SHM?

GO ON TO THE NEXT PAGE

MECHANICS III

An object of mass m is placed onto a dish that is attached to a spring (of spring constant k) hanging from the ceiling. Assume that the dish and the spring have negligible mass.

(a) What is the equilibrium extension of the spring?

(b) Suppose the dish (with the mass on it) is displaced from equilibrium by a small distance. What is the resulting frequency of oscillation?

(c) An experiment is designed to test the validity of the answer to part (b). Different masses are placed onto the same dish, connected to the same spring, of spring constant 100 N/m. The frequency of oscillation is measured with the following results:

Frequency (Hz)	1.29	1.12	1.05	0.91
Mass (kg)	1.5	2.0	2.5	3.0

What quantities could be graphed on horizontal and vertical axes to yield a straight line according to the answer to part (b)?

(d) Using your answer to part (c), plot the data on the grid below. Scale and label the axes, including units. Draw a straight line that best represents the data. The origin of your plot need not be (0, 0).

(e) Compare the slope of the line with the prediction of part (b).

GO ON TO THE NEXT PAGE

MULTIPLE-CHOICE QUESTIONS

Directions: Each multiple-choice question is followed by five answer choices. For each question, choose the best answer and fill in the corresponding circle on the answer sheet. You may refer to the formula sheet in the Appendix (pages 641–644).

Electricity and Magnetism

1. Consider the circuit shown, which contains two identical lightbulbs connected to a battery. The top bulb's intensity will

   (A) decrease slightly if either of the two switches is closed.
   (B) decrease slightly if S_1 is closed but not if S_2 is closed.
   (C) decrease slightly if S_2 is closed but not if S_1 is closed.
   (D) remain unchanged if S_2 is closed but go out completely if S_1 is closed.
   (E) remain unchanged if S_1 is closed but go out completely if S_2 is closed.

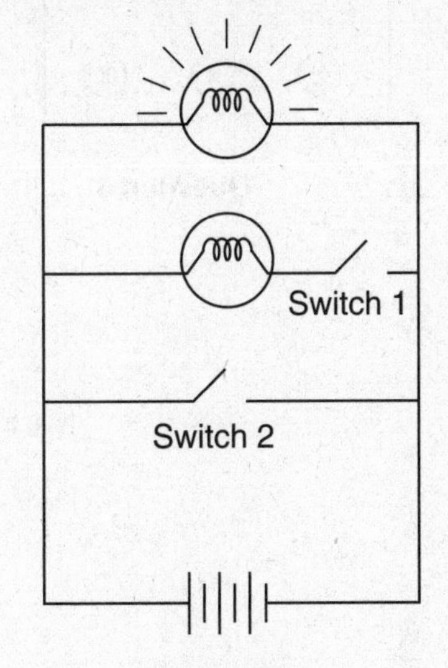

Question 1

2. What is the voltage V in the circuit shown in the figure?

   (A) 1 V
   (B) 2 V
   (C) 4 V
   (D) 6 V
   (E) 8 V

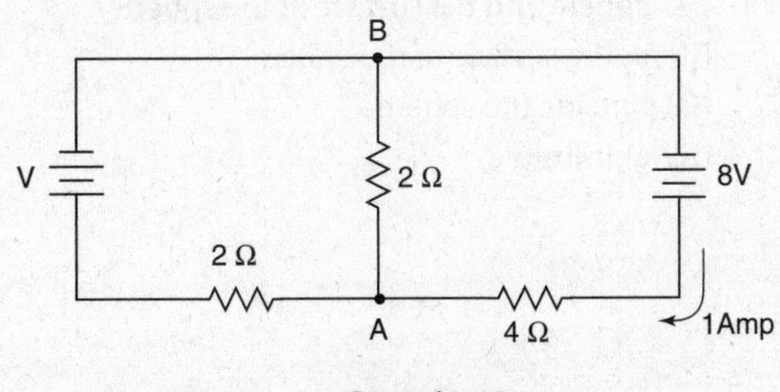

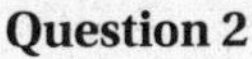

Question 2

3. A battery of voltage V is connected in series to a resistor of resistance R and an uncharged capacitor of capacitance C. What is the magnitude of the maximum charge on the capacitor?

   (A) $Q = \frac{CV}{2}$
   (B) $Q = \frac{CV}{4}$
   (C) $Q = CV^2$
   (D) $Q = CV$
   (E) Q grows without limit

GO ON TO THE NEXT PAGE

4. A sphere of radius R contains a negative charge with a charge density given by the function $\rho = -cr^4$, where r equals the distance from the center of the sphere. Where does the electric field have the smallest magnitude?

(A) at the center of the sphere
(B) at some point between the center of the sphere and the surface of the sphere
(C) at the surface of the sphere
(D) outside the sphere
(E) The electric field is constant in space.

5. Referring to the charge distribution described in question 4, where is the potential smallest in magnitude?

(A) at the center of the sphere
(B) at some point between the center of the sphere and the surface of the sphere
(C) at the surface of the sphere
(D) outside the sphere
(E) at infinity

6. A loop is moved out of a magnetic field at constant velocity as shown. Which of the following statements about the induced current in the loop and the applied force on the loop is correct? (The magnetic field points into the page.)

(A) The current is clockwise, and the applied force is to the right.
(B) The current is clockwise, and the applied force is to the left.
(C) The current is counterclockwise, and the applied force is to the right.
(D) The current is counterclockwise, and the applied force is to the left.
(E) The current is counterclockwise, and the applied force is into the page.

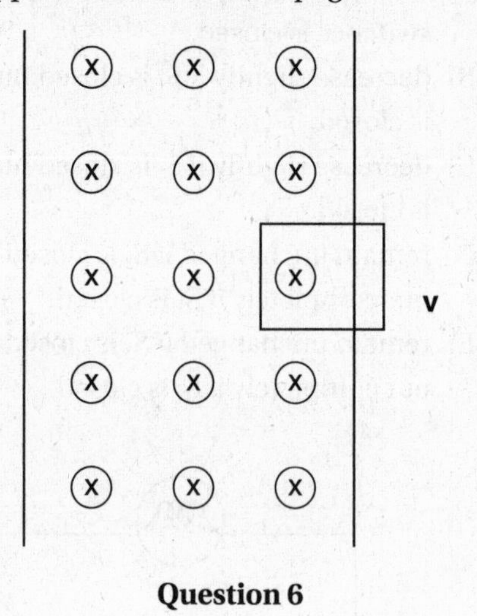

Question 6

GO ON TO THE NEXT PAGE

7. Which of the following statements about electric field lines and equipotential curves is *false*?

   (A) Equipotential lines are always perpendicular to electric field lines.
   (B) The electric field magnitude is greater in regions where the density of electric field lines is greater.
   (C) Electric field lines begin and end only at the locations of negative and positive charges, respectively.
   (D) The work required to move a charge along an equipotential line is zero.
   (E) Electric field lines never penetrate metals.

8. In an *LC* circuit composed of an inductor with inductance *L* and a capacitor of capacitance *C*, if the current through the capacitor is changing at a rate of dI/dt, what is the magnitude of the charge on the capacitor?

   (A) $CL(dI/dt)$
   (B) $(C/L)(dI/dt)$
   (C) $(CL^2/2)(dI/dt)$
   (D) $(L/C)(dI/dt)$
   (E) Information is not given.

9. A uniform cylindrical resistor has resistance *R*. Assume that the current density in this resistor is uniform. What is the resistance of a similar cylindrical resistor composed of the same material with half the radius and half the length?

   (A) $4R$
   (B) $2R$
   (C) R
   (D) $\frac{1}{2}R$
   (E) $\frac{1}{4}R$

10. A parallel plate capacitor of capacitance *C* has charge $+Q$ on one plate and $-Q$ on the other. It is discharged through a resistor of resistance *R*. The maximum value of the current flowing through the resistor is given by

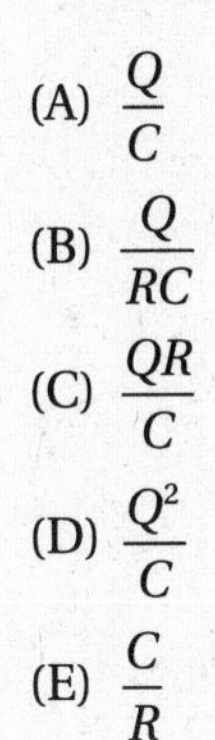

   (A) $\frac{Q}{C}$
   (B) $\frac{Q}{RC}$
   (C) $\frac{QR}{C}$
   (D) $\frac{Q^2}{C}$
   (E) $\frac{C}{R}$

11. The electric field lines due to a point charge suspended within a spherical conducting shell are shown. These field lines are consistent with which of the following charge distributions? (Q_p is the charge of the point charge, and Q_s is the net charge on the conducting shell.)

   (A) $Q_p < 0,\ Q_s < 0,\ |Q_p| > |Q_s|$
   (B) $Q_p < 0,\ Q_s > 0,\ |Q_p| > |Q_s|$
   (C) $Q_p < 0,\ Q_s > 0,\ |Q_p| < |Q_s|$
   (D) $Q_p > 0,\ Q_s < 0,\ |Q_p| > |Q_s|$
   (E) $Q_p > 0,\ Q_s < 0,\ |Q_p| < |Q_s|$

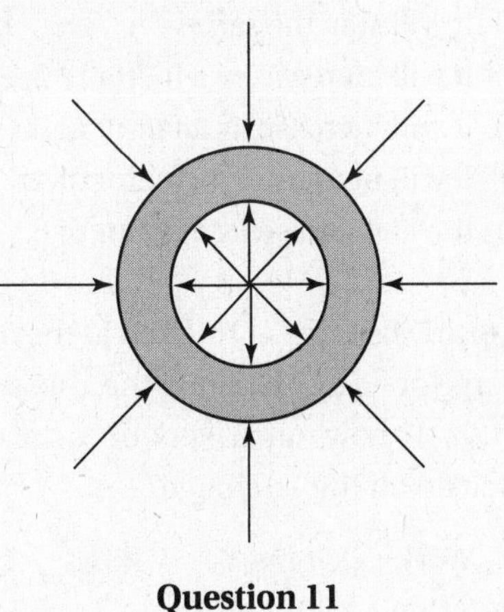

Question 11

GO ON TO THE NEXT PAGE

12. Three wires carry currents as shown. Which vector in the diagram shows the direction of the magnetic force on the top wire due to the other two?

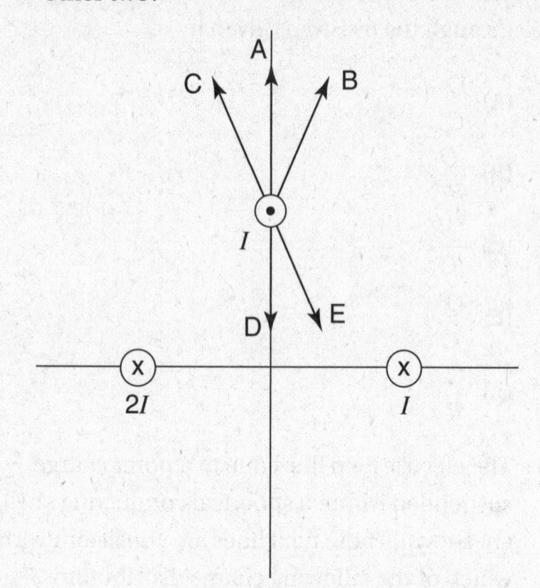

Question 12

13. Referring to the situation introduced in question 12, if the value of I is increased by a factor of 2 (doubling the current carried by each wire), how will the magnitude of the force on the top wire change?

(A) It will stay the same.
(B) It will increase by a factor of 2.
(C) It will increase by a factor of 3.
(D) It will increase by a factor of 4.
(E) It will increase by a factor of 6.

14. A positive charge $+Q$ is fixed at the origin. As a charge of $+q$ moves along the x-axis from $(d, 0)$ to $(2d, 0)$, how much work does the electric force do on the $+q$ charge?

(A) Work $= qQ/4\pi\varepsilon_0 d$
(B) Work $= qQ/8\pi\varepsilon_0 d$
(C) Work $= -qQ/4\pi\varepsilon_0 d$
(D) Work $= -qQ/8\pi\varepsilon_0 d$
(E) Work $= -Q/4\pi\varepsilon_0 d$

15. Which of the following statements is true about a discharging capacitor?

(A) The energy stored and the charge on the capacitor both increase with time.
(B) The energy stored and the charge on the capacitor both decrease with time.
(C) The energy stored decreases while the charge increases with time.
(D) The energy stored remains the same while the charge decreases with time.
(E) The energy stored increases while the charge decreases with time.

16. To calculate the resistance of an 80-W speaker, you measure the voltage across the speaker to be 20 V. Which of the following is the speaker's resistance?

(A) 0.2 Ω
(B) 0.25 Ω
(C) 4 Ω
(D) 5 Ω
(E) 1600 Ω

GO ON TO THE NEXT PAGE

17. A battery of voltage V is connected to two capacitors in series with capacitance C_1 and C_2. How much energy is stored in capacitor 1?

(A) $U = \dfrac{C_1 C_2^2 V^2}{2(C_1 + C_2)}$

(B) $U = \dfrac{C_1 C_2 V^2}{2(C_1 + C_2)^2}$

(C) $U = \dfrac{(C_1 + C_2) V^2}{2}$

(D) $U = \dfrac{C_2 C_1^2 V^2}{2(C_1 + C_2)^2}$

(E) $U = \dfrac{C_1 C_2^2 V^2}{2(C_1 + C_2)^2}$

18. Consider a spherical shell with a uniform surface charge density σ. If the radius of the shell is doubled while holding the charge density constant, what effect does this have on the magnitude of the electric field just outside the shell?

(A) The electric field decreases by a factor of 4.
(B) The electric field decreases by a factor of 2.
(C) The electric field remains the same.
(D) The electric field increases by a factor of 2.
(E) The electric field increases by a factor of 4.

19. A very long rod containing a charge density of $-\lambda$ coulombs/meter is moving along the $+x$-axis in the $+x$-direction with velocity v. The length of the rod is along the x-axis. What is the magnitude of the current, and in what direction is the conventional current flowing?

(A) $I = v\lambda$ along the $+x$-direction.
(B) $I = v\lambda$ along the $-x$-direction.
(C) $I = v\lambda^2$ along the $+x$-direction.
(D) $I = v\lambda^2$ along the $-x$-direction.
(E) $I = v/\lambda$ along the $+x$-direction.

20. Which of the following is a sketch of the electric field along the yz-plane due to the charged rod after the rod stops moving?

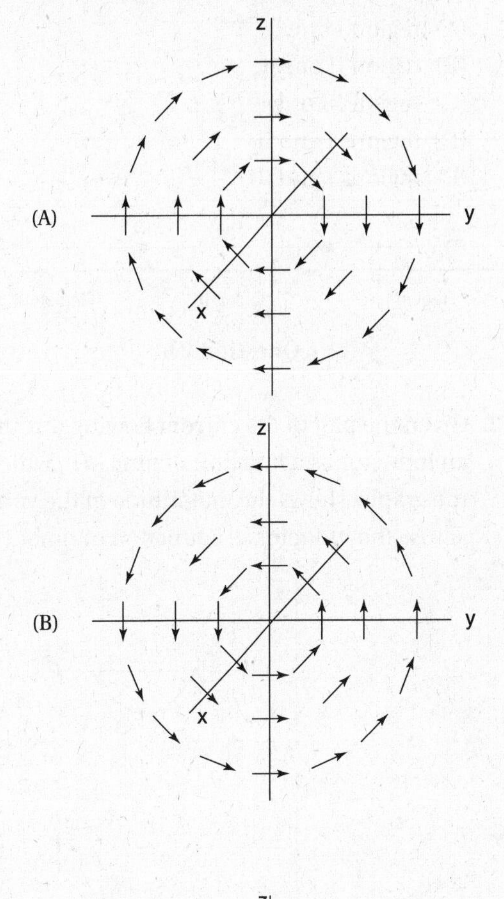

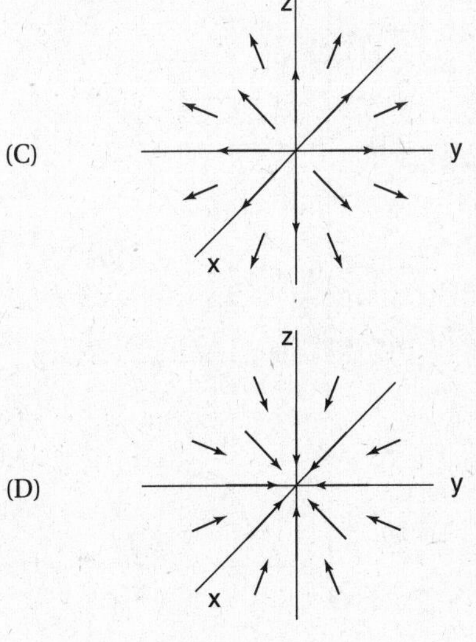

(E) The field is zero.

Question 20

GO ON TO THE NEXT PAGE

21. Consider the charge distribution shown. The potential is zero at points in which of the following regions (not including $\pm\infty$)?

(A) region I only
(B) region II only
(C) region III only
(D) regions I and II
(E) regions I and III

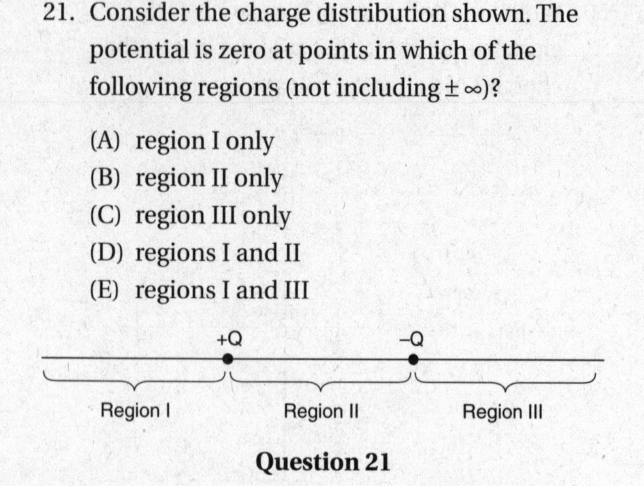

Question 21

22. Given the plot of the current flowing through an inductor as a function of time, $I(t)$, which of the graphs shows the magnitude of the voltage across the inductor as a function of time?

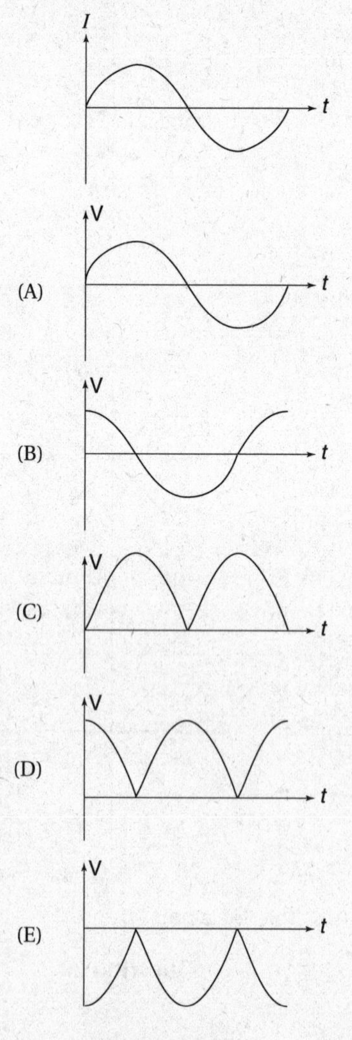

Question 22

23. Which of the following is most closely related to the brightness of a lightbulb? (Assume that all the energy dissipated in a lightbulb is emitted as light energy rather than thermal energy.)

(A) the current running through the lightbulb
(B) the voltage across the lightbulb
(C) the power dissipated by the lightbulb
(D) the resistance of the lightbulb
(E) the current density of the lightbulb

24. The figure shows the equipotential lines within a parallel plate capacitor. Which of the circuits could produce this set of equipotential lines?

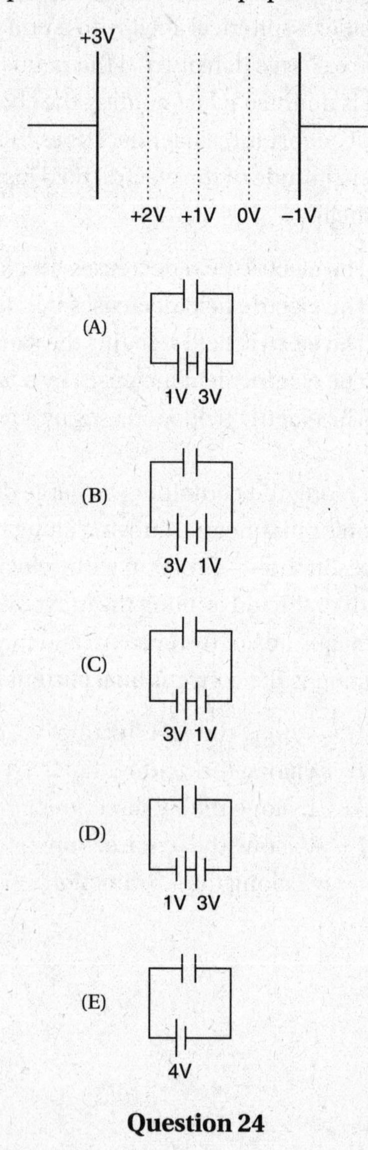

Question 24

GO ON TO THE NEXT PAGE

25. The electric field inside a sphere with uniform positive charge density has the following property:

(A) It is zero.
(B) It is a nonzero constant.
(C) It decreases as a function of radial distance from the center of the sphere.
(D) It increases as a function of radial distance from the center of the sphere.
(E) It is infinite.

26. A charged particle moving in the presence of both an electric and a magnetic field moves in a closed path (i.e., a path that begins and ends at the same point). If the energy of the particle at the end of the closed path is 2.0 J greater than its energy when it began the path, which of the following statements is correct?

(A) The magnetic field performed 2.0 J of work on the particle.
(B) The magnetic field performed –2.0 J of work on the particle.
(C) The electric field is conservative.
(D) The electric field is due to static charges.
(E) none of the above

27. Consider an electric dipole within a constant electric field as shown. (The dipole consists of two charges of equal magnitude and opposite sign rigidly connected to each other.) Describe the net force and net torque on the dipole.

(A) The net force is zero, and the net torque is clockwise.
(B) The net force is zero, and the net torque is counterclockwise.
(C) The net force is to the right, and the net torque is clockwise.
(D) The net force is to the right, and the net torque is counterclockwise.
(E) The net force is to the left, and the net torque is counterclockwise.

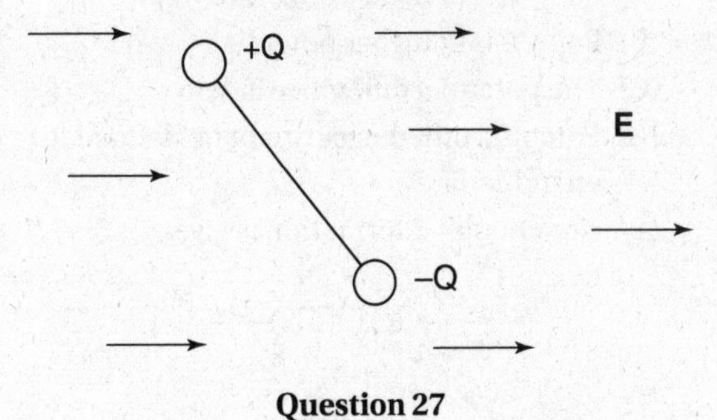

Question 27

28. A circular loop of wire of radius r lies in a plane perpendicular to a uniform magnetic field that is changing at a rate of a teslas/second. Which of the following statements correctly describes the power dissipated in the loop as a function of the loop's radius? (Recall that a wire's resistance is proportional to its length.)

(A) The power is independent of r.
(B) The power is proportional to r.
(C) The power is proportional to r^2.
(D) The power is proportional to r^3.
(E) The power is inversely proportional to r.

GO ON TO THE NEXT PAGE

29. If a potential difference of 3 V across a wire that is 2 m long and 0.5 mm in radius yields a current of 10 A, what is the resistivity of the wire?

(A) 1.2 Ω m
(B) 3.8 Ω m
(C) 1.2×10^{-7} Ω m
(D) 2.4×10^{-4} Ω m
(E) 1.2×10^{-18} Ω m

30. Consider the inductor shown in the figure below. If the current *I* is decreasing in time, what can be said about the potential difference between points *A* and *B*?

(A) Point *A* is at higher potential.
(B) Point *B* is at higher potential.
(C) The potential difference is zero.
(D) Potential difference cannot be defined for an inductor.
(E) Not enough information is given.

I
A *B*

Question 30

31. A dielectric material is inserted between the plates of an isolated charged capacitor. Which of the following is true?

(A) The potential energy of the capacitor increases.
(B) The charge of the capacitor decreases.
(C) The capacitance of the capacitor decreases.
(D) The voltage of the capacitor increases.
(E) none of the above

32. Consider charging a battery with an internal resistance of *R* and an EMF of V_0. (*Charging* a battery involves moving charge from the negative terminal to the positive terminal.) Which of the following is capable of charging a battery?

I. a potential difference greater than V_0
II. a potential difference equal to V_0
III. a potential difference smaller than V_0

(A) I only
(B) II only
(C) III only
(D) I and II
(E) I, II, and III

33. An electron is traveling through a region of space with electric and magnetic fields as shown (the velocity is in the plane of the **E** and **B** vectors). The force exerted on the electron has components in what directions?

(A) the +*y*-direction and the +*z*-direction
(B) the –*y*-direction and the –*z*-direction
(C) the –*y*-direction and the +*z*-direction
(D) the +*y*-direction and the –*z*-direction
(E) none of the above

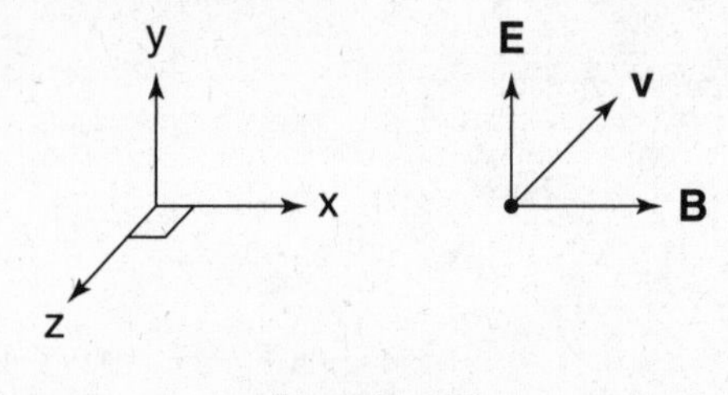

Question 33

GO ON TO THE NEXT PAGE

34. Consider an infinite line of positive charge that has a varying charge density. Very close to the line of charge, in which direction does the electric field approximately point?

(A) directly toward the line of charge
(B) directly away from the line of charge
(C) parallel to the line of charge, pointing in the direction of increasing charge density
(D) parallel to the line of charge, pointing in the direction of decreasing charge density
(E) The field is zero.

35. Consider a wire loop moving through a region of magnetic field pointing into the page as shown. Which is a graph of the magnetic flux through the loop as a function of position (designating inward flux as negative)?

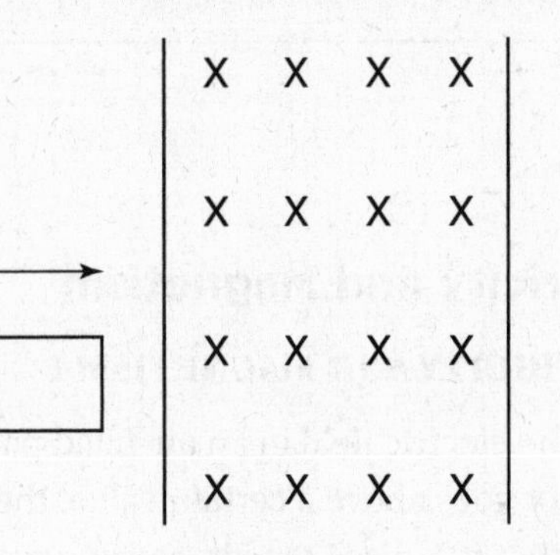

Question 35

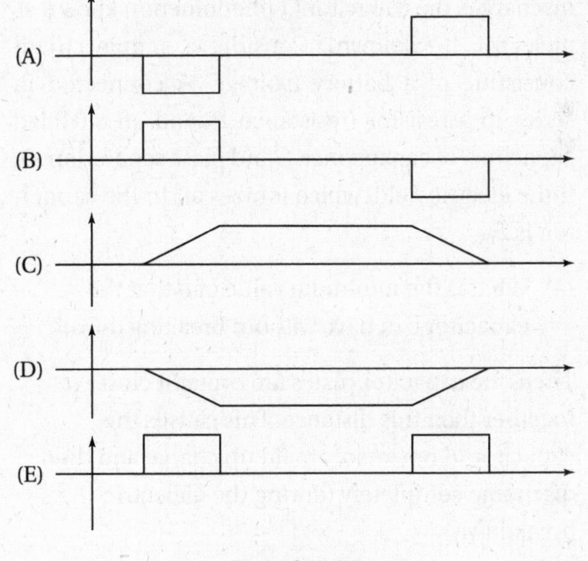

Question 35

GO ON TO THE NEXT PAGE

FREE-RESPONSE QUESTIONS

Directions: Use separate sheets of paper to write down your answers to the free-response questions. You may refer to the formula sheet in the Appendix (pages 641–644).

Electricity and Magnetism

ELECTRICITY AND MAGNETISM I

When the electric field in an air-filled parallel plate capacitor goes above a certain value, the field is so strong that it ionizes the air, causing a spark to fly across the capacitor that completely and rapidly discharges the capacitor (a phenomenon known as dielectric breakdown). Consider a simple circuit consisting of a battery (voltage V) connected in series to a resistor (resistance R) and an air-filled capacitor (of capacitance C and plate separation d). If the electric field, which ionizes air in the capacitor, is E_{BD},

(a) What is the minimum value of d that the capacitor can have without breaking down?

Then, the capacitor plates are brought closer together than this distance. This causes the capacitor to repeatedly build up charge and then discharge completely (during the dielectric breakdown).

(b) Given the new plate separation distance d, what is the voltage across the capacitor when dielectric breakdown occurs?

(c) Sketch the voltage across the capacitor as a function of time, $V(t)$.

(d) Calculate the amount of time that elapses between discharges.

(e) How much energy is lost during each dielectric breakdown?

Then, the capacitor plates are moved farther apart to d_1, so that dielectric breakdown does not occur. After the capacitor is completely charged, it is disconnected from the battery and the plates are brought together.

(f) Describe qualitatively what happens as the plates are brought together. In particular, does dielectric breakdown occur, and if so, at what plate separation?

ELECTRICITY AND MAGNETISM II

An insulating sphere of radius a has a negative charge distribution given by the function $\rho(r) = -c/r^2$ (where c is a positive constant with units of C/m).

(a) Calculate the total charge on the sphere.

(b) Use Gauss's law to calculate the electric field in the following regions:

(i) $r > a$

(ii) $r < a$

(iii) $r = a$

(c) Calculate the potential at $x = a$ by evaluating the appropriate line integral from $x = \infty$ to $x = a$.

(d) The field outside a spherically symmetric charge distribution is equal to the field that would be produced if the total charge were concentrated at the center of the distribution. Use this shortcut to confirm your result from part (c).

GO ON TO THE NEXT PAGE

ELECTRICITY AND MAGNETISM III

Consider the mass spectrometer shown in the figure. After particles are accelerated by a voltage V, they pass through three regions, each with a different uniform magnetic field. In order for a particle to reach the detector, it must travel along the trajectory indicated, moving through three semicircles of the indicated radii. Assume that the particles have the charge of an electron, $-e$.

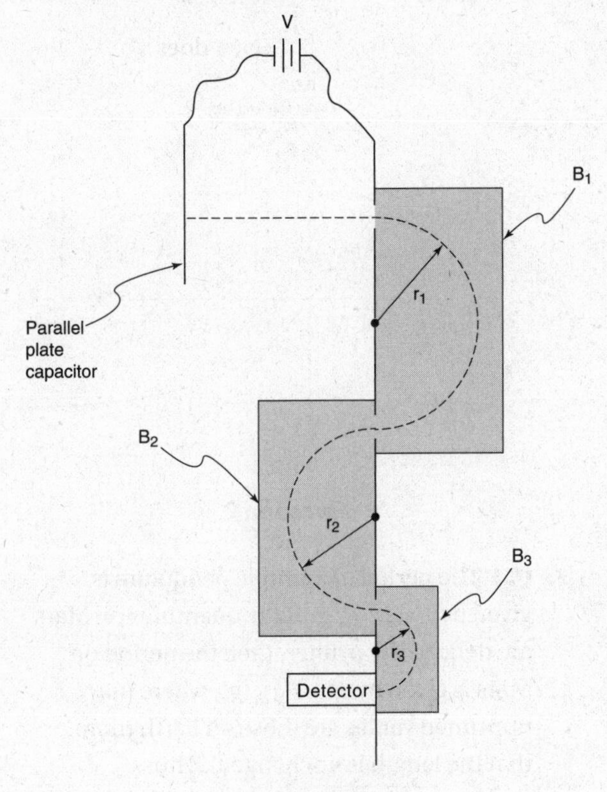

Question III

(a) What is the velocity of the electron as it enters region 1 in terms of V, e, and m (the mass of the charged particle)?

(b) What are the directions of the uniform magnetic fields $\mathbf{B}_1$, $\mathbf{B}_2$, and $\mathbf{B}_3$?

(c) Express r_1 in terms of m, e, V, and B_1.

(d) Calculate the values of r_2 and r_3 required to make it possible for a particle to make it to the detector. (Express your answer in terms of r_1, B_1, B_2, and B_3.)

(e) This setup (with only one semicircle) was used by J. J. Thomson to measure the ratio of the electron's charge to its mass. Rewrite your answer to part (c) so that it gives a prediction for e/m in terms of the voltage V, the radius r_1, and the magnetic field B_1. Assume that the radius is measured as a function of the voltage for fixed $B_1 = 0.01\,T$. Which quantities should be plotted in order to obtain a linear relationship?

(f) Assume that the measurements are carried out with the following results:

***V* (kV)**	2.0	2.5	3.0	3.5
***r* (cm)**	1.5	1.7	1.8	2.0

Plot the data using the quantities you determined in part (e). What value of e/m is obtained from these measurements? How does it compare with the known value?

MECHANICS

ANSWER KEY

1. **B**	8. **C**	15. **D**	22. **B**	29. **B**
2. **E**	9. **B**	16. **B**	23. **A**	30. **C**
3. **C**	10. **A**	17. **E**	24. **E**	31. **C**
4. **B**	11. **C**	18. **D**	25. **A**	32. **B**
5. **A**	12. **D**	19. **D**	26. **B**	33. **D**
6. **C**	13. **C**	20. **C**	27. **B**	34. **A**
7. **B**	14. **D**	21. **D**	28. **E**	35. **D**

ANSWERS EXPLAINED

Multiple-Choice

1. **(B)** Because the force is constant, the work is given by the dot product of the force and the displacement vectors:

$$W = \mathbf{F} \cdot \Delta\mathbf{r} = Fd\cos\theta$$

Choice (C) is the change in potential energy as the block is elevated. Choice (E) is the work that would have to be done by *F* to move the block up the incline at a constant speed; that is, the work done by *F* would be the negative of the work done by gravity and by friction (because the net work would have to be zero).

2. **(E)** See the figure. For large-amplitude oscillations, the restoring force is not linear, so the motion is not simple harmonic motion. Because the restoring force is smaller than its linear approximation about the origin (which would produce a period of *T*), it produces motion with a larger period. (Because the force is smaller, the particle oscillates more slowly and more time is required for each cycle.)

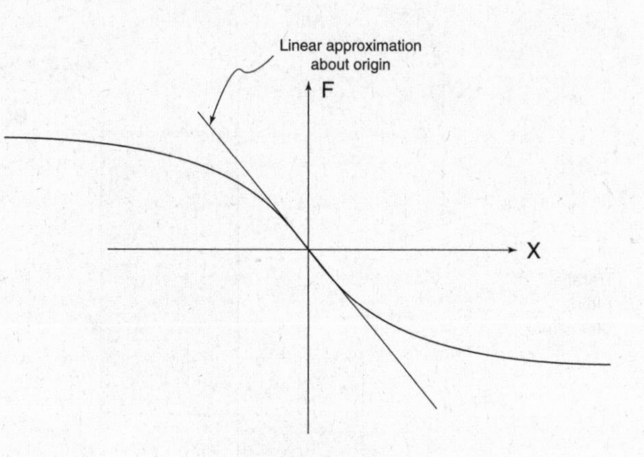

Question 2

3. **(C)** The period of a simple pendulum is given by $T = 2\pi\sqrt{l/g}$. If the quantities on Mars are denoted by primes, then the period on Mars is given by $T' = T\sqrt{g/g'}$, where the unprimed values are those on Earth (note that the length is unchanged). Thus, $T' = (2)\sqrt{9.8/3.71} = 3.25\text{ s}$.

4. **(B)** A simple harmonic oscillator is characterized by $a = -\omega^2 x$ (found by plugging Hooke's law into Newtons' second law). At equilibrium, $x = 0$ and thus the acceleration is also zero.

5. **(A)** According to the parallel axis theorem, the rotational inertia about an axis that does not pass through the center of mass (in this problem, the one that passes along the surface of the sphere) is equal to the rotational inertia about a parallel axis passing through the center of mass (in this problem, the axis passing through the sphere's center) plus MD^2, where *D* is the distance between the two axes (in this case, *r*):

$$\left[I_{\text{parallel axis}} = \frac{7}{5}Mr^2\right] = \left[MD^2 + I_{\text{CM}} = Mr^2 + I_{\text{CM}}\right]$$

Solving this equation for the rotational inertia about the axis that passes through the center of mass yields choice (A).

6. **(C)** Consider the figure. The pendulum is rotating in a circle, so it must experience a centripetal acceleration toward the center of the circle. The pendulum is also speeding up (because of conservation of energy, with gravitational potential energy being converted to kinetic energy), so it experiences a tangential acceleration parallel to its velocity. The sum of these accelerations produces a net acceleration that is neither purely radial nor purely tangential but rather points in the direction of choice (C).

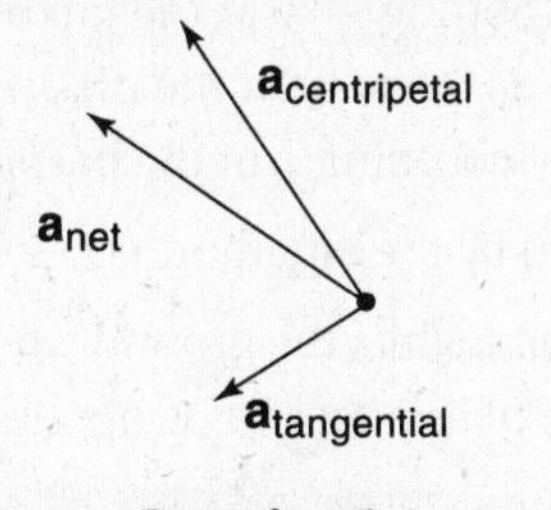

Question 6

7. **(B)** This is an inelastic collision, such that momentum is conserved and the final velocity of both masses is the same (because they stick together and move with a common velocity). Conservation of momentum yields

$$p_x = mv = (4m)\,v_{\text{final},\,x}$$

$$p_y = (3m)(2v) = (4m)\,v_{\text{final},\,y}$$

Solving these equations yields choice (B).

8. **(C)** $P = \mathbf{F} \cdot \mathbf{v}$. For the power to be negative, the force must have a component antiparallel to the velocity, as shown in the figure. Expressing the dot product as $P = \mathbf{F} \cdot \mathbf{v} = vF_{\parallel}$ and substituting in numbers indicates that the component of the force antiparallel to the velocity has a magnitude of $\sqrt{2}$ N directed as shown. Given that the magnitude of the force is 2 N, trigonometry indicates that the angle between the force and the velocity is equal to 135°.

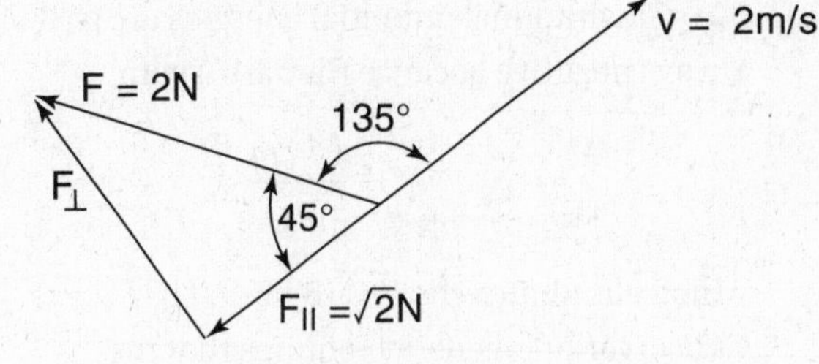

Question 8

9. **(B)** Since the gravitational potential energy is proportional to the height, a plot of potential energy vs position is a parabola similar to the $y(x)$-parabola. As in the case of a mass attached to a spring, whenever the plot of potential energy vs position is parabolic, there is a linear restoring force that gives rise to simple harmonic motion. Therefore, the position function is sinusoidal, initially having a positive value.

10. **(A)** As explained in the solution to question 9, the particle moves in simple harmonic motion. Therefore, the acceleration is sinusoidal, eliminating all choices except (A) and (B). Since the acceleration is initially negative (the object begins at rest and accelerates to the left), choice (A) is correct.

11. **(C)** $P = \mathbf{F} \cdot \mathbf{v}$. Therefore, the power is negative as the projectile moves away from Earth (because the velocity is antiparallel to the gravitational force), zero at the peak of the trajectory (when the velocity is zero), and positive as the projectile returns to Earth (because the velocity is parallel to Earth). The power is greatest the instant before the projectile hits Earth, at which point the velocity parallel to the force is greatest and the gravitational force is greatest.

12. **(D)** The magnitude of the acceleration is proportional to the magnitude of the force. The magnitude of the force, in turn, is maximized when the projectile is closest to Earth, which occurs both when the particle is launched and when the particle hits the ground.

13. **(C)** Gravitational potential energies are always negative because they are a sum:

$$\sum_{\text{all pairs of particles } i \text{ and } j} -\frac{Gm_i m_j}{r_{ij}}$$

which eliminates choices (B) and (D). Conservation of energy requires that the potential energy be larger at the peak of the trajectory (where the kinetic energy is minimized), so choice (C) is correct.

14. **(D)** Consider the equation $\omega_f^2 = \omega_0^2 + 2\alpha\Delta\theta$. In this case, the initial angular velocity is zero, so this equation simplifies to $\omega_f^2 = 2\alpha\Delta\theta$. Therefore, if $\Delta\theta$ decreases by a factor of 2, ω decreases by a factor of $\sqrt{2}$.

15. **(D)** There are two forces acting on the mass: tension pointing upward and gravity pointing downward. Therefore, $F_{net} = ma = T - mg$ (designating upward to be positive as indicated in the problem). Solving for the acceleration a yields choice (D). Note that the mass and the elevator have a common acceleration.

16. **(B)** Conservation of momentum requires that the speeds of m_1 and m_2 be related according to the equation $m_1v_1 = m_2v_2$, so $v_2 = m_1v_1/m_2$. (The total momentum of the system is zero.) The change in kinetic energy during the explosion is $\Delta\text{KE} = \frac{1}{2}m_1v_1^2 + \frac{1}{2}m_2v_2^2$ (recalling that initial kinetic energy is zero). Substituting $v_2 = m_1v_1/m_2$ and simplifying yields choice (B).

17. **(E)** Choice (A): The acceleration of a projectile is always g pointing downward. Choice (B): The power exerted by the gravitational force is maximized when the projectile's velocity parallel to the gravitational force (its downward velocity) is maximized, which occurs immediately before the projectile lands (based on the formula $P = \mathbf{F} \cdot \mathbf{v}$). Choice (C): The projectile's initial y-component of velocity is $v_0 \sin\theta$, and its final y-component of velocity (due to the symmetry of the parabolic trajectory) is $-v_0 \sin\theta$. Applying the equation $v_f = v_0 + at$, with the acceleration of $-g$, shows that the total time in the air is equal to $2v_0 \sin\theta/g$. Choice (D): Conservation of energy requires that the projectile has the same kinetic energy when it lands as when it was launched. Choice (E): The projectile has the same speed as when it was launched, but it is moving in a different direction (tilted downward rather than upward).

18. **(D)** Choices (A) and (C): The restoring force and acceleration are maximized when the displacement is maximized, which occurs at the turning points. Choice (B): The velocity is maximized at the equilibrium point (where all the energy is kinetic energy). Choice (D): The particle reverses direction at the point where its displacement is maximized.

19. **(D)** Setting $y = 0$ at the top of the cliff and $t = 0$ as the instant the mass slides off the cliff, the y-coordinate of the particle's position is given by the equation $y(t) = -\frac{1}{2}gt^2$. Because the mass hits the ground when $y = -h$, the time of impact can be obtained by solving the equation $y(t) = -\frac{1}{2}gt^2 = -h \Rightarrow t_{hit} = \sqrt{2h/g}$. The x-component of the velocity is a constant v_0, so the horizontal displacement during the fall is given by $\Delta x = v_0\Delta t = v_0\sqrt{2h/g}$.

20. **(C)** The torque is zero when the angular acceleration is zero. To calculate the angular acceleration from the angular position, we must take two derivatives:

$$\theta(t) = 2t^3 - 6t^2$$

$$\omega(t) = \frac{d}{dt}(2t^3 - 6t^2) = 6t^2 - 12t$$

$$\alpha(t) = \frac{d}{dt}(6t^2 - 12t) = 12t - 12$$

Therefore, $\alpha(t) = 0$ when $t = 1$.

21. **(D)** The tension force exerts no torque because it is parallel to the radial position vector. Therefore, the net torque is zero and angular momentum is conserved. When the radius decreases by a factor of 2, the rotational inertia, $I = mr^2$, decreases by a

factor of 4. In order for the angular momentum, $L = I\omega$, to remain constant, the angular velocity must then *increase* by a factor of 4. Given these changes in the rotational inertia and the angular velocity, the kinetic energy, $KE = \frac{1}{2}I\omega^2 = \frac{1}{2}(I\omega)(\omega) = (\text{constant})\omega$, increases by a factor of 4.

22. **(B)** Note that the *y*-coordinate of the top mass is irrelevant to answering this question. We could imagine that the top mass is also located at $(a, 0)$. We can use the equation $x_{CM} = \Sigma m_i x_{CM,i} / \Sigma m_i = (m \cdot 0 + m \cdot a + m \cdot a) / (m + m + m) = 2a/3$, choice (B). This answer makes sense because we have twice as much mass located at $x = a$ compared to $x = 0$, so the center of mass is located twice as close to $x = a$.

23. **(A)** When the mass m is displaced by an amount Δx, the net force on it is the sum of the two spring forces: $F_{net} = -k_1\Delta x - k_2\Delta x$. Setting this force equal to the force that would result from a single equivalent spring with spring constant k_{eq} yields

$$[F_{\text{equivalent spring}} = -k_{eq}\Delta x] = [F_{net} = -k_1\Delta x - k_2\Delta x]$$

Solving for the equivalent spring constant gives choice (A).

Note: Two springs in parallel have an equivalent spring constant of $k_{eq} = k_1 + k_2$. Two springs in series have an equivalent spring constant of

$$k_{eq} = \frac{1}{1/k_1 + 1/k_2} = \frac{k_1 k_2}{k_1 + k_2}$$

This is the same type of pattern seen with equivalent capacitances and resistances.

24. **(E)** Because of conservation of energy, the mass's speed is a function only of its height. Therefore, if the mass has enough energy to move over hill h_1 when moving to the right, it will also have enough energy to move over the hill when moving to the left. Therefore, it is impossible for the mass to become trapped in the valley between points *A* and *B* without some other force being present (e.g., a frictional force).

25. **(A)** The one-dimensional kinematics equation $v_f^2 = v_0^2 + 2a\Delta x$ can be solved for the acceleration, yielding $a = v_f^2/2\Delta x$. Applying Newton's second law, $F = ma$, indicates that $m = F/a = 2\Delta x F/v_f^2$.

26. **(B)** In order for the cylinders to rotate with a constant speed, they must experience zero net torque. Holding to the convention that positive torque is counterclockwise, this mathematically yields

$$\tau_{net} = RF + F_{unknown}(2R)\cos 45° - F(2R) = 0$$

Solving for the unknown force gives $F_{unknown} = F/\sqrt{2}$.

27. **(B)** Because the rod is fixed to a pivot (which can supply whatever force is necessary to make the net force zero), the only condition that must be satisfied for the rod to be in static equilibrium is that the net torque on the rod be zero. Calculating the torque about the pivot point (designating counterclockwise torques positive) yields

$$\tau_{net} = F_T\left(\frac{2l}{3}\right) + (Mg)\left(\frac{l}{6}\right) - (mg)\left(\frac{l}{3}\right) = 0$$

(The center of the rod is located a distance $l/6$ to the left of the pivot.) Solving for the tension force yields choice (B).

28. **(E)** The velocity of the ball with respect to the ground is equal to the vector sum of the velocity of the ball with respect to the car plus the velocity of the car with respect to the ground, as shown. Based on the vector geometry of this triangle, $\tan\theta = v/$speed of car, so that the speed of the car $= v/\tan\theta = v\cot\theta$.

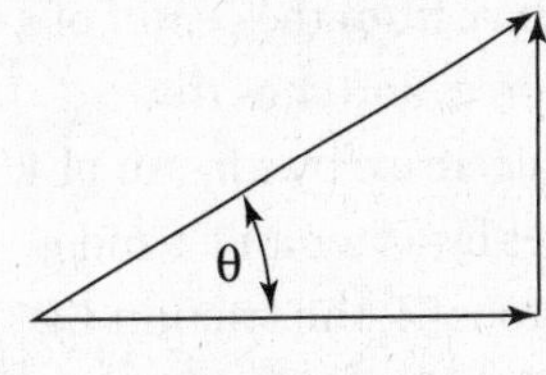

Question 28

29. **(B)** The frictional force provides as much centripetal force as is needed to keep the car moving in a circular path, up to the maximal static frictional force equal to $F_{fr} = \mu_k F_n = \mu_k mg$. Therefore, at the maximum speed, $mv^2/r = \mu_k mg$. Solving for v yields $v = \sqrt{\mu_k gr}$. Approximating g as 10 m/s^2,

$$v = \sqrt{\mu_k gr} = \sqrt{(0.5)(10\,\text{m/s}^2)(80\,\text{m})}$$
$$= \sqrt{400\,\text{m}^2/\text{s}^2} = 20\,\text{m/s}$$

30. **(C)** Dimensional analysis gives the answer. A will have dimensions $\left[\frac{L}{T^{5/2}}\right]$. In order for the kinetic energy to have the correct dimensions $\left[\frac{ML^2}{T^2}\right]$, the answer must be (C).

31. **(C)** Ignoring air friction and applying conservation of energy yields $\frac{1}{2}mv_0^2 + mgh = \frac{1}{2}mv_f^2$. Solving for the final speed gives $v_f = \sqrt{2gh + v_0^2}$. Because the frictional force always performs negative work, it decreases the final speed so that the final velocity is less than this value.

32. **(B)** Treating the three blocks as a single system, their common acceleration is

$$a = \frac{F_{\text{net}}}{m_{\text{total}}} = \frac{F}{m_1 + m_2 + m_3}$$

The net force on the leftmost block, which is equal to F_T, is then given by Newton's second law:

$$F_{\text{net}} = F_T = m_1 a = m_1\left(\frac{F}{m_1 + m_2 + m_3}\right)$$

33. **(D)** The formula for the period of a pendulum is $T = 2\pi\sqrt{l/g}$. Based on the law of universal gravitation ($F = Gm_1m_2/r^2$), doubling the distance from the center of Earth causes the force (and thus the acceleration g) to decrease by a factor of 4. Because l decreases by a factor of 2 and g decreases by a factor of 4, the fraction l/g *increases* by a factor of 2, so the period $T = 2\pi\sqrt{l/g}$ increases by a factor of $\sqrt{2}$.

34. **(A)** The rotational inertia of the system is equal to the rotational inertia of the rod plus the rotational inertia of the mass. The rotational inertia of the mass, mR^2, is maximized if the axis passes through point A (in order to maximize R, where R is the radius of the circular arc followed by m). The rotational inertia of the rod is given by the parallel axis theorem $I_{\text{rod}} = I_{\text{CM}} + MD^2$ and is maximized when the distance of the axis from the center of mass is maximized, at both points A and C. Therefore, the rotational inertia of the system is maximized at point A, where the rotational inertia due to both the rod and the mass are maximized.

35. **(D)** $\mathbf{a} = (\mathbf{v}_2 - \mathbf{v}_1)/\Delta t$, so it is parallel to $\mathbf{v}_2 - \mathbf{v}_1$. To obtain $\mathbf{v}_2 - \mathbf{v}_1$, we negate $\mathbf{v}_1$ (by making it point in the opposite direction) and add it to $\mathbf{v}_2$, as shown.

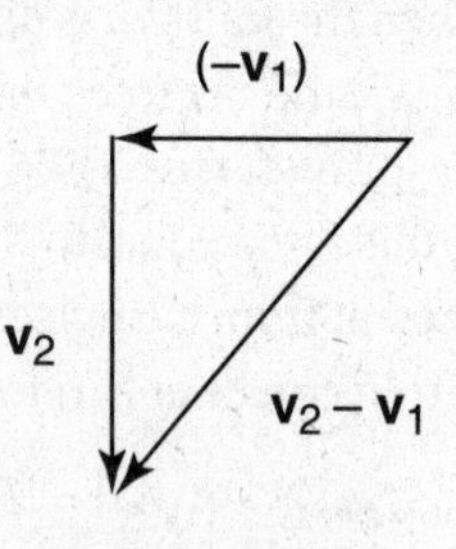

Question 35

Free-Response

MECHANICS I

(a) Rotational inertia is given by the integral $I = \int r^2 dm$. The cylinder must be divided into concentric cylindrical shells centered on the axle. The mass of each cylindrical shell is given by its volume multiplied by the density of the cylinder:

$$dm = \rho\, dV = (m/\pi R^2 l)(2\pi rl\, dr)$$

Substituting this differential mass into the integral for the rotational inertia, $I = \int r^2 dm$, yields

$$I = \left(\frac{m}{R^2 l}\right)\int_0^R (2\pi r^3 l\, dr) = \frac{mR^2}{2}$$

(b) This is an inelastic rotational collision: Energy is not conserved, but angular momentum is conserved (because any torques are exerted by internal forces).

$$mv_0R = \left(\frac{mR^2}{2} + mR^2\right)\omega$$

$$\omega = \frac{2v_0}{3R}$$

Multiplying by the radius to obtain the person's linear speed (rather than angular speed),

$$v = R\omega = \frac{2v_0}{3}$$

(c) The minimum required force occurs if the person exerts a constant force equal to the average force required. Any deviation below this minimum force for any length of time will require a larger than average force later. The average force can be calculated as follows (designating the $+x$-direction to point to the right).

$$\bar{F} = \frac{\Delta p}{\Delta t} = \frac{m\left[-2v_0/3\hat{i} - \left(-v_0\hat{i}\right)\right]}{\Delta t} = \frac{mv_0}{3\Delta t}\hat{i}$$

(d) Although energy is not conserved during the inelastic collision, energy *is* conserved as the person swings through the 180° rotation and rises a height $2r$, converting rotational kinetic energy to gravitational potential energy. Setting potential energy to zero at the bottom of the cylinder,

$$\frac{1}{2}\left(\frac{3}{2}mR^2\right)\left(\frac{2v_0}{3R}\right)^2 = \frac{1}{2}\left(\frac{3}{2}mR^2\right)\omega_f^2 + 2mgR$$

$$\omega_f = \sqrt{\frac{4v_0^2}{9R^2} - \frac{8g}{3R}}$$

(e) The net force required for circular motion is equal to the centripetal force. The downward force F exerted on the person is given by

$$F_{\text{net},y} = F + mg = F_{\text{cent}} = m\omega^2 R$$

$$F = m\omega^2 R - mg = mR\left(\frac{4v_0^2}{9R^2} - \frac{8g}{3R}\right) - mg$$

$$= -\frac{11}{3}mg + \frac{4mv_0^2}{9R}$$

Note: If the person's initial velocity is below a certain value, the answer above will be negative. This makes sense: As the person's initial speed decreases, her speed at the top decreases (according to the law of conservation of energy). Eventually, as the person's speed at the top decreases below $v^2/r < g$, the wheel actually has to exert an upward rather than a downward force on the person (rather than tending to fly off the wheel, she will tend to fall downward under the force of gravity).

MECHANICS II

(a) The total potential energy is the sum of the elastic and gravitational potential energies.

$$U(x) = U_{\text{elastic}}(x) + U_{\text{gravitational}}(x)$$
$$= \frac{1}{2}kx^2 - mgx\sin\theta$$

(b) $U(x)$, the quadratic function above, is graphed in the figure. The graph is a parabola with zeroes at $x = 0$ and $x = 2\,mg\sin\theta/k$.

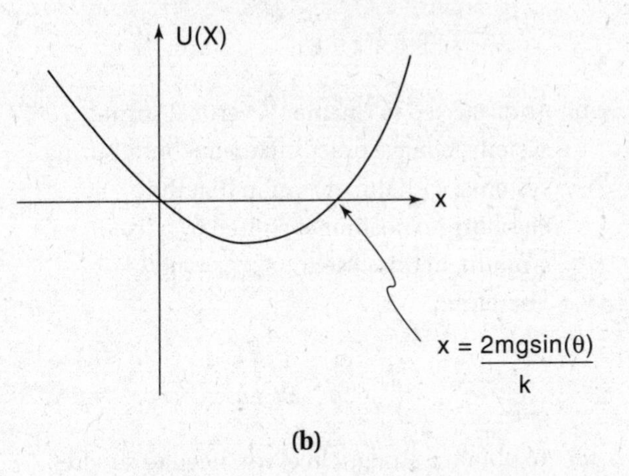

(b)

(c) The equilibrium point is located at the bottom of the $U(x)$ curve which, because of the symmetry of parabolas, lies exactly between the two zeroes, that is, at $x = mg\sin\theta/k$. (Imagine a ball rolling back and forth on top of the potential energy curve; its equilibrium position is at the bottom of the potential energy valley.)

(d) Based on Newton's second law, the system is in static equilibrium when

$$F_{x,\text{net}} = 0 = -kx_{\text{equilibrium}} + mg\sin\theta$$

Solving for $x_{\text{equilibrium}}$ yields the same answer as in part (c).

(e) $-dF/dx = d^2U/dx^2 = k = k_{\text{effective}}$. Therefore, as in vertical spring systems, the constant gravitational force does not affect the restoring force or the frequency (only the equilibrium point). Thus,

$$\omega = \sqrt{\frac{k_{\text{effective}}}{m}} = \sqrt{k/m}$$

MECHANICS III

(a) At equilibrium, the net force must be zero. (The vertical direction is chosen to be x, with $x = 0$ corresponding to the unstretched spring.)

$$F_{\text{net},x} = kx_{\text{equilibrium}} - mg = 0$$

$$\Rightarrow x_{\text{equilibrium}} = \frac{mg}{k}$$

(b) As discussed in Chapter 9, vertical spring systems behave exactly like horizontal spring systems, with the exception that the equilibrium position is shifted by a fixed amount, in this case: $x_{\text{equilibrium}} = mg/k$. Therefore,

$$f = \frac{\omega}{2\pi} = \frac{1}{2\pi}\sqrt{\frac{k}{m}}$$

(c) To obtain a straight line, you need to square the answer to part (b):

$$f^2 = \left(\frac{1}{2\pi}\right)^2 \frac{k}{m}$$

Then, you can choose either f^2 on the vertical axis and $1/m$ on the horizontal axis or vice versa.

(d) See the figure below, where f^2 is plotted on the vertical axis and $1/m$ is plotted on the horizontal axis. The axes are scaled so all of the data can be clearly plotted.

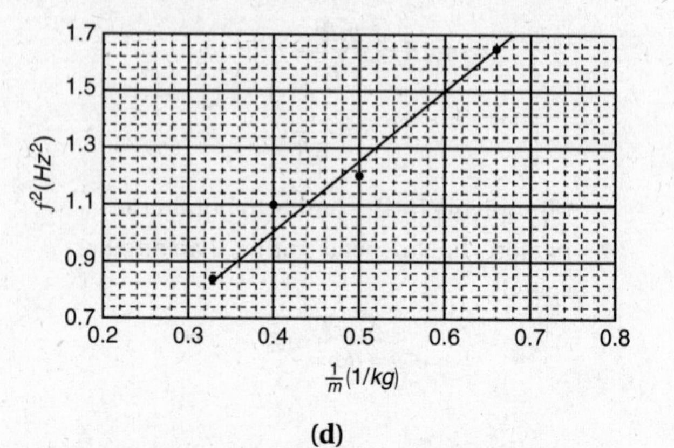

(d)

(e) By choosing the very top and very bottom points, you find:

$$\text{slope} = \frac{(1.66) - (0.83)}{(1/1.5) - (1/3.0)} = 2.49 \text{ kg/s}^2$$

whereas part (b) predicts a slope of $\frac{k}{4\pi^2} = 2.53$, a difference of about 2%.

ELECTRICITY AND MAGNETISM

ANSWER KEY

1. **E**	8. **A**	15. **B**	22. **D**	29. **C**
2. **D**	9. **B**	16. **D**	23. **C**	30. **B**
3. **D**	10. **B**	17. **E**	24. **E**	31. **E**
4. **A**	11. **E**	18. **C**	25. **D**	32. **A**
5. **E**	12. **B**	19. **B**	26. **E**	33. **C**
6. **A**	13. **D**	20. **D**	27. **A**	34. **B**
7. **C**	14. **B**	21. **B**	28. **D**	35. **D**

ANSWERS EXPLAINED

Multiple-Choice

1. **(E)** If S_1 is closed, a parallel circuit with an equivalent resistance of $R/2$ will be created. Thus the current drawn from the battery will double and half that current will go through the top bulb, yielding the same current that flowed before the switch S_1 was closed. If S_2 is closed, creating a path of zero resistance, no current will pass through either bulb.

2. **(D)** A Kirchhoff's loop equation for the right loop [clockwise: $8\text{ V} - (1\text{ A})(4\ \Omega) - (I)(2\ \Omega) = 0$] reveals that the current i moving from point A to point B must be 2.0 A (upward). A node equation at point A then shows the current through the left 2-Ω resistor must be 1.0 A flowing to the right. Finally, a Kirchhoff's loop equation for the left loop [clockwise: $V - (1\text{ A})(2\ \Omega) - (2\text{ A})(2\ \Omega) = 0$] allows us to calculate the voltage of the battery as 6 V.

3. **(D)** Capacitance is defined by the relationship $C = Q/V$. The battery's voltage determines the maximum potential difference across the capacitor plates.

4. **(A)** At the center of the sphere, as a result of symmetry, the electric field is zero. At every other point in space the electric field is nonzero.

5. **(E)** The electric field at all points in space is directed toward the center of the sphere because the sphere is negatively charged. Electric fields always point in the direction of decreasing potential. With the conventional choice of zero potential at infinity, the potential will be negative everywhere else and thus smallest in magnitude at infinity.

6. **(A)** To oppose the decrease in flux going into the page, the current flows clockwise. What magnetic force does this produce on the loop? The magnetic forces on the top and bottom of the loop, which point up and down, respectively, cancel each other out. The force on the left side of the loop points to the left, and there is zero force on the right side of the loop (because the magnetic field is zero in this region), so the net magnetic force points to the left. Because the applied force must cancel out this magnetic force in order for the loop to move at constant velocity, it then points to the right.

7. **(C)** Electric field lines begin and end at locations of positive and negative charges, respectively (not the other way around).

8. **(A)** Because the capacitor and the inductor are in series, they share a common current. Therefore, if the current through the capacitor is changing with a rate of dI/dt, the current through the inductor is changing at exactly the same rate. The voltage across the inductor is then $V = L(dI/dt)$, which is equal to the magnitude of the voltage across the capacitor (by Kirchhoff's loop rule). The charge on the capacitor is then given by the equation

$$Q = CV = C\left(L\frac{dI}{dt}\right)$$

9. **(B)** Recall the formula for the resistance of a uniform cylindrical resistor: $R = \rho l/A$, where l is the length of the resistor and A is its cross-sectional area. The second resistor has half the length of the first resistor, which alone would cause its resistance to decrease by a factor of 2 compared to R. The second resistor also has one-fourth the area of the first resistor (because $A = \pi r^2$, increasing r by a factor of 2 causes A to increase by a factor of 4), which alone would cause its resistance to increase by a factor of 4 compared to R. The combined results of these changes are an increase by a factor of 2.

10. **(B)** The potential difference across the fully charged capacitor is $V = Q/C$. The capacitor acts like a discharging battery whose maximum voltage is Q/C. Hence, the maximum current through the resistor is choice B, using Ohm's law. Note that using dimensional analysis also yields the correct answer.

11. **(E)** Inside the shell, the field lines point away from the point charge. Because field lines begin only at positive charges, this reveals that the point charge is positive, eliminating choices (A), (B), and (C).

    Outside the spherical shell, the field lines point toward the shell. Applying Gauss's law to any spherical surface outside the shell reveals that the net charge of the point charge and the shell must be negative. This can occur only if the net charge on the cylindrical shell is negative and greater in magnitude than the positive charge of the point charge, indicating that the correct answer is (E).

    Another way to rule out choice (D) is as follows. For the electric field within the metal to be zero, a net charge of $-Q_p$ must be distributed on the inside of the spherical shell. Conservation of charge then requires the charge on the outside of the spherical shell to be $Q_s + Q_p$. Because electric field lines terminate only at negative charges, the charge on the outside of the spherical shell must be negative. Because $Q_p > 0$ and $(Q_s + Q_p) < 0$, this requires Q_s to be negative and greater in magnitude than Q_p.

12. **(B)** Recall that wires carrying currents in opposite directions repel each other. (If you forget this, you can quickly figure it out by applying the right-hand rule twice to determine the direction of the magnetic field that one wire produces and the force exerted as the second wire carries current through this field.) Because the repulsive force between the top wire and the wire carrying current $2I$ has *twice* the magnitude of the repulsive force between the top wire and the wire carrying current I, the net force points upward and to the right as shown.

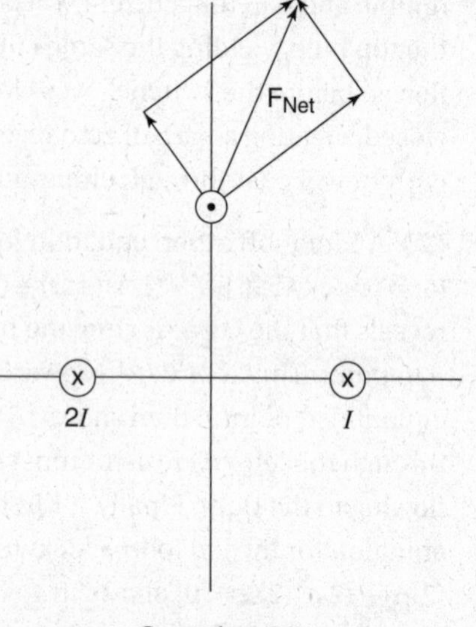

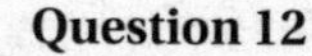

Question 12

13. **(D)** Recall that the force exerted by one wire on another parallel wire has a magnitude given by the formula force/length = $\mu_0 I_1 I_2 / 2\pi r$. Doubling I will double all the currents, so it will increase each of the forces by a factor of 4. Because each of the pairwise forces increases by a factor of 4, the net force (which is the sum of these forces) will also increase by a factor of 4.

14. **(B)** Choice (E) is incorrect as revealed by dimensional analysis. Work has the same units as energy, which has units of qV. However, comparison with the formula

$V = Q/4\pi\varepsilon_0 r$ indicates that choice (E) has the units of voltage, not of qV.

Because both the electrostatic force and the displacement point in the $+x$-direction, the dot product and the integral of work ($W = \int \mathbf{F} \cdot d\mathbf{r}$) must be positive, ruling out choices (C) and (D).

To actually solve the problem, note that the change in electric potential energy is

$$\Delta U = q\Delta V = q(V_f - V_0)$$
$$= q\left(\frac{Q}{4\pi\varepsilon_0(2d)} - \frac{Q}{4\pi\varepsilon_0 d}\right) = -\frac{qQ}{8\pi\varepsilon_0 d}$$

Because of conservation of energy, it is equal to the net external work (because the kinetic energy is constant), so that work = $qQ/8\pi\varepsilon_0 d$.

15. **(B)** A discharging capacitor certainly loses charge as it discharges. The decrease in charge leads to a decrease in the electric field between the plates; thus, the energy stored decreases as well.

16. **(D)** The most useful form of the formula for power here is $P = V^2/R$. Rearranging this yields $R = V^2/P = (20\text{ V})^2/80\text{ W} = 5\ \Omega$.

17. **(E)** One direct way to solve this problem is to start off by calculating the equivalent capacitance

$$C_{eq} = \frac{1}{1/C_1 + 1/C_2} = \frac{C_1C_2}{C_1 + C_2}$$

Then calculate the charge that flows through the circuit

$$Q = C_{eq}V = \frac{C_1C_2V}{C_1 + C_2}$$

and plug this charge into the equation for the energy stored in a capacitor:

$$U = \frac{Q^2}{2C_1} = \frac{1}{2C_1}\left(\frac{C_1C_2V}{C_1 + C_2}\right)^2 = \frac{C_1C_2^2V^2}{2(C_1 + C_2)^2}$$

Note: When compared with the equation $U = \frac{1}{2}CV^2$, choices (A) and (B) do not have the correct units and can be immediately eliminated.

18. **(C)** Gauss's law applied to a gaussian surface just outside the sphere indicates that $\Phi_E = EA = Q_{enc}/\varepsilon_0$. Doubling the radius increases the area of the shell, $A = 4\pi R^2$ (and thus the charge of the shell), by a factor of 4, but it also increases the area of the gaussian sphere by a factor of 4. These effects cancel, and the electric field remains the same.

Note: Very close to the surface of the sphere, the field is mainly due to the nearby charge on the surface of the sphere. Also, when very close to the sphere, the curvature can be ignored (just as we can ignore the curvature of Earth), so that the sphere can be approximated as a plane and the electric field approximated as $E = \sigma/2\varepsilon_0$. Therefore, it is not surprising that changing the radius doesn't affect the field near the sphere if the charge density is held constant.

19. **(B)** Carrying negative charge in the $+x$-direction causes the same net charge transport as carrying positive charge in the $-x$-direction. Therefore, the direction of the conventional current produced is in the $-x$-direction, which immediately eliminates choices (A), (C), and (E).

What is the magnitude of the current? The magnitude of current is defined as the amount of charge that passes a certain point per time. In a time increment dt, a length of rod $dx = vdt$ passes a certain point that carries a charge $dQ = \lambda dx = \lambda v\,dt$. The current is then $I = dQ/dt = \lambda v\,dt/dt = \lambda v$.

20. **(D)** The electric field lines point perpendicularly toward the line of negative charge, with greater magnitudes closer to the line of charge.

21. **(B)** Recall that the potential due to point charges is $V = \Sigma(Q_i/4\pi\varepsilon_0 r)$. For the net potential at a point to be zero, the magnitude of the positive potential due to $+Q$ must equal the magnitude of the negative potential due to $-Q$. This requires that the point be equidistant from the two point charges,

which occurs only at one place: exactly between the two charges in region II.

Note: In two-dimensional or three-dimensional space, the region where the potential is zero is a line or a plane, respectively.

22. **(D)** The magnitude of the voltage across an inductor is given by the equation $V = L(dI/dt)$. Because the current is a sine curve, the magnitude of the voltage must be proportional to the absolute value of a cosine curve with the same period.

23. **(C)** If all the energy dissipated in the lightbulb were emitted as light energy, the brightness of the lightbulb would directly reflect the amount of energy dissipated in the lightbulb per time, which is the power of the lightbulb.

24. **(E)** The key here is to remember that the value of potential at any point is meaningless—only differences in potential have significance. Therefore, sketching the capacitor with the left plate having a capacitor of +4 V and the right plate having a potential of 0 V would be physically equivalent to what is shown in the figure. Of all the battery arrays, the only one that produces a net potential difference of 4 V is choice (E). Choices (A) and (C) produce a net potential difference of +2 V with the positive terminal on the right, whereas choices (B) and (D) produce a net potential difference of +2 V with the positive terminal on the left. (To find the equivalent potential difference of any battery array, simply move from one end of the battery array to the other, summing potential jumps along the way.) Note that the equipotential lines are perpendicular to the electric field.

25. **(D)** Gauss's law tells us that the electric field is proportional to the enclosed charge. Drawing a gaussian sphere inside the sphere of charge shows us that the field will increase with increasing radial distance.

26. **(E)** The magnetic force is always perpendicular to the particle's displacement, so the magnetic force never can do any work on any particle. Therefore, choices (A) and (B) are incorrect. Because the magnetic field performs no work on the particle, the electric field must perform work on the particle. However, the fact that the electric field performs net work on the particle as it moves in a closed path indicates that the field is not conservative, ruling out choice (C). Because the electric fields due to static charges are always conservative, this eliminates choice (D). The solution therefore is (E).

27. **(A)** The force on $+Q$ is equal in magnitude and opposite in direction to the force on $-Q$, so the net force is zero. However, both forces produce a clockwise torque about the center of the dipole, so the net torque is clockwise.

28. **(D)** The EMF is proportional to the area of the loop, so it is proportional to r^2 (mathematically, $V = c_1 r^2$). The resistance is proportional to the length of the loop, so it is proportional to r (mathematically, $R = c_2 r$). Because we know the resistance and the voltage, the most useful formula for power to use here is

$$P = \frac{V^2}{R} = \frac{(c_1 r^2)^2}{c_2 r} = \frac{c_1^2}{c_2} r^3$$

29. **(C)** Resistivity is given by $\rho = \frac{AR}{L} = \frac{\pi r^2 V}{IL}$ for a wire of radius r. We have also used Ohm's law to write the resistance R in terms of the voltage and current.

30. **(B)** Point B is at higher potential. The potential difference between points A and B is given by $V_A - V_B = L\frac{dI}{dt}$ and $\frac{dI}{dt} < 0$ here.

31. **(E)** Choice (C): When a dielectric is added to a capacitor, the capacitance *increases* by a factor of k_D. Choice (B): Because the capacitor is isolated, there is nowhere for the charge to go and so it remains constant. Choice (D): Based on the equation $Q = CV$,

the charge remains constant and the capacitance increases, so the voltage decreases. Choice (A): The potential energy can be expressed as $Q^2/2C$. Because the charge remains constant while the capacitance increases, the potential energy decreases. (*Note*: As the dielectric is added, charges of opposite polarity are induced on the two surfaces of the dielectric adjacent to the capacitor plates. These induced charges attract the charge on the capacitor, drawing the dielectric toward it. As the dielectric is pulled into the capacitor, it can perform positive work on another object as it decreases the capacitor's potential energy.)

Therefore, by elimination the answer is (E).

32. **(A)** Consider the schematic in the figure, which shows a battery being charged. In order for the battery to be charged, the current must flow in the direction indicated. This requires the potential difference from A to B to be $V_0 + IR$ (with B at higher potential). Thus, to make current flow in the correct direction, a potential *greater* than V_0 is required.

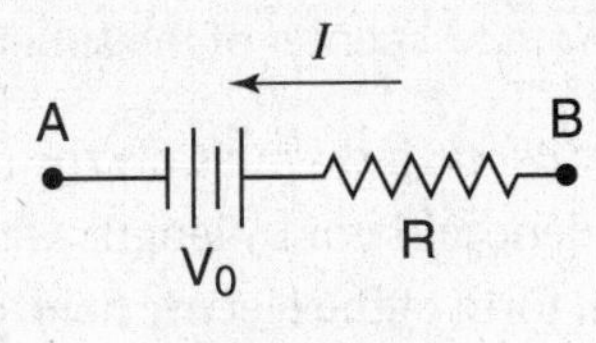

Question 32

33. **(C)** Because an electron has a negative charge, the electric force points antiparallel to the electric field, in the $-y$-direction. Similarly, because the particle is negative, the magnetic force points antiparallel to the direction indicated by the right-hand rule, in the $+z$-direction.

34. **(B)** Very close to the line of charge, the variation in the charge density with position is not apparent because the field is dominated by the charge located very close to the field point. The distance from the line of charge to the field point is small compared to the rate at which the charge density is changing. This effectively constant charge density produces a field pointing away from the line of charge (as does any uniform line of positive charge). This is analogous to the way the electric field is perpendicular to a surface charge distribution very close to an infinite surface even if the surface charge density is not constant.

35. **(D)** As the loop enters the field, the flux into the page (negative flux) steadily increases in magnitude, remains constant while the loop is completely within the magnetic field, and then steadily approaches zero as the loop exits the field. The discontinuities and plateaus featured in choices (A), (B), and (E) are similar to what the current or the EMF would look like as a function of position. Choice (C) is the same as choice (D) except that the sign is incorrect. (Flux out of the page is designated as positive, so the loop experiences negative flux.)

Free-Response

ELECTRICITY AND MAGNETISM I

(a) Using the relationship between voltage and electric field for a parallel plate capacitor, $V = Ed$, the minimum distance is

$$d = \frac{V}{E_{BD}}$$

(b) By the same equation, $V_{BD} = E_{BD}d$.

(c) See the figure. The capacitor-charging stage is exactly the same as in an RC circuit, except that when the breakdown voltage is reached, the capacitor rapidly discharges (as if it were shorted through a small resistance) and then repeats the charging step.

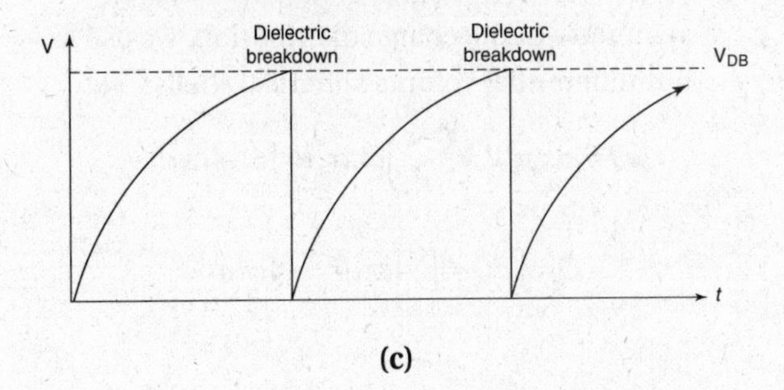

(c)

(d) Recall the equation for the charge on a charging capacitor:

$$Q = Q_f(1 - e^{-t/RC})$$

where Q_f is the charge on the capacitor when it is fully charged.

Using the relationship $Q = CV$, we can convert this to an equation expressing voltage as a function of time:

$$V_{(t)} = V_{\text{battery}}(1 - e^{-t/RC})$$

The time T between discharges is given by the duration of a single charging step (if we assume the discharge is essentially instantaneous), which is the amount of time it takes for the charging capacitor to reach the breakdown voltage:

$$V_{\text{BD}} = V_{\text{battery}}(1 - e^{-T/RC})$$

Solving for T yields

$$T = RC\ln\left(\frac{V_{\text{battery}}}{V_{\text{battery}} - V_{\text{BD}}}\right)$$

(e) The energy stored in the capacitor just before dielectric breakdown is given by the equation $U = \frac{1}{2}CV_{\text{BD}}^2 = \frac{1}{2}CE_{\text{BD}}^2d^2$.

(f) Because the battery is disconnected, the charge density on the plates is fixed (and thus, the charge density remains fixed). Since the electric field inside the capacitor depends only on the charge density of the plates, it is not affected and no dielectric breakdown occurs.

ELECTRICITY AND MAGNETISM II

(a) Because the charge density is not uniform, we must integrate. Mirroring the spherical symmetry of the charge distribution, we use as our differential volume spherical shells:

$$dQ = \rho(r)dV = \left(\frac{-c}{r^2}\right)(4\pi r^2 dr) = -4c\pi dr$$

$$Q = \int dQ = \int_0^a -4c\pi dr = -4c\pi a$$

(b) (i) We apply Gauss's law, noting that the enclosed charge is negative, which causes the electric field to point inward (antiparallel to the directed area vector) and yields a negative flux (note that E signifies the magnitude of the electric field)

$$\left[\oint \mathbf{E}\cdot d\mathbf{A} = -4\pi r^2 E\right] = \left[\frac{Q_{\text{enclosed}}}{\varepsilon_0} = \frac{-4c\pi a}{\varepsilon_0}\right]$$

$$E = \frac{ca}{\varepsilon_0 r^2} \quad \left(\begin{array}{l}\text{direction is toward the}\\ \text{center of the sphere}\end{array}\right)$$

Units check: The constant c has units of C/m, so the units of the numerator in the field expression are coulombs. We then recognize the usual dimensions of electric field: charge/ε_0(distance^2).

(b) (ii) At $r < a$, we have to integrate to obtain the enclosed charge:

$$Q = \int dQ = \int_0^r -4c\pi dr = -4c\pi r$$

Again applying Gauss's law,

$$\left[\oint \mathbf{E}\cdot d\mathbf{A} = -4\pi r^2 E\right] = \left[\frac{Q_{\text{enclosed}}}{\varepsilon_0} = \frac{-4c\pi r}{\varepsilon_0}\right]$$

$$E = \frac{c}{\varepsilon_0 r} \quad \left(\begin{array}{l}\text{direction is toward the}\\ \text{center of the sphere}\end{array}\right)$$

Units check: If we multiply the numerator and denominator by length, we recover the usual form of the electric field, charge/ε_0(distance^2), as in (b)(i).

(b) (iii) Both expressions for $r > a$ and $r < a$ agree that the magnitude of the field at $r = a$ is

$$E = \frac{c}{\varepsilon_0 a} \quad \left(\begin{array}{l}\text{direction is toward the}\\ \text{center of the sphere}\end{array}\right)$$

Recall that because there are no infinite charge densities (no point charges or surface charge densities), the electric field must be continuous, so the expressions for field in different regions must match up at the region's border.

(c) Taking the line integral of $V = -\int \mathbf{E} \cdot d\mathbf{r}$ along the x-axis from $x = \infty$ to $x = a$ yields

$$V(a) = V(a) - V(\infty)$$
$$= -\int_{\infty}^{a} E_x\, dx = -\int_{\infty}^{a} \left(\frac{-ca}{\varepsilon_0 x^2} \right) dx$$

(because the electric field points to the left, $E_x = -ca/\varepsilon_0 x^2$).

Evaluating the integral,

$$V(a) = -\frac{ca}{\varepsilon_0 x}\bigg|_{\infty}^{a} = -\frac{c}{\varepsilon_0}$$

Sign check: Based on the equation $V = \int dQ/4\pi\varepsilon_0 r$, because the charge is negative, the potential should be negative.

(d) Because the electric field outside the sphere is identical to that produced by a point charge located at the center of the sphere, the potential is also equal to that produced by a point charge:

$$V = \frac{Q}{4\pi\varepsilon_0 a}$$

Using the expression for Q calculated in part (*a*),

$$V = \frac{Q}{4\pi\varepsilon_0 a} = \frac{-4c\pi a}{4\pi\varepsilon_0 a} = -\frac{c}{\varepsilon_0}$$

ELECTRICITY AND MAGNETISM III

(a) Electric potential energy in the capacitor is converted to kinetic energy:

$$eV = \frac{1}{2}mv^2$$
$$v = \sqrt{\frac{2eV}{m}}$$

(b) B_1 and B_3 point into the page; B_2 points out of the page (remember that the charge is negative!).

(c) The magnetic force provides the centripetal force. (Here the $\sin\theta$ term in the cross-product is 1.)

$$\frac{mv^2}{r_1} = evB_1$$

$$r_1 = \frac{mv}{eB_1} = \frac{m}{eB_1}\sqrt{\frac{2eV}{m}} = \frac{1}{B_1}\sqrt{\frac{2mV}{e}}$$

(d) Given that the particle manages to make it through the first region, its mass must satisfy the equation above. As a particle traverses the semicircles, its speed does not change. (The magnetic field does no work because it is perpendicular to the displacement.) Therefore, equations with the same form can be written for the other two semicircles, with the results

$$r_2 = \frac{1}{B_2}\sqrt{\frac{2mV}{e}}$$
$$r_3 = \frac{1}{B_3}\sqrt{\frac{2mV}{e}}$$

Because the speed of the particle is constant throughout its trajectory, we can combine these equations:

$$\sqrt{\frac{2mV}{e}} = r_1 B_1 = r_2 B_2 = r_3 B_3$$

Therefore, $r_2 = r_1(B_1/B_2)$ and $r_3 = r_1(B_1/B_3)$.

(e) From part (c), you find that $\frac{e}{m} = \frac{2V}{r^2 B_1}$. Thus, to obtain a linear relationship, you should plot V versus r^2 or vice versa.

(f) Plot V versus r^2 (see the figure below). Fitting to a straight line gives the following slope:

$$\text{slope} = \frac{(3.5-2.0)\,\text{kV}}{(3.98-2.27)\,\text{cm}^2} = 8.77\times10^6\,\frac{\text{V}}{\text{m}^2}$$

You expect the slope of the line to be given by $\frac{eB_1}{2m}$. With $B_1 = 0.01\,T$, you find $\frac{e}{m} = 1.75\times10^{11}\,\frac{\text{C}}{\text{kg}}$. The accepted value is $1.76\times10^{11}\,\frac{\text{C}}{\text{kg}}$.

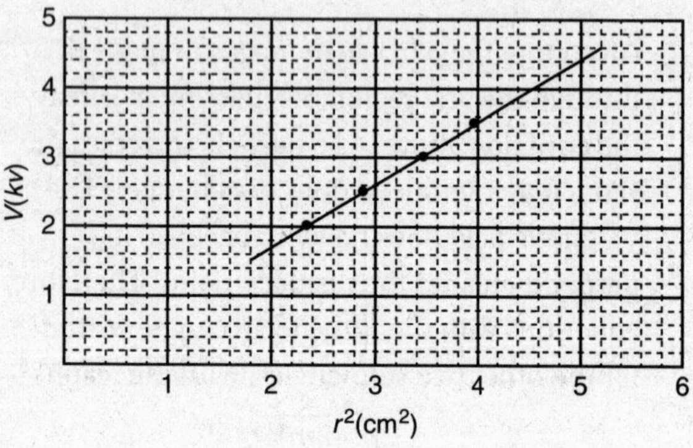

(f)

ANSWER SHEET
Practice Test 2

Mechanics

1.	Ⓐ Ⓑ Ⓒ Ⓓ Ⓔ	10.	Ⓐ Ⓑ Ⓒ Ⓓ Ⓔ	19.	Ⓐ Ⓑ Ⓒ Ⓓ Ⓔ	28.	Ⓐ Ⓑ Ⓒ Ⓓ Ⓔ
2.	Ⓐ Ⓑ Ⓒ Ⓓ Ⓔ	11.	Ⓐ Ⓑ Ⓒ Ⓓ Ⓔ	20.	Ⓐ Ⓑ Ⓒ Ⓓ Ⓔ	29.	Ⓐ Ⓑ Ⓒ Ⓓ Ⓔ
3.	Ⓐ Ⓑ Ⓒ Ⓓ Ⓔ	12.	Ⓐ Ⓑ Ⓒ Ⓓ Ⓔ	21.	Ⓐ Ⓑ Ⓒ Ⓓ Ⓔ	30.	Ⓐ Ⓑ Ⓒ Ⓓ Ⓔ
4.	Ⓐ Ⓑ Ⓒ Ⓓ Ⓔ	13.	Ⓐ Ⓑ Ⓒ Ⓓ Ⓔ	22.	Ⓐ Ⓑ Ⓒ Ⓓ Ⓔ	31.	Ⓐ Ⓑ Ⓒ Ⓓ Ⓔ
5.	Ⓐ Ⓑ Ⓒ Ⓓ Ⓔ	14.	Ⓐ Ⓑ Ⓒ Ⓓ Ⓔ	23.	Ⓐ Ⓑ Ⓒ Ⓓ Ⓔ	32.	Ⓐ Ⓑ Ⓒ Ⓓ Ⓔ
6.	Ⓐ Ⓑ Ⓒ Ⓓ Ⓔ	15.	Ⓐ Ⓑ Ⓒ Ⓓ Ⓔ	24.	Ⓐ Ⓑ Ⓒ Ⓓ Ⓔ	33.	Ⓐ Ⓑ Ⓒ Ⓓ Ⓔ
7.	Ⓐ Ⓑ Ⓒ Ⓓ Ⓔ	16.	Ⓐ Ⓑ Ⓒ Ⓓ Ⓔ	25.	Ⓐ Ⓑ Ⓒ Ⓓ Ⓔ	34.	Ⓐ Ⓑ Ⓒ Ⓓ Ⓔ
8.	Ⓐ Ⓑ Ⓒ Ⓓ Ⓔ	17.	Ⓐ Ⓑ Ⓒ Ⓓ Ⓔ	26.	Ⓐ Ⓑ Ⓒ Ⓓ Ⓔ	35.	Ⓐ Ⓑ Ⓒ Ⓓ Ⓔ
9.	Ⓐ Ⓑ Ⓒ Ⓓ Ⓔ	18.	Ⓐ Ⓑ Ⓒ Ⓓ Ⓔ	27.	Ⓐ Ⓑ Ⓒ Ⓓ Ⓔ		

Electricity and Magnetism

1.	Ⓐ Ⓑ Ⓒ Ⓓ Ⓔ	10.	Ⓐ Ⓑ Ⓒ Ⓓ Ⓔ	19.	Ⓐ Ⓑ Ⓒ Ⓓ Ⓔ	28.	Ⓐ Ⓑ Ⓒ Ⓓ Ⓔ
2.	Ⓐ Ⓑ Ⓒ Ⓓ Ⓔ	11.	Ⓐ Ⓑ Ⓒ Ⓓ Ⓔ	20.	Ⓐ Ⓑ Ⓒ Ⓓ Ⓔ	29.	Ⓐ Ⓑ Ⓒ Ⓓ Ⓔ
3.	Ⓐ Ⓑ Ⓒ Ⓓ Ⓔ	12.	Ⓐ Ⓑ Ⓒ Ⓓ Ⓔ	21.	Ⓐ Ⓑ Ⓒ Ⓓ Ⓔ	30.	Ⓐ Ⓑ Ⓒ Ⓓ Ⓔ
4.	Ⓐ Ⓑ Ⓒ Ⓓ Ⓔ	13.	Ⓐ Ⓑ Ⓒ Ⓓ Ⓔ	22.	Ⓐ Ⓑ Ⓒ Ⓓ Ⓔ	31.	Ⓐ Ⓑ Ⓒ Ⓓ Ⓔ
5.	Ⓐ Ⓑ Ⓒ Ⓓ Ⓔ	14.	Ⓐ Ⓑ Ⓒ Ⓓ Ⓔ	23.	Ⓐ Ⓑ Ⓒ Ⓓ Ⓔ	32.	Ⓐ Ⓑ Ⓒ Ⓓ Ⓔ
6.	Ⓐ Ⓑ Ⓒ Ⓓ Ⓔ	15.	Ⓐ Ⓑ Ⓒ Ⓓ Ⓔ	24.	Ⓐ Ⓑ Ⓒ Ⓓ Ⓔ	33.	Ⓐ Ⓑ Ⓒ Ⓓ Ⓔ
7.	Ⓐ Ⓑ Ⓒ Ⓓ Ⓔ	16.	Ⓐ Ⓑ Ⓒ Ⓓ Ⓔ	25.	Ⓐ Ⓑ Ⓒ Ⓓ Ⓔ	34.	Ⓐ Ⓑ Ⓒ Ⓓ Ⓔ
8.	Ⓐ Ⓑ Ⓒ Ⓓ Ⓔ	17.	Ⓐ Ⓑ Ⓒ Ⓓ Ⓔ	26.	Ⓐ Ⓑ Ⓒ Ⓓ Ⓔ	35.	Ⓐ Ⓑ Ⓒ Ⓓ Ⓔ
9.	Ⓐ Ⓑ Ⓒ Ⓓ Ⓔ	18.	Ⓐ Ⓑ Ⓒ Ⓓ Ⓔ	27.	Ⓐ Ⓑ Ⓒ Ⓓ Ⓔ		

Practice Test 2

MULTIPLE-CHOICE QUESTIONS

Directions: Each multiple-choice question is followed by five answer choices. For each question, choose the best answer and fill in the corresponding circle on the answer sheet. You may refer to the formula sheet in the Appendix (pages 641–644).

Mechanics

1. A variable force exerted during time interval Δt changes the velocity of mass m from $\mathbf{v}_0$ to $\mathbf{v}_f$. What is the average power exerted during this time interval?

   (A) $(\mathbf{v}_f - \mathbf{v}_0)/\Delta t$
   (B) $m\mathbf{v}_f - m\mathbf{v}_0$
   (C) $m(v_f^2 - v_0^2)/2$
   (D) $m(v_f^2 - v_0^2)/2\Delta t$
   (E) $m(\mathbf{v}_f - \mathbf{v}_0)/\Delta t$

2. A particle moves with constant speed v counterclockwise around a circle of radius r in the xy-plane. Which of the following correctly describes the x-component of the velocity?

   (A) It is constant and equal to v.
   (B) It is constant but not equal to v.
   (C) It varies between zero and v.
   (D) It varies between zero and $v/\sqrt{2}$.
   (E) It varies between $-v$ and v.

3. A baseball is falling toward Earth. If $\mathbf{a}_b$ is the acceleration of the baseball due to the gravitational pull of Earth and $\mathbf{a}_e$ is the acceleration of Earth due to the baseball, which of the following is true?

   (A) $\mathbf{a}_e = 0$
   (B) $\mathbf{a}_b = \mathbf{a}_e$
   (C) $a_b >> a_e$, and the accelerations are parallel.
   (D) $a_e >> a_b$, and the accelerations are parallel.
   (E) $a_b >> a_e$, and the accelerations are antiparallel.

4. If the angular position of a disk is given by the equation $\theta(t) = 2 + 4t - 3t^2$, what is the angular velocity of the disk as a function of time?

   (A) $4 - 6t$
   (B) $6t - 4$
   (C) $2t + 2t^2 - 3t^3$
   (D) $3t^3 - 2t^2 - 2t$
   (E) $3t^3 - 2t^2 - 2t + C$

GO ON TO THE NEXT PAGE

5. When an object moving in the circular track as shown has a velocity in the $+y$-direction, its acceleration has components in the $+x$- and $+y$-directions. Which of the following statements is true?

(A) The object is moving clockwise with increasing speed.
(B) The object is moving counterclockwise with increasing speed.
(C) The object is moving clockwise with decreasing speed.
(D) The object is moving counterclockwise with decreasing speed.
(E) The object is moving clockwise with constant speed.

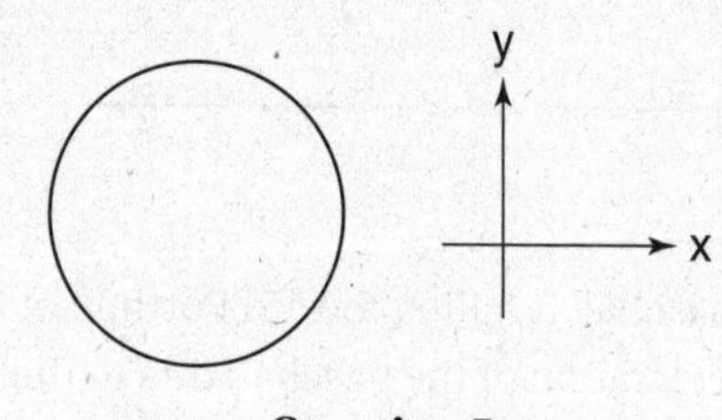

Question 5

6. A force F exerted on a block parallel to an inclined slope as shown causes the block's speed to increase from zero to v as it moves a distance d along the incline. If the coefficient of friction is μ_K, how much work does gravity do during this time interval?

(A) $\mu_K mg(\cos\theta)d$
(B) $\mu_K mg(\cos\theta)d + mg(\sin\theta)d$
(C) $\mu_K mg(\cos\theta)d + mg(\sin\theta)d + mv_f^2/2$
(D) $mgd\sin\theta$
(E) $-mgd\sin\theta$

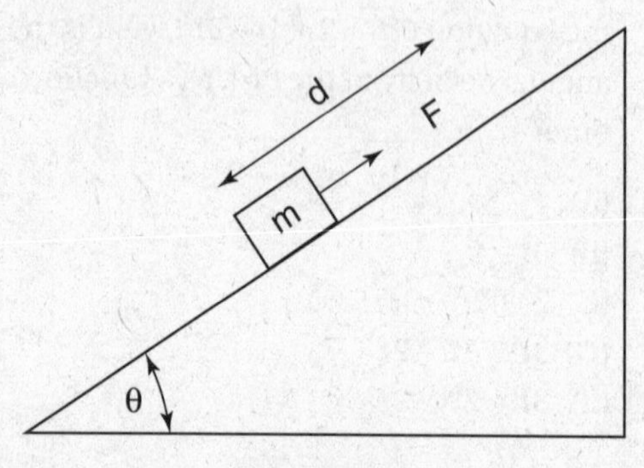

Question 6

7. A point mass is released from rest as shown. If the inclined planes are frictionless, which of the following statements is correct?

(A) The mass is undergoing simple harmonic motion with a period of $T = (2d/g\sin\theta)^{1/2}$.
(B) The mass is undergoing simple harmonic motion with a period of $T = 2(2d/g\sin\theta)^{1/2}$.
(C) The mass is undergoing simple harmonic motion with a period of $T = 4(2d/g\sin\theta)^{1/2}$.
(D) As the mass slides down the slope, the component of the momentum perpendicular to the slope is constant.
(E) The mass is in static equilibrium.

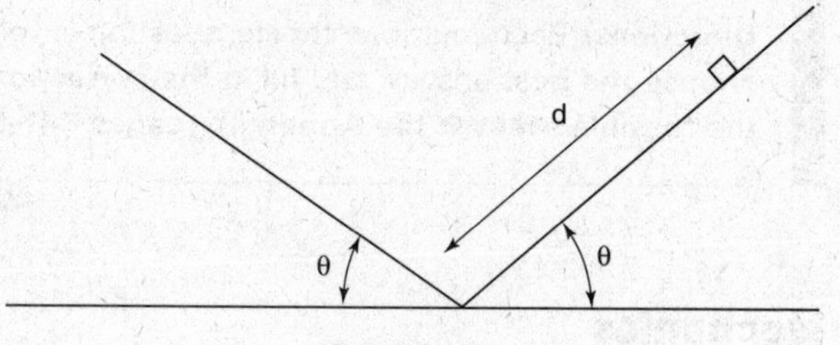

Question 7

8. An object of mass m has its position given by the equation $x(t) = at^4 - bt^3 + c$. What is the force on the object as a function of time?

(A) $F(t) = (4at^3 - 3bt^2)$
(B) $F(t) = (4at^3 - 3bt^2)m$
(C) $F(t) = (12at^2 - 6bt)$
(D) $F(t) = (12at^2 - 6bt)m$
(E) $F(t) = -(12at^2 - 6bt)m$

PRACTICE TEST 2

GO ON TO THE NEXT PAGE

9. A mass is suspended by two ropes as shown. What is the ratio of the tension forces in the two ropes?

(A) $T_1/T_2 = \sin\theta_1/\sin\theta_2$
(B) $T_1/T_2 = \sin\theta_2/\sin\theta_1$
(C) $T_1/T_2 = \cos\theta_1/\cos\theta_2$
(D) $T_1/T_2 = \cos\theta_2/\cos\theta_1$
(E) $T_1/T_2 = \tan\theta_1/\tan\theta_2$

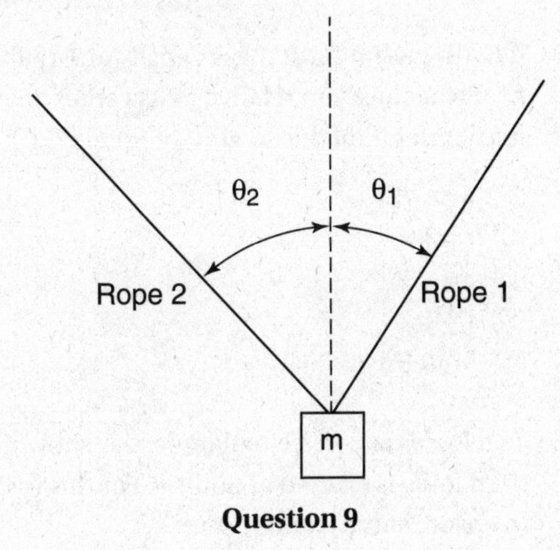

Question 9

10. Two projectiles are thrown at a vertical wall with initial velocities and positions as shown. Which of the following comparisons is correct regarding the height of the impact and the time the projectile spends in the air before it hits the wall?

	Height of impact	Time in the air
(A)	$h_1 > h_2$	$t_1 < t_2$
(B)	$h_1 = h_2$	$t_1 < t_2$
(C)	$h_1 > h_2$	$t_1 = t_2$
(D)	$h_1 < h_2$	$t_1 = t_2$

(E) The answer cannot be determined because it depends on whether the projectiles hit the wall before or after reaching the peaks of their trajectories.

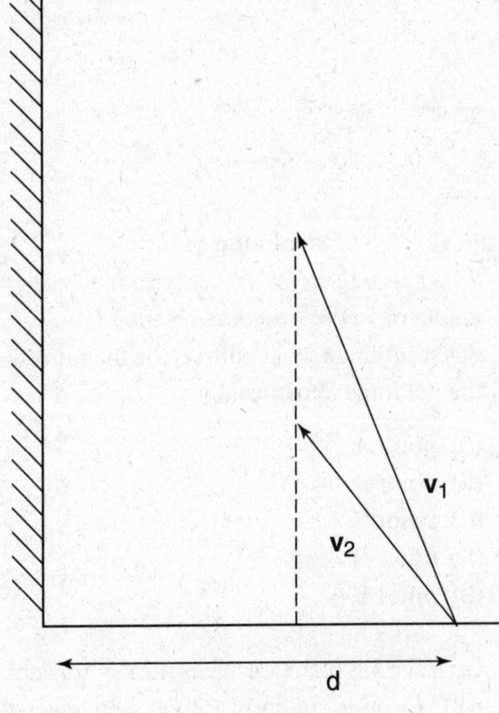

Question 10

PRACTICE TEST 2

GO ON TO THE NEXT PAGE

11. The figure shows the trajectory of a ball tossed off a cliff, subject to both gravity and air resistance. Which choice on the figure shows the direction of the net force on the ball at point C?

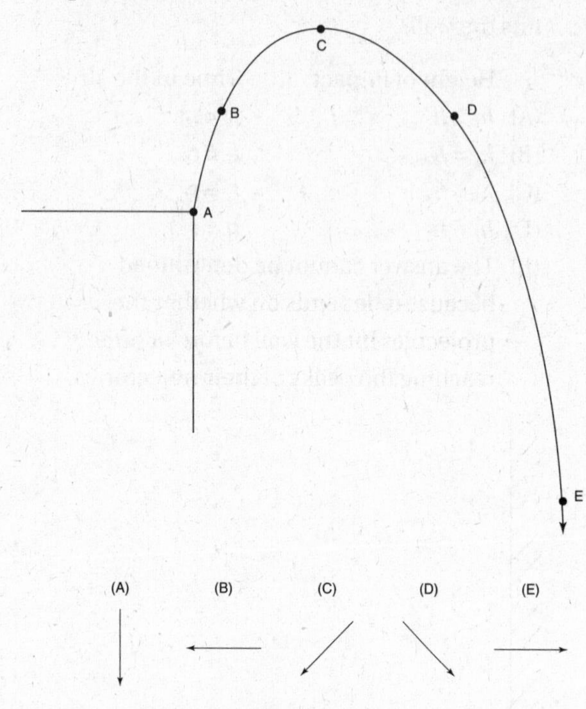

Question 11

12. Referring to the situation described in question 11, at what point is the magnitude of the net force maximized?

    (A) point A
    (B) point B
    (C) point C
    (D) point D
    (E) point E

13. On a certain planet, a ball is thrown upward with a kinetic energy of 100 J and an angle of elevation of 45°. If the ball has a kinetic energy of 75 J at a height of 10 m, what is the maximum height of its trajectory?

    (A) 10 m
    (B) 20 m
    (C) 40 m
    (D) 60 m
    (E) 80 m

14. A particle moves with speed v around a circle of radius r. What is the period of the particle's motion?

    (A) $T = 2\pi\sqrt{v/r}$
    (B) $T = 2\pi\sqrt{r/v}$
    (C) $T = 2\pi(v/r)$
    (D) $T = 2\pi(r/v)$
    (E) $T = r/2\pi v$

15. If Earth has uniform mass density and radius r_e, at what location relative to its center does the acceleration due to gravity equal $g/2$?

    (A) $r = r_e\sqrt{2}$
    (B) $r = 2r_e$
    (C) $r = r_e/2$
    (D) $r = 4r_e$
    (E) both (A) and (C)

16. Two forces act on a circular mass as shown. A third force is exerted at point P. For this force to ensure static equilibrium,

    (A) it must equal $F_2\hat{i} - F_1\hat{j}$
    (B) it must equal $-F_2\hat{i} + F_1\hat{j}$
    (C) it must equal $F_1\hat{i} - F_2\hat{j}$
    (D) it must equal $-F_2\hat{i} - F_1\hat{j}$
    (E) It is impossible for a force exerted at point P to cause static equilibrium.

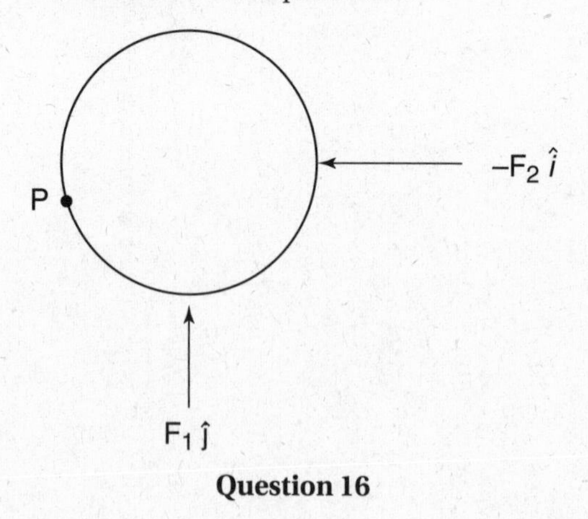

Question 16

GO ON TO THE NEXT PAGE

17. A railroad car is traveling to the right at a speed of 10 m/s relative to the ground. What is the velocity of an automobile relative to the railroad car if the velocity of the automobile relative to the ground is 2 m/s to the left?

(A) 8 m/s to the right
(B) 8 m/s to the left
(C) 12 m/s to the right
(D) 12 m/s to the left
(E) 2 m/s to the left

18. A block of mass 0.2 kg on a horizontal frictionless surface is attached to a spring of spring constant 200 N/m. The block is pulled 0.8 m from its equilibrium position and released from rest. It then executes simple harmonic motion. What is the maximum elastic potential energy of the system?

(A) 3 J
(B) 6.4 J
(C) 8,000 J
(D) 64 J
(E) 200,000 J

19. A ball attached to a string of length r swings in a vertical circle with the minimum energy needed to keep the ball in a circular path (so that the tension force is zero when the ball is at the top of the circle). What is the minimum speed of the ball?

(A) $v = \sqrt{gr}$
(B) $v = \sqrt{2gr}$
(C) $v = \sqrt{gr/2}$
(D) $v = 2\sqrt{gr}$
(E) $v = 4\sqrt{gr}$

20. Consider a spring with a restoring force given by $F(x) = -bx^3 + cx^5$. If the potential energy is chosen to be zero at the equilibrium position, which of the following functions gives the elastic potential energy as a function of the spring's displacement?

(A) $U(x) = 5cx^4 - 3bx^2$
(B) $U(x) = 3bx^2 - 5cx^4$
(C) $U(x) = (cx^6/6) - (bx^4/4)$
(D) $U(x) = (bx^4/4) - (cx^6/6)$
(E) none of the above

21. If the block shown moves with constant speed v, what form do the horizontal and vertical components of Newton's second law, respectively, take?

(A) $F\sin\theta = \mu_K mg$ for both components
(B) $F\cos\theta = \mu_K F_N$ and $F_N + F\sin\theta = mg$
(C) $F\sin\theta = \mu_K F_N$ and $F_N + F\cos\theta = mg$
(D) $F\cos\theta = \mu_K F_N$ and $F_N = mg + F\sin\theta$
(E) $F\sin\theta = \mu_K F_N$ and $F_N = mg + F\cos\theta$

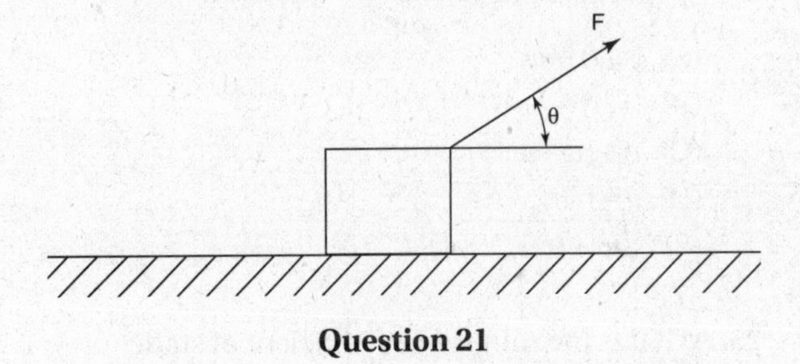

Question 21

22. What is the power exerted by the force F in question 21?

(A) $P = F(\sin\theta)v$
(B) $P = F(\cos\theta)v$
(C) $P = -F(\sin\theta)v$
(D) $P = -F(\cos\theta)v$
(E) $P = \mu_K mgv$

GO ON TO THE NEXT PAGE

Questions 23 through 25 refer to the same situation.

23. A horizontal bug of mass m stands at rest on the edge of a horizontal disk of mass M, radius r, and rotational inertia $\frac{1}{2}Mr^2$, which is also at rest. At a certain time, the bug starts walking counterclockwise around the rim (when viewed from above), so that it moves with a speed v with respect to the ground. If the disk is free to rotate about its center, what is the resulting angular velocity of the disk (with respect to the ground)?

(A) $\omega = mv/Mr$
(B) $\omega = 2mv/Mr$
(C) $\omega = Mv/2mr$
(D) $\omega = Mv/mr$
(E) none of the above

24. In terms of the speed of the bug, v (with respect to the ground), and the magnitude of the angular velocity of the disk, ω, what is the speed of the bug relative to the disk?

(A) $(v/r) + \omega$
(B) $(v/r) - \omega$
(C) $v + r\omega$
(D) $v - r\omega$
(E) $\sqrt{v^2 + (r\omega)^2}$

25. What is the minimum coefficient of static friction between the bug's feet and the disk required to keep the bug from sliding off the disk as it walks with constant speed v?

(A) mv^2/r
(B) v^2/rg
(C) $(v + \omega r)^2/rg$
(D) $(v - \omega r)^2/rg$
(E) No static friction is needed.

26. Two masses, m_1 and m_2, are moving in opposite directions with speeds v_1 and v_2, respectively. What is the speed of the center of mass of the two masses?

(A) $\dfrac{m_1v_1 + m_2v_2}{m_1 + m_2}$
(B) $\dfrac{m_1v_2 + m_2v_1}{m_1 + m_2}$
(C) $\dfrac{|m_1v_1 - m_2v_2|}{m_1 + m_2}$
(D) $\dfrac{|m_1v_2 - m_2v_1|}{m_1 + m_2}$
(E) Not enough information is given.

27. A force F pushes on two masses lying on a horizontal frictionless surface as shown, causing them to accelerate to the right with magnitude a. Which of the following statements is false?

(A) m_2 exerts a force of $-m_2a\hat{i}$ on m_1.
(B) The net force on m_1 is $F - m_2a$.
(C) The acceleration of the blocks is $a = F/(m_1 + m_2)$.
(D) The sum of the contact forces exerted by m_1 on m_2, and vice versa, produce net work on the two-block system.
(E) The net force on m_2 is m_2a.

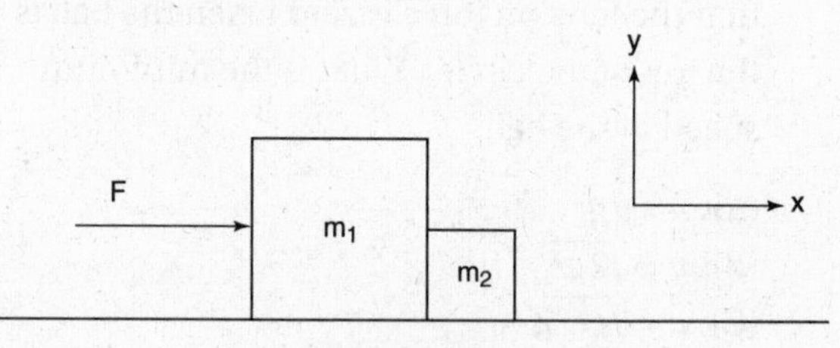

Question 27

GO ON TO THE NEXT PAGE

PRACTICE TEST 2

28. A particle is released from rest at point A. The figure shows the potential energy of the particle as a function of position. If no nonconservative forces are present, which of the following statements is true?

    (A) The particle oscillates between points B and F.
    (B) The particle cannot reach point G.
    (C) The particle moves in simple harmonic motion.
    (D) The particle experiences no force at points C, D, and E.
    (E) The particle's speed is maximized at point E.

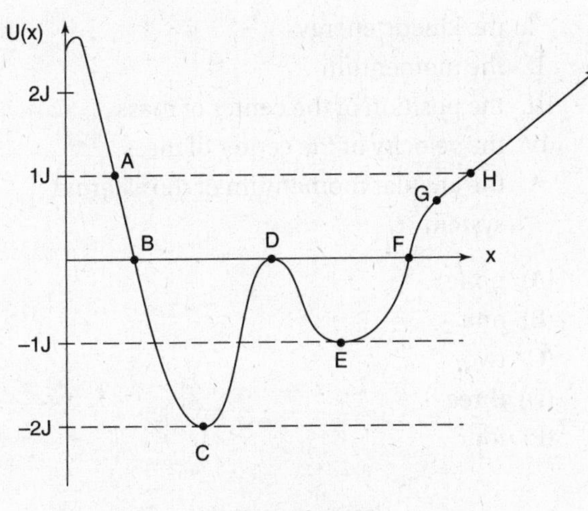

Question 28

29. If a rotating object accelerates from rest with an angular acceleration of 2 rad/s^2, through what angle will it rotate before a point on its edge a distance 0.5 m from the axle is moving with a speed of 2 m/s?

    (A) 0.5 rad
    (B) 1 rad
    (C) 2 rad
    (D) 4 rad
    (E) 8 rad

30. Consider the five figures, which show different forces as functions of position. How many of these forces can give rise to simple harmonic motion about the point $x = 0$?

    (A) none
    (B) one
    (C) two
    (D) three
    (E) four

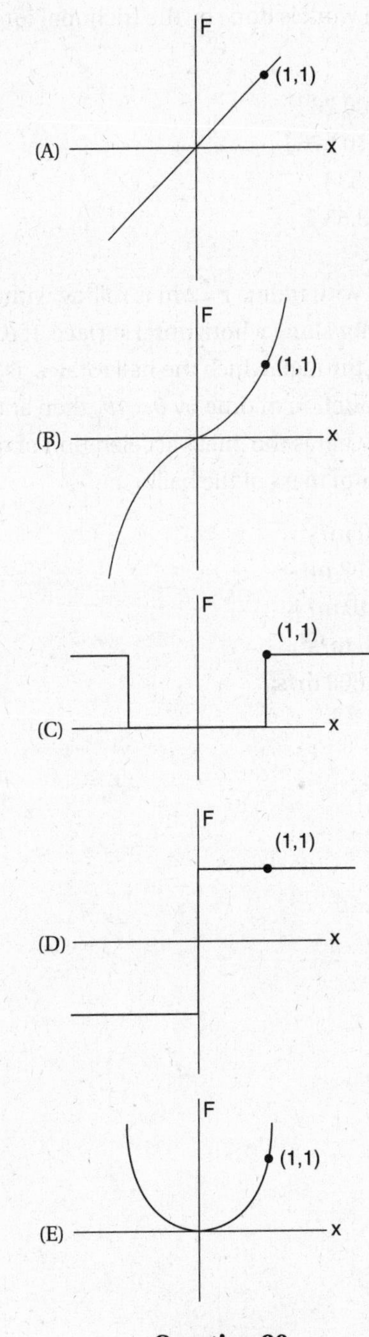

Question 30

GO ON TO THE NEXT PAGE

31. Referring to question 30, if the potential energy is set equal to zero at the origin, which force produces the greatest potential energy (in absolute value) at $x = 1$?

32. A block of mass 0.6 kg slides on a horizontal table where the coefficient of kinetic friction is 0.1. If the block is moved 3 m to the left and then 3 m back to its original position, how much work is done by the frictional force?

(A) 0
(B) 100.76 J
(C) –100.76 J
(D) 3.53 J
(E) –3.53 J

33. A ball with radius $r = 2$ m is rolling without slipping along a horizontal surface. If θ, the angle through which the ball rotates, is given as a function of time by $\theta = 4t^2$, then at time $t = 1$, what is the linear acceleration of the center of mass of the ball?

(A) 40 m/s^2
(B) 0.88 m/s^2
(C) 140 m/s^2
(D) 16 m/s^2
(E) 9,600 m/s^2

34. Referring to the ball introduced in question 33, what is the speed of a point on the very top of the ball at time $t = 1$ relative to the horizontal surface?

(A) 6 m/s
(B) 12 m/s
(C) 24 m/s
(D) 32 m/s
(E) 96 m/s

35. An asteroid moving at a constant velocity in space suddenly explodes. How many of the following quantities remain constant throughout and after the explosion?

I. the kinetic energy
II. the momentum
III. the position of the center of mass
IV. the velocity of the center of mass
V. the angular momentum of the asteroid system

(A) none
(B) one
(C) two
(D) three
(E) four

GO ON TO THE NEXT PAGE

FREE-RESPONSE QUESTIONS

Directions: Use separate sheets of paper to write down your answers to the free-response questions. You may refer to the formula sheet in the Appendix (pages 641–644).

Mechanics

MECHANICS I

A skier of mass m is standing on a slope with an angle of elevation θ. After being given a tiny nudge, the skier slides down the slope, grabs a rope at the bottom of the slope, and swings up to a height h, as shown. The coefficient of kinetic friction between the skier and the slope is μ_k.

(a) What is the minimum coefficient of static friction required so that the skier is initially at rest at the top of the slope without using his poles?

(b) How much energy is lost while the skier slides down the incline? Express your answer in terms of m, g, h, d, and θ.

(c) What is the coefficient of kinetic friction? Express your answer in terms of h, d, and θ.

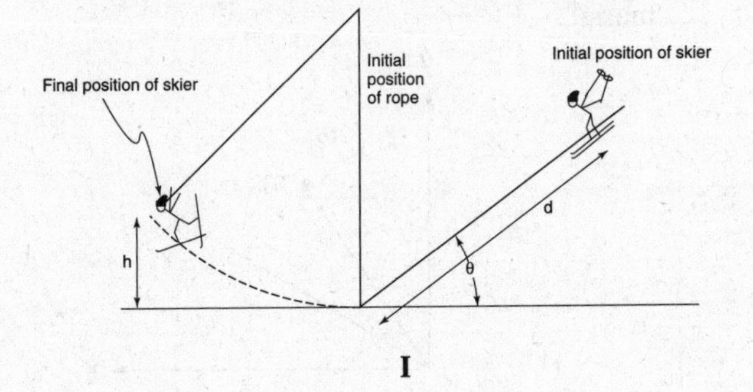

I

GO ON TO THE NEXT PAGE

MECHANICS II

A mass m is attached to a string of length l and released when the string is horizontal, causing the mass to fall in a circular path as shown. At the time that the mass has fallen through the angle θ, determine the following (in terms of m, l, g, and θ):

(a) the speed of the object
(b) the magnitude of the centripetal acceleration
(c) the magnitude of the tangential acceleration
(d) the magnitude of the tension force

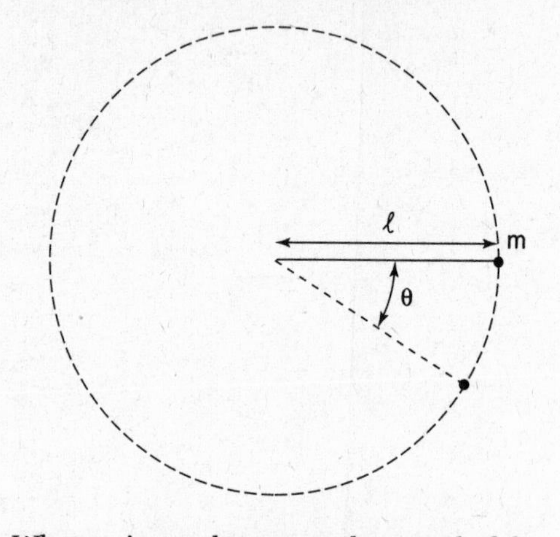

(e) When using a photogate, the speed of the mass is measured when it passes through the bottom of the circle. The experiment is repeated for strings of different lengths, and the following data are obtained:

Speed (m/s)	1.70	2.43	2.98	3.42	3.85
Length of string (cm)	15	30	45	60	75

What quantities should be plotted on the grid below to obtain a linear relation between the speed of the mass and the length of the string?

(f) Plot the data. Scale and label the axes, including units. After fitting the data to a straight line, check how well the experimental results agree with the theoretical prediction of part (a).

MECHANICS III

A mass m is connected to an axle by a massless rod of length l. The mass m feels a retarding force due to air resistance given by $\mathbf{F} = -b\omega$, where b is a positive constant with units of kg•m/s. The mass rotates in the xy-plane, with an initial angular speed ω_0 at time $t = 0$ as shown.

(a) What is the initial angular momentum of mass m about the z-axis?
(b) What is the torque due to air resistance?
(c) What is the initial angular acceleration of the mass?

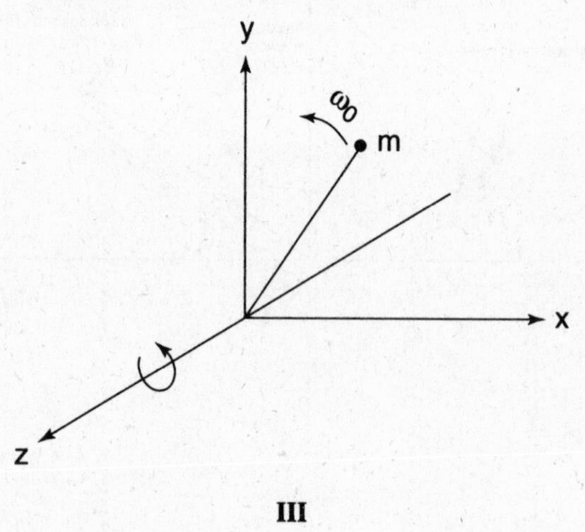

III

GO ON TO THE NEXT PAGE

MULTIPLE-CHOICE QUESTIONS

Directions: Each multiple-choice question is followed by five answer choices. For each question, choose the best answer and fill in the corresponding circle on the answer sheet. You may refer to the formula sheet in the Appendix (pages 641–644).

Electricity and Magnetism

1. A point charge $-q$ is released from rest at the center of a parallel plate capacitor, as shown. Note that the plates have small holes that allow the charge to exit the capacitor. Which of the following statements describes the charge's behavior a long time after it exits the capacitor?

   (A) It is moving to the right with a kinetic energy of KE = qV.
   (B) It is moving to the left with a kinetic energy of KE = qV.
   (C) It is moving to the right with a kinetic energy of KE = $qV/2$.
   (D) It is moving to the left with a kinetic energy of KE = $qV/2$.
   (E) It never reaches a constant velocity but instead continually oscillates in and out of the capacitor.

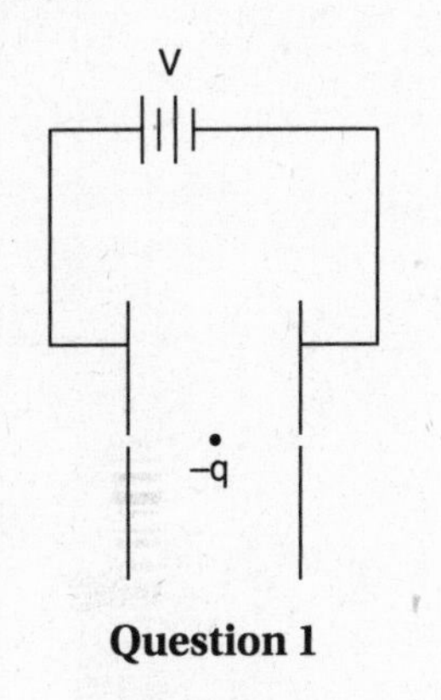

Question 1

2. Describe the potential V as a function of r inside a spherical shell with uniform surface charge density.

   (A) $V = 0$.
   (B) V is constant.
   (C) V has a constant slope.
   (D) V is proportional to $1/r$.
   (E) V is proportional to $1/r^2$.

3. While discharging a capacitor through a resistor, how many of the following functions are exponentially decreasing?

   I. the current through the circuit
   II. the energy stored in the capacitor
   III. the voltage across the resistor
   IV. the power dissipated in the resistor
   V. the electric field within the capacitor
   VI. the charge on the capacitor

   (A) two
   (B) three
   (C) four
   (D) five
   (E) six

PRACTICE TEST 2

GO ON TO THE NEXT PAGE

4. The figure shows two coaxial cylinders that have charge distributed uniformly on their surfaces, with a linear charge density of $-\lambda$ and $+\lambda$ coulombs per axial length on the inner and outer cylinders, respectively. Which of the graphs correctly shows the electric field as a function of radial position, $E(r)$?

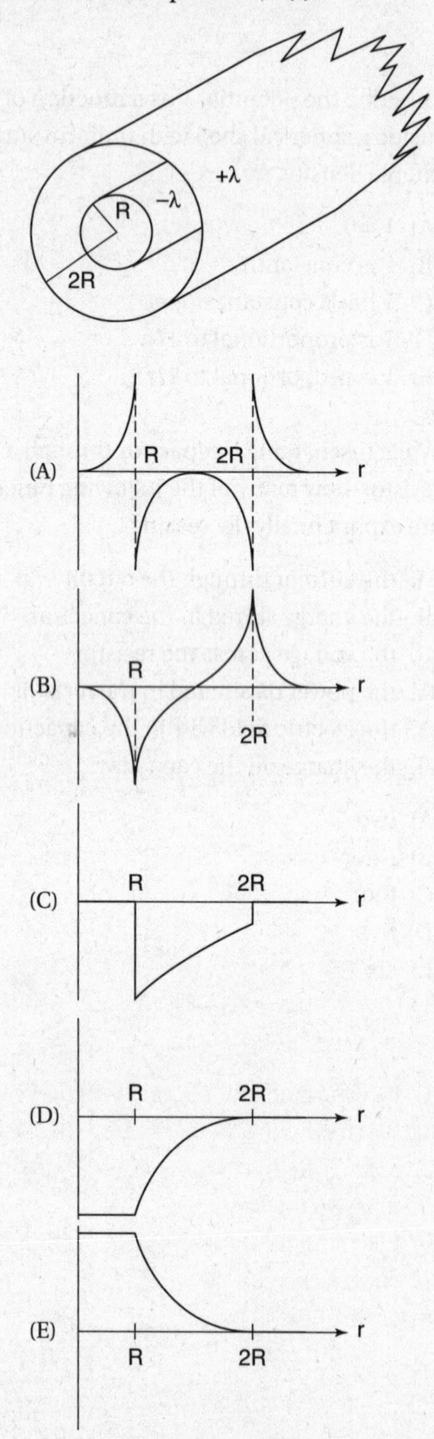

Question 4

5. Referring to the situation introduced in question 4, which of the graphs shows the voltage as a function of radial position, $V(r)$?

6. A loop of wire of radius r and resistance R is exposed to a magnetic field pointing into the page that steadily increases from zero to magnitude B in time t, as shown. Describe the induced current.

(A) $\pi rB/t$, clockwise
(B) $\pi rB/t$, counterclockwise
(C) $\pi r^2B/Rt$, clockwise
(D) $\pi r^2B/Rt$, counterclockwise
(E) $\pi r^2B/t$, counterclockwise

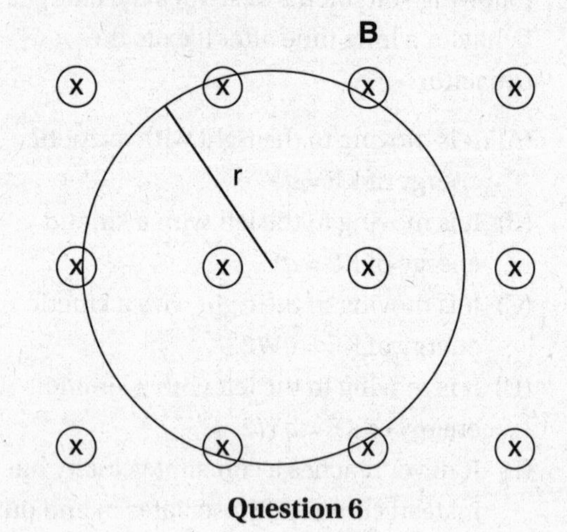

Question 6

GO ON TO THE NEXT PAGE

7. The potential in a region of space is given by the formula $V = ar$, where r is a radial coordinate and a is a positive constant. Which of the following statements is true of this region?

(A) The electric field has magnitude a and points away from the origin of the coordinates (in the $+\hat{r}$-direction).
(B) The electric field has a magnitude a and points toward the origin (in the $-\hat{r}$-direction).
(C) The electric field has a magnitude $r^2/2a$ and points away from the origin (in the $+\hat{r}$ direction).
(D) The electric field has a magnitude $r^2/2a$ and points toward the origin (in the $-\hat{r}$-direction).
(E) The electric force is nonconservative.

8. A capacitor of capacitance 25 μF is charged to 300 V and then connected in series with an inductor of inductance 10 mH. The maximum current flowing through the circuit will be

(A) 2.5 A
(B) 500 A
(C) 7,500 A
(D) 15 A
(E) 300 A

9. If it requires an input of E joules to move a charge Q from point A to point B at constant velocity, which of the following is true?

(A) $V_B - V_A = E/Q$
(B) $V_B - V_A = -E/Q$
(C) $V_B - V_A = QE$
(D) $V_B - V_A = -QE$
(E) $V_B - V_A = Q/E$

10. A parallel plate vacuum capacitor with capacitance C_0 is attached to a battery of voltage V. What is the change in the energy stored in the capacitor if the separation of the plates is doubled? (You can assume that the resistance in the circuit is very small so that the potential difference across the capacitor always equals V.)

(A) $\frac{1}{2}C_0V^2$
(B) $\frac{1}{4}C_0V^2$
(C) $-\frac{1}{2}C_0V^2$
(D) $-\frac{1}{4}C_0V^2$
(E) none of the above

11. A battery with EMF of V and an internal resistance of r_{int} is connected to a heating loop with resistance R. What is the correct expression for the rate at which energy is converted to heat by the heating loop?

(A) V^2/R
(B) $V^2r_{int}/(r_{int} + R)^2$
(C) $V/(r_{int} + R)$
(D) $V^2R^2/(r_{int} + R)^2$
(E) none of the above

12. A parallel plate capacitor connected to a battery contains a plastic dielectric material between the plates of the capacitor. Removing the dielectric has what effect on the electric field between the plates and the charge on the plates?

(A) Electric field and charge both decrease.
(B) Electric field and charge both increase.
(C) Electric field increases and the charge remains the same.
(D) Electric field remains the same and the charge decreases.
(E) none of the above

GO ON TO THE NEXT PAGE

13. An electron is released from rest at a distance of r_1 from a proton. What is the kinetic energy of the electron when it is a distance r_2 from the proton?

(A) 0

(B) $\frac{e^2}{4\pi\varepsilon_0}\left(\frac{1}{r_2^2}-\frac{1}{r_1^2}\right)$

(C) $\frac{e^2}{4\pi\varepsilon_0}\left(\frac{1}{r_1^2}-\frac{1}{r_2^2}\right)$

(D) $\frac{e^2}{4\pi\varepsilon_0}\left(\frac{1}{r_2}-\frac{1}{r_1}\right)$

(E) $\frac{e^2}{4\pi\varepsilon_0}\left(\frac{1}{r_1}-\frac{1}{r_2}\right)$

14. The magnetic field created by the two wires shown at point A in the figure is B_0 (pointing out of the page), and the two wires carry identical currents in the same direction. Which of the following is true?

(A) Both wires carry a current of $4\pi d/3\mu_0$ toward the top of the page.

(B) Both wires carry a current of $4\pi d/3\mu_0$ toward the bottom of the page.

(C) Both wires carry a current of $4\pi dB_0/\mu_0$ toward the top of the page.

(D) Both wires carry a current of $4\pi dB_0/\mu_0$ toward the bottom of the page.

(E) It is impossible for the wires to produce a magnetic field pointing perpendicular to the page.

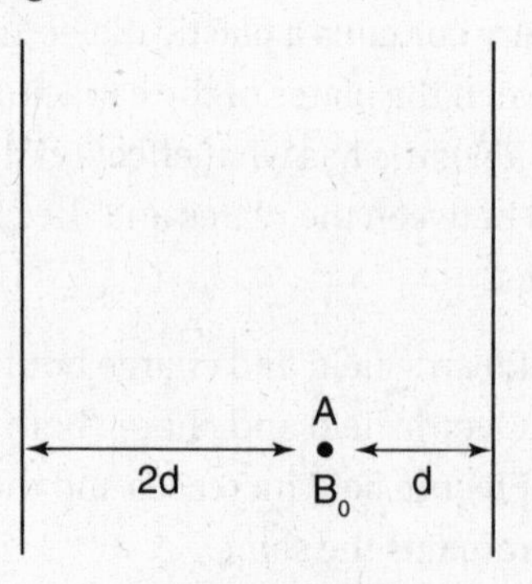

Question 14

15. Two insulating spheres of equal radii are oppositely charged with uniform charge densities of equal magnitude, as shown in the top half of the figure. The spheres are placed far apart from each other and can be considered completely isolated from each other. What is the effect of exchanging the top halves of these distributions with one another, as shown in the bottom half of the figure?

(A) The potential increases at point *A* and decreases at point *B*.

(B) The electric field magnitude increases at point *A* and decreases at point *B*.

(C) The potential increases at both points *A* and *B*.

(D) The electric field magnitude increases at both points *A* and *B*.

(E) After the exchange, positive work is required to move a positive charge from point *A* to point *B*.

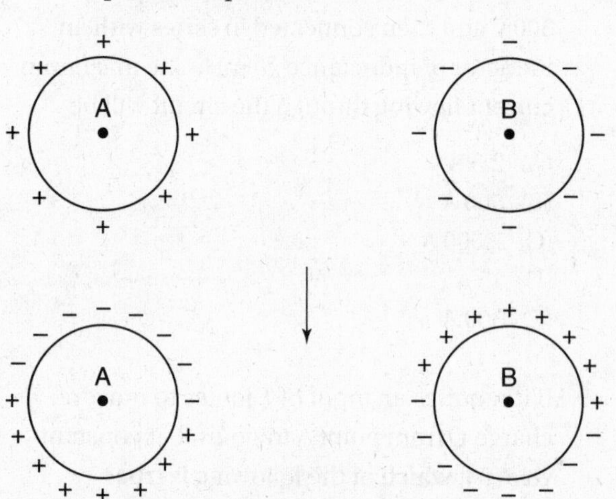

Question 15

GO ON TO THE NEXT PAGE

16. A loop of wire is pushed with constant velocity through a region of uniform magnetic field as shown. (The field points out of the page.) As the loop enters and exits the magnetic field, only one of the vertical segments of wire is within the magnetic field. What is the direction of the induced current within this vertical segment?

(A) upward when both entering and exiting the magnetic field
(B) downward when both entering and exiting the magnetic field
(C) upward when entering the magnetic field and downward when exiting the magnetic field
(D) downward when entering the magnetic field and upward when exiting the magnetic field
(E) zero, because the segment of wire is perpendicular to the magnetic field

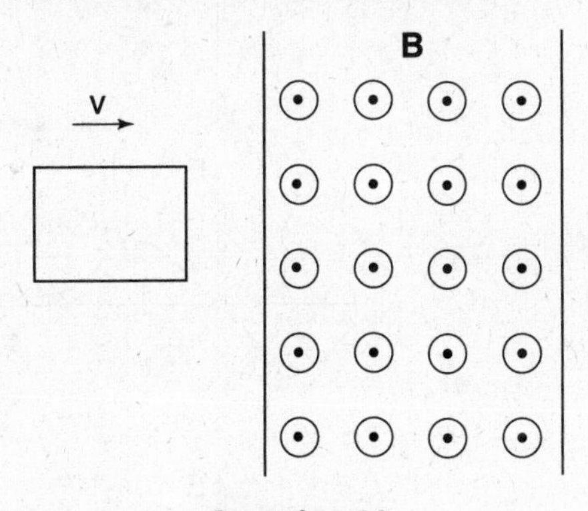

Question 16

17. What is the force on a positive charge $+Q$ located on a 10-V equipotential line?

(A) 10 Q in magnitude, pointing parallel to the line
(B) 10 Q in magnitude, pointing perpendicular to the line in the direction of decreasing potential
(C) 10 Q in magnitude, pointing perpendicular to the line in the direction of increasing potential
(D) of unknown magnitude, pointing perpendicular to the line in the direction of decreasing potential
(E) of unknown magnitude, pointing perpendicular to the line in the direction of increasing potential

18. The capacitor shown consists of two concentric metallic cylindrical shells. Assuming that the radius of the inner cylinder, r, is much greater than the plate separation distance, d, what is the approximate capacitance of the capacitor?

(A) $C = \varepsilon_0 \pi r^2 / d$
(B) $C = \varepsilon_0 \pi r l / d$
(C) $C = \varepsilon_0 \pi d l / r$
(D) $C = 2\varepsilon_0 \pi r l / d$
(E) $C = \varepsilon_0 d / 2\pi r l$

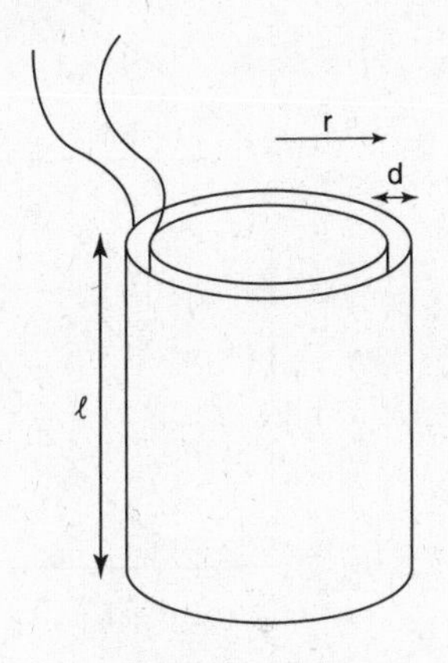

Question 18

19. A sphere has a uniform charge density distributed throughout its interior. What effect does doubling the radius of the sphere (while maintaining the same charge density) have on the electric field at a point outside the enlarged sphere?

(A) The field remains the same.
(B) The field increases by a factor of 2.
(C) The field increases by a factor of 4.
(D) The field increases by a factor of 8.
(E) The field increases by a factor of 16.

GO ON TO THE NEXT PAGE

20. A charged particle moves in a planar, circular orbit as a result of the presence of a uniform magnetic field perpendicular to the plane of its motion. Holding all other variables (such as the magnetic field, the particle's velocity, and the particle's charge) constant, which graph shows the relationship between the particle's mass and the radius of its path?

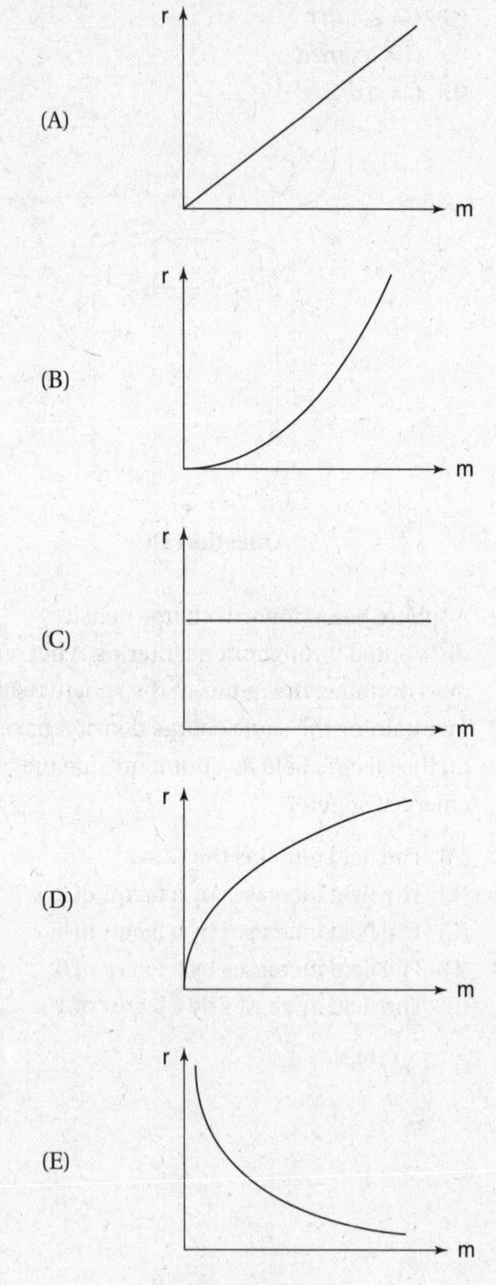

Question 20

21. A rod with charge density of $-\lambda$ C/m lies along the y-axis as shown. Which of the following plots correctly shows $E_x(x)$, the x-component of the electric field along the x-axis?

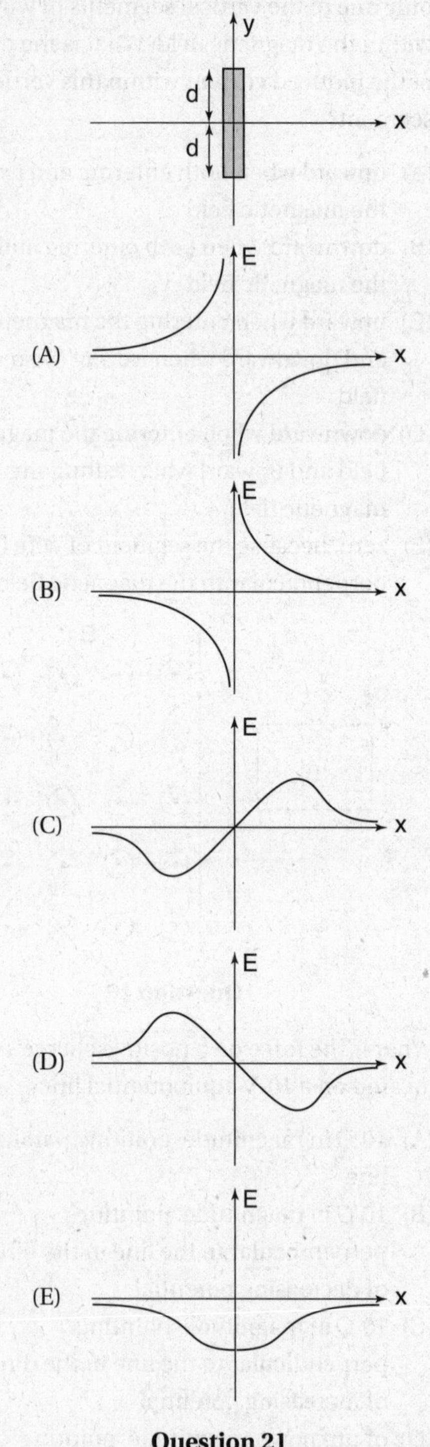

Question 21

GO ON TO THE NEXT PAGE

PRACTICE TEST 2

22. Consider a circuit consisting of the following elements in a single loop: a battery, a resistor, a switch, and a fourth unknown element E. A spark is most likely to occur at the switch if

(A) E is a capacitor and the switch is being closed for the first time.
(B) E is a capacitor and the switch is being opened after the circuit has reached equilibrium.
(C) E is an inductor and the switch is being closed for the first time.
(D) E is an inductor and the switch is being opened after the circuit has reached equilibrium.
(E) E is a battery arranged in series with the other battery in the circuit.

23. The two charge distributions shown each contain a net charge $+Q$ uniformly distributed along an arc of a circle with radius r. In distribution A this charge is spread along an arc twice the length of the arc in distribution B. Which of the following comparisons of the potential V and the magnitude E of the electric field at point A and point B are true?

(A) $V_A > V_B$ and $E_A > E_B$
(B) $V_A > V_B$ and $E_A < E_B$
(C) $V_A = V_B$ and $E_A > E_B$
(D) $V_A = V_B$ and $E_A < E_B$
(E) $V_A < V_B$ and $E_A > E_B$

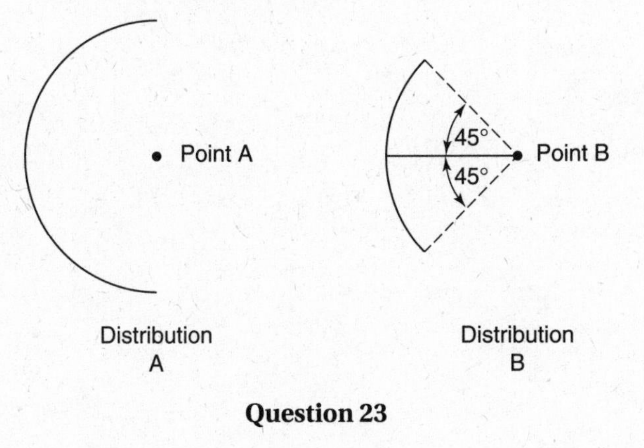

Question 23

24. Two cylindrical resistors are arranged in series and connected to a complicated circuit, as shown. Both resistors have the same resistivity, and the second resistor has twice the radius and twice the length of the first resistor. Which of the following statements is correct?

(A) The resistors have the same resistance.
(B) The resistors experience the same potential drop.
(C) The resistors dissipate the same amount of power.
(D) The resistors have the same current.
(E) The equivalent resistance of the two resistors is equal to $1.5R_2$.

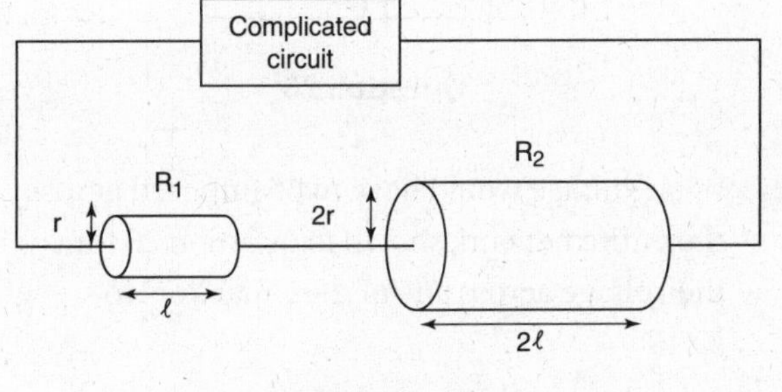

Question 24

25. In the circuit shown, if the potential at point A is equal to the potential at point B, what is the current passing through the unknown resistor R?

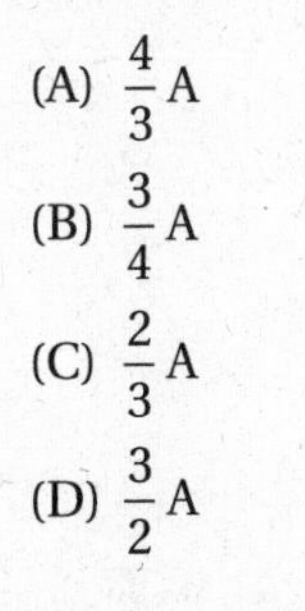

(A) $\frac{4}{3}$ A
(B) $\frac{3}{4}$ A
(C) $\frac{2}{3}$ A
(D) $\frac{3}{2}$ A
(E) none of the above

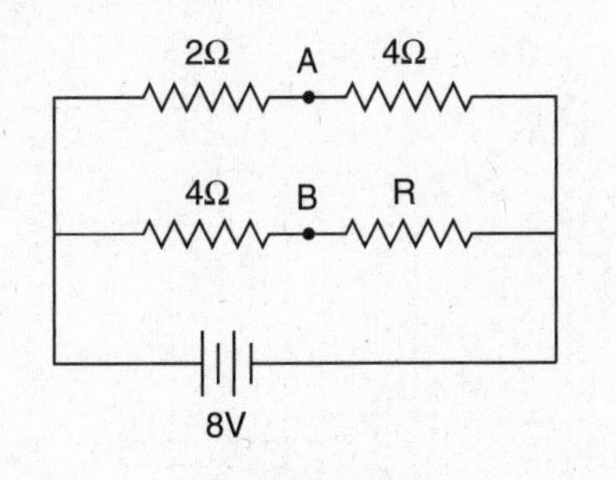

Question 25

GO ON TO THE NEXT PAGE

26. If each of the capacitors in the network shown has a capacitance of C, what is the equivalent capacitance of the network?

(A) $C/9$
(B) $C/3$
(C) C
(D) $3C$
(E) $9C$

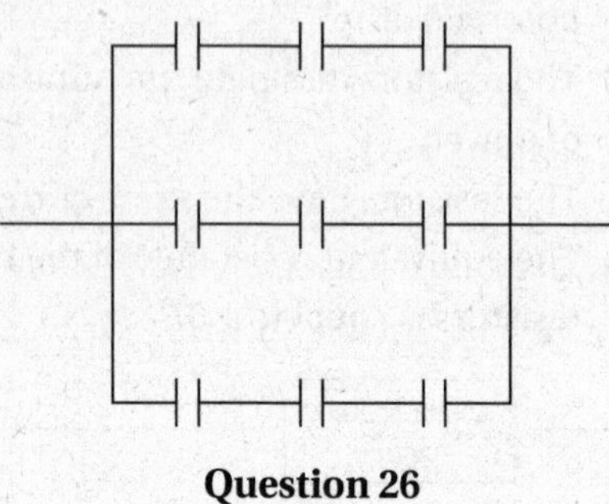

Question 26

27. What voltage would have to be imposed across the entire network shown in question 26 for the voltage across any of the capacitors to be V?

(A) $V/9$
(B) $V/3$
(C) C
(D) $3V$
(E) $9V$

28. The gaussian "unit cube" shown lies in a region where the electric field is constant and parallel to the z-axis: $\mathbf{E} = E\hat{k}$. Which of the following choices lists the outward flux through sides 1 through 6, respectively?

(A) 0, 0, $-E$, 0, $+E$, 0
(B) 0, 0, $+E$, 0, $-E$, 0
(C) 0, 0, $+E$, 0, $+E$, 0
(D) 0, 0, $-E$, 0, $-E$, 0
(E) none of the above

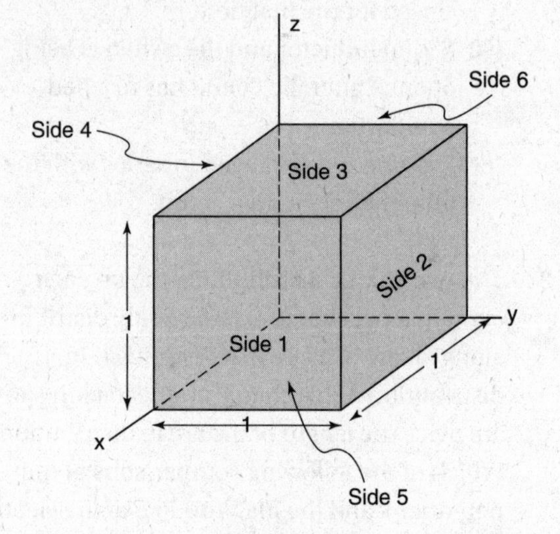

Question 28

PRACTICE TEST 2

GO ON TO THE NEXT PAGE

29. You have three identical capacitors you are using to design an equivalent capacitor that will be used to store energy for a camera flash, as shown. Charging time is defined as the amount of time required to reach 90% of the maximum energy. Rank the networks in order from least to greatest charging time.

(A) I, II, III, IV
(B) IV, III, II, I
(C) I, III, II, IV
(D) IV, II, III, I
(E) I, (II = III), IV

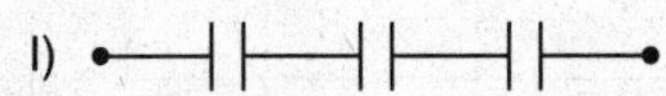

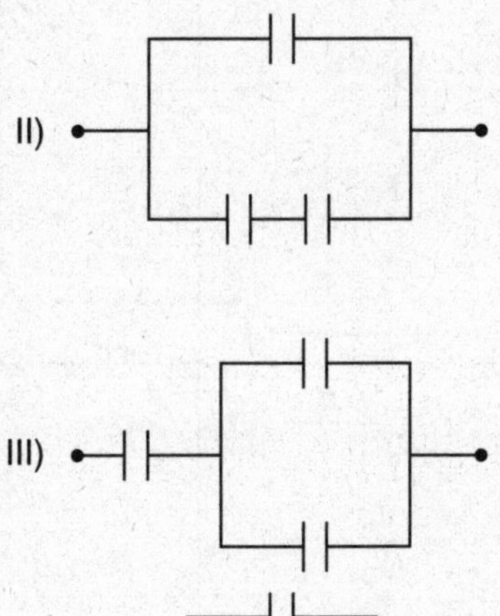

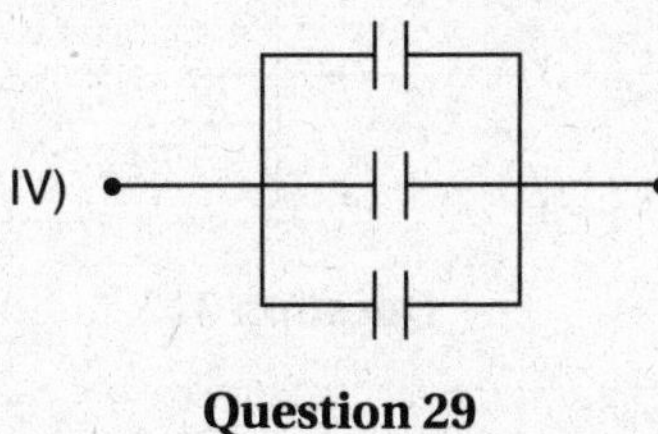

Question 29

30. Considering the situation introduced in the last problem, rank order the capacitor networks in terms of the energy stored when fully charged. (Assume that the capacitor networks are connected to the same circuit, with a fixed voltage.)

(A) I, II, III, IV
(B) IV, III, II, I
(C) I, III, II, IV
(D) IV, II, III, I
(E) I, (II = III), IV

31. What is the rate at which thermal energy is produced in the resistors shown?

(A) $\frac{1}{2}IV$
(B) IV
(C) $2IV$
(D) $4IV$
(E) none of the above

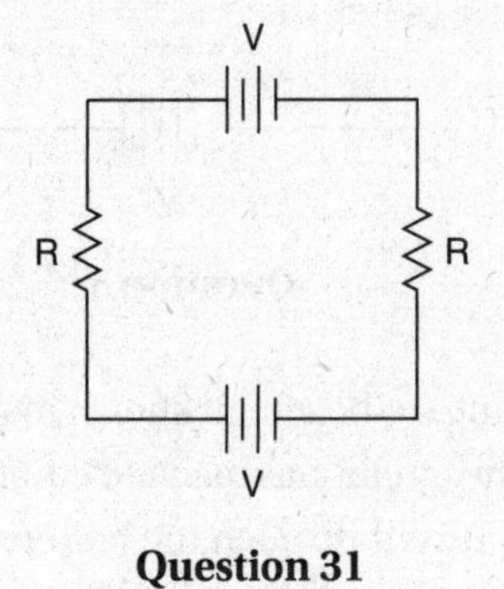

Question 31

GO ON TO THE NEXT PAGE

32. In the circuit shown, there is initially no current flowing through the inductor. The instant after the switch is closed, what is the current through the battery?

(A) V/R_1
(B) $V/(R_1 + R_2)$
(C) $V/(R_1 + R_3)$
(D) $\dfrac{V}{R_1 + \dfrac{R_2 R_3}{R_2 + R_3}}$
(E) $\dfrac{V}{R_1 + \dfrac{R_2 + R_3}{R_2 R_3}}$

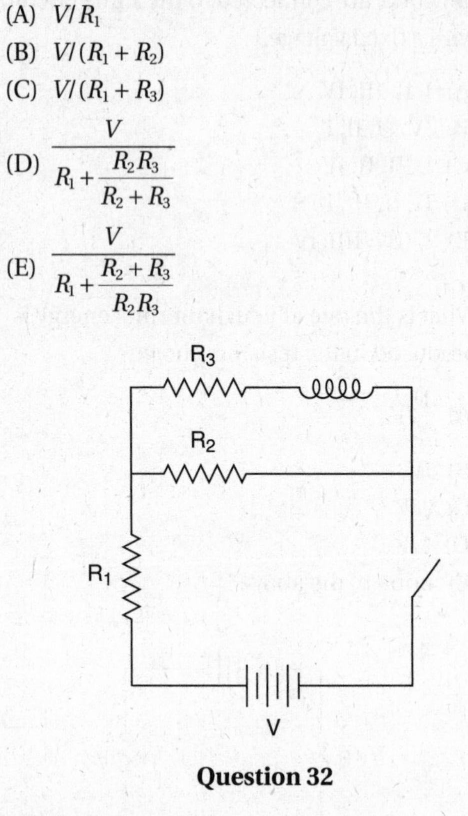

Question 32

33. Referring to the circuit shown in question 32, after the circuit has reached equilibrium, what is the current through the battery?

(A) V/R_1
(B) $V/(R_1 + R_2)$
(C) $V/(R_1 + R_3)$
(D) $\dfrac{V}{R_1 + \dfrac{R_2 R_3}{R_2 + R_3}}$
(E) $\dfrac{V}{R_1 + \dfrac{R_2 + R_3}{R_2 R_3}}$

34. Which of the sketches correctly shows potential as a function of position for the charge distribution shown?

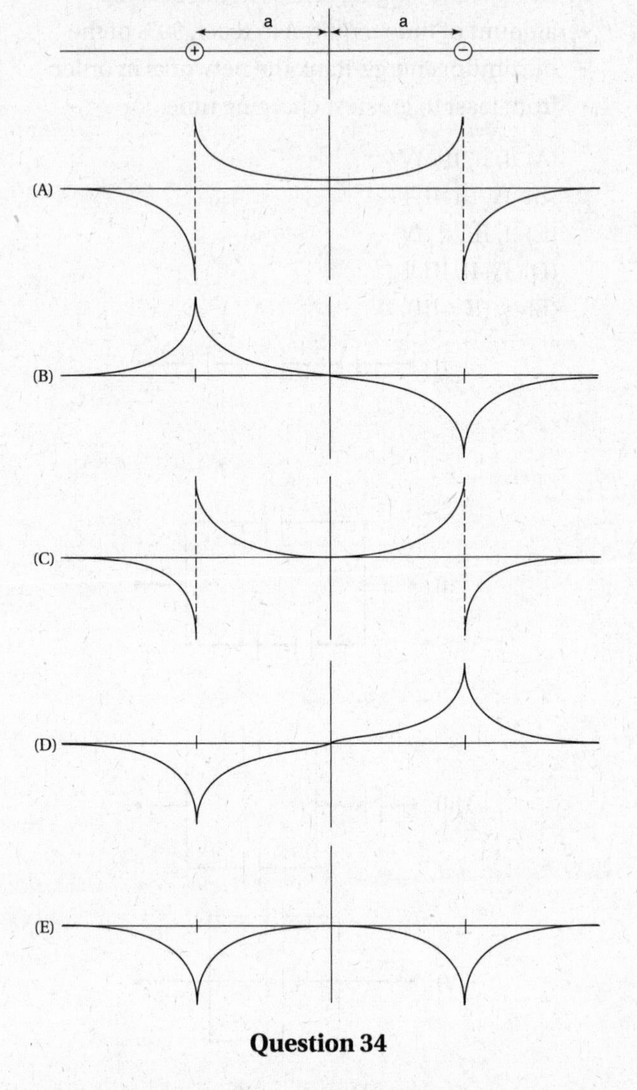

Question 34

35. Which graph in question 34 represents the electric field as a function of position?

GO ON TO THE NEXT PAGE

FREE-RESPONSE QUESTIONS

Directions: Use separate sheets of paper to write down your answers to the free-response questions. You may refer to the formula sheet in the Appendix (pages 641–644).

Electricity and Magnetism

ELECTRICITY AND MAGNETISM I

A semicircle of radius a and total positive charge Q (uniformly distributed along its length) lies on the xy-plane as shown.

(a) Calculate the potential at the origin.

(b) Calculate the electric field at the origin.

(c) If a charge of mass m and charge $+q$ is released from rest at the origin, what will its velocity and position be as time approaches infinity?

(d) How much work has the electric field done on the charge when the charge has moved a very small displacement Δx to the right of the origin?

I

ELECTRICITY AND MAGNETISM II

A cylinder of uniform material with resistivity ρ, length l, and area A is connected to an identical cylinder made of metal (with effectively zero resistance) of the same radius. If a current I runs through both cylinders with a uniform current density,

(a) What is the resistance of the two-cylinder apparatus?

(b) What is the potential difference between A and B? Between B and C? Which points are at higher potential?

(c) What is the electric field in the resistive material? In the metal?

(d) Calculate the amount of charge lying on the interface of the two cylinders by constructing a gaussian surface that contains the interface. (*Hint*: This gaussian surface behaves similarly to the surfaces used to calculate the electric fields of infinite planes of charge.)

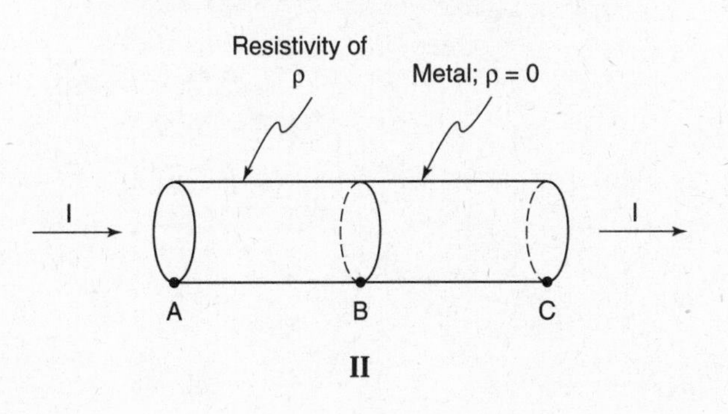

II

GO ON TO THE NEXT PAGE

(e) An experiment is carried out to measure the resistance of the cylinder. The following data are obtained for the potential difference between points A and C and the current flowing between those two points:

Potential difference (V)	2	4	6	8	10
Current (mA)	35	69	106	142	164

Plot the data on the grid below. Scale and label the axes, including units. After fitting the data to a straight line, determine the resistance.

ELECTRICITY AND MAGNETISM III

A charge $+Q$ is distributed on an isolated solid metal sphere of radius a.

(a) Use Gauss's law to calculate the electric field for $r < a$ and $r > a$.

(b) Integrate the electric field from $r = \infty$ to $r = a$ to calculate the potential on the surface of the sphere.

(c) If you consider the sphere to be a capacitor (with the opposite charge $-Q$ spread over a sphere an infinite distance away), what is its capacitance?

(d) You then place a very thin metal shell of radius $b > a$ concentric with the charged metal sphere. You can assume that the thickness of the shell (where the electric field is zero) is negligible, so that it does not affect the potential on the surface of the charged sphere. After connecting the sphere to the shell with a wire, how will the charge redistribute itself?

(e) What is the change in electrostatic potential energy as the charge moves from the sphere to the shell? If the energy change is not equal to zero, where does the energy disappear to or appear from?

MECHANICS

ANSWER KEY

1. **D**	8. **D**	15. **E**	22. **B**	29. **D**
2. **E**	9. **B**	16. **E**	23. **B**	30. **A**
3. **E**	10. **C**	17. **D**	24. **C**	31. **D**
4. **A**	11. **C**	18. **D**	25. **B**	32. **E**
5. **A**	12. **A**	19. **A**	26. **C**	33. **D**
6. **E**	13. **B**	20. **D**	27. **D**	34. **D**
7. **D**	14. **D**	21. **B**	28. **D**	35. **D**

ANSWERS EXPLAINED

Multiple-Choice

1. **(D)** Average power is given by the equation $\bar{P} = W/\Delta t$. What is the work done? Because there are no other forces acting on the block, $W = \Delta KE = m(v_f^2 - v_0^2)/2$ (according to the work-energy theorem). Dividing by the time increment to calculate the average power yields choice (D). Note that the correct answer can be obtained using only dimensional analysis.

2. **(E)** Consider the figure. At points *A* and *C*, the *x*-component of velocity is zero, at point *B* it is equal to $-v$, and at point *D* it is equal to $+v$. This range of values is consistent only with choice (E).

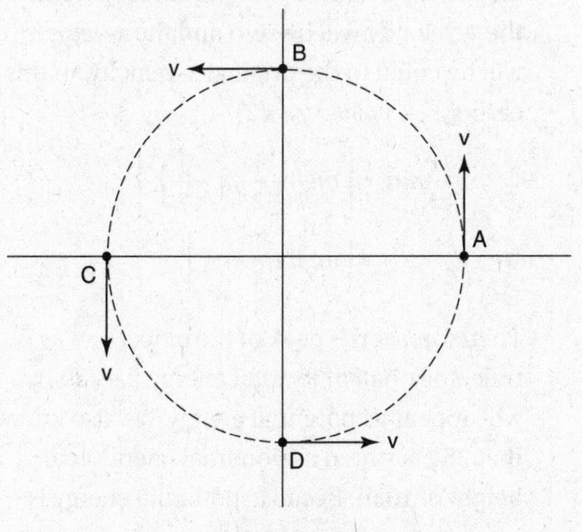

Question 2

3. **(E)** According to Newton's third law, the baseball and Earth exert equal and opposite forces on each other. Therefore, the accelerations they experience must be in opposite directions. Because $F = m_e a_e = m_b a_b$, the larger object (Earth) experiences a smaller acceleration (which is imperceptible).

4. **(A)** The angular velocity is simply the derivative of the angular position:

$$\omega(t) = \frac{d}{dt}[\theta(t)] = \frac{d}{dt}(2 + 4t - 3t^2) = 4 - 6t$$

5. **(A)** Because the object has a tangential acceleration parallel to its velocity, its speed is increasing. Because the centripetal acceleration (perpendicular to the velocity) always points toward the center of the circle, the object must move clockwise so that the centripetal acceleration points in the $+x$-direction while the object is instantaneously moving in the $+y$-direction.

6. **(E)** The gravitational force is constant, so the work it does is given by the equation $W = \mathbf{F} \cdot \Delta\mathbf{r} = (-mg\hat{j}) \cdot (\Delta x\hat{i} + \Delta y\hat{j}) = -mg\Delta y$. The *y*-coordinate increases by an amount $d \sin\theta$, so the work done by gravity is given by choice (E). (*Check*: The work done by gravity must come out negative because the force of gravity is antiparallel to the vertical displacement of the object.)

7. **(D)** Choices (A)–(C): The mass is not undergoing SHM because the restoring force is not proportional to the displacement (instead, it is constant and equal to $mg\sin\theta$). Choice (E): The mass would be in static equilibrium if placed at the bottom of the incline with zero initial velocity, but in the present case the object oscillates back and forth along the inclines. Choice (D): The component of the momentum perpendicular to the slope is zero as the mass initially slides down the slope, and it remains zero. (There is zero net force in this direction because the normal force cancels the component of the gravitational force perpendicular to the incline.)

8. **(D)** Differentiating twice to calculate the acceleration from the position yields $a(t) = (12at^2 - 6bt)$. Applying Newton's second law then gives $F = ma = ma(t) = m(12at^2 - 6bt)$.

9. **(B)** For the object to be in static equilibrium, the net force in the x-direction must be zero: $F_{net,x} = 0 = T_1 \sin\theta_1 - T_2 \sin\theta_2$. Rearranging this equation yields choice (B).

10. **(C)** The projectiles hit the wall when their x-positions are equal to the position of the wall. Because both projectiles have the same x-components of velocity, they require the same amount of time to hit the wall. Because projectile 1 has a greater y-component of velocity, $v_1(t) > v_2(t)$, it hits the surface at a greater height (noting that both projectiles hit the surface after the same amount of time t).

11. **(C)** The gravitational force points down while the frictional force points in a direction opposite that of the instantaneous velocity, that is, to the left. The net force, which is the sum of the gravitational force and the frictional force, has downward and leftward components.

12. **(A)** The frictional force of air resistance is proportional to the object's velocity and opposite in direction. Thus, the x-component of the frictional force (which in the x-direction equals the net force) is maximized when the x-component of the velocity is the greatest. This occurs at point A, before friction has begun to decrease the magnitude of the x-component of velocity.

    The y-component of the net force is equal to the force exerted by gravity plus the y-component of the frictional force. The magnitude of this sum is greatest when the two forces point in the same direction, which occurs for points to the left of point C (when the velocity has an upward component, causing a downward component in frictional force). The downward vertical component of the frictional force is greatest at point A (before the y-component of velocity has begun to decrease), so at point A the y-component of the net force is maximized.

    Because both components of the net force are maximized at point A, the magnitude of the net force is also maximized at point A.

13. **(B)** Consider the equation for conservation of energy between the instant the ball is launched and the instant it reaches its maximum height (at the maximum height, the y-velocity will be zero and the x-velocity will be equal to the original x-velocity, in this case $v_x = v_0\cos\theta = v_0/\sqrt{2}$).

$$\frac{1}{2}mv_0^2 = \left[mgh + \frac{1}{2}m\left(\frac{v_0}{\sqrt{2}}\right)^2\right]$$
$$= \left[mgh + \frac{1}{4}mv_0^2\right]$$

Therefore, at the peak of the object's trajectory, half of its total energy (i.e., 50 J) will appear as potential energy. We also know that 25 J is stored as potential energy at a height of 10 m. Because potential energy is proportional to height, the potential energy will equal 50 J, its value at the peak of the trajectory, at a height of 20 m.

14. **(D)** The only answers that have correct units are (D) and (E).

The period is the amount of time it takes for the particle to make one complete cycle, in this case one complete revolution around the circle. The total distance of a cycle is the circumference of the circle, $2\pi r$, and the speed is v, so the time required for one cycle is t = distance/rate = $2\pi r/v$.

15. **(E)** Outside Earth, the gravitational force is the same as if all the mass of Earth were concentrated at its center, and thus the acceleration due to gravity is given by $a = GM_{\text{Earth}}/r^2$. Therefore, the acceleration decreases by a factor of 2 when the distance increases by a factor of $\sqrt{2}$, making choice (A) correct.

Inside Earth, at a distance r from the center, the net gravitational force is due to the mass enclosed by a sphere of radius r. This enclosed mass is proportional to r^3. Therefore, the acceleration inside Earth is given by the equation

$$a = \frac{G\left[M_{\text{Earth}}\left(r^3/r_{\text{Earth}}^3\right)\right]}{r^2} = \frac{GM_{\text{Earth}}}{r_{\text{Earth}}^3} r$$

Because the gravitational force is proportional to r, it equals $g/2$ at $r = r_e/2$, making choice (C) correct as well.

For further explanation, consider Figure 10.1. Based on this figure, an acceleration smaller than g occurs at *two* locations, one within Earth and one outside the Earth.

16. **(E)** For the net force to be zero, the force at point P must be $F_2\hat{i} - F_1\hat{j}$ (pointing downward and to the right). However, this produces a net counterclockwise torque about the center of the object (note that the two forces in the figure exert zero torque about this point). Therefore, it is impossible for a force exerted at point B to produce static equilibrium.

17. **(D)** The velocity of the automobile relative to the ground obeys the formula

$$\mathbf{v}_{\text{auto with respect to ground}} = \mathbf{v}_{\text{auto with respect to railroad car}} + \mathbf{v}_{\text{railroad car with respect to ground}}$$

Thus $-2 = v_{\text{auto with respect to railroad car}} + 10$ and $v_{\text{auto with respect to railroad car}} = -12$ m/s.

18. **(D)** The elastic potential energy is given by $U = \frac{1}{2}kx^2 = \frac{1}{2}(200)(0.8)^2 = 64\ \text{J}$.

19. **(A)** According to conservation of energy, the ball's minimum speed occurs when the ball is at its maximum height. At the maximum height, when the ball is moving as slowly as possible without falling out of its circular path, the tension force is zero, so the gravitational force must provide the centripetal force. (If the ball were moving faster, the tension force would be positive; if the ball were moving slower, the gravitational force would be greater than the required centripetal force, so the ball would move in a *tighter* curve than the circular path.)

$$mg = \frac{mv^2}{r}$$

Solving this equation for v yields choice (A).

20. **(D)** $F_x = -dU/dx$, so

$$[U(x_1) - U(0) = U(x_1) - 0] = \left[-\int_0^{x_1} F(x)dx = -\int_0^{x_1}(-bx^3 + cx^5)dx = \frac{bx^4}{4} - \frac{cx^6}{6}\right]$$

21. **(B)** The free-body diagram is shown below.

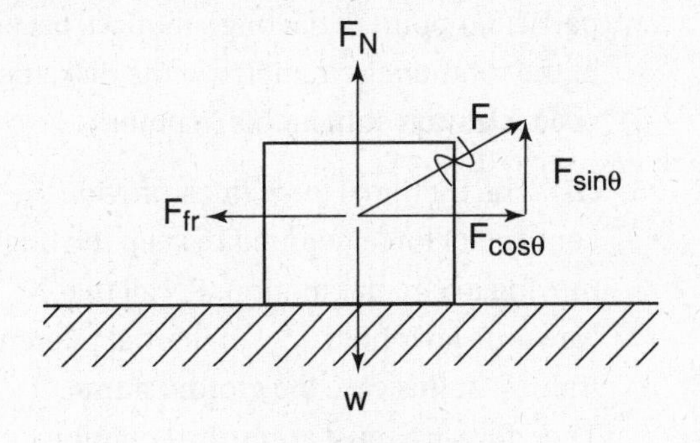

Question 21

22. **(B)** $P = \mathbf{F} \cdot \mathbf{v} = F_{\parallel} v = F(\cos\theta)v$. In words, the power is given by the component of the force parallel to the velocity, $+F\cos\theta$ multiplied by the magnitude of the velocity v.

23. **(B)** Angular momentum is conserved about the axle of the disk (assuming there is no friction in the axle). Because the initial angular momentum about this point is zero, the final angular momentum must also be zero. This requires that the disk rotate in the direction opposite the bug (clockwise) with an angular momentum of equal magnitude:

$$\left[I_{bug}\omega_{bug} = (mr^2)\left(\frac{v}{r}\right)\right] = \left[I_{disk}\omega_{disk} = \left(\frac{1}{2}Mr^2\right)\omega\right]$$

Solving for the magnitude of the angular velocity gives choice (B).

24. **(C)** Consider, for example, when the bug crosses the $+x$-axis (setting the origin of the coordinate system at the axle of the disk). The velocity of the bug with respect to the ground is $v\hat{j}$ because the bug is moving with speed v counterclockwise. The speed of the point on the disk where the bug stands is $-r\omega\hat{j}$ (relative to the ground) because the disk is rotating clockwise with angular velocity ω. The relative velocity of the bug with respect to the disk is given by

$$\mathbf{v}_{\text{bug relative to disk}} + \mathbf{v}_{\text{disk relative to lab}} = \mathbf{v}_{\text{bug relative to lab}}$$

$$\mathbf{v}_{\text{bug relative to disk}} = \mathbf{v}_{\text{bug relative to lab}} - \mathbf{v}_{\text{disk relative to lab}}$$

$$= v\hat{j} - (-r\omega\hat{j}) = (v + r\omega)\hat{j}$$

Although we solved this problem for a particular point in the bug's motion, because of the rotational symmetry of the disk, the speeds always sum in this manner.

25. **(B)** The frictional force must provide the centripetal force required to keep the bug moving in circular motion. Recall that Newton's laws hold only in inertial reference frames, in this case the ground frame. Therefore, we must apply the equation $F_{cent} = mv^2/r$ in the ground frame (note that v is the bug's speed in this frame of reference). Setting the centripetal force equal to the frictional force gives $mv^2/r = \mu_s mg$. Solving for the coefficient of friction yields choice (B).

Note: Why is this static rather than kinetic friction? Although the bug is moving, its feet are not moving with respect to the disk (i.e., the bug's feet are not slipping as it walks).

26. **(C)** If we set the $+x$-axis to point parallel to the velocity of m_1, the velocity of the center of mass is

$$v_{CM} = \frac{\sum m_i v_i}{\sum m_i} = \frac{m_1 v_1 - m_2 v_2}{m_1 + m_2}$$

The speed is then the absolute value of this velocity.

27. **(D)** Choice (D): The force that m_1 exerts on m_2 is an internal force, so it is balanced by another internal force so that together these internal forces make no contribution to the net force or the net work on the two-block system. (If we were considering the blocks separately, this force would no longer be an internal force.) See Example 4.6 on page 121 for further explanation.

28. **(D)** The particle oscillates between points A and H (the particle has a total energy of 1 J, so it cannot move beyond these points because that would require having negative kinetic energy). Since point G is within the boundaries of this oscillation, it will indeed reach and pass point G. Because there is no simple linear restoring force [which would be equivalent to an upward-parabolic $U(x)$ curve], the particle does not move in simple harmonic motion. The force, equal to $F(x) = -dU/dx$, is zero at points C, D, and E because the slope of the function $U(x)$ is zero at these points. The particle's speed is maximized at point C (not point E), as this is the place where the particle's potential energy is the lowest (and thus, its kinetic energy is the highest).

29. **(D)** The final angular velocity is $\omega = v/r$. Using the equation $\omega_f^2 = \omega_0^2 + 2\alpha\Delta\theta$ yields $\Delta\theta = \omega_f^2/2\alpha = v^2/2\alpha r^2$. Substituting in for all the values given in the problem results in choice (D).

30. **(A)** SHM requires a linear restoring force, which is graphically equivalent to the $F(x)$ curve passing through the x-axis with a negative slope (the x-intercept is the equilibrium of the SHM).

31. **(D)** Consider moving an object from the origin to $x = 1$. The force $F(x)$ performs an amount of work $W = \int \mathbf{F} \cdot d\mathbf{r} = \int F_x dx$, which is equal to the area under the $F(x)$ curve. Therefore, the force that produces the greatest potential energy at $x = 1$ must maximize the work and thus maximize the area under the curve. The force with the largest area under the $F(x)$ curve is force (D).

32. **(E)** The magnitude of the frictional force is given by $F_{fr} = \mu_k F_N = \mu_k mg = 0.59$ N. The work done by friction is $W = -F_{fr}d$, where d is the total displacement, i.e., 6 m. Thus, $W = -(0.59)(6) = -3.53$ J.

33. **(D)** The angular displacement is proportional to t^2; thus, the angular acceleration α is constant and given by the second time derivative of θ, i.e., $\alpha = 8$. Rolling without slipping implies that the acceleration a of the center of mass obeys $a = r\alpha = (2)(8) = 16\ \text{m/s}^2$.

34. **(D)** Differentiating once to calculate the angular velocity from the angular position gives $\omega = 8t$. The speed at the top of the ball is given by the equation $v = 2r\omega$ (for further explanation, see Chapter 9). Substituting in numbers yields choice (D).

35. **(D)** There is no external force, so the momentum and the velocity of the center of mass remain constant. Because the center of mass continues to move with its original velocity, the position of the center of mass does not remain constant but instead moves at constant velocity. Because there is no external torque, the angular momentum of the asteroid system (about any stationary reference point) remains constant. The kinetic energy does not remain constant during the explosion but rather increases dramatically as the asteroid explodes: Presumably some source of nonmechanical energy (e.g., chemical energy) was quickly converted to mechanical energy, resulting in the explosion.

Free-Response

MECHANICS I

(a) The free-body diagram for the skier at rest on the slope appears in the figure. If the skier is at rest, the acceleration is zero, and thus the net force in each of the directions of the tilted coordinate system is zero:

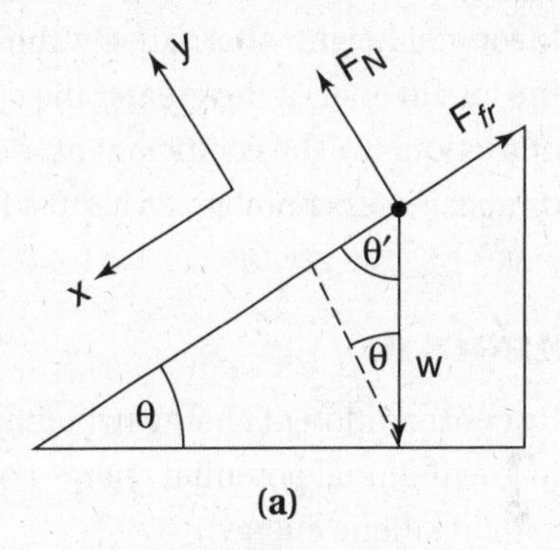

(a)

$$F_{\text{net},y} = 0 = F_N - mg\cos\theta \Rightarrow F_N = mg\cos\theta$$

$$F_{\text{net},x} = 0 = F_{fr} - mg\sin\theta \Rightarrow F_{fr} = mg\sin\theta$$

Finally, the minimum coefficient of static friction is

$$\mu_s = \frac{F_{fr}}{F_N} = \frac{mg\sin\theta}{mg\cos\theta} = \tan\theta$$

Note: If the coefficient of static friction were larger, the static frictional force would be the same (it would simply be less than the maximum possible frictional force for that coefficient of friction).

PRACTICE TEST 2

(b) We apply the law of conservation of energy. Setting $y = 0$ at the bottom of the incline and designating the final position where the skier has risen to a height h swinging on the rope,

$$E_{final} = KE_{final} + PE_{final} = 0 + mgh$$

$$E_{initial} = KE_{initial} + PE_{initial} = 0 + mg(d \sin \theta)$$

$$\text{Energy lost} = \Delta E = E_{final} - E_{initial} = mgh - mg(d \sin \theta)$$

(c) The change in the mechanical energy is equal to the work done by nonconservative forces, in this case kinetic friction:

$$\Delta E = \text{work}_{nonconservative} = \mathbf{F} \cdot \Delta \mathbf{r}$$

$$mg(h - d \sin \theta) = [\mathbf{F}_{fr} \cdot \Delta \mathbf{r} = -\mu_k mg \cos \theta(d)]$$

$$\mu_k = \frac{h - d\sin\theta}{-d\cos\theta} = \frac{d\sin\theta - h}{d\cos\theta}$$

Symbolic answer check: (1) If $d\sin \theta = h$, then no energy is lost because the skier rises to the initial height. Alternatively, the greater the loss in energy, the greater the coefficient of friction. (2) The coefficient of friction is a dimensionless number, as it must be.

MECHANICS II

(a) By conservation of energy (whereby the loss in gravitational potential energy equals the gain in kinetic energy):

$$mgl\sin\theta = \frac{1}{2}mv^2 \Rightarrow v = \sqrt{2gl\sin\theta}$$

Symbolic answer check: (1) The units are correct. (2) Velocity increases with increasing gravitational force, increasing length, and increasing θ, all of which make sense.

(b)

$$a_{cent} = \frac{v^2}{r} = \frac{2gl\sin\theta}{l} = 2g\sin\theta$$

(c) At this point we need the free-body diagram shown. The net force in the tangential direction is $mg \cos \theta$, so the acceleration in the tangential direction is $g \cos \theta$. Note that we do not have to know the value of the tension in order to compute the tangential acceleration or net force.

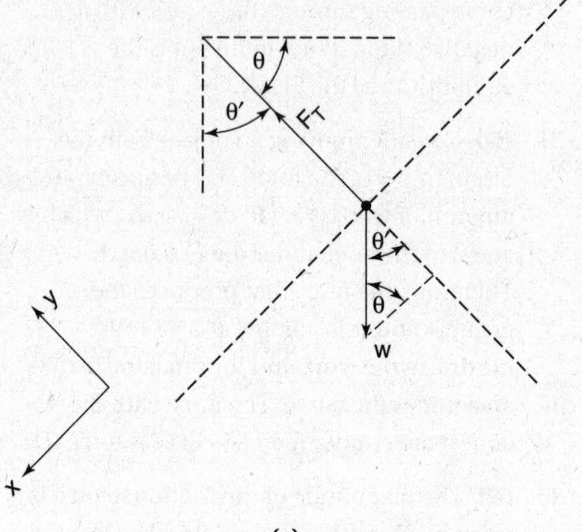

(c)

(d) Referring to the free-body diagram (and noting that the y-coordinate in this diagram is radial),

$$F_{net,y} = F_T - mg \sin \theta = F_{cent} = ma_{cent} = 2mg \sin \theta$$

$$F_T = 3mg \sin \theta$$

(e) From part (a), you know that $v = \sqrt{2g\ell\sin\theta}$. So you should plot v^2 versus ℓ. Note that at the bottom of the circle $\theta = \pi/2$ and $\sin\theta = 1$.

PRACTICE TEST 2

(f) The plot is shown below. Determine the slope of the fitted line:

$$\text{slope} = \frac{(3.85)^2 - (2.43)^2}{75-30}\ \text{cm/s}^2 = 19.8\ \text{m/s}^2$$

From part (a), you expect the slope to be $2g = 19.6\ \text{m/s}^2$.

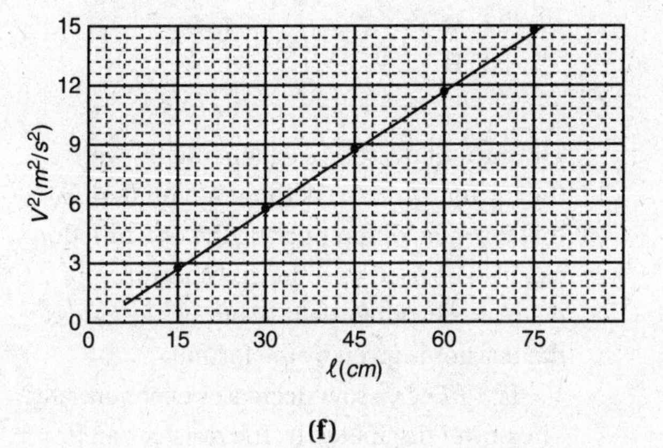

(f)

MECHANICS III

(a) The rotational inertia of a point mass is given by the equation $I = ml^2$. Its angular momentum is $\mathbf{L} = I\omega = ml^2\omega\hat{k}$ (using the right-hand rule to determine that the angular velocity is in the $+z$-direction).

(b) Because the retarding force is perpendicular to the position vector of the mass, the magnitude of the torque is equal to the product of the force and the distance from the axle: $\tau = rF_{\perp} = (l)(b\omega)$. The direction of the torque is in the $-z$-direction, opposing the angular velocity (because the force is retarding). Thus, vectorially, $\tau = -bl\omega\hat{k}$.

(c) The angular acceleration, α, obeys $\tau = I\alpha$. Thus, $\alpha = bl\omega / ml^2 = b\omega / ml$.

ELECTRICITY AND MAGNETISM

ANSWER KEY

1. **D**	8. **D**	15. **D**	22. **D**	29. **C**
2. **B**	9. **A**	16. **B**	23. **D**	30. **C**
3. **E**	10. **D**	17. **D**	24. **D**	31. **E**
4. **C**	11. **E**	18. **D**	25. **C**	32. **B**
5. **D**	12. **D**	19. **D**	26. **C**	33. **D**
6. **D**	13. **D**	20. **A**	27. **D**	34. **B**
7. **B**	14. **C**	21. **A**	28. **B**	35. **A**

ANSWERS EXPLAINED

Multiple-Choice

1. **(D)** The negative charge is attracted to the positively charged left plate and repelled by the negatively charged right plate, so it moves to the left. After it exits the capacitor, it continues on with constant velocity because the electric field is zero outside a capacitor. Its kinetic energy is KE = $qV/2$, half the kinetic energy that would be gained by moving the charge the entire way across the capacitor because it started in the middle of the capacitor.

2. **(B)** The electric field within a spherically symmetric shell of charge is zero as discussed in Chapter 13 (because the enclosed charge is zero), so the potential is constant, eliminating choices (C), (D), and (E). Consider the equation $V = \int dQ/(4\pi\varepsilon_0 r)$. Because all of the charge has the same sign, there is no cancellation and the potential should be nonzero, eliminating choice (A).

3. **(E)** The charge on the capacitor decreases exponentially according to the equation $Q = Q_0 e^{-t/RC}$. The energy stored in the capacitor is given by the formula

$$U = \frac{Q^2}{2C} = \frac{Q_0^2 e^{-2t/RC}}{2C}$$

so it too decreases exponentially. The electric field across the capacitor is equal to

$$E = \frac{V}{d} = \frac{Q/C}{d} = \frac{Q_0}{Cd} e^{-t/RC}$$

so it decreases exponentially.

The current decreases exponentially, with the form $I = I_0 e^{-t/RC}$. [Without remembering this formula, differentiating the $Q(t)$ function can also be used to show that current decreases exponentially.] The voltage across the resistor is given by the formula $V = IR = RI_0 e^{-t/RC}$, so it decreases exponentially. The power dissipated by the resistor can be written as $P = V^2/R = RI_0^2 e^{-2t/RC}$, so it decreases exponentially.

Note the way the energy in the capacitor decreases with the same time constant as the power dissipated by the resistor. This relationship is required by conservation of energy.

4. **(C)** This is a Gauss's law problem—calculating the field in each region involves drawing imaginary gaussian cylinders and applying Gauss's law. Such cylinders with radii $r > 2R$ and $r < R$ enclose zero net charge, leading to the conclusion that the electric field is zero for $r < R$ and $r < 2R$. The only graph that satisfies both these conditions is choice (C).

5. **(D)** Recall that $E_r = -dV/dr$. Based on this equation, because the electric field is zero within the regions $r < R$ and $r < 2R$, the potential must be constant, immediately eliminating choices (A) and (B). Because potential is the line integral of the electric field, $V = -\int \mathbf{E} \cdot d\mathbf{r}$, it must be continuous in the absence of infinite electric fields (e.g., due to a point charge), making choice (C) incorrect. To choose between choices (D) and (E), we simply decide what the sign of the potential should be for $r < R$. Because

points with $r < R$ are closer to the negative charge than the positive charge (and there are equal amounts of these charges), the potential must be negative [recall that potential is given by the integral $V = \int (dQ/4\pi\varepsilon_0 r)$]. The sign can also be determined by realizing that the electric field *between* the two cylinders points inward. Recalling that the electric field points from high to low potential, we see that choice (D) is correct.

6. **(D)** First, to oppose the increase in inward flux, the induced current must flow counterclockwise, producing an outward flux. What is the magnitude of the current? The induced EMF is

$$|\mathcal{E}| = \left|\frac{d\Phi_B}{dt}\right| = \left|\frac{d}{dt}(\mathbf{B}\cdot\mathbf{A})\right| = A\frac{dB}{dt} = \frac{AB}{t}$$

Because the magnetic field increases linearly, its slope dB/dt is simply

$$\frac{dB}{dt} = \frac{\Delta B}{\Delta t} = \frac{B-0}{t-0} = \frac{B}{t}$$

The resistance is R, so according to Ohm's law the current is $I = V/R = AB/Rt$. Finally, to express the result in terms of the variables in the problem, $A = \pi r^2$, so $I = \pi r^2 B/Rt$.

7. **(B)** The electric field equals the negative derivative of the potential along any line, and this holds for a line passing through the radius. Mathematically, $E_r = -dV/dr = -a$. Even without recalling this equation in its entirety, one could arrive at the correct answer by combining the information that (1) the magnitude of the electric field is related to the derivative of the potential, and (2) the electric field always points in the direction of decreasing potential.

   Choice (E) is false. Any field that has an associated potential function is conservative. (That is, any field for which a potential energy function can be defined is a conservative field.)

8. **(D)** The loop rule for an LC circuit is $L\frac{d^2Q}{dt^2} + \frac{Q}{C} = 0$. This looks just like a harmonic oscillator with $\omega = \frac{1}{\sqrt{LC}}$. The current is given by $I = \frac{dQ}{dt}$ and the maximum current obeys $I_{max} = \omega Q_{max}$. The maximum charge is the initial charge of the capacitor, $Q = CV = (25 \times 10^{-6})(300) = 7.5 \times 10^{-3}$C. Thus, $I_{max} = \omega Q_{max} = \frac{1}{\sqrt{(0.01)(25\times10^{-6})}} 7.5\times10^{-3} = 15\text{ A}$.

9. **(A)** Potential is equal to the potential energy per unit charge, so only choices (A) and (B) have the correct units. To figure out the signs, consider the case where Q and E are positive. If positive energy is required to move the charge from point A to point B, point B must be at higher potential energy (uphill), so $Q(V_B - V_A)$ must be positive. Because E is positive, only the equation $V_B - V_A = E/Q$ has signs that are consistent with this situation.

10. **(D)** The most useful form of the equation for the energy stored in a capacitor to use here is $U = \frac{1}{2}CV^2$. Because the voltage remains constant and the capacitance decreases by a factor of 2 (recall that $C = \varepsilon_0 A/d$), the energy stored in the capacitor decreases by a factor of 2, from $U_0 = \frac{1}{2}C_0V^2$ to $U_F = \frac{1}{4}C_0V^2$, and the change in energy is equal to $-\frac{1}{4}C_0V^2$.

11. **(E)** The current through the circuit is

$$I = \frac{V}{R_{circuit}} = \frac{V}{r_{int} + R}$$

   The power dissipated through the heating element is then

$$P = I^2R = \left(\frac{V}{r_{int}+R}\right)^2 R = \frac{V^2R}{(r_{int}+R)^2}$$

   Because this is not equivalent to any of the given choices, the answer is choice (E).

12. **(D)** The equation $V = -\int E_x dx$ always holds, so $V = Ed$ is valid even in the case of a dielectric. Therefore, because the capacitor is

PRACTICE TEST 2

always connected to the battery and the plate separation distance is unchanged, V and d are fixed and so E must remain the same.

What effect does this have on the plate charge? The key equation here is $Q = CV$. The capacitance decreases in the absence of the dielectric. The voltage is held fixed (because the capacitor is attached to the battery), so the charge on the plates must decrease.

How is it that the electric field can remain constant while the charge on the plates decreases? In the presence of the dielectric, the charge on the plates is partially screened by the induced charge on the surface of the dielectric. Once the dielectric is removed, the charge on the plates no longer needs to cancel this induced charge, so it decreases.

13. **(D)** By the work-energy theorem, we know that $\frac{1}{2}mv^2 = \frac{e^2}{4\pi\varepsilon_0}\left(\frac{1}{r_2} - \frac{1}{r_1}\right)$, where r_1 and r_2 are the initial and final separations of the electron and proton, respectively.

14. **(C)** A little experimentation with the right-hand rule shows that for points in between two wires carrying current in the same direction, the two wires produce magnetic fields in opposite directions. The wire closer to the point produces a larger magnetic field, so the net magnetic field points in the direction determined by this wire. In this case, because the magnetic field points out of the page, the current through the wire closest to point A (the right wire) must move toward the top of the page.

The magnitude of the magnetic field due to a current-carrying wire is $B = \mu_0 I/2\pi r$. Because the two wires produce fields in opposite directions, in this case the net magnetic field has magnitude

$$B_0 = \frac{\mu_0 I}{2\pi d} - \frac{\mu_0 I}{2\pi(2d)}$$

Solving this equation for the current yields $I = 4\pi d B_0/\mu_0$, so the correct answer is choice (C).

15. **(D)** Recall that potential is given by the integral $V = \int dQ/4\pi\varepsilon_0 r$. Therefore, moving negative charge closer to point A and positive charge farther away from point A *decreases* the potential at point A. The opposite occurs at point B, causing the potential to *increase* at point B. Therefore, choices (A) and (C) are false.

Because of symmetry, a sphere of uniform charge density produces a zero net field at its center. Therefore, the electric field at points A and B is initially zero. Exchanging the tops of the spheres leads to nonzero electric fields at points A and B (pointing up at point A and down at point B). Therefore, (B) is false, while (D) is true. As a result of the symmetry of the situation, points A and B have the same potential. Therefore, zero work is required to move any charge from point A to point B, making choice (E) false.

16. **(B)** While entering the magnetic field, the induced current flows clockwise, moving downward through the right segment of the loop that is exposed to the magnetic field. While leaving the magnetic field, the induced current flows counterclockwise, moving downward through the left segment of the loop that is exposed to the magnetic field.

17. **(D)** The force on the charge is proportional to the electric field according to the equation $\mathbf{F} = q\mathbf{E}$. The electric field is the negative derivative of the potential, but a single value of potential alone does not indicate anything about the electric field or the electric force.

However, we can still determine the direction of the force. Electric fields are always perpendicular to equipotential curves, pointing in the direction of decreasing potential. Because the charge is positive, the force is parallel to the electric field.

18. **(D)** Because the plate separation d is very small compared to the radius of the cylinder, it is reasonable to approximate the capacitor as a parallel plate capacitor with plate

PRACTICE TEST 2

separation d and plate area $2\pi rl$. (This makes sense: Because $d << r$, the electric field is relatively constant between the plates and is perpendicular to the plates.) Therefore, the approximate capacitance is

$$C = \frac{\varepsilon_0 A}{d} = \frac{\varepsilon_0(2\pi rl)}{d}$$

Even without making this observation, a few choices can be eliminated. Compared to the formula $C = \varepsilon_0 A/d$, choice (E) has incorrect units, so it can be eliminated. Choice (A) does not include the length of the capacitor. However, we know from the simpler case of a parallel plate capacitor that the dimension of the plates is relevant to the value of the capacitance, so we can safely eliminate choice (A).

19. **(D)** Imagine a spherical gaussian surface passing through any point outside the enlarged sphere. Doubling the radius of the sphere increases the enclosed charge by a factor of 8 (because the volume of a sphere is proportional to r^3 and the charge is proportional to the volume). Thus, by Gauss's law, the flux through the gaussian surface (whose surface area is fixed) increases by a factor of 8, and thus so does the electric field.

20. **(A)** The magnetic force provides the centripetal force: $mv^2/r = qvB$. Rearranging this equation to solve for r in terms of m yields $r = (v/qB)m$. Therefore, the graph is linear, passing through the origin.

21. **(A)** Because the rod is negatively charged, the field to the right of the rod is negative and the field to the left of the rod is positive, eliminating all choices except (A) and (D). What happens close to the origin? At points extremely close to the origin, the contributions to the field made by the rod near the point increase (based on the equation $dE = dQ/4\pi\varepsilon_0 r^2$), and the electric field due to points farther away decreases (because of vector cancellation; these points produce field mainly in the y-direction that cancels the field produced by matching pieces of charge an opposite distance from the origin). Therefore, the fact that the ends of the rod are missing doesn't significantly affect the electric field close to the rod, allowing us to approximate the rod as infinite and apply Gauss's law for small values of x. Applying Gauss's law with cylindrical symmetry indicates that the field increases without bound as x approaches zero, which is consistent only with choice (A).

22. **(D)** Recall that the current flowing through inductors is generally continuous because a discontinuous current would cause the time derivative of the current to be infinite and thus produce an infinite voltage. How is this theory applicable to the current problem? In choice (D), when the switch is opened, an abrupt change in the current occurs that produces a very large induced EMF, causing a spark to instantaneously connect the recently opened terminals of the switch. [In choice (C), the current is zero before and immediately after closing the switch, so there is no discontinuity and no spark.]

23. **(D)** All the charge in both distributions is at a fixed distance r from the points of interest. Therefore, the potential at both point A and point B is $V = Q/4\pi\varepsilon_0 r$ and these potentials are equal.

The electric field sums as vectors rather than scalars, so it is more complicated. In both cases, the net electric field points in the $+x$-direction because of symmetry about the x-axis. However, in distribution B, because the charge is concentrated closer to the horizontal axis, there is less cancellation of the differential electric field vectors compared to case A. (For further explanation, see Chapter 12.)

24. **(D)** Recalling that the same current flows through elements in series, you could recognize that choice (D) is correct without paying attention to the details of the problem.

The other choices are incorrect for the following reasons. The resistance of a cylindrical resistor is given by the equation $R = \rho l/A$. Because the second resistor has twice the length and four times the area compared to the first resistor, it has half the resistance of the first resistor, eliminating choice (A). The voltage across the resistors, given by $V = IR$, is different because the two resistors share the same current but have different resistances, eliminating choice (B). Similarly, power can be expressed as $P = I^2R$, and because the resistors have the same current and different resistances, they dissipate different amounts of power, eliminating choice (D). Finally, R_2 is half of R_1 (as discussed above), and because resistors in series add, the equivalent resistance is equal to $R_{eq} = 1.5R_1 = 3R_2$, so choice (E) is incorrect.

25. **(C)** The equivalent resistance of the top branch is 6 Ω, so the current running through the upper branch is $I = V/R = 8\,V/6\,\Omega = \frac{4}{3}$ A. Now the potential difference between points *A* and *B* is equal to the sum of the potential jumps as we move from point *A* to point *B*. Because the resistance *R* is unknown, we must proceed from point *A* through the 2-Ω resistor and the 4-Ω resistor. Equating the sum of these potential jumps to zero (and designating the current through the middle branch as *I*) yields

$$\left(\frac{4}{3}\text{ A}\right)(2\,\Omega) - I(4\,\Omega) = 0$$

This equation is easily solved for $I = 2/3$ A.

26. **(C)** Each set of three capacitors in series has an equivalent capacitance of $C/3$. Because capacitors in parallel add, the combination of these three equivalent capacitors (each with a capacitance of $C/3$) in parallel produces a network with an equivalent capacitance of *C*.

27. **(D)** Imagine charging the capacitor network. Equal amounts of charge would flow through the three branches of the network, causing equal amounts of charge to build up on every one of the nine capacitors. Therefore, for the voltage of one capacitor to be *V*, the voltage of every capacitor in the network must be *V*. When the voltage across each capacitor is *V*, what is the voltage across the entire network? To calculate the voltage across the network, we can imagine traversing one branch and summing the voltage jumps, as in a Kirchhoff's loop equation. Because the capacitors have the same polarity, their voltages sum, causing the voltage across the branch (and thus the network) to be 3 V.

28. **(B)** Sides 1, 2, 4, and 6 are parallel to the electric field, so the electric flux through them is zero. Sides 3 and 5 are perpendicular to the field, so the magnitude of the flux passing through each of these faces is equal to the area (which is 1) multiplied by the field, *E*. Because the field points in the positive *z*-direction, it enters the cube at side 5 (yielding *negative outward* flux) and leaves the cube at side 3 (yielding *positive outward* flux).

29. **(C)** The energy stored in a capacitor can be expressed in the form $U = Q^2/2C$. Therefore, the amount of time required for a capacitor to reach 90% of its maximum energy is equal to the amount of time required to reach 90% of its maximum Q^2 value (because *C* is fixed for any given capacitor). The circuit with the smaller time constant (smaller $\tau = RC$), and thus the smaller equivalent capacitance, reaches the 90% value more quickly. Therefore, ordering the networks from least to greatest charging time is simply a matter of ordering them from least to greatest equivalence capacitance. The equivalent capacitances of networks I through IV are $\frac{1}{3}C, \frac{3}{2}C, \frac{2}{3}C$, and $3C$, respectively.

30. **(C)** From the equation $U = \frac{1}{2}CV^2$, networks with higher equivalent capacitance store more energy when connected to the same voltage. Therefore, ordering the capacitors from least to greatest energy is equivalent to ordering them from least to greatest capacitance. This ordering is the same as in question 29, choice (C).

31. **(E)** What is the current through the circuit? A Kirchhoff's loop equation for the circuit (moving in the clockwise direction and choosing positive current clockwise) is $V - IR - V - IR = 0$, which simplifies to $2IR = 0$. Therefore, the current around the circuit is zero, which is not surprising because the two identical batteries are connected so as to oppose each other. Thus, the power dissipated in each of the resistors is zero, and absolutely no thermal energy is produced in the circuit.

32. **(B)** The instant after the switch is closed in an *RL* circuit, the inductor acts like an open circuit (because the current through the inductor must be continuous and thus initially zero). Therefore, the equivalent resistance is $R_1 + R_2$, and the current is calculated using Ohm's law with the result given by choice (B).

    Elimination approach: Even if you have forgotten about *RL* circuits, you could eliminate enough choices to make a good guess. First, because no circuit elements block the flow of current through resistors R_1 and R_2, these resistors determine the current and are expected to show up in the final answer, eliminating choices (A) and (C). The denominator of choice (E) has mixed units ($\Omega + \Omega^{-1}$), so it must be incorrect. This leaves us to guess between choices (B) and (D).

33. **(D)** At equilibrium the inductor acts like a short circuit, so to calculate the current we simply ignore the inductor, calculate the equivalent resistance, and apply Ohm's law. To calculate the equivalent resistance, note that R_2 and R_3 are in parallel with each other, so their equivalent resistance together is

$$\frac{1}{(1/R_2)+(1/R_3)} = \frac{R_2 R_3}{R_2 + R_3}$$

Because these resistors are in series with R_1, the equivalent resistance of the entire circuit is

$$R_{eq} = \frac{R_2 R_3}{R_2 + R_3} + R_1$$

Applying Ohm's law yields the current given in choice (D).

34. **(B)** Potential increases toward infinity very close to a positive charge (whether approaching from the right or the left) and decreases to negative infinity when very close to a negative charge (whether approaching from the right or the left) because of the equation $V = \Sigma(Q/4\pi\varepsilon_0 r)$. The only graph that satisfies these simple requirements is graph (B). (There are, of course, many other criteria that could be used to figure out which graph is correct.)

    Alternative solution: Sketch the potential-vs-position plots for both charges separately and superimpose them, as shown (valid because potential obeys the principles of superposition).

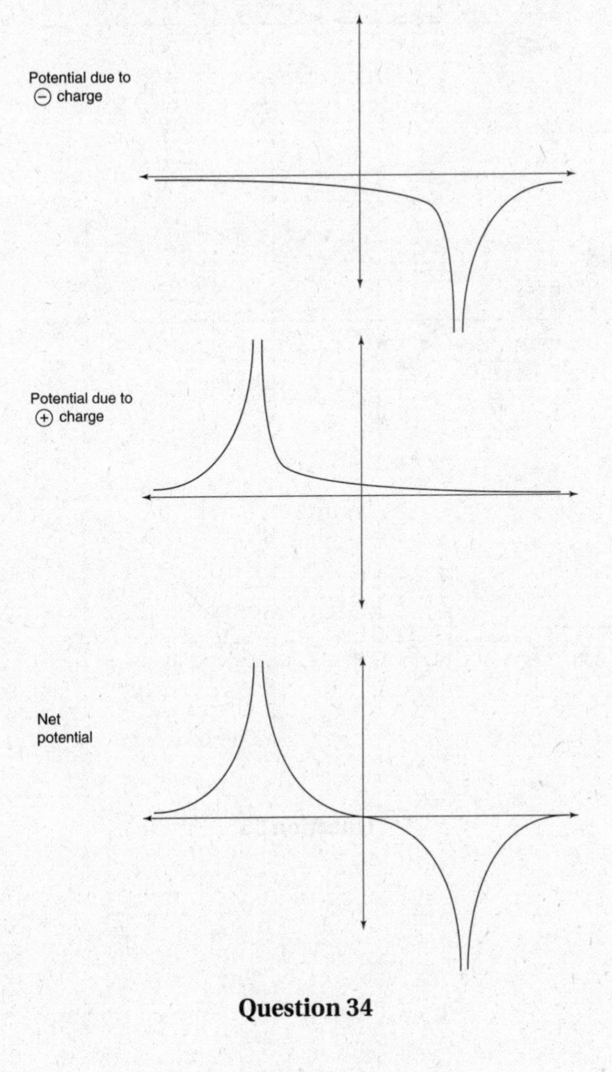

Question 34

PRACTICE TEST 2

35. **(A)** The electric field of a point charge (of either sign) jumps discontinuously from positive to negative infinity as we cross the position of the charge. This immediately eliminates choices (B), (D), and (E). The only difference between choices (A) and (C) is that the graph in choice (C) passes through the origin while the graph in choice (A) always lies above the origin. At the origin, the electric field for the charge distribution shown is nonzero and points to the right; thus, the answer is choice (A).

Alternative solution: Sketch the electric field-vs-position plots for both charges separately and superimpose them, as shown.

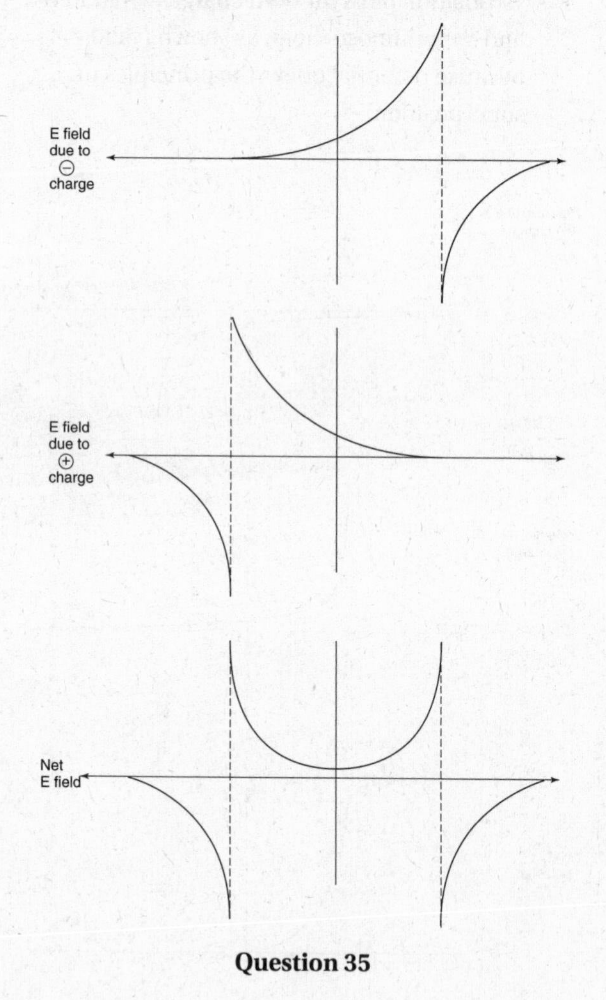

Question 35

ELECTRICITY AND MAGNETISM I

(a) All the differential charge elements making up the semicircle lie at a constant distance a from the origin, so the potential is simply

$$V=\frac{Q}{4\pi\varepsilon_0 a}$$

(b) Based on symmetry, the only well-defined direction at the origin is along the x-axis. Thus, the y-component of the electric field is zero. Therefore, we divide the semicircle into differential arc lengths ds and sum the contributions of each differential length to E_x. As shown in the figure, it is more useful to define the differential regions in terms of an angle θ, so that (recalling that arc length is equal to the radius multiplied by the angle of the arc in radians):

$$ds = a\, d\theta$$

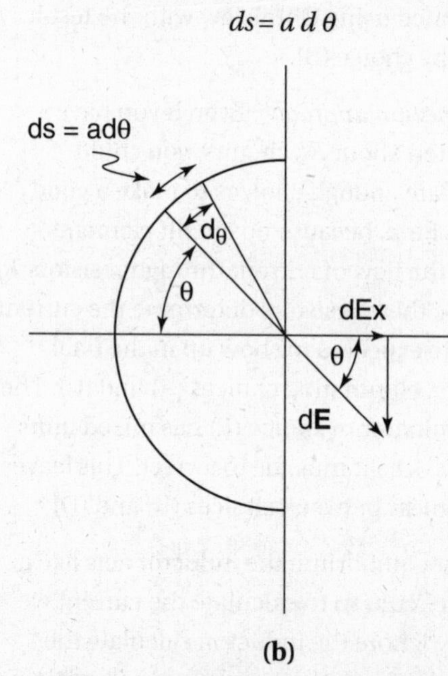

(b)

The amount of charge in each differential arc length is equal to the differential length multiplied by the linear charge density λ, which is given by λ = charge/length = $Q/\pi a$:

$$dQ=\lambda ds=\left(\frac{Q}{\pi a}\right)(a\, d\theta)=\frac{Q\, d\theta}{\pi}$$

The contribution of each differential length to the x-component of the electric field is

$$dE=\frac{dQ}{4\pi\varepsilon_0 a^2}\cos\theta=\frac{Q}{4\pi^2\varepsilon_0 a^2}\cos\theta\, d\theta$$

Integrating:

$$E_x = \int_{-\pi/2}^{\pi/2} \frac{Q}{4\pi^2 \varepsilon_0 a^2} \cos\theta \, d\theta = \frac{Q}{2\pi^2 \varepsilon_0 a^2}$$

(c) The positive charge will be repelled by the semicircle of positive charge. It moves in the +*x*-direction to infinity, converting electrical potential energy to kinetic energy. Therefore, as time approaches infinity, its position approaches (∞, 0) and its velocity can be obtained by conservation of energy:

$$qV = \frac{qQ}{4\pi\varepsilon_0 a} = \frac{1}{2}mv^2$$

$$v = \sqrt{\frac{qQ}{2\pi\varepsilon_0 am}}$$

(d) Recall that work is given by the equation $W = \int \mathbf{F} \cdot d\mathbf{r}$. For a very small increment, Δx, we can assume that the electric force is roughly constant, so that the work is

$$W = \int \mathbf{F} \cdot d\mathbf{r} \approx F\,\Delta x = (qE)\Delta x$$

$$= \left(\frac{qQ}{2\pi^2 \varepsilon_0 a^2}\right)\Delta x$$

ELECTRICITY AND MAGNETISM II

(a) The resistance of two elements in series is the sum of the two separate resistances. The resistance of the metal is approximately zero, and the resistance of the other cylinder is given by the formula $R = \rho l / A$. Thus, the sum of the two resistances, the resistance of the entire apparatus, is $R = \rho l / A$.

(b) Using Ohm's law, $V_{AB} = IR = I\rho l / A$, and $V_{BC} = IR = I(0) = 0$. Recall that current flows from high to low potential, so points upstream of the resistor (point A here) are at higher potential than downstream points (points B and C, which are at equal potential).

(c) Because the resistance/length is uniform throughout each cylinder, the electric field is constant and is given by the equation $|E| = V / l$. For the resistive cylinder, this yields $|E| = I\rho / A$. For the metal cylinder, this gives a value of $|E| = 0 / l = 0$.

(d) The interface is planar, so our gaussian surface is a cube (with faces of area *A*) bisected by the interface, with two faces parallel to the interface (see figure). The four faces perpendicular to the interface experience no flux because the electric field is parallel to the face and $\mathbf{E} \cdot d\mathbf{A} = 0$. Of the two faces parallel to the interface, the face immersed in the metal cylinder experiences zero flux because the electric field is zero in the metal. The last face, the face parallel to the interface and lying in the resistive material, experiences an inward flux as follows. (Keep in mind that the electric field is perpendicular to the surface and that inward flux is negative.)

$$\Phi = -EA = \frac{Q_{\text{enclosed}}}{\varepsilon_0}$$

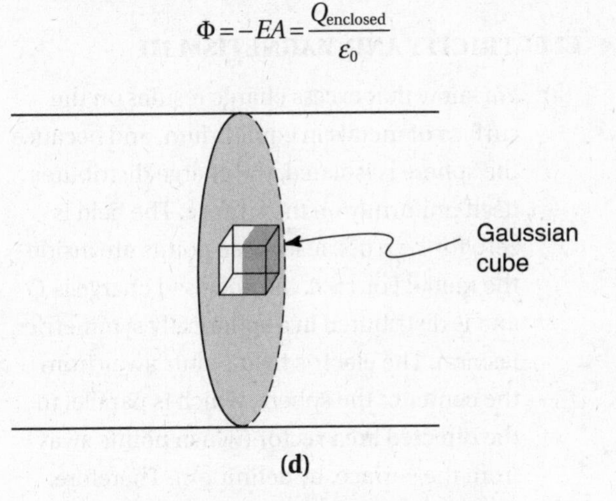

(d)

Therefore, the surface contains negative charge of surface density:

$$\sigma = \frac{Q_{\text{enclosed}}}{A} = -\varepsilon_0 E$$

Thus, the complete interface contains a negative charge equal to

$$Q_{\text{total}} = (\text{total area})\sigma = -A\varepsilon_0 E$$

$$= -A\varepsilon_0 \left(\frac{I\rho}{A}\right) = -I\rho\varepsilon_0$$

Check: The electric field lines in the resistive cylinder end on this negative charge.

(e) The plot is shown below. Calculate the slope of the fitted line:

$$\text{slope} = \frac{(172-35)\text{ V}}{(10-2)\text{ mA}} = 17{,}125\ \Omega$$

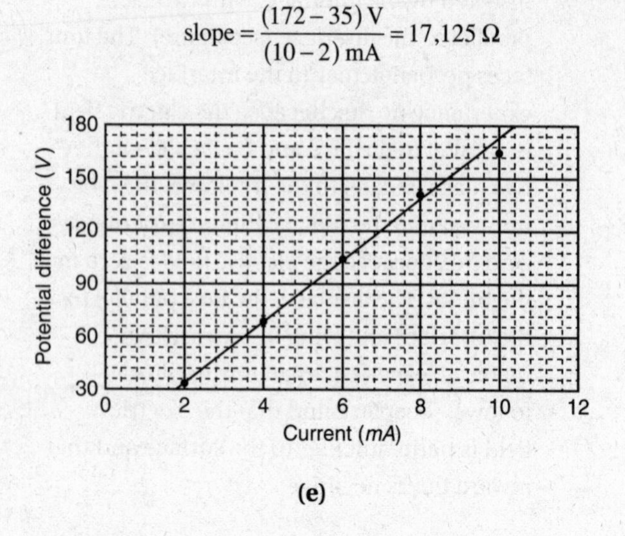

(e)

ELECTRICITY AND MAGNETISM III

(a) We know that excess charge resides on the surface of metals in equilibrium, and because the sphere is isolated, the charge distributes itself uniformly on the surface. The field is zero for $r < a$ because these points are inside the metal. For $r > a$, the enclosed charge is Q and is distributed in a spherically symmetric fashion. The electric field points away from the center of the sphere, which is parallel to the directed area vector (which points away from the surface, by definition). Therefore, applying Gauss's law to a Gaussian sphere of radius r,

$$\left[\oint \mathbf{E}\cdot d\mathbf{A} = E\oint dA = 4\pi r^2 E\right] = \frac{Q}{\varepsilon_0}$$

$$E = \frac{Q}{4\pi\varepsilon_0 r^2} \quad \left(\begin{array}{c}\text{the direction is away from}\\ \text{the center of the sphere}\end{array}\right)$$

This is the same magnitude that would be caused by a point charge $+Q$ at the center of the sphere, reflecting the fact that the electric field depends only on the value of the enclosed charge (not on its distribution) as long as the distribution is spherically symmetric.

(b) Because the electric field is the same as that of a point charge, the potential calculation is the same as that of a point charge. We choose to evaluate the potential integral along the x-axis (the direction of the electric field is in the $+x$-direction):

$$V(a) = V(a) - V(\infty)$$

$$= -\int_{\infty}^{a} E_x dx = -\int_{\infty}^{a} \frac{Q}{4\pi\varepsilon_0 x^2} dx = \frac{Q}{4\pi\varepsilon_0 a}$$

Note that this is the same answer we would obtain for a point charge Q located at the center of the sphere.

(c) From the definition of capacitance,

$$C = \frac{Q}{V} = \frac{Q}{Q/4\pi\varepsilon_0 a} = 4\pi\varepsilon_0 a$$

Reality check: This equation also makes sense in terms of thinking about the amount of charge required to achieve a certain voltage. For example, as $a \to 0$, the distribution becomes a point charge, the potential goes to infinity (an infinite amount of energy is stored by the charge when it is infinitesimally close), and the capacitance approaches zero.

(d) The charge moves from the surface of the inner sphere to the outer surface of the metal shell (like charge repels, so the positive charge tries to disperse itself as much as it can). Note that this distribution satisfies the requirement for the electric field to be zero within both the metal sphere and the metal shell.

(e) The key here is that the metal shell has the same charge distribution (a sphere with charge on its surface) as the metal sphere did before we introduced the shell. Thus, the shell acts like a capacitor with capacitance equal to $C_b = 4\pi\varepsilon_0 b$ [see part (c)]. This insight transforms the problem to a straightforward capacitance problem: A charge of $+Q$ is transferred from one capacitor of capacitance $C_a = 4\pi\varepsilon_0 a$ to a second capacitor of capacitance $C_b = 4\pi\varepsilon_0 b$. Because we are dealing with charges and capacitances, the most useful equation for the energy of a capacitor is $U = Q^2/2C$.
Therefore,

$$\begin{aligned}\Delta U &= U_{\text{final}} - U_{\text{initial}} \\ &= \frac{Q^2}{2(4\pi\varepsilon_0 b)} - \frac{Q^2}{2(4\pi\varepsilon_0 a)} \\ &= \frac{Q^2}{8\pi\varepsilon_0}\left(\frac{1}{b} - \frac{1}{a}\right)\end{aligned}$$

which satisfies $\Delta U < 0$. Thus, the potential energy decreases during the transfer. The energy is lost as heat while electrons flow through the wire (which acts as a resistor here) into the shell.

PRACTICE TEST 2

Appendix: Physics C Equations and Constants

MECHANICS

$v = v_0 + at$

$x = x_0 + v_0 t + \frac{1}{2}at^2$

$v^2 = v_0^2 + 2a(x - x_0)$

$\Sigma \mathbf{F} = \mathbf{F}_{net} = m\mathbf{a}$

$\mathbf{F} = \frac{d\mathbf{p}}{dt}$

$\mathbf{J} = \int \mathbf{F}\, dt = \Delta \mathbf{p}$

$\mathbf{p} = m\mathbf{v}$

$\mathbf{F}_{fric} \leq \mu N$

$W = \int \mathbf{F} \circ d\mathbf{r}$

$K = \frac{1}{2}mv^2$

$P = \frac{dW}{dt}$

$P = \mathbf{F} \cdot \mathbf{v}$

$\Delta U_g = mgh$

$a_c = \frac{v^2}{r} = \omega^2 r$

$\boldsymbol{\tau} = \mathbf{r} \times \mathbf{F}$

$\Sigma \tau = \tau_{net} = I\alpha$

$I = \int r^2 dm = \Sigma mr^2$

$\mathbf{r}_{CM} = \Sigma m\mathbf{r} / \Sigma m$

$\mathbf{a}$ = acceleration
$\mathbf{F}$ = force
f = frequency
h = height
I = rotational inertia
$\mathbf{J}$ = impulse
K = kinetic energy
k = spring constant
l = length
$\mathbf{L}$ = angular momentum
m = mass
$\mathbf{N}$ = normal force
P = power
p = momentum
r = radius or distance
$\mathbf{r}$ = position vector
T = period
t = time
U = potential energy
$\mathbf{v}$ = velocity
W = work done on a system
x = position
μ = coefficient of friction
θ = angle
τ = torque
$\boldsymbol{\omega}$ = angular speed
$\boldsymbol{\alpha}$ = angular acceleration

$v = r\omega$

$\mathbf{L} = \mathbf{r} \times \mathbf{p} = I\boldsymbol{\omega}$

$K = \frac{1}{2}I\omega^2$

$\omega = \omega_0 + \alpha t$

$\theta = \theta_0 + \omega_0 t + \frac{1}{2}\alpha t^2$

$\mathbf{F}_s = -k\mathbf{x}$

$U_s = \frac{1}{2}kx^2$

$T = \frac{2\pi}{\omega} = \frac{1}{f}$

$T_S = 2\pi\sqrt{\frac{m}{k}}$

$T_p = 2\pi\sqrt{\frac{\ell}{g}}$

$\mathbf{F}_G = -\frac{Gm_1m_2}{r^2}\hat{\mathbf{r}}$

$U_G = -\frac{Gm_1m_2}{r}$

ELECTRICITY AND MAGNETISM

$F = \frac{1}{4\pi\varepsilon_0}\frac{q_1q_2}{r^2}$

$\mathbf{E} = \frac{\mathbf{F}}{q}$

$\oint \mathbf{E} \cdot d\mathbf{A} = \frac{Q}{\varepsilon_0}$

$E = -\frac{dV}{dr}$

$V = \frac{1}{4\pi\varepsilon_0}\sum_i \frac{q_i}{r_i}$

$U_E = qV = \frac{1}{4\pi\varepsilon_0}\frac{q_1q_2}{r}$

$C = \frac{Q}{V}$

$C = \frac{\kappa\varepsilon_0 A}{d}$

$C_p = \sum_i C_i$

$\frac{1}{C_s} = \sum_i \frac{1}{C_i}$

$I = \frac{dQ}{dt}$

A = area
$\mathbf{B}$ = magnetic field
C = capacitance
d = distance
$\mathbf{E}$ = electric field
$\mathcal{E}$ = EMF
$\mathbf{F}$ = force
I = current
L = inductance
ℓ = length
n = number of loops of wire per unit length
P = power
Q = charge
q = point charge
R = resistance
r = distance
t = time
U = potential or stored energy
V = electric potential
$\mathbf{v}$ = velocity
ρ = resistivity
ϕ_m = magnetic flux
κ = dielectric constant

$$U_c = \frac{1}{2}QV = \frac{1}{2}CV^2$$

$$R = \frac{\rho\ell}{A}$$

$$V = IR$$

$$R_s = \sum_i R_i$$

$$\frac{1}{R_p} = \sum_i \frac{1}{R_i}$$

$$P = IV$$

$$\mathbf{F}_M = q\mathbf{v} \times \mathbf{B}$$

$$\oint \mathbf{B} \cdot d\ell = \mu_0 I$$

$$\mathbf{F} = \int I d\ell \times \mathbf{B}$$

$$B_s = \mu_0 nI$$

$$\phi_m = \int \mathbf{B} \circ d\mathbf{A}$$

$$\mathcal{E} = -\frac{d\phi_m}{dt}$$

$$\mathcal{E} = -L\frac{dI}{dt}$$

$$U_L = \frac{1}{2}LI^2$$

GEOMETRY AND TRIGONOMETRY

Rectangle

$A = bh$

Triangle

$A = \frac{1}{2}bh$

Circle

$A = \pi r^2$

$C = 2\pi r$

Parallelepiped

$V = lwh$

Cylinder

$V = \pi r^2 l$

$S = 2\pi rl + 2\pi r^2$

Sphere

$V = \frac{4}{3}\pi r^3$

$S = 4\pi r^2$

Right Triangle

$a^2 + b^2 = c^2$

$\sin\theta = \frac{a}{c}$

$\cos\theta = \frac{b}{c}$

$\tan\theta = \frac{a}{b}$

A = area
C = circumference
V = volume
S = surface area
b = base
h = height
l = length
w = width
r = radius

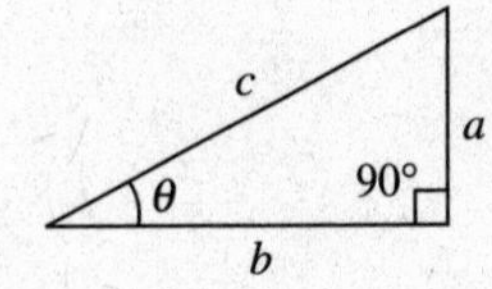

CALCULUS

$$\frac{df}{dx}=\frac{df}{du}\frac{du}{dx}$$

$$\frac{d}{dx}(x^n)=nx^{n-1}$$

$$\frac{d}{dx}(e^x)=e^x$$

$$\frac{d}{dx}(\ln x)=\frac{1}{x}$$

$$\frac{d}{dx}(\sin x)=\cos x$$

$$\frac{d}{dx}(\cos x)=-\sin x$$

$$\int x^n dx=\frac{1}{n+1}x^{n+1}, n\neq -1$$

$$\int e^x dx=e^x$$

$$\int \frac{dx}{x}=\ln|x|$$

$$\int \cos x\, dx=\sin x$$

$$\int \sin x\, dx=-\cos x$$

CONSTANTS

Proton mass, $m_p = 1.67 \times 10^{-27}$ kg
Neutron mass, $m_n = 1.67 \times 10^{-27}$ kg
Electron mass, $m_e = 9.11 \times 10^{-31}$ kg
Avogadro's number, $N_0 = 6.02 \times 10^{23}$ mol^{-1}
Electron charge magnitude, $e = 1.60 \times 10^{-19}$ C
1 electron volt, 1eV $= 1.60 \times 10^{-19}$ J
Speed of light, $c = 3.00 \times 10^8$ m/s
Universal gravitational constant, $G = 6.67 \times 10^{-11} (\text{N}\cdot\text{m}^2)/\text{kg}^2$
Acceleration due to gravity at Earth's surface, $g = 9.8$ m/s^2
Vacuum permittivity, $\varepsilon_0 = 8.85 \times 10^{-12}\ \text{C}^2/(\text{N}\cdot\text{m}^2)$
Coulomb's law constant, $k = 1/(4\pi\varepsilon_0) = 9.0 \times 10^9 (\text{N}\cdot\text{m}^2)/\text{C}^2$
Vacuum permeability, $\mu_0 = 4\pi \times 10^{-7}$ (T•m)/A
Magnetic constant, $\mu_0/(4\pi) = 1 \times 10^{-7}$ (T•m)/A

Index